T0201290

Foundations of Astrophysics

Foundations of Astrophysics provides a contemporary and complete introduction to astrophysics for astronomy and physics majors. With a logical presentation and conceptual and quantitative end-of-chapter problems, the material is accessible to introductory astrophysics students taking a two-semester survey course. Starting with the motions of the solar system and a discussion of the interaction of matter and light, the authors explore the physical nature of objects in the solar system, and the exciting new field of exoplanets. The second half of their text covers stellar, galactic, and extragalactic astronomy, followed by a brief discussion of cosmology. This is a reissue of the original 2010 edition, which has established itself as one of the market-leading astrophysics texts, well known for its clarity and simplicity. It has introduced thousands of physical science students to the breadth of astronomy, and helped prepare them for more advanced studies.

Barbara Ryden received her Ph.D. in astrophysical sciences from Princeton University. After postdocs at the Harvard–Smithsonian Center for Astrophysics and the Canadian Institute for Theoretical Astrophysics, she joined the astronomy faculty at The Ohio State University, where she is now a full professor. She has more than 25 years of experience in teaching, at levels ranging from introductory undergraduate courses to advanced graduate seminars. She won the Chambliss Astronomical Writing Award for her textbook *Introduction to Cosmology*.

Bradley M. Peterson received his Ph.D. from the University of Arizona. He was a member of the astronomy faculty at The Ohio State University from 1980 until his retirement in 2015, after serving as Department Chair for nine years. He received the Distinguished Scholar Award from Ohio State; the Outstanding Achievement Award from his alma mater, the University of Minnesota; and the NASA Exceptional Service Medal. He was a community co-chair for the Science and Technology Definition Team for the Large, Ultraviolet, Optical, Infrared Surveyor (LUVOIR), a large mission-concept study for NASA Astrophysics, and is the author of *An Introduction to Active Galactic Nuclei.*

For Nancy and Kent —B.R.

For Jan, Evan, Ethan, Erika, Lizzie, Ellyn,
Christopher, and Aden —B.M.P.

Foundations of Astrophysics

BARBARA RYDEN
The Ohio State University

BRADLEY M. PETERSON
The Ohio State University

Shaftesbury Road, Cambridge CB2 8EA, United Kingdom

One Liberty Plaza, 20th Floor, New York, NY 10006, USA

477 Williamstown Road, Port Melbourne, VIC 3207, Australia

314–321, 3rd Floor, Plot 3, Splendor Forum, Jasola District Centre, New Delhi – 110025, India

103 Penang Road, #05–06/07, Visioncrest Commercial, Singapore 238467

Cambridge University Press is part of Cambridge University Press & Assessment, a department of the University of Cambridge.

We share the University's mission to contribute to society through the pursuit of education, learning and research at the highest international levels of excellence.

www.cambridge.org
Information on this title: www.cambridge.org/9781108831956
DOI: 10.1017/9781108933001

© Barbara Ryden and Bradley M. Peterson 2021

This publication is in copyright. Subject to statutory exception and to the provisions of relevant collective licensing agreements, no reproduction of any part may take place without the written permission of Cambridge University Press & Assessment.

This book was previously published by Pearson Education, Inc., 2010
Reissued by Cambridge University Press & Assessment 2021 (version 6, March 2024)

Printed in the United Kingdom by TJ Books Limited, Padstow Cornwall, March 2024

A catalogue record for this publication is available from the British Library

ISBN 978-1-108-83195-6 Hardback

Additional resources for this publication at www.cambridge.org/rydenpeterson

Cambridge University Press & Assessment has no responsibility for the persistence or accuracy of URLs for external or third-party internet websites referred to in this publication and does not guarantee that any content on such websites is, or will remain, accurate or appropriate.

Contents

Preface

This book, like many textbooks, was inspired by teaching a class. The class in question was a two-quarter (5 hours per week) introductory survey course in astrophysics. The reader of this book, like the students in our course, is assumed to have studied a year of calculus (including differential and integral calculus, basic vector calculus, and a smattering of simple differential equations), as well as a year of calculus-based general physics. We assume that the reader has only a remote acquaintance, if any, with quantum physics, special relativity, or linear algebra.

Our fundamental goals for this book are twofold. First, we want to introduce students with a serious interest in physical science to the breadth of astronomy, preparing them for more advanced topical courses in the future. Second, we use astronomical examples to reinforce the physics that the students have already learned. To this end, we use SI (International System) units, which the students have already encountered in general physics class, rather than the cgs (centimeter, gram, second) units that are frequently encountered in the more advanced astronomical literature. Units that are peculiar to astronomers, such as parsecs, magnitudes, solar luminosities, and solar masses, are introduced as needed.

Our organization of the material is, in many respects, quite traditional. We start with the kinematics and dynamics of the solar system; then, after discussing the interaction of matter and light, we proceed to a discussion of the physical nature of objects in the solar system. We conclude our discussion of solar system astronomy with an examination of the solar system as illuminated by the exciting new field of exoplanets. The second half of the book covers stellar, galactic, and extragalactic astronomy, followed by a brief discussion of cosmology.

Our goals for the book, to some extent, dictate the relative emphasis placed on different fields of astronomy. Some particularly rich areas of astronomy, such as stellar populations, globular clusters, and the large-scale structure of the universe, are only briefly touched on. We regret the brevity with which we cover these and other fascinating topics in astronomy. However, we had to balance our desire to make the book of manageable size with our desire to cover thoroughly those topics that enhance understanding of important physical principles (such as blackbody radiation, physics of non-LTE gases, and gravitational accretion).

Our text benefited from criticism by many individuals. Most important, the book was shaped by several classes of undergraduate students at The Ohio State University, who provided detailed feedback on nearly every aspect of the book. In particular, most end-of-chapter problems in this book have been heavily field-tested; our students never hesitated to point out when a problem was clumsily or ambiguously worded. Many of the remaining end-of-chapter problems are classic problems that appear in somewhat similar form in earlier textbooks. The textbooks from which we have adopted and adapted problems are cited in the Bibliography at the end of the textbook.

We are grateful for reviews of individual chapters by instructors with experience in teaching astrophysics at this level, notably Byron D. Anderson, Phil Armitage, Don Bord, Tereasa Brainerd, David Cohen, John Cowan, Richard A. Crowe, Carsten Denker, George Djorgovski, Stephen Gottesman, Kim Griest, Peter H. Hauschildt, John Huchra, Philip A. Hughes, Steven Kawaler, Jeremy King, Chip Kobulnicky, Donald G. Luttermoser, Kevin MacKay, Michael P. Merilan, Stan Owocki, Eric S. Perlman, Lawrence S. Pinsky, Gary D. Schmidt, James Schombert, Horace Smith, Steven Stahler, Curtis J. Struck, Paula Szkody, Dan Wilkins, Jeff Wilkerson, Richard M. Williamon, Gerard Williger, Vincent Woolf, Kausar Yasmin, and Dennis Zaritsky, as well as a number of anonymous reviewers. We incorporated much of the advice received from these individuals.

We are especially grateful to friends and colleagues at The Ohio State University who provided invaluable assistance. Richard Pogge provided help with both scientific and technical issues. Jessica Orwig prepared many of the figures and tables. Marc Pinsonneault, David Weinberg, and Molly Peeples provided information for figures. Finally, the fact that this is a real book rather than a pile of incoherent notes and scrawled drawings is due to our diligent production team at Pearson Addison-Wesley.

1 Early Astronomy

The term **"astronomy"** is derived from the Greek words *astron*, meaning "star," and *nomos*, meaning "law." This reflects the discovery by ancient Greek astronomers that the motions of stars in the sky are not arbitrary but follow fixed laws. In modern times, astronomy is usually defined as the study of objects beyond the Earth's atmosphere, including not only stars but also celestial objects as small as interstellar dust grains and as large as superclusters of galaxies. The field of **cosmology**, which deals with the structure and evolution of the universe as a whole, is also regarded as part of astronomy.

In the late nineteenth century, the term **"astrophysics"** was invented, to describe specifically the field that studies how the properties of celestial objects are related to the underlying laws of physics. Thus, astrophysics could be regarded as both a subfield of physics and as a subfield of astronomy. However, because a knowledge of physics is crucial for any type of astronomical study, the terms "astronomy" and "astrophysics" are often used nearly interchangeably.

It is customary to start learning astronomy from a historical perspective. This is a natural way to learn about the universe; it permits our personal growth in knowledge to echo humanity's growth in knowledge, starting with relatively nearby and familiar objects, and then moving outward. Furthermore, as we will see, some of the most fundamental things we learn about the universe are based on simple, straightforward observations that don't require telescopes or space probes. Let us begin, therefore, by throwing away our telescopes and considering what we can see of the universe with our unaided eyes.

1.1 ▪ THE CELESTIAL SPHERE

When you look up at a cloudless night sky, you have little sense of depth. In Color Figure 1, for instance, it is not immediately obvious that the fuzzy streak in the upper part of the picture is a comet a few light-minutes away and that the fuzzy blob in the lower part is a galaxy two million light-years away. You can pick up a few clues about depth with your naked eyes (for instance, the Moon passes in front of stars, so it must

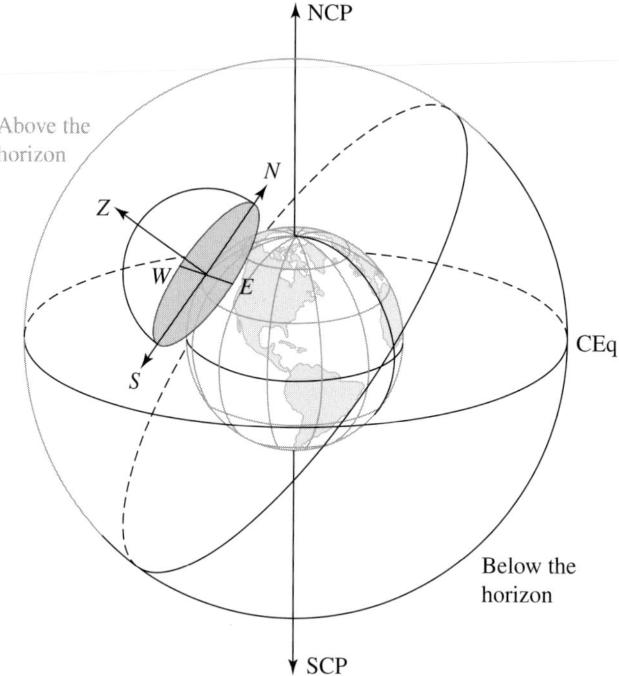

FIGURE 1.1 The celestial sphere surrounding the Earth. The Earth's north pole, south pole, and equator project onto the north celestial pole (NCP), south celestial pole (SCP), and celestial equator (CEq), respectively. For any observer, the horizon plane is tangent to the observer's location, and the zenith (Z) is directly overhead.

be closer to us than the stars are) but for the most part, determining distances to celestial objects requires sophisticated indirect methods.[1]

Although it is difficult to determine the distance to celestial objects, it is much easier to determine their position projected onto the **celestial sphere**. The celestial sphere is an imaginary spherical surface, centered on the Earth's center, with a radius immensely larger than the Earth's radius. (In Figure 1.1, the spherical Earth is exaggerated in size relative to the outer celestial sphere, for easy visibility.) Given the Earth's inconvenient opacity, an observer on the Earth's surface can see the sky only above the **horizon**, defined as a plane tangent to the idealized, perfectly spherical Earth at the observer's location (that is, it touches the Earth at the observer's feet and at no other place). The horizon is always defined locally, meaning that it moves with the observer. The horizon intersects the celestial sphere in a great circle called the **horizon circle**.[2] The horizon circle divides the celestial sphere into two hemispheres; only the hemisphere above the

[1] Some of these distance-measuring techniques will be discussed in Chapter 13.

[2] A "great circle" is a circle on the surface of a sphere whose center coincides with the sphere's center.

horizon is visible to the observer. The point directly above the observer's head, in the middle of the visible hemisphere of the celestial sphere, is called the **zenith** (point Z in Figure 1.1). The point directly below the observer's feet, opposite the zenith, is the **nadir**.

Since the celestial sphere is indeterminately large, distances between points on the celestial sphere are measured in angular units, as seen by an Earthly observer, rather than in physical units such as kilometers. Astronomers most frequently measure angles in degrees, arcminutes, and arcseconds, with 360 degrees (360°) in a circle, 60 arcminutes (60′) in a degree, and 60 arcseconds (60″) in an arcminute. When they measure angles smaller than an arcsecond, they revert to the decimal system and use milliarcseconds and microarcseconds.

When the Sun is above the horizon, it appears as a bright disk on the celestial sphere, 30 arcminutes across. The Moon, coincidentally, is also roughly 30 arcminutes in diameter as seen from Earth, but appears to change in shape as it waxes and wanes from new Moon to full and back again. When the Sun is below your horizon, you can see as many as 3000 stars with your unaided eyes.[3] The stars in the night sky appear as tiny lights, blurred by our imperfect human vision into blobs about an arcminute across. Starting in prehistoric times, astronomers have identified apparent groupings of stars called **constellations**. The stars in a constellation are not necessarily physically related, since they may be at very different distances from the Earth.

1.2 ▪ COORDINATE SYSTEMS ON A SPHERE

If we want to describe the approximate position of a star on the celestial sphere, we can say what constellation it lies within. However, since there are only 88 constellations on the entire celestial sphere, some of them quite large, merely knowing the constellation doesn't give a very precise location. For a more precise description of positions on the celestial sphere, we need to set up a coordinate system. On the two-dimensional celestial sphere, two coordinates will be needed to describe any position. Geographers have already shown us how to set up a coordinate system on a sphere; the system of **latitude** and **longitude** provides a coordinate system on the surface of the (approximately) spherical Earth.

On the Earth, the north and south poles represent the points where the Earth's rotation axis passes through the Earth's surface. The **equator** is the great circle midway between the north and south pole, dividing the Earth's surface into a northern hemisphere and a southern hemisphere. The latitude of a point on the Earth's surface is its angular distance from the equator, measured along a great circle perpendicular to the Earth's equator (Figure 1.2). Latitude is measured in degrees, arcminutes, and arcseconds, as is longitude. Thus, the use of latitude and longitude doesn't require knowing the size of

[3] This number assumes that you are in a dark location, far from the bright lights of the big city. In a populated area, you'll be lucky to see a few hundred stars.

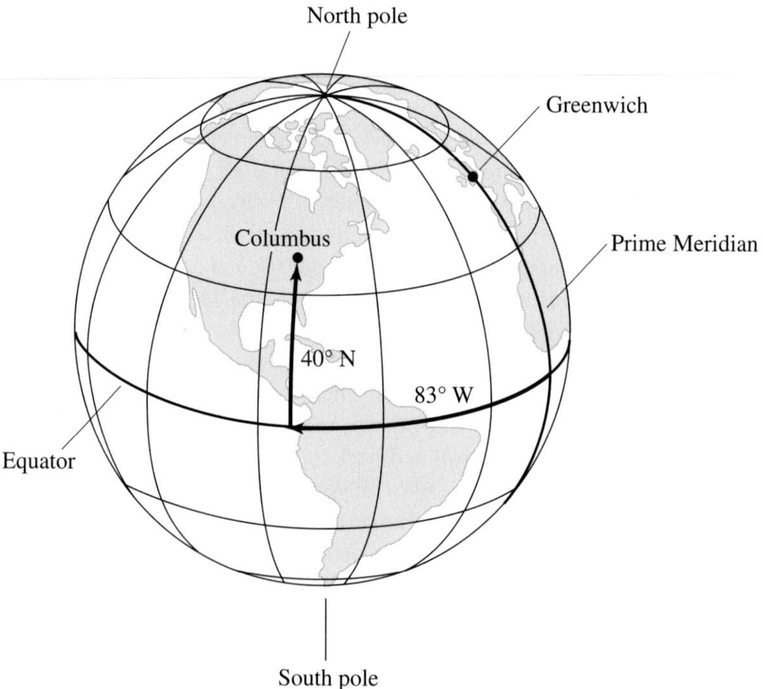

FIGURE 1.2 Latitude and longitude of a point on the Earth's surface.

the Earth in kilometers or any other unit of length.[4] In the example shown in Figure 1.2, the city of Columbus, Ohio, has a latitude of 40° N; that is, it's located 40° north of the equator.

Latitude alone doesn't uniquely specify a point on the Earth's surface. If you invited a friend to lunch at 40° N, he wouldn't know whether to expect hamburgers in Columbus, Peking duck in Beijing, or shish kebab in Ankara. The required second coordinate on the Earth's surface is the longitude. For each point on the Earth's surface, half a great circle can be drawn starting from the north pole, running through the point in question, and ending at the south pole. This half-circle, which intersects the equator at right angles, is called the point's **meridian of longitude**, or just "meridian" for short. The longitude of the point is the angle between the point's meridian and some other reference meridian. By international agreement, the reference meridian for the Earth, called the **Prime Meridian**, is the meridian that runs through the Royal Observatory at

[4] The use of latitude and longitude was successfully pioneered by the Greek astronomer Ptolemy in the second century AD, despite the fact that Ptolemy severely underestimated the size of the Earth. (Ptolemy's underestimate helped to encourage Christopher Columbus in his crazy plan to sail nonstop from the Canary Islands to Japan.)

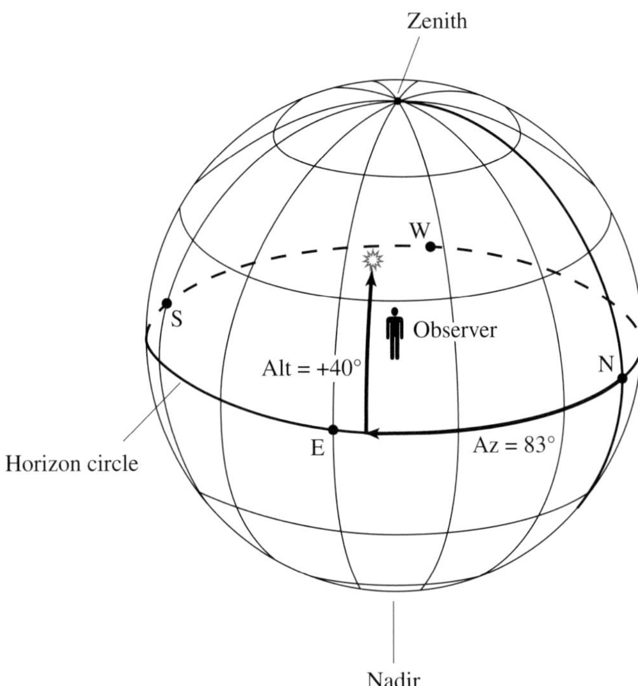

FIGURE 1.3 Altitude (Alt) and azimuth (Az) of a point on the celestial sphere, as seen by an observer on Earth.

Greenwich, England.[5] In Figure 1.2, the city of Columbus has a longitude of 83° W; that is, the meridian of Columbus is 83° west of the Prime Meridian.

The latitude–longitude coordinate system can be applied to other planets (and to spherical satellites as well). The rotation axis of the planet defines the poles and equator; the Prime Meridian is generally chosen to go through a readily identifiable landmark. The Martian Prime Meridian, for instance, runs through the center of a particular small crater called Airy-0. A coordinate system using latitude-like and longitude-like coordinates can also be applied to the celestial sphere. We just need to specify a great circle that can play the role of the equator on Earth, and a perpendicular meridian that can play the role of the prime meridian.

One such coordinate system on the celestial sphere is based on an observer's horizon, and hence is called the **horizon coordinate system**. In this system, illustrated in Figure 1.3, the latitude-like coordinate is the **altitude**, defined as the angle of a celestial object above the horizon circle. The zenith (the point directly overhead) is at an altitude of 90°. Points on the horizon circle are at an altitude of 0°. The nadir is at an altitude of

[5] Before the International Meridian Conference of 1884 agreed to adopt the Greenwich meridian as the Prime Meridian, different nations used different reference meridians.

$-90°$, but in practice, negative altitudes are seldom used, since they represent objects that are hidden by the Earth. The longitude-like coordinate in the horizon coordinate system is called the **azimuth**.[6]

For any point on the celestial sphere, half a great circle can be drawn from the zenith, through the point in question, to the nadir. The half-circle that runs through the north point on the horizon circle acts as the "prime meridian" in the horizon coordinate system. The azimuth is measured in degrees running from north through east. An object due north of an observer has an azimuth of 0°, an object due east has an azimuth of 90°, and so forth. If you know the altitude and azimuth of any object in your horizon coordinate system, you know where to point your telescope in order to see it. In the example shown in Figure 1.3, a star has an altitude of 40° and an azimuth of 83°; in other words, it's nearly halfway from the horizon to the zenith, off to the east of the observer.

One shortcoming of the horizon coordinate system is that every observer on Earth has a different, unique horizon and hence has a different, unique horizon coordinate system. A star that is near the zenith (altitude $\approx 90°$) for an observer in Buenos Aires will be near the nadir (altitude $\approx -90°$) for an observer in the antipodal city of Shanghai. To describe positions of objects on the celestial sphere, it is useful to have a coordinate system that all astronomers, regardless of location, can agree on, just as geographers all agree to use latitude and longitude to describe positions on the Earth.

To build a coordinate system useful for all Earthlings, we start by projecting the Earth's poles and equator outward onto the celestial sphere. The Earth's rotation axis, which passes through the north and south poles of the Earth, intersects the celestial sphere at the **north celestial pole** (labeled as NCP in Figure 1.1) and the **south celestial pole** (labeled as SCP). The north celestial pole is at the zenith for an observer at the Earth's north pole; more generally, for an observer at a latitude ℓ north of the equator, it will be at an altitude of ℓ and an azimuth of 0°.[7] The projection of the Earth's equator onto the celestial sphere is called the **celestial equator** (labeled as CEq in the figure). The celestial equator passes through the zenith for an observer on the Earth's equator.

On the Earth's surface, a point's latitude is its angular distance north or south of the equator. Similarly, on the celestial sphere, a point's **declination** (δ) is its angular distance north or south of the celestial equator. For points north of the celestial equator, the declination is positive ($0° < \delta \leq 90°$), and for points south of the celestial equator, the declination is negative ($-90° \leq \delta < 0°$).[8] However, the declination alone is insufficient

[6] The words "azimuth," "zenith," and "nadir," like many terms in astronomy, are derived from Arabic. ("Altitude" is from the Latin *altus*, meaning "high.")

[7] Similarly, the south celestial pole is at the zenith for an observer at the Earth's south pole; more generally, for an observer at a latitude ℓ south of the equator, it will be at an altitude ℓ and an azimuth of 180°.

[8] By analogy with the celestial poles and the celestial equator, a more logical term for declination might be "celestial latitude." However, the term "declination" has been in use for over six centuries; in Chaucer's *A Treatise on the Astrolabe* (ca. AD 1391), the poet included what he called "tables of the declinacions of the sonne."

to uniquely locate a point on the celestial sphere, just as latitude alone is insufficient to uniquely locate a point on the Earth. To determine the equivalent of longitude on the celestial sphere, it is necessary to choose a celestial "prime meridian" running from the north celestial pole to the south celestial pole. If we let the observer's zenith act as the celestial "Greenwich," then the **zenith meridian**, defined as the arc running from the north celestial pole through the zenith to the south celestial pole, will act as a celestial "prime meridian."[9] The longitude-like coordinate, measured westward from the zenith meridian, is called the **hour angle** (H). For a given observer at a given time, the declination (angular distance from the celestial equator) and hour angle (angular distance from the zenith meridian) uniquely specify the location of a star, or other object, on the celestial sphere.

One complication of using the hour angle to specify the location of a star is that observers at different longitudes will have different observer's meridians, and hence will measure different hour angles for the same star. If a star is on your zenith meridian, it will be 1° east of the zenith meridian for an observer 1° of longitude west of you. Another complication results from the fact that the Earth is rotating about the axis between its north and south poles, completing one rotation in about 24 hours. Although we know perfectly well, at an intellectual level, that the Earth is rotating from west to east, an observer pinned by gravity to the Earth's surface experiences a strong illusion that the Earth is stationary and the celestial sphere is rotating from east to west. Stars thus appear to follow circular paths called **diurnal circles** that are parallel to the celestial equator; that is, they stay a fixed angular distance from the celestial equator, and their declination remains constant. This situation is illustrated in Figure 1.4. Over the course of 24 hours, the hour angle of a star changes by 360° as it travels in its diurnal circle. Because of the constant rate of change of the hour angle (15° per hour), the hour angle is often measured in units of hours (h), minutes (m), and seconds (s) instead of degrees, arcminutes, and arcseconds, with $1^h = 15°$, $1^m = 15'$, and $1^s = 15''$. A star that is on the zenith meridian right now has hour angle $H = 0^h$; 6 hours from now it will be at $H = +6^h$, off to the observer's west; 12 hours from now it will be at $H = +12^h$, on the nadir meridian. Thus, the hour angle of a star can be thought of as the time that has elapsed since it was last on the zenith meridian.

The hour angle of a star is constantly changing because it is measured relative to an observer's meridian that is tied to the rotating Earth. If we want a longitude-like coordinate that is constant for a given star over the course of 24 hours, we need to measure it relative to a new meridian, one that is tied to the celestial sphere rather than to the Earth. In short, we need a point on the celestial sphere that acts as the astronomical equivalent of Greenwich, England. Instead of selecting one particular star to serve as a "Greenwich," astronomers have chosen a point on the celestial equator termed the "vernal equinox." (In Section 1.3, we give the technical definition of the vernal equinox; but remember,

[9] We can also define a complementary **nadir meridian** running from the north celestial pole through the nadir to the south celestial pole. The zenith meridian and the nadir meridian constitute the two halves of a great circle called the **observer's meridian**.

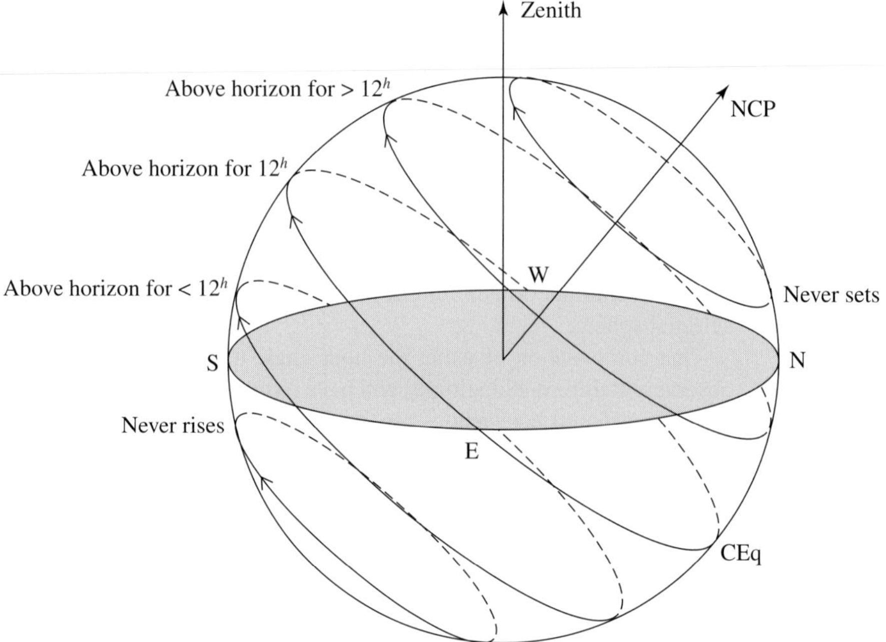

FIGURE 1.4 Diurnal circles of stars as seen by an observer in the northern hemisphere. Circumpolar stars near the north celestial pole never set; similarly, stars near the south celestial pole never rise. Stars on the celestial equator are above the horizon for 12 hours and below the horizon for 12 hours.

any point on the celestial sphere would work equally well, just as any point on the Earth would work just as well as Greenwich.)

Half a great circle drawn on the celestial sphere, from the north celestial pole, through the vernal equinox, to the south celestial pole, is the celestial equivalent of the Prime Meridian on Earth (Figure 1.5). The longitude-like coordinate measured *eastward* from this "Prime Meridian" is called the **right ascension** (α). The right ascension and declination of a star change slowly with time (just as the latitude and longitude of a city on Earth may change slowly thanks to plate tectonics), but they can be treated as constant over the course of a single night, unlike the inexorably changing hour angle. The right ascension of a celestial object, like its hour angle, is characteristically measured in hours, minutes, and seconds. The coordinate system using right ascension and declination is called the **equatorial coordinate system** and is widely used in astronomy; catalogs of stars, for instance, generally give their positions in terms of right ascension and declination. For the example shown in Figure 1.5, the star in question is at a right ascension $\alpha = 277° = 18^h28^m$ and a declination $\delta = +40°$. This is within the constellation Lyra, not far from the bright star Vega.

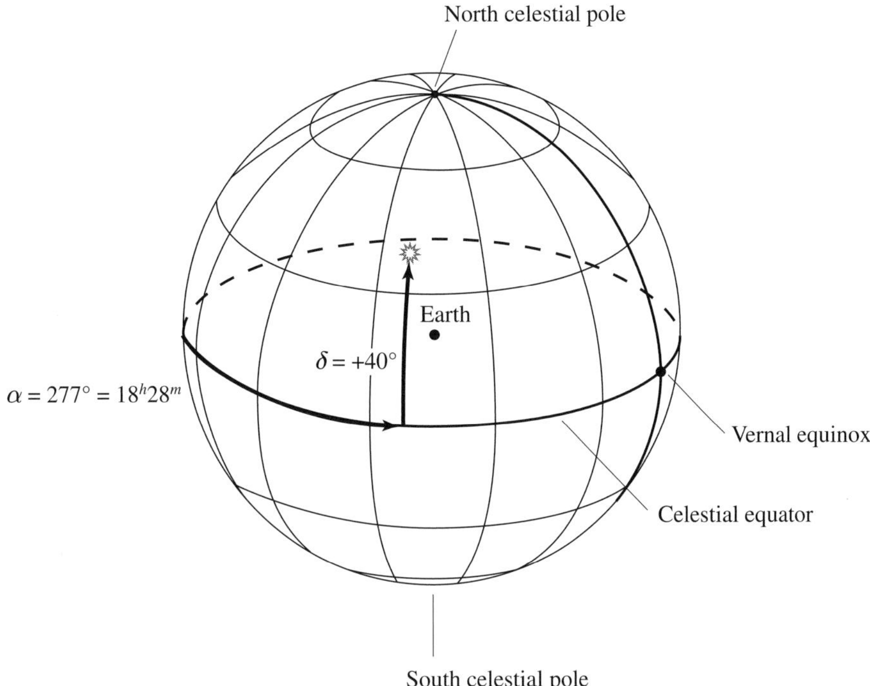

FIGURE 1.5 The right ascension (α) and declination (δ) of a point on the celestial sphere.

1.3 ▪ CELESTIAL MOTIONS

As mentioned above, and illustrated in Figure 1.4, an observer on the rotating Earth sees stars move in diurnal circles, just as if the Earth were stationary and the stars were glued to a rigid, rotating celestial sphere. The horizon plane of an observer bisects the celestial sphere, and thus also bisects the celestial equator (labeled "CEq" in Figure 1.4). Thus, stars on the celestial equator are above the horizon for 12 hours a day and below the horizon for 12 hours a day. The diurnal circles of stars not on the celestial equator are not bisected by the horizon (except in the special case when the observer is on the equator, when all diurnal circles are bisected). Consider an observer somewhere in the Earth's northern hemisphere, as shown in Figure 1.4.[10] For stars north of the celestial equator, more than half of their diurnal circles are above the horizon, so they spend more time above the horizon than below. For an observer at latitude ℓ, all stars within an angular distance ℓ of the north celestial pole (that is, with declination $\delta > 90° - \ell$)

[10] In our examples, we will practice blatant northern hemisphere chauvinism, rationalized by the fact that $\sim$90% of the human population lives in the northern hemisphere. Description of apparent motions for a southern hemisphere observer is left as an exercise for the reader.

FIGURE 1.6 Star trails over Mauna Kea, Hawaii, showing circumpolar stars around the north celestial pole.

will have diurnal circles that don't intersect the horizon plane at all. These stars, called **circumpolar stars**, never fall below the observer's horizon but can be seen to move in counterclockwise circles about the north celestial pole.

Figure 1.6 shows a long exposure of the night sky over Mauna Kea, Hawaii, at a latitude $\ell = 20°$; the star trails cover about 1/12 of a full circle, indicating the photographic exposure was ~ 2 hours long. By contrast with circumpolar stars, stars within an angular distance ℓ of the *south* celestial pole never rise above the horizon; again, the horizon plane never intersects their diurnal circles. For stars south of the celestial equator but farther than ℓ from the south celestial pole, less than half of their diurnal circles are above the horizon; these stars spend less than 12 hours per day above the northern observer's horizon, rising in the southeast and soon setting in the southwest.

As well as the stars, the Sun, Moon, and planets are seen to move in diurnal circles. However, if the Sun, Moon, and planets are observed for times much longer than a single night, additional motions are also seen. The most important motions are the following:

- The relative positions of **stars** can be approximated as constant, over human time scales. Although stars are in motion relative to each other and to the Sun, on time scales shorter than decades the motion cannot be detected without a telescope.
- The **Sun** moves eastward relative to the stars by about 1° per day. This is because the Earth is orbiting the Sun, and we see the Sun in projection against different background stars as we orbit around it. Because of the relative motion of the Sun and stars, the stars rise 4 minutes earlier each day relative to the Sun.

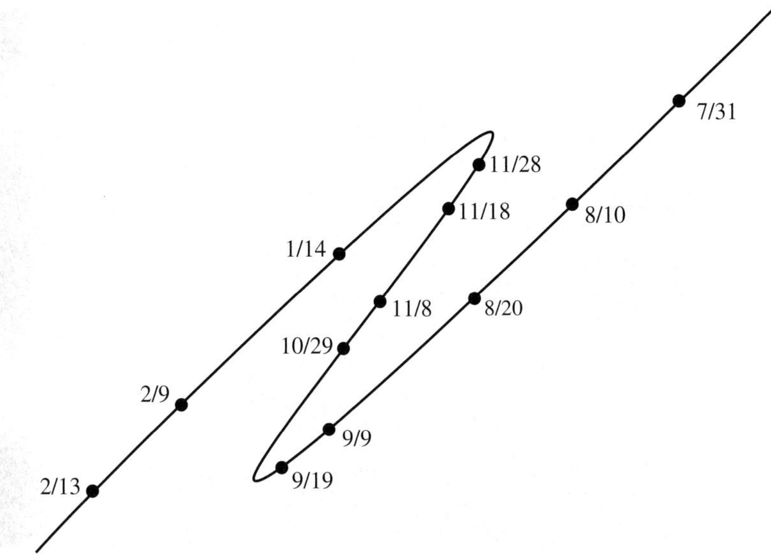

FIGURE 1.7 The apparent motion of Mars relative to the stars during late 2005 and early 2006. Mars was in retrograde motion from 2005 October 1 to 2005 December 9.

- The **Moon** also moves eastward relative to the stars, by about 13° per day. This is because the Moon orbits around the Earth in an eastward direction, taking 27.3 days for a complete orbit. The Moon's motion around the sky ($360^\circ/27.3$ days $\approx 13^\circ\ \mathrm{day}^{-1}$) is slow compared to the Earth's eastward rotation ($360^\circ\ \mathrm{day}^{-1}$), so we still see the Moon rise in the east and set in the west, just like the Sun. Relative to the Sun, the Moon moves eastward by about 12° per day, so it takes $360^\circ/12^\circ\ \mathrm{day}^{-1} \approx 30$ days for the Moon to "lap" the Sun. Because of the relative motion of Sun and Moon, the Moon rises about 50 minutes later each day.
- The **planets** known prior to the invention of the telescope were Mercury, Venus, Mars, Jupiter, and Saturn (in addition to the Earth, of course). Without a telescope, the planets look like unresolved stars. Early astronomers distinguished them from stars by the fact that planets move relative to the stars.[11] Ordinarily, planets move slowly eastward relative to the stars. On occasion, however, they reverse their motion and move westward relative to the stars for a short period. This reversed motion is called **retrograde motion**. Figure 1.7, illustrates, an example of retrograde motion for the planet Mars.

The great circle along which the Sun moves on the celestial sphere is called the **ecliptic**. The ecliptic represents the plane of the Earth's orbit around the Sun, projected onto the

[11] The word "planet" comes from the Greek word meaning "wanderer." The word "plankton" derives from the same root; plankton are tiny aquatic creatures condemned to wander where the ocean currents take them.

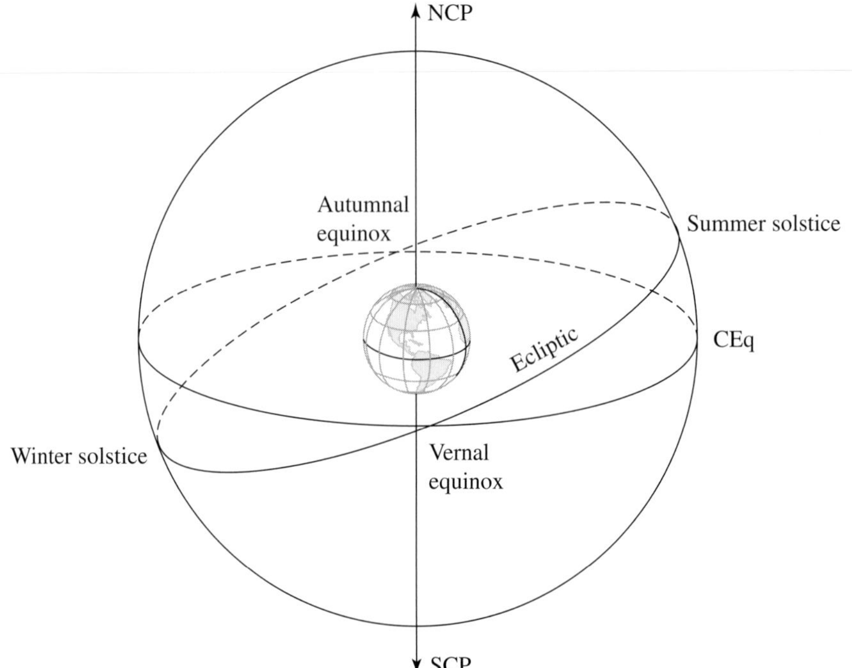

FIGURE 1.8 The relative positions of the ecliptic and the celestial equator on the celestial sphere. The equinoxes and solstices are indicated.

celestial sphere. The ecliptic, as shown in Figure 1.8, is inclined by 23.5° relative to the celestial equator. The tilt of 23.5° between the ecliptic and celestial equator is called the **obliquity of the ecliptic**. The obliquity is nonzero because the Earth's rotation axis is not exactly perpendicular to the orbit of the Earth around the Sun; instead, the axis is tilted by 23.5° from the perpendicular.

Since the ecliptic and celestial equator are two different great circles on a sphere, they intersect at two points, 180° apart. The two points of intersection are called the **equinoxes**. The point where the Sun moves from the northern celestial hemisphere to the southern is called the **autumnal equinox**; the Sun is at the autumnal equinox around September 21. The point where the Sun moves from the southern celestial hemisphere to the northern is called the **vernal equinox**; the Sun is at the vernal equinox around March 21. (Recall from Section 1.2 that the vernal equinox was chosen as the origin for the measurement of right ascension).

The point on the ecliptic that is farthest north of the celestial equator (it has declination $\delta = +23.5°$) is called the **summer solstice**; the Sun is at the summer solstice around June 21. The point on the ecliptic that is farthest south of the celestial equator ($\delta = -23.5°$) is called the **winter solstice**; the Sun is at the winter solstice around December 21. Astronomers usually use the terms "equinox" and "solstice" to refer to points on the celestial sphere; however, the terms can also refer to the *time* at which the Sun reaches those points.

The Sun's diurnal path varies during the year because its declination changes as it moves along the ecliptic. The time per day that the Sun is above the horizon depends on where it is relative to the celestial equator. At the equinoxes, the Sun is exactly on the celestial equator, and thus spends 12 hours above the horizon and 12 hours below the horizon.[12] When the Sun is north of the celestial equator, it is above the horizon for more than 12 hours for a northern hemisphere observer. When it's south of the celestial equator, it is above the horizon for less than 12 hours for a northern hemisphere observer. In the northern hemisphere, the shortest night of the year occurs when the Sun is at the summer solstice, its point farthest north of the celestial equator. Similarly, the longest night of the year in the northern hemisphere occurs when the Sun is at the winter solstice.[13]

As mentioned on page 10, stars with a declination $\delta > 90° - \ell$ are circumpolar stars for an observer at latitude ℓ north of the equator. This implies that a star in the northern celestial hemisphere, with declination $\delta > 0°$, will be a circumpolar star for all observers with latitude $\ell > 90° - \delta$. When the Sun is at the summer solstice, it has a declination $\delta = +23.5°$, and hence is a circumpolar star for observers north of latitude 66.5° N. Within this region, bounded by the **Arctic Circle**, observers experience the phenomenon of the midnight Sun around June 21; the Sun never sets but makes a complete circle in azimuth over 24 hours. At the same time, observers within the **Antarctic Circle**, at latitude 66.5° S, never see the Sun rise over the horizon during the course of 24 hours (see Figure 1.9). At the time of the winter solstice, around December 21, the situation is reversed; observers within the Arctic Circle have 24 hours of darkness while observers within the Antarctic Circle have 24 hours of sunlight.

Globes of the Earth usually have the Arctic and Antarctic Circles drawn on them (see Figure 1.9). They also display the Tropic of Cancer at 23.5° N and the Tropic of Capricorn at 23.5° S. At a latitude ℓ north of the equator, the zenith has a declination $+\ell$; thus, the Tropic of Cancer represents the latitude at which the Sun passes directly overhead when it's at the summer solstice. At a latitude ℓ *south* of the equator, the zenith has a declination $-\ell$; thus, the Tropic of Capricorn represents the latitude at which the Sun passes directly overhead when it's at the winter solstice. The region on Earth between the Tropic of Cancer and the Tropic of Capricorn is known as "the tropics."[14]

The Sun's annual motion along the ecliptic carries it through a group of constellations that comprise the **zodiac**. The 12 traditional members of the zodiac are Pisces (within which the vernal equinox is located), Aries, Taurus (where the summer solstice is located), Gemini, Cancer, Leo, Virgo (where the autumnal equinox is located), Libra,

[12] This equality accounts for the name "equinox," which comes from the Latin *equus* (equal) + *nox* (night). In other words, it's where the Sun is located when night is equal in length to day.

[13] The term "solstice" comes from the Latin *sol* (Sun) + *sistere* (to stand still). The solstices are the points where the Sun's declination reaches an extremum. Thus, although the Sun doesn't literally stand still relative to the background stars (its right ascension is continuously increasing), its declination is momentarily constant at a solstice.

[14] The words "tropic" and "tropical" derive from the Greek word *trope* (meaning "a turning"—as when the Sun, which has been moving away from the celestial equator, turns around and moves back toward the celestial equator).

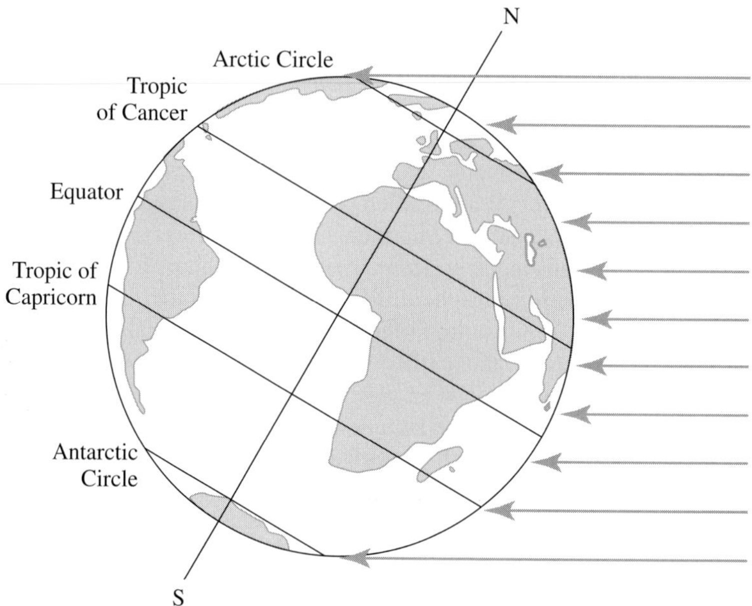

FIGURE 1.9 Sun's rays striking the Earth, around the time of the summer solstice. The Arctic and Antarctic Circles, as well as the Tropics of Capricorn and Cancer, are indicated.

Scorpius, Sagittarius (where the winter solstice is located), Capricornus, and Aquarius. However, using the constellation boundaries defined by the International Astronomical Union, the Sun also passes through the constellation Ophiuchus (from December 1 to December 18).

The vernal equinox has not always been in Pisces. In the second century BC, the Greek astronomer Hipparchus discovered that the equinoxes and solstices move westward along the ecliptic with respect to the fixed stars of the zodiac. This motion, called the **precession of the equinoxes**, occurs at a rate of $50.3''$ per year, completing a full circuit in 25,800 years. In the time of Hipparchus, the vernal equinox was located in the constellation Aries.[15] We know today, as Hipparchus did not, that the precession of the equinoxes is due to the precession, or "wobble," of the Earth's rotation axis, which moves in a cone of opening angle 47°, with a precession period of 25,800 years (Figure 1.10).

In Section 4.1, we examine the physical causes that make the Earth precess like a dying top. For the moment, however, we will focus on the practical implications of the precession. As the Earth's rotation axis precesses, the north and south celestial poles, which are the projections of that axis onto the celestial sphere, move in a circle of diameter

[15] Thus, the vernal equinox is sometimes referred to, anachronistically, as the "first point of Aries." Similarly, the Tropic of Cancer and Tropic of Capricorn gained their names when the summer and winter solstices were in the constellations Cancer and Capricornus, respectively.

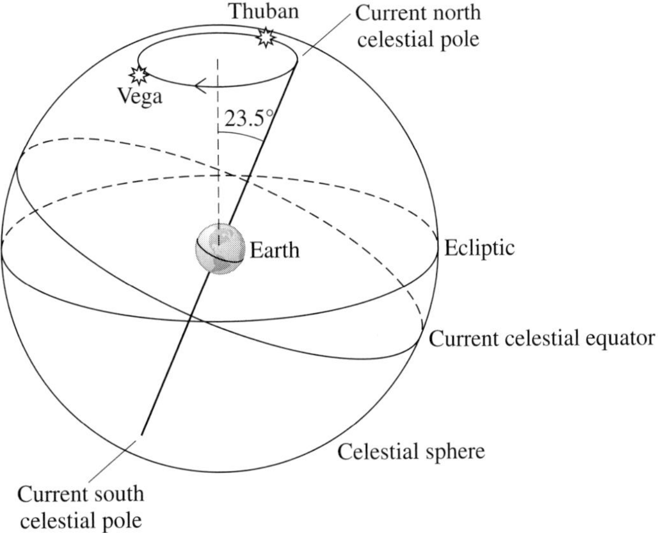

FIGURE 1.10 Precession of the Earth's rotation axis, with the resulting motion of the north celestial pole on the celestial sphere.

47°, taking 25,800 years for a complete circuit. The north celestial pole (at declination $\delta = 90°$) is currently near the bright star called Polaris, in the constellation Ursa Minor. In the year 2700 BC, the star Thuban, in the constellation Draco, was very close to the north celestial pole.[16] In the year AD 14,000, the bright star Vega, in the constellation Lyra, will be close to the north celestial pole (see Figure 1.10).

As the celestial poles and equator continuously move relative to the background stars, the declination of those stars must also continuously change. Also, since the vernal equinox continuously moves through the zodiac, the right ascension of stars (which is measured relative to the vernal equinox) must continuously change. Since the vernal equinox is moving westward across the celestial sphere, and right ascension is measured eastward from the vernal equinox, the right ascension of a fixed star will increase with time. Due to the time dependence of the coordinates, when a right ascension and declination are given, the epoch at which they are measured must also be specified. The most common standard used today is "equinox 2000.0," indicating right ascension and declination at the beginning of the year 2000. (Some older star charts and catalogs use 1950.0 or 1900.0 as their right ascension and declination standards.)

[16] When William Shakespeare put the words "I am as constant as the northern star" into the mouth of Julius Caesar, he was making an astronomical blunder. In the year 44 BC, when Caesar was assassinated, the closest bright star to the north celestial pole was Kochab, in the constellation Ursa Major; at a distance of 9° from the pole, its diurnal circle would have been blatantly obvious, and calling it "constant" would have been a real stretch.

1.4 ▪ BASIC TIMEKEEPING

Astronomy was initially developed largely for its practical applications, such as celestial navigation and timekeeping. Calendars are particularly important for agriculture; planting a crop at the correct time of year is vital. Thus, virtually all agrarian cultures developed astronomy to varying levels of sophistication. Some archeological sites have been shown to have a connection with astronomy. Stonehenge, on Salisbury Plain in the south of England, is a spectacular example of a prehistoric observatory, built in stages during the time span 2500–1700 BC. Various alignments of stones mark key events of the calendar; for instance, the direction in which the Sun rises at the time of the winter solstice and of the summer solstice. Other prehistoric structures throughout the world show similar alignments, giving concrete evidence for humanity's long-standing interest in astronomy.

With the invention of writing, astronomers began leaving systematic records of their observations of the sky. Chinese, Egyptian, and Mayan astronomers all made meticulous records, as did the ancient Babylonians. The Babylonian Empire was the dominant power in southern Mesopotamia (modern Iraq) from the reign of Hammurabi in the eighteenth century BC until it was absorbed into the Persian Empire in the sixth century BC. During that interval, careful observations of the Sun and Moon by Babylonian astronomers enabled an accurate determination of the length of the year and month. The Babylonians used a sexagesimal number system (base 60) rather than a decimal system (base 10); it is thanks to the Babylonians that there are 360 (6×60) degrees in a full circle, 60 arcminutes in a degree, and 60 arcseconds in an arcminute.

Astrology, which tracks the positions of planets in the belief that they influence human events, was also a major motivation for the development of astronomy in Babylonia and elsewhere. In fact, until relatively modern times, astronomy was not clearly distinguished from astrology at all. As late as the seventeenth century, for instance, the astronomer Johannes Kepler augmented his inadequate salary by casting horoscopes. "God provides for every animal his means of sustenance—for an astronomer he has provided astrology," Kepler wrote.

All common units of time are ultimately astronomical in origin. The **day** is based on (but is not identical to) the rotation period of the Earth. The **hour** is defined as a fraction of the day. Ancient cultures divided the day into 12 hours of daylight and 12 hours of darkness; thus, the daylight hours were longest near the time of the summer solstice and shortest near the time of the winter solstice. The division of the day into 24 hours of equal length didn't occur until the mechanical clock was invented in the thirteenth century. By the end of the Middle Ages, clocks were accurate enough to allow the subdivision of each hour into 60 **minutes**.[17] The measurement of **seconds**, defined as 1/60 of a minute, wasn't feasible until the invention of pendulum clocks in the seventeenth century.

The **month** and the **year** are based on (but are not identical to) the orbital period of the Moon around the Earth, and the Earth around the Sun, respectively. Even the **week** is tied, albeit loosely, to astronomy. The seven-day week currently in use is the merger

[17] The division of hours into 60 minutes was modeled on the much earlier division of degrees into 60 (arc)minutes. Thus, the 60 tick marks around the edge of a clock face ultimately trace back to the Babylonians.

TABLE 1.1 Days of the Week

Latin Name	Italian Name	English Name	Notes
dies Solis	domenica	Sunday	domenica = "Lord"
dies Lunae	lunedi	Monday	Moon = Luna
dies Martis	martedi	Tuesday	Mars ≈ Tiw
dies Mercurii	mercoledi	Wednesday	Mercury ≈ Woden
dies Iovis	giovedi	Thursday	Jupiter = Jove ≈ Thor
dies Veneris	venerdi	Friday	Venus ≈ Frigg
dies Saturni	sabato	Saturday	sabato = "Sabbath"

of two different cycles: the Jewish week, containing six work days plus the Sabbath, and the planetary week, in which each day is presided over by one of the seven wandering objects (or planets) known to ancient astronomers. In the planetary week, which may have originated among Egyptian astrologers, the days of the week are named, in order, after the Sun, the Moon, Mars, Mercury, Jupiter, Venus, and Saturn.[18] The Latin names for the days of the week, shown in Table 1.1, preserve this order. In Romance languages (Italian is shown as an example in the Table), the planetary names are retained for the workweek; however, Saturday is given a name derived from the Sabbath of the Jewish calendar, and Sunday is named the "Lord's Day." In the English names for the days of the week, the links to Saturn, the Sun, and the Moon are obvious in Saturday, Sunday, and Monday. The planetary associations are obscured for the other four days of the week, however, since the names of Roman deities have been replaced with their approximate Teutonic equivalents (see Table 1.1).

1.5 ▪ SOLAR AND SIDEREAL TIME

In Section 1.4, we noted that the length of the day, as it is most commonly defined, is not exactly equal to the rotation period of the Earth. Let's see why this is true. By convention, we define the "day" to be the interval between successive **upper transits** of a celestial object. Because of the Earth's rotation, a celestial object will cross, or **transit**, the observer's meridian twice a day. The upper transit occurs when the object crosses the zenith meridian, and the lower transit occurs half a day later, when it crosses the nadir meridian.[19] The time between two upper transits of a star is a **sidereal day**;

[18] To modern astronomers, the Sun is classified as a star, and the Moon is classified as a satellite. However, ancient astronomers lumped together the Sun and the Moon with the other "wanderers" they could see in the sky.

[19] For circumpolar objects, both transits are visible above the horizon, so it is particularly important to distinguish between them. The upper transit for a circumpolar object occurs when the object crosses the observer's meridian at a higher altitude traveling westward; the lower transit occurs when the object crosses the meridian at a lower altitude traveling eastward.

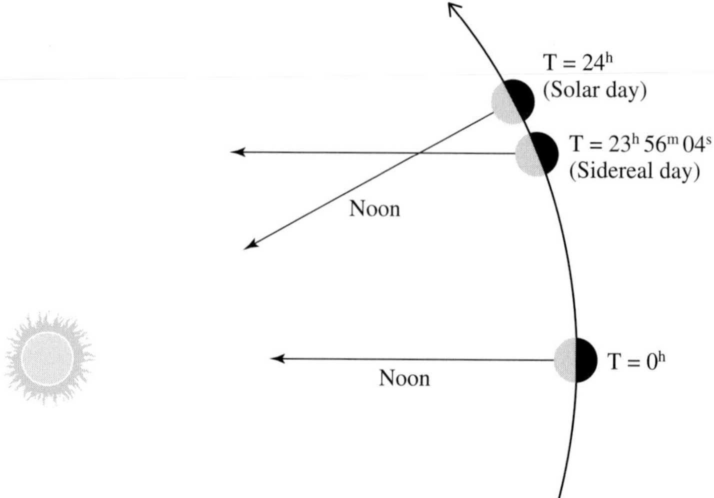

FIGURE 1.11 The relation between the solar and sidereal day; the solar day is slightly longer than the sidereal day because of the Earth's orbital motion around the Sun.

this represents the Earth's rotation period relative to the distant fixed stars.[20] The time between two upper transits of the *Sun* is a **solar day**, which is slightly longer than the sidereal day, as seen in Figure 1.11. The fundamental measure of time used by humans is solar time, since people find it more convenient to schedule their lives around how the Sun moves in the sky rather than how the inconspicuous nighttime stars move.

The difference in length between the sidereal and solar day is the result of a change in the observer's frame of reference. The sidereal day is the Earth's rotation period measured in the nonrotating frame of reference of the fixed stars, also known as the sidereal frame. The solar day is the Earth's rotation period measured in a reference frame that co-rotates with a line drawn from the Earth to the Sun. To examine the mathematical relation between the sidereal day and the solar day, let $\vec{\boldsymbol{\omega}}_{\mathbf{sid}}$ be the angular velocity of the Earth's rotation in the sidereal frame and let $\vec{\boldsymbol{\omega}}_{\mathbf{E}}$ be the angular velocity of the Earth's orbital motion in the same frame of reference. The difference between these is the angular velocity of the Earth's rotation in a reference frame that co-rotates with the Earth–Sun line; let's call this $\vec{\boldsymbol{\omega}}_{\mathbf{sol}}$. Specifically, we see that

$$\vec{\boldsymbol{\omega}}_{\mathbf{sid}}(t) = \vec{\boldsymbol{\omega}}_{\mathbf{sol}}(t) + \vec{\boldsymbol{\omega}}_{\mathbf{E}}(t). \tag{1.1}$$

If the angular velocity vectors are parallel, this can be rewritten as a scalar equation,

$$\omega_{\rm sid}(t) = \omega_{\rm sol}(t) + \omega_{\rm E}(t). \tag{1.2}$$

[20] The word "sidereal" is derived from the Latin word *sidereus*, meaning "starry."

For the Earth–Sun system, $\omega_{\rm sid}$ and $\omega_{\rm E}$ aren't exactly parallel, since they are tilted by 23.5° relative to each other. However, the parallel assumption gives a reasonable first approximation.

If, in addition, $\omega_{\rm sid}$ and $\omega_{\rm E}$ are constant, then equation (1.2) can be rewritten in terms of time rather than angular velocity. In that case, $|\omega| = 2\pi/P$, where P is the period of the circular motion in question. Thus, if $P_{\rm sid}$ is the length of the sidereal day, $P_{\rm sol}$ is the length of the solar day, and $P_{\rm E}$ is the Earth's orbital period around the Sun,

$$\begin{aligned} \frac{2\pi}{P_{\rm sid}} &= \frac{2\pi}{P_{\rm sol}} + \frac{2\pi}{P_{\rm E}} \\ \frac{1}{P_{\rm sid}} &= \frac{1}{P_{\rm sol}} + \frac{1}{P_{\rm E}}. \end{aligned} \tag{1.3}$$

If we define the solar day to be $P_{\rm sol} \equiv 1$ day, then we note that $P_{\rm E} \approx 365$ days $\gg P_{\rm sol}$. Thus, we may write

$$\begin{aligned} P_{\rm sid} = \left(\frac{1}{P_{\rm sol}} + \frac{1}{P_{\rm E}}\right)^{-1} &= P_{\rm sol}\left(1 + \frac{P_{\rm sol}}{P_{\rm E}}\right)^{-1} \\ &\approx P_{\rm sol}\left(1 - \frac{P_{\rm sol}}{P_{\rm E}}\right). \end{aligned} \tag{1.4}$$

The difference between the solar day and the sidereal day is then

$$\begin{aligned} P_{\rm sol} - P_{\rm sid} &\approx P_{\rm sol}\left(\frac{P_{\rm sol}}{P_{\rm E}}\right) \\ &\approx 1\text{ day}\left(\frac{1}{365}\right)\left(\frac{24\text{ hr}}{1\text{ day}}\right)\left(\frac{60\text{ min}}{1\text{ hr}}\right) \\ &\approx 4\text{ min}. \end{aligned} \tag{1.5}$$

Thus, the length of the sidereal day is $23^{\rm h}56^{\rm m}$. This means that, relative to the Sun, the stars rise 4 minutes earlier each day as the Sun moves slowly eastward along the ecliptic.

Although the Sun makes a convenient clock for terrestrial observers, and one that never needs winding, defining time in terms of the solar day has one major problem. The length of the apparent solar day, defined as the time between one upper transit of the Sun and the next, varies over the course of a year. The variations in the apparent solar day are not huge: the shortest apparent solar days, which occur in March and September, are less than a minute shorter than the longest apparent solar days, which occur in June and December. Nevertheless, the differences in the length of the apparent solar day were known to ancient Babylonian astronomers, thanks to their careful observations. From a purely empirical standpoint, astronomers circumvent the problem of the variable length of the apparent solar day by defining two types of time measurement:

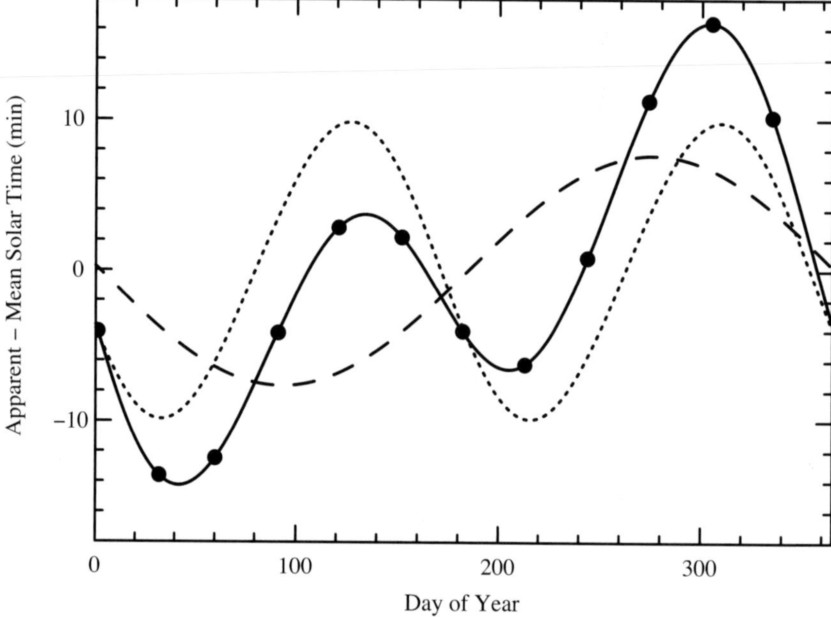

FIGURE 1.12 The solid line is the empirically determined equation of time; dots represent the first day of each calendar month. The dotted line is the contribution to the equation of time from the obliquity of the ecliptic; the dashed line is the contribution from the Earth's changing orbital speed.

- **Apparent solar time** is measured by the Sun's position relative to the local observer's meridian. Apparent solar time is the time measured by a sundial.[21]
- **Mean solar time** is the time kept by a fictitious "mean Sun" that travels eastward along the celestial equator at a constant rate, completing one circuit in one year. The mean solar day is thus equal to the average length of an apparent solar day. The mean solar day, which is constant over time, is the basis for the time kept by mechanical and electronic clocks.

These two measures of time are related by the **equation of time**. Specifically,

$$\text{Equation of Time} = \text{Apparent Solar Time} - \text{Mean Solar Time}. \tag{1.6}$$

The equation of time, as calculated from observations of the Sun, is shown in Figure 1.12. In mid-February, the accumulation of long apparent solar days causes apparent solar time to fall as much as 14 minutes behind mean solar time. Conversely, during early November, apparent solar time runs more than 16 minutes ahead of mean solar time.

[21] Brief reflection on how a sundial works, combined with the knowledge that mechanical clocks were invented in the northern hemisphere, should lead the reader to an understanding of why clocks run "clockwise."

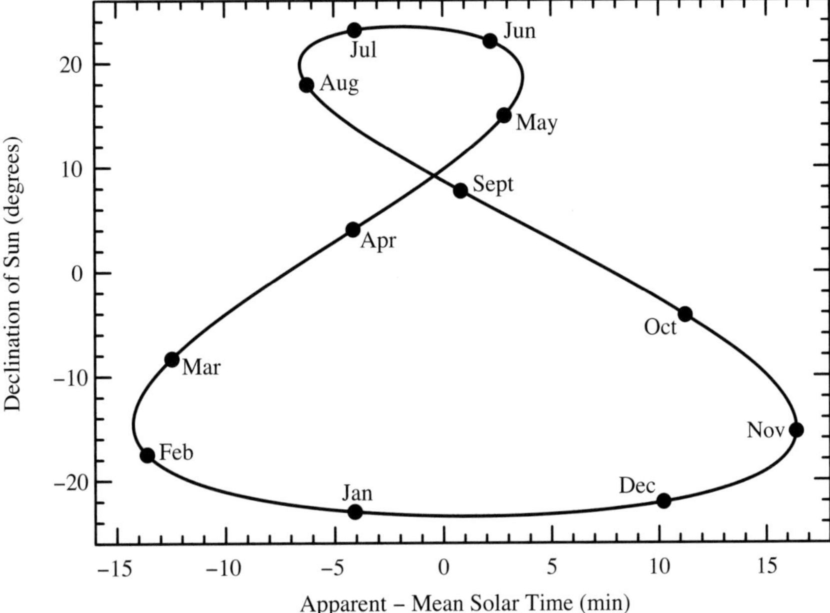

FIGURE 1.13 The analemma; that is, a plot of the Sun's declination as a function of the equation of time. The dots represent the Sun's position on the first day of each calendar month.

If the equation of time is plotted as a function of the Sun's declination rather than as a function of date, the result is a figure known as the **analemma** (Figure 1.13). The lopsided "figure eight" shape of the analemma is sometimes found printed on globes. Perhaps more striking, if you take a multiple exposure photograph of the Sun, taking an exposure at the same time each day (as measured by a clock) throughout the year, the resulting Sun images trace out the shape of an analemma. Such a photograph, taken from Arizona, is shown in Figure 1.14. Analemma photographs provide graphic evidence that the length of the apparent solar day is variable; if its length were constant, then the analemma would be a straight line segment, not a warped figure eight. The obvious next question is Why does the apparent solar day vary in duration?

The variation in length of the apparent solar day has two causes: the obliquity of the ecliptic (that is, the fact that $\vec{\boldsymbol{\omega}}_{\mathbf{sid}}$ and $\vec{\boldsymbol{\omega}}_{\mathbf{E}}$ are not parallel) and the nonuniform orbital speed of the Earth (that is, the fact that $\omega_{\mathrm{E}}(t)$ varies with time).[22] Even if the Sun moved at a perfectly constant rate along the ecliptic, the obliquity of the ecliptic would create a variable eastward motion of the Sun. The Sun's eastward motion (that is, the rate of increase of its right ascension) is its projected motion onto the celestial equator. The

[22] The angular velocity of the Earth's rotation, ω_{sid}, varies at a much, much slower rate than the angular velocity of the Earth's orbital motion.

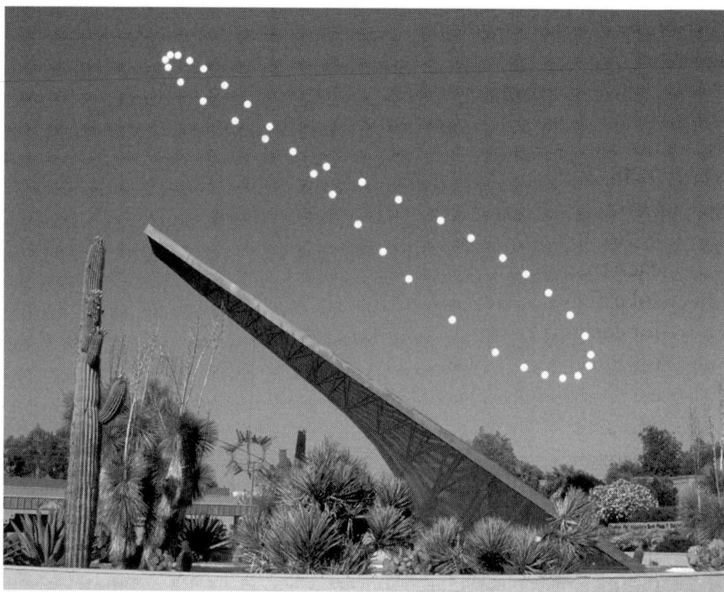

FIGURE 1.14 Analemma photographed over Carefree, Arizona, from 1990 September to 1991 August, at 8:00 am local standard time.

projected motion is greatest at the solstices, when the Sun's motion is parallel to the equator, and smallest at the equinoxes, when the eastward motion is reduced by a factor $\cos(23.5°) \approx 0.917$. As a result, the apparent solar day tends to be longer at the solstices (late June and late December) than at the equinoxes (late March and late September).

The contribution of the obliquity of the ecliptic to the equation of time is shown as the dashed line in Figure 1.12. The other contribution to the equation of time is due to the fact that the angular speed of the Earth on its orbit is not constant. As discussed in detail in Section 2.5, the Earth's orbit is not perfectly circular but is a mildly eccentric ellipse. The angular speed of the Earth is greatest when the Earth is at its closest approach to the Sun; this occurs near the beginning of January. The motion of the Sun, as seen from Earth, will thus be largest in January and smallest in July, six months later, when the Earth is at its greatest distance from the Sun. The contribution of the changing angular speed to the equation of time is shown as the dotted line in Figure 1.12. When the contributions of the obliquity of the ecliptic and the variable angular speed are added together, they produce the observed equation of time (the solid line in the figure).

Even after switching from apparent to mean solar time, a remaining difficulty is that time is defined locally, not globally. **Local noon**, defined as the instant when the center of the Sun makes an upper transit, is different for observers at different longitudes. For every degree of longitude that you travel westward, local noon occurs 4 minutes later. Prior to the nineteenth century, when it took far longer than 4 minutes to travel one degree in longitude, this was not a problem. However, the advent of high-speed communication (the telegraph) and high-speed transportation (the railways) raised a problem. To prevent railway conductors from having to reset their watch at each station, every railway company adopted a standard time, usually that of the company headquarters or the largest city serviced by the company. This resulted in unpleasant chaos at railroad stations served by more than one railway company.

In 1883, the major railway companies of the United States and Canada simplified matters by adopting **time zones**, within which all clocks would strike noon simultaneously. The adoption of time zones has since spread throughout the world. Each time zone is nominally 15° wide, but adjusted locally along political boundaries; the time within each time zone is called the "civil time" and can vary significantly from the local mean solar time, especially in broader time zones. Since the civil time increases by one hour for each time zone you travel to the east, there must be a boundary drawn from the north pole to the south at which the civil time jumps backward by 24 hours as you travel to the east. Otherwise, the civil time would be multiple-valued as it wound around and around and around the globe. This boundary is called the **International Date Line**. The International Date Line is based on the meridian opposite the Prime Meridian, but jogs back and forth to ensure that the division line doesn't pass through any nations.

Astronomers, and other scientists, frequently want to use a time measure that is independent of the observer's position on Earth. During the ninteenth century, when the meridian through Greenwich was adopted as the Prime Meridian, it was natural to use Greenwich Mean Time (GMT) as the universal standard, where GMT is defined as the mean solar time as measured at the Prime Meridian. Locally, mean solar time is then

$$\text{Mean Solar Time} = \text{GMT} + \ell_{\text{east}}, \tag{1.7}$$

where ℓ_{east} is the east longitude of the observer. Civil time, however, is

$$\text{Local Civil Time} = \text{GMT} + N_{\text{zone}} \times 1\,\text{hr}, \tag{1.8}$$

where N_{zone} is the integral number of time zones the observer is displaced eastward from the prime meridian.[23]

Until the twentieth century, the rotating Earth provided the ultimate basis for human measurements of time. However, the rotation rate of the Earth is not perfectly constant. The Moon's tidal effect (discussed in more detail in Section 4.2) slows the Earth's rotation rate by approximately 0.0016 s century^{-1}. In addition, the Earth's rotation rate varies seasonally because of changes in atmospheric and oceanic temperature and has irregular changes due to earthquakes, which make tiny changes to the Earth's moment of inertia. The Earth, in other words, is a clock that is "winding down," and is doing so at an irregular rate. In the twentieth century, atomic clocks were devised that measured time more accurately than the rotating Earth. The fundamental measure of time used today is therefore International Atomic Time (abbreviated TAI, from the French Temps Atomique International). In the SI system of units, the **second** is defined as 9,192,631,770 times the period of the radiation emitted by the hyperfine transition of the cesium-133 atom at absolute zero temperature. This definition was chosen so that the second was equal in length to 1/60 of 1/60 of 1/24 of a mean solar day, measured around the year AD 1900.

[23] The actual situation is complicated by the fact that some regions, such as India, Newfoundland, and central Australia, have time zones that are not offset by an integral number of hours.

Because of the slowing of the Earth's rotation, a mean solar day in AD 2000 was not $60 \times 60 \times 24 = 86{,}400$ s. Instead, it was 86,400.0016 s.

We thus have an inherent tension between time as measured by highly accurate atomic clocks and time as measured by the not-quite-as-accurate clock provided by the rotating Earth. To resolve this tension, scientists have adopted a reference time for Earth, called Coordinated Universal Time, or **UTC**.[24] In UTC, seconds are defined in accordance with the SI definition. UTC is synchronized with the (gradually slowing) mean Sun by occasionally interpolating a **leap second** when necessary to keep UTC within 0.9 seconds of the time measured by the mean Sun. Through the year 2008, a total of 34 leap seconds were required to align UTC with mean solar time.

One might ask how a spin-down rate of only 0.0016 s century^{-1} can lead to 34 leap seconds over a period of just over a century. This is because the effect of slowing is cumulative. Think of the Earth as being a clock that slows at a rate $\epsilon = 0.0016$ s century^{-1} = 4.4×10^{-8} s day^{-1}. If the length of a day is P_0, then during the first day we use the Earth as a clock, it loses a time $\epsilon P_0 = 4.4 \times 10^{-8}$ s. During the second day, however, it loses a time $2\epsilon P_0 = 8.8 \times 10^{-8}$ s, and in general, on the Nth day, it loses $N\epsilon P_0$. The total time lag after N days will be

$$\Delta t = \epsilon P_0 + 2\epsilon P_0 + \ldots + N\epsilon P_0 = \epsilon P_0 \sum_{i=1}^{N} i = \epsilon P_0 \frac{N(N+1)}{2}. \tag{1.9}$$

After a time t equal to many days has passed ($N \gg 1$), the time lag that must be filled in with leap seconds is

$$\Delta t \approx \epsilon P_0 \frac{N^2}{2} \approx \frac{\epsilon P_0}{2} \left(\frac{t}{1\text{ day}}\right)^2. \tag{1.10}$$

Thus, the time lag due to the Earth's spin-down is *quadratic* in t, not linear. Since there are 36,525 days in a century, the time lag expected, in seconds, is

$$\begin{aligned}\Delta t &\approx \frac{(4.4 \times 10^{-8}\text{ s day}^{-1})(1\text{ day})}{2} \left(\frac{36{,}525\text{ days}}{1\text{ century}}\right)^2 \left(\frac{t}{1\text{ century}}\right)^2 \\ &\approx 30\text{ s} \left(\frac{t}{1\text{ century}}\right)^2. \end{aligned} \tag{1.11}$$

This is only an approximate formula, because of the occasional small glitches in the Earth's rotation rate due to earthquakes. However, it gives the correct long-term trend: each century will require a greater number of leap seconds to keep the gradually lengthening mean solar day in synch with atomic clocks.

In addition to solar time, astronomers frequently find it useful to use an alternative time system, **sidereal time**. Because a sidereal day is the time between upper transits of a star other than the Sun, it represents the rotation period of the Earth relative to the

[24] English speakers wanted the abbreviation CUT; French speakers wanted TUC, for *temps universel coordiné*. UTC was chosen as the compromise.

distant fixed stars. A clock measuring sidereal time runs faster than a clock measuring mean solar time, by about 4 minutes per day.

Technically speaking, the **local sidereal time** (LST) is defined as the hour angle of the vernal equinox, which by definition has a right ascension $\alpha = 0$. Thus, when the vernal equinox makes an upper transit, the local sidereal time is 0^h. Local sidereal time is based on a 24-hour clock, running from 0^h to 24^h. If the vernal equinox is not above the horizon, the local sidereal time can be computed after measuring the hour angle (H) of a star with known right ascension (α):

$$\text{LST} = H + \alpha. \tag{1.12}$$

In practice, astronomers use this equation to compute the hour angle of a star with known right ascension at a particular local sidereal time.

1.6 ▪ CALENDARS

As mentioned in Section 1.4, having an accurate calendar is useful for an agrarian society. For a calendar to remain useful for agricultural purposes, it must remain in phase with the seasons of the year. That is, the Sun should return to the vernal equinox on the same calendar date each year. (In the calendar currently in use, that date happens to be March 21.) The interval of time that elapses between successive passes of the Sun through the vernal equinox is called the **tropical year**.[25] The length of the tropical year is 365.24219 mean solar days. Because of the precession of the equinoxes, the tropical year is slightly different in length from the **sidereal year**, which is the time it takes the Sun to make a complete circle of the ecliptic relative to the fixed background stars. The sidereal year, which is the orbital period of the Earth around the Sun, is 365.25636 days, or about 20 minutes longer than the tropical year.

The fact that the number of mean solar days in a tropical year, 365.24219, is not an integer led to a certain amount of difficulty when ancient cultures set up calendars. During the time of the Roman Republic, for instance, the Roman calendar contained 12 months adding up to only 355 days. It was the job of the board of pontifices (Roman priests) to interpolate an extra month when the calendar fell out of synchronization with the seasons. However, the priests were largely driven by nonastronomical considerations; they added the extra month when politicians friendly to them were in office, effectively extending their elected term, but omitted the month when their enemies were in power. By the time Julius Caesar became effective dictator of Rome, the Roman calendar was badly out of alignment with the seasons. In the year 46 BC, Caesar interpolated not one but three extra months to return the time of the vernal equinox to its traditional date in late March.[26] After consulting with an Alexandrian astronomer named Sosigenes, who

[25] It's called the "tropical" year because it is the time required for the Sun to go from being overhead at the Tropic of Cancer to being overhead at the Tropic of Capricorn and back again.

[26] Caesar called the year 46 BC, with its unusual length of 445 days, the *ultimus annus confusionis*, or "last year of confusion." Humorists in Rome emphasized the alternate meaning of the phrase: "the year of ultimate confusion."

was familiar with the 365-day calendar used by the Egyptians, Julius Caesar proclaimed a new calendar. In this **Julian calendar**, years ordinarily had 365 days; however, every fourth year, an extra day, called a **leap day**, was added. The Julian year thus has 365.25 days, on average; this is a fairly close approximation to the tropical year of 365.24219 days.

The initial small difference between the Julian year and the tropical year accumulated with time, amounting to one day every 128 years. By the sixteenth century, the vernal equinox fell on the date March 11, according to the Julian calendar. This caused problems for the Church, which computed the date of Easter using a formula devised in the fourth century that assumed that the vernal equinox occurred on March 21. Thus, the average date of Easter was gradually drifting later and later, relative to the true date of the equinox. Pope Gregory XIII foresaw, with displeasure, a future in which Easter fell during the summer, then during the fall, and eventually the winter. In the year 1582, therefore, Gregory issued a papal bull reforming the calendar. In October of that year, the calendar skipped 10 days, going straight from October 4 to October 15, and thus returning the date of the vernal equinox to March 21.

In addition, the papal bull announced a new algorithm for computing leap days; years evenly divisible by 4 would contain a leap day *unless* the year number was evenly divisible by 100 and not by 400. This means that the years 1600 and 2000 in the new **Gregorian calendar** were leap years, but that 1700, 1800, and 1900 were not. In 400 Gregorian years, there are 97 leap years and 303 regular years, totaling

$$N_{\text{Greg}} = 97 \times 366 + 303 \times 365 = 146{,}097 \text{ days}. \tag{1.13}$$

For comparison, 400 tropical years will contain

$$N_{\text{trop}} = 400 \times 365.24219 = 146{,}096.88 \text{ days}, \tag{1.14}$$

a difference that amounts to only 1 day in 3225 years (about 4% as large as the error in the Julian calendar). The accuracy of the Gregorian calendar eventually caused it to be adopted by members of all religions. Today all nations, even those that use other calendars for religious purposes, use the Gregorian calendar for business.

PROBLEMS

1.1 The Polynesian inhabitants of the Pacific reportedly held festivals whenever the Sun was at the zenith at local noon. How many times per year was such a festival held? At what time(s) of year was the festival held on Tahiti? At what time(s) of year was it held on Oahu? (Hint: any reputable world atlas will give you the latitude of Tahiti and Oahu. You may also find the information in Figure 1.13 useful.)

1.2 The right ascension and declination of the seven stars of the Big Dipper are given below.

Star	Right Ascension	Declination
Alkaid	13^h48^m	$+49°19'$
Mizar	13^h24^m	$+54°56'$
Alioth	12^h54^m	$+55°58'$
Megrez	12^h15^m	$+57°02'$
Phecda	11^h54^m	$+53°42'$
Merak	11^h02^m	$+56°23'$
Dubhe	11^h04^m	$+61°45'$

For what range of latitudes are all the stars of the Big Dipper circumpolar? What is the southernmost latitude from which all of the stars of the Big Dipper can be seen? For what range of latitudes are *none* of the stars of the Big Dipper ever seen above the horizon?

1.3 Columbus, Ohio, is in the Eastern Time Zone, for which the civil time is equal to the mean solar time along the 75° W meridian of longitude.

(a) Ignoring daylight saving time for the moment, are there any days of the year when civil noon (as shown by a clock) is the same as apparent local noon (as shown by the Sun) in the city of Columbus? If so, what day or days are they?
(b) Daylight saving time advances the clock by one hour from the second Sunday in March to the first Sunday in November ("Spring forward, fall back"). When daylight saving time is in effect, are there any days of the year when civil noon is the same as apparent local noon in the city of Columbus? If so, what day or days are they?

1.4 Suppose you've been granted access to a large telescope during the last week in September. One of the two objects you want to observe is in the constellation Virgo; the other is in the constellation Pisces. You only have time to observe one object: which should you choose? Please explain your answer.

1.5 In *The Old Man and the Sea*, Hemingway described the old man lying in his boat off the coast of Cuba, looking up at the sky just after sunset: "It was dark now as it becomes dark quickly after the Sun sets in September. He lay against the worn wood of the bow and rested all that he could. The first stars were out. He did not know the name of Rigel but he saw it and knew soon they would all be out and he would have all his distant friends." Explain what is astronomically incorrect about this passage. (Hint: what are the celestial coordinates of the star Rigel?)

1.6 (a) Consider two points on the Earth's surface that are separated by 1 arcsecond as seen from the center of the (assumed to be transparent) Earth. What is the physical distance between the two points?
(b) Consider two points on the Earth's equator that are separated by 1 second of time. What is the physical distance between the two points?

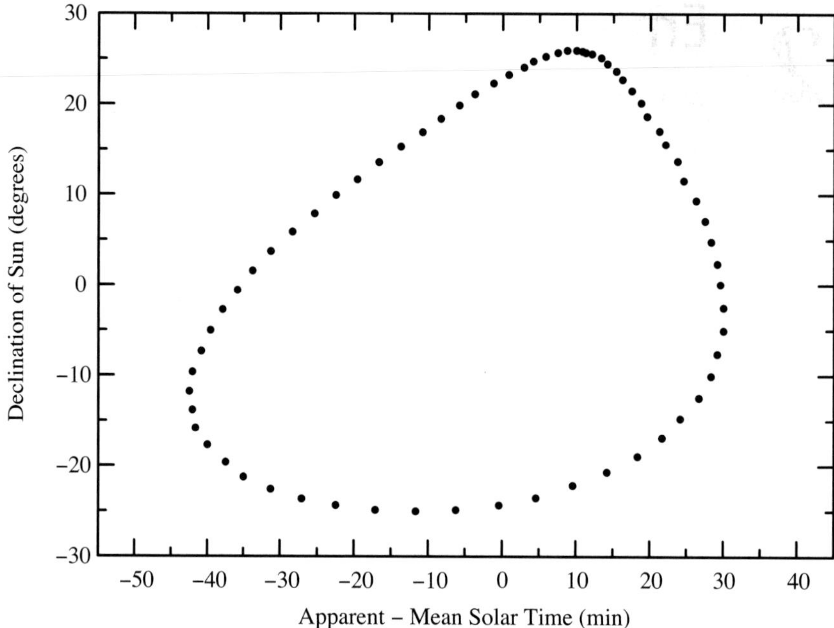

FIGURE 1.15 The analemma of the planet Mars.

1.7 The bright star Mintaka (also known as δ Orionis, the westernmost star of Orion's belt) is extremely close to the celestial equator. Amateur astronomers can determine the field of view of their telescope (that is, the angular width of the region that they can see through the telescope) by timing how long it takes Mintaka to drift through the field of view when the telescope is held stationary in hour angle. How long does it take Mintaka to drift through a 1° field of view?

1.8 (a) Imagine that technologically advanced, but highly mischievous, space aliens have reduced the tilt of the Earth's rotation axis from 23.5° to 0°, while leaving the Earth's orbit unchanged. Sketch the analemma in this case.

(b) Now imagine that the aliens have restored the axial tilt to its previous value of 23.5° but have changed the Earth's orbit to a perfect circle, with the Earth's orbital speed being constant over the course of a year. Sketch the analemma in this case.

(c) The martian analemma is shown in Figure 1.15. What is the tilt of the rotation axis of Mars?

1.9 How many square degrees are on the complete celestial sphere?

2 Emergence of Modern Astronomy

Modern astronomy has deep historical roots. The main path of development for astronomy begins with the ancient Babylonians. Greek astronomers built on the observations of the Babylonians, creating a science of astronomy that was mathematical and deductive in nature. Ancient knowledge about the heavens was preserved and expanded during medieval times by Arabic scientists. During the Renaissance, the heliocentric theory of Copernicus led to additional advances by scientists such as Galileo and Kepler. This lineage, Babylonians to Greeks to Arabs to Europeans, is a great oversimplification of the rich history of astronomy. However, in a single chapter, we have only enough space for a broad overview of how modern astronomy evolved.

2.1 ▪ EARLY GREEK ASTRONOMY

Of the nine muses of classical mythology, eight dealt with various forms of music, dance, and poetry; the ninth muse, Urania, was the Muse of Astronomy. This is indicative of the ancient Greek approach to astronomy: the motions of Sun, Moon, and planets were regarded as a type of cosmic dance, revealing an underlying rhythm and harmony. A main goal of ancient Greek astronomers was to build, using deductive reasoning and mathematical computations, a conceptual model for the universe that explained the (sometimes complicated) motions of celestial bodies. To provide a bit of clarification, when historians of science talk about "ancient Greek astronomy," they aren't talking solely about developments in the geographical region currently called Greece. Rather, they embrace the entire Greek-speaking world, which in Hellenistic times, after the conquests of Alexander the Great, embraced much of the Mediterranean basin and the Near East.

Our knowledge of Greek astronomy, particularly in the time prior to Aristotle, is sadly fragmentary, due to the incompleteness of the written record. Many early Greek astronomical works are lost and are known to have existed only because they were cited by later writers. Some general aspects of Greek astronomy are well established, however. For instance, the Greeks were the first known culture to realize that the sky is three-dimensional; that is, it has a significant depth. Earlier societies, such as the Babylonians and Egyptians, thought that the sky was a thin, solid dome, arching over a flat Earth. The most famous written description of such a domed universe is in the first book of Genesis:

"God made the firmament, and divided the waters which were under the firmament from the waters which were above the firmament; and it was so. And God called the firmament heaven."[1] Greek astronomers, however, realized that the Sun and Moon, instead of being disks stuck to a domed sky, were spherical objects, at different distances from the Earth.

The realization that space was three-dimensional led Greek astronomers to an understanding of various celestial effects. For instance, they correctly explained the causes of the **phases of the Moon**. During the course of 29.5 days, the Moon appears to change in shape against the sky (see Figure 4.10b, for instance). The Moon wanes from a full circle on the sky (the full Moon) through gibbous and crescent phases until it seems to disappear (the new Moon). It then waxes through the crescent and gibbous phases until it reaches full Moon again, 29.5 days after the previous full Moon. The ancient Greeks realized that the phases occur because the Moon is an opaque sphere illuminated by the Sun. As the Moon orbits the Earth, we see different fractions of the illuminated hemisphere of the Moon, causing the apparent change in shape.[2]

The Greeks also realized the cause of **eclipses**. During a lunar eclipse, the Moon darkens dramatically; this is because the Moon passes through the Earth's shadow, depriving it of the sunlight that usually illuminates the Moon's surface. During a solar eclipse, the Sun darkens dramatically; this is because the opaque Moon passes between the Earth and the Sun, blocking the sunlight that usually reaches the Earth's surface. Thus, Greek astronomers realized that the Sun is farther away from us than the Moon is.[3]

Aristotle (384–322 BC) was one of the great philosophers and scientists of the Greek world. In his work *On the Heavens,* written around 350 BC, he turned his attention to astronomy. In this work, Aristotle pointed out that the Earth was spherical and gave four physical reasons, based on observation, why this must be true. His first reason was based on his observations of how gravity works: since gravity draws dense materials toward the center of the Earth, the resulting compression must squeeze the Earth's substance into as compact a form as possible—which is a sphere. His second reason was based on observations of partial lunar eclipses: when the edge of the Earth's shadow falls on the Moon, it always forms an arc of a circle. The only object that *always* casts a circular shadow is a sphere; thus, the Earth must be spherical.

His third reason was based on observing that new stars appear above the horizon when you head south toward the equator: on a spherical Earth, observers at a latitude ℓ north of the equator cannot see stars with declination $\delta < -90^\circ + \ell$. To take an example known in ancient times, the star Canopus ($\delta \approx -53^\circ$) is invisible from Athens ($\ell \approx 38^\circ$ N) but is visible from Alexandria, in Egypt ($\ell \approx 31^\circ$ N). This showed that the Earth was curved in the north–south direction, as a sphere would be. His fourth reason was based on observing elephants: since elephants existed both in Morocco, the westernmost region known to Aristotle, and in India, the easternmost region known to him, the two regions

[1] The image portrayed is more graphic in the original Hebrew; the word translated as "firmament" in the King James translation is *raqia*, which means a metal sheet or bowl that has been hammered out of a solid ingot.

[2] The Moon's phases are discussed in more detail in Section 4.4.

[3] Eclipses are discussed in more detail in Section 4.6.

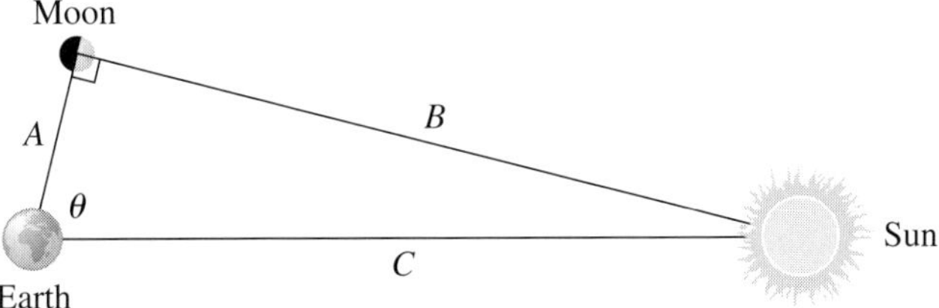

FIGURE 2.1 The geometrical method of Aristarchus for determining the relative distances to the Moon and to the Sun. (Not to scale.)

must actually be adjacent to each other on the spherical surface of the Earth. (This last, elephant-based reason sounds absurd to modern ears, but it's actually an illustration of how you can arrive at the right answer for the wrong reasons.)

The astronomer Aristarchus (ca. 310–230 BC) was notorious in his day for his unprecedented belief that the Earth orbits the Sun, rather than vice versa. The only surviving book of Aristarchus, *On the Sizes and Distances of the Sun and Moon,* doesn't explicitly mention his **heliocentric** (Sun-centered) model for the universe; instead, it puts forward geometric methods for determining the relative distances to the Sun and Moon, and their relative sizes. Aristarchus realized that when we, on the Earth, see half the Moon's disk illuminated, then the Earth–Moon–Sun angle must be exactly 90°, as seen in Figure 2.1. When the Earth–Moon–Sun angle is 90°, then the ratio of the Earth–Moon distance A to the Earth–Sun distance C is

$$\frac{A}{C} = \cos\theta, \tag{2.1}$$

where θ is the measurable angle between the Sun and Moon as seen from the Earth. Unfortunately, the angle θ is difficult to measure with sufficient accuracy, since the difference between θ and 90° is tiny. Aristarchus thought the angle was $\theta = 87°$, which would give

$$C = A/\cos 87° = 19A. \tag{2.2}$$

However, the actual value of the angle is $\theta = 89.853°$, much closer to a right angle, which gives

$$C = A/\cos 89.853° = 390A. \tag{2.3}$$

Because of the difficulty of measuring θ with sufficiently high accuracy, Aristarchus underestimated the distance to the Sun, relative to that of the Moon, by a factor of 20.

Nevertheless, Aristarchus did correctly deduce that the Sun is much farther away than the Moon is. Since the Sun and the Moon are the same angular size as seen from Earth, we know from similar triangles that the ratio of their diameters is the same as the ratio of their distances from Earth. That is, Aristarchus thought that the Sun was 19 times bigger than the Moon in diameter (whereas, the Sun is actually 390 times bigger than the Moon). Aristarchus knew that the Moon was smaller than the Earth, since it fits inside the Earth's shadow during a total lunar eclipse. Moreover, he calculated, by further geometric arguments, that the diameters of Moon, Earth, and Sun had the approximate relative values 1:3:19. Again, although the exact numbers are wrong (they are actually closer to 1:4:390), Aristarchus correctly deduced that the Sun is much larger than the Earth, thus lending support to, or perhaps even inspiring, his heliocentric model for the universe. It seemed sensible to Aristarchus that the small Earth should go around the large Sun rather than the reverse.

Aristarchus deduced the relative sizes of the Moon, Earth, and Sun; absolute values for their sizes, in physical units, were provided by the work of Eratosthenes (276–195 BC), who served as the head librarian of the famous Library of Alexandria. Although the original works of Eratosthenes have been lost, a later textbook by the astronomer Cleomenes records the method by which Eratosthenes determined the circumference of the Earth.[4] Eratosthenes was aware that exactly at noon at the time of the summer solstice, the Sun was at the zenith as seen from the town of Syene (the modern city of Aswan, in upper Egypt).

On the same date, however, the Sun doesn't pass through the zenith as seen from Alexandria; instead, as shown in Figure 2.2, it is an angle α south of the zenith at noon. Eratosthenes measured the angle α and found it to be 1/50 of a full circle, or $\alpha = 7°12'$. At this point, Eratosthenes assumed that the Earth is spherical (he had read his Aristotle) and that the Sun is far enough away that the Alexandria–Sun line is effectively parallel to the Syene–Sun line. In that case, angle β in Figure 2.2 must be equal to angle α.[5] Since β, the angular distance between Alexandria and Syene, is equal to 1/50 of a full circle, the physical distance D between Alexandria and Syene must be 1/50 of the circumference of the Earth. That is,

$$C = 50D, \tag{2.4}$$

where C is the circumference of the Earth. The distance between Alexandria and Syene was known to be 5000 stades; the *stade* was a Greek unit of length, based on the length of the stadium in which foot races were held. This meant that the Earth's circumference was

$$C = 50 \times 5000 \text{ stades} = 250{,}000 \text{ stades}. \tag{2.5}$$

The length of the stade was not uniform throughout the ancient world, and historians of science have had a grand time debating the exact length of the stade used by Eratosthenes. Perhaps the most widely used stade at the time of Eratosthenes was the Attic, or Athenian, stade, equal in length to 185 meters. If we assume that Eratosthenes used the Attic stade, then the circumference that he computed was

$$C = 250{,}000 \text{ stades} \left(\frac{185 \text{ m}}{1 \text{ stade}}\right) = 4.6 \times 10^7 \text{ m} = 46{,}000 \text{ km}. \tag{2.6}$$

[4] Like most textbook writers, Cleomenes labored in humble obscurity; in fact, so obscure was Cleomenes that estimates of when he wrote his text range from 100 BC to AD 470.

[5] This equality is proved as Proposition 29 in Book I of Euclid's *Elements*, written ca. 300 BC.

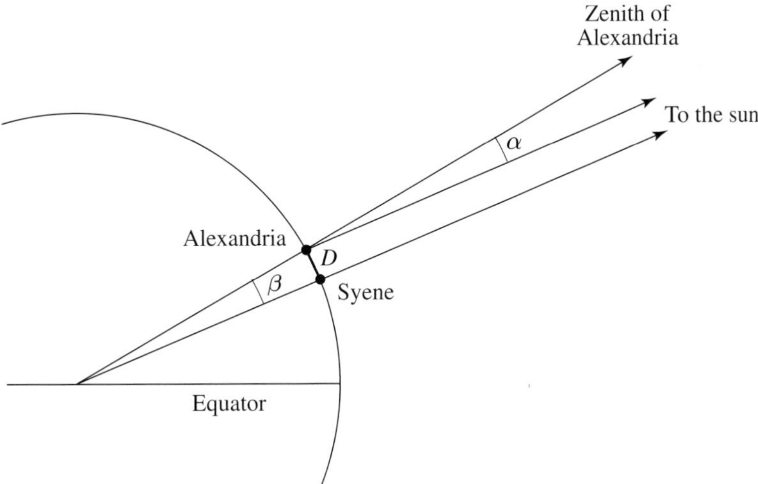

FIGURE 2.2 The geometrical method of Eratosthenes for determining the circumference of the Earth.

This is only 15% bigger than the correct value of 40,000 km. Thus, by the time of Eratosthenes, Greek astronomers not only knew the Earth is spherical but also had a reasonably correct idea of its size.

Hipparchus (ca. 190–120 BC) was perhaps the greatest astronomical observer during antiquity. Hipparchus is credited with a number of accomplishments:

- He produced an accurate catalog of hundreds of star positions. It was his careful observations that led Hipparchus to the discovery of the precession of the equinoxes, mentioned on page 14. He noted that the bright star Spica, which lies close to the ecliptic, was 6° west of the autumnal equinox. However, a star catalog made 150 years earlier had described Spica as being 8° west of the autumnal equinox. Hence, Hipparchus deduced that the equinoxes were slipping westward relative to Spica and the other stars at a rate of 2° per 150 years, equivalent to $48'' \text{ yr}^{-1}$; this is close to the accurate modern value of $50.3'' \text{ yr}^{-1}$.
- He established the **magnitude** system for describing the brightness of stars. He called the brightest stars "first magnitude," and then worked downward through second, third, fourth, and fifth magnitudes, all the way down to the faintest stars he could see, which were labeled "sixth magnitude." The more quantitative magnitude system that is used by astronomers today (described in more detail in Section 13.2) is based on that of Hipparchus.
- He computed a more accurate distance to the Moon. Although the original work of Hipparchus is lost, like so many works of Greek astronomy, a later commentator stated that Hipparchus found the average Earth–Moon distance to be roughly 70 times the Earth's radius. The actual average separation is 60.5 Earth radii.

- He measured the length of the tropical year with an error of less than 7 minutes. (Despite having such an accurate measure of the length of the year available, the Roman pontifices *still* botched their calendar!)

The observations of Hipparchus were the basis of the Ptolemaic model for the universe, which dominated Western astronomy for more than 14 centuries.

2.2 ▪ PTOLEMAIC ASTRONOMY

Claudius Ptolemaeus (called "Ptolemy" for short) lived and worked in Alexandria, Egypt, during the mid-second century AD. The scanty details that we know about his life come from his surviving astronomical books. His main work, which Ptolemy called *Mathematike Syntaxis* ("Mathematical Treatise") is better known by the name applied to it in the middle ages: the *Almagest*, a name that comes from an Arabic phrase meaning "the best." As you might guess from its flattering nickname, the *Almagest* was the most highly regarded astronomical work in the Western world from the time it was written until the sixteenth century.[6] The main portion of the *Almagest* is devoted to a geometrical model that describes the motions of the stars, Sun, Moon, and planets as seen from Earth. Before going into detail about Ptolemy's model, let's briefly review the motions of celestial bodies that he was attempting to explain.

- Stars move in diurnal circles about the celestial poles, with one complete circuit requiring one sidereal day. The stars are fixed in position relative to each other (this is only approximately true, but the relative motions of the stars are too gradual for the Greeks to have discovered).
- The Sun moves eastward relative to the stars along the ecliptic, which is tilted at 23.5° relative to the celestial equator. The average rate of motion is roughly 1° per day, but this varies over the course of a year.
- The Moon moves eastward relative to the stars along a path close to, but not identical with, the ecliptic. The average rate of motion is roughly 13° per day, but this varies over the course of a month.
- The planets Mercury, Venus, Mars, Jupiter, and Saturn usually move eastward relative to the stars, along a path close to the ecliptic; sometimes, however, they reverse course and move westward. An example of the **prograde** (eastward) and **retrograde** (westward) motion of Mars is shown in Figure 1.7.

Ptolemy's job was made unnecessarily complicated by the erroneous assumptions that he made. First, he assumed that the Earth was stationary at the center of the universe. In other words, the Ptolemaic model was **geocentric** (Earth-centered) rather than heliocentric (Sun-centered). Second, he assumed that celestial bodies moved in perfect circles at constant speed. This doctrine of **uniform circular motion** can be traced to early Greek

[6] From now on, all dates in this textbook will be AD, unless otherwise indicated.

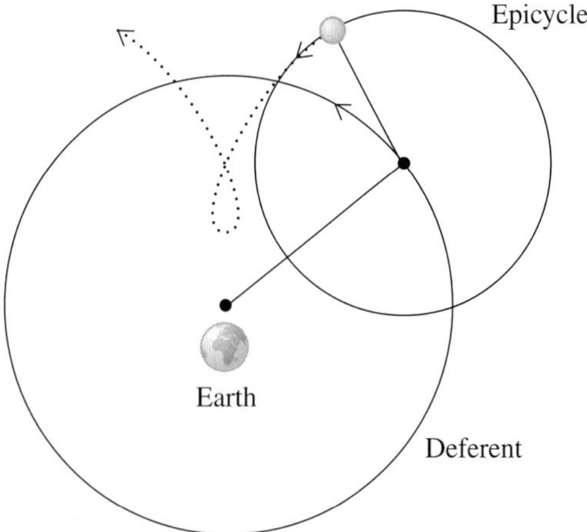

FIGURE 2.3 A planet moves at constant speed around the center of its epicycle, while the center of the epicycle moves at a constant speed along the deferent. The combination causes a model planet to move in a "loop-the-loop" motion.

philosophers such as Pythagoras and Plato. They believed that the heavens were perfect, in contrast to the obviously imperfect Earth, and that heavenly bodies must therefore move in circles (which were regarded as a perfect shape) at a perfectly constant speed.

Given these assumptions, explaining the apparent motions of the "fixed stars" was easy; Ptolemy assumed they were affixed to a rigid spherical shell, which rotated from east to west about the celestial poles, completing one rotation every sidereal day. Explaining the apparent motion of the Sun was more difficult. How could the nonuniform motion of the Sun along the ecliptic be reconciled with the dogma of uniform circular motion? Ptolemy followed the example of his predecessors by using a concept known as the *eccentric*. The Sun, Ptolemy assumed, moved along a circular orbit at a constant speed; however, the Earth was offset from the orbital center by a short distance. This small offset was referred to as the orbit's eccentric.[7] As the Sun moves along the orbit at a constant physical speed, its angular speed as seen from Earth is greatest when it's closest to the Earth, and smallest when it's farthest from the Earth. Ptolemy found that when he displaced the Earth from the orbital center by roughly 4% of the orbital radius, he could reproduce the observed motions of the Sun with fair accuracy.

Although the eccentric can describe an angular speed that varies with time, it cannot describe retrograde motion, in which the angular speed of a planet actually changes sign, rather than simply slowing down and speeding up. Ptolemy explained retrograde motion of a planet by using an **epicycle**, illustrated in Figure 2.3. In the epicyclic model, a planet travels at a constant speed around a circular path called an epicycle. At the

[7] The word "eccentric" literally means "away from the center"; thus, when you call a friend's behavior eccentric, that's another way of saying that he's a few standard deviations away from the mean.

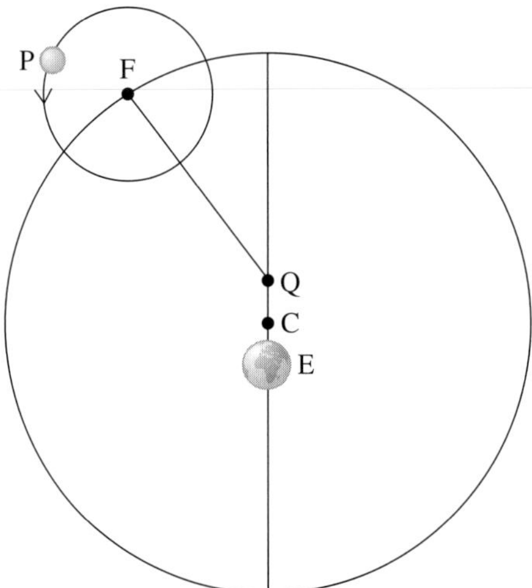

FIGURE 2.4 The complete Ptolemaic model for a planet's motion, including the equant.

same time, the center of the epicycle moves at a constant speed around the center of a larger circle called the **deferent**. The combination of an epicycle and a deferent can produce retrograde motion. Suppose the planet moves counterclockwise at a speed v around its epicycle, while the center of the epicycle moves counterclockwise at a speed w around its deferent, as shown in Figure 2.3. When the planet is at the outside of its epicycle, its speed relative to the center of the deferent is $v + w$; when it's at the *inside* of its epicycle, its speed is $v - w$. Thus, if $w > v$, the planet is actually moving backward (or in retrograde) when it is closest to the center of the deferent. A typical path traced out by a planet on an epicycle is shown in Figure 2.3. By fiddling with the sizes of eccentrics and epicycles, and by playing with the relative orbital speeds of epicycles and deferents, Ptolemy could get a fairly good fit to the observed motions of planets on the celestial sphere, but not quite good enough. His models were unable to match the observations exactly. Eccentrics, deferents, and epicycles were ideas that Ptolemy had inherited from previous Greek astronomers. However, in order to match the observations with the necessary accuracy, Ptolemy introduced a new device called the **equant**, illustrated in Figure 2.4.

In Ptolemy's new construction, the Earth (labeled E in the figure) is offset from the center of the planet's deferent (labeled C) by a small distance. Ptolemy dictated, however, that the center of the epicycle (labeled F) moved along the deferent at a changing physical speed, such that its angular speed would be constant as seen from the **equant point** (labeled Q). The equant (Q), orbital center (C), and Earth (E) lie along a straight line and are spaced so that the distance Q–C is equal to the distance C–E. The concept of the equant stretched the doctrine of uniform circular motion to the absolute limit; according to Ptolemy's critics, it stretched it beyond the limit. Many medieval astronomers were dissatisfied by the rather contrived notion of the equant.

Nevertheless, Ptolemy's complete model for a planetary orbit, including a deferent, epicycle, and equant, had enough adjustable parameters to enable Ptolemy to make quite accurate predictions of the motions of planets as seen from Earth. It is not clear that Ptolemy intended his complicated geocentric model to be an actual physical model of the cosmos. It worked adequately as a mathematical model, which accounted for its popularity during medieval times; people wanted reasonably accurate predictions of the locations of the Sun, Moon, and planets, which the Ptolemaic model provided. The fact that Ptolemy's model was geocentric also made it conceptually acceptable. There were a number of plausible arguments, during Ptolemy's time and later, why a geocentric model seemed correct:

- We cannot feel the motion of the Earth. A circumference of 250,000 stades implies a rotation speed at the Earth's equator of roughly 3 stades per second, or about 50 times the speed of the fastest sprinter. It seemed inconceivable that such a rapid speed should be imperceptible.
- The Earth's centrality and importance was somehow gratifying. (The Earth must be important; after all, *we* live on it.)
- **Stellar parallax** is not observed. This is the most serious scientific objection to a heliocentric model and deserves a fuller discussion, which is given below.

In general, the term **parallax** refers to the shift in apparent position of an object when seen from two different locations. For instance, if you hold up your thumb at arm's length and view it first through your right eye and then through your left, you will see your thumb's image jump from left to right by roughly 5° relative to objects in the background. In astronomy, the term **geocentric parallax** refers to the shift in apparent position of a relatively nearby object, such as the Moon or a planet, when seen from two different points on the Earth's surface. Geocentric parallax, illustrated in Figure 2.5a, is also referred to as **diurnal parallax**. If you want to observe geocentric parallax, you don't have to go on an expedition; during the course of 12 hours, the daily (or diurnal) rotation of the Earth will carry you through a distance $d = D \cos \ell$, where $D \approx 12{,}700$ km is the Earth's diameter and ℓ is your latitude. The closer an object is to the Earth, the larger its geocentric parallax will be. The Moon shifts in apparent position by as much as 2° when viewed from antipodal points on the Earth; however, the Sun's corresponding shift in apparent position is smaller by a factor of 390, since the Sun is 390 times farther away than the Moon is. Thus, the geocentric parallax of the Moon was easily measured by ancient astronomers (it's how Hipparchus measured the distance to the Moon, in fact), but the diurnal parallax of the Sun, and of the yet more distant stars, is too small to be measured by the naked eye.

The daily rotation of the Earth causes a change in position of an observer on the Earth; so does the annual revolution of the Earth around the Sun. **Heliocentric parallax** is the shift in apparent position of a relatively nearby star when seen from two different points on the Earth's orbit. Heliocentric parallax, illustrated in Figure 2.5b, is also referred to as **annual parallax**. If you want to observe heliocentric parallax, you don't have to launch

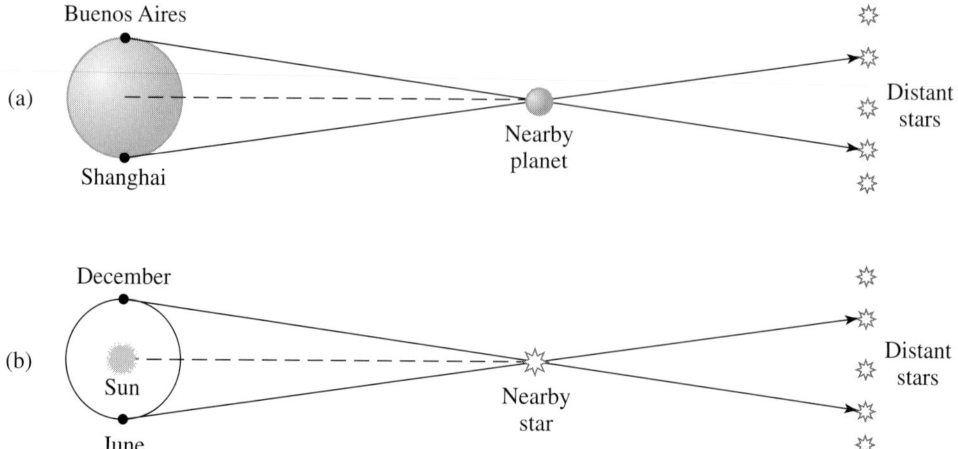

FIGURE 2.5 (a) Geocentric, or diurnal, parallax due to a change in position relative to the Earth's center. (b) Heliocentric, or annual, parallax due to a change in position relative to the Sun.

a spacecraft; during the course of half a year, the annual revolution of the Earth will carry you through a distance equal to the diameter of the Earth's orbit.

Before the invention of the telescope, astronomers attempted to measure the annual parallax of nearby stars but were unsuccessful. They recognized two possible explanations for the lack of detectable annual parallax: either the Earth was stationary or the stars were so far away that the annual parallax, like the diurnal parallax, was too small to be measured. Given the accuracy with which stellar positions could be measured in antiquity, Ptolemy and others deduced that if the solar system were heliocentric, then the nearest stars would have to be at a distance of *at least* a few thousand times the Earth–Sun distance. Such a large amount of empty space made astronomers uneasy. They preferred the more compact geocentric model. As we discuss further in Chapter 13, stellar parallax was not measured until long after the invention of the telescope. Even the Sun's nearest neighbors among the stars are at a distance of 270,000 times the Earth–Sun distance. The small, tidy Ptolemaic universe may have been psychologically comforting, but the universe is under no obligation to make us comfortable.

2.3 ▪ COPERNICAN ASTRONOMY

The Polish astronomer Nicolaus Copernicus (1473–1543) was the first scientist since antiquity to advance a heliocentric model for the universe. Copernicus was a Renaissance man metaphorically as well as chronologically; in addition to studying astronomy and mathematics, he also traveled to Italy in order to study medicine and law. After taking minor orders in the Church, he served in a variety of administrative positions. His work for the Church left Copernicus with enough time to make astronomical observations and work out his heliocentric model in detail. By the year 1514, Copernicus was

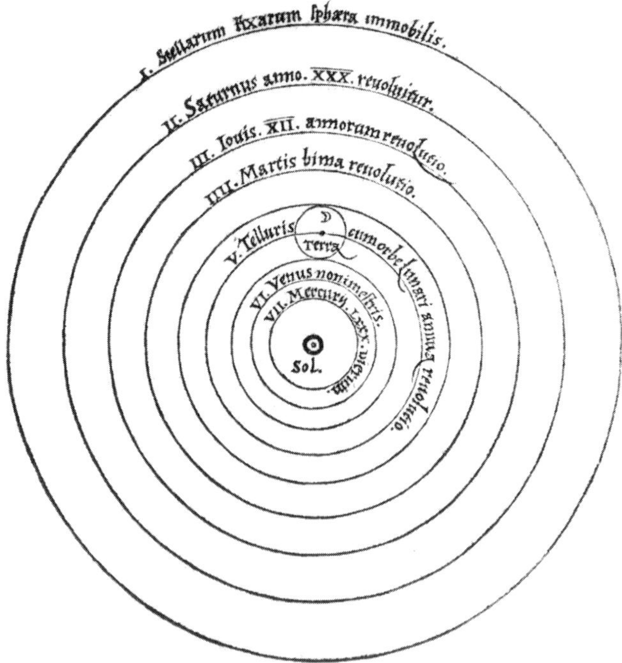

FIGURE 2.6 A schematic diagram of the heliocentric model, drawn by Copernicus (note that *Sol*, the Sun, is at center).

circulating a brief manuscript about his ideas among his friends; the grand summary of his work, the book *De Revolutionibus Orbium Coelestium* ("On the Revolutions of the Heavenly Spheres"), was not published until Copernicus was on his deathbed, in the year 1543.

The most radical aspect of the Copernican model was its insistence that the Sun, not the Earth, was at the center of the solar system (Figure 2.6), and that the Earth was both rotating about its axis and revolving about the Sun. The Copernican model, however, also had conservative aspects. For instance, Copernicus wholeheartedly embraced the dogma of uniform circular motion. One of his proudest claims for his heliocentric model was that it eliminated the need for equants (however, to match the observations, it still needed eccentrics and epicycles).

The Copernican model, although it retained eccentrics and epicycles, was conceptually simpler than the Ptolemaic model in many respects. In the Copernican model, retrograde motion of the planets is accounted for by the fact that inner planets move faster along their orbits than the outer planets do. Thus, as an inner planet, such as the Earth, overtakes an outer planet, such as Mars, the outer planet undergoes retrograde motion as seen from the inner planet. This is demonstrated graphically in Figure 2.7. In a heliocentric model, with the Earth being one of many planets orbiting the Sun, it

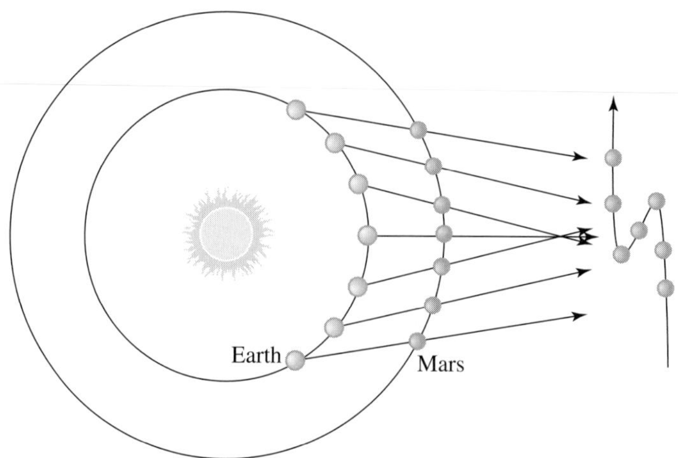

FIGURE 2.7 Explanation of the retrograde motion of Mars in a heliocentric system.

is useful to divide the planets into two groups, based on their distance from the Sun compared to that of the Earth:

- **Inferior planets** are those with orbits smaller than the Earth's orbit, that is, Mercury and Venus.
- **Superior planets** are those with orbits larger than the Earth's orbit. Mars, Jupiter, and Saturn were the superior planets known at the time of Copernicus; the planets Uranus and Neptune and the dwarf planets Ceres, Pluto, Haumea, Makemake, and Eris were not discovered until after the invention of the telescope.

In the Copernican model, the Earth is in motion around the Sun. Thus, for an Earthly observer, the positions of planets are measured from a reference frame that is co-rotating with the Earth–Sun line. It is particularly useful, as we shall see, to measure the position of planets on the celestial sphere relative to the Sun.

Some special positions of the superior planets relative to the Sun are shown in Figure 2.8. Names have been given to these special positions:

- **Opposition** occurs when the Earth lies between the Sun and the superior planet. That is, the Sun and planet are 180° apart on the celestial sphere as seen from the Earth.
- **Conjunction** occurs when the Sun lies between the Earth and the superior planet. That is, the Sun and planet are 0° apart as seen from the Earth.
- **Quadrature** occurs when the Sun and the superior planet are 90° apart as seen from the Earth. The quadrature can be either eastern, when the planet appears 90° east of the Sun on the sky, or western, when the planet appears 90° west of the Sun.

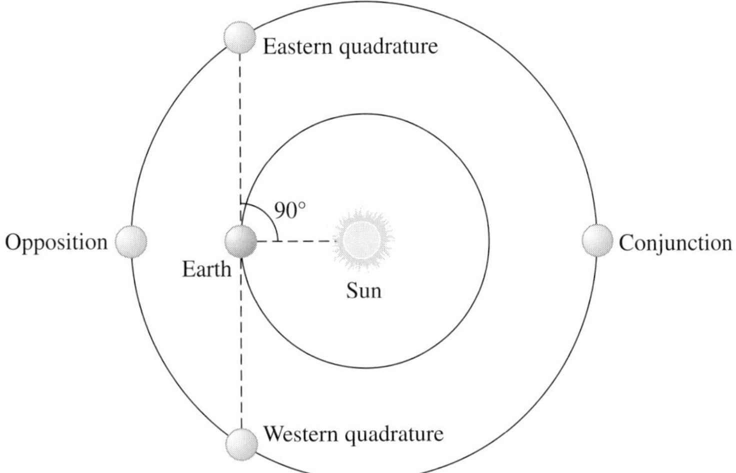

FIGURE 2.8 Configurations of superior planets. In this and following diagrams, we adopt a convention of looking down on the solar system from above the Earth's north pole.

Although inferior planets cannot be seen in opposition or in quadrature, they do have two different conjunctions, as shown in Figure 2.9:

- **Inferior conjunction** occurs when the inferior planet lies between the Earth and the Sun.
- **Superior conjunction** occurs when the Sun lies between the Earth and the inferior planet.

When a planet is not in conjunction, it is separated from the Sun on the celestial sphere by an angle θ referred to as the planet's **elongation**. Note from Figure 2.9 that an inferior planet can have the same elongation θ at two different distances from the Earth.

One of the happy results of the Copernican model is that it enabled Copernicus to compute the orbital periods of the planets, relative to the Earth's orbital period, and compute the size of planetary orbits, relative to the size of the Earth's orbit. Let's first see how Copernicus computed orbital periods, and then how he computed orbital sizes.

As seen from the Earth, planets undergo motion that can be described as periodic; that is, there is a fixed time interval between consecutive appearances of a particular planetary configuration. This time interval, known as the **synodic period** of the planet, can be found by measuring the time elapsed between successive conjunctions (for a superior planet) or the time elapsed between successive inferior conjunctions (for an inferior planet).[8] The synodic period is different from the **sidereal period** of the planet, which is the time

[8] The term "synodic" comes from the Greek word *synodos*, meaning a "coming together"—in this case, a coming together of the Sun and the planet when the planet is at conjunction.

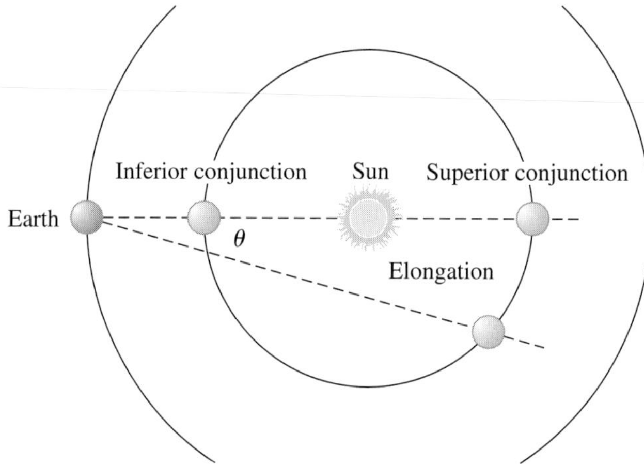

FIGURE 2.9 Configurations of inferior planets. When the planet is not in conjunction, the angle θ between the Sun and the planet, as seen from Earth, is the planet's elongation.

it takes the planet to complete one full circuit of the sky relative to the fixed stars. The synodic period of a planet is longer than its sidereal period for much the same reason that the solar day is longer than the sidereal day (as discussed in Section 1.5). As a reminder, the sidereal day is the Earth's rotation period in the nonrotating frame of reference of the distant stars (the sidereal frame); the solar day is the Earth's rotation period in a frame of reference co-rotating with the Earth–Sun line. Similarly, the sidereal period of a planet is the planet's orbital period in the nonrotating sidereal frame; the synodic period is its orbital period in a frame of reference co-rotating with the Earth–Sun line.

As in equation (1.1), let $\vec{\omega}_E$ be the angular velocity of the Earth's orbital motion in the sidereal frame; let $\vec{\omega}_P$ be the angular velocity of the planet's orbital motion in the same frame. Figure 2.10 shows the orbital motions of the Earth and an inferior planet; for an inferior planet, $\omega_P > \omega_E$. The difference between these two angular velocities is $\vec{\omega}_{syn}$, the angular velocity of the planet's orbital motion in the frame co-rotating with the Earth–Sun line. Specifically, we see that

$$\vec{\omega}_P = \vec{\omega}_E + \vec{\omega}_{syn}. \tag{2.7}$$

If $\vec{\omega}_P$ and $\vec{\omega}_E$ are parallel (that is, if the orbits of the Earth and the planet are coplanar and they orbit in the same direction about the Sun), we may write, for an inferior planet,

$$\begin{aligned}
\omega_P &= \omega_E + \omega_{syn} \\
\frac{2\pi}{P_P} &= \frac{2\pi}{P_E} + \frac{2\pi}{P_{syn}} \\
\frac{1}{P_P} &= \frac{1}{P_E} + \frac{1}{P_{syn}}.
\end{aligned} \tag{2.8}$$

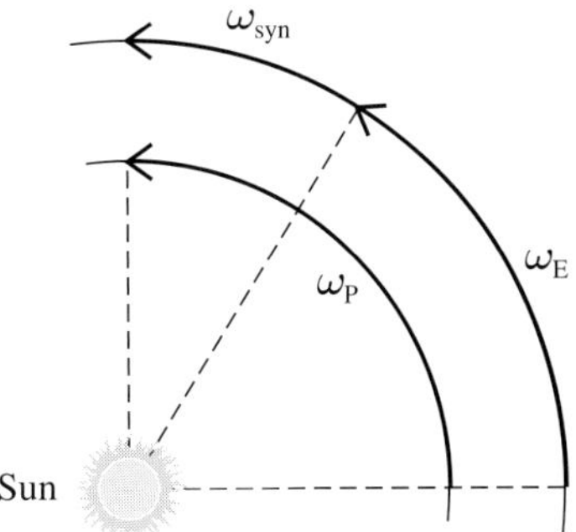

FIGURE 2.10 The angular speed of the Earth is ω_{E} and the angular speed of an inferior planet is ω_{P}. The difference between them is ω_{syn}, the angular speed of the planet in a reference frame that co-rotates with the Earth–Sun line.

In equation (2.8), P_{E} is the sidereal orbital period of the Earth, P_{P} is the sidereal orbital period of the inferior planet, and P_{syn} is the synodic orbital period of the inferior planet, as seen from Earth. As an example of an inferior planet, consider Venus. The synodic period of Venus is measured to be $P_{\text{syn}} = 583.92$ days. The Earth's sidereal orbital period is $P_{\text{E}} = 365.256$ days.[9] We can then compute the sidereal period of Venus:

$$P_{\text{Venus}} = \left[\frac{1}{365.256\text{ days}} + \frac{1}{583.92\text{ days}}\right]^{-1} = 224.70\text{ days}. \qquad (2.9)$$

In the case of a superior planet, $\omega_{\text{P}} < \omega_{\text{E}}$. If we refer to Figure 2.11, we see that $\vec{\boldsymbol{\omega}}_{\text{syn}}$ is in the opposite sense to $\vec{\boldsymbol{\omega}}_{\text{E}}$ and $\vec{\boldsymbol{\omega}}_{\text{P}}$. Equation (2.8) then becomes, for a superior planet,

$$\begin{aligned} \omega_{\text{P}} &= \omega_{\text{E}} - \omega_{\text{syn}} \\ \frac{1}{P_{\text{P}}} &= \frac{1}{P_{\text{E}}} - \frac{1}{P_{\text{syn}}}. \end{aligned} \qquad (2.10)$$

As an example of a superior planet, consider Mars. The synodic period of Mars is measured to be $P_{\text{syn}} = 779.95$ days. Given the length of the sidereal period of Earth, $P_{\text{E}} = 365.256$ days, we compute the sidereal period of Mars to be

$$P_{\text{Mars}} = \left[\frac{1}{365.256\text{ days}} - \frac{1}{779.95\text{ days}}\right]^{-1} = 686.98\text{ days}. \qquad (2.11)$$

In addition to permitting a determination of a planet's sidereal orbital period, the Copernican model also permits us to compute the distance of each planet from the Sun. For an inferior planet, this computation is straightforward. We need only measure the

[9] Remember that due to the precession of the equinoxes, the sidereal year is slightly longer than the tropical year of $P = 365.24219$ days.

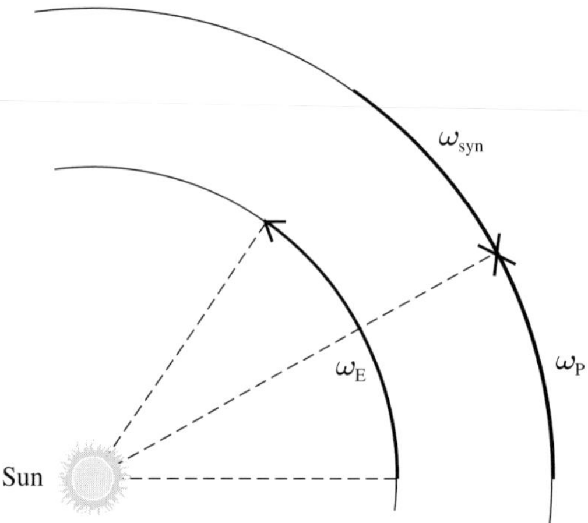

FIGURE 2.11 The angular speed of the Earth is ω_E and the angular speed of a superior planet is ω_P. The difference between them is ω_{syn}. To an observer on Earth, the angular velocity ω_{syn} of a superior planet is negative.

inferior planet's **greatest elongation**, that is, the maximum angular separation between the planet and Sun as seen from the Earth. As shown in Figure 2.12, if we approximate the orbit of the inferior planet as a perfect circle, then greatest elongation occurs when the line of sight from the Earth to the planet is exactly tangent to the planet's orbit. When that happens, the angle Earth–planet–Sun is a right angle, as the figure shows. The distance B from the planet to the Sun is then given by the relation

$$B/C = \sin\theta, \tag{2.12}$$

where θ is the angle of greatest elongation and C is the Earth–Sun distance. This method, therefore, only gives the radius of the planet's orbit in units of the Earth–Sun distance. The average distance from the Earth to the Sun is of such importance to astronomers that it is given the name **astronomical unit**, or **AU** for short. Copernicus, like the Greek astronomers before him, did not have an accurate knowledge of the absolute length of the astronomical unit.[10] However, he did know the *relative* sizes of the planets' orbits. For instance, the greatest elongation of Venus is $\theta = 46°$, so its orbital radius is

$$B = (\sin 46°)(1\ \text{AU}) = 0.72\ \text{AU}. \tag{2.13}$$

The size of the orbits of superior planets can be determined by a similar but slightly more complicated method. First, we must measure the time interval τ between opposition and eastern quadrature of the superior planet. As shown in Figure 2.13, the angle swept out by the Earth during the time interval τ is $\omega_E\tau$, where ω_E is the angular speed of the Earth's orbital motion. Over the same time interval, the superior planet (assumed to

[10] We now know that 1 AU = 149,597,870.7 km.

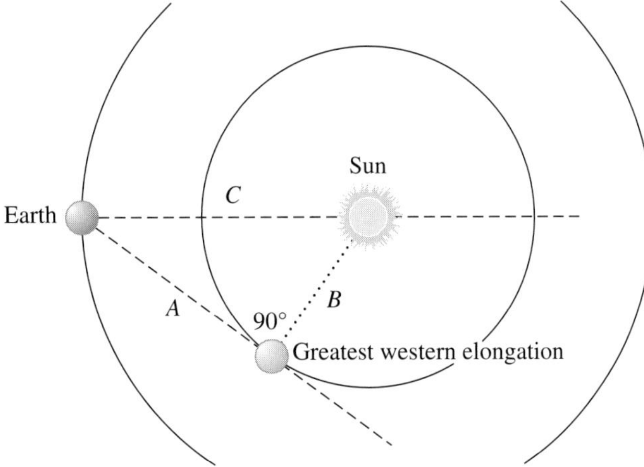

FIGURE 2.12 Measurement of the greatest elongation θ of an inferior planet allows determination of its distance B from the Sun.

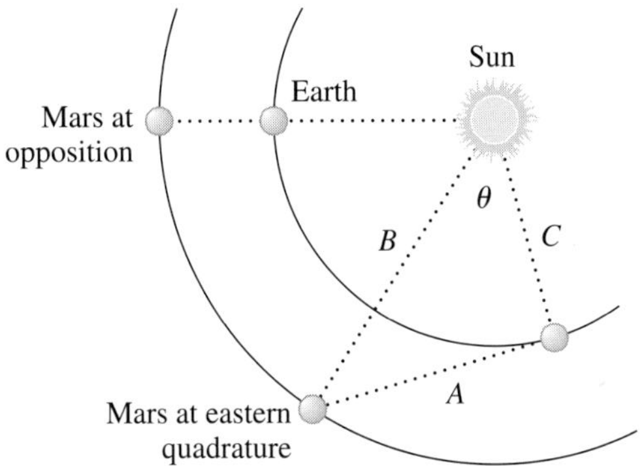

FIGURE 2.13 In the time τ between opposition and eastern quadrature, Mars sweeps out an angle $\omega_{\text{Mars}}\tau$ and the Earth sweeps out an angle $\omega_{\text{E}}\tau$. The difference between these angles is θ, with $\cos\theta = C/B$.

be Mars in the figure) sweeps out an angle $\omega_{\text{P}}\tau$, where ω_{P} is the angular speed of the planet's orbital motion. The difference between these angles is the angle $\theta = (\omega_{\text{E}} - \omega_{\text{P}})\tau$ shown in the figure. When Mars is at quadrature, the angle Mars–Earth–Sun is a right angle, so we have the relation

$$C/B = \cos\theta, \tag{2.14}$$

where C is the Earth–Sun distance and B is the Mars–Sun distance. In the case of Mars, the time from opposition to eastern quadrature is $\tau = 107$ days. Thus, the angle θ is

TABLE 2.1 Planetary Orbits

Planet[a]	Sidereal Period (years)	Orbital Radius (AU)
Mercury	0.2408	0.3871
Venus	0.6152	0.7233
Earth	1.000	1.000
Mars	1.881	1.524
Ceres	4.599	2.766
Jupiter	11.863	5.203
Saturn	29.447	9.537
Uranus	84.017	19.189
Neptune	164.79	30.070
Pluto	247.92	39.482
Haumea	283.28	43.133
Makemake	306.17	45.426
Eris	559.55	67.903

a. Dwarf planets in italics.

$$\begin{aligned}\theta &= \left(\frac{2\pi}{P_{\rm E}} - \frac{2\pi}{P_{\rm Mars}}\right)\tau \\ &= 2\pi\left(\frac{1}{365.256\ \text{days}} - \frac{1}{686.98\ \text{days}}\right)(107\ \text{days}) \\ &= 0.862\ \text{rad}\left(\frac{180^\circ}{\pi\ \text{rad}}\right) = 49^\circ. \end{aligned} \tag{2.15}$$

The distance from Mars to the Sun is then

$$B = \frac{C}{\cos\theta} = \frac{1\ \text{AU}}{\cos 49^\circ} = 1.52\ \text{AU}. \tag{2.16}$$

Table 2.1 shows the sidereal orbital period and the orbital radius for each of the planets and dwarf planets in the solar system (including those that were unknown at the time of Copernicus).[11]

[11] Truth in advertising: the simple calculations we have done in this section assume that planetary orbits are perfectly circular. Although this is a good first approximation, the orbits are actually ellipses, and what we call the "orbital radius" in Table 2.1 is actually the semimajor axis of the ellipse.

2.4 ▪ GALILEO: THE FIRST MODERN SCIENTIST

Both the Ptolemaic and Copernican models could explain the observed motions of the Sun, Moon, and planets on the celestial sphere. Why, then, should one believe that the Earth is in motion rather than the Sun? We know now that the Earth does orbit the Sun rather than vice versa, but direct experimental proof of the Earth's orbital motion was not provided until the eighteenth century, nearly two centuries after the death of Copernicus. Nevertheless, the heliocentric model came to be accepted without direct proof. This was partly because of its elegant simplicity; the motions of the planets are less complicated in a heliocentric model than in a geocentric model. This is an application of the general principle often referred to as **Occam's Razor.**[12] In its typically quoted form, Occam's Razor states that "the simplest description of Nature is most likely to be most nearly correct." In other words, unnecessary complications should be "shaved away" from a theory. Of course, when using a razor, it is important not to cut too deep; Albert Einstein is said to have rephrased Occam's Razor in the form "Everything should be made as simple as possible . . . but not simpler."

In addition to the aesthetic appeal of the heliocentric model's relative simplicity, compelling indirect evidence for heliocentrism was provided by the telescopic observations of Galileo Galilei (1564–1642). Galileo is sometimes called the first modern experimental physicist. Instead of relying purely on the pronouncements of Aristotle, Galileo tried to understand how nature works by carrying out experiments, such as swinging pendulums back and forth, and sliding weights down inclined planes. Although Galileo didn't invent the telescope, he was among the first individuals to use a telescope as a scientific instrument. The actual inventor of the telescope may possibly have been a Dutch optician called Hans Lippershey. In October 1608, Lippershey applied for a patent on what he called a *kijker*, or "looker" in English. The patent was denied by the Dutch government, however, on the grounds that "many other persons had a knowledge of the invention." Indeed, news of the telescope reached Galileo in Italy as early as May 1609; soon thereafter, he built several telescopes, each superior to the one before.

Although Galileo's telescopes had apertures of only an inch or two, they provided Galileo with many important observations. Galileo, knowing the potentially revolutionary impact of his discoveries, rushed into print in March 1610 with a pamphlet entitled *Sidereus Nuncius* ("Starry Messenger"). Many of Galileo's observations were startling to his contemporaries:

- The Moon is not smooth and perfect. Instead, as Galileo wrote, it is "uneven, rough, and crowded with depressions and bulges. And it is like the face of the Earth itself, which is marked here and there with chains of mountains and depths of valleys." In other words, there is not a vast difference between the Earth's surface and that of a celestial object, namely the Moon.

[12] Occam's Razor is named after William of Occam, a fourteenth-century friar and logician.

FIGURE 2.14 Galileo's illustration of the four bright satellites of Jupiter (the four asterisks), shown relative to Jupiter itself (the central disk).

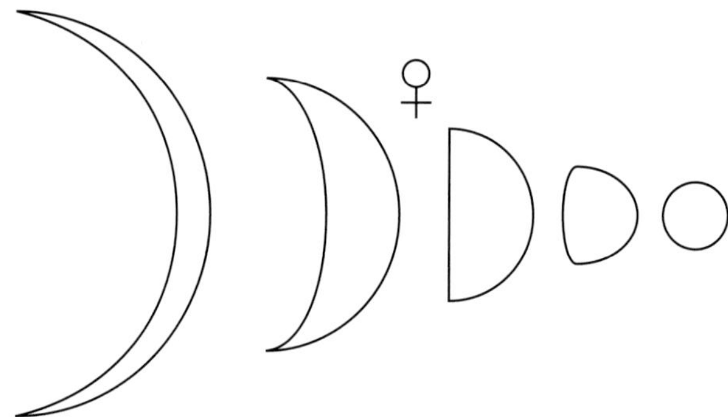

FIGURE 2.15 The phases and relative angular size of Venus, from crescent to full.

- The Milky Way, the nebulous band of light that extends around the sky, actually consists of numerous faint stars. "To whatever region of it you direct your spyglass," Galileo wrote, "an immense number of stars immediately offer themselves to view."
- Through a telescope, stars remain unresolved points, but planets show as disks. As Galileo put it, "the planets present entirely smooth and exactly circular globes that appear as little moons." (Unfortunately for astronomers, even the nearest stars are too distant to be resolved with conventional telescopes, even telescopes much larger than Galileo's.)
- The planet Jupiter has four large, bright satellites. Although Galileo called these satellites the "Medicean Stars," in honor of Cosimo de Medici, Grand Duke of Tuscany, later astronomers named them the **Galilean satellites**. The individual names of the four Galilean satellites are Io, Europa, Ganymede, and Callisto.

The Galilean satellites of Jupiter, shown in Figure 2.14 were an indirect piece of support for the Copernican system. One objection to a heliocentric model was that it required multiple centers of motion; the Earth went around the Sun while the Moon went around the Earth. This was regarded as more complex than a geocentric model in which everything goes around the Earth. However, Galileo provided clear evidence that there *had* to be multiple centers of motion; obviously, the Galilean satellites were going around Jupiter, regardless of whether Jupiter was going around the Sun or around the Earth.

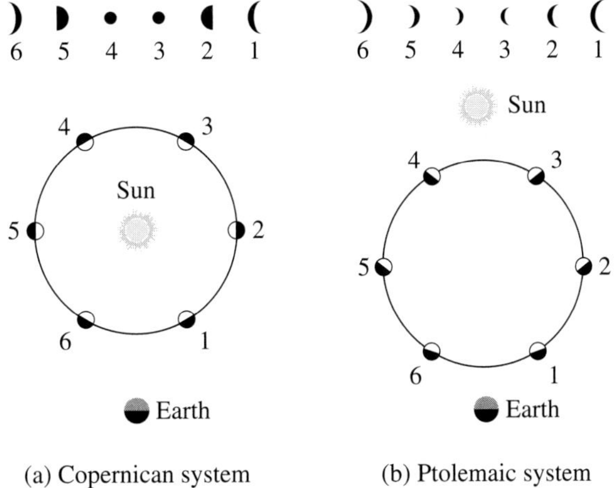

FIGURE 2.16 (a) The phases of Venus in the Copernican model. (b) The phases of Venus in the Ptolemaic model.

By the end of the year 1610, Galileo made another telescopic discovery that further undermined the Ptolemaic model. He found that Venus went through all the phases that the Moon did, from full to new. Moreover, he found that the angular size of Venus was smallest when it was full and largest when it was a thin crescent. The phases of Venus, illustrated by Galileo in his later work *Il Saggiatore*, are shown in Figure 2.15. Ptolemy, in his geocentric system, had the task of explaining why Venus should always lie within 46° of the Sun if the two bodies were on independent orbits around the Earth. Ptolemy managed it by saying that the center of Venus's epicycle always lies directly between the Earth and the Sun (as shown in Figure 2.16b) and that the epicycle is big enough to subtend an angle of 92° as seen from the Earth. The geometry of this situation requires that we see primarily the nighttime side of Venus, that is, the side away from the Sun. In the Ptolemaic system, then, we would always see a new or crescent phase for Venus, as illustrated in Figure 2.16b, top.

Galileo demonstrated, however, that we see gibbous and full Venuses, as well as crescent and new Venuses. This is easily explained in the Copernican system, as shown in Figure 2.16a. In the Copernican model, the sunlit side of Venus is turned toward us when Venus is at superior conjunction; this is when Venus is at its greatest distance from Earth, and hence has its smallest angular size. Conversely, the nighttime side of Venus is turned toward us when it is at inferior conjunction, when it is closest to Earth and has its largest angular size.[13] This is in accord with the observations of Galileo.

[13] Tantalizingly, when Venus is in its crescent phase, it is just under an arcminute across and thus can *almost* be resolved by the human eye. If our eyes were a bit better, or Venus were a bit larger, the phases of Venus would have been seen before the invention of the telescope, thus altering the course of astronomical history.

2.5 ▪ KEPLER'S LAWS OF PLANETARY MOTION

As increasingly accurate observations of planetary motions were made, the flaws of both the Ptolemaic and Copernican models became more evident. Tycho Brahe (1546–1601) was probably the greatest astronomical observer prior to the invention of the telescope; it was his observations of planetary motions that both revealed the inadequacy of the Copernican system and provided the necessary data for calculating the true nature of planetary orbits around the Sun. Tycho was a Danish aristocrat and received large sums of money from the King of Denmark to set up an elaborate observatory on the island of Hven, near Copenhagen. For more than 20 years, Tycho observed the positions of planets and stars with an accuracy of 1 arcminute. Interestingly, Tycho did not believe that the heliocentric model was correct. He noted, as did the Greeks before him, that the stars do not show parallax. The absence of parallaxes larger than 1 arcminute implies that the nearest stars must be farther away than a few thousand AU, given a heliocentric solar system. Tycho thought this distance was implausibly large and thus devised a compound system in which all the planets other than the Earth went around the Sun, while the Sun orbited the Earth, carrying its entourage of planets along with it.

In the year 1599, after a major falling-out with the Danish king, Tycho accepted a post as Imperial Mathematician to the Holy Roman Emperor in Prague. There he hired a new assistant named Johannes Kepler (1571–1630). Initially, Kepler was frustrated by Tycho's reluctance to share his data. However, Kepler soon had complete access to Tycho's observations; in October 1601, less than two years after Kepler arrived in Prague, Tycho died, and Kepler was appointed his successor as Imperial Mathematician. By using Tycho's observations of the planet Mars, and by doing several years' worth of calculations, Kepler was able to formulate a mathematical description of its orbit, and by extension, the orbits of other planets. His basic findings are encapsulated in **Kepler's laws of planetary motion**.

1. **Kepler's first law:** *Planets travel on elliptical orbits with the Sun at one focus.* The properties of the closed curve known as an **ellipse** are best described by explaining how to draw one (Figure 2.17). Take a piece of string and tie each end to a pin. Stick the pins into a piece of paper, separated by a distance less than the string's length. Use a pencil to stretch the string taut and draw a complete, closed curve; this is an ellipse. The two pins are located at the **foci** of the ellipse.[14] Expressed mathematically, the ellipse is the locus of points for which the sum of the distances to the foci is a constant (equal to the length of the string, in our graphic example). If the pins are moved closer together, for a given length of string, the ellipse becomes more nearly circular; if they are moved farther apart, the ellipse becomes more flattened.

   The longest distance across the ellipse (which passes through both foci) is called the **major axis**. The shortest distance across the ellipse, passing through the ellipse's center, is called the **minor axis**. The **semimajor axis** is half the major axis, and the **semiminor axis** is half the minor axis. The **eccentricity** of the ellipse

[14] "Foci" is the plural of the word "focus."

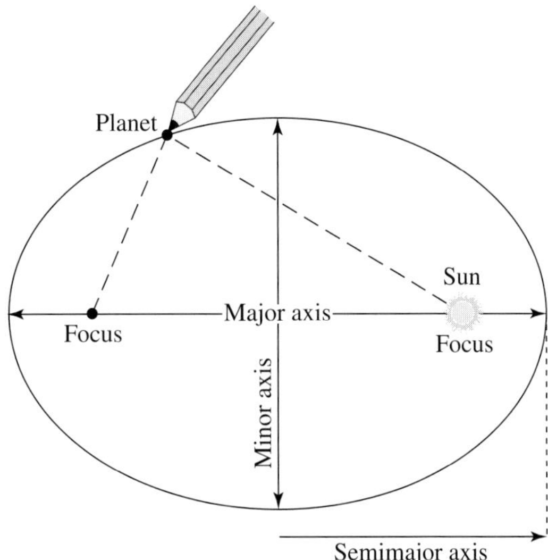

FIGURE 2.17 The properties of an ellipse.

is the distance between the foci divided by the length of the major axis. If the foci coincide, then $e = 0$, and the ellipse is a circle. The other limiting case, $e = 1$, represents the case in which the foci are separated by the full length of the string. It was quite a feat for Kepler to discover the elliptical shape of planetary orbits, since most planets have orbits with small eccentricity. Of the planets known to Kepler, Mercury had the largest eccentricity, $e = 0.21$; all the others had $e < 0.1$.

2. **Kepler's second law:** *A line drawn from the Sun to a planet sweeps out equal areas in equal time intervals.* This law provides a quantitative description of how the orbital speed of planets changes with their distance from the Sun; not only is motion not circular, Kepler discovered, it doesn't have uniform speed, either. The second law is graphically demonstrated in Figure 2.18. A mythical planet has its motion plotted during two time intervals, each 10 days long, separated by half the planet's orbital period. The two wedge-shaped areas swept out by the planet–Sun line are of equal area, even though they are of different shape. Kepler's second law implies that planets move most rapidly at **perihelion**, the point on their orbit closest to the Sun, and least rapidly at **aphelion**, the point farthest from the Sun.[15] As we show in Section 3.1, Kepler's second law is a simple consequence of the conservation of angular momentum.

3. **Kepler's third law:** *The squares of the sidereal orbital periods of the planets are proportional to the cubes of the semimajor axis of their orbits.* Kepler's third law

[15] Sometimes you hear "aphelion" pronounced as "ap-helion," sometimes as "af-felion." Both pronunciations can be found in reputable dictionaries.

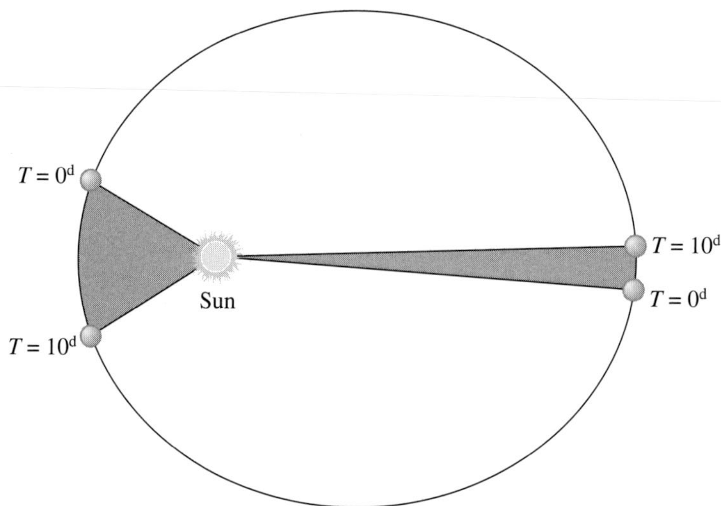

FIGURE 2.18 The area swept out by the planet–Sun line in each 10-day interval is identical.

can be expressed more compactly in mathematical notation:

$$P^2 = Ka^3, \tag{2.17}$$

where P is a planet's sidereal orbital period, a is the length of the semimajor axis of its orbit, and K is a constant. For objects orbiting the Sun,

$$K = 1\,\mathrm{yr}^2\,\mathrm{AU}^{-3}. \tag{2.18}$$

A plot of orbital period versus semimajor axis (like that of Figure 2.19) shows that all planets in the solar system, even those unknown to Kepler, follow his third law. In addition, Figure 2.19 shows that the Galilean satellites of Jupiter also obey equation (2.17), but with $K \approx 1050\,\mathrm{yr}^2\,\mathrm{AU}^{-3}$ rather than $K \approx 1\,\mathrm{yr}^2\,\mathrm{AU}^{-3}$.

2.6 ▪ PROOF OF THE EARTH'S MOTION

Although Galileo's discoveries convinced many individuals that the heliocentric model was correct, definitive proof that the Earth revolves around the Sun and rotates on its axis wasn't provided until much later. The rotation of the Earth about its axis was proved by detecting the Coriolis effect; this was done most famously by Jean Foucault, using what is now called a Foucault pendulum. The revolution of the Earth about the Sun was proved by detecting the effect known as aberration of starlight; later confirmation came from measuring the annual parallax of nearby stars.

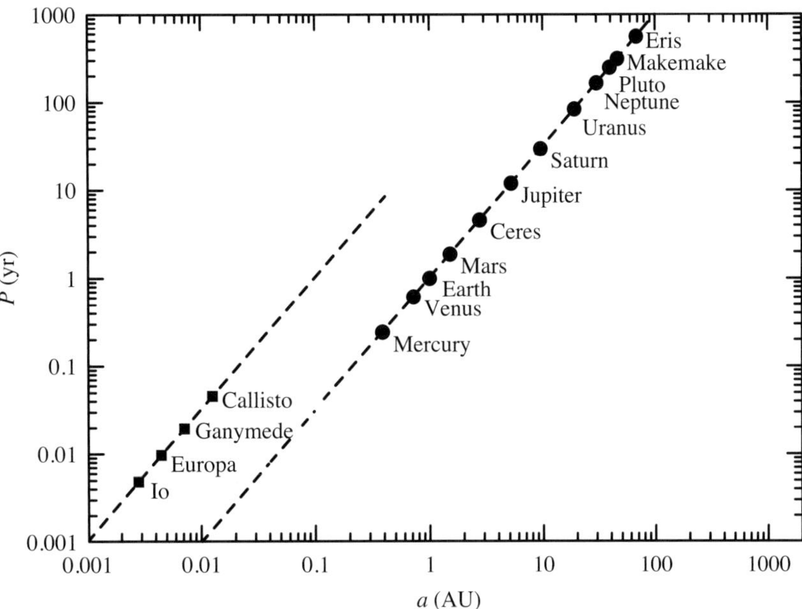

FIGURE 2.19 Orbital period P versus semimajor axis a for planets and dwarf planets orbiting the Sun (circular dots) and for the Galilean satellites orbiting Jupiter (square dots).

2.6.1 Rotation of the Earth

When we measure the trajectory of a projectile (such as a bullet or a thrown ball), we are measuring the trajectory relative to the Earth's surface. However, because of the Earth's rotation, any set of coordinates fixed to the Earth's surface is rotating with an angular velocity $\vec{\omega}$. The magnitude of $\vec{\omega}$ is $\omega \approx 2\pi \text{ day}^{-1} \approx 7.3 \times 10^{-5} \text{ s}^{-1}$, and the direction of $\vec{\omega}$ is pointing from south to north, parallel to the Earth's rotation axis. By watching the motion of the projectile, we can detect the Earth's rotation; its trajectory in the Earth's rotating frame of reference is subtly different from what it would be in a nonrotating frame of reference.

To quantify the difference in trajectories, let's start by writing down the relevant equations of motion. In a nonrotating frame, the motion of an object is famously given by Newton's second law of motion:

$$\vec{\mathbf{a}} = \vec{\mathbf{F}}/m, \tag{2.19}$$

where $\vec{\mathbf{a}}$ is the measured acceleration of the object, $\vec{\mathbf{F}}$ is the net force applied, and m is the object's mass. However, the equation of motion is different when the acceleration $\vec{\mathbf{a}}$ is measured in a frame of reference rotating with angular velocity $\vec{\omega}$:

$$\vec{\mathbf{a}} = \vec{\mathbf{F}}/m + 2(\vec{\mathbf{v}} \times \vec{\omega}) - \vec{\omega} \times (\vec{\omega} \times \vec{\mathbf{r}}), \tag{2.20}$$

where $\vec{\mathbf{v}}$ is the object's velocity and $\vec{\mathbf{r}}$ is the object's position, both measured in the rotating frame of reference.

The last term on the right-hand side of equation (2.20) is called the **centrifugal** acceleration. The centrifugal acceleration points away from the rotation axis, and has a magnitude

$$a_{\text{cent}} = |\vec{\omega} \times (\vec{\omega} \times \vec{\mathbf{r}})| = \omega^2 R, \tag{2.21}$$

where R is the distance of the object from the rotation axis of the frame of reference. In other words, when we rotate with the Earth, we see objects at a distance R from the Earth's rotation axis move in diurnal circles of physical radius R; motion in a circle of radius R with uniform angular speed ω requires an acceleration $a = \omega^2 R$. For objects near the Earth's surface, the centrifugal acceleration is greatest at the equator, where $R \approx 6.4 \times 10^6$ m is equal to the Earth's radius. This implies a centrifugal acceleration near the equator of

$$a_{\text{cent}} = \omega^2 R \approx (7.3 \times 10^{-5}\,\text{s}^{-1})^2 (6.4 \times 10^6\,\text{m}) \approx 0.034\,\text{m s}^{-2}. \tag{2.22}$$

This is not a large acceleration. In the jargon of auto advertisements, it would take you from "zero to sixty mph" in 13 minutes. More relevantly in this context, a_{cent} is small compared to the gravitational acceleration at the Earth's surface, $g = 9.8\,\text{m s}^{-2}$. In principle, traveling from the poles to the equator should reduce your acceleration toward the Earth's center, and thus reduce your weight. However, the fractional weight loss will be only $a_{\text{cent}}/g \approx 0.003$.

The middle term on the right-hand side of equation (2.20) is called the Coriolis acceleration, or the **Coriolis effect**, after a French scientist named Gustave Coriolis, who published the equations of motion for a rotating frame in the year 1835. It is sometimes computationally convenient to think of the Coriolis acceleration,

$$\vec{\mathbf{a}}_{\text{cor}} = 2(\vec{\mathbf{v}} \times \vec{\omega}), \tag{2.23}$$

as being due to a fictitious "Coriolis force" equal to $2m(\vec{\mathbf{v}} \times \vec{\omega})$. In truth, however, no physical force is being applied to the particle; the Coriolis acceleration results from the fact that the particle is being observed from a rotating, and hence accelerated, reference frame. The cross-product in equation (2.23) tells us that the Coriolis acceleration is always perpendicular to the direction of motion of the particle. When the cross-product is worked out in detail, it is seen that a moving particle is deflected to its right in the northern hemisphere and to its left in the southern hemisphere as Figure 2.20 demonstrates.

The magnitude of the Coriolis acceleration is

$$a_{\text{cor}} = 2v\omega \sin \Theta, \tag{2.24}$$

where Θ is the angle between $\vec{\mathbf{v}}$ and $\vec{\omega}$. Thus, the Coriolis effect is maximized when the particle's motion is perpendicular to the Earth's rotation axis; it vanishes when the particle's motion is parallel to the rotation axis. For other directions of motion, we may make the rough approximation

$$a_{\text{cor}} \sim v\omega. \tag{2.25}$$

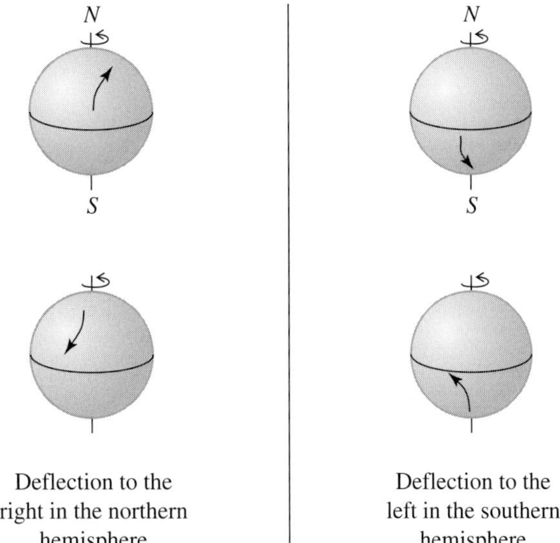

FIGURE 2.20 In a reference frame co-rotating with the Earth, moving particles are deflected to the right in the northern hemisphere, and to the left in the southern hemisphere.

If a particle is in flight for a time Δt, its velocity will be altered by a fractional amount

$$\frac{\Delta v}{v} \sim \frac{a_{\rm cor}\Delta t}{v} \sim \omega \Delta t. \tag{2.26}$$

Thus, the change in the particle's direction of motion will be small as long as its time of flight is much shorter than

$$\omega^{-1} \sim \frac{1}{2\pi}\ \text{days} \sim 4\ \text{hr} \sim 14{,}000\ \text{s}. \tag{2.27}$$

Usually, when a ball is thrown or a bullet is fired, it reaches its target within a few seconds, so the Coriolis effect is negligible. However, the Coriolis acceleration can significantly affect the ballistic trajectory of projectiles when the time of flight is sufficiently long. During the projectile's flight, it will be deflected by a distance

$$\Delta d \sim \frac{1}{2} a_{\rm cor} (\Delta t)^2 \sim \frac{1}{2} v\omega (\Delta t)^2, \tag{2.28}$$

to the right of its initial trajectory in the northern hemisphere and to the left in the southern hemisphere. During World War I, for instance, the German army used an immense artillery gun to bombard Paris from a distance of ~ 120 km. The Paris Gun had a muzzle velocity $v \sim 1.6\ \text{km s}^{-1}$; shells were sent on a parabolic trajectory with a maximum altitude of ~ 40 km and a time of flight $\Delta t \sim 170$ s. This led to a deflection

FIGURE 2.21 A Foucault pendulum at the Earth's north pole.

$$\Delta d \sim \frac{1}{2} v\omega(\Delta t)^2 \sim 2\,\text{km}, \tag{2.29}$$

to the right of where the gun was aimed.

The Coriolis acceleration also affects wind patterns. As air moves inward toward an area of low pressure, the Coriolis acceleration causes it to swerve to the right (in the northern hemisphere of Earth), and sets up a counterclockwise circulation. As a consequence, hurricanes in the northern hemisphere rotate counterclockwise; conversely, circular storms in the southern hemisphere rotate clockwise. Urban legend to the contrary, water draining from a sink doesn't invariably spiral counterclockwise in the northern hemisphere and clockwise in the southern hemisphere. Draining a sink takes much less time than forming a hurricane; during the time it takes a sink to empty, the Δv caused by the Coriolis effect remains small compared to the speed of the eddies that form as you fill the sink and wash your hands.[16]

A celebrated demonstration of the Coriolis effect is the **Foucault pendulum**, first demonstrated in the year 1851 by a French scientist named Jean Foucault. A Foucault pendulum is nothing more than a long pendulum suspended from a ball-and-socket joint overhead, so it is free to swing in any direction. Although Foucault set up his own pendulum in Paris, it is easier to visualize the principle behind the Foucault pendulum if we imagine one installed at the Earth's north pole (Figure 2.21). If we set the pendulum oscillating, it will continue to oscillate back and forth in the same plane, as viewed by a nonrotating observer. Thus, a sidereal nonrotating observer would report, "The

[16] In a classic experiment, A. H. Shapiro of MIT (latitude 42° N) managed to detect the Coriolis effect by filling a 6-foot diameter tank with water, letting it sit covered with a plastic sheet for 24 hours at constant temperature, then carefully pulling out the small, centrally located drain plug. Under such controlled conditions, the water did indeed spiral counterclockwise down the drain. (*Nature*, 1962, vol. 196, p. 1080).

Earth rotates counterclockwise (viewed from above the Earth's north pole), completing one rotation in a sidereal day; the plane of the pendulum's oscillation is not rotating." However, an observer co-rotating with the Earth would report, "The Earth is not rotating with respect to my frame of reference; the plane of the pendulum's oscillation is rotating clockwise (viewed from above the Earth's north pole), completing one rotation in a sidereal day."

Analyzing the rotation of a Foucault pendulum at locations other than the north or south pole requires a more detailed analysis of the Coriolis acceleration of the pendulum bob; the result found is that the pendulum's plane of oscillation rotates at a rate $2\pi \sin \ell$ radians per sidereal day, where ℓ is the latitude at which the Foucault pendulum is located. (This accounts for the popularity of Foucault pendulums at high-latitude science museums; near the equator, the excruciatingly slow rotation of a Foucault pendulum is a less visually exciting demonstration of the Earth's rotation.)

2.6.2 Revolution of the Earth

The **aberration of starlight** was first detected by Jean Picard in 1680, but it wasn't explained until 1729, by the astronomer James Bradley. The aberration of starlight is an effect that causes the apparent positions of stars on the celestial sphere to be deflected in the direction of the observer's motion. The common analogy to explain the aberration of starlight involves running through a rainshower with an umbrella; even if the rain is falling straight down, you have to tilt your umbrella in the direction of motion in order to keep your head dry. Similarly, in order to catch photons from a distant star, you have to tilt your telescope in the direction of motion (Figure 2.22). Photons travel at a large but finite speed, $c = 3.0 \times 10^5$ km s^{-1}. The orbital speed of the Earth averages

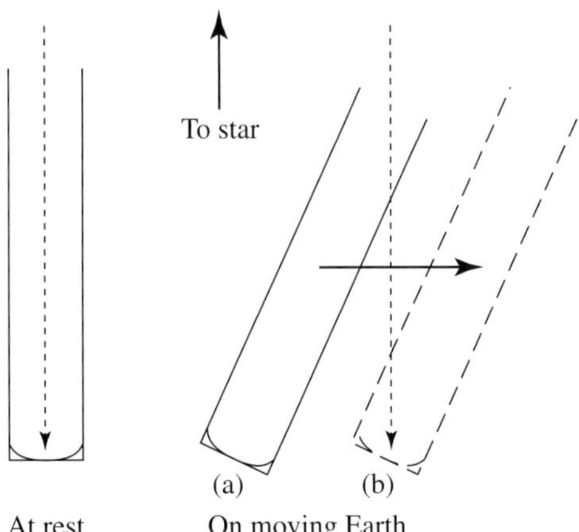

FIGURE 2.22 Telescopes must be tilted in the direction of the Earth's motion by an angle $\theta \approx v/c$ to assure that photons arrive at point P at the same time as the bottom of the telescope.

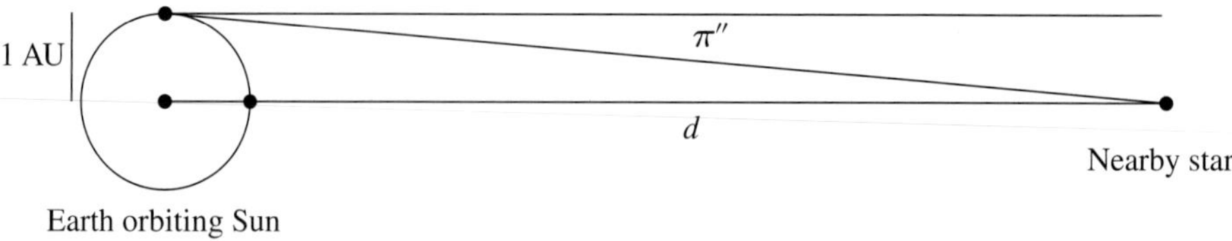

FIGURE 2.23 Definition of the parallax π'' of a star.

$v = 29.8\text{ km s}^{-1} \approx 10^{-4}\,c$. If your telescope is 1 m long, then during the time it takes light to pass through the telescope, the Earth's motion will have translated the telescope through a distance of 0.1 mm. Figure 2.22 shows that the angle through which the telescope must be tilted is given by the relation

$$\tan\theta = v/c. \tag{2.30}$$

Since the Earth's speed is so much smaller than the speed of light, we may use the small-angle approximation:

$$\theta \approx \frac{v}{c} \approx \frac{29.8\text{ km s}^{-1}}{3.0\times 10^{5}\text{ km s}^{-1}}\left(\frac{180^\circ}{\pi\text{ rad}}\right)\left(\frac{3600''}{1^\circ}\right) \approx 20.5''. \tag{2.31}$$

Aberration of starlight causes the positions of stars in the sky to follow an annual path that is the projection of the Earth's motion onto the sky: an ellipse of semimajor axis $20.5''$ and a semiminor axis $20.5''\beta$, where β is the angular distance of the star from the ecliptic.

Stellar parallax was introduced earlier, on page 37, when we emphasized the fact that observers couldn't detect it prior to the invention of the telescope. In fact, even after the invention of the telescope, it took a long time before stellar parallax was first measured. It wasn't until 1838, more than two centuries after the first telescopes, that the astronomer Friedrich Wilhelm Bessel announced that he had finally measured the annual parallax of a star. Formally, astronomers define parallax π'' (Figure 2.23) as the apparent displacement of a star, in arcseconds, due to a change in the position of the observer by 1 AU perpendicular to the line of sight to the star.[17] Although parallaxes are defined in terms of a 1 AU displacement, the actual baseline used for parallax measurements can be as large as 2 AU, by using observations six months apart at the appropriate times of year. From Figure 2.23, we see that the distance d from the Sun to another star is simply related to the star's parallax:

$$d = \frac{a}{\tan\pi''}. \tag{2.32}$$

[17] We avoid confusion with the irrational number $\pi = 3.14159265\ldots$ by using the double prime, gently reminding us that parallaxes are generally measured in units of arcseconds.

Using the small-angle approximation, and converting the parallax from radians to arcseconds, we find that

$$d = \frac{a}{\pi''[\text{arcsec}]}\left(\frac{180^\circ}{\pi\ \text{rad}}\right)\left(\frac{3600''}{1^\circ}\right) = \frac{206{,}265\ \text{AU}}{\pi''}. \tag{2.33}$$

The distance at which a star has a parallax of exactly $1''$ is known as the **parsec**, short for "**par**allax of one arc**sec**." The number of AU in one parsec is equal to the number of arcseconds in a radian: 206,265. The nearest star to the Sun, Proxima Centauri, has a parallax $\pi'' = 0.76''$, and hence is at a distance $d = 270{,}000$ AU $= 1.3$ parsecs. Stellar parallax causes the positions of stars on the celestial sphere to follow a path that is the projection of the Earth's orbit onto the sky: an ellipse with semimajor axis π'' and semiminor axis $\pi''\beta$, where β is the angular distance of the star from the ecliptic.[18] It took a while for stellar parallax to be measured, but when it was, it confirmed two initially controversial assertions made by Copernicus. First, the Earth goes around the Sun, rather than vice versa. Second, space is big (really big).

PROBLEMS

2.1 Over the course of the year, which gets more hours of daylight, the Earth's north pole or south pole? (Hint: the Earth is at perihelion in January.)

2.2 On 2003 August 27, Mars was in opposition as seen from the Earth. On 2005 July 14 (687 days later), Mars was in western quadrature as seen from the Earth. What was the distance of Mars from the Sun on these dates, measured in astronomical units (AU)? Is this greater than or less than the semimajor axis length of the Martian orbit? You may assume the Earth's orbit is a perfect circle. (Hint: the sidereal period of Mars is also 687 days.)

2.3 In the 1670s, the astronomer Ole Rømer observed eclipses of the Galilean satellite Io as it plunged through Jupiter's shadow once per orbit. He noticed that the time between observed eclipses became shorter as Jupiter came closer to the Earth and longer as Jupiter moved away. Rømer calculated that the eclipses were observed 17 minutes earlier when Jupiter was in opposition compared to when it was close to conjunction. This was attributed by Rømer to the finite speed of light. From Rømer's data, compute the speed of light, first in AU min^{-1}, then in m s^{-1}.

[18] Note that the aberrational shift of $20.5''$, which is independent of the star's distance, is much greater than the parallactic shift for even the nearest stars, which have $\pi'' \leq 0.75''$.

2.4 In addition to aberration of starlight due to the Earth's orbital motion around the Sun, there should also be diurnal aberration due to the Earth's rotation. Where on the Earth is this effect the largest, and what is its amplitude?

2.5 A light-year is defined as the distance traveled by light in a vacuum during one tropical year. How many light-years are in a parsec?

2.6 The planets of the solar system all orbit the Sun in the same sense: counterclockwise as seen from above the Earth's north pole. Imagine a "wrong-way" planet orbiting the Sun in the opposite (clockwise) sense, on an orbit of semimajor axis length $a = 1.3$ AU. What would the sidereal period of this planet be? What would its synodic period be as seen from the Earth? What would its synodic period be as seen from Mars?

2.7 Consider a football thrown directly northward at a latitude 40° N. The distance of the quarterback from the receiver is 20 yards (18.5 m), and the speed of the thrown ball is $25\ \mathrm{m\ s^{-1}}$. Does the Coriolis effect deflect the ball to the right or to the left? By what amount (in meters) is the ball deflected? Does the receiver need to worry about correcting for the deflection, or should he be more worried about being nailed by the free safety? (Hint: the angular velocity $\vec{\omega}$ of the Earth's rotation is parallel to the rotation axis.)

3 Orbital Mechanics

Isaac Newton (1642/3–1727) was born in rural England; his birth date was 1642 December 25 according to the Julian calendar (still in use in England at the time), but 1643 January 4 according to the Gregorian calendar. When young Newton proved to be incompetent at managing his family's farm, he was sent to Cambridge University and started to thrive as a scholar. In 1665, the year in which Newton earned his bachelor's degree, an outbreak of the plague closed down the university, and Newton retreated to his family's farm and began to think—very hard. The period when the university was closed was Newton's *annus mirabilis*, during which he discovered calculus, formulated his three **laws of motion** and his **law of universal gravitation**, and performed ground-breaking experiments in optics. Much of the remainder of Newton's long life was dedicated to developing the ideas he had in this burst of youthful creativity.[1]

Newton didn't publish his laws of motion and law of universal gravitation until 1687, when his book *Philosophiae Naturalis Principia Mathematica* ("Mathematical Principles of Natural Philosophy") was published. The laws of motion can be summarized as follows:

1. An object's velocity remains constant unless a net outside force acts upon it.
2. If a net outside force acts on an object, its acceleration is directly proportional to the force and inversely proportional to the mass of the object. In short, $\vec{\mathbf{F}} = m\vec{\mathbf{a}}$, where $\vec{\mathbf{F}}$ is the outside force, m is the mass, and $\vec{\mathbf{a}}$ is the acceleration.
3. Forces come in pairs, equal in magnitude and opposite in direction. (As Newton put it: *Actioni contrariam semper et aequalem esse reactionem*, or "Every action has an equal and opposite reaction.")

Newton's law of universal gravitation can be concisely expressed in mathematical form. Suppose that two spherical objects, of mass M and m, are separated by a distance r.

[1] He also performed many alchemical experiments while trying to systematize chemistry in the way he did physics, not to mention writing reams of theological works, becoming Master of the Royal Mint, and serving as president of the Royal Society for nearly a quarter-century.

(The distance r is measured between the centers of the two objects.) Newton's law tells us that the gravitational attraction between the two objects is

$$F = -\frac{GMm}{r^2}, \tag{3.1}$$

where G, called the **gravitational constant**, is a universal constant whose value is $G = 6.67 \times 10^{-11}\,\mathrm{Nm^2\,kg^{-2}}$ (where N stands for newton).[2] The negative sign in equation (3.1) tells us that gravity is always an attractive force.

3.1 ▪ DERIVING KEPLER'S LAWS

Newton derived the form of equation (3.1) by requiring that the force of gravity result in planetary orbits that obey Kepler's laws of planetary motion. Newton was solving the problem in the difficult direction: he deduced the form of the law of gravitation starting from the observations. Since we aren't as smart as Newton, we will take the easier direction in the following section; starting with Newton's law of universal gravitation, we'll show that Kepler's laws follow as a consequence. Although it may seem numerically incongruous, the derivations will flow more smoothly if we begin by deriving Kepler's second law, then go on to the first and third laws.

3.1.1 Kepler's Second Law

Gravity is an example of a **central force**, defined as a force directed straight toward or away from some central point, with a magnitude that depends only on the distance r from that point. The gravitational force qualifies as a central force because the force $\vec{\mathbf{F}}$ acting on the mass m always points toward the mass M (the central point of the force), and the magnitude of the gravitational force is $\propto 1/r^2$, where r is the separation of the two masses.[3] While analyzing the motion of a particle responding to a central force, it is convenient to be able to switch from Cartesian coordinates to polar coordinates.

In a Cartesian coordinate system (Figure 3.1), the unit vectors along the x, y, and z axes are $\hat{\imath}$, $\hat{\jmath}$, and $\hat{\mathbf{k}}$, respectively. Suppose we choose our Cartesian coordinate axes such that the larger mass M lies at the origin, and the position $\vec{\mathbf{r}}$ and velocity $\vec{\mathbf{v}}$ of the smaller mass m lie in the xy plane. (For the sake of concreteness, let's call mass M the Sun, and mass m a planet, although the situation applies in general to any system of two spherical masses: a planet and a moon, a planet and an artificial satellite, a supermassive black hole and a star—you name it.) The planet's position (x, y) can now be expressed in polar coordinates, where the polar coordinates (r, θ) are related to the Cartesian coordinates (x, y) by the relations $x = r\cos\theta$ and $y = r\sin\theta$. In polar coordinates, as illustrated in

[2] The newton (N)—the force required to accelerate 1 kilogram at one meter per second per second—is equivalent to 3.6 ounces, or about the weight of a small apple.

[3] The electrostatic repulsion or attraction between two charged particles is another example of a central force.

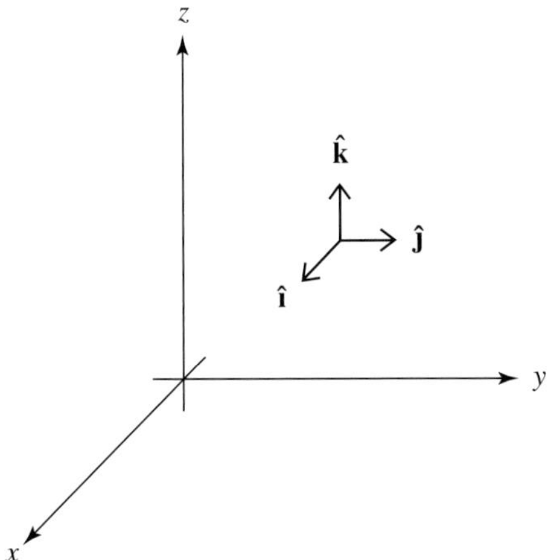

FIGURE 3.1 Axes and unit vectors in a Cartesian coordinate system.

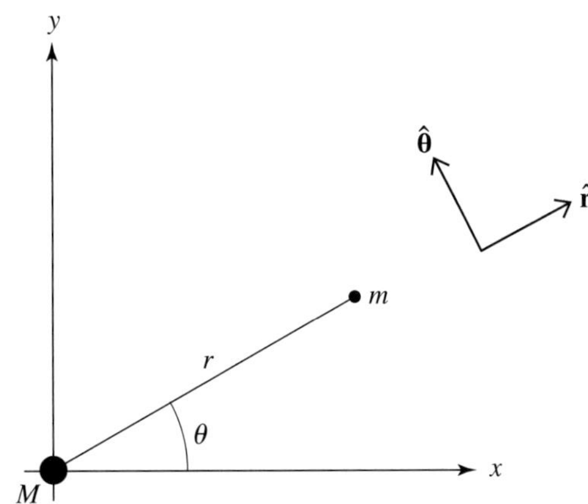

FIGURE 3.2 Axes and unit vectors in a polar coordinate system.

Figure 3.2, the unit vectors $\hat{\mathbf{r}}$ and $\hat{\boldsymbol{\theta}}$ are

$$\hat{\mathbf{r}} = \hat{\imath}\cos\theta + \hat{\jmath}\sin\theta \tag{3.2}$$

and

$$\hat{\boldsymbol{\theta}} = -\hat{\imath}\sin\theta + \hat{\jmath}\cos\theta. \tag{3.3}$$

The dot product (or scalar product) of these unit vectors is

$$\hat{\mathbf{r}} \cdot \hat{\boldsymbol{\theta}} = -\cos\theta\,\sin\theta + \sin\theta\,\cos\theta = 0, \tag{3.4}$$

and their cross product (or vector product) is

$$\hat{\mathbf{r}} \times \hat{\boldsymbol{\theta}} = \begin{vmatrix} \hat{\imath} & \hat{\jmath} & \hat{\mathbf{k}} \\ \cos\theta & \sin\theta & 0 \\ -\sin\theta & \cos\theta & 0 \end{vmatrix} = \hat{\mathbf{k}}(\cos^2\theta + \sin^2\theta) = \hat{\mathbf{k}}, \tag{3.5}$$

thus demonstrating that $\hat{\mathbf{r}}$ and $\hat{\boldsymbol{\theta}}$ are mutually orthogonal as well as being orthogonal to $\hat{\mathbf{k}}$, the unit vector in the z direction.

From equations (3.2) and (3.3), we see that

$$\frac{d\hat{\mathbf{r}}}{d\theta} = \frac{d}{d\theta}(\hat{\imath}\cos\theta + \hat{\jmath}\sin\theta) = -\hat{\imath}\sin\theta + \hat{\jmath}\cos\theta = \hat{\boldsymbol{\theta}} \tag{3.6}$$

and

$$\frac{d\hat{\boldsymbol{\theta}}}{d\theta} = \frac{d}{d\theta}(-\hat{\imath}\sin\theta + \hat{\jmath}\cos\theta) = -\hat{\imath}\cos\theta - \hat{\jmath}\sin\theta = -\hat{\mathbf{r}}. \tag{3.7}$$

We can then apply the chain rule to find the rate of change of the unit vectors $\hat{\mathbf{r}}$ and $\hat{\boldsymbol{\theta}}$:

$$\frac{d\hat{\mathbf{r}}}{dt} = \frac{d\hat{\mathbf{r}}}{d\theta}\frac{d\theta}{dt} = \hat{\boldsymbol{\theta}}\frac{d\theta}{dt} \tag{3.8}$$

and

$$\frac{d\hat{\boldsymbol{\theta}}}{dt} = \frac{d\hat{\boldsymbol{\theta}}}{d\theta}\frac{d\theta}{dt} = -\hat{\mathbf{r}}\frac{d\theta}{dt}. \tag{3.9}$$

Note that since $\hat{\mathbf{r}}$ and $\hat{\boldsymbol{\theta}}$ are unit vectors, they change only in direction, not in magnitude.

The velocity of the planet can be expressed in polar coordinates as

$$\vec{\mathbf{v}} \equiv \frac{d\vec{\mathbf{r}}}{dt} = \frac{d(r\hat{\mathbf{r}})}{dt} = \frac{dr}{dt}\hat{\mathbf{r}} + r\frac{d\hat{\mathbf{r}}}{dt} = v_r\hat{\mathbf{r}} + v_t\hat{\boldsymbol{\theta}}, \tag{3.10}$$

where

$$v_r = \frac{dr}{dt} \tag{3.11}$$

is the **radial velocity** and

$$v_t = r\frac{d\theta}{dt} \tag{3.12}$$

is the **tangential velocity**.

The angular momentum of the planet is defined as

$$\vec{\mathbf{L}} \equiv \vec{\mathbf{r}} \times \vec{\mathbf{p}}, \tag{3.13}$$

where $\vec{\mathbf{p}} = m\vec{\mathbf{v}}$ is the linear momentum. The rate of change of the angular momentum is then

$$\frac{d\vec{\mathbf{L}}}{dt} = \frac{d\vec{\mathbf{r}}}{dt} \times \vec{\mathbf{p}} + \vec{\mathbf{r}} \times \frac{d\vec{\mathbf{p}}}{dt} = \vec{\mathbf{v}} \times m\vec{\mathbf{v}} + \vec{\mathbf{r}} \times m\frac{d\vec{\mathbf{v}}}{dt}. \tag{3.14}$$

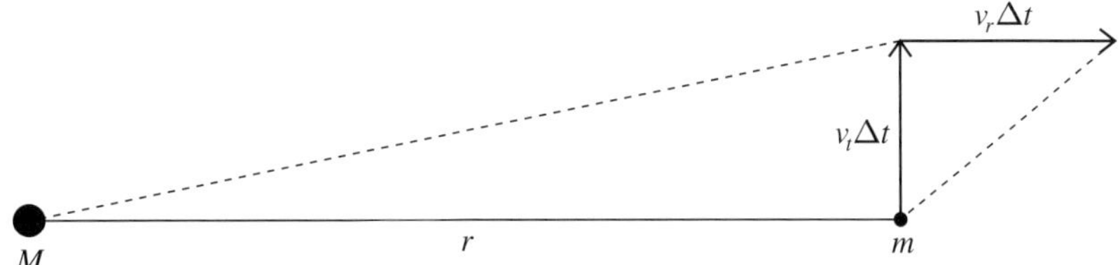

FIGURE 3.3 The motions of a planet during a short time interval Δt.

From Newton's second law of motion, we know that $m d\vec{\mathbf{v}}/dt = \vec{\mathbf{F}}$. Thus, equation (3.14) can be rewritten as

$$\frac{d\vec{\mathbf{L}}}{dt} = m(\vec{\mathbf{v}} \times \vec{\mathbf{v}}) + \vec{\mathbf{r}} \times \vec{\mathbf{F}}. \tag{3.15}$$

However, $\vec{\mathbf{v}} \times \vec{\mathbf{v}} = 0$ (that's just a vector identity), and for a central force, $\vec{\mathbf{F}}$ is parallel to $\vec{\mathbf{r}}$ and thus $\vec{\mathbf{F}} \times \vec{\mathbf{r}} \propto \vec{\mathbf{r}} \times \vec{\mathbf{r}} = 0$. We conclude that for gravity or any other central force, angular momentum is conserved:

$$\frac{d\vec{\mathbf{L}}}{dt} = 0. \tag{3.16}$$

Note that the direction as well as the magnitude of $\vec{\mathbf{L}}$ is constant; this tells us that the motion of an object moving under the influence of a central force is confined to a plane.

The conservation of angular momentum is equivalent to Kepler's second law; to demonstrate that this is true, we use equation (3.10) to write the angular momentum explicitly as

$$\vec{\mathbf{L}} = \vec{\mathbf{r}} \times m\vec{\mathbf{v}} = mrv_t\hat{\mathbf{k}} = L\hat{\mathbf{k}}, \tag{3.17}$$

where v_t is the tangential velocity. Referring to Figure 3.3, consider a planet of mass m; at a time t, it is at a distance r from the Sun, which has mass M. During a brief time interval Δt, the planet moves a distance $v_t\Delta t$ in the tangential direction and a distance $v_r\Delta t$ in the radial direction. The area ΔA swept out by the planet–Sun line during this brief interval can be approximated as the sum of two triangles:

$$\Delta A \approx \frac{1}{2} r(v_t\Delta t) + \frac{1}{2}(v_r\Delta t)(v_t\Delta t), \tag{3.18}$$

where the two terms represent the left-hand triangle and the right-hand triangle in Figure 3.3.[4] In the limit $v_r\Delta t \ll r$, the right-hand triangle is vanishingly small compared to the left-hand triangle, and the area swept out can be further simplified as

$$\Delta A \approx \frac{1}{2} r(v_t\Delta t). \tag{3.19}$$

[4] In Figure 3.3, we are looking at the specific case $v_r > 0$, but performing a time reversal will yield the case $v_r < 0$.

The rate at which the planet–Sun line sweeps out area can then be written

$$\lim_{\Delta t \to 0} \frac{\Delta A}{\Delta t} = \frac{dA}{dt} = \frac{1}{2} r v_t. \tag{3.20}$$

However, since we know that $L = mrv_t$, from equation (3.17), we can rewrite equation (3.20) in the form

$$\frac{dA}{dt} = \frac{1}{2} \frac{L}{m}. \tag{3.21}$$

Since L and m are constant, so is the rate dA/dt at which the planet–Sun line sweeps out area. In other words, we have demonstrated that Kepler's second law will be true for a body acting under any central force, not just the force of gravity.

3.1.2 Kepler's First Law

To demonstrate that Kepler's first law follows from Newton's law of universal gravitation, we will have to demonstrate that the trajectory $r(\theta)$ of the mass m (the planet) constitutes an ellipse with the larger mass M (the Sun) at one focus. Using equations (3.12) and (3.17), we can write the angular momentum per unit mass of the orbiting body as

$$\frac{L}{m} = r^2 \frac{d\theta}{dt}, \tag{3.22}$$

which is constant for any central force. If the force acting on the mass m is gravitational, then from Newton's law of universal gravitation and second law of motion,

$$\vec{\mathbf{F}} = -\frac{GMm}{r^2} \hat{\mathbf{r}} = m \frac{d\vec{\mathbf{v}}}{dt}. \tag{3.23}$$

The orbital acceleration under the influence of gravity is then

$$\frac{d\vec{\mathbf{v}}}{dt} = -\frac{GM}{r^2} \hat{\mathbf{r}}. \tag{3.24}$$

From equation (3.9), we know that

$$\hat{\mathbf{r}} = -\left(\frac{d\theta}{dt}\right)^{-1} \frac{d\hat{\boldsymbol{\theta}}}{dt}. \tag{3.25}$$

By combining equations (3.24) and (3.25), we find that the acceleration of the planet is

$$\frac{d\vec{\mathbf{v}}}{dt} = \frac{GM}{r^2} \left(\frac{d\theta}{dt}\right)^{-1} \frac{d\hat{\boldsymbol{\theta}}}{dt}. \tag{3.26}$$

Combining this equation with equation (3.22), we see

$$\frac{L}{GMm} \frac{d\vec{\mathbf{v}}}{dt} = \frac{d\hat{\boldsymbol{\theta}}}{dt}. \tag{3.27}$$

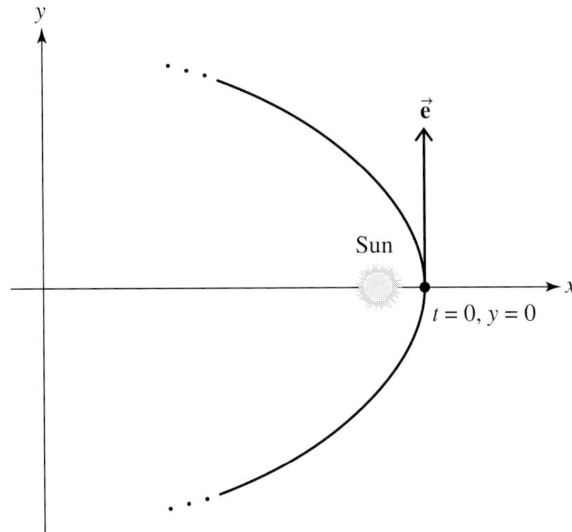

FIGURE 3.4 Time $t = 0$ corresponds to perihelion passage, with the planet crossing the x axis with its velocity in the positive y direction.

Integration of this simple differential equation yields

$$\frac{L}{GMm}\vec{\mathbf{v}} = \hat{\boldsymbol{\theta}} + \vec{\mathbf{e}}, \tag{3.28}$$

where $\vec{\mathbf{e}}$ is a constant of integration that depends on the initial conditions of the orbiting planet. We may choose the initial conditions for our own convenience. Let's choose the time $t = 0$ to correspond to a perihelion passage of the planet, and orient the axes so that perihelion passage occurs on the positive x axis (Figure 3.4). With this choice of coordinates, $\vec{\mathbf{v}}$ and $\hat{\boldsymbol{\theta}}$ both point in the y direction at $t = 0$; thus, we may write $\vec{\mathbf{e}} = e\hat{\mathbf{j}}$, where e is a constant. Equation (3.28) is then

$$\frac{L}{GMm}\vec{\mathbf{v}} = \hat{\boldsymbol{\theta}} + e\hat{\mathbf{j}}. \tag{3.29}$$

We now take the dot product of this equation and the unit vector $\hat{\boldsymbol{\theta}}$:

$$\frac{L}{GMm}\vec{\mathbf{v}} \cdot \hat{\boldsymbol{\theta}} = \hat{\boldsymbol{\theta}} \cdot \hat{\boldsymbol{\theta}} + e\hat{\mathbf{j}} \cdot \hat{\boldsymbol{\theta}}. \tag{3.30}$$

To simplify the right-hand side of equation (3.30), we use equation (3.3) to find that $\hat{\mathbf{j}} \cdot \hat{\boldsymbol{\theta}} = \cos\theta$. To simplify the left-hand side, we write

$$\vec{\mathbf{v}} \cdot \hat{\boldsymbol{\theta}} = \left[v_r\hat{\mathbf{r}} + v_t\hat{\boldsymbol{\theta}}\right] \cdot \hat{\boldsymbol{\theta}} = v_t. \tag{3.31}$$

But, since equation (3.17) tells us that $mrv_t = L$, we may write

$$\vec{\mathbf{v}} \cdot \hat{\boldsymbol{\theta}} = v_t = \frac{L}{mr}. \tag{3.32}$$

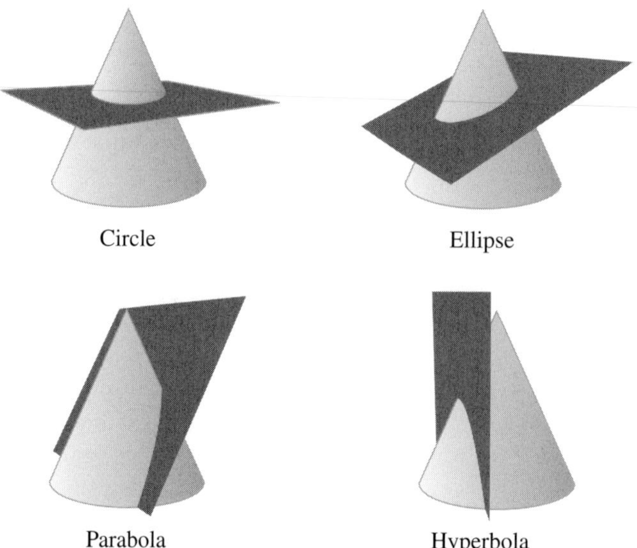

FIGURE 3.5 Conic sections demonstrated by slicing a cone.

Substituting equation (3.32) back into equation (3.30), we find a relationship between r and θ for fixed values of M, m, L, and e:

$$\frac{L^2}{GMm^2r} = 1 + e\cos\theta, \tag{3.33}$$

which can also be written in the form

$$r = \frac{L^2}{GMm^2(1 + e\cos\theta)}. \tag{3.34}$$

Equation (3.34) is the equation of a **conic section** in polar coordinates; as such, it provides a generalization of Kepler's first law.

Conic sections can be obtained by slicing a cone with a plane, as illustrated in Figure 3.5. If the plane is perpendicular to the cone's axis, then the conic section is a **circle**; from equation (3.34), we see that a circle corresponds to the special case $e = 0$, and hence $r = L^2/(GMm^2) =$ constant. If the slicing plane is tilted from the perpendicular by an angle less than the half-opening angle of the cone, the conic section obtained is an **ellipse**; this corresponds to the special case $0 < e < 1$.[5] When the slicing plane is tilted from the perpendicular by an angle exactly equal to the half-opening angle of the cone, the conic section resulting is a **parabola**; this is the special case $e = 1$. Finally, when the slicing plane is tilted by a larger angle, the conic section that results is a **hyperbola**,

[5] Yes, the parameter e in equation (3.34) is the same as the eccentricity e that we encountered while discussing elliptical orbits in Section 2.5, that is, the distance between foci divided by the major axis length.

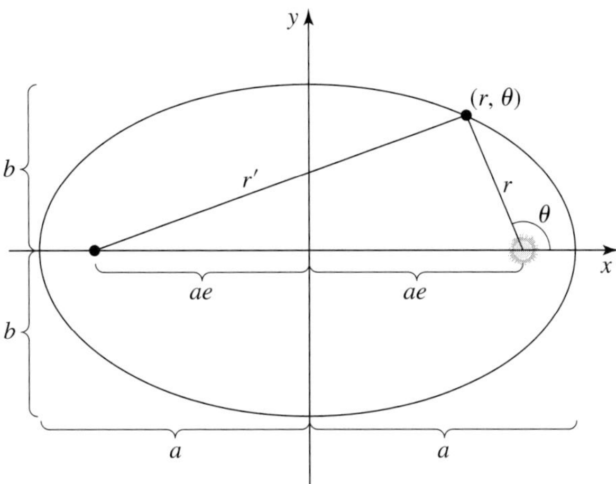

FIGURE 3.6 An ellipse of semimajor axis a and semiminor axis b.

which has $e > 1$. Kepler's first law is thus a special case that deals with **closed orbits**; that is, orbits with $e < 1$, which form closed curves (ellipses or circles). The basic physics of gravitation, however, permits **open orbits** as well, that is, parabolic or hyperbolic orbits with $e \geq 1$.

We have blithely asserted that the parameter e in equation (3.34), when it lies in the range $0 \leq e < 1$, is precisely the same as the eccentricity of an ellipse, defined as the distance between the foci divided by the length of the major axis. It is time to support that assertion by looking at the properties of ellipses in more depth. In Figure 3.6, an ellipse is shown along with a set of Cartesian coordinates; the origin of the coordinates is the center of the ellipse; the x axis lies along the major axis of the ellipse; and the y axis lies along the minor axis. We also define a system of polar coordinates centered on one of the foci. Let's call the focus at the origin the **principal focus** and require that it be the focus where the Sun is located, if the ellipse is regarded as a planetary orbit. The angular coordinate θ is measured counterclockwise from the x axis in the manner shown in Figure 3.6. The semimajor axis has length a and the semiminor axis has length b; each of the foci is displaced from the origin of the Cartesian coordinates by a distance ae. An arbitrary point on the ellipse is displaced by a distance r from the principal focus and a distance r' from the other focus; the basic property of an ellipse is that $r + r'$ is constant. By considering the two points of the ellipse lying on the x axis ($x = \pm a$, $y = 0$), we find that $r + r' = 2a$. It also follows that the perihelion distance, if the ellipse is regarded as a planetary orbit, is $q = a(1 - e)$ and the aphelion distance is $Q = a(1 + e)$.

Consider the point of the ellipse that lies on the positive y axis, where $r = r' = a$ as shown in Figure 3.7. From the Pythagorean theorem, as applied to the right triangle drawn in the figure, we find that $b^2 + (ae)^2 = r^2$, or since $r = a$,

$$b^2 = a^2(1 - e^2). \tag{3.35}$$

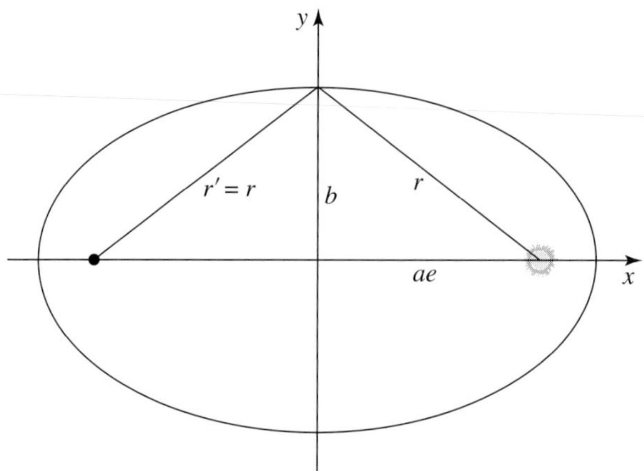

FIGURE 3.7 The relationship among the semimajor axis a, the semiminor axis b, and the eccentricity e.

This enables us to translate between the axis ratio of an ellipse, b/a, and its eccentricity,

$$e = (1 - b^2/a^2)^{1/2}. \tag{3.36}$$

It can also be shown that the average distance of all points on the ellipse from either focus is equal to the semimajor axis length a. To prove this, consider an arbitrary point on the ellipse, $P(x, y)$, and its reflection across the y axis, $P'(-x, y)$, as shown in Figure 3.8. The distance from point P to the focus on the positive x axis is r. By symmetry, the distance from the complementary point P' to the focus on the positive x axis is r', where r' is the distance from point P to the focus on the negative x axis. The average distance of the two points from the focus on the positive x axis is then

$$\langle r \rangle = \frac{r + r'}{2} = \frac{2a}{2} = a. \tag{3.37}$$

Since this relation holds for all (P, P') pairs, regardless of the choice of P, it is true that the average distance $\langle r \rangle$ from the focus over the entire ellipse is a.

Let us now describe the ellipse in terms of the polar coordinates (r, θ), where r is the distance from the principal focus and θ is the polar angle measured counterclockwise from the positive x axis, as shown in Figure 3.9. (When the ellipse represents an orbit, the angle θ is called the **true anomaly**.) Note in the figure that we can draw a triangle from the principal focus at $r = 0$, to an arbitrary point (r, θ) on the ellipse, to the other focus, then back to the principal focus. The internal angle of the vertex at the principal focus (as shown in Figure 2.17) is $\pi - \theta$. We can thus use the law of cosines to write

$$r'^2 = r^2 + (2ae)^2 - 2(2ae)r\cos(\pi - \theta). \tag{3.38}$$

Using the trigonometric identity $\cos(\pi - \theta) = -\cos\theta$, this becomes

$$r'^2 = r^2 + 4a^2e^2 + 4aer\cos\theta. \tag{3.39}$$

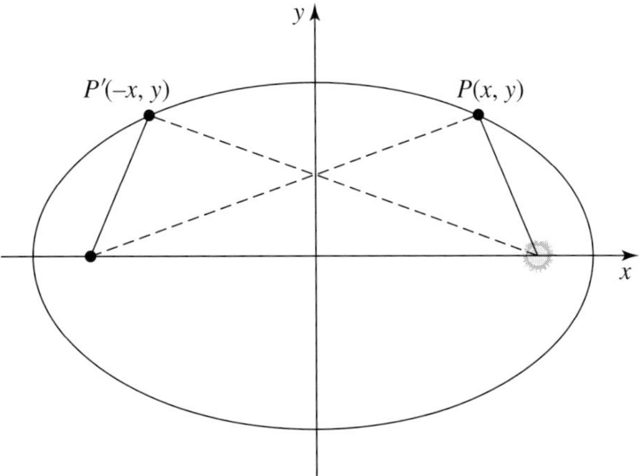

FIGURE 3.8 The point $P(x, y)$ is at a distance r from the focus on the positive x axis and a distance r' from the other focus. The complementary point $P'(-x, y)$ is at a distance r' from the focus on the positive x axis and a distance r from the other focus.

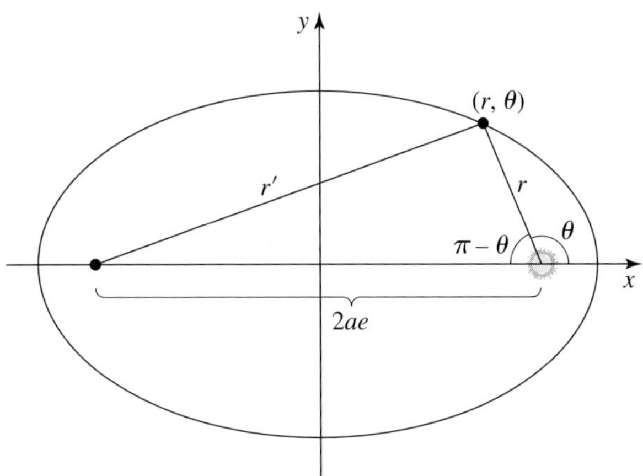

FIGURE 3.9 An ellipse in polar coordinates.

However, from the definition of the ellipse, we know that $r' = 2a - r$, which yields (squaring each side of the equation)

$$r'^2 = 4a^2 - 4ar + r^2. \tag{3.40}$$

Since the right-hand sides of equations (3.39) and (3.40) are equal, this tells us

$$4a^2e^2 + 4aer\cos\theta = 4a^2 - 4ar. \tag{3.41}$$

After dividing by $4a$ and doing a bit of rearranging, we find

$$r = \frac{a(1 - e^2)}{1 + e\cos\theta}. \tag{3.42}$$

This equation for r as a function of θ is the equation for an ellipse in polar coordinates, with the origin at one focus. This is equivalent in form to equation (3.34), which gives the shape of an orbit if Newton's law of universal gravitation holds true. Comparison of equations (3.34) and (3.42) tells us that the angular momentum L of a planet's orbital motion is related to the size and shape of its orbit by the relation

$$\frac{L^2}{m^2} = GMa(1 - e^2). \tag{3.43}$$

Since $L = mrv_t$, this relation can also be written in the form

$$r^2 v_t^2 = GMa(1 - e^2). \tag{3.44}$$

When a planet is at perihelion, its velocity is entirely tangential ($v_{\rm pe} = v_t$), and its distance from the Sun is $q = a(1 - e)$. This implies that for a planet at perihelion,

$$v_{\rm pe}^2 a^2 (1 - e)^2 = GMa(1 - e^2), \tag{3.45}$$

or

$$v_{\rm pe} = \left[\frac{GM}{a}\frac{1 + e}{1 - e}\right]^{1/2}. \tag{3.46}$$

A similar analysis of the planet's speed at aphelion, where its velocity is also entirely tangential ($v_{\rm ap} = v_t$), tells us that

$$v_{\rm ap} = \left[\frac{GM}{a}\frac{1 - e}{1 + e}\right]^{1/2}. \tag{3.47}$$

3.1.3 Kepler's Third Law

Kepler's second law (equation 3.21) tells us that the area swept out per unit time by the planet–Sun line is a constant, $L/(2m)$. The area swept out in one orbital period, P, is the area of the ellipse, given by the standard formula $A = \pi ab$. For one complete orbital period, then, we may write

$$\frac{\pi ab}{P} = \frac{L}{2m}. \tag{3.48}$$

By squaring this equation and making the substitution $b^2 = a^2(1 - e^2)$, we have

$$\frac{\pi^2 a^4 (1 - e^2)}{P^2} = \frac{L^2}{4m^2}. \tag{3.49}$$

Since equation (3.43) gives us a relation among L, a, and e, namely,

$$\frac{L^2}{m^2} = GMa(1 - e^2), \tag{3.50}$$

we can substitute back into equation (3.49) to find

$$\frac{\pi^2 a^4 (1 - e^2)}{P^2} = \frac{GMa(1 - e^2)}{4}, \tag{3.51}$$

or

$$P^2 = \frac{4\pi^2}{GM} a^3, \tag{3.52}$$

which we recognize as Kepler's third law, $P^2 = Ka^3$, with the proportionality constant $K \propto 1/M$. With somewhat more exertion, taking into account the acceleration of the Sun (mass M) as well as the lower-mass planet (mass m), it is possible to reach the more general form

$$P^2 = \frac{4\pi^2}{G(M + m)} a^3. \tag{3.53}$$

Within the solar system, however, even the most massive of the planets, Jupiter, has a mass only $1/1000$ that of the Sun, so the approximation $M + m \approx M$ is adequate.

The masses of celestial bodies are measured by how they accelerate nearby masses. In particular, we can use the orbital periods and semimajor axes of the planets to determine the mass of the Sun:

$$M = \frac{4\pi^2 a^3}{GP^2}. \tag{3.54}$$

The orbital period of the Earth, for instance, is 365.256 days $\times$ 86,400 s day^{-1} = 3.16×10^7 s.[6] The semimajor axis of the Earth's orbit is $a = 1\ \text{AU} = 1.496 \times 10^{11}$ m. Thus, we can compute the mass of the Sun as

$$\begin{aligned} M &= \frac{4\pi^2 (1.496 \times 10^{11}\ \text{m})^3}{6.67 \times 10^{-11}\ \text{m}^3\ \text{s}^{-2}\ \text{kg}^{-1} (3.16 \times 10^7\ \text{s})^2} \\ &= 1.98 \times 10^{30}\ \text{kg} \equiv 1 M_\odot. \end{aligned} \tag{3.55}$$

Later in this book, we will find that the solar mass ($M_\odot$) is a useful unit for expressing the masses of stars (and larger objects).[7]

[6] A useful approximation is that the length of the year is $\pi \times 10^7$ s.

[7] The "dot in a circle" symbol $\odot$ is the standard astronomical symbol for the Sun. It is of great antiquity, being identical to the Egyptian hieroglyph for the Sun god Ra, seen here, for instance, as the first syllable in the name of the pharaoh Ramses the Great: 𓍹𓇳𓄟𓋴𓋴𓍺

3.2 ▪ ORBITAL ENERGETICS

Suppose you place a particle of mass m at a location $\vec{\mathbf{r}}$ relative to an object of mass M; you give it a kick so that it is initially moving at a velocity $\vec{\mathbf{v}}$. What determines whether its orbit is closed (a circle or ellipse, with $e < 1$) or open (a parabola or hyperbola, with $e \geq 1$)? In a sense, it's all about the energy. The particle will have an energy E that is the sum of its kinetic energy K and its gravitational potential energy U:

$$E = K + U = \frac{1}{2}mv^2 - \frac{GMm}{r}. \tag{3.56}$$

The square of the velocity can be determined by squaring equation (3.28):

$$\left(\frac{L}{GMm}\right)^2 \vec{\mathbf{v}} \cdot \vec{\mathbf{v}} = \hat{\boldsymbol{\theta}} \cdot \hat{\boldsymbol{\theta}} + 2e\hat{\boldsymbol{\theta}} \cdot \hat{\mathbf{j}} + e^2\hat{\mathbf{j}} \cdot \hat{\mathbf{j}}$$

$$\left(\frac{L}{GMm}\right)^2 v^2 = 1 + 2e\hat{\boldsymbol{\theta}} \cdot \hat{\mathbf{j}} + e^2. \tag{3.57}$$

Since $\hat{\boldsymbol{\theta}} \cdot \hat{\mathbf{j}} = \cos\theta$, from equation (3.3), we may now write the kinetic energy as

$$K = \frac{1}{2}mv^2 = \frac{1}{2}m\left(\frac{GMm}{L}\right)^2 (1 + e^2 + 2e\cos\theta). \tag{3.58}$$

The kinetic energy is greatest at perihelion ($\theta = 0$), which is as it should be, since that's when the particle is moving fastest. Now using equation (3.34) for r as a function of θ, we can write the potential energy as

$$U = -\frac{GMm}{r} = -\frac{(GM)^2m^3}{L^2}(1 + e\cos\theta). \tag{3.59}$$

The amplitude of the potential energy, $|U|$, is greatest at perihelion ($\theta = 0$), which is as it should be, since that's when the particle is closest to the mass M. By adding together the kinetic energy (equation 3.58) and the potential energy (equation 3.59), and doing a bit of rearranging, we find

$$E = \left(\frac{GMm}{L}\right)^2 \frac{m}{2}(e^2 - 1). \tag{3.60}$$

This is constant, which is as it should be, since energy is conserved for this isolated two-body system. We can also, if we so choose, write the orbital eccentricity as a function of energy E and angular momentum L:

$$e = \left(1 + \frac{2EL^2}{G^2M^2m^3}\right)^{1/2}. \tag{3.61}$$

We can readily identify three distinct cases:

1. **Hyperbolic orbits:** As we recall from our discussion of conic sections (page 68), the case $e > 1$ represents a hyperbola. Equation (3.60) shows that $e > 1$ corresponds

to a total energy $E > 0$; that is, $K > |U|$. This is an open orbit; the mass m is not gravitationally bound to the mass M. The mass m makes a single perihelion passage at $\theta = 0$ and does not return—its value of r, the distance from the mass M, continues to increase monotonically after perihelion passage.

2. **Parabolic orbits:** In the case where $e = 1$ exactly, the mass m is marginally unbound to M; that is, its velocity approaches zero asymptotically as r approaches infinity. In the case of a parabolic orbit, equation (3.60) shows that $e = 1$ corresponds to $E = 0$, or $K = |U|$. Equation (3.56) reveals that a particle will be on a parabolic orbit if its speed is equal to the **escape speed**:

$$v_{\text{esc}}(r) = \left(\frac{2GM}{r}\right)^{1/2}. \tag{3.62}$$

If its velocity is greater than v_{esc}, it will be on a hyperbolic orbit.

3. **Elliptical orbits:** In the case where $e < 1$, the mass m is gravitationally bound; it goes around the mass M on an elliptical orbit. The total energy, when $e < 1$, is $E < 0$, corresponding to $K < |U|$. The special case $e = 0$ corresponds to a perfectly circular orbit. Equation (3.60) shows that a circular orbit is the orbit that minimizes the energy E for a given angular momentum L.

3.3 ▪ ORBITAL SPEED

It is not possible in general to obtain a simple equation that gives the time dependence of a planet's distance from the Sun, $r(t)$, or orbital speed, $v(t)$.[8] However, it is possible to find the orbital speed v as a simple function of r, which can be useful. We start with the equation for a conic section (equation 3.42), which we write in the form

$$e \cos\theta = \frac{a(1-e^2) - r}{r}. \tag{3.63}$$

The orbital speed as a function of θ is given by equation (3.58):

$$v^2 = \frac{2K}{m} = \left(\frac{GMm}{L}\right)^2 (1 + e^2 + 2e\cos\theta). \tag{3.64}$$

Thus, by combining equations (3.63) and (3.64), we find an equation that gives the orbital speed as a function of r:

$$v^2 = \frac{G^2M^2m^2}{L^2}\left(1 + e^2 + \frac{2}{r}[a(1-e^2) - r]\right). \tag{3.65}$$

Using equation (3.43), which tells us $L^2/m^2 = GMa(1-e^2)$, we find

[8] This also implies that there is no simple equation for $\theta(t)$, since if we had one, we could use the conic section equation for $r(\theta)$ to find $r(t)$.

$$v^2 = \frac{G^2M^2}{GMa(1-e^2)}\left(\frac{r+e^2r+2a(1-e^2)-2r}{r}\right)$$
$$= \frac{GM}{a(1-e^2)}\left(\frac{2a(1-e^2)-r(1-e^2)}{r}\right)$$
$$= \frac{GM}{a}\left(\frac{2a}{r}-1\right) = GM\left(\frac{2}{r}-\frac{1}{a}\right). \tag{3.66}$$

The resulting equation

$$v^2 = GM\left(\frac{2}{r}-\frac{1}{a}\right) \tag{3.67}$$

is called the **vis viva** equation. The Latin term *vis viva*, which translates literally to "living force," is an archaic bit of scientific terminology that was first employed by Gottfried Leibniz (best known as the other discoverer of calculus). Leibniz used the term *vis viva* to refer to the quantity mv^2, what we would now call $2K$, or twice the kinetic energy. The *vis viva* equation is a statement of how the kinetic energy of an orbiting object changes as a function of r. By using Kepler's third law (equation 3.52), we can also write the *vis viva* equation in the form

$$v(r) = \frac{2\pi a}{P}\left(2\frac{a}{r}-1\right)^{1/2}. \tag{3.68}$$

This implies that the orbital angular speed $\omega = v/r$ of a planet is

$$\omega(r) = \frac{2\pi}{P}\frac{a}{r}\left(2\frac{a}{r}-1\right)^{1/2}. \tag{3.69}$$

At perihelion, where $r = q = a(1-e)$, the angular speed of the planet is

$$\omega_{\rm pe} = \frac{2\pi}{P}\frac{(1+e)^{1/2}}{(1-e)^{3/2}}, \tag{3.70}$$

and at aphelion, where $r = Q = a(1+e)$, the angular speed is

$$\omega_{\rm ap} = \frac{2\pi}{P}\frac{(1-e)^{1/2}}{(1+e)^{3/2}}. \tag{3.71}$$

Here on Earth, for instance, the observed average angular speed of the Sun along the ecliptic is equal to 2π radians per sidereal year, or $\omega = 0.986^\circ/\text{day}$. However, since the Earth's orbit has an eccentricity $e = 0.017$, the observed angular speed is greatest at the time of perihelion (early January), when $\omega_{\rm pe} = 1.020^\circ/\text{day}$, and smallest at the time of aphelion (early July), when $\omega_{\rm ap} = 0.953^\circ/\text{day}$.

An interesting application of the *vis viva* equation (eq. 3.68) addresses the problem of the **transfer orbit**. In traveling from the Earth to another planet, the transfer orbit is the route you would take from the Earth to the other planet's orbit. The **Hohmann transfer orbit**, illustrated in Figure 3.10, is an ellipse whose perihelion is at the orbit of

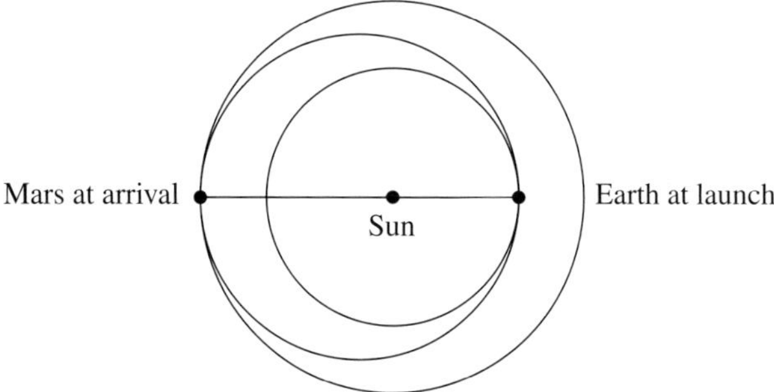

FIGURE 3.10 A Hohmann transfer orbit for interplanetary travel (here from Earth to Mars). The transfer orbit is an ellipse with its perihelion at Earth and its aphelion at the orbit of Mars.

the inner planet and whose aphelion is at the orbit of the outer planet. As the German engineer Walter Hohmann pointed out in the 1920s, the Hohmann transfer orbit has two desirable properties. First, it requires only two engine burns when done properly: one when leaving Earth and one when the destination planet is reached. The rest of the time, the spacecraft is "coasting" on a Newtonian orbit. Second, it is economical in its fuel use; launching your spacecraft on a hyperbolic orbit will cause it to reach its destination faster but requires more energy.

As a concrete example, suppose you want to send a spacecraft to Mars. As a first approximation, we can assume that the orbit of the Earth is a circle of radius $a_\oplus = 1\text{ AU} = 1.50 \times 10^8$ km, with orbital period $P_\oplus = 1\text{ yr} = 3.16 \times 10^7$ s.[9] We further assume that the orbit of Mars is a larger circle, of radius $a_{\text{Mars}} = 1.52 a_\oplus = 2.27 \times 10^8$ km, with orbital period $P_{\text{Mars}} = 1.88\text{ yr} = 5.94 \times 10^7$ s. The semimajor axis of the Hohmann transfer orbit from Earth to Mars is

$$a_{\text{to}} = \frac{a_\oplus + a_{\text{Mars}}}{2} = \frac{1\text{ AU} + 1.52\text{ AU}}{2} = 1.26\text{ AU}. \tag{3.72}$$

The orbital period for the transfer orbit is then

$$P_{\text{to}}[\text{ yr}] = (a[\text{ AU}])^{3/2} = (1.26)^{3/2} = 1.41. \tag{3.73}$$

Traveling from Earth to Mars requires half an orbit, or a time $t = P_{\text{to}}/2 = 0.71\text{ yr} \approx 260$ days.

The average speed of the Earth on its orbit is

$$v_\oplus = \frac{2\pi a_\oplus}{P_\oplus} = \frac{2\pi(1.50 \times 10^8\text{ km})}{3.16 \times 10^7\text{ s}} = 29.8\text{ km s}^{-1}. \tag{3.74}$$

[9] The "cross in a circle" symbol $\oplus$ is the standard astronomical symbol for the Earth.

The average speed of Mars is slower:

$$v_{\rm Mars} = \frac{2\pi a_{\rm Mars}}{P_{\rm Mars}} = \frac{2\pi(2.27 \times 10^8\ {\rm km})}{5.94 \times 10^7\ {\rm s}} = 24.0\ {\rm km\ s^{-1}}. \tag{3.75}$$

When the spacecraft has just left the Earth, it is at the perihelion of the Hohmann transfer orbit. Its speed, from the *vis viva* equation (eq. 3.68), is

$$\begin{aligned} v_{\rm pe} &= \frac{2\pi a_{\rm to}}{P_{\rm to}}\left(\frac{2a_{\rm to}}{a_\oplus} - 1\right)^{1/2} \\ &= \frac{2\pi(1.26\ {\rm AU})(1.50 \times 10^8\ {\rm km\ AU^{-1}})}{(1.41\ {\rm yr})(3.16 \times 10^7\ {\rm s\ yr^{-1}})}\left[\frac{2(1.26\ {\rm AU})}{1.00\ {\rm AU}} - 1\right]^{1/2} \\ &= 26.7\ {\rm km\ s^{-1}}(1.52)^{1/2} = 32.9\ {\rm km\ s^{-1}}. \end{aligned} \tag{3.76}$$

Thus, at the perihelion of the Hohmann transfer orbit, the spacecraft must be going *faster* than the Earth by an amount $\Delta v = v_{\rm pe} - v_\oplus = 3.1\ {\rm km\ s^{-1}}$. When the spacecraft is just reaching Mars, it is at the aphelion of the Hohmann transfer orbit. Its speed, from equation (3.68), is then

$$v_{\rm ap} = \frac{2\pi a_{\rm to}}{P_{\rm to}}\left(\frac{2a_{\rm to}}{a_{\rm Mars}} - 1\right)^{1/2} = 26.7\ {\rm km\ s^{-1}}\left[\frac{2(1.26\ {\rm AU})}{1.52\ {\rm AU}} - 1\right]^{1/2} = 21.7\ {\rm km\ s^{-1}}. \tag{3.77}$$

Thus, in order to match its velocity to that of Mars, the spacecraft must increase its speed by $\Delta v = v_{\rm Mars} - v_{\rm ap} = 2.3\ {\rm km\ s^{-1}}$. (If you want your spacecraft to go into orbit around Mars, like the *Mars Reconnaissance Orbiter*, the time, direction, and duration of your engine burn depend on the orbital parameters you want to attain.)

Use of a Hohmann transfer orbit requires careful timing. If you are sending a spacecraft to Mars, for instance, the craft must reach the aphelion of its orbit just as Mars reaches that point. This restricts launches to certain times, known as **launch windows**. If you fail to launch during one launch window, you could wait for one synodic period of the target planet before launching again. For a mission to Mars, whose synodic period is 2.1 years, this could be a frustrating wait.

3.4 ▪ THE VIRIAL THEOREM

If a system contains only two spherical bodies, such as a star and planet, there is a simple analytic solution (first seen in Section 2.5) for the planet's trajectory, $r(\theta)$. Similarly, Section 3.2 yields simple formulas for the planet's kinetic energy $K(\theta)$ and potential energy $U(\theta)$, while Section 3.3 gives the *vis viva* equation for v as a function of r. In a system containing more than two bodies, however, there are no longer any simple analytic solutions for the bodies' properties. Thus, when astronomers study large stellar systems such as star clusters and galaxies, they generally use numerical techniques to compute the stellar orbits using a computer. However, despite the complexity of many-body systems such as star clusters, it is possible to find useful statistical results that describe the average global properties of the system. One such result is the **virial theorem**, which relates the total kinetic energy of a system to its total potential energy.

To derive the virial theorem, let's suppose we have a system containing N stars (or planets, or other compact massive bodies). The mass of the ith star is m_i, and its location is $\vec{\mathbf{r}}_i$. We can define a function

$$A \equiv \sum_{i=1}^{N} m_i \frac{d\vec{\mathbf{r}}_i}{dt} \cdot \vec{\mathbf{r}}_i. \tag{3.78}$$

The reason for defining this function starts to become a bit more obvious when we take the time derivative of A:

$$\frac{dA}{dt} = \sum_{i=1}^{N} \left(m_i \frac{d\vec{\mathbf{r}}_i}{dt} \cdot \frac{d\vec{\mathbf{r}}_i}{dt} + m_i \frac{d^2\vec{\mathbf{r}}_i}{dt^2} \cdot \vec{\mathbf{r}}_i \right). \tag{3.79}$$

The first term on the right-hand side of equation (3.79) is twice the kinetic energy, and the second term can be transformed using Newton's second law,

$$m_i \frac{d^2\vec{\mathbf{r}}_i}{dt^2} = \vec{\mathbf{F}}_i, \tag{3.80}$$

where $\vec{\mathbf{F}}_i$ is the net force acting on the ith star. Thus, we may write

$$\frac{dA}{dt} = 2K + \sum_{i=1}^{N} \vec{\mathbf{F}}_i \cdot \vec{\mathbf{r}}_i, \tag{3.81}$$

where K is the sum of the kinetic energies of all the stars in the system. The term $\sum \vec{\mathbf{F}}_i \cdot \vec{\mathbf{r}}_i$ was named the **virial** by the physicist Rudolf Clausius.[10]

Equation (3.81) is the most general form of the virial theorem. It applies to any system of bodies that follow Newton's second law, regardless of the forces $\vec{\mathbf{F}}_i$ acting on them. A more useful form of the virial theorem can be found by taking the time average of equation (3.81). If we average over the time interval $t = 0 \to t = \tau$, we find

$$\begin{aligned} 2\langle K\rangle + \langle \sum_{i=1}^{N} \vec{\mathbf{F}}_i \cdot \vec{\mathbf{r}}_i \rangle &= \langle \frac{dA}{dt} \rangle \\ &= \frac{1}{\tau} \int_0^{\tau} \frac{dA}{dt} dt \\ &= \frac{A(\tau) - A(0)}{\tau}. \end{aligned} \tag{3.82}$$

If the system is bound, then the velocity of each particle, as well as its displacement from the origin, remains finite. In that case, $A(t)$, as given by equation (3.78), is finite at all

[10] Clausius also coined the term "entropy," probably his most memorable contribution to the scientific vocabulary.

times, and the right-hand side of equation (3.82) goes to zero in the limit $\tau \to \infty$. Thus, for any bound system of particles, the time-averaged virial theorem has the form

$$2\langle K\rangle + \langle\sum_{i=1}^{N} \vec{\mathbf{F}}_i \cdot \vec{\mathbf{r}}_i\rangle = 0. \tag{3.83}$$

The virial theorem as expressed in equation (3.83) can be applied to any bound system, for instance, to a gas of molecules enclosed within a box. However, as astronomers, we are interested in the specific case of an isolated bound stellar system, in which the force acting on the ith star is the sum of the gravitational forces exerted by the other $N - 1$ stars in the system:

$$\vec{\mathbf{F}}_i = \sum_{j\neq i} \frac{Gm_i m_j(\vec{\mathbf{r}}_j - \vec{\mathbf{r}}_i)}{|\vec{\mathbf{r}}_j - \vec{\mathbf{r}}_i|^3}. \tag{3.84}$$

For such a system, what is the value of the virial, $\sum \vec{\mathbf{F}}_i \cdot \vec{\mathbf{r}}_i$? Let's start with a simple system containing only two stars. For this system, the virial will be

$$\begin{aligned}\vec{\mathbf{F}}_1 \cdot \vec{\mathbf{r}}_1 + \vec{\mathbf{F}}_2 \cdot \vec{\mathbf{r}}_2 &= \frac{Gm_1m_2(\vec{\mathbf{r}}_2 - \vec{\mathbf{r}}_1)\cdot\vec{\mathbf{r}}_1}{|\vec{\mathbf{r}}_2 - \vec{\mathbf{r}}_1|^3} + \frac{Gm_2m_1(\vec{\mathbf{r}}_1 - \vec{\mathbf{r}}_2)\cdot\vec{\mathbf{r}}_2}{|\vec{\mathbf{r}}_1 - \vec{\mathbf{r}}_2|^3}\\ &= -\frac{Gm_1m_2|\vec{\mathbf{r}}_2 - \vec{\mathbf{r}}_1|^2}{|\vec{\mathbf{r}}_2 - \vec{\mathbf{r}}_1|^3}\\ &= -\frac{Gm_1m_2}{|\vec{\mathbf{r}}_2 - \vec{\mathbf{r}}_1|}.\end{aligned} \tag{3.85}$$

The right-hand side of equation (3.85) is simply the potential energy U of the two-star system. By extension, for a three-star system, the virial will be equal to the sum of the potential energies of all three pairs: (1,2), (2,3), and (3,1). For a system containing N stars, the virial will be equal to the sum of the potential energies of all $N_{\text{pair}} = N(N-1)/2$ pairs of stars that can be drawn from the system. We can thus write

$$\sum_{i=1}^{N} \vec{\mathbf{F}}_i \cdot \vec{\mathbf{r}}_i = U = \sum_{i=1}^{N}\sum_{j>i} -\frac{Gm_i m_j}{|\vec{\mathbf{r}}_i - \vec{\mathbf{r}}_j|}, \tag{3.86}$$

and the virial equation (eq. 3.83) becomes

$$2\langle K\rangle + \langle U\rangle = 0. \tag{3.87}$$

The virial theorem is useful to astronomers, as we find in Section 20.2, when it enables us to estimate the mass of distant galaxies.

PROBLEMS

3.1 Comet Hale-Bopp has an orbit about the Sun with eccentricity $e = 0.9951$ and semimajor axis length $a = 186.5$ AU. What is the sidereal orbital period of Comet Hale-Bopp? What is Comet Hale-Bopp's distance from the Sun at perihelion? What is its distance from the Sun at aphelion? Comet Hale-Bopp passed through perihelion on 1997 April 1; did the previous perihelion passage of Comet Hale-Bopp occur before or after the birth of Aristotle?

3.2 The asteroid Eros is seen in opposition from the Earth once every 847 days. What is the sidereal orbital period of Eros? What is the length a of the semimajor axis of Eros' orbit? The eccentricity of the orbit of Eros is $e = 0.223$. Does Eros ever come within 1 AU of the Sun?

3.3 Consider a satellite in a circular, low-Earth orbit; that is, the satellite's elevation above the Earth's surface is $h \ll R_\oplus$. Show that the orbital period P for such a satellite is approximately

$$P = C\left(1 + \frac{3h}{2R_\oplus}\right).$$

What is the numerical value of the constant C in minutes? When Puck, in *A Midsummer Night's Dream*, boasted, "I'll put a girdle round about the Earth in forty minutes" (Act 2, Scene 1), could he have done so by traveling on a circular orbit, accelerated by the Earth's gravity alone? If so, what would be his elevation h?

3.4 What is the orbital period for a low-*lunar* orbit (as was used by the Apollo command modules)?

3.5 (a) Io is the innermost Galilean satellite of Jupiter. The orbital period of Io is $P = 1.769$ days; the semimajor axis of its orbit is $a = 421{,}600$ km (slightly larger than the Moon's orbit about the Earth). Given this information, find the mass of Jupiter.
(b) Phobos is the inner moon of Mars. The orbital period of Phobos is $P = 0.32$ days; the semimajor axis of its orbit is $a = 9370$ km. Find the mass of Mars. (Hint: you may assume the masses of Io and Phobos are negligible compared to those of their parent planets.)

3.6 Communications and weather satellites are often placed in *geosynchronous* orbits. A geosynchronous orbit is an orbit about the Earth with orbital period P exactly equal to one sidereal day. What is the semimajor axis a_{gs} of a geosynchronous orbit? What is the orbital velocity v_{gs} of a satellite on a circular geosynchronous orbit?

3.7 Starting with the equation for an ellipse in polar coordinates (eq. 3.42), derive the more familiar Cartesian form,

$$\frac{x^2}{a^2} + \frac{y^2}{b^2} = 1.$$

3.8 The *Hubble Space Telescope (HST)* is on a circular, low-Earth orbit, at an elevation $h = 600$ km above the Earth's surface. What is its orbital period? For an observer who sees *HST* pass through the zenith, how long is *HST* above the horizon during each orbit?

3.9 One way of lifting a satellite into geosynchronous orbit is to use the space shuttle to lift it into a circular, low-Earth orbit (with $h = 300$ km above the Earth's surface), and then use a booster rocket to place the satellite on a Hohmann transfer orbit (see Section 3.3) up to a circular geosynchronous orbit. What is the orbital velocity v_{ss} of the satellite while it is still in low-Earth orbit? What is the orbital velocity at pericenter, v_{pe}, of the appropriate Hohmann transfer orbit? What is the orbital velocity at apocenter, v_{ap}, of the Hohmann transfer orbit? How long does it take the satellite to travel from the low-Earth orbit to the geosynchronous orbit?

3.10 A small particle of mass m is on a circular orbit of radius R around a much larger mass M. Suppose that we suddenly increase the speed at which the mass m is moving, by a factor α (that is, $v_{\rm final} = \alpha v_{\rm initial}$, with $\alpha > 1$). Compute the major axis, minor axis, pericenter distance, and apocenter distance for the new orbit; express your answers in terms of R and α alone.

4 The Earth–Moon System

In the previous chapter, we treated all massive bodies as if they were point masses. This provides a good first-order approximation of how bodies in the solar system interact. However, a better description requires that we take into account the fact that stars, planets, and satellites are extended bodies that do not have a perfectly spherical shape. In addition, when computing the orbits of planets around the Sun, we must take into account the fact that there are many sources of gravitational force besides the Sun. These cause perturbations of what would otherwise be a perfectly elliptical orbit. In this chapter, we will see how these effects influence the dynamics of the Earth–Moon system in particular. As we will see, the more subtle, second-order effects of gravity, such as precession and tides, have significant long-term effects.

4.1 ▪ PRECESSION

A rapidly rotating plastic body will distort from a spherical shape.[1] The rotation causes the body to take the shape of an **oblate spheroid**, with its short axis coinciding with the rotation axis. A pliable, rapidly rotating lump of pizza dough will become a thin disk; the Earth is more rigid than pizza dough, but it still has a slight equatorial bulge, shown grossly exaggerated in Figure 4.1. In reality, the Earth's equatorial radius (the average distance from the Earth's center to the equator) is $R_{\rm eq} = 6378.14$ km; its polar radius (the distance from the Earth's center to the north or south pole) is $R_{\rm pol} = 6356.75$ km, smaller by about 21.4 km, or 0.3%.

The equatorial bulge of the Earth provides an extra "spare tire" of material at the Earth's equator on which the Moon and Sun can pull. Neither the Sun nor the Moon lies in the Earth's equatorial plane; thus, the pull of the Sun and Moon on the Earth's equatorial bulge works to align the equatorial and ecliptic planes. If we consider a small bit of mass m within the Earth's equatorial bulge, the force exerted on it by the Moon will be

[1] We are using "plastic" here in its original sense, meaning "able to flow slowly." We don't mean to imply that the planets are made of polyethelene or polyvinyl chloride.

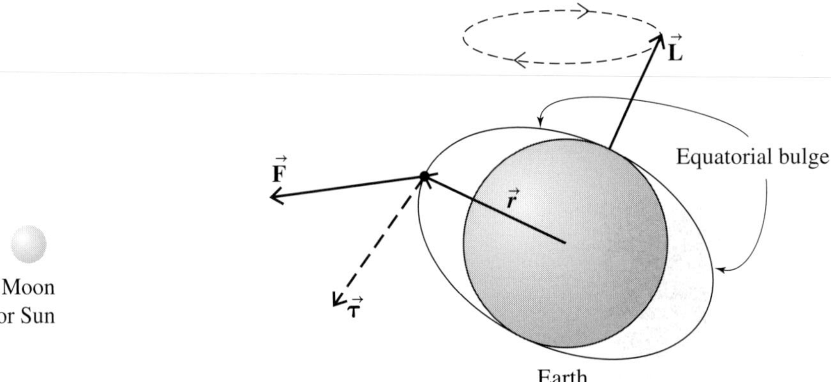

FIGURE 4.1 Gravitational torque on the Earth's equatorial bulge results in precession of the rotation axis. The torque vector is pointing out of the page.

$$F = -\frac{GM_{\text{Moon}}m}{r^2_{\text{Moon}}}. \tag{4.1}$$

Because of the tilt of the Earth's axis relative to the Moon's orbit, the direction of this force will not be exactly parallel to the vector $\vec{\mathbf{r}}$ from the Earth's center to the mass bit m. The mass m will thus experience a torque $\vec{\tau}$ given by the relation (see equation 3.15)

$$\vec{\tau} = \vec{\mathbf{r}} \times \vec{\mathbf{F}} \neq 0. \tag{4.2}$$

If the Earth were perfectly spherical, then the net torque on the entire Earth would vanish, by symmetry arguments. However, the "spare tire" of the equatorial bulge has a net torque exerted on it. As seen from above the Earth's north pole, this torque causes the Earth's rotation axis to precess clockwise. The combined torque of the Sun and the Moon acting on the Earth's equatorial bulge causes the observed precession period of 25,800 years.

4.2 ▪ TIDES

Around an isolated spherical body, the equipotential surfaces (contours of constant gravitational potential) will be perfect spheres. However, if another massive object is brought close, the equipotential surfaces will become distorted. To the extent that the initially spherical body is fluid, it will change its shape to fill a nonspherical equipotential surface. The distortions of the equipotential surfaces, and of the body within them, are called **tides**. Because the Earth has a large, nearby Moon, its shape is measurably distorted by the tidal effect of the Moon.[2] The most spectacular manifestation of tides for

[2] The gravitational effects of the Sun also have a significant tidal effect on the Earth, as we'll compute later, but let's use the lunar tides as our example.

(a) (b)

FIGURE 4.2 (a) High tide in the Bay of Fundy. (b) Low tide at the same location.

an inhabitant of the Earth is the rise and fall of the water level at the seashore; Figure 4.2 shows "high tide" and "low tide" in the Bay of Fundy. The fluid water is more easily distorted by the differential tidal forces than is the rigid rock. However, even the solid crust of the Earth is significantly distorted in shape by the tidal forces.

To compute the differential tidal force, let's start by considering the gravitational force of the Moon acting on the matter of which the Earth is made. A small bit of matter (perhaps a drop of ocean water, or perhaps a pebble) has mass m. The gravitational force exerted on it by the Moon is

$$F_{\text{Moon}}(r) = -\frac{GM_{\text{Moon}}m}{r^2}, \tag{4.3}$$

where $M_{\text{Moon}} = 7.4 \times 10^{22}$ kg is the Moon's mass, and r is the distance of the bit of matter from the Moon's center. The average distance r_0 between the center of the Earth and the center of the Moon is 384,000 km, about 60 times the Earth's mean radius of $R_\oplus = 6370$ km. Thus, the force of gravity at a distance r from the Moon's center can be profitably expanded as a Taylor series for locations within the Earth or on the Earth's surface:

$$F_{\text{Moon}}(r) \approx F_{\text{Moon}}(r_0) + (r - r_0)\frac{dF_{\text{Moon}}}{dr}\bigg|_{r=r_0}. \tag{4.4}$$

Since, from equation (4.3), we know that

$$\frac{dF_{\text{Moon}}}{dr} = \frac{2GM_{\text{Moon}}m}{r^3}, \tag{4.5}$$

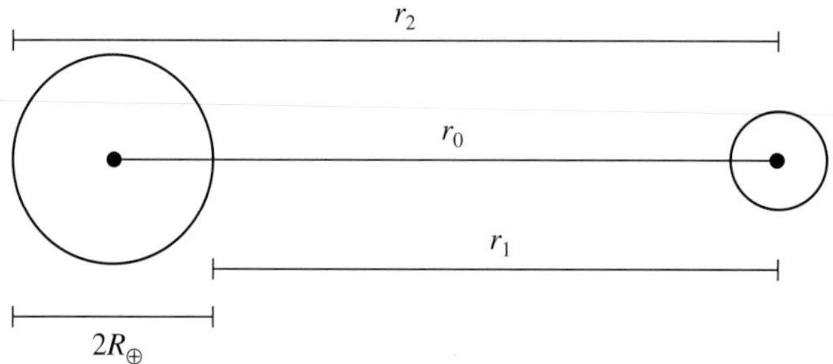

FIGURE 4.3 The differential gravitational force of the Moon (right) on the Earth (left).

we can write the *differential* gravitational force near the Earth's center as

$$\Delta F_{\rm Moon} \equiv F_{\rm Moon}(r) - F_{\rm Moon}(r_0) = (r - r_0)\frac{2GM_{\rm Moon}m}{r_0^3}. \quad (4.6)$$

The point on the Earth's surface nearest the Moon has $r = r_1 = r_0 - R_\oplus$, as shown in Figure 4.3. This means that the differential force at this near point is

$$\Delta F_{\rm Moon}(r_1) = -\frac{2GM_{\rm Moon}mR_\oplus}{r_0^3}. \quad (4.7)$$

Similarly, we can compute the differential force at the point on the Earth's surface *farthest* from the Moon, $r = r_2 = r_0 + R_\oplus$; it is

$$\Delta F_{\rm Moon}(r_2) = \frac{2GM_{\rm Moon}mR_\oplus}{r_0^3}, \quad (4.8)$$

equal in magnitude but opposite in sign to the differential force at the near point.

Note that the differential force associated with tides falls off as the cube of the Earth–Moon distance. The differential force due to the Sun can similarly be written as

$$\Delta F_\odot = \frac{2GM_\odot mR_\oplus}{a_\oplus^3}, \quad (4.9)$$

where $a_\oplus = 1\,\text{AU} = 1.50 \times 10^8$ km is the average Earth–Sun distance. The ratio of the differential forces due to the Sun and the Moon is

$$\frac{\Delta F_\odot}{\Delta F_{\rm Moon}} = \frac{M_\odot}{M_{\rm Moon}}\left(\frac{r_0}{a_\oplus}\right)^3 = \frac{2.0 \times 10^{30}\,\text{kg}}{7.4 \times 10^{22}\,\text{kg}}\left(\frac{3.8 \times 10^5\,\text{km}}{1.5 \times 10^8\,\text{km}}\right)^3 \approx 0.44. \quad (4.10)$$

We thus see that the *differential* effects of the Moon's gravity are about twice as large as those due to the Sun's gravity. Although the mass of the Sun is 27 million times the Moon's mass, the cube of an astronomical unit is $(390)^3 \approx 59$ million times the cube of the Earth–Moon distance.

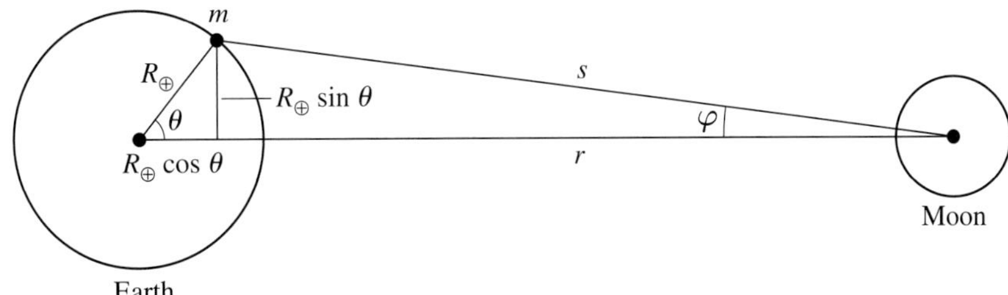

FIGURE 4.4 Geometry for computing tidal distortion of the shape of the Earth.

Thus, a back-of-envelope calculation tells us that Moon-related tides should have roughly twice the amplitude of Sun-related tides. We can consider tidal effects in more detail, over the entire surface of the Earth, by using Figure 4.4 as a starting point.

At the center of the Earth, the force on a test mass m due to the Moon's mass M_{Moon} is

$$\vec{\mathbf{F}}_c = \frac{GM_{\text{Moon}}m}{r^2}\hat{\mathbf{\imath}}, \tag{4.11}$$

where r is the distance from the Earth's center to the Moon's center. At a point P on the Earth's surface (where mass m lies in the figure), the force on the test mass m is

$$\vec{\mathbf{F}}_P = \frac{GM_{\text{Moon}}m}{s^2}(\hat{\mathbf{\imath}}\cos\phi - \hat{\mathbf{\jmath}}\sin\phi). \tag{4.12}$$

The difference in force between these two locations is

$$\Delta\vec{\mathbf{F}} = \vec{\mathbf{F}}_P - \vec{\mathbf{F}}_c = GM_{\text{Moon}}m\left[\hat{\mathbf{\imath}}\left(\frac{\cos\phi}{s^2} - \frac{1}{r^2}\right) - \hat{\mathbf{\jmath}}\frac{\sin\phi}{s^2}\right]. \tag{4.13}$$

Inspection of Figure 4.4 tells us that the relation between the distance s, the Earth's radius $R_\oplus$, and the Earth–Moon distance r is

$$\begin{aligned} s^2 &= (r - R_\oplus\cos\phi)^2 + (R_\oplus\sin\phi)^2 \\ &= r^2 - 2rR_\oplus\cos\phi + R_\oplus^2\cos^2\phi + R_\oplus^2\sin^2\phi \\ &= r^2\left[1 - \frac{2R_\oplus}{r}\cos\phi + \frac{R_\oplus^2}{r^2}\right]. \end{aligned} \tag{4.14}$$

Since $R_\oplus/r \ll 1$, we can do an expansion, ignoring terms of order $(R_\oplus/r)^2$ and higher:

$$\begin{aligned} \frac{1}{s^2} &\approx \frac{1}{r^2}\left(1 - \frac{2R_\oplus}{r}\cos\phi\right)^{-1} \\ &\approx \frac{1}{r^2}\left(1 + \frac{2R_\oplus}{r}\cos\phi\right). \end{aligned} \tag{4.15}$$

Referring again to Figure 4.4, we can also write the equation

$$\begin{aligned}\sin\phi &= \frac{R_\oplus \sin\theta}{s} \approx \frac{R_\oplus \sin\theta}{r}\left(1 - \frac{2R_\oplus}{r}\cos\theta\right)^{-1/2} \\ &\approx \frac{R_\oplus \sin\theta}{r}\left(1 + \frac{R_\oplus}{r}\cos\theta\right) \\ &\approx \frac{R_\oplus \sin\theta}{r},\end{aligned} \tag{4.16}$$

discarding, as before, terms of order $(R_\oplus/r)^2$ and higher. Similarly, we may write

$$\begin{aligned}\cos\phi &= \frac{r - R_\oplus \cos\theta}{s} \\ &= \frac{r - R_\oplus \cos\theta}{r}\left(1 - \frac{2R_\oplus}{r}\cos\theta\right)^{-1/2} \\ &\approx \left(1 - \frac{R_\oplus}{r}\cos\theta\right)\left(1 + \frac{R_\oplus}{r}\cos\theta\right) \approx 1.\end{aligned} \tag{4.17}$$

By substituting equations (4.15), (4.16), and (4.17) back into equation (4.13), we find the differential force as a function of r and θ:

$$\Delta\vec{\mathbf{F}} \approx \frac{GM_{\text{Moon}} m R_\oplus}{r^3}\left(\hat{\imath}\, 2\cos\theta - \hat{\jmath}\,\sin\theta\right). \tag{4.18}$$

Figure 4.5 shows the differential force vectors for several locations on the surface of the Earth. Notice that in addition to producing extension along the x axis, the differential forces also produce compression along the y axis. Since the Earth–Moon system possesses rotational symmetry about the Earth–Moon line, there will also be compression

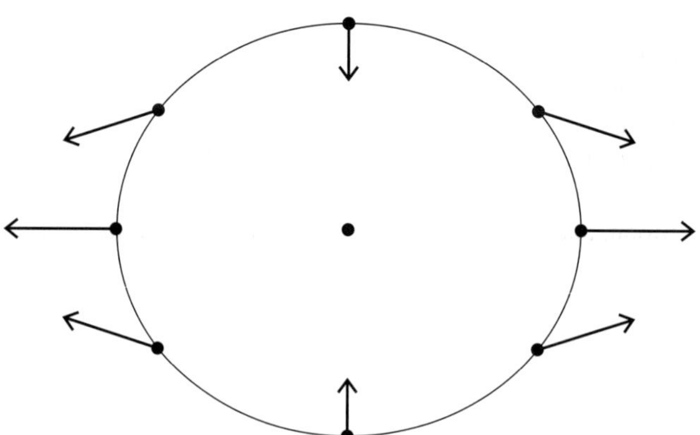

FIGURE 4.5 A plot of tidal distortion vectors at different locations on the Earth.

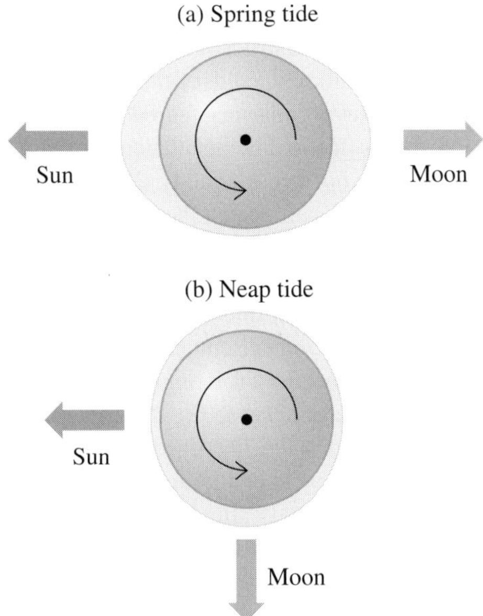

FIGURE 4.6 (a) Spring tides occur when the tidal bulges produced by the Sun and the Moon are aligned. (b) Neap tides occur when the tidal bulges are at right angles.

along the z axis, and the differential forces will mold the Earth into a prolate shape, elongated along the x axis. Thus, the Earth has two tidal bulges, one on the side of the Earth facing the Moon, and the other on the opposite side. Since the Moon makes an upper transit once every 24^h50^m, observers on Earth experience high tide every 12^h25^m, as one of the tidal bulges reaches their location. The height of the tidal bulges raised by the Moon is typically 1 m in the open ocean and only 0.2 m on land. Irregularities of the ocean floor, however, can cause dramatic local variations in the height of tides. For instance, the Bay of Fundy, between New Brunswick and Nova Scotia, is a long bay gradually narrowing toward its head. As the tidal bulges are funneled into the Bay of Fundy, they can produce high tides as much as 12 m above the low tides.[3]

The tidal bulges created by the Sun are about half as high as those created by the Moon, and thus are not negligible when computing the actual height of tides. When the Sun and Moon are separated by either 180° or 0°, as seen from the Earth, their tidal bulges add constructively, causing relatively high-amplitude tides called **spring tides**

[3] Please don't fall into the beginner's error of confusing the Earth's tidal bulges with its equatorial bulge. The equatorial bulge is an oblate distortion due to the Earth's rotation, with an amplitude of several kilometers. The tidal bulges are a prolate distortion due to the effect of the Moon or Sun, with an amplitude of only a meter or so.

(Figure 4.6).[4] When the Sun and Moon are separated by 90°, as seen from the Earth, their tidal bulges sum destructively, causing relatively low-amplitude tides called **neap tides**.[5] For sailors, a knowledge of the tides is essential. This was especially true in the age of sail, when every mariner feared being "neaped"; that is, running aground at the height of a spring tide and being stuck during the subsequent neap tides.[6]

High tide does not occur exactly at the instant when the Moon makes an upper or lower transit. In the open ocean, there is an average delay of ~ 40 minutes between the Moon's transit and the following high tide.[7] This lag is caused by the effects of friction on the Earth's tidal bulges. The Earth rotates once per sidereal day. However, the Moon-induced tidal bulges go around the Earth's center at the same angular rate as the Moon's orbital motion, that is, one circuit per month. This means that friction between the tidal bulges and the more rapidly rotating body of the Earth tends to drag the bulges forward of where they would otherwise be; the angle by which they are dragged forward turns out to be 10° (Figure 4.7). The observable consequence is that high tide occurs when the Moon is slightly west of the upper meridian, that is, shortly *after* its upper transit (and also, of course, shortly after its lower transit, given the two tidal bulges on opposite sides of the Earth). Since the Moon pulls slightly more strongly on the nearer tidal bulge than the farther tidal bulge, there is a net torque acting to slow the rotation of the Earth; this decrease of the Earth's rotational angular momentum is referred to as **tidal braking**. Conversely, because the nearer tidal bulge pulls slightly more strongly on the Moon than does the farther tidal bulge, the Moon is pulled forward in the direction of its orbital motion, increasing its orbital angular momentum.

If we approximate the Moon's orbit as a circle of radius r, the Moon's orbital angular momentum is

$$L_{\rm orb} = M_{\rm Moon} v r = M_{\rm Moon} \left(\frac{G M_\oplus}{r} \right)^{1/2} r = (G M_\oplus M_{\rm Moon}^2 r)^{1/2}. \tag{4.19}$$

If the rate of increase of the Moon's orbital angular momentum is small (so that the orbit is always well approximated as Keplerian) then an increase of the orbital angular momentum results in an increased orbit size:

$$\frac{dL_{\rm orb}}{dt} = \left(G M_\oplus M_{\rm Moon}^2 \right)^{1/2} \frac{1}{2} r^{-1/2} \frac{dr}{dt}. \tag{4.20}$$

[4] The name "spring tides" has nothing to do with the season of spring; they can occur at any time of year. The name refers to the fact that high tides spring higher into the air when the Sun and Moon are aligned.

[5] The word "neap" is of obscure etymology; scholars of Old English surmise that its original meaning was "lacking power."

[6] Captain James Cook, for instance, had the misfortune to run aground on the Great Barrier Reef on 1770 June 11, just at high tide. Thus, he had to wait until the next high tide, over 12 hours later, to be floated free. Unfortunately, he had the further misfortune to run aground during a high *spring* tide, and the next high tide was significantly lower in amplitude. The ship's crew, as Cook's log recorded, had to jettison "Guns, Stone & Iron Ballast, Casks, Hoop Staves, Oil Jars, decay'd Stores, etc." before the ship floated.

[7] On your next moonlit vacation at the seashore, you are likely to notice a delay larger or smaller than 40 minutes. This is because continents, acting as obstacles to the tidal flow, significantly alter the times of high tide.

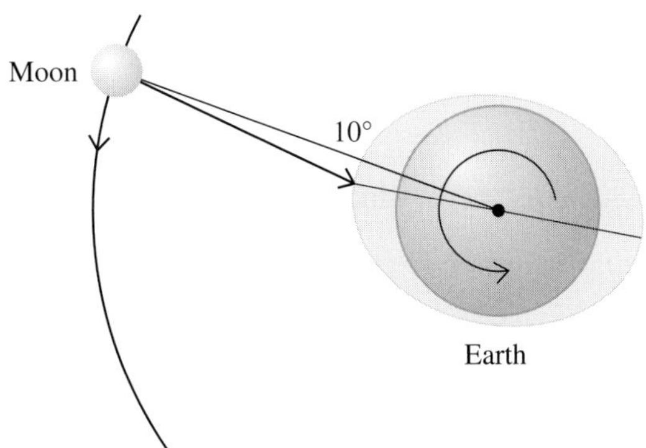

FIGURE 4.7 The Moon, by pulling on the Earth's tidal bulges, decreases the rotation speed of the Earth and increases the orbital angular momentum of the Moon.

If the Earth–Moon system is regarded as an isolated two-body system, then the angular momentum gained by the Moon must be equal to that lost by the Earth:

$$\frac{dL_{\rm orb}}{dt} = -\frac{dL_{\rm rot}}{dt}, \tag{4.21}$$

where $L_{\rm rot}$ is the angular momentum of the Earth's rotation. In general, for a rotating body, the angular momentum can be written as $L_{\rm rot} = I\omega$, where ω is the angular speed of rotation and I is the moment of inertia of the rotating body. If we approximate the Earth as a constant-density sphere,[8] its moment of inertia is

$$I = \frac{2}{5} M_\oplus R_\oplus^2. \tag{4.22}$$

With the approximation of constant density, the Earth's rotational angular momentum is

$$L_{\rm rot} = \frac{2}{5} M_\oplus R_\oplus^2 \left(\frac{2\pi}{P_{\rm rot}} \right), \tag{4.23}$$

and the rate of change of the rotational angular momentum is

$$\frac{dL_{\rm rot}}{dt} = \frac{4\pi M_\oplus R_\oplus^2}{5} \left(-\frac{1}{P_{\rm rot}^2} \frac{dP_{\rm rot}}{dt} \right). \tag{4.24}$$

Because angular momentum is conserved (equation 4.21), we can equate the Earth's loss of rotational angular momentum (equation 4.24) to the Moon's gain of orbital angular

[8] Since the Earth is differentiated, becoming denser near the center, the constant-density approximation actually overestimates the Earth's true moment of inertia by about 20%.

momentum (equation 4.20):

$$\frac{(GM_\oplus M_{\rm Moon}^2)^{1/2}}{2r^{1/2}}\frac{dr}{dt} = \frac{4\pi}{5}\frac{M_\oplus R_\oplus^2}{P_{\rm rot}^2}\frac{dP_{\rm rot}}{dt}. \tag{4.25}$$

Solving for the rate at which the Moon is receding from the Earth, we find

$$\frac{dr}{dt} = \frac{8\pi}{5}\left(\frac{M_\oplus r}{G}\right)^{1/2}\frac{R_\oplus^2}{M_{\rm Moon}P_{\rm rot}^2}\frac{dP_{\rm rot}}{dt}. \tag{4.26}$$

As we discussed in Section 1.5, the rate of slowing of the Earth's rotation is measurable and amounts to

$$\frac{dP_{\rm rot}}{dt} = 0.0016\ {\rm s\ century}^{-1} = 5.2\times10^{-13}\ {\rm s\ s}^{-1}. \tag{4.27}$$

That is, the length of the sidereal day increases by 0.0016 s every century. Inserting this measured quantity into equation (4.26), and using the known masses of the Earth and Moon, the known radius of the Earth, and the known (present) Earth–Moon distance r, we find that the Moon must be receding from the Earth at the rate

$$\frac{dr}{dt} \approx 4\ {\rm cm\ yr}^{-1}. \tag{4.28}$$

For comparison, this is roughly the rate at which London and New York are moving apart due to continental drift, which in turn is approximately equal to the rate at which your fingernails grow. The rate at which the Moon is moving away from the Earth has been confirmed experimentally by lunar laser ranging experiments. In these experiments, powerful lasers are aimed at reflectors left on the Moon's surface by the Apollo astronauts; the round-trip travel time t is measured with exquisite accuracy, permitting a calculation of the one-way distance $d = ct/2$.[9]

4.3 ▪ LIMITS ON THE SIZE OF ORBITS

Just as the Moon produces tidal bulges on the Earth, the Earth produces tidal bulges on the Moon. Since the differential gravitational force acting on the Moon is inversely proportional to the cube of the Earth–Moon distance, if we could bring the Moon closer to the Earth, the tidal distortions of the Moon would increase dramatically. At some critical distance, the differential gravitational forces would tear the Moon apart. Thus, tidal effects impose a minimum permissible orbit size. Conversely, there is a maximum possible size for the orbit of the Moon, beyond which the differential gravitational acceleration caused by the Sun would exceed the acceleration of the Moon by the Earth. The Moon would then escape into a solar orbit. This section will discuss first the minimum orbit size (the Roche limit), then the maximum orbit size (the Hill radius).

[9] The rate at which your fingernails grow can be confirmed experimentally by letting them grow unclipped for a year; more practically, you can save up a year's worth of fingernail clippings and place them side by side.

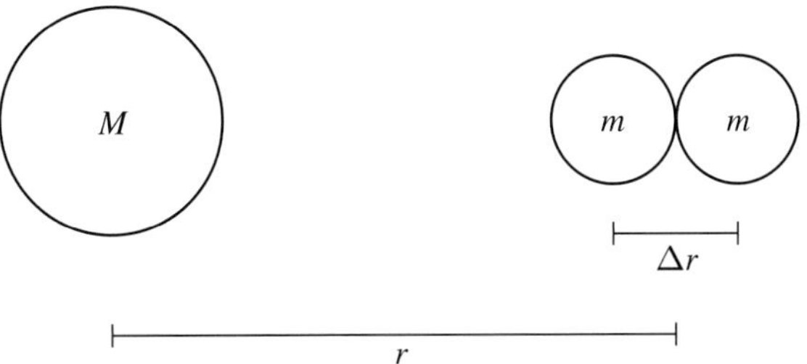

FIGURE 4.8 The two small spheres, which have center-to-center separation Δr, are a self-gravitating system experiencing a differential gravitational force from the larger mass M at a distance r.

4.3.1 Minimum Orbit Size: Roche Limit

Every object has a self-gravity that tends to hold it together and mold it into a spherical body. However, if a small mass is sufficiently close to a larger mass, the differential gravitational force on the smaller mass can exceed its self-gravity, causing the small mass to be ripped apart. How close does the smaller mass have to come to the larger one for this tidal disruption to occur? To answer this question, let's first consider the artificial yet illustrative case of two spherical masses, each of mass m and diameter Δr, that are just touching each other, as shown in Figure 4.8. The centers of the two masses are thus separated by a distance Δr. This two-body system is separated from a larger mass M by a distance $r > \Delta r$. The differential force on the two masses is

$$\Delta F = \frac{dF}{dr}\Delta r = \frac{2GMm}{r^3}\Delta r. \tag{4.29}$$

The self-gravity holding the two masses together is

$$F = -\frac{Gmm}{(\Delta r)^2}. \tag{4.30}$$

As we make r smaller, there is a critical distance known as the **Roche limit**, r_{R}, at which the differential tidal force, which tends to pull the masses apart, is larger in magnitude than the self-gravity force, which tends to bring the masses together. By comparing equations (4.29) and (4.30), we find that the Roche limit r_{R} is given by the relation

$$\frac{2GMm}{r_{\mathrm{R}}^3}\Delta r = \frac{Gmm}{(\Delta r)^2}. \tag{4.31}$$

Solving for r_{R}, we find

$$r_{\mathrm{R}} = \left(\frac{2M}{m}\right)^{1/3}\Delta r. \tag{4.32}$$

It is frequently more convenient to express the Roche limit in terms of the mass density of the objects involved, rather than their masses. Since we have assumed the objects are spherical, we can write

$$M = \frac{4\pi}{3} R^3 \rho_{\rm M}, \tag{4.33}$$

where R is the radius of the larger mass, and $\rho_{\rm M}$ is its mass density. Similarly, for each of the two smaller objects,

$$m = \frac{4\pi}{3} \left(\frac{\Delta r}{2} \right)^3 \rho_{\rm m}, \tag{4.34}$$

where $\rho_{\rm m}$ is the mean mass density. Substituting these two relations into equation (4.32), we obtain

$$r_{\rm R} = \left(\frac{16\rho_{\rm M}}{\rho_{\rm m}} \right)^{1/3} R \approx 2.5 \left(\frac{\rho_{\rm M}}{\rho_{\rm m}} \right)^{1/3} R. \tag{4.35}$$

Admittedly, approximating the Moon as a pair of spheres touching each other is an extraordinarily crude approximation. A more exact treatment of the problem treats the smaller-mass object not as a pair of hard spheres but as a deformable fluid. This more exact approach changes only the numerical coefficient; the Roche limit then becomes

$$r_{\rm R} = 2.44 \left(\frac{\rho_{\rm M}}{\rho_{\rm m}} \right)^{1/3} R. \tag{4.36}$$

Thus, if a planet and its satellite are of comparable density, once the satellite comes within ~ 2.4 planetary radii of its parent planet, the differential gravitational forces will be greater than the satellite's self-gravity.

Let's consider how close the Moon could come to the Earth before it would be tidally disrupted. The mean density of the Earth is

$$\rho_\oplus = \frac{3M_\oplus}{4\pi R_\oplus^3} = \frac{3(5.97 \times 10^{24}\,{\rm kg})}{4\pi(6.38 \times 10^6\,{\rm m})^3} \approx 5500\,{\rm kg\,m^{-3}}, \tag{4.37}$$

about 5.5 times the density of water. Similarly, the density of the Moon is

$$\rho_{\rm Moon} = \frac{3M_{\rm Moon}}{4\pi R_{\rm Moon}^3} = \frac{3(7.35 \times 10^{22}\,{\rm kg})}{4\pi(1.74 \times 10^6\,{\rm m})^3} \approx 3300\,{\rm kg\,m^{-3}}. \tag{4.38}$$

The Roche limit for the Moon is therefore

$$r_{\rm R} = 2.44 \left(\frac{5500}{3300} \right)^{1/3} R_\oplus \approx 2.9 R_\oplus. \tag{4.39}$$

Thus, the Moon is quite safe from tidal disruption; as mentioned in Section 4.2, its actual distance from the Earth is $r \approx 60 R_\oplus$.

It should be noted, however, that if a satellite ventures inside the Roche limit, it is not necessarily torn apart instantly. For instance, Phobos is the inner satellite of Mars; it's an irregularly shaped body with a density $\rho_{\rm Phobos} \approx 1900\,{\rm kg\,m^{-3}}$. Its parent planet Mars

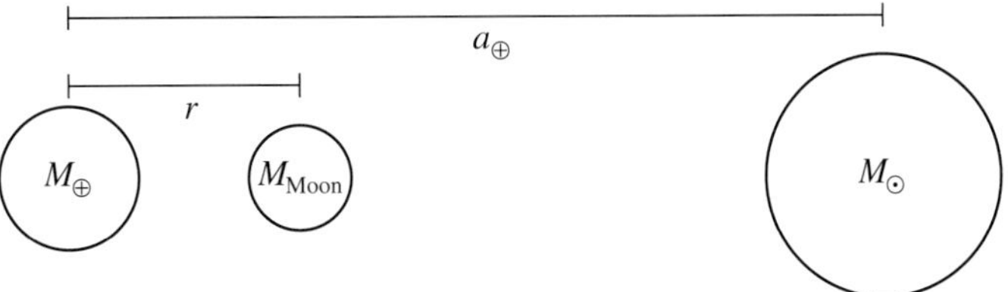

FIGURE 4.9 The Moon and the Earth, with masses $M_{\rm Moon}$ and $M_\oplus$, respectively, form a gravitationally bound system with separation r. The Sun, with mass $M_\odot$ at a distance $a_\oplus$ from the Earth, perturbs the Earth–Moon system. (Not to scale.)

has a density $\rho_{\rm Mars} \approx 3900\,{\rm kg\,m^{-3}}$ and a radius $R_{\rm Mars} = 3397\,{\rm km}$. The Roche limit for Phobos is therefore

$$r_{\rm R} = 2.44\left(\frac{3900}{1900}\right)^{1/3} R_{\rm Mars} = 3.10 R_{\rm Mars}, \tag{4.40}$$

or about 10,500 km. However, the orbit of Phobos, which is nearly circular, has a semimajor axis of only $a_{\rm Phobos} = 9400\,{\rm km} = 2.76 R_{\rm Mars}$. Phobos is inside its Roche limit. It hasn't been torn apart by tides because it is not held together purely by self-gravity. Instead, the electrical forces binding together the material of which it is made are sufficient to keep it intact.[10]

4.3.2 Maximum Orbit Size: Hill Radius

The maximum distance at which the Moon could orbit the Earth is the distance at which the differential acceleration of the Moon away from the Earth, due to the Sun's gravity, is equal to the acceleration of the Moon toward the Earth, due to the Earth's gravity.

Consider the three masses illustrated in Figure 4.9, which may be thought of as the Earth, the Moon, and the Sun. The Earth, of mass $M_\oplus$, and the Moon, of mass $M_{\rm Moon}$, are separated by a distance r. The much larger mass of the Sun, $M_\odot$, is separated from the Earth by a distance $a_\oplus \gg r$. The difference between the Earth's acceleration due to the Sun, $g_\oplus$, and the Moon's acceleration due to the Sun, $g_{\rm Moon}$, is

$$\Delta g = g_{\rm Moon} - g_\oplus = \frac{GM_\odot}{(a_\oplus - r)^2} - \frac{GM_\odot}{a^2}. \tag{4.41}$$

In the case of the Earth–Moon system, $r \ll a_\oplus$, so we may make the approximation

$$(a_\oplus - r)^{-2} = a_\oplus^{-2}(1 - r/a_\oplus)^{-2} \approx a_\oplus^{-2}\left(1 + \frac{2r}{a_\oplus}\right). \tag{4.42}$$

[10] Similarly, you are not torn apart by tides, despite being inside the Roche limit; you are held together by strong electrical forces that prevent you from being disrupted by the relatively feeble differential gravitational force between your head and feet.

This means that the differential acceleration Δg is roughly

$$\Delta g \approx \frac{GM_\odot}{a_\oplus^2}(1+\frac{2r}{a_\oplus}) - \frac{GM_\odot}{a_\oplus^2} \approx \frac{2GM_\odot r}{a_\oplus^3}. \quad (4.43)$$

(Note the now-familiar inverse cube law for a differential gravitational effect. Now, however, instead of a differential force stretching out the Earth, we have a differential force stretching out the Earth–Moon system.)

As we make the Earth–Moon separation r larger, the differential acceleration Δg increases. At a critical distance known as the **Hill radius**, r_H, the differential acceleration Δg is equal in magnitude to the Moon's acceleration due to the Earth. Using equation (4.43), we find

$$\frac{2GM_\odot r_\mathrm{H}}{a_\oplus^3} = \frac{GM_\oplus}{r_\mathrm{H}^2}. \quad (4.44)$$

Solving for the Hill radius r_H, we find

$$r_\mathrm{H} = \left(\frac{M_\oplus}{2M_\odot}\right)^{1/3} a_\oplus. \quad (4.45)$$

The similarity in appearance between the equation for the Roche limit (eq. 4.32) and the equation for the Hill radius (eq. 4.45) is not coincidental. In each case, we were computing the radius at which a differential gravitational force was sufficient to pull apart something bound by gravity. When computing the Roche limit, the "something" was a single extended object, such as a satellite; when computing the Hill radius, the "something" was a two-body system, such as a planet and its satellite.

We can now compute a numerical value for the Hill radius of the Earth–Moon system while it is being perturbed by the Sun's gravity:

$$r_\mathrm{H} = \left(\frac{M_\oplus}{2M_\odot}\right)^{1/3} a_\oplus = \left(\frac{6.0 \times 10^{24}\ \mathrm{kg}}{2(2.0 \times 10^{30}\ \mathrm{kg})}\right)^{1/3} \times (1\ \mathrm{AU})$$
$$= 0.011\ \mathrm{AU} \approx 1.7 \times 10^6\ \mathrm{km}, \quad (4.46)$$

or about 4.5 times the Moon's present average distance from the Earth. Although the Moon is well inside the Hill radius, and will not be lost in a tug-of-war with the Sun, solar tidal forces are sufficient to distort the Moon's orbit from a true elliptical shape. The Sun also causes eastward precession of the line of apsides (the major axis of the Moon's orbit) with a period of 8.85 years. Finally, the Sun's tidal forces cause westward precession of the line of nodes (the intersection of the Moon's orbital plane with the ecliptic) with a period of 18.6 years.[11] If you find it difficult to see where these two types

[11] You may have noted that although we carefully described where the ecliptic lies on the celestial sphere, we didn't provide a similarly careful description of the great circle along which the Moon travels. That's because of the short period of the precession of the line of nodes; a description that would be valid today would be obsolete after only a few years.

of precession come from, you might take some comfort in knowing that Isaac Newton once stated that "his head never ached but when he was studying that subject [the motions of the Moon]."

4.4 ▪ PHASES OF THE MOON

As the Moon orbits the Earth, an Earthly observer sees the Moon go through its complete set of **phases**, as different fractions of the sunlit hemisphere of the Moon are visible from the Earth. Figure 4.10 shows the relative orientation of Sun, Moon, and Earth for each Moon phase. At new Moon, the Moon is almost directly between the Earth and the Sun, presenting its dark side to the Earth. As the Moon orbits the Earth, an increasing fraction of the sunlit face is visible from the Earth as a waxing crescent Moon. About a week after new Moon, approximately half of the illuminated face is presented toward us, at the phase we call first quarter Moon; the name "first quarter" comes from the fact that this phase occurs one-quarter of the way through the complete cycle of phases, starting at new Moon. In subsequent days, the lunar phase becomes waxing gibbous.[12] About two weeks after new Moon, we see full Moon, when the Sun is opposite the Moon in the sky and we see the complete illuminated face of the Moon. Following full Moon, the lunar phase proceeds through the waning (decreasing) gibbous Moon, to the third quarter (sometimes called "last quarter") Moon, to the waning crescent Moon, ultimately returning to the new phase, one synodic month after the previous new Moon.

The orbit of the Moon about the Earth is tilted by 5.1° relative to the Earth's orbit around the Sun. This means that the declination of the Moon can never be greater than $\delta = 23.5° + 5.1° = 28.6°$, and can never be less than $\delta = -23.5° - 5.1° = -28.6°$. Because the Moon never strays far from the celestial equator as seen from Earth, most observers on Earth can see the Moon in the sky for about half of each day (ignoring the technicality that it is difficult to see the Moon when its phase is nearly new, because of its proximity to the Sun). However, the time of day when the Moon is visible varies with its phase. For instance, we see from Figure 4.10 that an observer on Earth would see the first quarter Moon rise above the eastern horizon about the time of local noon; the Moon would then transit the local meridian around sunset, and set around midnight. The waxing crescent Moon rises shortly after the Sun does, and sets shortly after sunset; conversely, the waning crescent Moon rises shortly before the Sun does, and sets shortly before sunset.

Because the Earth is orbiting the Sun at the same time the Moon is orbiting the Earth, the **sidereal month** (another name for the sidereal orbital period of the Moon) is shorter than the **synodic month** (the time that elapses between one new Moon and the next). Expressed mathematically, the relation between the sidereal month and the synodic month is similar to that between the sidereal period and the synodic period of a planet, as described in Section 2.3. In both cases, you need to convert between an orbital period as

[12] The adjective "gibbous" comes from a Latin word meaning "hunchbacked"; the lopsided appearance of the gibbous Moon (seen as images D and F in Figure 4.10b) gave rise to this analogy.

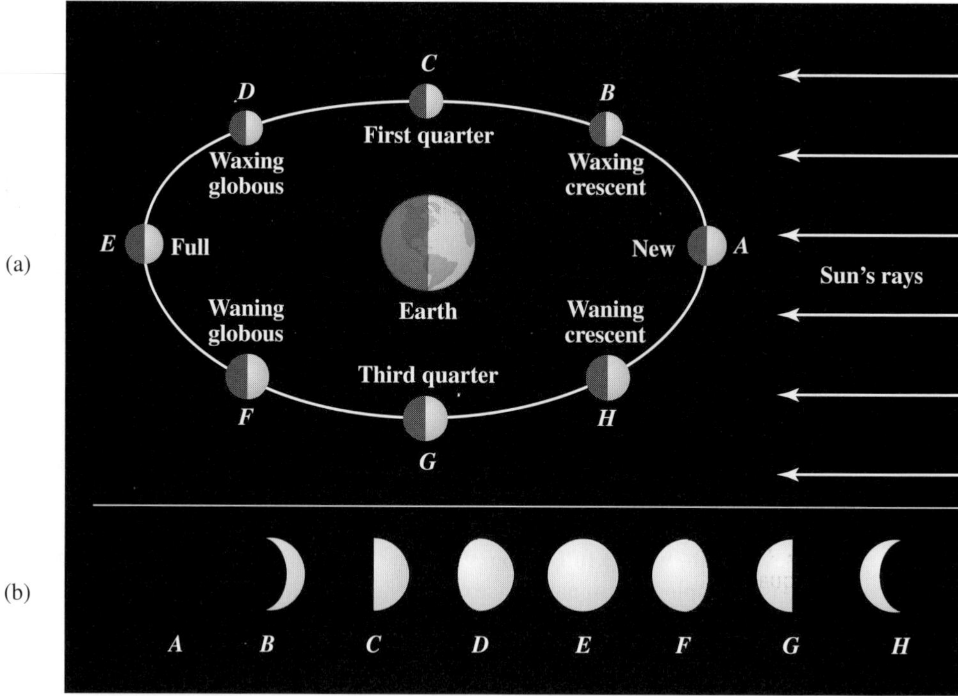

FIGURE 4.10 Phases of the Moon. (a) What an outside observer sees. (b) What an Earthly observer sees.

measured in the sidereal frame of reference to an orbital period as measured in a frame of reference co-rotating with the Earth–Sun line. As before, let $\vec{\omega}_E$ be the angular velocity of the Earth's orbital motion about the Sun, measured in the sidereal frame. Let $\vec{\omega}_{\mathbf{sid}}$ be the angular velocity of the Moon's orbital motion about the Earth, also measured in the sidereal frame. If $\vec{\omega}_{\mathbf{syn}}$ is the angular velocity of the Moon's orbital motion, measured in a frame of reference co-rotating with the Earth–Sun line, then (compare with equation 2.7)

$$\vec{\omega}_{\mathbf{sid}} = \vec{\omega}_E + \vec{\omega}_{\mathbf{syn}}. \tag{4.47}$$

If we ignore the 5.1° tilt between the Moon's orbit and the Earth's orbit, we can reduce the problem to a scalar equation:

$$\omega_{\text{sid}} = \omega_E + \omega_{\text{syn}}. \tag{4.48}$$

If we further ignore the eccentricities of the Moon's orbit and the Earth's orbit, we can write

$$\frac{2\pi}{P_{\text{sid}}} = \frac{2\pi}{P_E} + \frac{2\pi}{P_{\text{syn}}}, \tag{4.49}$$

where P_{sid} is the length of the sidereal month, P_{E} is the length of the sidereal year ($P_{\text{E}} = 365.256$ days), and P_{syn} is the length of the synodic month. The synodic month has been carefully measured over the course of human history, since it is the unit of time on which lunar calendars, such as the Muslim calendar, are based. The length of the synodic month is $P_{\text{syn}} = 29.531$ days. We can now compute the length of the sidereal month:

$$P_{\text{sid}} = \left(\frac{1}{P_{\text{E}}} + \frac{1}{P_{\text{syn}}} \right)^{-1} \tag{4.50}$$

$$= \left(\frac{1}{365.256 \text{ days}} + \frac{1}{29.531 \text{ days}} \right)^{-1} = 27.322 \text{ days}.$$

4.5 ▪ ROTATION OF THE MOON

The Moon raises tidal bulges on the Earth; similarly, the Earth raises tidal bulges on the Moon. The differential gravitational acceleration across the Earth due to the Moon is (see equation 4.7)

$$\Delta g = \frac{\Delta F}{m} \propto \frac{M_{\text{Moon}} R_{\oplus}}{r^3}, \tag{4.51}$$

where r is the distance between the Earth's center and the Moon's center. Conversely, the differential gravitational acceleration across the *Moon* due to the *Earth* is

$$\Delta g \propto \frac{M_{\oplus} R_{\text{Moon}}}{r^3}. \tag{4.52}$$

Since the Earth's mass is roughly 80 times that of the Moon, and the Earth's radius is roughly four times that of the Moon, we see that $M_{\oplus} R_{\text{Moon}} \sim 20 M_{\text{Moon}} R_{\oplus}$. Tidal bulges on the Moon are thus larger than those on the Earth, and tidal braking is more effective on the Moon than on the Earth. In fact, tidal braking has been so effective that the Moon is now locked into **synchronous rotation**; that is, its sidereal period of rotation is equal to its sidereal period of revolution, $P_{\text{sid}} = 27.322$ days. Since the rotation period of the Moon is equal to its orbital period, the Moon always presents the same hemisphere toward the Earth. For much of human history, the "far side of the Moon" was symbolic of all that is mysterious and unknown; in October 1959, the Soviet spacecraft Luna 3 took the first photographs of the Moon's far side.

Synchronous rotation has resulted in permanent tidal bulges on the Moon that are aligned with the Earth–Moon line, making the synchronous rotation stable against perturbations. If, for example, the Moon's rotation rate speeded up, the tidal bulges would be dragged ahead of the Earth–Moon line, and tidal braking would slow the Moon's rotation until the tidal bulges were aligned again. Conversely, if the Moon's rotation rate slowed down, the tidal bulges would fall behind the Earth–Moon line, and the tidal braking effect would work in reverse, adding angular momentum to the Moon's rotation until the tidal bulges were once again aligned.

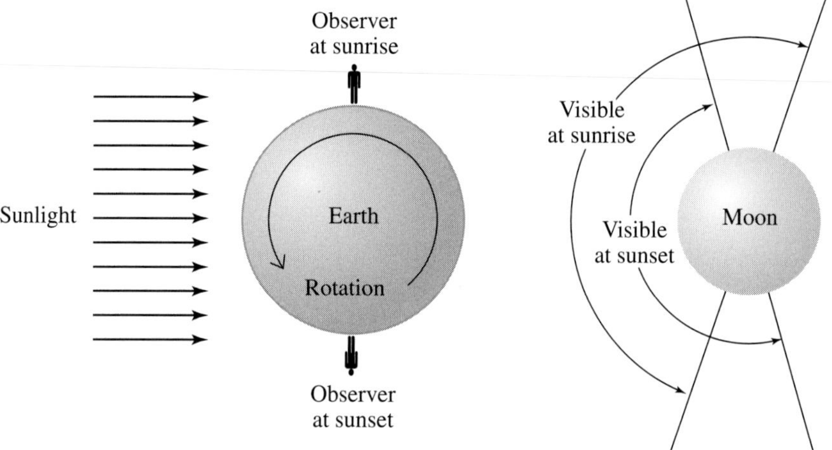

FIGURE 4.11 Diurnal libration is caused by motion of the observer due to the Earth's rotation between moonrise and moonset.

During the course of one sidereal month, somewhat more than 50% of the Moon's surface is visible from the Earth. As seen by an observer on Earth, the Moon "wobbles" in the east–west direction and "nods" in the north–south direction. These motions are known as **lunar librations**, of which there are three distinct types.[13]

First, there is the effect known as **diurnal libration**. This is essentially an effect due to diurnal parallax, as shown in Figure 4.11, which depicts an observer looking at the full Moon at moonrise (when the Sun is setting) and at moonset (when the Sun is rising). The Earth's rotation carries an observer a distance of nearly 13,000 km between moonrise and moonset. The change in viewpoint means that at moonrise an observer can see 1° farther east in lunar longitude than average; at moonset, the observer can see 1° farther west in lunar longitude.

Second, there is the physical effect known as **libration in longitude**. Although the Moon's rotation rate is constant, at exactly 2π radians per sidereal month, its orbital angular speed varies because of the eccentricity of the Moon's orbit. Thanks to the conservation of angular momentum (Kepler's second law), the orbital angular speed is greatest when the Moon is at perigee (closest to the Earth) and lowest when the Moon is at apogee (farthest from the Earth); this is shown schematically in Figure 4.12. Immediately after perigee, when the Moon's orbital angular speed is greater than average, the Moon travels through more than one-quarter of an orbit during the time it takes to make one-quarter of a rotation (going from point 1 to point 2 in Figure 4.12). This enables us, here on Earth, to see a little more of the Moon's right-hand limb than usual when it's at point 2 on its orbit. Conversely, the Moon's slower orbital speed just after apogee enables us

[13] The word "libration" comes from the Latin word *libra* meaning "balance" or "scales" (as in the name of the zodiacal constellation). When you think of how an old-fashioned two-arm balance wobbles back and forth before coming to an equilibrium, you can understand the origins of the word "libration."

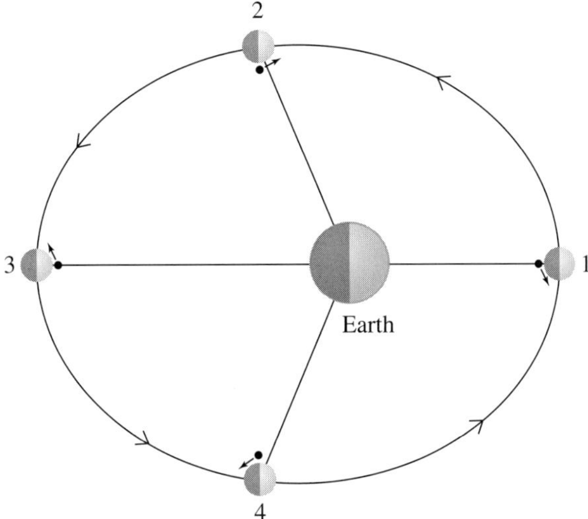

FIGURE 4.12 Libration in longitude is caused by the variable orbital speed of the Moon along its elliptical orbit.

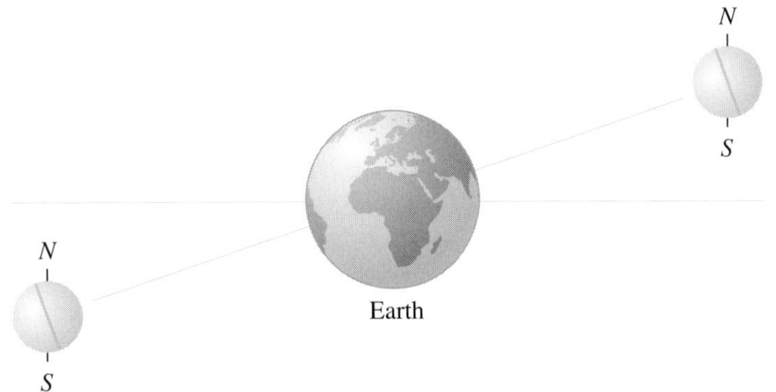

FIGURE 4.13 Libration in latitude is caused by the tilt of the Moon's rotation axis relative to the Moon's orbit around the Earth.

to see a little more of the Moon's left-hand limb than usual when it's at point 4 on its orbit. The Moon's orbit has an eccentricity of $e = 0.055$, which provides a libration in longitude of about 6°.

Third and last is the effect known as **libration in latitude**. As shown in Figure 4.13, the Moon's rotation axis is not perpendicular to the plane of its orbit around the Earth. Instead, the Moon's rotation axis is tilted by $\sim 6.5°$; intriguingly, the direction of the tilt is such that the Moon's rotation axis is nearly perpendicular to the ecliptic plane. As a consequence of the axial tilt, observers on the Earth can see slightly beyond the Moon's north pole when the Moon is south of the ecliptic, and slightly beyond the Moon's south pole when the Moon is north of the ecliptic. Adding together the effects of all three types of libration, it is possible over time to see 59% of the Moon's surface without the necessity of leaving the Earth.

4.6 ▪ ECLIPSES

In general, an **eclipse** can be defined as the passage of one body through the shadow of another. Here on Earth, we experience a **lunar eclipse** when the Earth casts a shadow on the Moon, and a **solar eclipse** when the Moon casts a shadow on the Earth. For the Earth to cast a shadow on the Moon, it must be directly between the Sun and Moon; thus, lunar eclipses occur during full Moon (see Figure 4.10). For the Moon to cast a shadow on the Earth, it must be directly between the Sun and Earth; thus solar eclipses occur during new Moon. The reason there is not a lunar eclipse every full Moon and a solar eclipse every new Moon is that the Moon's orbit is not coplanar with the Earth's orbit. The tilt of 5.1° between the two means that during most full Moons, the Earth's shadow passes north or south of the Moon; similarly, during most new Moons, the Moon's shadow passes north or south of the Earth.

As seen on the celestial sphere, the ecliptic intersects the Moon's path at two points, called the **nodes**. These two points are actually projections onto the celestial sphere of the line of intersection, or **line of nodes**, of the ecliptic plane and the plane of the Moon's orbit. For an eclipse to occur, both the Sun and the Moon must be near a node; a solar eclipse happens when they're near the same node, and a lunar eclipse happens when they're near opposite nodes.[14] Because the Sun and Moon are not point sources but subtend angles of $\sim 0.5^\circ$ as seen from Earth, an exact alignment of Sun and Moon is not required for an eclipse to occur.

The geometry of solar eclipses is shown in Figure 4.14. If the Sun were a geometric point of light, the Moon would cast a simple, sharply defined shadow. Since the Sun is an extended source, however, the Moon's shadow has two parts. The **umbra** is the inner region of the shadow, shaped like a long, tapering cone stretching away from the Moon; within the umbra, the Sun is completely hidden from view and objects are in total shadow.[15] The **penumbra** is the outer region of the shadow, shaped like a widening cone, stretching out to infinity; within the penumbra, the Sun is only partially hidden from view and objects still receive some light from the Sun. Observers in the Moon's penumbra will experience a **partial solar eclipse**; they will see the Sun's disk partially obscured by the Moon. Observers in the Moon's umbra will experience a **total solar eclipse**; the Sun's disk will be completely obscured by the Moon. During a total solar eclipse, observers can see the faint, tenuous outer atmosphere of the Sun, called the corona.

[14] Put another way, *eclipses* occur when the Moon is near the *ecliptic*. The word "eclipse" comes from a Greek word meaning "to fail to appear"; the word "ecliptic" is a truncation of the Latin phrase *linea ecliptica*, meaning "line of eclipses."

[15] "Umbra" is the Latin word for "shadow" (from which we deduce that umbrellas were originally used as sunshades rather than as protection from the rain).

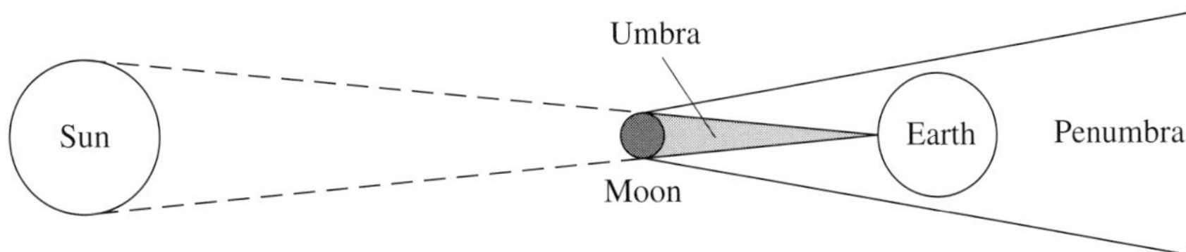

FIGURE 4.14 The geometry of a solar eclipse, showing the Earth's central shadow cone (umbra) and outer partial shadow (penumbra).

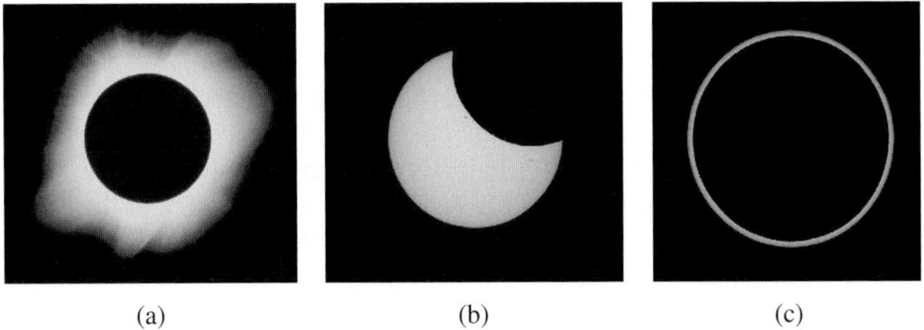

FIGURE 4.15 (a)–(c): total solar eclipse, partial solar eclipse, and annular eclipse.

Figure 4.14 is not to scale, of course. The Sun–Moon distance is actually very long compared to the Sun's radius, and hence the Moon's umbra is very long compared to the Moon's radius. The actual length of the Moon's umbra is $\ell \approx 380{,}000$ km; by a striking coincidence, this is nearly equal to the semimajor axis of the Moon's orbit, $a = 384{,}000$ km.[16] The Moon's orbit is significantly eccentric, though, with $e = 0.055$; this means that the actual Earth–Moon distance varies from $q = (1 - e)a = 363{,}000$ km at perigee to $Q = (1 + e)a = 405{,}000$ km at apogee. As a consequence, observers on the Earth can see a total solar eclipse only when the Moon is close to perigee. When the Moon is near apogee, the tip of the Moon's umbra falls short of the Earth's surface; in that case, observers on Earth can see an **annular eclipse**. During an annular eclipse, the Moon is too small in angular size to blot out the entire disk of the Sun, so observers see a ring, or annulus, of Sun surrounding the Moon. Figure 4.15 shows images of a total solar eclipse (with the Sun's corona prominent), a partial solar eclipse, and an annular eclipse.

Because the Earth's diameter is nearly four times that of the Moon, the Earth's umbra is nearly four times longer than the Moon's. At the average distance of the Moon, the Earth's umbra has a width of about 9000 km; this is more than twice the diameter of the Moon, so the Moon is easily able to fit within the Earth's umbra. The geometry of lunar

[16] The coincidence is a temporary one; remember that the Moon's orbit is increasing in size by 40 km per million years.

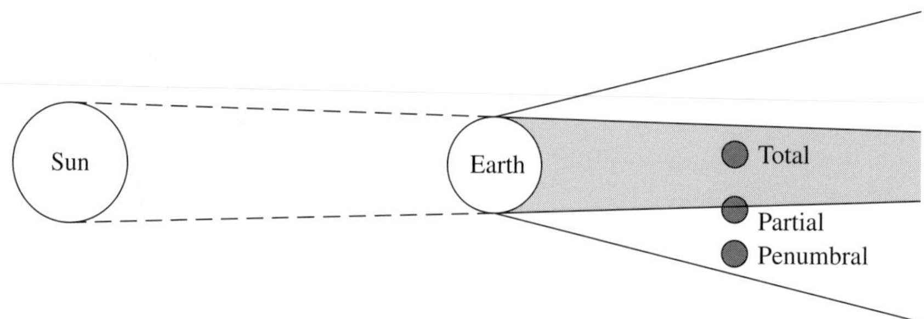

FIGURE 4.16 The geometry for lunar eclipses: total, partial, and penumbral.

eclipses is shown in Figure 4.16. A **total lunar eclipse** occurs when the Moon passes entirely into the Earth's umbra. A **partial lunar eclipse** occurs when only part of the Moon enters the Earth's umbra. A **penumbral lunar eclipse**, which happens when the Moon passes through the penumbra but not the umbra of the Earth, is a fairly frequent occurrence but is barely noticeable from the Earth.

Because the Sun and Moon are of finite angular size, they don't have to be *exactly* at a node for an eclipse to occur. However, computing how far away they can be from a node while still undergoing an eclipse is a bit complicated. Let's consider, for concreteness, the case of a solar eclipse. The average angular radius of the Sun, as seen from the Earth, is

$$\theta_\odot = \frac{R_\odot}{a_\oplus} = \frac{6.96 \times 10^5\,\text{km}}{1.50 \times 10^8\,\text{km}} = 0.00465\,\text{rad} = 0.27^\circ. \tag{4.53}$$

The average angular radius of the Moon is

$$\theta_{\text{Moon}} = \frac{R_{\text{Moon}}}{r_0} = \frac{1.74 \times 10^3\,\text{km}}{3.84 \times 10^5\,\text{km}} = 0.00453\,\text{rad} = 0.26^\circ. \tag{4.54}$$

If an observer on the Earth's surface sees the centers of the Sun and Moon separated by an angular distance $\theta \lesssim \theta_\odot + \theta_{\text{Moon}} = 0.53^\circ$, then that particular observer will experience a solar eclipse. However, the Moon, since it's so nearby, has a significant amount of diurnal parallax when viewed from different points on the Earth. An observer at one location might see a total solar eclipse, while an observer at a second location sees a partial eclipse, and an observer at a third location sees no eclipse at all.[17]

Because of the effects of diurnal parallax on the Moon's apparent location on the celestial sphere, when astronomers talk about *the* angular distance between the Sun's center and the Moon's center, they use the angular distance Δ as measured by an idealized observer at the center of a perfectly transparent Earth. This has the advantage of removing the diurnal parallax effects due to observing the Sun and Moon from different locations

[17] When Hipparchus measured the distance to the Moon (see Section 2.1), he started with the observation that when a total solar eclipse was seen from the Hellespont, the Moon's disk covered only 80% of the Sun's diameter as seen from Alexandria, 10° of latitude away.

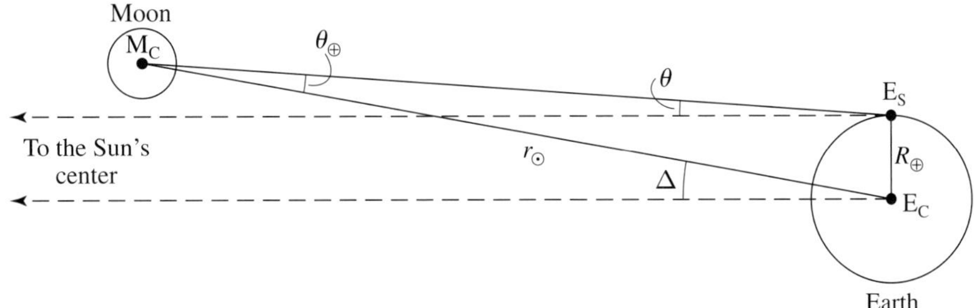

FIGURE 4.17 Geometric arrangement of Earth, Moon, and Sun (offstage left) when the Moon just grazes the Sun as seen from point E_S on Earth. (Not to scale.)

on the Earth. Let's now compute the maximum possible angular separation Δ between the Sun's center and the Moon's center (as measured from the Earth's center) that would still allow a minimal partial solar eclipse to be seen from somewhere on the Earth's surface. The geometry of this bare-minimum solar eclipse is shown in Figure 4.17. In this configuration, an observer at point E_S on the Earth will see the limb of the Moon just barely grazing the limb of the Sun. This means that for the observer at E_S, the centers of the Sun and Moon are separated by an angle $\theta = \theta_\odot + \theta_{\text{Moon}} = 0.53°$. What, then, will be the angle Δ between the centers of the Sun and Moon as seen from the transparent Earth's center?

To simplify matters, we can ignore the effects of diurnal parallax on the distant Sun and assume that the direction of the Sun's center as seen from point E_S is the same as the direction as seen from the Earth's center.[18] In this case, when we look back at Figure 4.17, we see that the triangle formed by point E_S, point E_C (the Earth's center), and point M_C (the Moon's center) has vertices with angles $90° + \theta$ (at E_S), $90° - \Delta$ (at E_C), and $\theta_\oplus$ (at M_C). The angle $\theta_\oplus$ is the shift in the Moon's apparent location as a result of moving from the Earth's center to point E_S on its surface. From the law of sines applied to the triangle $E_S E_C M_C$, we may write

$$\frac{R_\oplus}{\sin\theta_\oplus} = \frac{r_0}{\sin(90° + \theta)}, \tag{4.55}$$

where r_0 is the distance from the Earth's center to the Moon's center. However, since $\theta \ll 90°$, it's safe to make the approximation $\sin(90° + \theta) \approx 1$, and thus

$$\theta_\oplus \approx \frac{R_\oplus}{r_0} = \frac{6.37 \times 10^3\ \text{km}}{3.84 \times 10^5\ \text{km}} = 0.0166\ \text{rad} = 0.95°. \tag{4.56}$$

[18] This is very similar to the assumption made by Eratosthenes (see Section 2.1) when he stated that the direction from Alexandria to the Sun was the same as the direction from Syene to the Sun. Because the Earth is much smaller than the Earth–Sun distance, no matter where we stand on (or in) the Earth, the shift in the Sun's apparent position will be tiny.

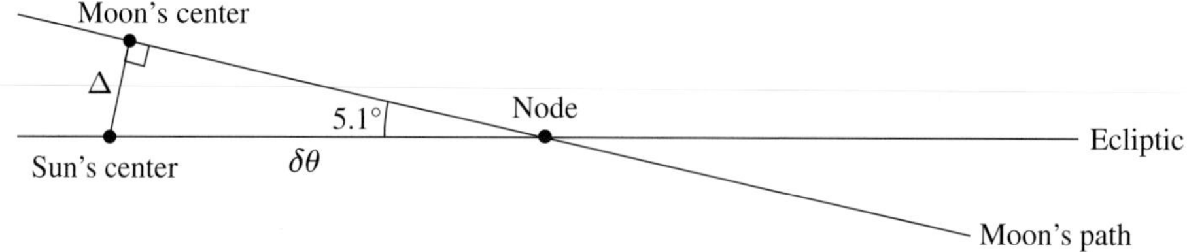

FIGURE 4.18 As seen in projection on the celestial sphere, the Moon and Sun are just barely close enough to the node for an eclipse to occur.

In other words, $\theta_\oplus$ is approximately equal to the angular radius of the Earth as seen from the Moon. Since the angles at the vertices of a triangle add up to 180°, the triangle $E_S E_C M_C$ must satisfy the equation

$$(90^\circ + \theta) + (90^\circ - \Delta) + \theta_\oplus = 180^\circ. \tag{4.57}$$

Solving for Δ, we find

$$\begin{aligned} \Delta &= \theta_\oplus + \theta \\ &\approx \theta_\oplus + \theta_\odot + \theta_{\text{Moon}} \approx 0.95^\circ + 0.27^\circ + 0.26^\circ \approx 1.48^\circ. \end{aligned} \tag{4.58}$$

Thus, even if the Sun's center and the Moon's center are nearly 1.5 degrees apart as seen from the Earth's center, one lucky observer at point E_S on the Earth's surface will glimpse a minimal partial solar eclipse.

When the centers of the Sun and the Moon are separated by an angular distance $\Delta = 1.48^\circ$ on the celestial sphere, what is their maximum possible angular distance from a node? Figure 4.18 shows the configuration in which the Sun is at the maximum possible distance $\delta\theta$ from a node while still being within a center-to-center angular distance Δ of the Moon. From the right triangle in Figure 4.18, we find that

$$\delta\theta_{\text{ps}} = \frac{\Delta}{\sin(5.1^\circ)} = 11.3\Delta = 16.65^\circ. \tag{4.59}$$

Thus, there exists an "eclipse window" of width $2\delta\theta_{\text{ps}} = 33.3^\circ$ centered on each of the two nodes, within which the Sun can undergo a partial solar eclipse. Geometric arguments similar to those presented in Figure 4.17 tell us that the value of Δ required for a central (total or annular) solar eclipse is

$$\Delta \approx \theta_\oplus + \theta_\odot - \theta_{\text{Moon}} \approx 0.96^\circ, \tag{4.60}$$

creating an eclipse window for central solar eclipses of width

$$2\delta\theta_{\text{cs}} = \frac{2\Delta}{\sin(5.1^\circ)} \approx 21.6^\circ. \tag{4.61}$$

The width of the eclipse window for lunar eclipses (both partial and total) can also be computed. For partial lunar eclipses, the width of the eclipse window is $2\delta\theta_{\rm pl} \approx 21.2^\circ$; for total lunar eclipses, the width of the eclipse window is $2\delta\theta_{\rm tl} \approx 9.6^\circ$.[19]

The rate at which the Sun moves along the ecliptic is

$$\left(\frac{360^\circ}{365.26\ \text{days}}\right) \times \left(\frac{29.53\ \text{days}}{1\ \text{synodic month}}\right) = 29.1^\circ/\ \text{synodic month}. \tag{4.62}$$

When we compare this rate to the width of the eclipse window for partial solar eclipses, $2\delta\theta_{\rm ps} = 33.3^\circ$, we note that the Sun cannot pass entirely through this window during one synodic month. Thus, there must be at least one and at most two solar eclipses each time the Sun passes through a node. However, the distance moved by the Sun during one synodic month is greater than the width of the eclipse window for central solar eclipses or the width of the eclipse window for lunar eclipses (either partial or total). Thus, there will be no more than one central solar eclipse and no more than one lunar eclipse each time the Sun passes through a node.

Because of the westward precession of the line of nodes, the interval between nodal passages by the Sun is about nine days less than half a year. Thus, up to three opportunities for eclipses can present themselves in a single calendar year. The minimum number of eclipses during a calendar year is two; if this is the case, both eclipses will be solar eclipses. The number of central solar eclipses in one calendar year ranges from zero to three. However, the Moon's umbral shadow on the Earth is small compared to the Earth's radius, and central eclipses are therefore seen only along a narrow path on the Earth's surface. The number of lunar eclipses per year is also in the range zero to three, but these are seen by many more people, since they are visible to everyone on the night side of the Earth.

Total solar eclipses, as mentioned above, are seen over a limited region of Earth, the path of totality. The Moon orbits the Earth at a speed $v_{\rm Moon} \sim 3400\ \text{km hr}^{-1}$, so its umbra, tied to the Moon, moves at this speed from the west to the east. The Earth also rotates in an eastward direction, but at a speed of only $v_{\rm rot} = 1670 \cos \ell\ \text{km hr}^{-1}$, where ℓ is the latitude. Thus, the Earth's rotation cannot keep up with the speed of the Moon's shadow, and the tip of the umbra sweeps along a path of totality from west to east. Depending on the relative separations of the Sun, Earth, and Moon, the path of totality can be infinitesimally narrow, or as wide as 270 km (such a wide path requires that the Moon be at perigee). Similarly, the duration of totality at any point on the path can be instantaneously short, or as long as 7.5 minutes.

Some eclipses occur in series when specific conditions repeat; the eclipses in the series are thus nearly identical. The most famous eclipse cycle is the **Saros cycle**, which was known to the ancient Babylonians. They noticed, from their observations, that eclipses with similar durations and paths of totality occurred, separated by intervals of 6585-1/3

[19] The numbers quoted here for the width of the eclipse windows are averages only. Since the Earth–Moon distance and the Earth–Sun distance both vary, the width of the eclipse windows around the nodes vary with time.

days (about 18 years, 11 days). The length of the Saros interval is almost exactly 223 synodic months, which comes to 6585.32 days. It is also almost the same as 242 nodical months, where the "nodical month" is the time it takes the Moon to make a complete orbit relative to the precessing line of nodes; 242 nodical months amount to 6585.36 days.

Finally, the Saros interval is almost equal to 239 anomalistic months, where the "anomalistic month" is the time it takes the Moon to make a complete orbit relative to its precessing perigee; 239 anomalistic months amount to 6585.54 days. Moreover, since the Saros interval is nearly an integral number of years, the Sun will be at nearly the same location on the ecliptic at times separated by one Saros interval. If an eclipse takes place at some time t, then at time $t + 6585.333$ days, the relative positions of the Earth, Moon, and Sun will be nearly identical, and a very similar eclipse will occur. As an example, the total solar eclipse of 1999 August 11 was part of a Saros cycle labeled by astronomers as "Saros 145" (Figure 4.19). The previous eclipse of this Saros cycle took place on 1981 July 31; the next eclipse of the cycle will happen on 2017 August 21. The number of days in a Saros interval is greater than an integer by about 1/3; thus, you will notice in Figure 4.19 paths of successive solar eclipses in a Saros are rotated westward by about 120° in longitude.

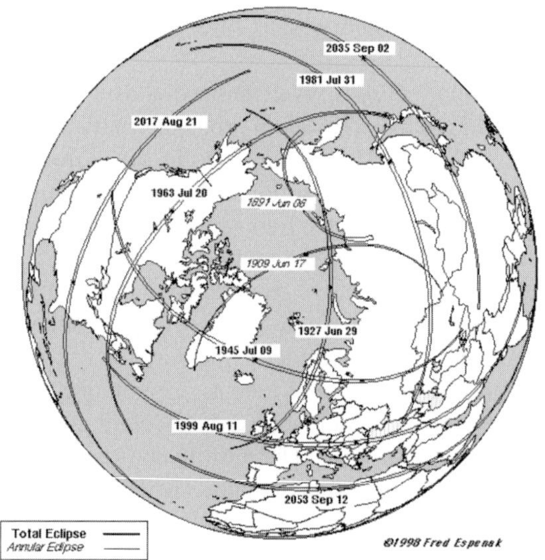

FIGURE 4.19 Central solar eclipses associated with Saros 145.

PROBLEMS

4.1 What is the largest angular distance possible between the center of the Moon's disk and the ecliptic? What is the largest angular distance possible between the center of the Moon's disk and the celestial equator? (Give your answers to the nearest tenth of a degree.)

4.2 How close to the Sun could the planet Jupiter come without suffering tidal disruption?

4.3 Compute the differential tidal force ΔF exerted on the Earth by Mars when it's at opposition. Express your result as a numerical fraction of the differential tidal force exerted by the Moon,

$$\Delta F_{\rm Moon} = \frac{2GM_{\rm Moon} m R_\oplus}{r_0^3},$$

where $r_0 = 384{,}000$ km $= 0.00257$ AU is the Earth–Moon distance and $M_{\rm Moon} = 7.2 \times 10^{22}$ kg is the mass of the Moon. Repeat to find the differential tidal force ΔF exerted by Jupiter at opposition, also expressed as a fraction of $\Delta F_{\rm Moon}$. (Assume that the Moon, Earth, Mars, and Jupiter are on circular coplanar orbits.)

4.4 Imagine a test particle of mass m at the point on the Earth's surface closest to the Moon. Compute the ratio of the differential tidal force $|\Delta \vec{\mathbf{F}}|$ acting on this particle at spring tide to the differential tidal force acting at neap tide.

4.5 A satellite orbits a planet; at the same time, the satellite–planet system orbits a star. Show that if the satellite–planet distance is less than the Hill radius, the sidereal period of the satellite about the planet must be shorter than the sidereal period of the satellite–planet system about the star. (You may assume circular coplanar orbits.)

4.6 Given the amplitudes of lunar librations given in the text, demonstrate that over time 59% of the lunar surface can be seen from the Earth.

4.7 In the timeline used by geologists, the Cambrian period began 542 million years ago. What was the length of the apparent solar day at the beginning of the Cambrian period? (Assume that the slowing of the Earth's rotation, $dP_{\rm rot}/dt$, has been constant.)

4.8 The Earth will be in synchronous rotation with the Moon once its rotation period has increased to 47 days.

(a) How far away will the Moon be from the Earth when this happens?

(b) How long will it be until the Earth attains this synchronous rotation, assuming that $dP_{\rm rot}/dt$ is approximately constant?

4.9 Standing at the Kennedy Space Center (latitude 28° N), you notice the third quarter Moon at your zenith.

(a) Approximately what time of day is it?
(b) Approximately what time of year is it?
(c) Half a synodic month later, what will be the altitude of the first quarter Moon when it makes its upper transit?

5 Interaction of Radiation and Matter

Much of what we know about the universe comes from collecting and analyzing electromagnetic radiation, otherwise known as "light." Thanks to wave–particle duality, light can be thought of either as electromagnetic waves or as a stream of massless particles, called **photons**. Electromagnetic waves are characterized by their wavelength λ, or frequency $\nu = c/\lambda$, where c is the speed of light. Photons are characterized by their energy $E = h\nu$, where h is the Planck constant, $h = 6.626 \times 10^{-34}$ J s. In studying atomic structure, a handy unit of energy is the electron volt (eV), defined as the change in energy of an electron when the electrical potential drops by one volt. When expressed in terms of joules, the electron volt is seen to be a small amount of energy: $1\,\text{eV} = 1.602 \times 10^{-19}$ J. Using electron volts as our unit of energy, the Planck constant is $h = 4.135 \times 10^{-15}$ eV s.

Early astronomers could detect only visible light, that is, light that stimulates a response in the retina of the human eye. Visible light lies in the wavelength range $4 \times 10^{-7}\,\text{m} < \lambda < 7 \times 10^{-7}$ m, corresponding to photons in the energy range $1.8\,\text{eV} < E < 3.1\,\text{eV}$. Modern astronomers, as we see in Chapter 6, have instruments that enable them to detect photons over a much broader range of energies. It has proved convenient for scientists to subdivide the continuous electromagnetic spectrum into different wavelength ranges, from radio waves, which have the longest wavelength and smallest photon energy, to gamma rays, which have the shortest wavelength and highest photon energy. A summary of the main subdivisions of the full spectrum, with approximate ranges in wavelength and photon energy, is given in Table 5.1.

In this chapter, we will study how light and matter interact at the level of individual particles. To begin, we'll explore the nature of atomic structure by examining Niels Bohr's model of the hydrogen atom. Although this derivation is semiclassical rather than fully quantum mechanical, it will give us the insight we need to understand basic atomic structure and its relation to the emission and absorption of light.

5.1 ▪ ATOMIC STRUCTURE

The Bohr model for a hydrogen-like atom begins with a nucleus consisting of Z protons, each with positive electric charge $+e$ and mass m_p, and $A - Z$ neutrons, each with zero electric charge and mass $m_n \sim m_p$. The number A, called the mass number, is the total number of nucleons (protons and neutrons) in the atomic nucleus. The mass of the nucleus

TABLE 5.1 Electromagnetic Spectrum

Type	λ (meters)	$h\nu$ (eV)
radio	$1 \rightarrow \infty$	$0 \rightarrow 10^{-6}$
microwave	$10^{-3} \rightarrow 1$	$10^{-6} \rightarrow 10^{-3}$
infrared	$7 \times 10^{-7} \rightarrow 10^{-3}$	$10^{-3} \rightarrow 1.8$
visible	$4 \times 10^{-7} \rightarrow 7 \times 10^{-7}$	$1.8 \rightarrow 3.1$
ultraviolet	$10^{-8} \rightarrow 4 \times 10^{-7}$	$3.1 \rightarrow 100$
X-ray	$10^{-10} \rightarrow 10^{-8}$	$100 \rightarrow 10^4$
gamma ray	$0 \rightarrow 10^{-10}$	$10^4 \rightarrow \infty$

is $\sim Am_p$, and its charge is Ze. Orbiting the nucleus of the Bohr atom is a single electron with negative charge $-e$ and mass $m_e \approx m_p/1836$. If we treat the electron as a classical particle traveling on a circular orbit of radius r about the atomic nucleus, its acceleration must be $-v^2/r$, where v is the electron's orbital speed. Equating the electromagnetic attraction between the nucleus and electron with the force necessary to keep it on its circular orbit, we have

$$-\frac{(Ze)e}{4\pi\epsilon_0 r^2} = -\frac{m_e v^2}{r}, \tag{5.1}$$

where ϵ_0 is the vacuum permittivity (sometimes called the permittivity of free space); in the SI system, $\epsilon_0 = 8.854 \times 10^{-12}\,\mathrm{C^2\,J^{-1}\,m^{-1}}$.[1] The kinetic energy of the orbiting electron is

$$K = \frac{1}{2}m_e v^2 = \frac{Ze^2}{8\pi\epsilon_0 r}, \tag{5.2}$$

and its potential energy is

$$U = -\frac{Ze^2}{4\pi\epsilon_0 r}. \tag{5.3}$$

Thus, the total energy of the electron is

$$E = K + U = \frac{Ze^2}{8\pi\epsilon_0 r} - \frac{Ze^2}{4\pi\epsilon_0 r} = -\frac{Ze^2}{8\pi\epsilon_0 r}. \tag{5.4}$$

So far, the electron, moving on a circular orbit under the influence of an attractive force that follows an inverse square law, is strongly analogous to a planet moving around a star. Bohr's key insight, though, was that unlike gravitational orbits, the orbital angular momentum of an electron is *quantized*. That is, it can't have an arbitrary value. Only

[1] The coulomb (C) is the SI unit of electric charge; the charge of a single proton is $e = 1.602 \times 10^{-19}$ C.

angular momenta that are integer multiples of some constant value are allowed; more specifically, the orbital angular momentum of an electron must be

$$L = m_e v r = \frac{nh}{2\pi} = n\hbar, \tag{5.5}$$

where the **quantum number** n must be an integer. In equation (5.5), h is the Planck constant and $\hbar$ is the reduced Planck constant, $\hbar \equiv h/(2\pi) = 1.052 \times 10^{-34}$ J s.

By squaring the angular momentum (given by equation 5.5), we find that

$$m_e^2 v^2 r^2 = n^2 \hbar^2. \tag{5.6}$$

However, the kinetic energy relation (equation 5.2) tells us that

$$m_e v^2 = \frac{Ze^2}{4\pi\epsilon_0 r}. \tag{5.7}$$

By substituting equation (5.7) into equation (5.6), and solving for r, we find that the orbital radius for an electron with quantum number n is

$$r_n = \frac{4\pi\epsilon_0\hbar^2}{Ze^2 m_e} n^2 = \frac{5.29 \times 10^{-11}\,\text{m}}{Z} n^2. \tag{5.8}$$

We are now, obviously, entering the realm of the small. When discussing atomic structure, a common unit of length is the **angstrom** (Å), defined as 1 Å $\equiv 10^{-10}$ m. In these units, the orbital radius of the Bohr atom is

$$r_n = \frac{0.529\,\text{Å}}{Z} n^2. \tag{5.9}$$

A schematic drawing of the first three orbits of a Bohr atom is shown in Figure 5.1. Using the relation for the orbital radius (equation 5.8) in the equation for the total electron energy (equation 5.4), we find the energy as a function of the quantum number n:

$$E_n = -\frac{Ze^2}{8\pi\epsilon_0}\left(\frac{Ze^2 m_e}{4\pi\epsilon_0\hbar^2}\right)\frac{1}{n^2} = -\left(\frac{e^2}{4\pi\epsilon_0}\right)^2 \frac{m_e}{2\hbar^2}\frac{Z^2}{n^2}. \tag{5.10}$$

A useful dimensionless number, in this context, is the **fine-structure constant**, defined as

$$\alpha \equiv \frac{1}{4\pi\epsilon_0}\frac{e^2}{\hbar c} \approx 7.30 \times 10^{-3} \approx \frac{1}{137}. \tag{5.11}$$

Using the fine-structure constant α, the electron orbital energy can be written in the more convenient form

$$E_n = -\frac{m_e c^2}{2}\alpha^2 \frac{Z^2}{n^2}, \tag{5.12}$$

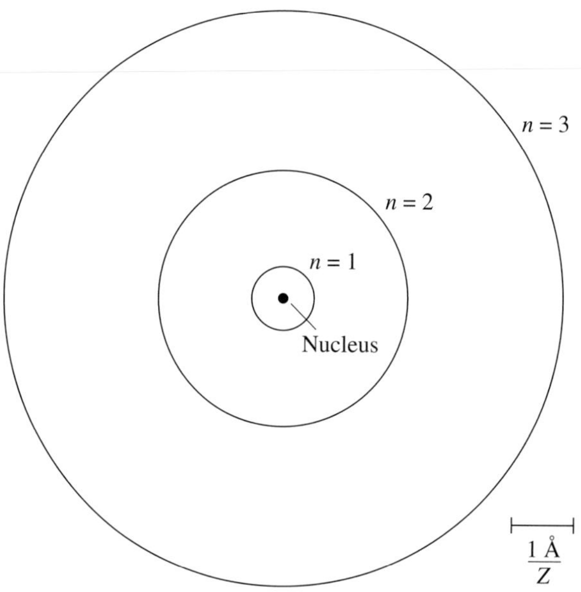

FIGURE 5.1 The $n = 1$, $n = 2$, and $n = 3$ orbits of a Bohr atom depicting the relative size of the orbits; the scale bar in the lower right has a length of 1 Å divided by the charge Z of the central nucleus.

which expresses the orbital energy in terms of the rest energy of the electron, $m_e c^2 = 8.19 \times 10^{-14}\,\mathrm{J} = 0.511\,\mathrm{MeV}$. Since $\alpha^2/2 \approx 3 \times 10^{-5}$, the orbital energy is small compared to the rest energy. Expressed in units of electron volts, the orbital energy is

$$E_n \approx -\frac{0.511\,\mathrm{MeV}}{2}\left(\frac{1}{137}\right)^2 \frac{Z^2}{n^2} \approx -13.6\,\mathrm{eV}\frac{Z^2}{n^2}. \tag{5.13}$$

The orbital radius can also be written conveniently using α:

$$r_n = \frac{\hbar}{m_e c}\frac{1}{\alpha}\frac{n^2}{Z} = \frac{\lambda_c}{2\pi}\frac{1}{\alpha}\frac{n^2}{Z}, \tag{5.14}$$

where $\lambda_c \equiv h/(m_e c) = 0.0243$ Å is the Compton wavelength of the electron.

According to classical physics, since the electron is accelerated, it should be radiating at the rate

$$P = \frac{2}{3}\frac{e^2}{4\pi\epsilon_0}\frac{a^2}{c^3} = \frac{2}{3}\hbar\alpha\frac{a^2}{c^3}, \tag{5.15}$$

where a is the acceleration of the electron.[2] For the electron moving on its circular orbit,

$$a = \frac{v^2}{r} = \frac{Ze^2}{4\pi\epsilon_0}\frac{1}{m_e r^2} = \frac{Z\alpha\hbar c}{m_e}\frac{1}{r^2}, \tag{5.16}$$

and using equation (5.14) to give us the appropriate radius r_n for each orbit,

$$a = \frac{Z\alpha\hbar c}{m_e}\frac{m_e^2c^2\alpha^2Z^2}{\hbar^2n^4} = \frac{Z^3\alpha^3m_ec^3}{\hbar n^4}. \tag{5.17}$$

Substituting the acceleration back into the formula for power radiated, we find

$$P = \frac{2}{3}\alpha^7\frac{(m_ec^2)^2}{\hbar}\frac{Z^6}{n^8}. \tag{5.18}$$

If the orbiting electron actually radiated at this classical rate, the lifetime of the orbit would be

$$\begin{aligned}\tau &= \frac{E}{dE/dt} = \frac{E_n}{P}\\ &= \frac{3}{4}\frac{1}{\alpha^5}\frac{\hbar}{m_ec^2}\frac{n^6}{Z^4}\\ &\approx 4.7\times10^{-11}\,\text{s}\left(\frac{n^6}{Z^4}\right).\end{aligned} \tag{5.19}$$

The time for an electron in the $n = 2$ orbit of hydrogen ($Z = 1$) to decay to the $n = 1$ orbit is then

$$\tau \approx (4.7\times10^{-11}\,\text{s})(2^6) \approx 3\times10^{-9}\,\text{s}. \tag{5.20}$$

The actual lifetime of the $n = 2$ state in hydrogen is $\tau = 1.6\times10^{-9}$ s, so the semiclassical Bohr analysis gives a result correct to within a factor of 2.

The only truly stable state of a Bohr atom is the **ground state**, with $n = 1$. Electrons in **excited states**, those with $n \geq 2$, will decay to states with lower n, losing energy by emitting photons whose energy is equal to the energy difference ΔE between the two states. If an electron transfers from level n to level n', where $n > n'$, the emitted photon will have energy

$$\begin{aligned}\Delta E = E_n - E_{n'} &= \frac{m_ec^2}{2}\alpha^2Z^2\left[\frac{1}{(n')^2} - \frac{1}{n^2}\right]\\ &= 13.6\,\text{eV}\,Z^2\left[\frac{1}{(n')^2} - \frac{1}{n^2}\right].\end{aligned} \tag{5.21}$$

[2] This is just the Larmor formula for the power radiated by an accelerated charged particle; it's a standard formula from classical electromagnetic theory.

The photon energy can be translated into a frequency ν or wavelength λ for the emitted light:

$$\Delta E = h\nu = \frac{hc}{\lambda}. \tag{5.22}$$

In terms of wavelength,

$$\begin{aligned}\lambda = \frac{hc}{\Delta E} &= \left(\frac{2h}{m_e c\alpha^2 Z^2}\right)\left[\frac{1}{(n')^2} - \frac{1}{n^2}\right]^{-1} \\ &= \frac{911.6\ \text{Å}}{Z^2}\left[\frac{1}{(n')^2} - \frac{1}{n^2}\right]^{-1}.\end{aligned} \tag{5.23}$$

Equations (5.21) and (5.23) enable us to compute the photon energies or wavelengths that are emitted or absorbed by hydrogen gas as the electrons undergo **atomic transitions** between one permitted orbit and another. Transitions from higher-energy states to lower-energy states result in emission of photons; transitions from lower to higher energy result from absorption of photons.

The energy level differences in hydrogen follow distinct and recognizable patterns. For example, all the downward transitions that end in the $n = 1$ ground state correspond to wavelengths in the ultraviolet range of the spectrum. This group of transitions is known as the **Lyman series**, named after the physicist Theodore Lyman. In Figure 5.2, we show an energy level diagram for hydrogen based on equation (5.21) and specifying the transitions of the Lyman series. The first line in the Lyman series is the one with the lowest energy (that is, $n = 2 \rightarrow n' = 1$) and is called Lyman α, or Lyα λ1216. The number 1216 is the wavelength, in angstroms, of the light emitted by the transition, as computed in equation (5.23). The next highest energy transition is Lyβ λ1026 ($n = 3 \rightarrow n' = 1$), the next is Lyγ λ972 ($n = 4 \rightarrow n' = 1$), and so forth. The **series limit** for the Lyman series corresponds to $n = \infty$, with a wavelength $\lambda = 912$ Å. When an electron in the ground state ($n = 1$) absorbs a photon with wavelength $\lambda < 912$ Å, the electron becomes unbound and has an energy $E > 0$ that is unquantized.

A second series of transitions in the hydrogen atom is called the **Balmer series**, named after the mathematician Johann Balmer. The Balmer series consists of transitions in which the lower-energy level is the $n = 2$ state, not the $n = 1$ ground state. The first line in the Balmer series corresponds to the $n = 3 \rightarrow n' = 2$ transition; this is the Hα λ6563 line. A wavelength of 6563 Å lies in the red range of the visible spectrum. The Balmer β line is Hβ λ4861, corresponding to a blue-green color. The Balmer series limit is at 3650 Å, in the near ultraviolet.

The **Paschen series**, named after the physicist Friedrich Paschen, consists of the transitions for which the lower-energy level is the $n = 3$ state. In the Paschen series, the lowest energy transition is Paα λ1.87 μ m and the series limit is at 8220 Å, in the near infrared. There are also the Brackett series (to and from the $n = 4$ level), the Pfund series (to and from $n = 5$), and the Humphreys series (to and from $n = 6$). Those series for which the lower energy level is at $n > 6$ don't have special names.

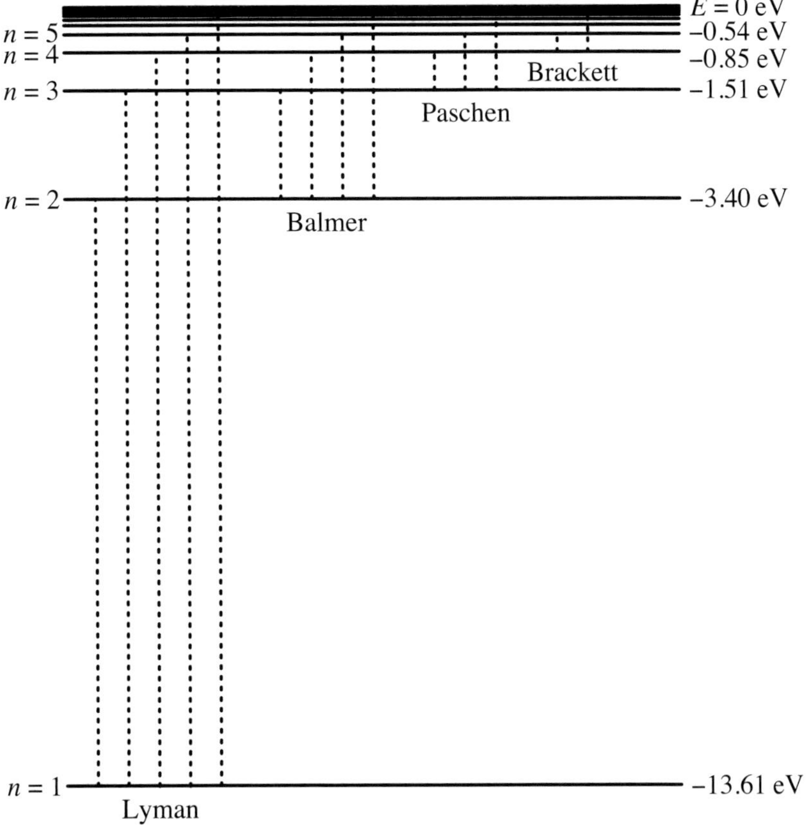

FIGURE 5.2 The energy level diagram for hydrogen, showing the lower-order lines of the Lyman, Balmer, Paschen, and Brackett series. The transitions of the Lyman series are labeled on the left.

The Bohr model of the atom is a useful starting point for studying atomic structure; it helps us to understand some basic principles involving energy and angular momentum, and it accurately predicts the energy levels for a hydrogen-like atom with a single electron. However, for multielectron atoms, the semiclassical approach of the Bohr model fails miserably, and we must compute the energy levels using the full power of quantum mechanics. Although a complete review is beyond the scope of this book, a few key results from quantum mechanics should be remembered.

- Electrons are **fermions**, particles with half-integer spin. As such, they obey the **Pauli exclusion principle**, which states that any given atomic state can be occupied by no more than one electron.

- Atomic systems spontaneously tend toward their lowest permissible energy state. That is, electrons in high-energy states will emit photons as they cascade down to the lowest unoccupied electron energy state.
- The quantum number n in the Bohr model is called the **principal quantum number**. In actuality, there are several quantum numbers that specify the orbital angular momentum and spin angular momentum of the bound electrons, as well as their distribution about the nucleus. Electron orbits, in the quantum mechanical picture, are best described in terms of probability distributions that indicate where the electron might be found.
- For most excited states, there are a number of paths that might be taken to the lowest unoccupied state. The relative probability of an electron making the transition for any particular lower state can be predicted quantum mechanically; it depends on the quantum numbers associated with the upper state and the lower state. The probabilities can be described in terms of **selection rules**. Transitions that do not violate selection rules are called **permitted transitions**. Those that do violate selection rules are called **forbidden transitions**; this is something of a misnomer, since these transitions are not absolutely forbidden, but merely of much lower probability than the permitted transitions.

5.2 ▪ ATOMIC PROCESSES

Now let us consider, at a schematic level, how a single photon interacts with a single atom. In Figure 5.3a, we show the energy level diagram for a simple (and imaginary) two-level atom. The energy difference between the upper level, or excited state (energy E_2), and the lower level, or ground state (energy E_1), is $\Delta E = E_2 - E_1$. If the electron is in the ground state, then the atom can absorb a photon if the photon energy $h\nu$ matches the energy difference ΔE. Absorption of a photon with $h\nu = \Delta E$ leaves the electron in the excited state. From the atom's point of view, we call this process **photoexcitation**; from the photon's point of view, we call it **absorption**. In any case, the energy formerly carried by the photon has been transferred to the atom as internal energy. We can describe the process of photoexcitation symbolically as

$$X + h\nu \rightarrow X^*, \tag{5.24}$$

where X represents the atom in its ground state, and X^* represents the atom in its excited state.

Excitation can also be brought about by a collision with a particle, usually a free electron, as shown in Figure 5.3b. In this case, some of the free electron's kinetic energy is transferred to the internal energy of the atom through **collisional excitation**. We can describe the process of collisional excitation symbolically as

$$X + \frac{1}{2}m_e v^2 \rightarrow X^* + \frac{1}{2}m_e(v')^2, \tag{5.25}$$

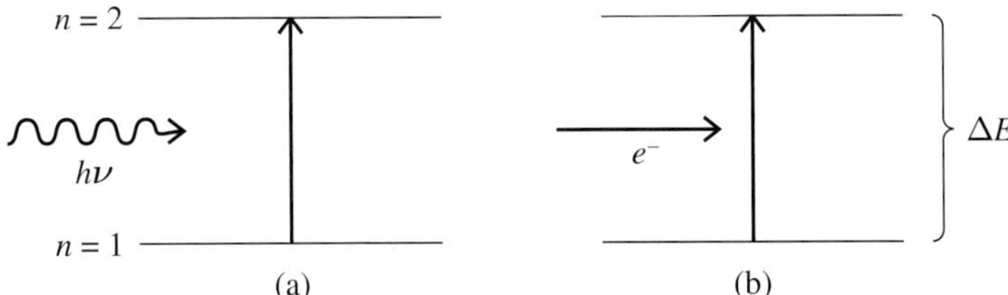

FIGURE 5.3 A schematic two-level atom with states separated by ΔE. (a) Photoexcitation, in which the electron is moved to the upper level by absorption of a photon of energy $h\nu = \Delta E$. (b) Collisional excitation, in which the electron is moved to the upper level by collision with a free electron.

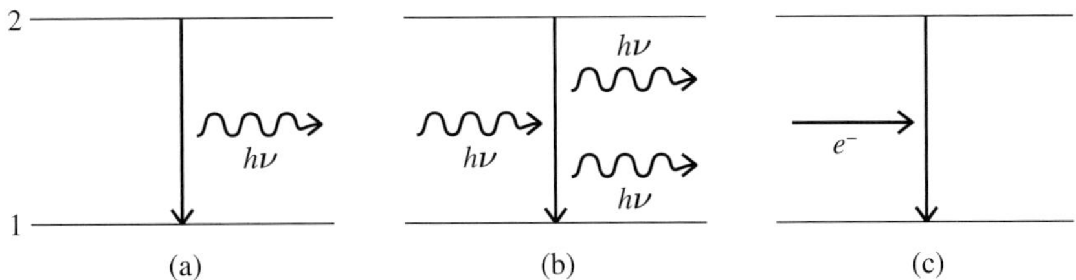

FIGURE 5.4 A schematic two-level atom as in Figure 5.3. (a) Spontaneous emission of a photon. (b) Interaction with a photon resulting in stimulated emission. (c) Collisional de-excitation, which does not result in the emission of a photon.

where v and v' are the speed of the electron before and after the collision; since the electron loses kinetic energy in the process of collisional excitation, $v > v' \geq 0$. Since the electron must initially have at least enough kinetic energy to raise the electron to the excited state, the threshold criterion for collisional excitation is

$$\frac{1}{2}m_e v^2 \geq \Delta E, \tag{5.26}$$

leaving the outgoing electron with a kinetic energy $m_e(v')^2/2 = m_e v^2/2 - \Delta E$.

As illustrated in Figure 5.4, the reciprocal processes to photoexcitation and collisional excitation can also occur. If an atom starts in an excited state, photodeexcitation yields a photon that carries away some of the internal energy of the atom; this can happen by either spontaneous or stimulated emission. **Spontaneous emission**, illustrated in Figure 5.4a, is a result of the inherent instability of excited states; eventually Humpty Dumpty will fall to the ground, even if he isn't pushed. If vacant lower levels exist, eventually an electron in an excited state will spontaneously decay to a lower energy state, producing a photon of energy $h\nu = \Delta E$. Symbolically, the process of spontaneous emission can be written as

$$X^* \to X + h\nu. \tag{5.27}$$

In the case of **stimulated emission**, the downward transition of the electron in the excited state is triggered, or stimulated, by interaction with a photon of energy $h\nu = \Delta E$. An interesting feature of stimulated emission is that the new photon has the same direction and phase as the photon that triggered the transition. The process of stimulated emission can be written as

$$X^* + h\nu \rightarrow X + h\nu + h\nu. \tag{5.28}$$

The rate of stimulated emission is proportional to the intensity of the radiation field at the relevant frequency $\nu = \Delta E/h$. Since stimulated emission increases the number of photons at this frequency, the process of stimulated emission, under the correct conditions, can greatly amplify the intensity of light with frequency $\nu = \Delta E/h$. This is the process that is utilized in lasers.[3]

Figure 5.4c shows that **collisional de-excitation** can occur. In this case, the energy ΔE is transferred to the free electron:

$$X^* + \frac{1}{2}m_e v^2 \rightarrow X + \frac{1}{2}m_e (v')^2, \tag{5.29}$$

where in this case $v' > v$. During the process of collisional de-excitation, a photon is not emitted; all the energy is transferred to the kinetic energy of the outgoing free electron.

When a sufficiently energetic photon is absorbed by a bound electron, the electron can become unbound, as shown schematically in Figure 5.5a. This process is called **photoionization**, and requires a photon of energy $h\nu > \chi$, where χ is the ionization potential; unbound states are unquantized, so any photon with $h\nu > \chi$ can be absorbed. The photoionization process releases an electron whose kinetic energy equals $h\nu - \chi$. Written in symbolic terms, photoionization is the reaction

$$X + h\nu \rightarrow X^+ + \frac{1}{2}m_e v^2, \tag{5.30}$$

where X^+ is a positively charged ion.

Collisional ionization can occur when a free electron with kinetic energy greater than the ionization potential χ collides with the atom (see Figure 5.5b). In this case,

$$X + \frac{1}{2}m_e v^2 \rightarrow X^+ + \frac{1}{2}m_e (v')^2 + \frac{1}{2}m_e (v'')^2, \tag{5.31}$$

where v and v' are the precollision and postcollision speed of the ionizing free electron, and v'' is the postcollision speed of the liberated electron. The speeds v' and v'' are determined by conservation of energy and momentum.

Finally, a free electron can be captured by an ion, with a photon carrying away the excess energy (Figure 5.6); this process is known as **recombination**. During recombination, the electron's kinetic energy contributes to the energy of the photon produced:

$$X^+ + \frac{1}{2}m_e v^2 \rightarrow X + h\nu. \tag{5.32}$$

[3] The word "laser" is an acronym for **l**ight **a**mplification by **s**timulated **e**mission of **r**adiation.

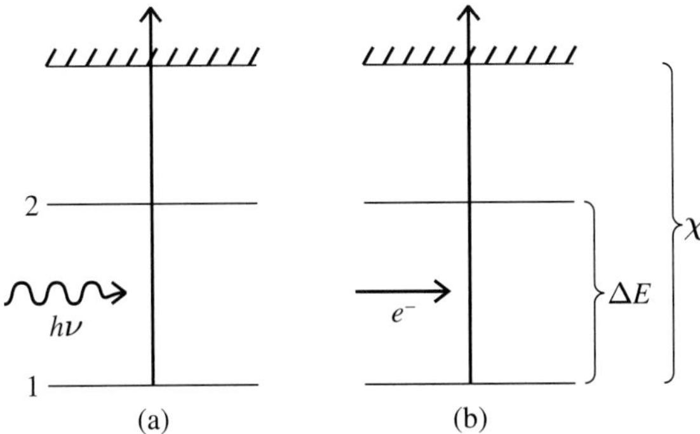

FIGURE 5.5 A schematic two-level atom with ionization potential χ. (a) Photoionization from the ground state by a photon with energy $h\nu > \chi$. (b) Collisional ionization from the ground state by a high-energy electron.

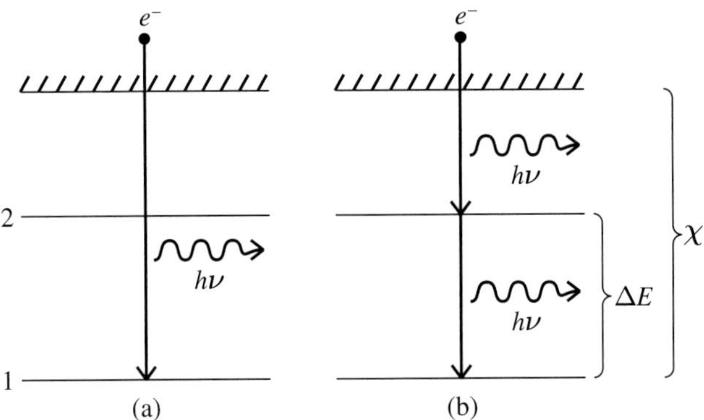

FIGURE 5.6 A schematic two-level atom as in Figure 5.5. (a) A free electron undergoes radiative recombination to the ground state. (b) A free electron undergoes radiative recombination to the upper level, followed by transition to the ground state.

Recombination can occur directly to the ground state, as shown in Figure 5.6a; this produces a photon with energy $h\nu = \chi + m_e v^2/2$. However, recombination can also occur to excited states, as shown in Figure 5.6b; this results in a downward cascade that produces a series of photons whose energies sum to $\chi + m_e v^2/2$.

5.3 ▪ EMISSION AND ABSORPTION SPECTRA

The **spectrum** of an object is its distribution of photons as a function of wavelength or energy. The processes described in the previous section produce identifiable spectra. De-excitations produce photons only at energies corresponding to energy level differences in the atoms in which they originate. Thus, de-excitations produce a distinctive **emission spectrum**. Color Figure 2 shows the emission spectra of the elements hydrogen, helium, oxygen, and neon. Note the Balmer lines in the hydrogen spectrum, and notice how the emission lines in the neon spectrum cluster toward the red end of the visible band; this accounts for the characteristic red color of neon lights. Each emission line is due to a transition between a specific pair of energy levels. Since the energy-level differences are a unique feature of each ion, the pattern of emission lines leads to a unique identification of the ion that produced the photons.

Similarly, if a continuous spectrum (one with light at all wavelengths) is passed through a gas, an **absorption spectrum** is produced. That is, individual absorption lines are produced at energies that correspond to atomic transitions in the ions that make up the gas. Photons at other energies pass freely through the gas. Since the atomic transitions that produce the absorption lines are the same as those that produce emission lines, the distribution of lines in wavelength again uniquely identifies the ion responsible. The basic results for emission and absorption are summarized in empirical principles known as **Kirchhoff's laws:**

- A solid, liquid, or dense gas produces a continuous spectrum (discussed in more detail in Section 5.7).
- A tenuous gas seen against a hot, glowing background produces an absorption spectrum.
- A tenuous gas seen against a cool, dark background produces an emission spectrum.

A schematic summary of Kirchhoff's laws is shown in Color Figure 3; the solid filament of an incandescent lightbulb produces a continuous spectrum; a hydrogen cloud seen in projection against the bulb produces an absorption spectrum; and the hydrogen cloud seen in projection against a dark background produces an emission spectrum.

Real emission and absorption lines obviously have a finite width in wavelength, $\Delta\lambda$, about their central wavelength, λ. To determine what this width will be, let's go back to the case of spontaneous emission, illustrated in Figure 5.4. The excited state has a finite lifetime τ before it suddenly undergoes a transition to a lower state. The probability, per second, that this transition will occur can be computed quantum mechanically and is known as the **Einstein A coefficient.**[4] If there are n_2 ions per unit volume in the $n = 2$ excited state, the number of photons expected per second per unit volume from spontaneous emission will be

$$\frac{dN_{\text{phot}}}{dt} = n_2 A_{21}, \tag{5.33}$$

where A_{21} is the Einstein coefficient for transitions from the $n = 2$ state to the $n = 1$ state. The numerical value of A depends strongly on whether the transition in question

[4] Yes, that Einstein. Although he wasn't tremendously fond of the probabilistic aspects of quantum physics, Einstein was quite adept at quantum mechanical calculations.

is permitted or forbidden. Typical values are

$$A_{21} \sim 10^8\,\mathrm{s}^{-1} \quad \text{(permitted lines)}$$
$$\sim 1\,\mathrm{s}^{-1} \quad \text{(forbidden lines)}. \tag{5.34}$$

Thus, an electron that must make a forbidden transition to jump to a lower energy level will dawdle for $\tau \sim 1$ s in the higher energy level, but an electron that can make a permitted transition will leap to the lower level in $\tau \sim 10$ nanoseconds.

Since the Einstein coefficients simply give transition probabilities, the lifetime of an excited state is uncertain. Thus, the **Heisenberg uncertainty principle** tells us that the energy of the state is also uncertain, which leads in turn to an uncertainty in the energy of the photon produced by the transition from the excited state to a lower state. More specifically, the Heisenberg uncertainty principle states that a particle's position x and momentum p have uncertainties Δx and Δp such that

$$\Delta x \cdot \Delta p \gtrsim \hbar. \tag{5.35}$$

For a photon, with speed c, energy $E = h\nu$, and momentum $p = E/c$, the Heisenberg uncertainty principle can also be written as

$$\left(\frac{\Delta x}{c}\right)(c\Delta p) = \Delta t \cdot \Delta E \gtrsim \hbar, \tag{5.36}$$

where ΔE is the uncertainty in the photon energy and Δt is the uncertainty in its time of creation; we equate Δt with the lifetime τ associated with the transition that produces the photon. Thus, permitted transitions, with $\tau \sim 10^{-8}$ s, produce broader emission (and absorption) lines than forbidden transitions, with $\tau \sim 1$ s.

Thanks to the Heisenberg uncertainty principle, uncertainty in the lifetimes of energy states leads to **natural broadening** of spectral lines. An excited state with principal quantum number n can make a transition to any state with $n' < n$. It thereby becomes useful to define a **damping constant**

$$\gamma_n = \sum_{n'<n} A_{nn'}. \tag{5.37}$$

The **line profile** associated with a particular transition is the observed flux of light per unit frequency (or wavelength) as a function of frequency (or wavelength). For an emission line that shows only natural broadening, the line profile is given by a function called the **Lorentz distribution**:[5]

$$\phi(\nu)d\nu = \frac{(\gamma_n/4\pi)}{(\nu-\nu_0)^2 + (\gamma_n/4\pi)^2}\frac{d\nu}{\pi}, \tag{5.38}$$

where ν_0 is the frequency at the line center. The Lorentz distribution is compared with a Gaussian distribution in Figure 5.7. A notable feature of the Lorentz distribution is that for $|\nu - \nu_0| \gg \gamma_n$, the distribution falls off only as $\phi \propto \gamma_n(\nu - \nu_0)^{-2}$. The wings of the

[5] In probability theory, this function is also called the Cauchy distribution.

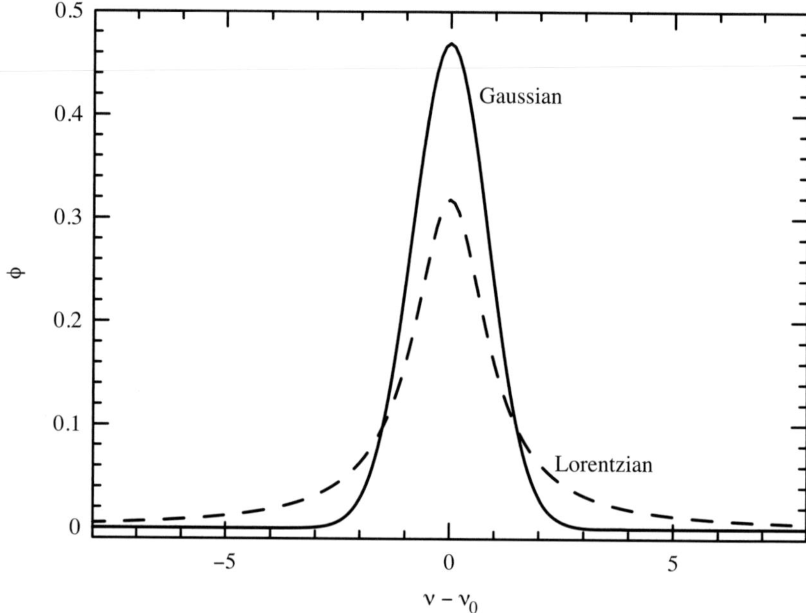

FIGURE 5.7 A Lorentz distribution (dotted line) and a Gaussian distribution (solid line). For each distribution, the function falls to half its maximum value at $\nu - \nu_0 = \pm 1$. The amplitudes are normalized so that the area under each curve is equal to 1.

Lorentz distribution become important when $|\nu - \nu_0| = \Delta\nu \sim \gamma_n/4\pi$. For the Ly$\alpha$ line of hydrogen, for instance, the damping constant is measured to be $\gamma_2 = 6.26 \times 10^8\,\mathrm{s}^{-1}$ (from which we deduce that the Lyman α line is permitted). The width of the Lyα line core is

$$\frac{\Delta\lambda}{\lambda} = \frac{\Delta\nu}{\nu} \approx \frac{\gamma_n/4\pi}{\lambda/c} \approx \frac{(6.26 \times 10^8\,\mathrm{s}^{-1})(1215\,\text{Å})}{4\pi(3 \times 10^8\,\mathrm{m\,s}^{-1})(10^{10}\,\text{Å}\,\mathrm{m}^{-1})} \approx 2 \times 10^{-8}. \tag{5.39}$$

Thus, natural broadening smears out the Lyman α line by only 20 parts per billion.

There are, however, physical mechanisms that can produce much broader emission and absorption lines. Random thermal motions of particles in a gas will produce a temperature-dependent distribution of line-of-sight velocities. Photons emitted or absorbed by a gas at temperature $T > 0$ will thus have a distribution of Doppler shifts, effectively broadening the emission or absorption lines that arise in the gas.

From thermodynamics, we know that the equilibrium distribution of particle speeds in an ideal gas is given by the **Maxwell-Boltzmann distribution**,

$$F(v)dv = 4\pi\left(\frac{m}{2\pi kT}\right)^{3/2} v^2 \exp\left(-\frac{mv^2}{2kT}\right) dv, \tag{5.40}$$

where m is the particle mass, T is the gas temperature, and k is the Boltzmann constant, $k = 1.38 \times 10^{-23}\,\mathrm{J\,K^{-1}} = 8.62 \times 10^{-5}\,\mathrm{eV\,K^{-1}}$. Since $F(v)$ is a probability distribution, the constants in front are chosen to normalize the function such that

$$\int_0^\infty F(v)dv = 1. \tag{5.41}$$

The most probable speed v_p for a particle is the speed v for which $F(v)$ is a maximum. This speed is

$$v_p = \left(\frac{2kT}{m}\right)^{1/2} \approx 1.41 \left(\frac{kT}{m}\right)^{1/2}. \tag{5.42}$$

The average speed $\langle v \rangle$ of particles in the gas is given by

$$\langle v \rangle = \int vF(v)dv. \tag{5.43}$$

Performing this integration yields

$$\langle v \rangle = \left(\frac{8kT}{\pi m}\right)^{1/2} \approx 1.60 \left(\frac{kT}{m}\right)^{1/2}. \tag{5.44}$$

The Maxwell-Boltzmann distribution can also be expressed as a distribution of particle kinetic energy E, rather than particle speed v, using the relation $E = mv^2/2$. Expressed as a distribution of kinetic energy, the Maxwell-Boltzmann distribution is

$$\begin{aligned} F(E) &= F(v)\frac{dv}{dE} \\ &= \frac{2}{\sqrt{\pi} kT}\left(\frac{E}{kT}\right)^{1/2} \exp\left(-\frac{E}{kT}\right). \end{aligned} \tag{5.45}$$

From this equation, we can compute the mean kinetic energy per particle:

$$\langle E \rangle = \int_0^\infty EF(E)dE = \frac{3}{2}kT, \tag{5.46}$$

independent of the particle mass. This implies a mean square speed

$$\langle v^2 \rangle = \frac{2\langle E \rangle}{m} = \frac{3kT}{m} \tag{5.47}$$

and a root mean square (rms) speed of

$$\langle v^2 \rangle^{1/2} = \left(\frac{3kT}{m}\right)^{1/2} \approx 1.73 \left(\frac{kT}{m}\right)^{1/2}. \tag{5.48}$$

If we set up a Cartesian coordinate system, the distribution of speeds along any one axis, given a Maxwell-Boltzmann velocity distribution, is

$$\phi(v_z)dv_z = \left(\frac{1}{2\pi\sigma_z^2}\right)^{1/2} \exp\left(-\frac{v_z^2}{2\sigma_z^2}\right) dv_z, \tag{5.49}$$

where the one-dimensional velocity dispersion is given by the relation $\sigma_z^2 = kT/m$. Knowing the distribution of one-dimensional speeds is useful because when you measure Doppler shifts, in the nonrelativistic regime, you detect only the (one-dimensional) speed along the line of sight.

Because the mass m of a gas particle is a very small number when expressed in kilograms, it is frequently useful to use the dimensionless **molecular mass** μ of the gas particle, defined as its mass m divided by the atomic mass unit u. Technically, the atomic mass unit is defined as 1/12 the mass of a carbon-12 nucleus, or $u = 1.6605 \times 10^{-27}$ kg. In practice, the atomic mass unit differs from the mass of a proton by less than 1% ($m_p = 1.6726 \times 10^{-27}$ kg $= 1.0073u$); thus, we usually write the molecular mass as $\mu = m/m_p$.[6] In terms of the molecular mass μ, the line-of-sight velocity dispersion is

$$\sigma_z = \left(\frac{kT}{\mu m_p}\right)^{1/2} \approx 100\ \mathrm{m\,s^{-1}} \left(\frac{T}{1\,\mathrm{K}}\right)^{1/2} \mu^{-1/2}. \tag{5.50}$$

For a gas of atomic hydrogen, which has $\mu = 1$, a temperature $T = 100$ K corresponds to $\sigma_z = 1\ \mathrm{km\,s^{-1}}$. If helium ($\mu = 4$) is present at the same temperature, its velocity dispersion will be only half as great.

The spread in Doppler shift introduced by thermal motions is

$$\frac{\Delta\lambda}{\lambda} \approx \frac{\sigma_z}{c} \approx 3 \times 10^{-7} \left(\frac{T}{1\,\mathrm{K}}\right)^{1/2} \mu^{-1/2}. \tag{5.51}$$

Comparison of this value with the width of the Lorentz distribution due to natural broadening (equation 5.39) shows that even at low temperatures and high molecular masses, the Doppler broadening is greater than the natural broadening.

In addition to natural broadening and thermal Doppler broadening, there are other physical mechanisms that can broaden spectral lines:

- **Turbulent Doppler broadening** is similar to thermal broadening, except that it involves chaotic bulk motions of gas rather than random motions of individual atoms or ions. In principle, turbulent broadening is distinguishable from thermal broadening by the fact that turbulent broadening doesn't depend on the molecular mass μ of individual particles. We are also likely to suspect that turbulent broadening is present if the line widths are larger than expected given the temperature of the gas (as happens, for instance, in supernova remnants and the broad line regions of active galactic nuclei).
- **Rotational Doppler broadening** occurs in rotating stars, since the line-of-sight velocity varies across the visible hemisphere of the star (unless the rotational axis

[6] The name "molecular mass" is used (in an admittedly sloppy manner) even when the gas particles are atoms, or ions and free electrons.

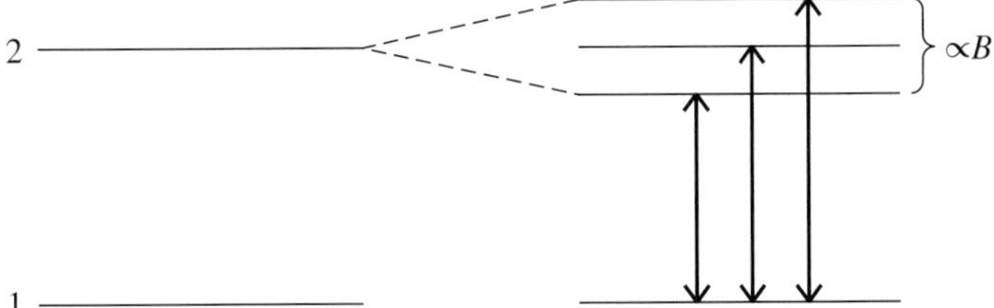

FIGURE 5.8 A schematic two-level atom illustrating Zeeman splitting of the upper level by a magnetic field B.

of the star happens to be pointing straight toward us). Again, rotational broadening is distinguishable from thermal broadening by the fact that rotational broadening is independent of μ; thus, all the star's absorption lines will be rotationally broadened by the same amount. In addition, rotational broadening produces lines that are not Gaussian in shape.

- **Pressure broadening**, also called collisional broadening, occurs when the number density of atoms and ions is sufficiently high that they cannot be regarded as isolated systems. When an atom, for instance, undergoes a close encounter with another atom, or an ion, or a free electron, the electric field of the intruding particle causes an upward or downward shift of the atom's energy levels. For an ensemble of atoms, the random upward and downward shifts in energy levels have the effect of broadening the observed emission or absorption lines. When this "pressure broadening" is a significant effect, the line profile can be written in the form

$$\phi(\nu)d\nu = \frac{(\Gamma/4\pi)}{(\nu - \nu_0)^2 + (\Gamma/4\pi)^2} \frac{d\nu}{\pi}, \tag{5.52}$$

  where $\Gamma = \gamma_n + 2N_{\rm col}$; here, γ_n is the usual damping constant given in equation (5.37) and $N_{\rm col}$ is the rate at which atoms or ions undergo encounters close enough to significantly shift their energy levels.

- **Zeeman broadening** or **Zeeman splitting** occurs in the presence of a magnetic field. The magnetic field breaks the degeneracy of certain atomic states. In the simplest case, known as the normal Zeeman effect, the initial energy E of the affected atomic state is split into three levels, $E - \Delta E_B$, E, and $E + \Delta E_B$, where $\Delta E_B \propto B$, as shown in Figure 5.8. With a sufficiently high-resolution spectrograph, the individual lines can be seen, and the effect is known as Zeeman "splitting." With a lower-resolution spectrograph, the individual lines blur together into a single broadened line, and the effect is called Zeeman "broadening."

5.4 ▪ THE EQUATION OF RADIATIVE TRANSFER

We are now ready to consider how photons interact with matter on scales larger than the size of an atom. We begin with a simple case: photons pass through a gas in which

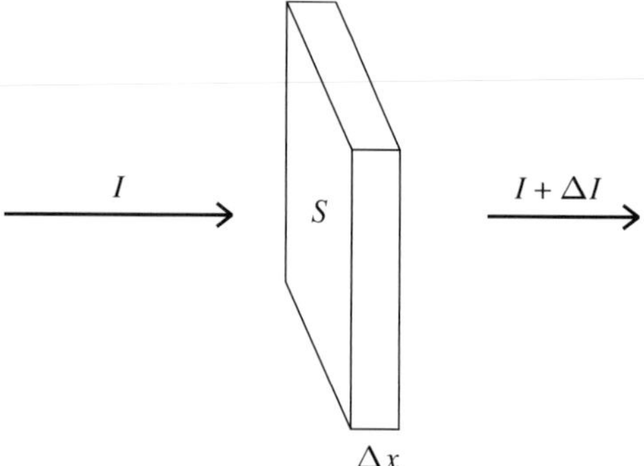

FIGURE 5.9 A volume element of surface area S and thickness Δx. The number density of absorbing particles is n; each particle has a cross-section σ for absorption.

some of the photons are absorbed by atoms. In Figure 5.9, we consider light entering a medium with some intensity I, where the intensity specifies the number of photons passing through a unit area in a particular direction per unit time.[7] We consider the interactions occurring in a volume element of surface area S and thickness Δx, where Δx is sufficiently small that the individual atoms (or other absorbing particles) within the volume do not overlap or shadow each other. The intensity of the stream of light emerging from our box of gas is $I + \Delta I$; if there are no sources of light within the box, then we expect $\Delta I \leq 0$.

It is a useful convention to express the probability that a photon will be absorbed by an absorbing particle in terms of a **cross-section** σ. Although the cross-section has units of area, it isn't necessarily equal to the geometric cross-section of the absorbing particle. In fact, the cross-section is usually a strong function of wavelength if the absorption is due to atomic processes such as photoexcitation. If we assume that the volume element in Figure 5.9 has only a single absorbing particle within it, then the fraction of light lost in passing through is

$$\frac{\Delta I}{I} = -\frac{\sigma}{S}, \tag{5.53}$$

where the right-hand side is simply the fraction of the surface area occupied by a single particle. If there are N nonoverlapping, absorbing particles within the volume element, then the fraction of light lost is N times greater. The number of absorbing particles within the volume element is

$$N = nS\Delta x, \tag{5.54}$$

[7] As we will see later, we can also define the intensity in terms of the *energy* carried by the photons. For the moment, however, we will simply count the photons instead of undertaking the more arduous bookkeeping task of adding up their energies.

where n is the number density of particles. Thus, equation (5.53) becomes

$$\frac{\Delta I}{I} = -(nS\Delta x)\frac{\sigma}{S} = -n\sigma\Delta x. \quad (5.55)$$

In the limit where the volume element becomes infinitesimally thin ($\Delta x \to 0$), this becomes the simple differential equation

$$\frac{dI}{I} = -n\sigma dx. \quad (5.56)$$

If the number density n and cross-section σ of absorbing particles are independent of x, equation (5.56) is easily integrated to yield

$$\ln I = -n\sigma x + C, \quad (5.57)$$

where C is a constant of integration. If we impose the boundary condition that $I = I_0$ at $x = 0$, this equation becomes

$$I(x) = I_0 e^{-n\sigma x}. \quad (5.58)$$

Equation (5.58) is the simplest form of the **equation of radiative transfer**. The only physical process going on is absorption; we see, in this case, that the light becomes attenuated exponentially as it shines through the gas.

We can generalize equation (5.58) slightly to the case where the number density n is a function of x, but the cross-section σ is not. Specifically, we define the **column density** of the gas as

$$N(x) \equiv \int_0^x n(x')dx'. \quad (5.59)$$

The column density has units of m^{-2} and represents the total number of absorbing particles in a column with a cross-sectional area of $1\,\text{m}^2$ and a length of x. The column density of the Earth's atmosphere, for instance, integrated from sea level upward, is $N \sim 2 \times 10^{29}$ molecules m^{-2}. We can also define a dimensionless number called the **optical depth**:

$$\tau(x) = \sigma \int_0^x n(x')dx' = \sigma N(x). \quad (5.60)$$

The optical depth is a measure of how much the intensity of light is attenuated by traveling through the gas; if n is constant, then $\tau(x) = n\sigma x$. In terms of the optical depth, the intensity $I(x)$ can be written

$$I(x) = I_0 e^{-\tau(x)}. \quad (5.61)$$

If a particular blob of gas (or other material) has $\tau \ll 1$, it is referred to as "optically thin," or "transparent"; if it has $\tau \gg 1$, it is called "optically thick," or "opaque." The average distance that a photon will travel through a gas before being absorbed is the **mean free**

path. For a gas with constant n and σ, the mean free path can be computed as

$$\langle x \rangle = \frac{\int_0^\infty x e^{-n\sigma x} dx}{\int_0^\infty e^{-n\sigma x} dx} = \frac{1}{n\sigma}. \tag{5.62}$$

The mean free path is thus the distance over which the optical depth grows from $\tau = 0$ to $\tau = 1$.

The cross-section σ_ν is, in general, a function of frequency. Thus, the equation of radiative transfer (eq. 5.61) should more properly be written in the form

$$I_\nu = I_{\nu,0} e^{-\tau_\nu(x)} = I_{\nu,0} e^{-\sigma_\nu N(x)}, \tag{5.63}$$

where the **specific intensity** is a frequency-dependent quantity. For a single absorption line, $\sigma_\nu = \sigma_0 \phi(\nu)$, where σ_0 is the total cross-section for that particular absorption line and $\phi(\nu)$ is the profile function, which for a single absorbing atom is the Lorentz distribution (equation 5.38).

5.5 ▪ THE CURVE OF GROWTH

Consider the absorption line due to one particular atomic transition in one particular type of atom or ion. We expect the shape of that absorption line to depend on the optical depth at the line center, τ_0. At low optical depth, $\tau_0 \lesssim 1$, the shape and width of the absorption line is usually determined by thermal Doppler broadening; the shape in this case will be Gaussian, and the width will depend on the temperature of the gas and the molecular mass of the absorbing particles (equation 5.51).

At larger optical depth, $1 \lesssim \tau_0 \lesssim 10^4$, the line core **saturates**; that is, the line is black at the line center, because none of the photons with a wavelength near the line center make it through the absorbing gas. The observed width of the line, however, grows only very slowly with optical depth. For example, Figure 5.10 shows the Lyman α absorption line for $\tau_0 = 1, 3, 10, 30$, and 100. The wings of the Gaussian are produced by those very few absorbing atoms that are traveling much faster than σ_z toward us or away from us along the line of sight; even when you increase the number of absorbing atoms along the line of sight, there aren't going to be many of those anomalously fast atoms.

At large optical depth, $\tau_0 \gtrsim 10^4$, the cumulative effect of the $(\nu - \nu_0)^{-2}$ wings of the Lorentz profile (see Figure 5.7) becomes important. While the wings for a line with small optical depth are weak, the column density of atoms is now so large that even photons far displaced from the line center ($|\nu - \nu_0| \gg \gamma_n$) have a fair probability of being absorbed.

In many astronomical applications, the line profiles are not fully resolved. The principal challenge of astronomical observations is that the observed sources tend to be very faint. It is thus impractical to disperse light into a spectrum that has such high resolution in wavelength that line profiles can be studied in great detail. Fortunately, as we will show, high resolution spectra are not necessary for many purposes.

Even at fairly low spectral resolution, absorption lines are still easily detected, and the amount of light missing from the spectrum due to absorption is a measurable quantity.

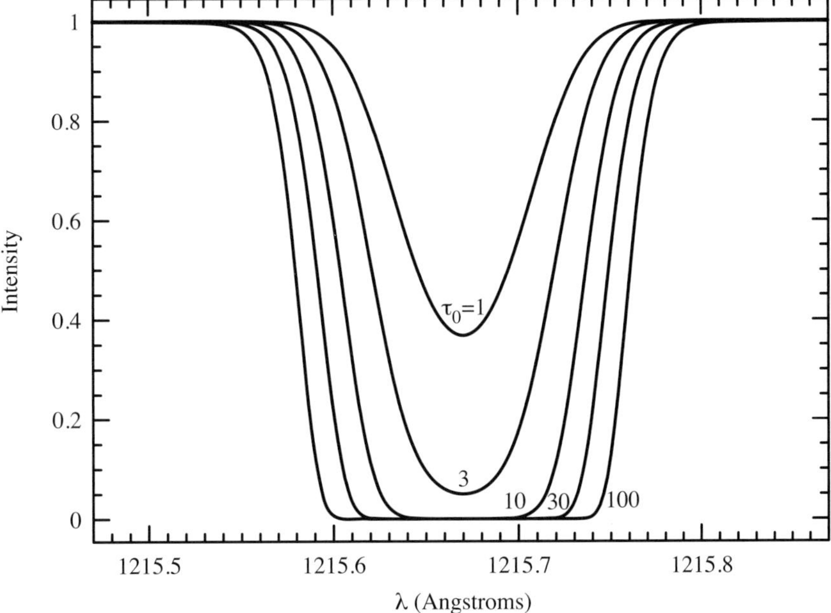

FIGURE 5.10 Lyman α absorption for τ_0 ranging from 1 to 100. Note that as τ_0 more than triples from 30 to 100, the line width increases by only a small amount.

As shown on the right in Figure 5.11, we can define an artificial absorption line that is completely black at its bottom ($I_\lambda = 0$) and has perfectly vertical sides. The width W of the artificial absorption line can be varied until the area of the artificial line is the same as that of the real absorption line. When this is the case, we refer to the parameter W as the **equivalent width** of the line. The equivalent width is usually found by plotting I_λ versus λ (as opposed to I_ν versus ν) and thus has wavelength units. A formal definition of the equivalent width, based on Figure 5.11, is

$$W = \int \frac{I_{\text{continuum}} - I_{\text{line}}}{I_{\text{continuum}}} d\lambda = \int \left[1 - e^{-\tau_\lambda}\right] d\lambda. \tag{5.64}$$

The equivalent width is a useful parameter for astronomers who are doing spectroscopy. It is easily measured even in low-resolution spectra, and it is a function of the not-so-easily measured parameter τ_0, the optical depth at the center of the absorption line.

A plot of an absorption line's equivalent width W as a function of τ_0 (or alternatively, as a function of the column density N of the absorbing atoms) is called the **curve of growth**. A curve of growth for a particular absorption line is shown in Figure 5.12, where the dependence of W upon τ_0 is seen to have three different regimes:

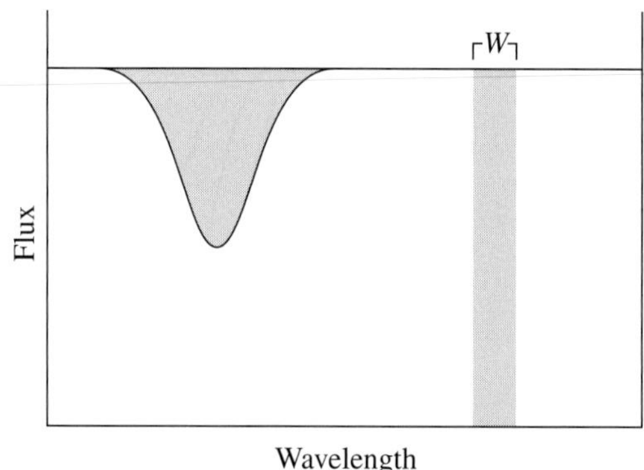

FIGURE 5.11 Determining the equivalent width W of an absorption line; the two shaded regions are of equal area.

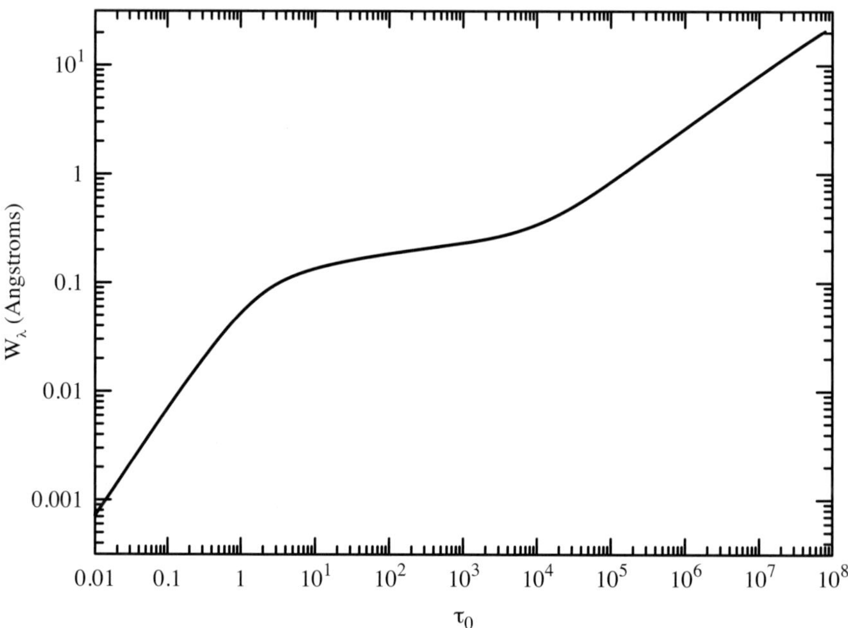

FIGURE 5.12 Curve of growth for the Lyman α absorption line, showing equivalent width W as a function of τ_0.

1. When $\tau_0 \lesssim 1$, $W \propto \tau_0$. In this regime, the absorption line has not saturated, and its strength is directly dependent on the number of absorbers along the line of sight. This is the "linear part" of the curve of growth.

2. When $1 \lesssim \tau_0 \lesssim 10^4$, the center of the line is saturated, and the equivalent width grows only slowly with increasing τ_0 as the optical depth grows in the weak Doppler wings of the absorption lines. In this "Doppler part" of the curve of growth, $W \propto \sqrt{\ln \tau_0}$. This function grows so slowly with increasing τ_0 that it is sometimes called the "flat part" of the curve of growth.

3. When $\tau_0 \gtrsim 10^4$, even the Doppler wings have saturated. Now, though, such a large number of absorbing atoms are along the photon path that the damping wings of the Lorentz distribution become important. In this regime, $W \propto \tau_0^{1/2}$; this is called the "square root part," or the "damping part," of the curve of growth.

5.6 ▪ LOCAL THERMODYNAMIC EQUILIBRIUM

In everyday life, when we talk about the temperature of an object, we are referring to what thermodynamicists technically call the **kinetic temperature**, which is a measure of the average kinetic energy per particle associated with random thermal motions. For instance, suppose that the air around you has a temperature of 68° Fahrenheit, equivalent to 20° Celsius, or 293 Kelvin. This means that the random speeds of the air molecules reflects a Maxwell–Boltzmann distribution (equation 5.40):

$$F(v) \propto v^2 \exp\left(-\frac{mv^2}{2kT}\right), \tag{5.65}$$

where m is the mass of a molecule, T is the kinetic temperature of the gas, and $k \sim 10^{-4}$ eV K is the Boltzmann constant. As we have already seen (from equation 5.46), the average kinetic energy per molecule is

$$\frac{1}{2}m\langle v^2\rangle = \frac{3}{2}kT. \tag{5.66}$$

Because of the frequent collisions between molecules, all the molecules in the air—nitrogen, oxygen, carbon dioxide, and so forth—all have the same average kinetic energy $(3/2)kT = 0.038$ keV, and thus have the same kinetic temperature $T = 293$ K.[8]

If every component of a system is characterized by the same temperature T, that system is in **thermodynamic equilibrium**. Isolated systems have a natural tendency to approach thermodynamic equilibrium, as heat flows from high-temperature regions to low-temperature regions; this is just a result of the second law of thermodynamics. However, in real systems, absolute thermodynamic equilibrium is rare. Consider the air

[8] This implies that the more massive molecules are moving more slowly, on average, than the less massive molecules. At room temperature, nitrogen molecules, with molecular mass $\mu = 28$, have an average random speed $v \approx 490$ m s^{-1}. Carbon dioxide molecules, with $\mu = 44$, have an average speed of only $v \approx 390$ m s^{-1} at the same temperature.

inside the room where you are reading this book.[9] The temperature of the air will not be identical throughout the room. (For one thing, *you* provide a heat source, raising the temperature in your immediate vicinity.) In addition, at any given location in the room, not all the particles will be in thermodynamic equilibrium with each other. The air molecules have an average kinetic energy $\langle E \rangle \approx 0.04\,\text{eV}$. However, the photons of visible light that stream through the air and enable you to read the book have an average energy $h\nu \approx 3\,\text{eV}$. Although the molecules and photons coexist spatially, they have a low probability of interaction; in other words, the mean free path for the photons is much larger than the size of the room. This means that the molecules and photons fail to come into equilibrium.

Although absolute thermodynamic equilibrium is rare, it is useful to employ the concept of **local thermodynamic equilibrium** (LTE). Consider a cube of air (or any other material) with a side of length L. If the length L is much larger than the mean free path λ for particle collisions, and much smaller than the length scale over which temperature varies significantly, then we state that the particles in the cube are in local thermodynamic equilibrium. For instance, at sea level on Earth, the number density of molecules in the atmosphere is $n \approx 2.5 \times 10^{25}\,\text{m}^{-3}$; for the most common molecules in the atmosphere, the cross-section for molecule–molecule collisions is $\sigma \approx 1 \times 10^{-18}\,\text{m}^2$, or approximately 100 square angstroms. The mean free path for molecules is then

$$\lambda \approx \frac{1}{n\sigma} \approx 4 \times 10^{-8}\,\text{m} \approx 400\,\text{Å}. \tag{5.67}$$

An air molecule will travel just 400 Å, on average, before colliding with another molecule. If a cube is much larger than 400 Å on a side, then the molecules within it will collide many times and will be able to come to temperature equilibrium. In general, a macroscopic cube will have a temperature difference ΔT between its hottest neighborhood and its coolest. However, as long as $\Delta T \ll T$, where T is the average temperature in the cube, we may safely make the approximation that the colliding particles in the cube are in LTE at a temperature T.

Since the Earth's atmosphere is transparent to many wavelengths of light, the molecules in the air are in LTE with each other, but not with the photons passing through. Gaseous interstellar nebulae and the atmospheres of stars are also examples of transparent systems in which the photons are not in LTE with the massive particles present.[10] The interior of a star, which is composed of dense, ionized gas, is an example of a system where the mean free path for photons is short compared to the total system; thus, it is possible to carve out regions of a stellar interior where the assumption of LTE holds true for photons as well as for the ions and free electrons present. For local thermodynamic equilibrium to apply to photons as well as to the massive molecules, atoms, and ions in a system, the following conditions suffice:

1. Both photons and massive particles have a high number density.
2. The system is optically thick ($\tau \gg 1$) at all wavelengths.

[9] If you are doing your reading *al fresco*, imagine yourself inside a room.

[10] Although LTE is not formally true in the atmospheres of stars, we will still find that it is not a wretchedly bad first approximation.

If the system is optically thin, photons will not be able to come into equilibrium with the massive particles. If the number density of particles is low, their mean free path for collisions will be long, and they may not be able to come into equilibrium with each other.

The assumption of local thermodynamic equilibrium is a very powerful one, since it allows us to bring up the big guns of statistical mechanics and thermodynamics. For instance, in LTE, the relative populations of different states of an atom or ion are described by the **Boltzmann equation**. For the simple two-level atom from Figure 5.3, the Boltzmann equation is

$$\frac{n_2}{n_1} = \frac{g_2}{g_1} \exp\left(-\frac{\Delta E}{kT}\right), \tag{5.68}$$

where n_1 and n_2 are the number density of atoms with electrons in level 1 and level 2; g_1 and g_2 are the statistical weights of the first and second levels, accounting for degeneracy (multiple states with the same energy); and ΔE is the energy gap between the two levels. For a system in LTE, the temperature T in equation (5.68) is just the local kinetic temperature. The Boltzmann equation is the reflection of a statistical equilibrium; while any individual electron will undergo a round of excitations, de-excitations, ionizations, and recombinations, a large ensemble of electrons will satisfy the Boltzmann equation in a statistical sense.

Similarly, for a system in LTE, the **Saha equation** gives the relative population of different ions of a particular element:

$$\frac{n^{i+1} n_e}{n^i} = 2\frac{Q^{i+1}}{Q^i} \left(\frac{m_e kT}{2\pi \hbar^2}\right)^{3/2} \exp\left(-\frac{\chi_i}{kT}\right), \tag{5.69}$$

where n_e is the number density of free electrons, n^i is the number density of particles in ionization state i (with i electrons stripped away), and χ_i is the ionization potential from state i to state $i+1$. The numbers Q^i and Q^{i+1} are **partition functions**, which can be thought of as a sum of the statistical weights of all possible excitation states of an ion, weighted by the relative probability that the ion is in that state. For our purposes, we need merely note that the factor $2Q^{i+1}/Q^i$ in equation (5.69) is a dimensionless number that is usually of order unity.

By combining the Boltzmann equation and Saha equation, we can determine the population for any particular atomic state, and thus predict the strength of the absorption lines produced by a gas. Stellar absorption lines are produced in the atmospheres of stars, where LTE is only a rough approximation, but the assumption of LTE is a useful place to start. As an example, consider the population of the $n = 2$ level of atomic hydrogen; this is a useful example because the Balmer absorption lines, much used by astronomers, arise from this state. Hydrogen has only two ionization states, neutral and singly ionized, so the total number density of hydrogen is

$$n = n^0 + n^1, \tag{5.70}$$

where n^0 is the number density of neutral hydrogen atoms and n^1 is the number density of positive hydrogen ions—in other words, bare protons. A neutral hydrogen atom has a large number of energy levels (technically, an infinite number), so the total number density of neutral hydrogen atoms is

$$n^0 = \sum_i n_i^0. \tag{5.71}$$

(Please remember that we are using superscripts to designate ionization states, and subscripts to designate excitation states of an ion.) To determine the strength of Balmer absorption, we need to know the number density of neutral hydrogen atoms with electrons in the $n = 2$ level. Relative to the total number of hydrogen atoms and ions, this is

$$\begin{aligned} \frac{n_2^0}{n} = \frac{n_2^0}{n^0 + n^1} &= \frac{n_2^0}{n_1^0}\left(\frac{n_1^0}{n^0 + n^1}\right) \\ &= \frac{n_2^0}{n_1^0}\left(\frac{1}{(n^0/n_1^0) + (n^1/n_1^0)}\right). \end{aligned} \tag{5.72}$$

The temperature in typical stellar atmospheres is $T \lesssim 50{,}000$ K, corresponding to thermal energy $kT \lesssim 4$ eV. This means that the Boltzmann factor $\exp(-\Delta E/kT)$ in equation (5.68) is small, since it takes an energy $\Delta E = 10.2$ eV to lift an electron to the $n = 2$ energy level. Thus, most of the neutral hydrogen atoms in a stellar atmosphere are in the ground state: $n^0 \approx n_1^0 \gg n_2^0$. Accordingly, in equation (5.72) we may safely make the substitution $n_1^0 = n^0$, yielding

$$\frac{n_2^0}{n} \approx \frac{n_2^0}{n_1^0}\left(\frac{1}{1 + (n^1/n^0)}\right). \tag{5.73}$$

The first term on the right-hand side of this equation is given by the Boltzmann equation (eq. 5.68); the second term can be found from the Saha equation (eq. 5.69). Figure 5.13 shows a plot of n_2^0/n_1^0 from the Boltzmann equation, a plot of n^1/n^0 from the Saha equation, and a plot of n_2^0/n from equation (5.73). We see that the fraction of hydrogen that has a bound electron in the $n = 2$ state peaks at a temperature $T \approx 10{,}000$ K. At temperatures much cooler than 10,000 K, most of the bound electrons are in the ground state ($n = 1$), because collisions are not sufficiently energetic to populate the higher levels. At temperatures much hotter than 10,000 K, few of the electrons are bound to hydrogen atoms at all, since collisions are sufficiently energetic to ionize the hydrogen. Consequently, stars with atmospheric temperatures around 10,000 K are those that produce the strongest Balmer absorption lines.

We can repeat this analysis for all atomic species, finding that different absorption lines of different ionization states of different atoms have maximum strength at different temperatures. Thus, the relative strength of the absorption lines in a stellar atmosphere can be used to determine the star's temperature. This is the physical underpinning for stellar spectral classification, which we discuss in Chapter 14.

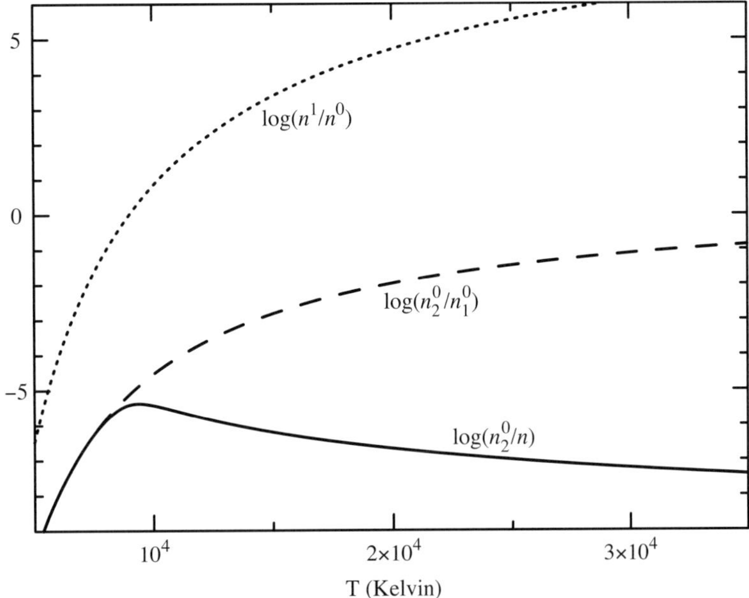

FIGURE 5.13 Dotted line: the ratio n^1/n^0 from the Saha equation (a free electron density $n_e = 10^{20}$ m^{-3} is assumed). Dashed line: the ratio n_2^0/n_1^0 from the Boltzmann equation. Solid line: the ratio n_2^0/n from equation (5.73), that is, the fraction of hydrogen atoms that have a bound electron in the $n = 2$ state.

5.7 ▪ BLACKBODY RADIATION

So far, we have discussed the production of emission and absorption spectra by low-density gas. It is time to consider why dense gases, liquids, and solids produce a continuous spectrum, as described empirically in the first of Kirchhoff's laws. Let's start by considering the simple case of an ideal gas in LTE.

The technical definition of the **specific intensity** I_ν can be illustrated by considering a small, perfectly transparent window of area dA. We create a unit vector $\hat{\mathbf{n}}$ that is normal to the surface of the window, and select a small angular area $d\Omega$ around the point on the celestial sphere at which $\hat{\mathbf{n}}$ is pointing. We then ask how much energy dE is carried through the window during a time dt, carried by photons coming from the solid angle $d\Omega$ with frequencies in the range $\nu \to \nu + d\nu$. The answer to our question is

$$dE = I_\nu \, dt \, dA \, d\Omega \, d\nu. \tag{5.74}$$

The intensity I_ν thus has units of J s^{-1} m^{-2} steradian^{-1} Hz^{-1}.[11] The specific intensity can also be expressed as a function of wavelength rather than frequency, I_λ rather than

[11] 1 joule per second is widely known as the "watt," abbreviated W.

I_ν. The intensity I is simply the integral of the specific intensity over all frequencies (or wavelengths):

$$I = \int_0^\infty I_\nu d\nu. \tag{5.75}$$

In general, the specific intensity I_ν is a function of the orientation $\hat{\mathbf{n}}$ of our window as well as its location $\vec{\mathbf{r}}$ in the universe. In many cases, though, the anisotropy of the photons' motion is irrelevant; for instance, when we simply want the number of photons that are available to excite an electron to some specific state, the direction of motion of the photons doesn't matter. When direction is unimportant, we are really interested in the **mean specific intensity** J_ν, averaged over all directions (that is, averaged over all possible orientations of our little window):

$$J_\nu = \frac{1}{4\pi} \int I_\nu d\Omega. \tag{5.76}$$

We want to know why dense gases, liquids, and solids produce a continuous spectrum, and what the shape of that spectrum is. In other words, what is I_ν as a function of ν? Dense bodies are distinguished from tenuous gas clouds by the fact that they satisfy the sufficient conditions for being in LTE (page 134): their atoms are packed closely together, and they are optically thick.

To begin our investigation of radiation from dense bodies, let's consider a simple gas made of our familiar two-level atoms, each of which has energy states separated by ΔE, as illustrated in Figure 5.3. The atomic gas coexists with an isotropic radiation field with intensity J_ν. The photons that have frequency $\nu = \Delta E/h$ are able to interact with the two-level atoms, causing photoexcitation and stimulated emission. In addition, photons with $\nu = \Delta E/h$ are created when excited atoms undergo spontaneous emission. The rate at which level 1 (the ground state) is depopulated is the radiative excitation rate:

$$\frac{dn_1}{dt} = -n_1 B_{12} J_\nu, \tag{5.77}$$

where n_1 is the number density of atoms in the $n = 1$ state, J_ν is the mean specific intensity at $\nu = \Delta E/h$, and B_{12} is the **Einstein absorption coefficient**, which can be computed quantum mechanically, and is proportional to the cross-section for photon absorption.

Similarly, the rate at which level 2 (the excited state) is depopulated is

$$\frac{dn_2}{dt} = -n_2 B_{21} J_\nu - n_2 A_{21}, \tag{5.78}$$

where the first term is due to stimulated emission and the second term is due to spontaneous emission. In equation (5.78), n_2 is the number density of atoms in the $n = 2$ state, and B_{21} is the **Einstein stimulated emission coefficient**, which can also be calculated quantum mechanically.

Now let us assume that the population of two-level atoms is in **statistical equilibrium**; this is a weaker assumption than LTE, since it merely assumes that the net populations of the two atomic levels are constant with time. In this case, the rate at which electrons go from level 1 to level 2 (equation 5.77) must be equal to the rate at which they go from level 2 to level 1 (equation 5.78). Thus,

$$n_1 B_{12} J_\nu = n_2 B_{21} J_\nu + n_2 A_{21} \tag{5.79}$$

and the ratio of excited atoms to atoms in the ground state must be

$$\frac{n_2}{n_1} = \frac{B_{12} J_\nu}{B_{21} J_\nu + A_{21}}. \tag{5.80}$$

This looks like a complicated relation, involving three different Einstein coefficients, but the laws of quantum mechanics actually provide a simple relation among the Einstein coefficients (here cited without proof):

$$A_{21} = \frac{2h\nu^3}{c^2} B_{21}, \tag{5.81}$$

where $\nu = \Delta E/h$, and

$$B_{12} = \frac{g_2}{g_1} B_{21}, \tag{5.82}$$

where g_1 and g_2 are the statistical weights of the two levels, as in the Boltzmann equation (eq. 5.68). Substituting these two relations back into equation (5.80), we find that when the levels are in statistical equilibrium,

$$\frac{n_2}{n_1} = \frac{(g_2/g_1) J_\nu}{J_\nu + (2h\nu^3/c^2)}. \tag{5.83}$$

Thus, if we knew n_2/n_1, the ratio of excited to ground-state atoms, we would know the mean intensity J_ν required to keep that ratio constant with time.

If we make the stronger assumption that the mix of atoms and photons is in local thermodynamic equilibrium at a temperature T, then the ratio n_2/n_1 is given by the Boltzmann equation (eq. 5.68):

$$\frac{n_2}{n_1} = \frac{g_2}{g_1} \exp\left(-\frac{h\nu}{kT}\right), \tag{5.84}$$

where we are now writing $h\nu$ instead of ΔE. So now, by equating the Boltzmann equation, which assumes LTE, with equation (5.83), which makes the weaker assumption of statistical equilibrium, we find that

$$\frac{(g_2/g_1) J_\nu}{J_\nu + (2h\nu^3/c^2)} = (g_2/g_1) e^{-h\nu/kT}. \tag{5.85}$$

By canceling the factors of g_1/g_2 and solving for J_ν, the intensity of light required to keep the system in LTE at temperature T, we reach our result:

$$J_\nu(T) = \frac{2h\nu^3}{c^2} \frac{1}{e^{h\nu/kT} - 1}. \tag{5.86}$$

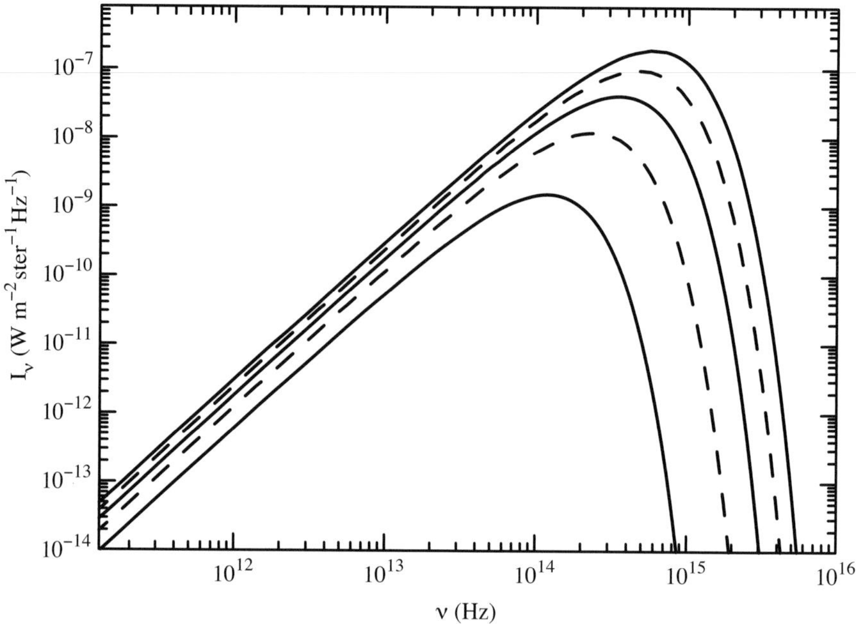

FIGURE 5.14 The Planck function for the temperatures $T = 10{,}000$ K, 8000 K, 6000 K, 4000 K, and 2000 K (top to bottom).

This is the **Planck function** for radiation, first derived by Max Planck in the year 1900. In LTE, the radiation field is isotropic, so we can write $I_\nu = J_\nu$. The Planck function for three different temperatures is plotted in Figure 5.14.

We derived the Planck function for a two-level atom with a totally arbitrary energy level difference ΔE and photon frequency $\nu = \Delta E/h$. In a dense gas, pressure broadening becomes so significant that the medium becomes optically thick at all frequencies, and photons of all energies can be absorbed and emitted. Thus, the particular elements of which the gas is made becomes irrelevant; the dense gas radiates a spectrum that is equal to the Planck function (equation 5.86) at all frequencies. The Planck function thus is the spectrum produced by a perfect emitter (a body that can emit photons at all frequencies); since a perfect emitter is also a perfect absorber, we often refer to light that has a Planck spectrum as **blackbody radiation**. Objects made of dense gas, or of opaque liquid or solid material, produce radiation that is, to a first approximation, blackbody radiation.

In the limit that $h\nu \ll kT$, representing photon energies much smaller than the thermal energy per particle, the Planck function (equation 5.86) reduces to the simple power law

$$I_\nu \approx \frac{2kT}{c^2}\nu^2. \tag{5.87}$$

This limit of low photon energy is known as the **Rayleigh–Jeans limit**. In the opposite limit, where $h\nu \gg kT$, the Planck function reduces to the form

$$I_\nu \propto \nu^3 e^{-h\nu/kT}. \tag{5.88}$$

In this limit, known as the **Wien limit**, the Boltzmann equation imposes an exponential cutoff on the flux of high-energy photons.

It is sometimes useful to express the Planck function in terms of wavelength rather than frequency. The intensity $I_\nu d\nu$ in the frequency interval $\nu \to \nu + d\nu$ must be the same as the intensity $I_\lambda d\lambda$ in the corresponding wavelength interval $\lambda \to \lambda + d\lambda$. Thus, we can write

$$I_\lambda d\lambda = I_\nu d\nu = I_\nu \left| \frac{d\nu}{d\lambda} \right| d\lambda, \tag{5.89}$$

Since $\nu = c/\lambda$, $|d\nu/d\lambda| = c/\lambda^2$, we can write

$$\begin{aligned} I_\lambda d\lambda &= \left[\frac{2h}{c^2} \left(\frac{c}{\lambda} \right)^3 \frac{1}{e^{hc/\lambda kT} - 1} \right] \frac{c}{\lambda^2} d\lambda \\ &= \frac{2hc^2}{\lambda^5} \frac{1}{e^{hc/\lambda kT} - 1} d\lambda. \end{aligned} \tag{5.90}$$

(A plot of the Planck function I_λ for a variety of temperatures is shown in Figure 5.15).

Since the main body of a star is made of dense gas, we can approximate the radiation emitted by a star as having a Planck spectrum. Let's consider the energy carried away by photons from a small area ΔA on the surface of the star, as shown in Figure 5.16. The escaping energy ΔE, integrated over all directions, is

$$\frac{\Delta E}{\Delta t \Delta \nu} = \int I_\nu \Delta A \cos\theta d\Omega, \tag{5.91}$$

where $\Delta A \cos\theta$ is the projected area on the surface for radiation headed at an angle θ relative to the normal (see Figure 5.16 for the geometry). Thus, the **specific flux** through the star's surface, in watts per square meter per Hertz, is

$$F_\nu = \frac{\Delta E}{\Delta t \Delta A \Delta \nu} = \int I_\nu \cos\theta d\Omega. \tag{5.92}$$

Since we are at the star's surface, the intensity is highly anisotropic; we can assume that the intensity is given by the Planck formula for radiation coming from within the star ($0 \leq \theta \leq \pi/2$) and is zero for radiation coming from the darkness of interstellar space outside the star ($\theta > \pi/2$). With this approximation, the specific flux passing through the surface is

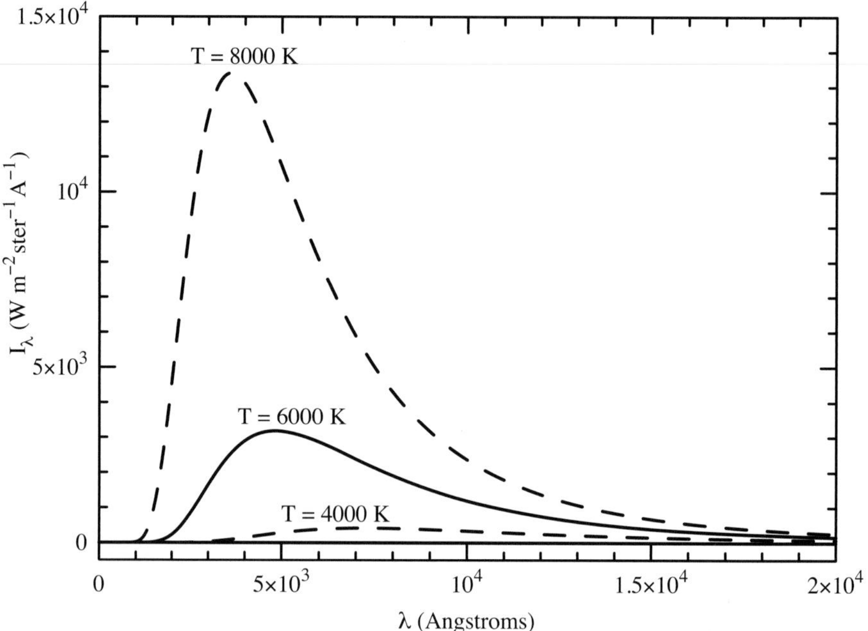

FIGURE 5.15 The Planck function, as a function of wavelength λ, for temperatures $T = 8000$ K, 6000 K, and 4000 K (top to bottom).

$$\begin{aligned} F_\nu &= \int_{\phi=0}^{2\pi} d\phi \int_{\theta=0}^{\pi/2} I_\nu \cos\theta \sin\theta d\theta \\ &= 2\pi I_\nu \int_{\theta=0}^{\pi/2} \sin\theta d(\sin\theta) \\ &= \pi I_\nu. \end{aligned} \tag{5.93}$$

To obtain the total flux emitted per unit area by the star, we integrate over all frequencies:

$$\begin{aligned} F &= \int_0^\infty F_\nu d\nu = \pi \int_0^\infty I_\nu d\nu \\ &= \frac{2\pi h}{c^2} \int_0^\infty \frac{\nu^3 d\nu}{\exp(h\nu/kT) - 1}. \end{aligned} \tag{5.94}$$

The integral over frequency can be transformed by setting $x \equiv h\nu/kT$, and thus

$$F = \left(\frac{2\pi h}{c^2}\right)\left(\frac{kT}{h}\right)^4 \int_0^\infty \frac{x^3 dx}{e^x - 1}. \tag{5.95}$$

The definite integral in the above equation is simply a numerical factor; it turns out to be $\pi^4/15 \approx 6.5$, yielding a flux

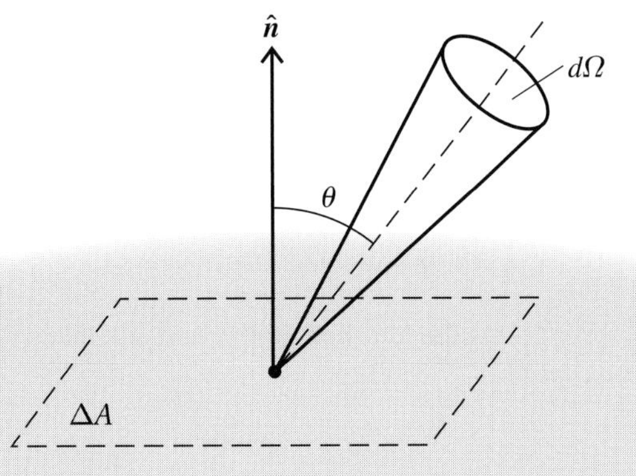

FIGURE 5.16 Radiation emerging from a star at an angle θ to the perpendicular.

$$F = \frac{2\pi^5}{15}\frac{k^4}{c^2h^3}T^4 = \sigma_{SB}T^4, \tag{5.96}$$

where the **Stefan-Boltzmann constant** is

$$\sigma_{SB} \equiv \frac{2\pi^5}{15}\frac{k^4}{c^2h^3} = 5.67\times10^{-8}\,\mathrm{J\,s^{-1}\,m^{-2}\,K^{-4}}. \tag{5.97}$$

The Sun, for instance, has a surface temperature $T_\odot \approx 5780$ K; thus, it has a flux of about 63 megawatts for every square meter of its surface.[12]

If a star is approximated as a spherical blackbody with temperature T and radius R, its total **luminosity**, in watts, is

$$L = 4\pi R^2\sigma_{SB}T^4. \tag{5.98}$$

For the Sun, with a radius $R_\odot = 6.96\times10^8$ m and a surface temperature $T_\odot \approx 5780$ K, the total luminosity is

$$L_\odot = 4\pi R_\odot^2\sigma_{SB}T_\odot^4 = 3.8\times10^{26}\,\mathrm{J\,s^{-1}} = 3.8\times10^{26}\,\mathrm{W}, \tag{5.99}$$

and of course, the luminosity of other stars can be scaled to that of the Sun:

$$L = 1L_\odot\left(\frac{R}{R_\odot}\right)^2\left(\frac{T}{T_\odot}\right)^4. \tag{5.100}$$

[12] You have a surface temperature $T \approx 310$ K, so you produce about 520 watts for every square meter of your surface.

PROBLEMS

5.1 Verify that the Maxwell-Boltzmann distribution (equation 5.40) has its maximum at a speed

$$v_p = \left(\frac{2kT}{m}\right)^{1/2}.$$

5.2 Verify that for the Maxwell-Boltzmann distribution (equation 5.40), the average speed is

$$\langle v \rangle = \left(\frac{8kT}{\pi m}\right)^{1/2}.$$

(Hint: you may find it useful to know the definite integral $\int_0^\infty x^3 \exp(-x^2)dx = 1/2$.)

5.3 Verify that for the Maxwell-Boltzmann distribution (eq. 5.46), the mean kinetic energy per particle is

$$\langle E \rangle = \frac{3}{2}kT.$$

(Hint: you may find it useful to know the definite integral $\int_0^\infty x^{3/2}e^{-x}dx = 3\sqrt{\pi}/4$.)

5.4 Molecules have additional degrees of freedom that atoms don't possess, namely, rotation and vibration. The energies associated with molecular rotation and vibration are quantized, and photons can be emitted or absorbed by molecules making transitions from one rotational or vibrational state to another.

(a) Show that the *rotational* energy of a system can be written as

$$E_{\rm rot} = \frac{L^2}{2I},$$

where L is the angular momentum and I is the moment of inertia.

(b) Suppose that angular momentum is quantized according to Bohr's hypothesis: $L = j\hbar$, with j being a positive integer. Consider the case of a diatomic molecule where the two atoms have equal mass M (for instance, H_2, O_2, or N_2). Derive an expression for the rotational energy $E_{\rm rot}$ in terms of j, $\hbar$, M, and r_0, the separation between the two atomic nuclei in the molecule.

(c) In the case of molecular hydrogen (H_2), which has $r_0 \approx 1$ Å, estimate the wavelength of light produced by the $j = 2 \rightarrow 1$ rotational transition. Is this longer or shorter than the wavelength of visible light?

5.5 (a) A neutral sodium atom has an ionization potential of $\chi = 5.1\,\mathrm{eV}$. What is the speed of a free electron that has just barely enough kinetic energy to collisionally ionize a sodium atom in its ground state? What is the speed of a free *proton* with just enough kinetic energy to collisionally ionize this atom?
(b) What is the temperature T of a gas in which the average particle kinetic energy is just barely sufficient to ionize a sodium atom in its ground state?
(c) At the temperature T computed in part (b), what is the expected thermal Doppler broadening, $\Delta\lambda/\lambda$, of a sodium spectral line? (Hint: the only stable isotope of sodium has mass number $A = 23$.)

5.6 For the Planck function $I_\nu(T)$ (see equation 5.86), what is the most probable frequency ν_p at a given temperature T? For the Planck function expressed as a function of wavelength, $I_\lambda(T)$ (see equation 5.90), what is the most probable wavelength λ_p at a given temperature T? For what range of temperatures does λ_p fall in the visible range of the electromagnetic spectrum?

5.7 A slab of glass 0.2 m thick absorbs 50% of the light passing through it. How thick must a slab of identical glass be in order to absorb 90% of the light passing through it? How thick must it be to absorb 99% of the light? How thick to absorb 99.9% of the light?

5.8 If an incandescent light bulb has a luminosity $L = 60$ W and a filament temperature of $T = 2900$ K, what must be the surface area of its filament? If the filament consists of a cylindrical wire with diameter $d = 4.6 \times 10^{-5}$ m (as in a standard incandescent 60 watt, 120 volt bulb), what is the length of the wire?

5.9 Demonstrate that the Lorentz distribution as given in equation (5.38) is correctly normalized so that

$$\int_0^\infty \phi(\nu)d\nu = 1.$$

5.10 Show that for an ensemble of particles with temperature T and particle mass μm_p, the line profile from thermal Doppler broadening will be

$$\phi(\nu)d\nu = \frac{c}{\nu_0}\sqrt{\frac{\mu m_p}{2\pi kT}}\exp\left[-\frac{\mu m_p c^2(\nu-\nu_0)^2}{2kT\nu_0^2}\right]d\nu,$$

where ν_0 is the frequency at the line center.

6 Astronomical Detection of Light

As the previous chapter revealed, there is a wealth of information to be gained from observing the stars. In particular, spectroscopy of stars yields temperatures, elemental abundances, stellar rotation rates, and magnetic field strengths, among other information. While stars are intrinsically luminous, they are (except for the Sun) at extremely large distances, and hence appear to be very faint. The challenge facing astronomers over the centuries has been to collect the faint light from distant stars and other astronomical objects, and preserve and analyze the information it contains. Large telescopes and sophisticated instrumentation are required.

6.1 ▪ THE TELESCOPE AS A CAMERA

In many ways, a telescope and its associated instrumentation can be thought of as a camera, albeit one with a very large and unwieldy telephoto lens. A review of how cameras work will thus be useful for understanding the basics of imaging science. The word "camera" is the Latin word for "room," and is a shortening of the term "camera obscura," or "darkened room." A camera obscura, the earliest and simplest of all cameras, was an unlit room with a tiny hole cut in one wall; Figure 6.1 is an illustration of a camera obscura from the sixteenth century, before the invention of the telescope. Because the hole in the wall is small, only light rays headed in a specific direction can reach the far wall of the room. Thus, there is a one-to-one mapping between points on the **object** (the source of light) and the **image**.

A compact version of the camera obscura is the pinhole camera (Figure 6.2), an opaque box with a tiny pinhole in the middle of one wall. If a permanent record of the projected image is required, an electronic detector or a piece of photographic film can be placed on the wall where the image is located. From Figure 6.2, we see that the image is inverted, and its size is proportional to the length F of the box, called the **focal length**. Because the pinhole is small, it admits photons at a slow rate. A detector such as a piece of film requires a certain number of photons per unit area to yield a detectable signal; thus, the **exposure time** t required to produce a detectable image using a pinhole camera can be very long. For a fixed pinhole size, the exposure time is directly proportional to the area of the image; thus, $t \propto F^2$, where F is the focal length of the pinhole camera. Because of

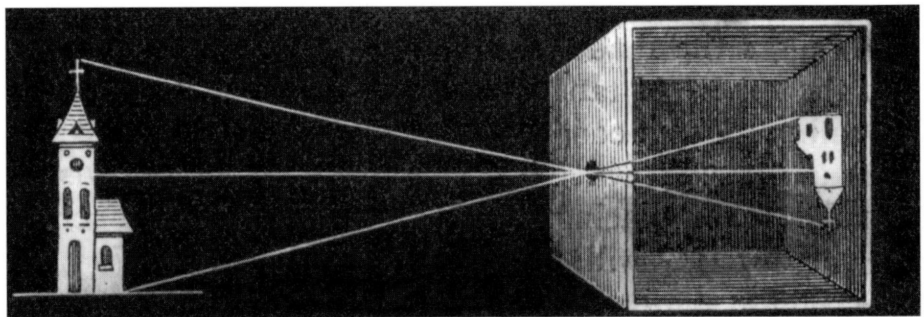

FIGURE 6.1 A camera obscura, as used to project the image of a nearby church.

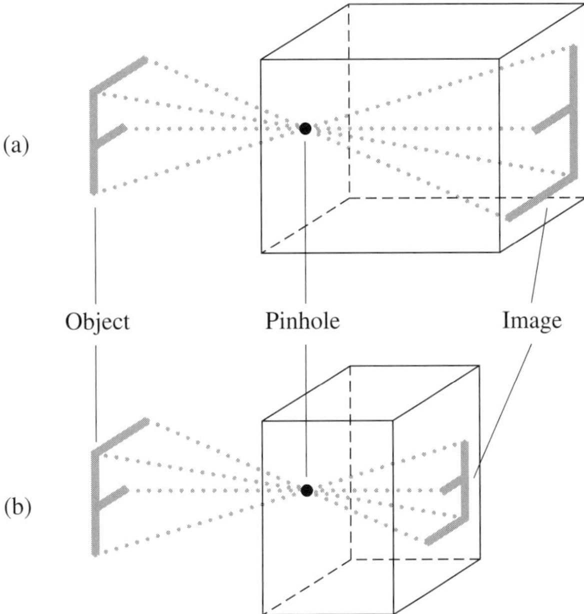

FIGURE 6.2 A pinhole camera, which is simply a miniature camera obscura. Points on the object, to the left, map onto the image plane on the right. A long camera (a) produces a larger image than a short camera (b).

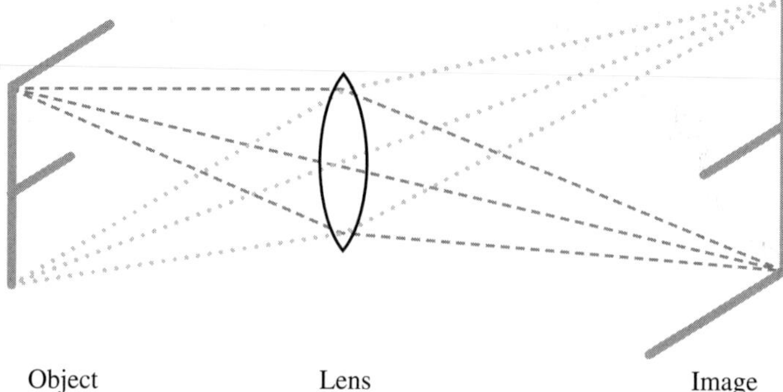

FIGURE 6.3 By replacing the pinhole in Figure 6.2 with a convex lens, we can admit more light. The image plane is now fixed, and its location depends on the shape of the lens.

the relation between focal length and exposure time, a camera with a short focal length is called a *fast* system, and a camera with a long focal length is called a *slow* system.

In order to reduce exposure times while keeping the image size large, photons must be admitted into the camera at a faster rate. The easiest way to do this is to increase the size of the pinhole. This has the unfortunate consequence of destroying the one-to-one mapping between the object and the image; now many light rays from different parts of the object can reach the same point on the image plane. This yields a blurry image. To restore the one-to-one mapping, we place a convex **lens** in the aperture where the pinhole used to be, as illustrated in Figure 6.3.

A lens is made from a transparent material (often glass) whose **refractive index** is different from that of the air surrounding it. The refractive index of a material is defined as $n \equiv c/v_m$, where c is the speed of light in a vacuum and v_m is the speed of light in the material. The refractive index of air is $n = 1.0003$. Most types of glass have $n \approx 1.5$; flint glass, which contains lead oxide and is therefore quite dense, can have a refractive index as large as $n \approx 2$. As light travels from one medium to another—from air to glass, for instance, or from glass to air—its path will be bent, or refracted, unless it happens to strike the air/glass interface exactly at a right angle.

A properly shaped lens will refract the light rays from a luminous object in such a way that all the light from a particular point on the object is directed to a single point on the image plane. One consequence of inserting a lens into the camera is that the focal length of the system is now fixed. In a pinhole camera, the image plane can be placed at an arbitrary distance from the pinhole; in a camera with a lens, the image will be in focus only at a fixed distance F from the lens. The distance F to the **focal plane** depends on the shape of the lens, as well as on its refractive index. A highly curved lens made of material with a large refractive index n will have a short focal length; a gently curved lens with smaller n will have a longer focal length. For lenses, a useful parameter is the focal ratio $f = F/D$, where D is the diameter of the lens. Among photographers, it is conventional to write the focal ratio as the "f-number": for instance, $f/8$ for a lens with

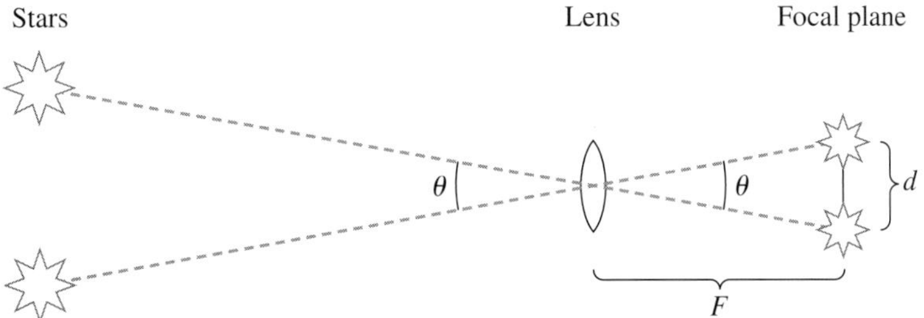

FIGURE 6.4 Two stars separated by a small angle θ on the sky have images that are separated by a physical distance d on the focal plane.

$F/D = 8$.[1] The size of the image produced is not affected by the diameter D of the lens but only by the focal length F.

Another useful parameter, in addition to the focal length, is the scale of the image on the focal plane, known for historical reasons as the **plate scale**.[2] Specifically, an angular distance θ on the celestial sphere is related to a physical distance d on the image plane by the plate scale s:

$$\theta[\text{arcsec}] = s[\text{arcsec/mm}] \cdot d[\text{mm}]. \tag{6.1}$$

From Figure 6.4, we can compute the separation d between a pair of images when the pair of objects we are observing (let's call them stars) are separated by an angular distance θ on the celestial sphere. If the angular distance θ between the two stars is small, we can use the small-angle approximation:

$$\theta[\text{radians}] = \frac{d}{F}, \tag{6.2}$$

where F is the focal length, or

$$\theta[\text{arcsec}] = \theta[\text{radians}]\left(\frac{180^\circ}{\pi\text{ radians}}\right)\left(\frac{3600\text{ arcsec}}{1^\circ}\right) = 206{,}265\left(\frac{d}{F}\right). \tag{6.3}$$

Combining equations (6.1) and (6.3), we have a relationship between the plate scale s and the focal length F:

$$s[\text{arcsec/mm}] = \frac{206{,}265}{F[\text{mm}]}. \tag{6.4}$$

[1] In the human eye, the distance F from the lens to the retina is nearly constant, but the diameter D of the pupil changes significantly as it constricts and dilates. In bright surroundings, the focal ratio of the eye is approximately $f/8$; in dark surroundings, it's roughly $f/2$.

[2] In the past, it was common for astronomical images to be taken using photographic emulsions deposited on glass plates, hence the term "plate scale." Glass was used rather than plastic film because glass is more rigid.

The human eye, for instance, has a focal length $F \approx 17$ mm, and hence a "plate scale" $s \approx 12{,}100$ arcsec/mm, or $s \approx 3.4°$/mm; when you look at the full Moon, its image covers an area of your retina less than 0.15 mm across.[3]

Large astronomical telescopes have focal lengths that are more conveniently expressed in meters than in millimeters. For these big telescopes, we may write

$$s[\text{arcsec/mm}] = \frac{206.265}{F[\text{m}]} = \frac{206.265}{f D[\text{m}]}, \tag{6.5}$$

where f is the focal ratio, and D is the diameter of the telescope's aperture. As an example, the famous "forty-inch" Yerkes Telescope (at Williams Bay, Wisconsin) has an aperture $D = 1.02$ m and a focal ratio $f = 19$. The plate scale of the Yerkes Telescope is thus

$$s = \frac{206.265}{19(1.02)} \text{ arcsec/ mm} = 10.6 \text{ arcsec/ mm}. \tag{6.6}$$

An image of the full Moon produced by the Yerkes Telescope is therefore 170 mm across, about the size of a salad plate.[4]

For a given aperture size D, we can increase the size of the image simply by increasing the focal ratio f of the lens. However, it is important to realize that by increasing the focal ratio f, we do not necessarily produce images with more detail. At the level of fundamental physics, the image quality is limited by diffraction. When light from a point source passes through a circular aperture of finite size, diffraction produces an image that looks like a central bright disk surrounded by a series of alternately dark and bright rings (illustrated in Figure 6.5). The diameter of the central bright disk, known as the "Airy disk" after its discoverer George Airy, is determined by the aperture size D and the wavelength λ of the observed light. Two point sources can be resolved as separate objects when the center of the Airy disk of one source falls into the dark ring surrounding the Airy disk of the other source. This happens when the centers of the two light sources are separated by a distance

$$\theta_{\text{min}}[\text{rad}] = 1.22\frac{\lambda}{D}. \tag{6.7}$$

If we observe at visible wavelengths, $\lambda \approx 5000$ Å $\approx 5 \times 10^{-7}$ m, then resolving two stars separated by an angular distance $\theta = 1'' = 4.8 \times 10^{-6}$ rad requires a telescope of minimum diameter

$$D = \frac{1.22\lambda}{\theta} \approx \frac{1.22(5 \times 10^{-7}\text{ m})}{4.8 \times 10^{-6}} \approx 0.13 \text{ m}. \tag{6.8}$$

If diffraction were the only limit to angular resolution, we could see astronomical objects with finer detail simply by building larger telescopes. However, for those of us living

[3] The image is also, as shown in Figure 6.3, inverted. Your brain processes the signal coming from the optic nerve to return the image right-side-up.

[4] Which gives another meaning to the phrase "plate scale."

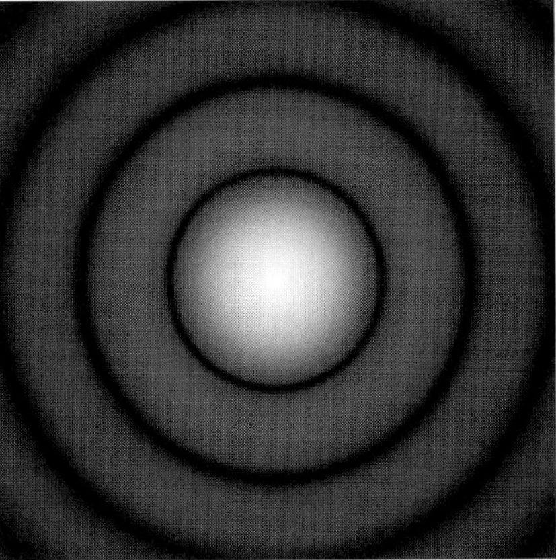

FIGURE 6.5 The diffraction pattern produced by monochromatic light from a point source passing through a circular aperture of finite size.

near the surface of the Earth, there is an additional factor to consider. Image quality for telescopes on Earth is generally limited by what is referred to as atmospheric **seeing**. That is, turbulence in the Earth's atmosphere blurs the images of stars, and causes them to jitter back and forth (the process that produces "twinkling" of stars). The blurring caused by turbulence in the Earth's atmosphere is typically about $1''$. At the best observing sites, where turbulence in the air flow is minimized, the atmospheric seeing is rarely better than $\sim 0.25''$.[5]

Thus, for an ordinary ground-based telescope, making the aperture larger than $D \sim 0.5$ m does not improve the resolution of the image. Nevertheless, there are advantages to an aperture many meters across. The primary advantage of a large-aperture telescope is that it gathers more photons. The light-gathering power of a telescope is proportional to D^2, so a larger aperture enables you to see fainter objects during a given exposure time.

6.2 ▪ REFRACTING AND REFLECTING TELESCOPES

So far, we have assumed that the primary light-gathering element of the telescope (called the **primary** for short) is a lens. Telescopes that use lenses to gather light and bend it toward the focal plane are called **refracting telescopes**, or **refractors**. Although small refractors are commonly used, there are serious intrinsic problems in making very large refracting telescopes. First, the weight of a lens rapidly becomes unwieldy, since the

[5] The technique termed "adaptive optics," described in Section 6.7, affords an opportunity to achieve better resolution.

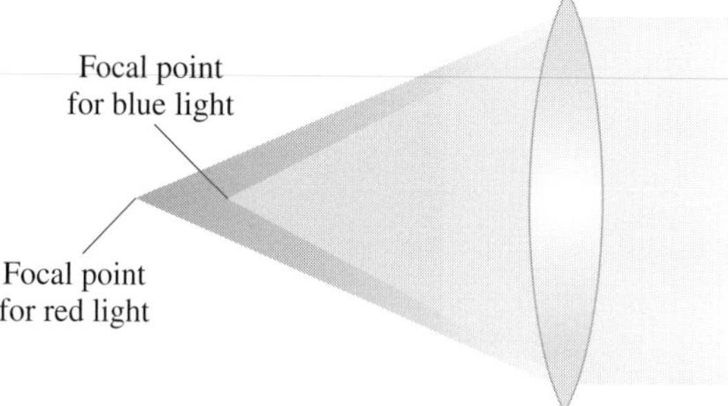

FIGURE 6.6 Chromatic aberration in a refractor; the focal length for blue light is shorter than for red light.

volume, and hence the weight, of a lens scales as D^3. Since light must pass freely through the lens, it can only be supported around its edges; this means that a large heavy lens tends to sag in the middle, distorting the images that it produces.

Matters are made worse by the fact that the heavy lens is at the top of a long tube that must be extremely stiff; otherwise, the tube will flex and wobble as it swings around to point in different directions on the sky. In addition, lenses suffer from the problem known as **chromatic aberration**; since the refractive index decreases with wavelength, short wavelengths of light are bent through a greater angle than long wavelengths. This means that the focal length for shorter wavelengths of visible light (violet and blue) is shorter than the focal length for longer wavelengths (orange and red). As illustrated in Figure 6.6, this means that if the red light from a star is in optimum focus, the blue light will be blurred, and vice versa. The effect of chromatic aberration can be reduced by making compound lenses out of two pieces of glass with different refractive indices. However, this does not completely eliminate chromatic aberration.[6]

An alternative to using lenses is to bring the light to a focus by using a concave mirror. A telescope that uses a mirror as its primary is called a **reflecting telescope**, or a **reflector**. Like a correctly shaped lens, a correctly shaped mirror takes the light from a distant object and creates an image on a focal plane (Figure 6.7). Like a refractor, a reflector can be described in terms of its focal length F and its aperture diameter D—in this case, the diameter of the primary mirror. Reflectors have a number of advantages over refractors, particularly at large apertures. For instance, reflecting telescopes do not suffer from chromatic aberration. In addition, a mirror, unlike a lens, can be supported over its entire back surface, which prevents sagging and the consequent distortion of images.

[6] In the mid-sixteenth century, Huygens tried to minimize chromatic aberration by building telescopes of extremely high focal ratio; one of his telescopes had a lens with $D = 2$ inches and $F = 123$ feet, yielding a telescope with $f/738$. He eliminated the flexing, wobbling tube altogether, which had the unfortunate side effect of making the lens and the eyepiece nearly impossible to keep aligned.

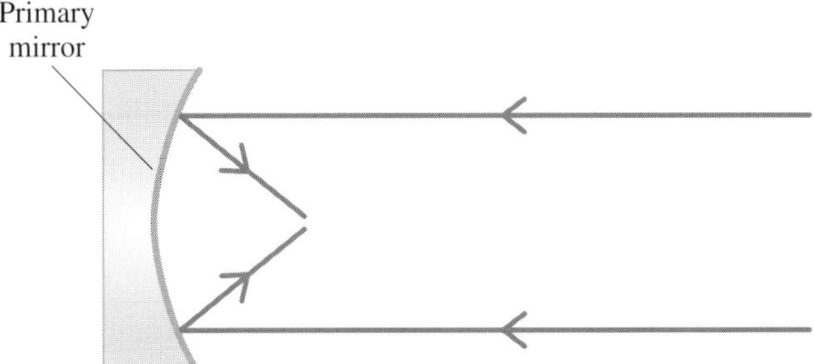

FIGURE 6.7 The primary mirror of a reflecting telescope brings light to a focus.

The preferred material for mirrors is glass, because it can be shaped to extremely high accuracy; if you are observing at a wavelength λ, you want any bumps and valleys on the mirror surface to be smaller than $\sim \lambda/4$ in height. The front surface of the mirror is coated with a thin layer of metal, usually aluminum. Silver is actually more reflective than aluminum at visible wavelengths, but it tarnishes rapidly. Mirrors intended for observations at infrared wavelengths are frequently coated with gold, which is highly reflective in the infrared, but not at the blue/violet end of the visible spectrum.[7] A gold-plated mirror might seem extravagant, but the layer of gold can be very thin; coating a $D = 2$ m mirror with a layer of gold 1 micron thick requires less than a hundred grams of gold. The entire purpose of the large, massive glass mirror, if you think about it, is to keep the extremely thin layer of metal on its surface in the right shape to bring light to a focus.

An obvious problem with reflectors is that the focal plane lies between the mirror and the light source. For a large reflecting telescope using a relatively small detector, the detector can be placed directly at the **prime focus** of the telescope, as shown in Figure 6.8a. If the detector is much smaller in cross-sectional area than the primary mirror, the amount of light it blocks is negligible.[8] When Isaac Newton constructed a reflecting telescope in the year 1668, he diverted the converging beam of light from the primary mirror by using a flat secondary mirror, as shown in Figure 6.8b. Most small reflectors use this **Newtonian focus**, but it is unwieldy for large telescopes with heavy detectors, which would have to be attached to the side of the telescope tube, throwing it off-balance.

The most commonly used focus in moderate-size research telescopes, a few meters in aperture, is the **Cassegrain focus** (Figure 6.8c). In a Cassegrain telescope, the converging beam from an approximately parabolic primary mirror is directed back toward the primary by an approximately hyperbolic secondary mirror; a hole in the primary mirror

[7] Poor reflectivity in the blue is actually desirable for infrared observations, since this provides short-wavelength moonlight suppression.

[8] Placing an opaque object within the light path doesn't cause a "hole" in the image on the focal plane; instead, it results in an overall dimming of the image.

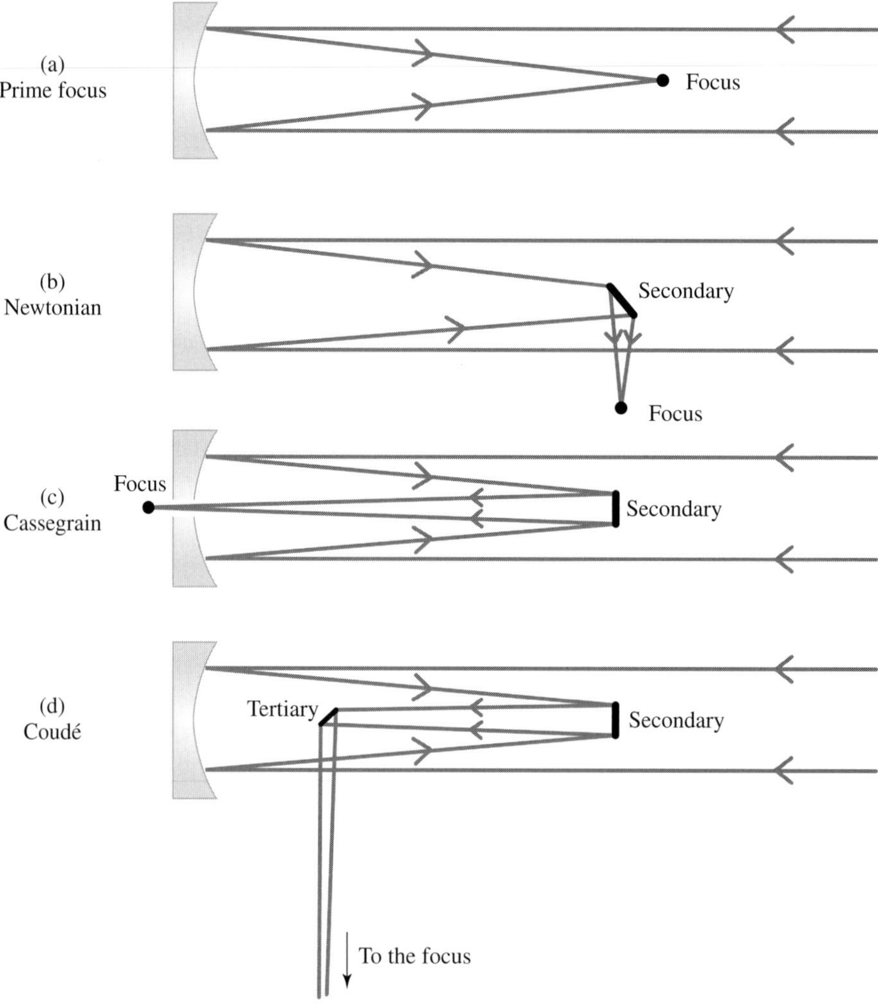

FIGURE 6.8 (a)–(d) Four different configurations of a reflecting telescope, observing a light source off to the right.

allows the light to come to a focus behind the primary mirror. The Cassegrain telescope enables us to make a telescope with a long focal length that is still reasonably compact in size. It also permits the use of large spectrographs and cameras, since the heaviest parts of the telescope, the mirror and the detectors, are close together at the base of the telescope; this reduces balance problems. The **Coudé focus** (Figure 6.8d) is useful for instruments that either are too large to be attached to the telescope or require a great deal of mechanical stability. The Coudé telescope relies on multiple mirrors to direct the light beam to a location distant from the telescope.

In general, multiple reflections in a telescope are undesirable. Even under good conditions, no more than 90% of the light incident on a mirror will be reflected. This fraction can become much lower with time as the reflective coating degrades with exposure to the elements.[9] In a Coudé telescope requiring four reflections, the fraction of incident light that survives the reflections is only $(0.9)^4 \approx 0.66$. Throwing away a third of the scarce photons from a distant star is not something to be done frivolously; the Coudé focus of a telescope is used only when absolutely necessary.

6.3 ▪ QUALITY OF IMAGES

All optical systems suffer from different types of aberrations, some of which are inherent to the design of the system, and some of which arise due to errors in manufacture or the wear and tear of everynight use. We have already noted the existence of chromatic aberration, which is unique to refracting telescopes. Five other types of aberration can appear in both refractors and reflectors.

1. **Spherical aberration**, illustrated in Figure 6.9, occurs when different annuli of a lens or mirror have different focal lengths. In this case, there is not a unique focal plane; wherever you place your detector, light from some or all of the annuli will be out of focus. The name "spherical" aberration comes from the fact that this type of aberration is seen when the surfaces of a lens or mirror are sections of a sphere. The nonspherical surfaces that create images free of spherical aberration are more difficult, and hence more expensive, to make than spherical surfaces.
2. **Coma**, illustrated in Figure 6.10, is the aberration that occurs when light rays near the edge of the lens or mirror come to a focus at a larger distance from the optical axis than light rays passing through the center of the mirror.[10] Coma is characteristic of the off-axis images formed by parabolic mirrors. *Coma* is the Latin word for "hair," and describes the fuzzy-looking appearance of images that suffer from this type of aberration. Figure 6.11 shows the image of a star displaying coma.
3. **Astigmatism**, illustrated in Figure 6.12, is the aberration that occurs when the radius of curvature of the lens or mirror is different along different axes. This means that the focal length is different in the vertical plane, for instance, than in the horizontal plane. Astigmatism causes a point source of light to produce an elongated, distorted image, as shown in Figure 6.13. Astigmatism is a commonly found aberration in the human eye.

[9] Telescope mirrors are periodically stripped and recoated, but this is a process that consumes time and money; at most research telescopes, it happens every two or three years.

[10] The "optical axis" of a simple telescope is the axis passing through the middle of the primary lens or mirror, perpendicular to the primary's surface at that point.

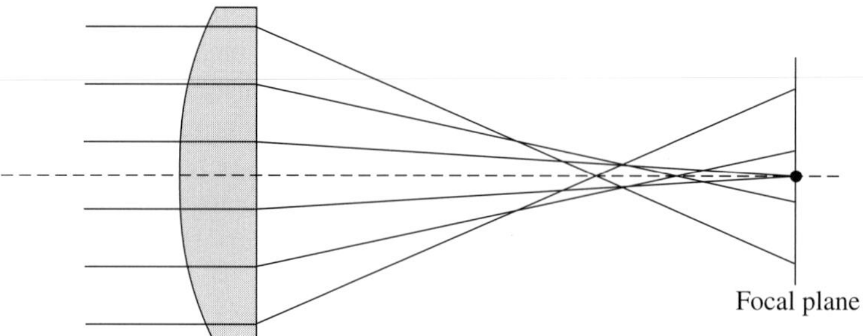

FIGURE 6.9 Spherical aberration in a refracting telescope.

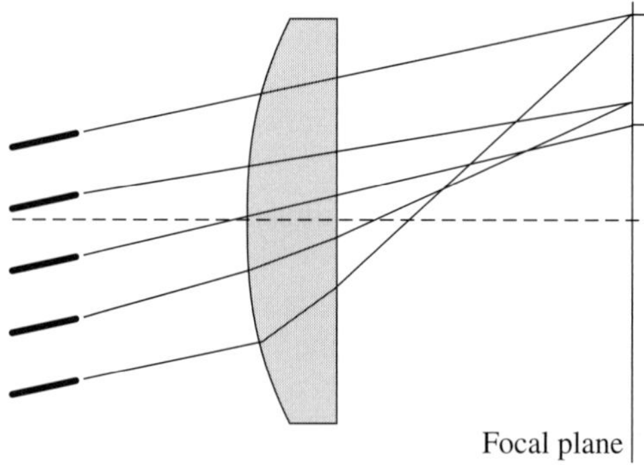

FIGURE 6.10 Coma for an off-axis image in a refracting telescope.

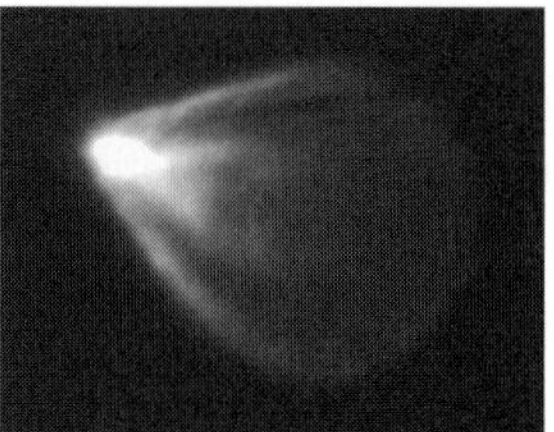

FIGURE 6.11 An example of coma, reminding us that "coma" and "comet" come from the same Latin root.

4. **Curvature of field**, illustrated in Figure 6.14, is the aberration that occurs when the focal surface is not a flat plane but is curved. In fact, curvature of field is so common that it's regarded as a benign feature rather than a bug. Your eye, for instance, has curvature of field; this is necessary to produce a focused image on your curved retina.

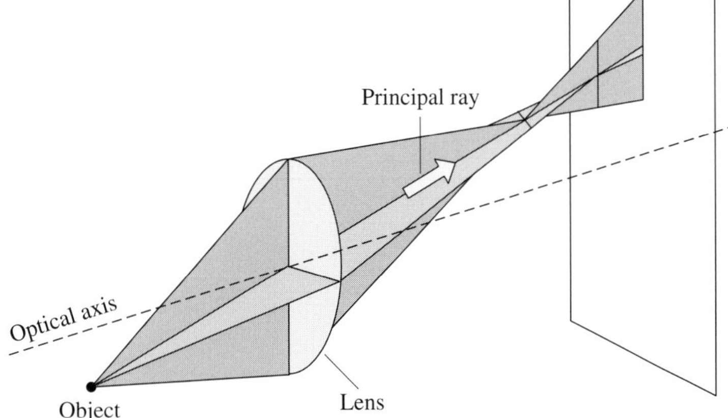

FIGURE 6.12 Astigmatism in a refracting telescope.

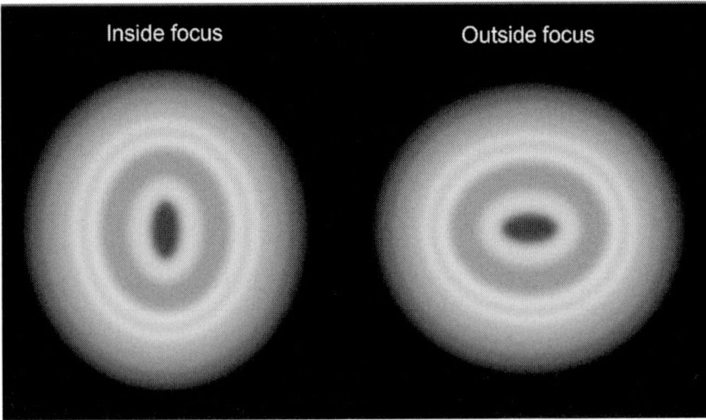

FIGURE 6.13 Astigmatic images of a star.

5. **Distortion**, illustrated in Figure 6.15, is the aberration that occurs when the mapping from object to image does not have a constant plate scale. Distortion comes in two types: pincushion distortion, also called positive distortion, and barrel distortion, also called negative distortion. Both types of distortion do not lower the angular resolution of the image; they simply change its shape.

There exist two special types of telescope that are designed to circumvent some of these aberrations. The **Schmidt telescope** uses a spherical primary mirror, which affords a very large field of view. Spherical aberration, which would ordinarily be present for a spherical primary, is suppressed by using a correcting lens in front of the primary mirror. The **Ritchey-Chrétien telescope** uses primary and secondary mirrors that are both hyperboloids. This design, intended for use with the Cassegrain focus, produces

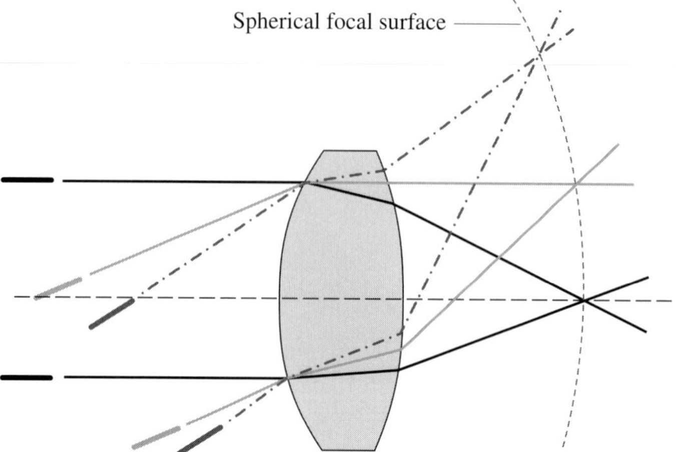

FIGURE 6.14 Curvature of field in a refracting telescope.

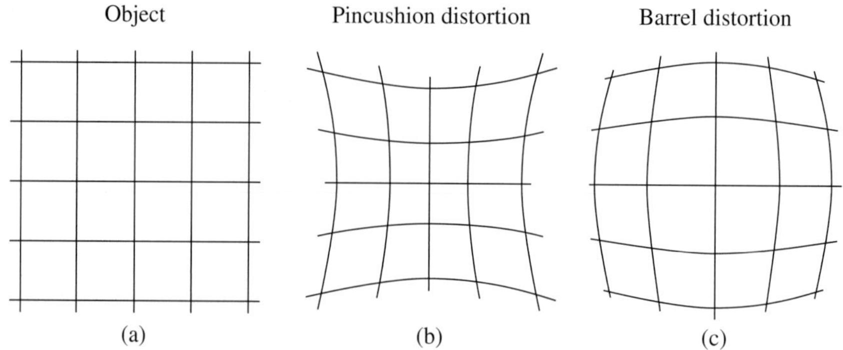

FIGURE 6.15 An object covered with a Cartesian grid (a) is mapped to an image with pincushion distortion (b) and an image with barrel distortion (c).

round images of stars, free of coma and spherical aberration. Most modern telescopes, including the *Hubble Space Telescope*, are of the Ritchey-Chrétien design.[11]

Astronomical images of point-like sources, such as distant stars, are never observed to be true mathematical points. This is because of diffraction limits, atmospheric seeing, or the types of aberration listed above. It is common to describe the observed angular distribution of light from a point source in terms of a **point spread function**, or **PSF**. Depending on the particular optical system and on the types of aberration present, the

[11] The *Hubble Space Telescope* notoriously suffered from spherical aberration when it was first deployed. This was due to an error in manufacturing the primary mirror, however. It wasn't the fault of the telescope's design.

PSF need not be circularly symmetric, and it may have considerable structure (such as the Airy disk and rings shown in Figure 6.5). It is customary to characterize the PSF either by the angular size (often the full width at half maximum) of the image of a point source, or by the angular radius that encircles some specified percentage of the total light. The PSF at a given telescope generally changes with time; a common lament of astronomers on an observing run is "The seeing has just gone to hell!"

6.4 ▪ ASTRONOMICAL INSTRUMENTS AND DETECTORS

Astronomical instruments fall into two basic categories: **imaging cameras** and **spectrographs**. Most instruments are one or the other, but some are a hybrid of the two.

Imaging cameras are designed to produce high-quality images over a suitably large field of view. Most of them have interchangeable filters that can be put into the light beam to restrict the wavelengths observed to some desired range. The pretty colored images of nebulae that you see in coffee-table astronomy books, for instance, are made by combining images taken through filters admitting different wavelengths. As we see in Chapter 13, images taken through different filters are not merely an aid to creating pretty pictures; they also permit estimates of stellar temperatures and provide other useful information.

Spectrographs are designed to disperse light by wavelength, in order to produce spectrograms of astronomical sources. The basic elements in an astronomical spectrograph are illustrated in Figure 6.16. The dispersing element can be a prism, such as Isaac Newton used to disperse sunlight into a spectrum. However, the dispersion is more frequently performed by a reflection grating, as shown in the figure.[12] Sometimes a transmission grating or a "grism" (a combination of a transmission grating and a prism) is used instead. The dispersing elements are usually interchangeable, to allow observation of different regions of the spectrum at different spectral resolutions. If a spectrograph can barely distinguish two spectral features at wavelengths λ and $\lambda + d\lambda$, its spectral resolution is $R = \lambda/d\lambda$.

Many types of detector technologies are used in astronomical observations; we will restrict ourselves to making a few general observations. The **quantum efficiency** of a detector describes how effectively it responds to light. In general, the quantum efficiency of a device is the number of photons detected divided by the number of photons that strike the detector. In photographic emulsions, photons are detected when they trigger a chemical reaction (in classic black-and-white film, the reaction is the conversion of a silver salt to metallic silver). The quantum efficiency of a photographic plate is typically only $\sim 1\%$. In the dark-adapted human retina, photons are detected when they trigger a structural change in the rhodopsin molecule. Under ideal conditions, the quantum efficiency of the dark-adapted human eye is $\sim 10\%$.

[12] The finely spaced pits on a DVD act as a reflection grating, as do the regularly spaced spheres of silicon dioxide within an opal and the parallel rods of melanin within a peacock feather.

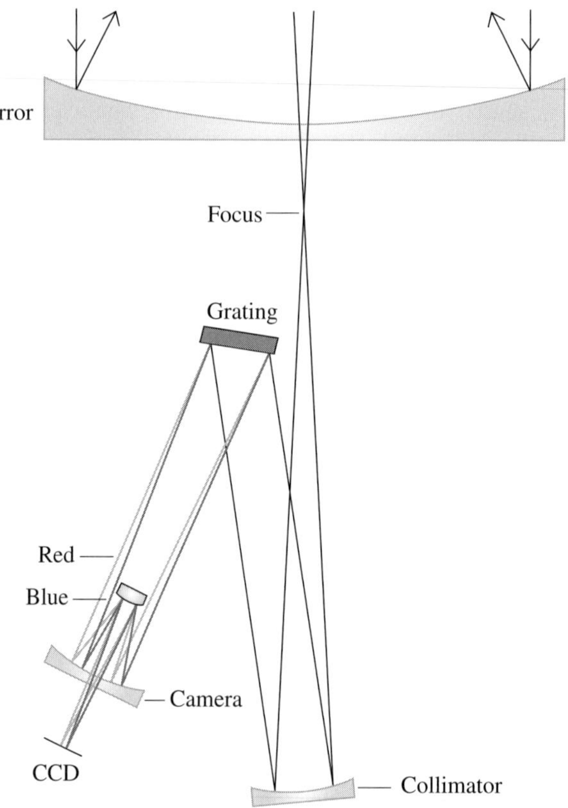

FIGURE 6.16 Schematic diagram of a spectrograph. The collimator creates a parallel beam of light which falls on a reflection grating. The grating disperses the light by wavelength, and a camera brings the dispersed light to a focus on the detector, in this case a charge-coupled device (CCD).

Most modern astronomical detectors fall into one of two categories: photoemissive or photoconductive. **Photoemissive** detectors make use of the photoelectric effect; a photon striking the surface of the detector liberates an electron. An electric field can be used to accelerate the free electrons and ultimately convert them into a measurable current, proportional to the incident photon flux. A photomultiplier tube is an example of a photoemissive detector. The typical peak quantum efficiency for a photomultiplier is about 10%.

Photoconductive detectors are solid state devices that accumulate local charges as photons strike them. The most common photoconductive detector is the charge-coupled device (CCD), used in commercial digital cameras as well as in astronomical imaging cameras and spectrographs. A CCD is an integrated circuit that can be thought of as a rectangular grid of tiny capacitors coupled together. A photon striking the detector can free an electron, causing each capacitor to accumulate a charge proportional to the number of photons striking its immediate vicinity. The peak quantum efficiency of a CCD is $\sim 80\%$; this high quantum efficiency makes CCDs the detector of choice in a wide range of applications. A drawback to CCDs is that they are also sensitive to cosmic rays (high-speed, charged particles), so strategies for minimizing their effects must be employed.

CCDs come in various shapes, generally either a square or an elongated rectangle (useful for spectroscopy). Each electron-accumulating capacitor in a CCD corresponds to a picture element, or **pixel**. The size of pixels should be carefully matched to the desired application. Using pixels larger than the point spread function (PSF) of your image causes a loss of information. However, using a pixel size that is very much smaller than the PSF is sheer waste. A useful approximation is that the pixel size should be 1/2 times the full-width half-maximum (FWHM) of the PSF for the best seeing conditions expected at the telescope. For a telescope with a plate scale $s = 10$ arcsec/mm and seeing with a FWHM of 0.5 arcsec, this would require pixels of a width

$$w \approx \frac{0.25 \text{ arcsec}}{10 \text{ arcsec/ mm}} \approx 0.025 \text{ mm}. \tag{6.9}$$

A CCD with a 4096×4096 ($4\text{k} \times 4\text{k}$) array of pixels on this scale would then be about 10 centimeters on a side.

6.5 ▪ OBSERVATIONS AND PHOTON COUNTING

One practical problem that an observer faces is determining how long to observe a particular source in the sky. How long does it take to obtain data of acceptable quality? In this section, we address this problem at a fundamental statistical level.

In an astronomical observation, we detect an integral number of photons from a particular source. Given a source of known brightness, we might expect to observe μ photons in some given amount of time. However, because of random statistical fluctuations, each time we observe the object, we will count a slightly different number of photons. The number of photons observed will follow a **Poisson probability distribution**:

$$P(x, \mu) = \frac{\mu^x}{x!} e^{-\mu}, \tag{6.10}$$

where μ is the expected number of photons (which isn't necessarily an integer) and x is the actual number of photons observed (which *is* necessarily an integer). Since $P(x, \mu)$ is a probability distribution, it is normalized so that

$$\sum_{x=0}^{\infty} P(x, \mu) = 1, \tag{6.11}$$

which is simple to see, given the mathematical relation

$$\sum_{x=0}^{\infty} \frac{\mu^x}{x!} = e^{\mu}. \tag{6.12}$$

If we perform the experiment of observing the astronomical object numerous times, the mean number of photons we observe will be

$$\langle x \rangle = \sum_{x=0}^{\infty} x P(x, \mu) = \sum_{x=1}^{\infty} \frac{x \mu^x e^{-\mu}}{x!}, \tag{6.13}$$

where we have eliminated the $x = 0$ term, since it equals zero. Using the relation $x! = x(x-1)!$, we may write

$$\langle x \rangle = \sum_{x=1}^{\infty} \frac{\mu^x e^{-\mu}}{(x-1)!}. \tag{6.14}$$

Making the substitution $z = x - 1$, so the sum goes from $z = 0$ to infinity, we see that

$$\langle x \rangle = \sum_{z=0}^{\infty} \mu \frac{\mu^z e^{-\mu}}{z!} = \mu, \tag{6.15}$$

as expected.

We would also like to know the width of the distribution $P(x, \mu)$; in other words, if we perform the experiment only once, how far from μ is x likely to be? A useful measure of the width of the distribution is the variance,

$$\sigma^2 \equiv \langle (x-\mu)^2 \rangle = \sum_{x=0}^{\infty} (x-\mu)^2 P(x, \mu). \tag{6.16}$$

We may write the variance as

$$\begin{aligned} \sigma^2 &= \sum_{x=0}^{\infty} x^2 P(x, \mu) - 2\mu \sum_{x=0}^{\infty} x P(x, \mu) + \mu^2 \sum_{x=0}^{\infty} P(x, \mu) \\ &= \sum_{x=0}^{\infty} x^2 P(x, \mu) - \mu^2, \end{aligned} \tag{6.17}$$

making use of equations (6.11) and (6.15). By a process similar to that used to compute $\langle x \rangle$, it can be shown that

$$\langle x^2 \rangle = \sum_{x=0}^{\infty} x^2 P(x, \mu) = \mu^2 + \mu, \tag{6.18}$$

and hence, substituting into equation (6.17),

$$\sigma^2 = \mu. \tag{6.19}$$

The **standard deviation** of the Poisson distribution is $\sigma = \sqrt{\sigma^2} = \mu^{1/2}$. When we make a measurement x, there is a probability of $\sim 2/3$ that it will be between $\mu - \sigma$ and $\mu + \sigma$.

As the expected number of photons μ increases (by increasing the exposure time, for instance), the uncertainty increases more slowly, like $\mu^{1/2}$. The uncertainty in the measurement is often referred to as the "noise" in the measurement; it represents random fluctuations in the quantity we want to measure, the "signal." This leads us to define a figure of merit for the measurement, which we call the **signal-to-noise** ratio:

$$S/N = \frac{\text{measurement}}{\text{uncertainty}} = \frac{\mu}{\sigma} = \frac{\mu}{\mu^{1/2}} = \mu^{1/2}. \tag{6.20}$$

The signal-to-noise ratio is the reciprocal of the fractional uncertainty in the measurement. For example, if we want to measure an intensity to within 1% accuracy, we require that $S/N = 1/0.01 = 100$, which requires $\mu = 10^4$.

The signal-to-noise ratio is $S/N = \mu^{1/2}$ if fluctuations in the photon count are the only source of uncertainty. In most applications, there are additional sources of noise. For example, measurements of faint objects will include a significant number of background photons as well.[13] A separate measurement is usually required to estimate the background contribution. You observe the astronomical source in question, you observe an apparently blank patch of sky next to it (the "background measurement"), and then you subtract the background measurement from your observation of source + background. Since the background photons are also Poisson distributed, uncertainty in the background measurement also adds to the noise. The uncertainties of two Poisson distributions add in quadrature; that is,

$$\sigma_{\text{total}}^2 = \sigma_{\text{s}}^2 + \sigma_{\text{b}}^2, \tag{6.21}$$

where the contributing factors are the source counts (μ_{s}, with associated uncertainty $\sigma_{\text{s}} = \mu_{\text{s}}^{1/2}$) and the background counts (μ_{b}, with associated uncertainty $\sigma_{\text{b}} = \mu_{\text{b}}^{1/2}$). Thus, the signal-to-noise ratio for an astronomical source contaminated with a background is

$$S/N = \frac{\mu_{\text{s}}}{(\sigma_{\text{s}}^2 + \sigma_{\text{b}}^2)^{1/2}} = \frac{\mu_{\text{s}}}{(\mu_{\text{s}} + \mu_{\text{b}})^{1/2}}. \tag{6.22}$$

Astronomers frequently find themselves doing faint-object astronomy; this is the case where the source counts are outnumbered by the background counts ($\mu_{\text{b}} > \mu_{\text{s}}$). In this case, the signal-to-noise ratio is

$$S/N \approx \frac{\mu_{\text{s}}}{\mu_{\text{b}}^{1/2}}, \tag{6.23}$$

and the observation is referred to as **background limited**. Let's consider the case of background limited observations in more detail. First, let's assume that we are observing a star of luminosity L (in watts) and specific luminosity L_λ (in watts per angstrom). The star is at a distance r from the Earth. The **specific flux** received at the top of the Earth's atmosphere will be[14]

$$F_\lambda = \frac{L_\lambda}{4\pi r^2}. \tag{6.24}$$

[13] The term "background" is a bit misleading, since the background photons can come from in front of the object observed as well as behind it. Contributions to the background can come from the Earth's atmosphere and from sunlight scattered by interplanetary dust, as well as from distant, unresolved stars and galaxies.

[14] As we see later in the book, stars radiate nearly isotropically, and interstellar space is usually optically thin.

We can convert from *energy* flux to *photon* flux by dividing by the energy per photon: $F_\lambda / h\nu = \lambda F_\lambda / hc$. The rate $\dot{n}_s$ at which photons are measured by our detector will be

$$\dot{n}_s = \frac{\lambda F_\lambda}{hc} \cdot \Delta\lambda \cdot \frac{\pi D^2}{4} \cdot \phi_a \cdot \phi_t \cdot q, \tag{6.25}$$

where $\Delta\lambda$ is the bandwidth of the observation (that is, the range of wavelengths observed), D is the aperture of the telescope, ϕ_a is the fraction of the light that makes it through the Earth's atmosphere,[15] ϕ_t is the fraction of the light that makes it through the telescope to the detector, and q is the quantum efficiency of the detector. The total number of photons measured during an exposure time t will then be $\mu_s = \dot{n}_s t$.

In addition to photons from the star itself, we also detect background photons. The night sky background will be described in terms of a **surface brightness** S_λ, which is the specific flux F_λ produced per unit solid angle on the sky; thus, the units of S_λ will be watts per square meter per angstrom per square arcsecond of sky. The rate $\dot{n}_b$ at which background photons are measured by our detector will be

$$\dot{n}_b = \frac{\lambda S_\lambda}{hc} \cdot \frac{\pi \theta^2}{4} \cdot \Delta\lambda \cdot \frac{\pi D^2}{4} \cdot \phi_a \cdot \phi_t \cdot q, \tag{6.26}$$

where $\pi\theta^2/4$ is the area of sky that we are observing. If we are looking at a point source, the relevant value of θ for computing the amount of contaminating background light is the width of the point spread function. A broader point spread function leads to larger images of point sources and thus more background light underlying the image.

In the background limited case, the signal-to-noise ratio is thus (combining equations 6.23, 6.25, and 6.26)

$$S/N \approx \frac{\mu_s}{\mu_b^{1/2}} \approx \frac{\dot{n}_s t}{(\dot{n}_b t)^{1/2}} \propto \frac{F_\lambda D \phi_a}{\theta} \left(\frac{\Delta\lambda \phi_t q}{S_\lambda} t \right)^{1/2}. \tag{6.27}$$

Frequently, an astronomer observing a star, or other astronomical source, will want to reach a particular signal-to-noise ratio. The integration time t required to reach a given value of S/N is

$$t \propto \left(\frac{\theta}{F_\lambda D \phi_a} \right)^2 \frac{S_\lambda}{\Delta\lambda \phi_t q}. \tag{6.28}$$

This equation tells us, first of all, that the aperture size of the telescope is crucial; the observing time is inversely proportional to the collecting area of the telescope. However, it is a bit more surprising to realize that the size of the image θ is as important as the size of the telescope D. Data of the same quality can be achieved in equal time with a 1 m telescope that has $0.5''$ seeing as with a 4 m telescope that has $2''$ seeing.

If you observe a region of the sky containing many stars, then the **limiting flux** is the flux of the faintest star that you can detect at a specified signal-to-noise ratio S/N during

[15] Typically, $\phi_a \approx 0.8$ in the visible portion of the spectrum.

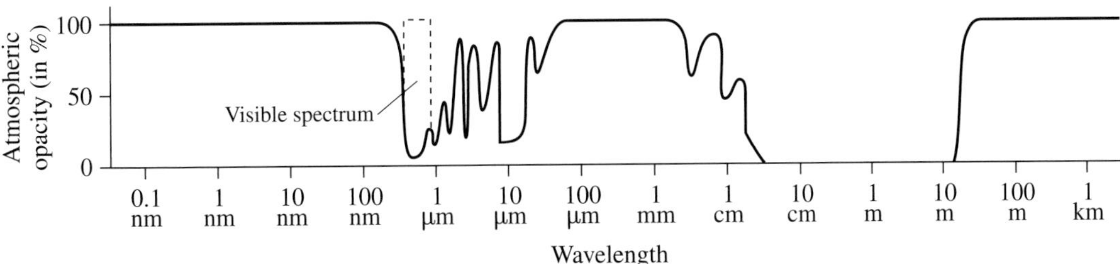

FIGURE 6.17 Atmospheric opacity as a function of the wavelength of light. The "optical window" ($\lambda \sim 0.5\,\mu$m) and the "radio window" ($\lambda \sim 3$ cm $\rightarrow$ 10 m) occur where the atmosphere is most nearly transparent.

an integration time t. In the background limited case, the limiting flux will be

$$F_\lambda(t) \propto \frac{\theta}{D\phi_a}\left(\frac{S_\lambda}{\Delta\lambda\phi_t q}\frac{1}{t}\right)^{1/2}. \tag{6.29}$$

As you integrate for a longer and longer period of time, you will detect fainter and fainter stars, but the limiting flux is inversely proportional to the square root of time.

6.6 ▪ OBSERVATIONS AT OTHER WAVELENGTHS

In the previous sections, we implicitly assumed that we were observing at visible wavelengths. Initially, of course, astronomers were compelled to observe at visible wavelengths because the only detectors available were their eyes. Even now, however, observations at visible wavelengths are of primary importance to astronomers for several reasons. First, most atomic transitions have ΔE on the order of a few eV; this means that the emitted and absorbed photons are in the visible or ultraviolet (UV) region of the spectrum. Further, the surface temperatures of stars are in the range $T \approx 2000$ K $\rightarrow$ 100,000 K; at these temperatures, they emit most of their radiation in the UV, visible, and infrared (IR) regions of the spectrum. Finally, the Earth's atmosphere is nearly transparent from wavelengths of $\lambda \approx 3100$ Å, in the near UV, to $\lambda \approx 1.1\,\mu$m, in the near IR; this range of wavelengths is known as the "optical window" (Figure 6.17). At wavelengths between $\sim 1.1\,\mu$m and ~ 3 cm, the Earth's atmosphere is only partially transparent, mostly because of absorption by water vapor. Observations in the infrared can be made only in a few "windows" where the opacity is relatively low. At wavelengths shorter than ~ 3100 Å, the Earth's atmosphere is almost completely opaque. This means that observations in most of the ultraviolet, and in all of the X-ray and gamma-ray regions of the electromagnetic spectrum, cannot be made from the ground.

The other transparent "window" through the Earth's atmosphere is in the radio portion of the spectrum. **Radio astronomy** began in earnest shortly after World War II, driven in part by technical advances made during the development of radar.[16] Within two decades, radio astronomy had led to the discovery of exotic sources such as pulsars (to be discussed in Section 18.2) and quasars (Section 21.1.2). Interstellar dust, which absorbs and scatters light, leaves much of our own galaxy unobservable to us at visible wavelengths. However, the transparency of dust at radio wavelengths allows us to map the structure of our galaxy. In addition, complex molecules have low-energy transitions seen at radio wavelengths; we see these molecules in dark, dusty, cold, interstellar clouds.[17]

Radio telescopes are generally much larger than optical telescopes, primarily to obtain adequate angular resolution. The Earth's atmosphere does not significantly distort radio waves; thus, the quality of images produced by a radio telescope is determined by the diffraction limit rather than by atmospheric seeing. Equation (6.7) tells us that the angular resolution for a diffraction-limited telescope is proportional to the wavelength of observation divided by the aperture of the telescope. If we observe at a wavelength of $\lambda \sim 5$ cm (roughly 100,000 times longer than visible wavelengths), then to obtain an angular resolution of $\sim 1''$ (routinely obtained at visible wavelengths), we must have an aperture of $D \sim 13$ km. Thus, very-large-aperture radio telescopes enable us to obtain reasonable angular resolution as well as enabling us to detect intrinsically faint radio sources. Fortunately, it is feasible to build radio telescopes that are much larger than optical telescopes. The accuracy with which a primary mirror must be manufactured is $\sim \lambda/4$, where λ is the wavelength being observed. Thus, a radio telescope operating at $\lambda \sim 5$ cm can have irregularities as much as a centimeter in height and still produce high-quality images.

The largest single-dish radio telescope in operation is the Arecibo Telescope, in Puerto Rico. Its aperture is $D = 305$ m, which yields an angular resolution of $\sim 40''$ when observing at a wavelength of $\lambda \sim 5$ cm, and proportionally poorer resolution when observing at longer wavelengths. It is impractical to make a single-dish telescope large enough to produce images with subarcsecond resolution. However, it is possible to use multiple radio telescopes for **interferometry**; the signals from the radio telescopes can be combined, if sufficient care is taken, to give diffraction-limited performance with resolution equivalent to that of a single dish as large as the maximum separation between the linked radio telescopes. The Very Large Array (VLA) in New Mexico consists of 27 radio telescopes mounted on rails so that the distance between them can be varied. The maximum possible separation between telescopes is $D = 36$ km, giving a resolution of $\sim 0.05''$ at $\lambda \sim 0.7$ cm, the shortest wavelength at which the VLA can observe. When 10 additional radio telescopes, spread from Hawaii to the Virgin Islands, are electronically tied to the 27 telescopes of the VLA, the resulting Very Long Baseline Array (VLBA) has a maximum separation between telescopes of $D = 8600$ km, giving angular resolution of less than a milliarcsecond.

[16] "Radar" is an acronym for **ra**dio **d**etection **a**nd **r**anging. Tracking ships and aircraft by bouncing radio signals off them required the ability to detect faint reflected radio waves.

[17] These "molecular clouds," as they are called, are particularly interesting to astronomers because they are where stars form.

At wavelengths that do not correspond to an atmospheric "window," observations must be made from above the Earth's atmosphere. Indeed, the principal reason for putting astronomical telescopes in space is to observe at wavelengths where the atmosphere is opaque. Each wavelength regime of the electromagnetic spectrum requires different technologies for photon detection, and each has its own challenges. One challenge in building X-ray telescopes, for instance, is the difficulty in building lenses or mirrors that are effective at focusing X-rays. Most materials have a refractive index n that is very close to one at X-ray energies; this means that X-ray lenses tend to have inconveniently long focal lengths. Making X-ray mirrors is not easy either, since X-ray photons are energetic enough to penetrate metal surfaces rather than reflect from them.

The current generation of X-ray telescopes uses "grazing incidence" mirrors; if X-ray photons strike a metal surface at an angle less than a small critical angle (typically 2° or less), it will reflect off rather than be absorbed. Gamma-ray photons are even more energetic ($h\nu > 10^4$ eV) than X-ray photons. In practice, they are detected with the same types of detectors that are used in particle accelerators here on Earth. Scintillation detectors, for instance, convert the energy of a high-energy particle, such as a gamma ray, into a burst of lower-energy photons. One major challenge in gamma-ray astronomy is distinguishing the actual gamma rays (high-energy photons) from cosmic rays (high-energy charged particles such as protons and helium nuclei).

The biggest challenge for space telescopes operating at infrared wavelengths is separating the signal from distant sources from the thermal background emitted by the telescope and its detectors. Thermal emission can be reduced by cooling with liquid nitrogen ($T \sim 77$ K) or by liquid helium ($T \sim 4$ K). For CCDs and similar detectors, cooling also reduces the "dark current"; that is, the self-generated electronic signal that is produced even when the detector is not exposed to photons.[18] At wavelengths longer than $\sim 5\,\mu$m, liquid helium temperatures are required to suppress the thermal background.

Even at visible wavelengths, there is a major advantage to going above the Earth's atmosphere; telescopes in orbit are diffraction-limited instead of seeing-limited. The higher angular resolution this permits largely accounts for the success of the *Hubble Space Telescope*. Although the *Hubble* is a modest-size telescope by research standards, with $D = 2.4$ m, it produces images of width $\theta = 0.05''$ at $\lambda = 5000$ Å. Equation (6.29) then tells us that the *Hubble Space Telescope* can go deeper (that is, see objects of lower flux) than any ground-based telescope currently in existence, as well as producing sharper images.

There are distinct disadvantages, as well as the obvious advantages, to placing astronomical telescopes in space. The most important disadvantage is cost. Space telescopes are extremely expensive to build, launch, and operate. A telescope in space costs roughly 100 to 1000 times as much as a comparably sized telescope on the ground. Moreover, neither the telescope nor its instruments can be easily serviced; indeed, the *Hubble Space*

[18] Indeed, most electronic detectors, even at visible and ultraviolet wavelengths, are cooled at least to liquid nitrogen temperature to suppress dark current.

Telescope is the only space-based observatory that can have its instruments changed and telescope refurbished by regular manned missions.

There are a number of misconceptions about the supposed advantages of space astronomy:

- Misconception: *In space, you can observe a celestial object continuously, not just at night.* For a telescope in low Earth orbit, nearly half the celestial sphere is blocked by the Earth at any given time. Thus, most celestial objects can be viewed for only half an orbit at a time (45 minutes on, 45 minutes off).
- Misconception: *The sky seen from orbit is dramatically darker than the night sky on Earth.* Although the night sky is darker as seen from orbit, the improvement is not as great as you might expect. At blue wavelengths, the sky is about half as bright as from the ground. Much of the sky background is due to sunlight scattered from interplanetary dust and distant unresolved stars and galaxies.
- Misconception: *Observations from space are not affected by weather.* In fact, space observations have their own environmental difficulties; instruments and detectors deteriorate rapidly due to the harsh radiation environment and the broad temperature range through which they are cycled. In addition, there is "weather" of a sort in space, largely due to solar magnetic activity (described in Section 7.2).

6.7 ▪ MODERN TELESCOPES

During the 1960s, 70s, and 80s, the major advances in observational astronomy were due to improvements in detector technology and the opening of new spectral windows through space observations. It became apparent, however, that certain limitations had been reached; in particular, any further advance in detector technology would be only incremental, since the quantum efficiencies of detectors were approaching the theoretical limit. Equation (6.29) tells us that the key elements in detecting fainter objects, once the quantum efficiency has reached its limit, is minimizing the image size θ and maximizing the telescope aperture D.

To minimize the image size, we must first select sites that are capable of delivering good seeing; these are generally at high altitudes (so there is less air above the telescope) in regions where the air flow is smooth and laminar. We then need to control seeing quality to the fullest extent possible; this really means making sure that the telescope and its immediate vicinity are not contributing to turbulence in the atmosphere. Finally, after we have exhausted such "passive control" measures, we need to pursue *active* control of the image quality. This can be done through **adaptive optics**; in this technique, we sense wavefront distortions (by observing a bright star, for instance) and compensate for these measured distortions with a deformable mirror. Such systems, although technically difficult, are beginning to come into regular use.

Maximizing the collecting area of telescopes means making larger mirrors. Unfortunately, conventional telescope mirrors are limited to $D \lesssim 5$ m. The problem with larger mirrors is **thermal inertia**. To perform well, the mirror must be at the same temperature as the air surrounding it. If it is warmer than its surroundings, the mirror will radiate

FIGURE 6.18 The Large Binocular Telescope on Mount Graham in Arizona.

infrared light, heating the air above the mirror and causing turbulence;[19] this makes the seeing very poor. The thermal inertia is the measure of an object's ability to store heat during the day and radiate it away at night. Thermal inertia scales with mass, and the mass of a conventional mirror scales as D^3, since the thickness must increase proportionately with its diameter in order to keep the mirror stiff. There are a number of ways around the thermal inertia problem, and all of them have met with success:

- **Multiple mirrors.** A large effective collecting area can be achieved by combining the light collected by smaller individual mirrors. This is one reason why the Large Binocular Telescope (LBT), which has two 8.4 m mirrors on a single mount (Figure 6.18), was not designed as a "Very Large Monocular Telescope," with a solitary 11.9 m mirror.
- **Segmented mirrors.** Instead of a monolithic mirror made from a single slab of glass, a mirror can be made of individual segments kept aligned through computer-controlled active supports. This has been done successfully on the two 10 m Keck telescopes, each of which consists of thirty-six 1 m segments.
- **Honeycombed mirrors.** Instead of starting with a solid disk of glass, we can melt the glass for our mirror over heat-resistant ceramic forms that leave the back of the mirror with a honeycomb structure that is stiff but lightweight.[20] It is also possible to control the temperature of the mirror by circulating coolant through the structure behind the reflecting surface. The two mirrors of the LBT are made with this honeycombed structure.
- **Active mirror support.** A large telescope mirror can be made very thin, and consequently flexible, if it is actively supported along its back to retain the correct

[19] This is the same effect that causes shimmering mirages above an asphalt highway on a hot day.

[20] Much the same honeycomb structure is seen inside bird bones, which also must be stiff but lightweight.

shape as the telescope swings back and forth. The mirrors of the four Very Large Telescope units at the European Southern Observatory use this technology, as do the mirrors of the two Gemini telescopes, one on Mauna Kea (Hawaii) and one on Cerro Pachon (Chile).

PROBLEMS

6.1 Assume that your vision is diffraction limited at $\lambda = 5000$ Å and that the diameter of the pupil of your eye is $D = 8$ mm. What angular resolution can you achieve with your unaided eye? How does this compare with the maximum angular size of Venus and Jupiter as seen from the Earth?

6.2 (a) The Hiltner Telescope at the MDM Observatory (on Kitt Peak, Arizona) has an aperture $D = 2.4$ m. Its Cassegrain focus has an f-number $f/7$. What is the focal length F and plate scale s?

(b) The Mayall Telescope at the Kitt Peak National Observatory (also on Kitt Peak) has an aperture $D = 4.0$ m. Its prime focus has an f-number $f/2.7$, its Cassegrain focus has $f/8$, and its Coudé focus has $f/160$. What is the focal length and plate scale for *each* of these three foci?

(c) The Keck Telescope (on Mauna Kea, Hawaii) has an aperture $D = 10.0$ m. Its Cassegrain focus has $f/15$. What is the focal length and plate scale?

6.3 With the $D = 2.4$ m telescope at the MDM Observatory, I can obtain a spectrum of a particular star with signal-to-noise ratio $S/N = 100$ in $t = 20$ minutes when the atmospheric seeing is average ($\theta = 1''$). How long would it take me to obtain the same data with the Keck Telescope ($D = 10.0$ m) with excellent seeing ($\theta = 0.4''$)?

6.4 A charge-coupled device (CCD) detector is mounted at the focus of an $f/7$ reflecting telescope with a $D = 50$ cm mirror. The CCD chip contains 1024×1024 pixels, with each square pixel being 10 μm on a side.

(a) What is the area (in square arcseconds) of the sky that is imaged on a single pixel?

(b) What is the area (in square arcminutes) of the sky that is imaged on the entire chip? Would the image of the full Moon fit into the chip?

(c) How many separate exposures would be required to cover the entire celestial sphere (4π steradians)?

6.5 Suppose that you want to see stars that are as faint as possible in the background-limited case. The Astronomy Fairy gives you a choice: *either* she can increase the quantum efficiency of your retina from $q = 0.1$ to $q = 1$, *or* she can double the maximum pupil size of your eye while guaranteeing diffraction-limited angular resolution. Which of these choices would produce a lower limiting flux F_λ? Explain your choice.

6.6 The Atacama Large Millimeter/Submillimeter Array (ALMA) is designed to operate over the wavelength range $\lambda = 0.3 \rightarrow 9.6$ mm. It will consist of 80 independent 12 m telescopes with a maximum baseline of 18 km.

(a) What is the highest angular resolution achievable with ALMA?

(b) How large would a single-dish antenna have to be to have the same collecting area as ALMA?

6.7 Prove that equation (6.18) is correct for a Poisson probability distribution.

7 The Sun

Although the Sun is more massive and more luminous than most of the stars in its neighborhood, it is by no means freakishly bright. Thus, by studying the Sun in particular, we can learn a great deal about stars in general. A study of stellar interiors (including the solar interior) will be deferred until Chapter 15, when we will have covered the observable properties of stars in more detail. At this point, however, some discussion of the Sun's outer layers is merited. This is partly because there are aspects of solar physics we need to understand in order to understand the evolution of the solar system as a whole. However, it is also true that the Sun is the only star whose surface has been studied in detail, and thus it deserves some extra attention.

7.1 ▪ OBSERVABLE LAYERS OF THE SUN

An image of the Sun at visible wavelengths (Figure 7.1) has a sharp, well-defined boundary, implying that the Sun has a well-defined surface. The observed surface of the Sun at visible wavelengths is the **photosphere**. More exactly, the photosphere is defined as the layer of the Sun's atmosphere from which nearly all of the observed photons escape. The optical depth τ increases rapidly with depth in the photosphere; as a result, the photosphere is not very thick when compared to the size of the Sun as a whole. The vast majority of the light we observe from the Sun comes from a photosphere only ~ 400 km thick. The base of the photosphere is at a distance $R_\odot = 696{,}000$ km from the Sun's center. It's the thinness of the photosphere that gives the Sun its sharp-edged appearance. The temperature T beneath the photosphere increases with depth, as does the degree of ionization; the interior of the Sun is a hot plasma of free electrons and positively charged ions.

Because the top of the photosphere is cooler than the layers beneath, the photosphere produces absorption lines in the spectrum of the Sun. Detailed analysis of high resolution spectra allows us to deduce the physical properties of the solar photosphere. Elemental abundances, for instance, are very different from those found on Earth. By mass, hydrogen constitutes 73.4% of the photosphere, helium constitutes 25.0%, and

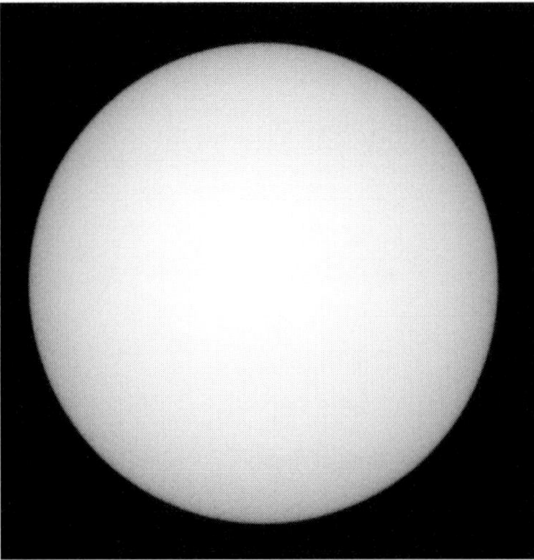

FIGURE 7.1 The photosphere is the region of the solar atmosphere from which most of the visible light is emitted. Note the effect of "limb darkening"; the solar disk has a higher surface brightness at the center than at the edges.

all the remaining elements of the periodic table contribute only 1.6% of the mass of the photosphere.[1]

In the photosphere, the principal source of opacity is the **H^- ion**, that is, a hydrogen atom with an additional electron. In a gas containing both neutral hydrogen atoms and free electrons, H^- ions can form by the reaction

$$H + e^- \rightarrow H^- + \gamma. \tag{7.1}$$

Many metals have low ionization potentials and thus are partially ionized at the temperature of the Sun's photosphere; this provides the primary source of free electrons for the creation of H^- ions. However, the second electron in the H^- ion is quite loosely bound, with an ionization energy $\chi = 0.75$ eV. The fragility of the H^- ion implies that it is abundant enough to affect the opacity only under special conditions; the density of gas must be fairly high, and the temperature must be in the range $2500\text{ K} \lesssim T \lesssim 10{,}000\text{ K}$. At lower temperatures, there are essentially no free electrons available, and at higher temperatures, the H^- ions are blasted apart by photons as soon as they form. The energy $\chi = 0.75$ eV required to remove the second electron corresponds to a wavelength $\lambda = 1.7\ \mu$m. This means that when H^- is present, it can absorb ultraviolet photons, visible photons, and infrared photons out to a wavelength of $1.7\ \mu$m. Since the density of particles in the photosphere decreases with height, eventually the density drops to the

[1] Astronomers sometimes lump together all elements heavier than helium as "metals." This puts them, we realize, on the same level as primitive tribes who count "one, two, many," but sometimes the ability to count beyond two is overrated.

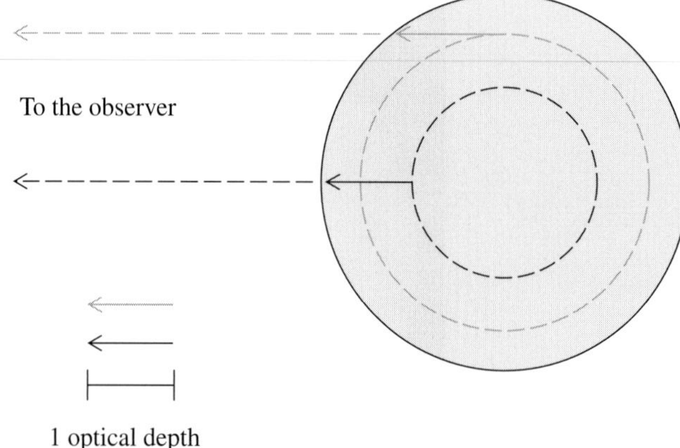

FIGURE 7.2 An observer far to the left observes limb darkening of the Sun. Photons from the center of the Sun's disk come, on average, from deeper, hotter layers of the photosphere.

point where the collisions between free electrons and hydrogen atoms that lead to formation of H^- occur too slowly to keep up with the photodissociation rate. Thus, at the top of the photosphere, the H^- has disappeared, the opacity due to H^- has disappeared, and photons can escape.

The photons that we observe from the photosphere come from a variety of depths, and thus, given the temperature gradient across the photosphere, represent blackbody emission at a range of temperatures. The average depth from which the observed photons originate is determined by the column density of H^- along our line of sight. At the center of the solar disk, the physical depth corresponding to an optical depth $\tau = 1$ is larger than the physical depth at the limb, or edge, of the Sun. This effect is illustrated in Figure 7.2.

As a consequence of this effect, the Sun displays **limb darkening**; the surface brightness is greater at the center of the Sun's disk because the photons we see come from deeper, hotter layers of the photosphere, on average. By comparison, the limb of the Sun's disk is lower in surface brightness because the photons come from higher, cooler layers of the photosphere, on average. The average temperature (that is, the one that gives the best-fitting Planck spectrum) at the center of the disk is $T \approx 6100$ K. However, when photons from the entire disk are pooled together, the best-fitting temperature is $T \approx 5700$ K, thanks to the contribution from the cooler photons from the limb.

When the Sun's disk is viewed at high angular resolution, the photosphere is seen to be broken up into **granules** (Figure 7.3). The granules are convection cells in the photosphere. Hot gas rises at the center of the granule; after the hot gas cools by radiation, the cooler gas sinks back down at the edges of the granules. The typical size of granules is $d \sim 1000$ km, and the typical lifetime of a granule before it breaks up is only $t \sim 10$ minutes. A time-lapse movie of the photosphere looks like a seething vat of soup.

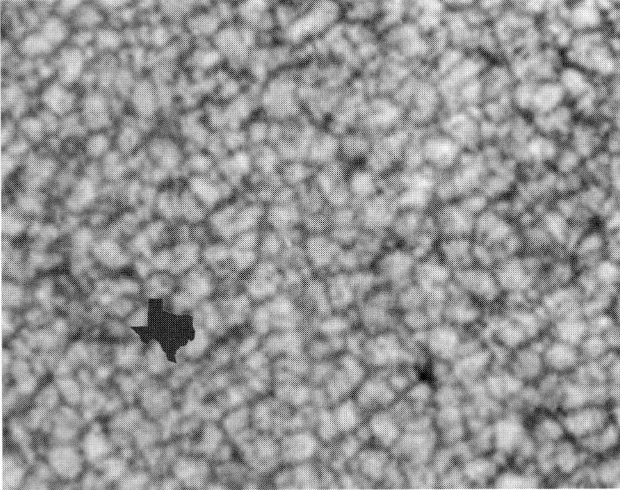

FIGURE 7.3 Detail of the photosphere showing granules; typically, they are the size of Texas ($d \sim 1000$ km).

The **chromosphere** is the layer of the Sun's atmosphere immediately above the photosphere. The chromosphere, as illustrated in Color Figure 4, is most easily seen during a total solar eclipse, when the Moon blocks the light from the much-brighter (by definition) photosphere. The chromosphere produces an emission spectrum, as expected from Kirchhoff's laws; it consists of hot, tenuous gas seen, during an eclipse, against a dark background. The characteristic red color that gives the chromosphere its name[2] is due to strong emission from the Hα λ6563 line. When the emission spectrum of the chromosphere was measured during an eclipse on 1868 August 18, astronomers were surprised to detect a yellow emission line ($\lambda = 5875$ Å) that didn't correspond to any known element. The English scientist Norman Lockyer decided that the line was due to a previously unknown element that he called "helium," after the Sun god Helios. Chemists did not isolate the element helium in their laboratories until 1895, when its emission spectrum was verified.

The temperature of the chromosphere increases with distance from the Sun's center (unlike the temperature structure of the photosphere, where the temperature drops with increasing distance). At the top of the photosphere, which constitutes the base of the chromosphere, the temperature is $T \approx 4400$ K; at the top of the chromosphere, at a height of ~ 2500 km above its base, the temperature has risen sharply to $T \approx 9000$ K.[3]

[2] *Chromo-* is a Greek root meaning "color."

[3] The reason why helium absorption lines are not seen in the Sun's photospheric spectrum is that the cooler temperatures there result in photons that are insufficiently energetic to raise electrons above the ground state of helium.

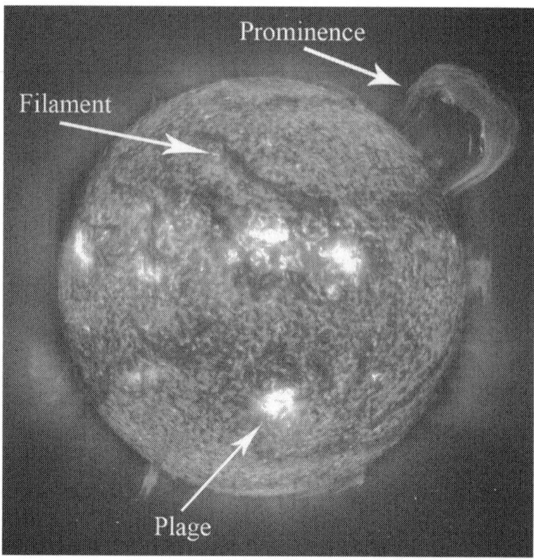

FIGURE 7.4 A spectroheliogram, showing features of the chromosphere.

If you are insufficiently patient to wait for a total solar eclipse, you can also observe the chromosphere by using a **spectroheliogram**. These are simply images taken through filters with a narrow bandwidth, centered on the wavelengths of strong photospheric absorption lines. At the wavelengths of the strongest absorption lines, such as the Hα λ6563 line, the photosphere is essentially black, and we can detect the chromospheric emission lines against the black background. Figure 7.4 is a spectroheliogram of the Sun, showing extensive chromospheric structure. The structure of the chromosphere is irregular and variable with time. Bright extended regions in the chromosphere are called **plages**.[4] Plages are regions where the magnetic field is particularly strong. The chromosphere also shows dark, linear features called **filaments**; these are long clouds of relatively cool gas that are lifted above the chromosphere by the Sun's magnetic field. Although filaments are cooler than the chromosphere below them, they are still hot enough to be seen as bright **prominences** when they extend beyond the Sun's limb and are seen against the relative darkness of the sky.

On finer scales, the chromosphere shows features called **spicules**, which are relatively narrow columns of gas moving vertically at speeds of $\sim 10\,\mathrm{km\,s^{-1}}$. For comparison, these speeds are higher than the speed of $\sim 1\,\mathrm{km\,s^{-1}}$ with which gas circulates in granules. Like plages, and like the filaments and prominences, the spicules are related to the Sun's magnetic field. The spicules consist of charged particles moving along magnetic field lines. In Figure 7.5, a spectroheliogram taken during the transit of Venus on 2004 June 8 (which explains the black dot), the spicules are seen as short, dark lines against the

[4] The word "plage" is the French term for a beach or, more generally, an open area. Because of its French origin, "plage" rhymes with "mirage" and not with "rage."

FIGURE 7.5 A spectroheliogram, showing dark spicules (as well as Venus in the process of transiting).

hotter body of the chromosphere. However, like the larger filaments, the short spicules are bright when they extend beyond the limb of the Sun and are seen in projection against the sky.

The **corona** is the outer layer of the Sun's atmosphere. The corona emits about 1 million times less light than the photosphere, which makes it roughly as bright as the full Moon. Ordinarily, the coronal light is lost in the glare of the garish photosphere; as a consequence, the corona is most easily observed during a total solar eclipse. Viewing the coronal spectrum during an eclipse permits us to identify two separate components of the corona, each with a distinctive spectrum. The **K corona**, which dominates near the Sun (within $r \sim 2.5R_\odot$), has a continuum spectrum with emission lines superimposed on it, but no detectable absorption lines.[5] The **F corona**, which dominates farther from the Sun, has a continuum spectrum with absorption lines, similar in appearance to the photospheric spectrum.[6]

When the spectrum of the K corona was first studied in the nineteenth century, it was found to have a green emission line ($\lambda = 5303$ Å) that didn't correspond to any known element. Astronomers, following the example of Norman Lockyer, the discoverer of helium, declared that the emission line was due to a previously unknown element, called "coronium." It wasn't until the year 1942 that the mystery emission line was discovered to correspond to a forbidden transition of Fe XIV, that is, an iron atom with 13 of its 26 electrons stripped away. It takes a very high temperature to collisionally ionize iron to such an extent; the highest temperatures in the corona are $T \approx 2 \times 10^6$ K. The

[5] The "K" stands for *Kontinuum*; the name was devised by a German scientist.

[6] The "F" stands for *Fraunhofer*, the spectroscopist who first studied the Sun's absorption lines.

continuum light emitted by the K corona consists of light from the photosphere that has scattered from the free electrons in the coronal plasma. The large thermal velocity of the hot electrons causes enough Doppler broadening to smear out the strong photospheric absorption lines, and leaves the continuum nearly featureless aside from the emission lines that arise in the K corona itself.

The light emitted by the F corona, farther from the photosphere, consists of light from the photosphere that has scattered from dust grains.[7] Since the dust grains are moving slowly compared to the free electrons, the absorption lines in the photospheric spectrum are preserved when they scatter from dust in the F corona.

Detailed study of the structure of the solar atmosphere reveals that the temperature jumps upward from $T \sim 10^4$ K to $T \sim 10^6$ K in a thin transition region between the chromosphere and the corona (Figure 7.6). One question that has puzzled astronomers is, What gives the corona its extremely high temperature? Since the density of the corona is low, it doesn't take an extraordinarily high amount of thermal energy to raise it to $T \gtrsim 10^6$ K, but you still need a way of transporting that energy from lower levels of the Sun out to the corona. Early hypotheses proposed that sound waves, generated by the Sun's convective cells, traveled outward from the photosphere until they steepened into shock waves ("sonic booms") in the corona. When a shock wave passes through a gas, it causes a rise in temperature. More recent hypotheses have proposed that the changing magnetic fields of the Sun cause electric currents to run through the corona; the electrical resistance of the coronal gas would then cause it to heat up. The main source of coronal heating is not yet clear.

The rms speed of the protons in the hot corona is (equation 5.48):

$$v_{\rm rms} = \left(\frac{3kT}{m_p}\right)^{1/2} \approx 160\ {\rm km\ s^{-1}} \left(\frac{T}{10^6\ {\rm K}}\right)^{1/2} . \tag{7.2}$$

This can be compared to the escape speed at a distance r from the Sun's center (equation 3.62):

$$v_{\rm esc} = \left(\frac{2GM_\odot}{r}\right)^{1/2} \approx 620\ {\rm km\ s^{-1}} \left(\frac{r}{R_\odot}\right)^{-1/2} . \tag{7.3}$$

As the escape speed decreases with distance from the Sun's center and the coronal temperature increases, an increasing fraction of the particles in the corona exceed the escape speed. The outward flow of escaping particles is called the **solar wind**.

The solar wind has been measured by interplanetary spacecraft; at $r = 1$ AU from the Sun, the speed of the wind is $v \sim 400$ km s^{-1}, on average, while the number density of particles is $n \sim 10^7$ m^{-3}. For the mix of electrons, protons, and helium nuclei that make up the solar wind, this provides a mass density $\rho \sim 10^{-21}$ kg m^{-3} at $r = 1$ AU. If we make the assumption that the solar wind is in a steady state, we can compute the rate at

[7] Dust grains are ubiquitous in the solar system, but tend to be concentrated in the ecliptic plane; it's these dust grains, as we see in Chapter 11, that give rise to the zodiacal light.

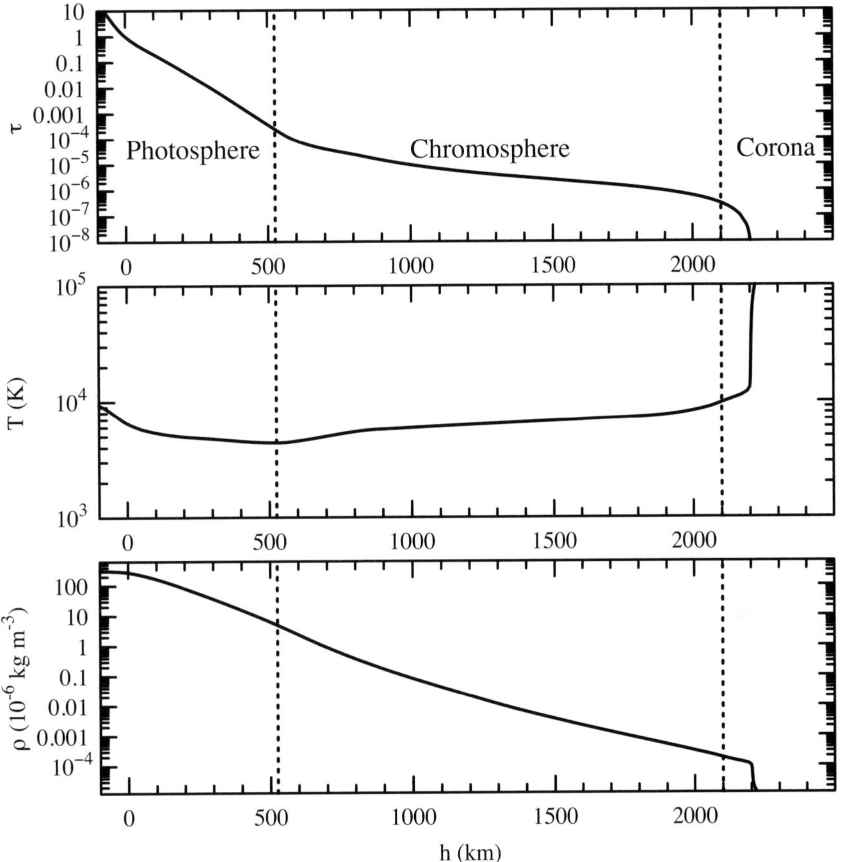

FIGURE 7.6 The density and temperature of the solar atmosphere (photosphere, chromosphere, and corona) as a function of height above the base of the photosphere.

which the Sun is losing mass. (The steady-state assumption isn't strictly true—there are "gusts" in the solar wind—but it's sufficient for a rough approximation.) Consider a thin spherical shell of radius r and thickness Δr centered on the Sun, as shown in Figure 7.7. The mass of the solar wind in the shell at any given instant is just the shell's volume times the density:

$$\Delta M = (4\pi r^2 \Delta r)\rho. \tag{7.4}$$

The mass flux through the shell can be obtained by dividing by the time Δt it takes a solar wind particle to move outward a distance Δr:

$$\frac{\Delta M}{\Delta t} = 4\pi r^2 \frac{\Delta r}{\Delta t}\rho. \tag{7.5}$$

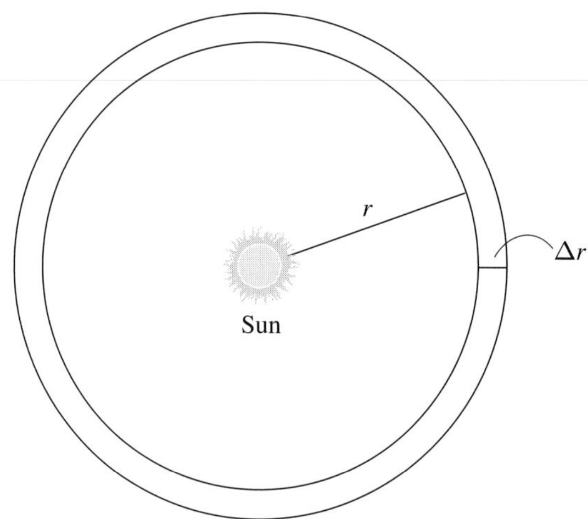

FIGURE 7.7 A thin spherical shell is centered on the Sun.

As we let $\Delta t \to 0$, we can write this as a mass continuity equation for a steady-state spherical flow:

$$\frac{dM}{dt} = 4\pi r^2 \frac{dr}{dt} \rho, \tag{7.6}$$

where we can identify dM/dt as the mass loss rate of the Sun, $\dot{M}_\odot$, since it is assumed to be constant at all radii. We can also identify dr/dt with the wind speed v, which will be constant as long as nothing is acting to accelerate or decelerate the wind particles after they leave the corona. Thus, we may write

$$\dot{M}_\odot = 4\pi r^2 v \rho. \tag{7.7}$$

Using $v \sim 400\text{ km s}^{-1}$, $r = 1$ AU, and $\rho \sim 10^{-21}\text{ kg m}^{-3}$, we find

$$\dot{M}_\odot = 4\pi (1.5 \times 10^{11}\text{ m})^2 (4 \times 10^5\text{ m s}^{-1})(10^{-21}\text{ kg m}^{-3}) \sim 10^8\text{ kg s}^{-1}. \tag{7.8}$$

This mass loss rate of 100,000 tons per second sounds big until you realize that it is equivalent to $\dot{M}_\odot \sim 10^{-14} M_\odot\text{ yr}^{-1}$. The timescale for mass loss is then

$$t_M = M_\odot / \dot{M}_\odot \sim 10^{14}\text{ yr}, \tag{7.9}$$

whereas the age of the Sun is only $\sim 5 \times 10^9$ yr. If the rate of mass loss has been approximately constant over the age of the Sun, then the solar wind has not significantly pared down the Sun's mass.

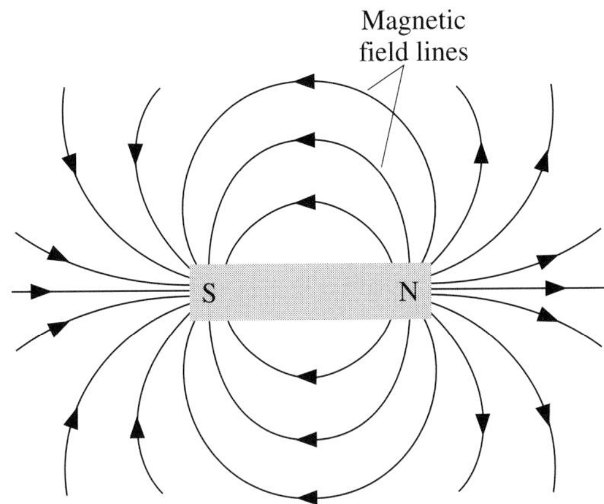

FIGURE 7.8 Magnetic field lines of a common bar magnet.

7.2 ▪ SOLAR ACTIVITY

Violent and variable phenomena in the Sun's atmosphere are collectively referred to as **solar activity**. All forms of solar activity are tied to the complex magnetic field of the Sun. To understand solar activity, we must begin with a discussion of the interaction between charged particles, such as electrons and protons, and magnetic fields. A magnetic field is a vector field $\vec{\mathbf{B}}$; that is, it has a direction as well as a magnitude at every position in space. In SI units, magnetic fields are measured in teslas (T), where $1\,\mathrm{T} = 1\,\mathrm{kg\,s^{-1}\,C^{-1}}$.[8] Magnetic fields are usefully visualized by drawing magnetic field lines; these are curves such that $\vec{\mathbf{B}}$ at every point of the curve is exactly tangent to the curve. Figure 7.8, for example, shows magnetic field lines of a bar magnet. Regions where the field lines converge and come closer together are regions of stronger magnetic field.

A particle with electric charge q is moving with a velocity $\vec{\mathbf{v}}$ through a magnetic field. The charged particle will be accelerated by the Lorentz force,

$$\vec{\mathbf{F}} = q\vec{\mathbf{v}} \times \vec{\mathbf{B}}, \tag{7.10}$$

where $\vec{\mathbf{B}}$ is the magnetic field strength, in teslas. The sense of the acceleration is shown in Figure 7.9. The particle velocity $\vec{\mathbf{v}}$ can be decomposed into a component $v_\parallel$ that is parallel to $\vec{\mathbf{B}}$ and a component $v_\perp$ that is perpendicular to $\vec{\mathbf{B}}$. The speed $v_\parallel$ is unchanged by the Lorentz force. However, if $v_\perp$ is nonzero, the Lorentz force will cause the particle to execute a circular motion, with constant speed $v_\perp$, perpendicular to the magnetic field.

[8] The Earth's magnetic field has a strength of $B \sim 3 \times 10^{-5}\,\mathrm{T}$ at the Earth's surface. An MRI (magnetic resonance imaging) machine uses an electromagnet with $B \sim 1\,\mathrm{T}$; this is a strong magnetic field by terrestrial standards.

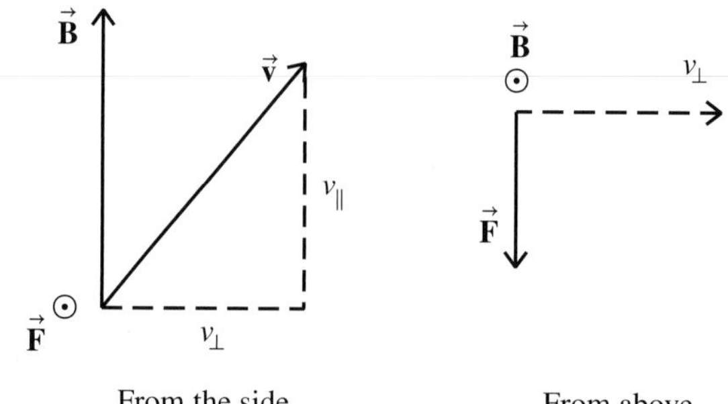

FIGURE 7.9 The Lorentz force accelerates a charged particle in a direction perpendicular to both the magnetic field lines and the instantaneous particle velocity. (The symbol ⊙ represents a vector coming out of the page.)

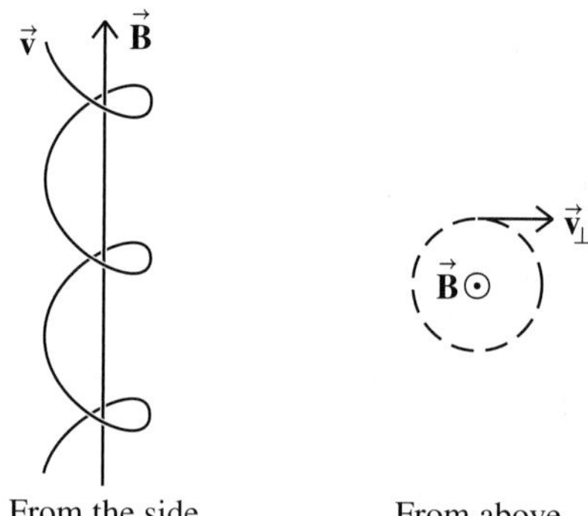

FIGURE 7.10 The Lorentz force causes charged particles to travel in a helix around magnetic field lines.

The constant motion in the direction parallel to $\vec{\mathbf{B}}$, combined with the circular motion perpendicular to $\vec{\mathbf{B}}$, causes charged particles to move in a helix, as shown in Figure 7.10. Newton's second law tells us that the acceleration, for a particle of charge q and mass m, will be

$$a = \frac{q}{m} v_\perp B. \tag{7.11}$$

On a circular orbit, $a = v_\perp^2 / r$, so the radius of the particle's circular path will be given by the relation

$$\frac{v_\perp^2}{r} = \frac{q}{m} v_\perp B. \tag{7.12}$$

The radius of the particle's orbit will then be

$$r_c = \frac{mv_\perp}{qB}, \tag{7.13}$$

known as the **Larmor radius** of the charged particle.

Magnetic field lines thus tend to guide charged particles; the helix along which a particle moves is wrapped around a magnetic field line like a vine wrapped around a beanpole. As long as the gas pressure is low, the motion of charged particles will be dictated by the magnetic field. At high gas pressure, however, any bulk motions of the gas will tend to drag the magnetic field along with them. Whether the gas controls the motion of magnetic field lines, or the magnetic field controls the motion of gas particles, depends on the ratio of the energy density of the magnetic field to the pressure of the gas.[9] The magnetic energy density (or magnetic pressure) is

$$P_B = \frac{B^2}{2\mu_0}, \tag{7.14}$$

where μ_0 is the **permeability constant**. In SI units, $\mu_0 = 4\pi \times 10^{-7}\,\mathrm{kg\,m\,C^{-2}}$. The magnetic pressure is then

$$P_B = 4.0 \times 10^5\,\mathrm{N\,m^{-2}} \left(\frac{B}{1\,\mathrm{T}}\right)^2. \tag{7.15}$$

Within the Sun, the gas pressure is given by the ideal gas law, $P_{\mathrm{gas}} = nkT$, where n is the particle density, T is the temperature, and k is Boltzmann's constant. In terms of the mass density of the gas, $\rho = \mu m_p n$, the gas pressure is

$$P_{\mathrm{gas}} = \frac{\rho kT}{\mu m_p} \approx \frac{5000\,\mathrm{N\,m^{-2}}}{\mu} \left(\frac{\rho}{10^{-4}\,\mathrm{kg\,m^{-3}}}\right)\left(\frac{T}{6000\,\mathrm{K}}\right), \tag{7.16}$$

where we've scaled the equation to densities and temperatures appropriate for the solar photosphere.[10]

Many observable properties vary along with the Sun's time-variable magnetic field. The most obvious solar phenomena linked to the magnetic field are **sunspots**. Figure 7.11 illustrates a pair of particularly large sunspot groups. Sunspots are regions in the photosphere where the magnetic field is locally enhanced and the gas temperature is relatively cool. At the center of the solar disk, as mentioned earlier, the photospheric temperature is ~ 6100 K; however, in a sunspot, the temperature is ~ 4300 K. Thus, sunspots are still hot—much hotter than the filament of an incandescent lightbulb—and

[9] Note that energy density ($\mathrm{J\,m^{-3}}$) and pressure ($\mathrm{N\,m^{-2}}$) have the same units ($\mathrm{kg\,m^{-1}\,s^{-2}}$).

[10] For purposes of comparison, the pressure of the Earth's atmosphere at sea level is $P \sim 10^5\,\mathrm{N\,m^{-2}}$. The Earth's atmosphere is cooler than the Sun's photosphere, and has a higher mean molecular mass μ, but at sea level it is much denser, with $\rho \sim 1\,\mathrm{kg\,m^{-3}}$.

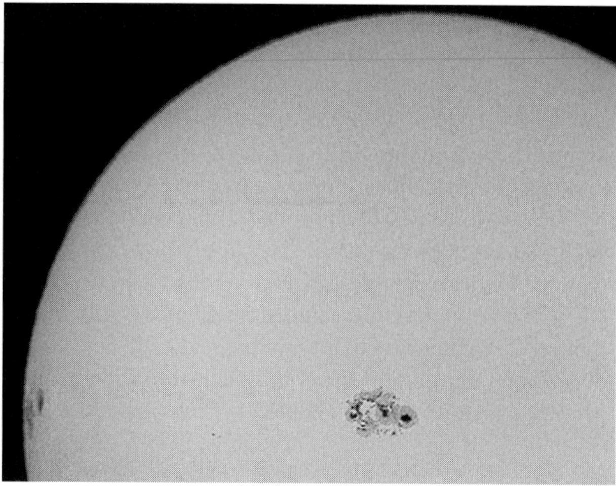

FIGURE 7.11 Sunspot groups 10484 (near the center of the Sun's disk) and 10486 (near the limb), seen in autumn 2003.

radiate like blackbodies. However, they look dark relative to the surrounding photosphere because the surface flux F is strongly dependent on temperature: $F \propto T^4$, and $(6100/4300)^4 \approx 4$.

Sunspots were first seen by Chinese astronomers over 2000 years ago; at sunrise and sunset, large sunspot groups (similar to those in Figure 7.11) can be seen with the naked eye. With the advantage of a telescope, Galileo and other seventeenth-century astronomers were able to track sunspots as they moved across the Sun's disk. The movement of sunspots demonstrated that the Sun rotates differentially; the rotation period near the equator is $P = 25.4$ days, while near the poles it is around $P = 35$ days. (This differential rotation is permissible because the Sun is a fluid body rather than a rigid body.) Individual sunspots are transient features, and have lifetimes ranging from a few hours to a few months. That sunspots are associated with strong magnetic fields is revealed by the Zeeman effect. As discussed in Section 5.3, there exist atomic energy levels that are degenerate with no magnetic field present, but which split into an upper level and a lower level in the presence of a magnetic field, with the amount of splitting proportional to the magnetic field strength B. As seen in Figure 7.12, there exist absorption lines that display Zeeman splitting within a sunspot, but no perceptible splitting outside the spot.

The strong magnetic fields within sunspots explain why they are cooler than their surroundings. The total pressure in the sunspot (magnetic pressure plus gas pressure) must equal the pressure outside the sunspot. If they were not equal, then the sunspot would expand or contract until the pressures were equalized. If we assume that the magnetic pressure outside the spot is negligibly small, we see that

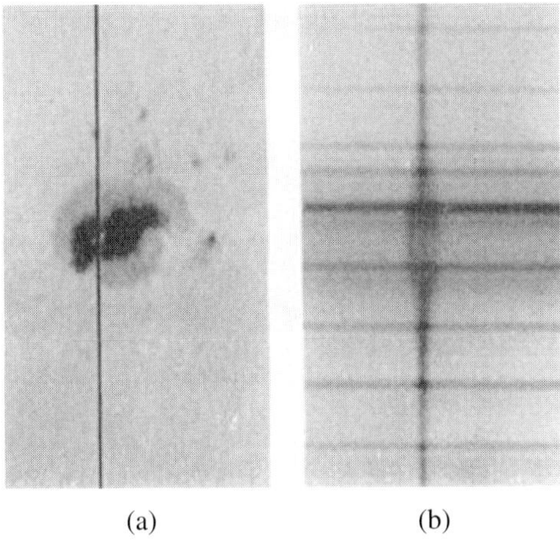

FIGURE 7.12 (a) A sunspot, with the vertical line representing the slit of a spectrograph. (b) The resulting spectrogram. For each point on the vertical slit, the spectrum is dispersed in the horizontal direction. The prominent absorption line shows Zeeman splitting where the slit crosses the sunspot.

$$\frac{\rho k T_s}{m_p} + \frac{B^2}{2\mu_0} = \frac{\rho k T_p}{m_p}, \tag{7.17}$$

where T_s is the sunspot temperature and T_p is the temperature of the surrounding photosphere.[11] Solving for the magnetic field strength, we find

$$B = \left[\frac{2\mu_0 \rho k (T_p - T_s)}{m_p} \right]^{1/2}. \tag{7.18}$$

By referring to a detailed model of the solar atmosphere, we find that, at an optical depth $\tau = 1$ (in the middle of the photosphere), the mass density is $\rho \approx 3.5 \times 10^{-4}\,\mathrm{kg\,m^{-3}}$. With this mass density, and with a temperature difference $T_p - T_s = 6100\,\mathrm{K} - 4300\,\mathrm{K} = 1800\,\mathrm{K}$, we find that

$$\begin{aligned} B &\approx \left[\frac{2(4\pi \times 10^{-7})(3.5 \times 10^{-4})(1.38 \times 10^{-23})(1800)}{1.67 \times 10^{-27}} \right]^{1/2} \\ &\approx 0.1\,\mathrm{T}. \end{aligned} \tag{7.19}$$

[11] We have assumed a mean molecular mass $\mu = 1$ in both the sunspot and the surrounding photosphere; this is approximately correct and has the benefit of eliminating confusion between μ and μ_0 (the permeability constant).

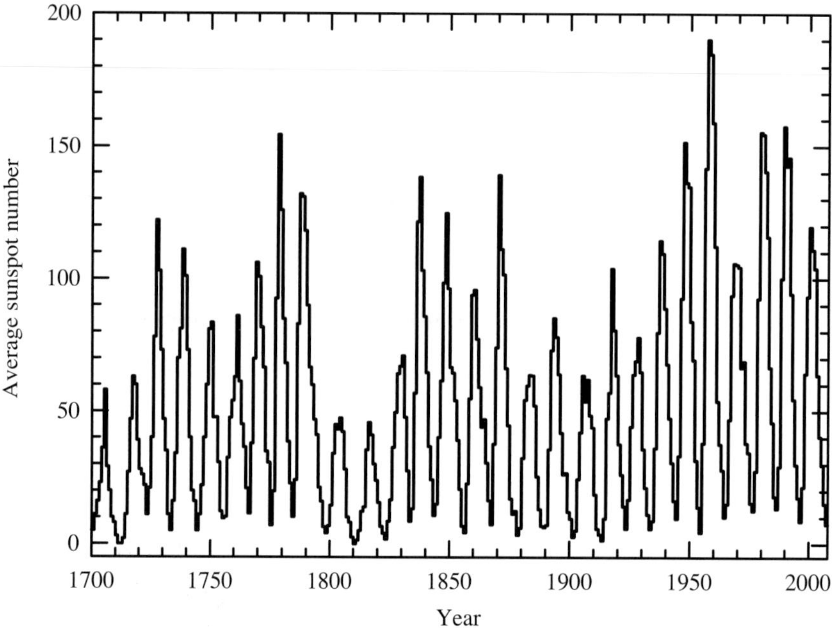

FIGURE 7.13 Historical sunspot counts, from 1700 to 2007.

By contrast, the average magnetic field strength on the surface of the Sun is in the range $0.01 \rightarrow 0.001$ T. The magnetic field in a sunspot effectively acts to cool its surroundings by decreasing the local gas pressure.

The number of sunspots visible at any given time varies over an 11-year cycle, as shown in Figure 7.13. It is not a perfectly repeatable cycle, since the peak number of sunspots can vary dramatically from one cycle to the next. In fact, during the **Maunder minimum**, between the years 1645 and 1710, sunspots seemed to disappear almost entirely. During the course of the solar cycle, the location of sunspots on the solar disk also changes. Early in the solar cycle, sunspots first appear at relatively high latitudes (typically at solar latitudes $\pm 30°$), but later in the cycle, they migrate closer to the equator. A plot of sunspot latitude as a function of the time of appearance of the sunspot is shown as Figure 7.14. This plot is known as a **butterfly diagram**, from its fancied resemblance to butterflies with their wings spread.

Sunspots tend to occur in pairs of opposite magnetic polarity; that is, the magnetic field lines run upward out of one sunspot and downward into the other sunspot. However, the sense of the polarity is reversed between the Sun's northern and southern hemispheres; if in the northern hemisphere the leading (eastern) sunspot has a positive polarity, then in the southern hemisphere the leading sunspot will have a *negative* polarity. This switch in polarity is illustrated in Figure 7.15, which is a magnetogram of the Sun. A magnetogram is a plot of magnetic field strength (determined by Zeeman splitting) over the disk of the Sun; white and black represent different magnetic polarities. In the figure, there is one

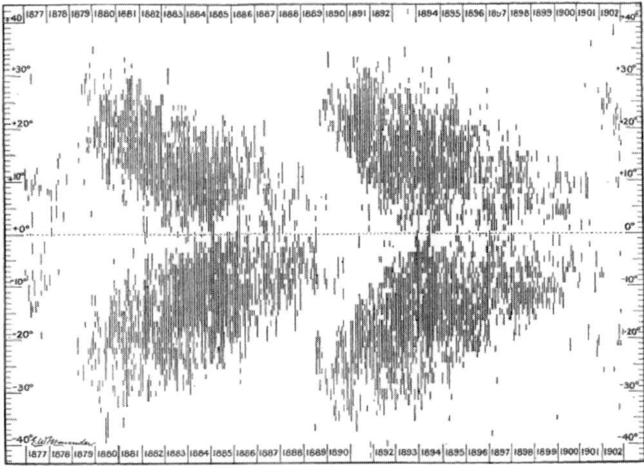

FIGURE 7.14 Sunspot solar latitude as a function of time, as plotted by Edward Maunder (after whom the Maunder minimum is named) for the two solar cycles from 1879 to 1901.

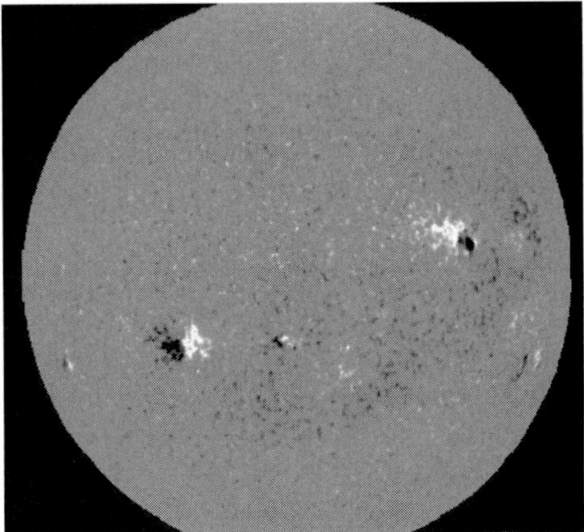

FIGURE 7.15 A magnetogram of the Sun, in which white and black represent strong magnetic fields of opposite polarity. Gray represents weak magnetic fields.

sunspot pair in each hemisphere; in one hemisphere, the "white" spot leads, and in the other, the "black" spot leads. In addition to the reversed polarity of sunspot pairs in opposite hemispheres, the polarities switch between one cycle and the next. That is, in one 11-year cycle, the *positive* polarity sunspot will lead in the northern hemisphere; in the next 11-year cycle, the *negative* polarity sunspot will lead in the same hemisphere. Thus, the complete solar cycle actually has a period of 22 years rather than 11 years.

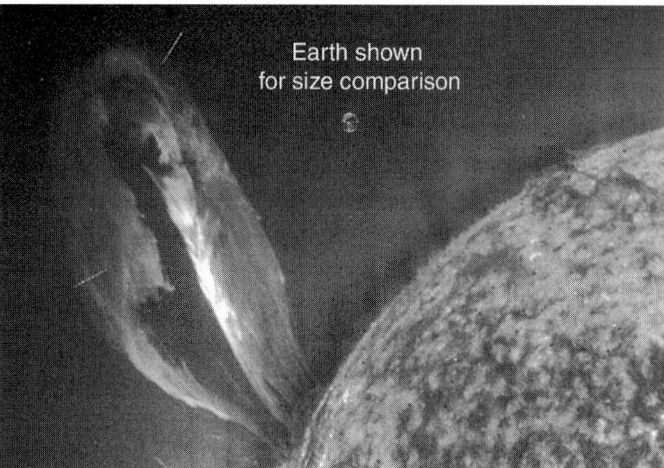

FIGURE 7.16 This image taken by the SOHO satellite shows a particularly large prominence.

In the photosphere, the gas pressure exceeds the magnetic pressure (although only by a relatively small margin in sunspots). Thus, in the photosphere, bulk motions of ionized gas carry the magnetic field lines with them. The turbulent convection in the photosphere randomly pushes and bends the magnetic field lines, which produces kinks in the field lines. A kinked magnetic field can develop loops that erupt outward through the photosphere. These outward loops produce a **bipolar magnetic region**, which can be a sunspot pair, as described above, or a larger complex of sunspots. The magnetic loop can be removed by merger of the magnetic poles. Ionized gas trapped between merging magnetic poles will be accelerated along the magnetic field lines, forming a prominence. An example of a large prominence is shown in Figure 7.16. (When a prominence is seen in absorption against the bright photosphere, it is called a filament.)

Large prominences can be highly energetic, with gas moving upward at speeds higher than the Sun's escape speed. An erupting prominence—that is, one that sends matter upward at speeds greater than the escape speed—can cause a **coronal mass ejection**. A coronal mass ejection can be simply defined as a big blob of ionized gas ejected from the Sun. Coronal mass ejections occur a few times a day when the Sun is active, and once every few days during the quiet part of the solar cycle. A large coronal mass ejection typically contains $\sim 10^{13}$ kg of gas. The average speed of a coronal mass ejection is $\sim 500 \text{ km s}^{-1}$ once it is far from the Sun, implying a kinetic energy $\sim 10^{24}$ J for a large coronal mass ejection.

Solar flares are events that can be even more energetic than coronal mass ejections. Solar flares are chromospheric eruptions, triggered by sudden releases of energy stored in the Sun's magnetic field. A solar flare produces a bright burst of electromagnetic radiation, much of it in the form of high-energy ultraviolet, X-ray, and gamma-ray photons. In addition, the flare accelerates charged particles such as electrons and protons to relativistic speeds. Solar flares occur several times a day when the Sun is active, and once a week or so when the Sun is quiet. The energy of a solar flare can be as much as $\sim 10^{25}$ J, and the temperatures within a flare can reach 10^7 K. Although the exact mechanism of solar flares is not understood, they can apparently be explained as a consequence of magnetic reconnection events in bipolar magnetic regions.

Both coronal mass ejections and solar flares produce charged particles that can interact with the Earth's magnetic field. The slower coronal mass ejections take a few days to reach the Earth from the Sun, while the relativistic charged particles associated with a solar flare can reach the Earth in two hours or less. In either case, the charged particles are accelerated by the Earth's magnetic field toward the magnetic poles. There, they interact with atoms in the Earth's upper atmosphere, producing the **aurora borealis** (in the Earth's northern hemisphere) and the **aurora australis** (in the southern hemisphere). More colloquially, they are known as the northern lights and southern lights. A fine specimen of an aurora is shown in Color Figure 6.

7.3 ▪ ANGULAR MOMENTUM OF THE SUN

Although the Sun contains 99.8% of the mass of the solar system, Jupiter contains most of the angular momentum. From equation (3.49), we know that the angular momentum of Jupiter's orbital motion is

$$L_{\rm Jup} = m_{\rm Jup} a_{\rm Jup}^2 \omega_{\rm Jup} \sqrt{1-e^2}, \tag{7.20}$$

where $a_{\rm Jup} = 5.2$ AU is the semimajor axis of Jupiter's orbit, $e = 0.048$ is its eccentricity, and $\omega_{\rm Jup} \equiv 2\pi/P_{\rm Jup} = 1.5 \times 10^{-3}\ {\rm day}^{-1}$ is the average angular speed of Jupiter on its orbit. By comparison, the Sun's rotational angular momentum can be approximated as

$$L_\odot \approx \frac{2}{5} M_\odot R_\odot^2 \omega_\odot, \tag{7.21}$$

where $\omega_\odot$ is the average angular speed of the Sun (since the Sun is not a rigid body, ω varies with position in the Sun). Since $P_\odot \approx 28$ days, then $\omega_\odot = 2\pi/P_\odot \approx 0.22\ {\rm day}^{-1}$.

The ratio of Jupiter's orbital angular momentum to the Sun's rotational angular momentum is

$$\frac{L_{\rm Jup}}{L_\odot} \approx \frac{5}{2}\left(\frac{m_{\rm Jup}}{M_\odot}\right)\left(\frac{a_{\rm Jup}}{R_\odot}\right)^2\left(\frac{\omega_{\rm Jup}}{\omega_\odot}\right)\sqrt{1-e^2}. \tag{7.22}$$

Although Jupiter is much less massive than the Sun ($M_\odot \approx 1040 m_{\rm Jup}$), and its orbital angular speed is less than the Sun's rotational angular speed ($\omega_\odot \approx 150\omega_{\rm Jup}$), Jupiter's orbit is much larger than the Sun's radius ($a_{\rm Jup} \approx 1100 R_\odot$). This means that

$$\frac{L_{\rm Jup}}{L_\odot} \approx \frac{5}{2}\left(\frac{1}{1040}\right)(1100)^2\left(\frac{1}{150}\right)\sqrt{0.998} \approx 20. \tag{7.23}$$

The Sun's small angular momentum is puzzling, particularly since young stars comparable in mass to the Sun are seen to rotate at a much higher rate.[12] The puzzle can be solved if the Sun has lost a significant fraction of its angular momentum.

One way in which the Sun can lose angular momentum is to have it carried off by the solar wind. Consider a particle of mass m leaving the Sun's atmosphere at radius $R_\odot$; it will have an angular momentum $L = mR_\odot^2\omega$, where ω is the angular speed of rotation at the point where the particle leaves. Taking $\omega \approx \omega_\odot \approx 0.22\ \mathrm{day}^{-1}$, the rate at which angular momentum leaves the Sun is

$$\frac{dL}{dt} = \frac{dm}{dt} r^2\omega = \dot{M}_\odot R_\odot^2 \omega_\odot, \tag{7.24}$$

where $\dot{M}_\odot$, the Sun's average rate of mass loss, is $\sim 10^{-14} M_\odot$ yr. Given the Sun's rotational angular momentum,

$$L_\odot \approx \frac{2}{5} M_\odot R_\odot^2 \omega_\odot, \tag{7.25}$$

the timescale over which the Sun loses its angular momentum through the solar wind should be

$$\begin{aligned} \tau = \frac{L_\odot}{dL/dt} &\approx \frac{2M_\odot R_\odot^2 \omega_\odot}{5\dot{M}_\odot R_\odot^2 \omega_\odot} \\ &\approx \frac{2}{5}\frac{M_\odot}{\dot{M}_\odot} \approx \frac{2}{5}\frac{1 M_\odot}{10^{-14} M_\odot\ \mathrm{yr}^{-1}} \approx 4 \times 10^{13}\ \mathrm{yr}. \end{aligned} \tag{7.26}$$

This is much longer than the age of the Sun, $t \sim 5 \times 10^9$ yr, and argues that the solar wind is as inefficient at paring away the Sun's angular momentum as it is paring away the Sun's mass.

However, the above calculation neglects the effect of the Sun's magnetic field. At sufficiently large radii, the Sun's magnetic field lines can be approximated as stretching radially outward (see Figure 7.17 for one simulation of the Sun's magnetic field). Since the magnetic field lines are rooted in the dense ionized gas of the Sun's interior, they co-rotate with the Sun. Close to the Sun, the magnetic energy density is greater than the energy density of the solar wind. Thus, at small radii, the charged particles of the solar wind are compelled to move along the (approximately) radial magnetic field lines, and they too co-rotate with the Sun. Since their angular speed $\omega \approx \omega_\odot$ remains constant while their radius r increases, their angular momentum, $L \propto r^2\omega$, must increase. However, at some point the magnetic energy density drops below the energy density of the solar wind. The **Alfvén point** is the radius r_A at which the magnetic energy density is equal to the kinetic energy density of the solar wind:

[12] Rotation speeds are estimated from the rotational Doppler broadening of the stars' absorption lines; the stars are known to be young because they are associated with types of stars that have short lifetimes.

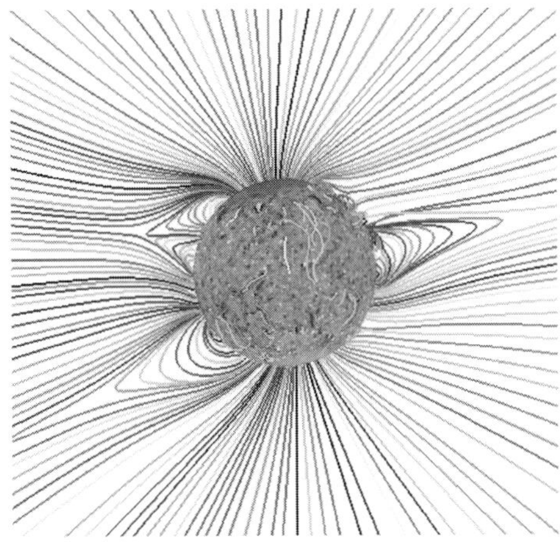

FIGURE 7.17 The Sun's magnetic field lines at a time of low solar activity.

$$\frac{B^2}{2\mu_0} = \frac{1}{2}\rho v^2. \tag{7.27}$$

Outside the Alfvén radius, the solar wind is no longer torqued by the magnetic field, and the angular momentum of each solar wind particle is constant.

Since the magnetic flux is conserved, we can write for an arbitrary distance r from the Sun's center

$$4\pi r^2 B(r) = 4\pi R_\odot^2 B_\odot, \tag{7.28}$$

assuming that the magnetic field $\vec{\mathbf{B}}$ is purely radial. Thus, the magnetic field strength falls as the inverse square of distance:

$$B(r) = B_\odot \left(\frac{R_\odot}{r}\right)^2, \tag{7.29}$$

and the magnetic energy density falls off even more rapidly:

$$\frac{B(r)^2}{2\mu_0} = \frac{B_\odot^2}{2\mu_0}\left(\frac{R_\odot}{r}\right)^4. \tag{7.30}$$

However, for a steady-state wind, we have found that the density falls off as (equation 7.7)

$$\rho(r) = \frac{\dot{M}_\odot}{4\pi v}\frac{1}{r^2}, \tag{7.31}$$

and thus the kinetic energy density of the solar wind is

$$\frac{1}{2}\rho(r)v^2 = \frac{\dot{M}_\odot v}{8\pi}\frac{1}{r^2}. \tag{7.32}$$

By equating the magnetic energy density (equation 7.30) and the kinetic energy density of the solar wind (equation 7.32), we find the Alfvén radius r_A:

$$\frac{B_\odot^2}{2\mu_0}\frac{R_\odot^4}{r_A^4} = \frac{\dot{M}_\odot v}{8\pi}\frac{1}{r_A^2}, \tag{7.33}$$

yielding

$$\frac{r_A}{R_\odot} = \left(\frac{4\pi B_\odot^2 R_\odot^2}{\mu_0 \dot{M}_\odot v}\right)^{1/2}. \tag{7.34}$$

Using $\dot{M}_\odot \approx 10^{-14}\,M_\odot\,\mathrm{yr}^{-1}$, $v \approx 400\,\mathrm{km\,s}^{-1}$, and $B_\odot \approx 10^{-3}\,\mathrm{T}$, we can compute

$$\frac{r_A}{R_\odot} \approx 140. \tag{7.35}$$

That is, the outward-streaming solar wind co-rotates with the Sun out to a distance of $r_A \approx 140 R_\odot \approx 0.6$ AU. Thus, the angular momentum carried away by each particle is larger, by a factor $\sim (140)^2 \sim 2 \times 10^4$, than it would be in the absence of a magnetic field. This decreases the timescale for angular momentum loss from $\tau \approx 4 \times 10^{13}$ yr to $\tau \approx 2 \times 10^9$ yr, shorter than the age of the Sun. Thus, when the Sun's magnetic field is taken into account, there is no problem explaining the low angular momentum of the Sun. Much of the Sun's initially large angular momentum has been carried away by the solar wind.

PROBLEMS

7.1 Using data in this chapter, compute the Larmor radius r_c for a typical electron in the K corona.

7.2 The thermal energy density of a gas is equal to its number density of particles times the mean kinetic energy $\langle E \rangle$ of each particle due to random thermal motions.

(a) What is the thermal energy density at the base of the Sun's photosphere?
(b) What is the thermal energy density of the Earth's atmosphere at sea level ($n = 2.5 \times 10^{25}\,\mathrm{m}^{-3}$ and $T = 290$ K)?

7.3 At what rate does the solar wind carry kinetic energy away from the Sun? Give your result first in watts, then as a fraction of the Sun's luminosity in photons, $L_\odot = 3.8 \times 10^{26}$ W.

7.4 How many rotations (and how much time) does it take for the equatorial regions of the Sun to "lap" the polar regions by one full rotation?

7.5 The normal Zeeman effect splits a spectral line at frequency ν_0 into three components: a central line at ν_0 and two satellite lines at $\nu_0 \pm eB/(4\pi m_e)$. By what amount (in angstroms) are the satellite lines of the hydrogen Balmer α line ($\lambda_0 = 6562.81$ Å) split from the central component in a typical sunspot?

7.6 A solar flare erupts in a region where the average magnetic field strength is $B = 0.03$ T; the flare releases an energy $E = 2 \times 10^{24}$ J.

(a) What was the magnetic energy density in the region prior to the eruption?
(b) What was the minimum volume V required to supply enough magnetic energy to fuel the flare?
(c) If the volume V is spherical, what is its radius? Is this greater than or less than the typical radius $r \approx 10^4$ km of a sunspot?

7.7 If the entire photosphere of the Sun had $B_\odot = 0.1$ T, what would the Alfvén radius of the Sun be? (Hint: assume the properties of the solar wind would be unchanged.) What would be the timescale τ for the loss of the Sun's angular momentum?

7.8 Vertical motions of gas in photospheric granules typically have speeds $v \sim 2\ \mathrm{km\ s^{-1}}$.

(a) What angular resolution (in arcseconds) is required to see an individual granule in the Sun's photosphere?
(b) Neutral sodium has a pair of absorption lines at rest wavelength $\lambda_0 = 5889.973$ Å and 5895.940 Å. What is the ratio of the thermal Doppler broadening of these lines to the Doppler shift expected from the vertical motion of the granules?

7.9 Imagine a sphere of gas with a uniform number density n of gas particles; each particle has a cross-section σ. Consider the optical depth τ along a path parallel to a line through the sphere's center, displaced from the center by some distance z. Compute $\tau(z)$ for such paths, then compute $d\tau/dz$. Explain, given this calculation, why a gaseous sphere can appear to have a very sharp limb.

8 Overview of the Solar System

Our aim over the next four chapters is to understand the basic characteristics of the solar system, and to integrate this information into a self-consistent picture of how the solar system formed and how it has evolved with time. Some basic facts should be kept in mind as we pursue our goal:

- The Sun contains 99.8% of the mass in the solar system.
- Most of the remaining 0.2% of the mass is confined to a flattened disk. Within this disk, all the planets revolve in the same direction, most (but not all) of the planets rotate in the same direction, and all objects have similar ages ($t \sim 4.6$ billion years, when measurable).

Figure 8.1 shows a plot of mass versus orbital semimajor axis for the largest objects known to be orbiting the Sun (in the case of the smaller objects, the mass is a rough estimate). The eight largest objects orbiting the Sun have been given the collective name **planets**. The rocky and metallic objects in the asteroid belt, lying primarily between the planets Mars and Jupiter, are called **asteroids**; the largest asteroid, Ceres, is also a **dwarf planet**, according to the definition approved by the International Astronomical Union.[1] The icy objects beyond the orbit of Neptune are called **trans-Neptunian objects** (TNOs); many of the known TNOs are in a region just beyond Neptune called the Kuiper belt. The largest known TNOs—Eris, Pluto, Haumea, and Makemake—are also dwarf planets.

8.1 ▪ TWO TYPES OF PLANETS

The planets can be divided into two major types, or families, each named after their largest member. The smaller **terrestrial** planets, named after the Earth (alias Terra), are closer to the Sun; the larger **Jovian** planets, named after Jupiter (alias Jove), are farther

[1] A dwarf planet must be large enough to be squeezed by its own self-gravity into a spherical shape, but not large enough to be gravitationally dominant in the region near its orbit. We discuss the definition of "planet" and "dwarf planet" in more detail in Section 11.2.

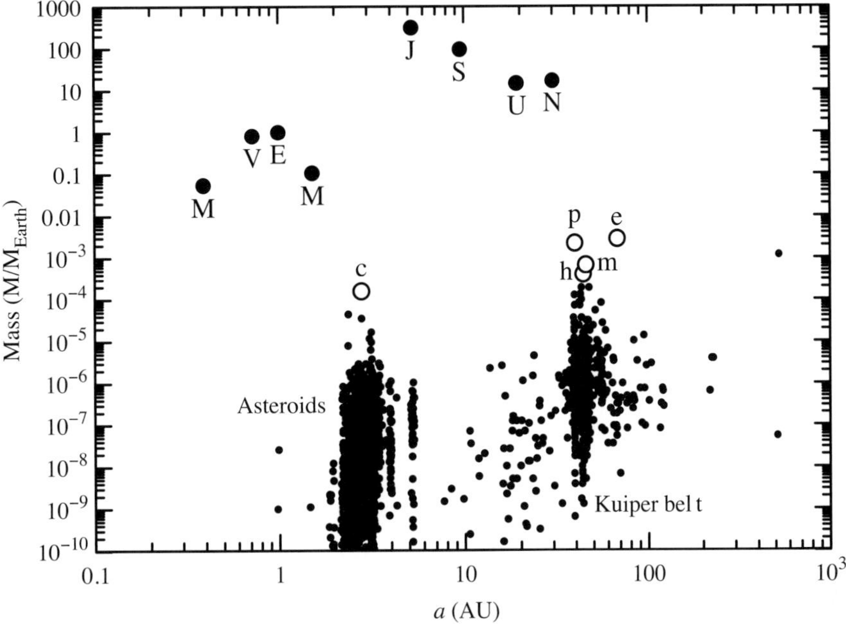

FIGURE 8.1 Estimated mass (in units of the Earth's mass) as a function of semimajor axis for bodies orbiting the Sun. Planets are labeled with their initials. (The dwarf planets Ceres, Pluto, Makemake, Haumea, and Eris are labeled with lowercase initials.)

from the Sun.[2] The characteristics of the two families are compared and contrasted in Table 8.1.

Any useful theory for the origin of the solar system must explain the observed differences between terrestrial and Jovian planets. In addition, the following questions should be addressed:

- Why are planetary orbits nearly circular?
- What is the nature and origin of the small "debris," such as comets and asteroids?
- What is the origin of the planetary satellites?
- Why are there differences in chemical composition among bodies in the solar system?

[2] The dwarf planets Ceres, Pluto, Makemake, Haumea, and Eris don't fit into this scheme. Ceres resembles the rocky satellites of the solar system, while Pluto resembles the icy satellites. The properties of Makemake, Haumea, and Eris are as yet poorly known, but their high albedo, or reflectivity, indicates that they have icy surfaces.

TABLE 8.1 Characteristics of Planetary Types

Characteristic	Terrestrial	Jovian
Mass	Low ($\leq 1\,M_\oplus$)	High ($> 10\,M_\oplus$)
Composition	Rocky/metallic ($\rho \gtrsim 3000$ kg m^{-3})	Gaseous/icy ($\rho \lesssim 2000$ kg m^{-3})
Rotation	Slow ($P \geq 24$ hr)	Fast ($P < 18$ hr)
Satellites	Few	Many
Distance from Sun	$a < 2$ AU	$a > 5$ AU

- Why are there rings around the Jovian planets (and not around the terrestrial planets)?
- Why are the terrestrial planets chemically differentiated, with a rocky outer layer wrapped around a metallic core?

We will first describe the physical characteristics of the constituents of the solar system, introducing new physical concepts as necessary. We will then try to put these elements into a single coherent picture of the solar system.

8.2 ▪ PHYSICAL PROPERTIES OF PLANETS

The masses of astronomical objects are measured by looking at how they accelerate neighboring objects. It is particularly easy to measure the mass of a planet when it has a nearby small satellite; in that case, it is a simple matter of applying Kepler's third law (equation 3.54):

$$M_{\text{planet}} \approx \frac{4\pi^2 a^3}{GP^2}, \tag{8.1}$$

where a and P are the semimajor axis and period of the satellite's orbit. Planets that do not have satellites (that is, Mercury and Venus) pose a more difficult problem. Before the advent of interplanetary space probes, the masses of Venus and Mercury were determined from their (tiny) perturbations to the orbits of other planets. Much more accurate masses are now available from space probes that have flown by them or orbited them.

The radius R of a planet is computed from its measured angular radius and its distance.[3] The mean density can then be computed from the known radius and mass:

$$\rho = \frac{3M}{4\pi R^3}. \tag{8.2}$$

Densities of $\rho \sim 700 \to 2000$ kg m^{-3} are typical of Jovian planets, indicating a composition that is mostly gas or ice, where "ice," in the language of astronomers, can

[3] Distances within the solar system can be measured accurately using radar, as described in Section 13.1.

refer to frozen water, but also to frozen carbon dioxide ("dry ice"), frozen methane, and frozen ammonia. The lowest-density Jovian planet, Saturn, has a mean density $\rho \approx 710\,\mathrm{kg\,m^{-3}}$, which is less than that of water ($\rho = 1000\,\mathrm{kg\,m^{-3}}$).[4] Densities of $\rho \sim 3000 \to 5500\,\mathrm{kg\,m^{-3}}$ are typical of terrestrial planets, indicating a composition that is mostly rock and metal.

The surface temperature T of objects in the solar system depends on a number of factors. Since the Sun is an important source of energy, the temperature of an object depends on its distance from the Sun and on its **albedo**, or reflectivity.[5] However, the surface temperature also depends on whether the object has any internal heat sources, and whether it has an atmosphere that can act as an insulating blanket wrapped around its surface.

The Sun, to a rough approximation, radiates like a blackbody. The Sun's luminosity is thus

$$L_\odot = 4\pi R_\odot^2 \sigma_{\mathrm{SB}} T_\odot^4, \tag{8.3}$$

where the Sun's effective temperature is $T_\odot \approx 5800\,\mathrm{K}$. It was left as an exercise in Chapter 5 (Problem 5.6) to show that the peak of the Planck function I_λ occurs at a wavelength

$$\lambda_{\mathrm{p}} \approx 0.20 \frac{hc}{kT} \approx \frac{2.9 \times 10^7\,\text{Å K}}{T} \approx \frac{2900\,\mu\mathrm{m\,K}}{T}, \tag{8.4}$$

a relation that is known as **Wien's law**, after the physicist who discovered it empirically. For $T_\odot \approx 5800\,\mathrm{K}$, $\lambda_{\mathrm{p}} \approx 5000\,$Å, so the solar spectrum, expressed in terms of wavelength, peaks in the visible part of the spectrum.

The flux of energy received by a planet at a distance r from the Sun is

$$F(r) = \frac{L_\odot}{4\pi r^2}. \tag{8.5}$$

The energy that the planet *absorbs* per second is the flux F times the cross-section of the planet (πR^2) times the fraction of the incident light that is absorbed rather than reflected. Since the albedo A of the planet is the fraction of the light energy that is reflected, the fraction that is absorbed is $1 - A$. Thus, the rate at which the planet absorbs energy is

$$W_{\mathrm{p}} = \left(\frac{L_\odot}{4\pi r^2}\right)(\pi R^2)(1 - A). \tag{8.6}$$

The energy absorbed by the planet will raise its surface temperature to some value T_{p}. If we approximate the planet as a blackbody, it will radiate energy at a rate

$$L_{\mathrm{p}} = 4\pi R^2 \sigma_{\mathrm{SB}} T_{\mathrm{p}}^4, \tag{8.7}$$

assuming a uniform surface temperature T_{p} for the planet.

[4] The standard Bad Astronomical Joke is that Saturn would float in a large enough bathtub . . . but that it would leave a ring behind.

[5] The word "albedo" comes from the Latin word *albus*, meaning "white." A white object has a high albedo since it reflects most of the light that strikes it; a black object has a low albedo.

When the planet is in equilibrium, the rate at which energy is emitted by the planet, $L_{\rm p}$, is equal to the rate at which energy is absorbed by the planet, $W_{\rm p}$. In equilibrium, then,

$$4\pi R^2 \sigma_{\rm SB} T_{\rm p}^4 = \frac{L_\odot}{4\pi r^2} \pi R^2 (1 - A). \tag{8.8}$$

Solving for the equilibrium surface temperature $T_{\rm p}$, we find

$$T_{\rm p} = \left(\frac{R_\odot}{r}\right)^{1/2} \left(\frac{1-A}{4}\right)^{1/4} T_\odot. \tag{8.9}$$

Inserting numerical values for $R_\odot$ and $T_\odot$, and expressing the distance from the Sun in astronomical units, we find

$$T_{\rm p} \approx 279\ {\rm K}(1 - A)^{1/4} \left(\frac{r}{1\ {\rm AU}}\right)^{-1/2}. \tag{8.10}$$

If the object being heated by the Sun is a blackbody, then $A = 0$ and

$$T_{\rm bb} \approx 279\ {\rm K} \left(\frac{r}{1\ {\rm AU}}\right)^{-1/2}, \tag{8.11}$$

which is the **equilibrium blackbody temperature** for a spherical blackbody of uniform temperature at a distance r from the Sun.

The assumption of uniform surface temperature $T_{\rm p}$ is a good approximation for a planet that is rotating rapidly, like a chicken on a spit, or has efficient atmospheric circulation. However, we should also consider the case of a planet that is rotating slowly and is not a good conductor of heat. In that case, the energy that is absorbed by a small patch of area Σ must be re-radiated by the same patch. The equilibrium condition then becomes

$$\Sigma \sigma_{\rm SB} T_{\rm p}^4 = \frac{L_\odot}{4\pi r^2} \Sigma (1 - A), \tag{8.12}$$

where $T_{\rm p}$ is now the temperature of a small patch on the planet's surface for which the Sun is at the zenith. Solving for $T_{\rm p}$, the equilibrium temperature of this patch, we find

$$T_{\rm p} = \left(\frac{R_\odot}{r}\right)^{1/2} (1 - A)^{1/4} T_\odot, \tag{8.13}$$

or

$$T_{\rm p} \approx 395\ {\rm K}(1 - A)^{1/4} \left(\frac{r}{1\ {\rm AU}}\right)^{-1/2}. \tag{8.14}$$

If the slowly rotating body is a blackbody, with $A = 0$, then the resulting temperature is the **subsolar blackbody temperature**

$$T_{\rm ss} \approx 395\ {\rm K} \left(\frac{r}{1\ {\rm AU}}\right)^{-1/2}, \tag{8.15}$$

TABLE 8.2 Planetary Albedos

Rocky Surfaces	Mercury	0.06
	Moon	0.07
	Mars	0.16
Complex Mix	Earth	0.40
Gases and Ices	Saturn	0.50
	Jupiter	0.51
	Neptune	0.62
	Uranus	0.66
	Venus	0.76

TABLE 8.3 Computed vs. Observed Temperatures

Planet	Albedo	r (AU)	Uniform T_p $T_{bb}(1-A)^{1/4}$	Slow Rotator $T_{ss}(1-A)^{1/4}$	T_{obs}
Venus	0.76	0.72	230 K	325 K	740 K
Earth	~ 0.4	1	246 K	347 K	290 K
Neptune	0.62	30.1	40 K	57 K	59 K

which represents the highest temperature a body can reach at a distance r from the Sun, if sunlight is the only source of heat.

Real objects within the solar system are neither perfectly white ($A = 1$) nor perfectly black ($A = 0$). Strictly speaking, the albedo is a function of wavelength; however, within the solar system, it's a reasonable approximation to use the albedo at visible wavelengths, since that's where most of the Sun's radiated energy is located. Table 8.2 summarizes the approximate albedos of several solar system objects, listed from darkest to brightest. The Earth, thanks to its changing cloud cover, has a highly variable albedo, ranging from a low of $A \approx 0.3$ to a high of $A \approx 0.5$.

In Table 8.3, we give the computed surface temperatures, T_p, for a sample of three planets; we give both the case of uniform temperature and that of slow rotation. For purposes of comparison, we also give the average observed temperature, T_{obs}, for each of these planets. Naïvely, we would expect all these planets to approximate the "uniform temperature" case: the Earth and Neptune are rapid rotators, and although Venus is a slow rotator, it has a dense atmosphere with global mixing. We see, however, that all the planets are warmer than expected. Neptune is warmer than expected because, like all the Jovian planets, it started its existence much hotter than it is today and has not had time to radiate away all its internal heat. Both the Earth and Venus are warmer than expected because of the **greenhouse effect**.

The greenhouse effect occurs when a planet's atmosphere is transparent at visible wavelengths but opaque at infrared wavelengths. To see why that influences temperature at the planet's surface, consider the energy radiated by the warm surface. For the Earth, with $T_p \approx 290$ K, the Planck spectrum peaks at $\lambda_p \approx 10\ \mu$m. For Venus, with $T_p \approx 740$ K, the peak is at $\lambda_p \approx 4\ \mu$m. These wavelengths are in the infrared range of the spectrum. If the atmosphere is opaque at infrared wavelengths, the radiated light will not freely escape from the atmosphere. Instead, it will be absorbed and re-radiated, until the light energy makes its way to the upper atmosphere, where the optical depth at infrared wavelengths drops to $\tau \sim 1$. Then, at last, the light can escape. Thus, the photosphere of the Earth and Venus—the layer from which their radiated infrared light escapes—is higher than, and cooler than, the solid surface of the planet. Naturally occurring greenhouse gases include water vapor (H_2O), carbon dioxide (CO_2), and methane (CH_4). The greenhouse effect is particularly strong on Venus because its dense atmosphere is roughly 95% carbon dioxide.

The flux of light we observe from a planet depends on wavelength. Consider, as an example, the planet Neptune, with a measured temperature of $T = 59$ K. Using Wien's law (equation 8.4), its spectrum should peak at a wavelength $\lambda_p = 50\ \mu$m, in the far infrared. The light we see from Neptune at visible wavelengths is reflected sunlight, modified by absorption in the atmosphere of Neptune. (Uranus and Neptune both appear bluer than the Sun at visible wavelengths, as shown in Color Figure 5, because their atmospheres contain methane, which strongly absorbs red wavelengths of light.) The spectrum of a planet can thus be roughly approximated as the sum of two blackbody spectra. One spectrum is that of reflected sunlight, containing both the Fraunhofer absorption lines of sunlight and any absorption lines contributed by the planet itself. The other spectrum is that of the planet's thermal emission. The reflected sunlight has an integrated energy proportional to the albedo A; the thermal emission has an energy proportional to $1 - A$ (plus any internal source of energy). For instance, Figure 8.2 shows the spectrum of Mars at UV, visible, and IR wavelengths. Iron oxide in the soil of Mars strongly absorbs blue light and thus makes Mars appear reddish at visible wavelengths (in other words, Mars looks red because it's rusty). The thermal emission peaks at $\lambda_p \sim 13\ \mu$m, corresponding to a temperature $T \sim 225\text{ K} \sim -48°$C, about average for Mars.

An underlying reason for many of the observed differences between terrestrial and Jovian planets is that the Jovian planets have a chemical composition that closely resembles that of the Sun; that is, they are mostly hydrogen and helium. The terrestrial planets are deficient in hydrogen and helium because they are too hot to retain such light elements.

Let's consider the conditions under which planets can retain different atmospheric gases. The hotter a planet's atmosphere, the higher the random thermal speeds of the individual gas particles. In a dense atmosphere, though, any individual particle will travel only a short distance before colliding with another particle and changing its velocity. In the Earth's atmosphere at sea level, for instance, we computed the mean free path of a typical molecule to be only ~ 400 Å (see equation 5.67). However, at higher levels in the atmosphere, the density of molecules decreases, and the mean free path therefore increases. At some height in the atmosphere, the mean free path increases to the point

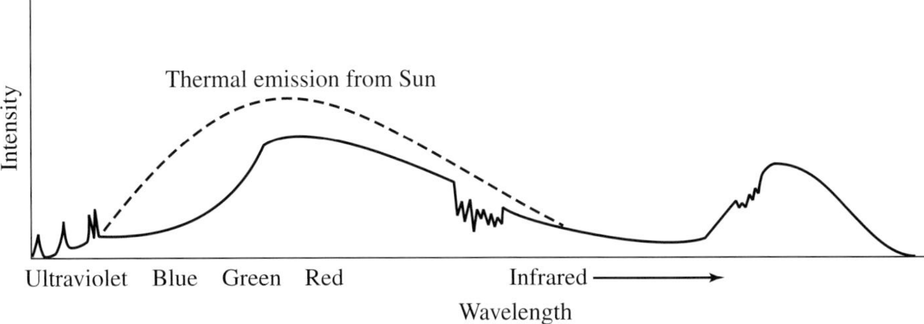

FIGURE 8.2 The spectrum of Mars at ultraviolet, visible, and infrared wavelengths. The peak on the left is reflected sunlight; the peak on the right is thermal emission at $T \sim 225$ K, the temperature of the surface of Mars.

where a gas particle moving upward faster than the escape speed is able to escape from the atmosphere before colliding with another particle. This height is called the **exobase**, and the layer of the atmosphere above the exobase is called the **exosphere**.

To compute where the exobase lies, let's assume that gas particles in the atmosphere have an average cross-section σ for collisions. (For small molecules, like N_2, O_2, and CO_2, we expect $\sigma \sim 10^{-18}\,\text{m}^2$.) The number density of gas particles in the atmosphere is $n(z)$, where z is the height above the planet's surface, or above some convenient reference level, if the planet has no solid or liquid surface. The height z_{ex} of the exobase is given by the relation

$$\int_{z_{\text{ex}}}^{\infty} \sigma n(z) dz = \sigma N(z_{\text{ex}}) = 1. \tag{8.16}$$

Here $N(z_{\text{ex}})$ is the column density of gas particles above the exobase. Thus, for typical molecule sizes, the exobase is located where the column density falls to $N \sim 1/\sigma \sim 10^{18}\,\text{m}^{-2}$. In the Earth's atmosphere, the exobase is at a height $z_{\text{ex}} \sim 500$ km above sea level.

At the exobase, the rms speed of gas particles with mass m will be

$$v_{\text{rms}} = \left(\frac{3kT_{\text{ex}}}{m}\right)^{1/2}, \tag{8.17}$$

where T_{ex} is the temperature at the exobase. We can compare this to the escape speed at the distance R_{ex} of the exobase from the planet's center:

$$v_{\text{esc}} = \left(\frac{2GM}{R_{\text{ex}}}\right)^{1/2}, \tag{8.18}$$

where M is the planet's mass. If the typical particle speed v_{rms} is comparable to the escape speed, the atmosphere quickly escapes. Even if $v_{\text{rms}} < v_{\text{esc}}$, there will be some fraction of particles in the high-speed exponential tail of the Maxwell–Boltzmann distribution (equation 5.40) that will exceed the escape speed. A useful rule of thumb[6] is that for a planet to retain a particular gas in its atmosphere for the age of the solar system, the particles of the gas must have $v_{\rm rms} \lesssim v_{\rm esc}/6$. Squaring both sides, we can write this condition as

$$v_{\rm rms}^2 \lesssim \frac{1}{36} v_{\rm esc}^2$$

$$\frac{3kT_{\rm ex}}{m} \lesssim \frac{GM}{18R_{\rm ex}}. \tag{8.19}$$

In other words, retaining gas particles of mass m requires a temperature

$$T_{\rm ex} \lesssim \frac{GMm}{54kR_{\rm ex}}. \tag{8.20}$$

It is useful to scale this result to the Earth's mass, $M_\odot = 5.97 \times 10^{24}$ kg, and the Earth's radius, $R_\odot = 6.37 \times 10^6$ m, while expressing the particle mass in terms of the molecular mass $\mu = m/m_p$. With this scaling,

$$T_{\rm ex} \lesssim \frac{m_p}{54k} \mu g R_{\rm ex} \tag{8.21}$$

$$\lesssim 140\ {\rm K} \left(\frac{M}{M_\oplus}\right) \left(\frac{R_{\rm ex}}{R_\oplus}\right)^{-1} \mu. \tag{8.22}$$

For a planet with exobase temperature $T_{\rm ex}$, we can write the condition for retaining a gas in the form

$$\mu \gtrsim \frac{54kT_{\rm ex}}{gR_{\rm ex}m_p}. \tag{8.23}$$

The Earth's exobase has a higher temperature than the atmosphere at sea level, because the upper atmosphere is heated by interactions with high-energy solar wind particles; during daytime, the exobase temperature is $T_{{\rm ex},\oplus} \approx 1000$ K. Scaling to the temperature of the Earth's exobase, we find that the condition for retaining a gas is

$$\mu \gtrsim 7.1 \left(\frac{T_{\rm ex}}{1000\ {\rm K}}\right) \left(\frac{M}{M_\oplus}\right)^{-1} \left(\frac{R_{\rm ex}}{R_\oplus}\right). \tag{8.24}$$

Thus, the Earth, where $R_{\rm ex} = 1.08R_\oplus$, can retain gases with $\mu \gtrsim 8$; H_2 ($\mu = 2$) and He ($\mu = 4$) can escape, but N_2 ($\mu = 28$), O_2 ($\mu = 32$), and CO_2 ($\mu = 44$) are retained. However, let's now look at the planet Mercury, which has so few molecules in its atmosphere that its exobase is located at its solid surface. The mass and radius of Mercury are $M = 0.055M_\oplus$ and radius $R = 0.38R_\oplus$, and the daytime temperature is $T = 700$ K.

[6] The phrase "rule of thumb" apparently derives from brewers who estimated the temperature of their mash by sticking in their thumb. Hence, it means an approximate (as opposed to precise) guideline.

Thus, during daytime on Mercury, the retention condition for gas particles is

$$\mu_{\rm Merc} \gtrsim \frac{7.1 \cdot 0.7 \cdot 0.38}{0.055} \approx 34. \tag{8.25}$$

Thus, Mercury would be able to retain only highly massive molecules in its atmosphere.

8.3 ▪ FORMATION OF THE SOLAR SYSTEM

In this section, we give a broad overview of current ideas about how the solar system formed. We return to this topic in Chapter 12 and fill in some of the details.

The modern view of planetary formation is that it is a natural consequence of star formation. Put simply, formation of planets is a way for a collapsing gas cloud to rid itself of angular momentum so that it can shrink to the size of a star. Star formation occurs when a large gas cloud collapses under its own gravity; since this can happen only if the self-gravity exceeds the support provided by gas pressure, star formation occurs only in dense, cold ($T \lesssim 10$ K) regions of interstellar gas. If the collapsing gas cloud has a net angular momentum, it will collapse until it forms a rotationally supported disk. The dense central region of the disk ultimately becomes a star, while smaller condensations that occur within the disk ultimately grow into planets. Such rotationally supported, dusty, gaseous disks, called **protoplanetary disks**, are seen around young stars in our galaxy. For instance, Figure 8.3 shows a protoplanetary disk in the Orion Nebula. The edge-on protoplanetary disk resembles an edge-on hamburger: the dark "patty" represents a thin, dust-rich disk, and the bright "buns" consist of light from the central protostar scattered by a more diffuse distribution of dust.

The first step in planet formation is **condensation**, the formation of solid particles within the gaseous disk as it cools. Different materials condense into solids at different temperatures; metals condense at high temperatures, rocky materials at intermediate temperatures, and ices at low temperatures. Materials that condense into a solid (or liquid)

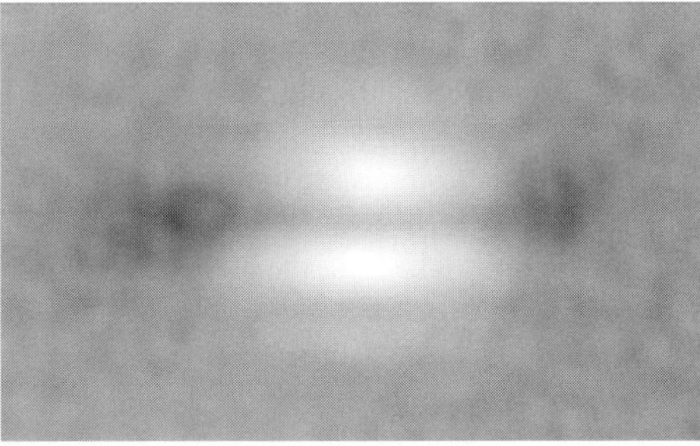

FIGURE 8.3 An image, taken at $\lambda \approx 2\ \mu$m, of the edge-on protoplanetary disk Orion 114-426. The size of the region shown is 1300 AU by 800 AU.

TABLE 8.4 Simplified Condensation Sequence

T (K)	Condensate	Planet
1500	Metal oxides	Mercury
1300	Fe, Ni	
1200	Silicates	
700	FeS (iron sulfide)	Venus
200	H_2O	Earth, Mars
150	NH_3	Jovian planets
120	CH_4	Pluto, Eris
65	Ar, Ne	

only at very low temperatures are called **volatile** materials.[7] The gaseous disk from which the planets ultimately form is hottest at the center, where the protosun is located, and becomes cooler with increasing distance. In the central parts of the protoplanetary disk, only the most refractory materials can condense to form solid particles; a **refractory** material is one that condenses into a solid (or liquid) at relatively high temperatures. In the outer parts of the disk, by contrast, the temperature is sufficiently low that even volatile materials can condense. Table 8.4 gives approximate condensation temperatures for different substances; the right-hand column lists the planets that will ultimately form at the given temperature.

Condensation, more specifically, is the process by which solids grow molecule by molecule, as individual molecules (or atoms) adhere to the solid body. As an example, snowflakes grow by condensation. Eventually, however, the major mode of growth switches over from condensation to **accretion**, in which solid condensates come together and are held together by weak electrical forces. For instance, individual snowflakes can accrete together to form a snowball.[8] In the early solar system, the collisions between individual condensates, or "snowflakes," are gentle, since they are on similar orbits. Objects grow by accretion until they are roughly 1 km across; these intermediate-sized bodies are known as **planetesimals**.[9] Near the Sun, planetesimals are made of the least volatile materials: metal oxides and metals. Farther from the Sun, where the temperatures are cooler, the planetesimals are made of a mix of materials with different degrees of volatility.

Eventually, planetesimals are drawn toward each other by gravity and merge to form larger bodies; this process is known as **coalescence**. Since only the least volatile materials can condense close to the Sun, the planets built up by coalescence within 1.5 AU of the Sun are rich in low-volatility elements, even though such elements are relatively rare

[7] The adjective "volatile" comes from the Latin word *volare*, meaning "to fly." A volatile solid is one in whose atoms fly apart to form a gas at a low temperature.

[8] In other subfields of astronomy, "accretion" can refer to a flow of gas onto a compact object such as a black hole. Such dual definitions are annoying, but it's hard to change entrenched usage.

[9] "Planetesimal" = "planet" + "infinitesimal." It's a portmanteau word, as Lewis Carroll would say.

TABLE 8.5 Uncompressed Densities of Terrestrial Planets

Planet	Density ($kg\ m^{-3}$)
Mercury	5400
Venus	4200
Earth	4200
Mars	3300

in the universe at large. These metal-rich and silicate-rich bodies close to the Sun are the terrestrial planets. Since materials with low volatility tend to be dense, the terrestrial planets all have high densities. Table 8.5 lists the uncompressed densities of the terrestrial planets; in other words, this is the density the planet would have if it were not squeezed to a smaller volume and higher density by its own gravity.[10] Note that the uncompressed density of the terrestrial planets decreases with increasing distance from the Sun. (It is interesting, however, that the uncompressed density of the Moon is only $3350\ kg\ m^{-3}$, much less than the Earth's uncompressed density, despite the fact that they are at the same distance from the Sun. The unexpectedly low density of the Moon is a consequence of its unique formation history, discussed in Section 9.5.)

In the outer part of the protoplanetary disk, containing planetesimals of both high-volatility and low-volatility materials, larger bodies are able to form. Once these "protoplanets" reach a mass of $M \sim 15M_\oplus$, they are massive enough, given the lower temperatures in the outer disk, to retain hydrogen and helium. Since hydrogen and helium are the most abundant elements, being able to hang onto them permits the Jovian planets to grow to much larger masses than the terrestrial planets.

Protoplanets grow gradually by the coalescence of planetesimals until they reach the masses of the planets as we know them today. During and immediately following this early phase of coalescence, the interior and surface structure of terrestrial planets are changed by several processes, among them chemical differentiation, cratering, volcanic flooding, and slower processes of surface evolution.

Chemical differentiation is the process by which dense elements sink to the center of terrestrial planets while the lower-density elements float up to the surface. Since planets form by a jumbled coalescence of planetesimals of differing chemical composition, we might expect that planets would show the same jumble at their center as at their surface. However, study of the interiors of terrestrial planets shows that they are chemically differentiated; that is, their chemical composition varies with distance from the center. The cores of terrestrial planets are rich in dense elements like iron and nickel, with uncompressed densities of $8000 \rightarrow 9000\ kg\ m^{-3}$. However, their outer layers are primarily made of silicate rocks, with typical uncompressed densities of $\sim 3000\ kg\ m^{-3}$.

[10] The Earth is the most massive terrestrial planet and has the greatest gravitational compression. The measured mean density of the Earth is $\rho_\oplus = 5500\ kg\ m^{-3}$, significantly higher than its uncompressed density of $4200\ kg\ m^{-3}$.

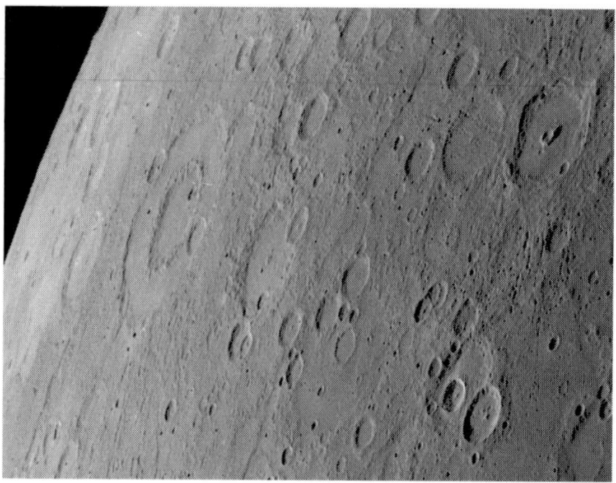

FIGURE 8.4 Mercury as imaged by the *MESSENGER* spacecraft.

Differentiation occurs naturally in a fluid body; the denser material sinks while the lower-density material rises to the top. The fact that terrestrial planets are differentiated is an indication that they were once hot enough to be fully molten. In the Earth, the radioactive decay of elements such as uranium, thorium, and potassium-40 releases enough heat to melt iron. The iron, as it sinks toward the Earth's center, converts gravitational potential energy into heat, leading to an "iron catastrophe," that is, a runaway heating process that doesn't cease until the interior is highly differentiated.

Cratering begins as the newly formed planets sweep up the remaining planetesimals. The process of cratering modifies the outer crust of the terrestrial planets. For instance, cratering has given Mercury (Figure 8.4) its characteristically pock-marked appearance. Cratering was an important process until about 3.3 billion years ago; at that time, the early "period of heavy bombardment" ended because the planetestimals and other small objects had been largely cleared out of the inner solar system.

Volcanic flooding of planetary surfaces occurred at the same time as cratering. Fracturing a planet's crust by large impacts leads to flooding of the surface by lava welling up from below. This obliterates the older surface features; craters are paved over by lava, just as old potholes in a road are paved over by a new layer of asphalt. Surface features on a terrestrial planet can also be worn away by slower processes, such as erosion by wind and water. At the same time, surface features can be built up by other slow processes, such as plate tectonics and volcanism. On a geologically active body such as the Earth, surface features are in a constant state of flux.

The surface of a terrestrial planet can change with time; so can its atmosphere. The primeval atmosphere of a planet is whatever is left behind by the formation process. The atmospheres of terrestrial planets evolve for several reasons:

1. Gases can escape from planets if the individual gas particles are moving sufficiently rapidly, as discussed in Section 8.2. Planets with cooler temperatures and higher surface gravity will retain more of their primeval atmosphere.

2. Outgassing from the planet's interior releases gases that were trapped in the interior during the formation process. These planetary belches occur as part of the differentiation process and continue through ongoing volcanic activity.
3. Chemical interactions between the atmosphere and the surface can alter the atmosphere's composition. For instance, the interaction of gaseous CO_2 with liquid water removed most of the CO_2 from the Earth's atmosphere and dissolved it in the Earth's oceans.

The mechanism by which planetary magnetic fields are generated is poorly understood. In general terms, it is thought to result from dynamo action in the planetary core; hot, partially ionized matter wells up by convection in planetary interiors and is deflected by the Coriolis effect. Our expectation, then, is that larger magnetic fields will be found to be associated with larger bodies (which will have larger liquid cores) and faster rotators (which will have a larger Coriolis effect). In fact, as we will see, the planet Jupiter, which is both the largest and the fastest rotating planet in the solar system, has the strongest magnetic field.

PROBLEMS

8.1 What is the mean mass density $\overline{\rho}$ of Saturn's largest satellite, Titan? What does this suggest about the composition of Titan?

8.2 Radioactive decay of elements in the Earth's interior results in a mean heat flux through the Earth's surface of 5×10^{-2} W m^{-2}. What is this flux expressed as a fraction of the energy flux due to thermal re-radiation of absorbed solar energy? If radioactive decay were the *only* heat source for the Earth, what would the Earth's surface temperature be?

8.3 Mercury has an orbit with semimajor axis $a = 0.387$ AU and eccentricity $e = 0.206$. Mercury is a slowly rotating planet with no atmosphere. What is the temperature of the subsolar point on Mercury at aphelion? What is the temperature of the subsolar point on Mercury at perihelion? (The "subsolar point" is the location on the planet's surface where the Sun is at the zenith.)

8.4 Pure, solid water ice has an albedo $A \approx 0.35$. What is the minimum distance from the Sun at which a rapidly rotating ice cube would remain frozen? Between which two planets does this distance lie?

8.5 Suppose that Uranus were moved to the location of Jupiter; would Uranus then retain its hydrogen-rich atmosphere?

8.6 Because Venus has a very feeble magnetic field, the solar wind collides with its atmosphere, instead of being deflected by magnetic forces. Suppose that if a solar wind particle strikes the atmosphere of Venus, all its kinetic energy will be absorbed.

(a) What is the rate, in watts, at which Venus absorbs energy from the solar wind? Assume that the energy density of the solar wind is

$$\rho v^2/2 = 2 \times 10^{-9}\,\mathrm{J\,m^{-3}},$$

and that the solar wind speed is $v = 400\,\mathrm{km\,s^{-1}}$.

(b) What is the rate, in watts, at which Venus absorbs energy from sunlight? Is the solar wind a significant heat source for Venus?

8.7 Jupiter's moon Callisto is slowly rotating and has a low albedo ($A \approx 0.2$). What is the temperature of Callisto's subsolar point? Would you expect Callisto to retain an atmosphere of N_2? What about an atmosphere of He? (Hint: you may assume that the exobase lies at the surface of Callisto.)

9 Earth and Moon

In this chapter, we will concentrate on the large-scale characteristics of the Earth (the planet about which we have the most information) and the Moon (the satellite about which we have the most information). We will later use our knowledge of the Earth and Moon for purposes of comparison to other terrestrial planets and rocky satellites.

9.1 ▪ THE EARTH'S INTERIOR

The interior of the Earth is highly differentiated, with the densest material in the center. The layers into which the Earth is divided are shown in Figure 9.1. The Earth has a central core, made primarily of iron and nickel, with a radius of ~ 3500 km. The core is further divided into a solid inner core, of radius ~ 1300 km, and a liquid outer core, of thickness ~ 2200 km. The core, which contributes only 16% of the Earth's volume, provides 31% of its mass. The core is surrounded by a rocky mantle of thickness ~ 2900 km. Atop the mantle is a thin crust, which is on average ~ 30 km thick. The crust differs from the mantle in its chemical composition; the rocks making up the crust are poorer in heavy elements like iron, and richer in light elements like silicon and aluminum. Above the crust is the gaseous atmosphere.

The deepest holes drilled by humans have reached only 12 km into the crust. Thus, we are forced to infer the properties of the mantle and core from indirect means. The most fruitful method for studying the Earth's interior is the analysis of **seismic waves**. A seismic wave is a disturbance, usually generated by an earthquake, that travels through the Earth's interior. There are two main types of seismic wave, as illustrated in Figure 9.2. **P-waves**, or pressure waves, are compressional waves; they are regions of alternating high and low pressure traveling through the Earth. Individual particles move back and forth in a direction parallel to the direction of wave propagation. In short, a P-wave is a sound wave.[1] By contrast, **S-waves**, or shear waves, are transverse waves; that is, the individual particles move back and forth in a direction perpendicular to the direction of wave propagation. S-waves are thus analogous to the waves that travel along a rope when you wiggle one end back and forth.

[1] Most P-waves, however, have a frequency $\lesssim 20$ Hz, in the "infrasound" range that human ears cannot detect.

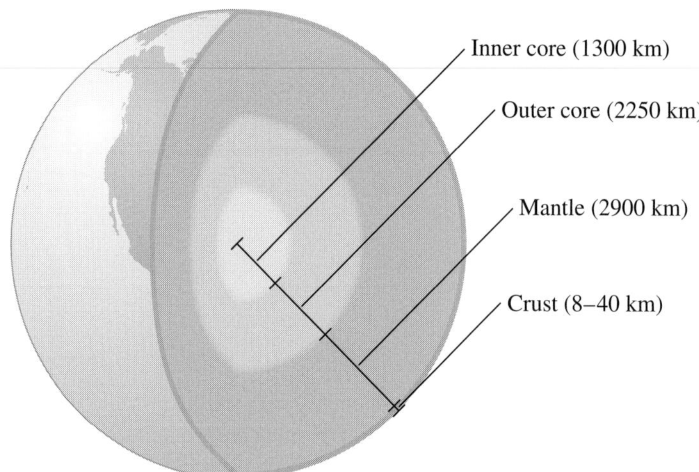

FIGURE 9.1 The differentiated structure of the Earth: inner core, outer core, mantle, and crust.

When an earthquake occurs, most of the energy is released within a small region known as the hypocenter of the earthquake.[2] Both P-waves and S-waves radiate away from the hypocenter; however, the speed of P-waves is greater. As a consequence, by measuring the time elapsed between the arrival of P-waves and S-waves at a given seismometer, geologists can compute the distance from the seismometer to the hypocenter.[3] With data from multiple observers, the position of the hypocenter can be uniquely triangulated.

Computing the position of the hypocenter is complicated by the fact that seismic waves, as illustrated in Figure 9.3, do not generally travel along straight lines. As you go deeper into the Earth's interior, the pressure increases, and the rock and metal is squeezed until it is very stiff.[4] Seismic waves travel more rapidly through a stiffer medium; the increasing wave speed with increasing depth causes seismic waves to be refracted upward.

Observations of seismic waves indicate that S-waves are detected only within 103° of the epicenter. At a greater distance from the epicenter, S-waves are not seen, because they cannot travel through the Earth's liquid outer core. In addition, no direct P-waves are detected between 103° and 142° of the epicenter, as shown in Figure 9.3. This is

[2] The epicenter is the point on the Earth's surface directly above the hypocenter. Since most people live on the Earth's surface, rather than burrowed underground, they are more concerned with where the epicenter is than where the hypocenter is.

[3] In fact, the terms "P-wave" and "S-wave" are short for primary wave and secondary wave (not pressure wave and shear wave), acknowledging the order in which they reach seismometers.

[4] In the jargon of materials scientists, the "bulk modulus" and "shear modulus" increase with increasing depth in the Earth.

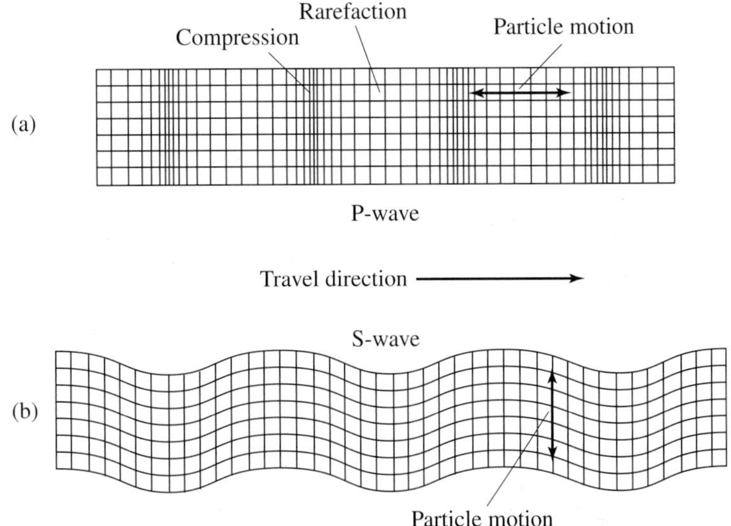

FIGURE 9.2 Seismic waves can be either compressional P-waves (a) or transverse S-waves (b).

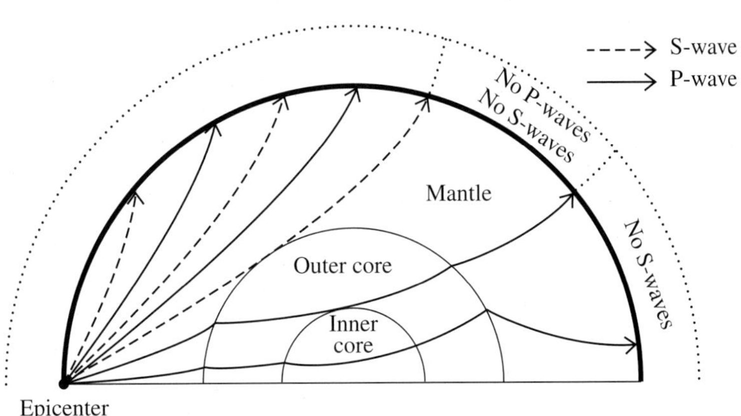

FIGURE 9.3 The propagation of seismic waves through the Earth's interior. Note that S-waves cannot travel through the liquid outer core. All seismic waves are refracted as they move to stiffer material at greater depth.

the result of the increase in refraction as the P-waves travel from the mantle to the outer core. Finally, P-waves arriving at the antipodal point to the epicenter arrive slightly earlier than they would if the core were entirely liquid; this indicates that P-waves that travel through the center of the Earth are speeded up as they pass through a solid inner core.

These observations lead us to conclude that the Earth has a liquid outer core and solid inner core.

The solid surface of the Earth, known as the **lithosphere**, consists of the solid crust plus the upper solid part of the mantle, down to a depth of $\sim$ 100 km. The lithosphere is fragmented into roughly a dozen plates that move quasi-rigidly, floating on the partially molten upper mantle, a region known as the **asthenosphere**.[5] **Plate tectonics**, the relative motion and interaction of these plates, accounts for many earthquakes and drives the formation of many geological features. For instance, **rift zones** are regions where plates are moving away from each other. This allows molten rock to rise upward from the mantle and cool to form a new solid surface. The Mid-Atlantic Ridge, where the North American and Eurasian plates are moving apart from each other, is an example of a rift zone. Iceland, which straddles the Mid-Atlantic Ridge, is noted for its extreme volcanic activity.[6] By contrast, **subduction zones** are regions where plates are moving toward each other, with one plate being driven underneath the other. The friction between the plates results in partial melting of the lithosphere, and an abundance of volcanic activity. Plate tectonics and volcanism are constantly renewing the surface of the Earth, which has the youngest surface of any of the terrestrial planets.

The Earth's core is at a temperature of $T \sim 5000$ K; because metals are good conductors of heat, the temperature within the core is nearly constant with radius. This high temperature is maintained by the radioactive decay of unstable elements such as uranium-238, thorium-232, and potassium-40, which have half-lives of approximately 10^9 yr. Within the outer core, the temperature of $T \sim 5000$ K is sufficient to keep the iron and nickel liquid; however, the higher pressures in the inner core force the iron and nickel into a solid form.

The measured geothermal heat flux through the surface of the Earth is $F_{\rm geo} \approx 5 \times 10^{-2}\ {\rm W\,m^{-2}}$, on average (although it's much higher in Iceland). Thus, the rate at which the Earth is radiating away its internal energy is

$$\frac{dE}{dt} = 4\pi R_\oplus^2 F_{\rm geo} \approx 2.6 \times 10^{13}\ {\rm W}. \tag{9.1}$$

For purposes of comparison, the average power consumption by the world's human population is $\sim 1.4 \times 10^{13}$ W. The stored thermal energy of the Earth can be estimated from the energy per atom ($E = 3kT$ for solids), divided by the mass per atom (about $m \approx 40m_p$ for the mix of elements in the Earth), multiplied by the mass of the Earth:

$$E_{\rm therm} \approx \frac{3kT}{40m_p} M_\oplus \approx 1.1 \times 10^{31}\ {\rm J}, \tag{9.2}$$

[5] The term "asthenosphere" is derived from the Greek word *asthenes*, meaning "not strong." The asthenosphere is softened by heat to the point where it flows like a thick liquid when subjected to a gradual shear but still behaves like a solid when subjected to a sharp blow. (Silly Putty® behaves in much the same way; if you hit a lump of Silly Putty with a hammer, it will fracture instead of flowing.)

[6] As a consolation for the possibility of being inundated with lava, the inhabitants of Iceland get more than half of their energy needs from geothermal sources.

where we have estimated the current average temperature of the Earth's interior to be $T \approx 3000$ K. Thus, even if radioactive decay stopped altogether, it would take a long time for the Earth to cool down; the cooling time would be

$$\tau = \frac{E_{\text{therm}}}{dE/dt} \approx \frac{1.1 \times 10^{31}\,\text{J}}{2.6 \times 10^{13}\,\text{J s}^{-1}}$$

$$\approx 4.2 \times 10^{17}\,\text{s} \approx 1.4 \times 10^{10}\,\text{yr}. \tag{9.3}$$

This is larger than the age of the Earth, $t_\oplus \approx 4.6 \times 10^9$ yr. Thus, the Earth is geologically active, and it will remain so for billions of years.

9.2 ▪ THE EARTH'S ATMOSPHERE

The Earth's atmosphere has evolved through a number of stages. The **primeval atmosphere** of the Earth was comprised mostly of hydrogen, helium, methane, and ammonia accumulated during formation. The hydrogen and helium were rapidly lost (as described in Section 8.2). Other compounds left over from the primeval atmosphere, such as methane and ammonia, were dissociated by solar ultraviolet radiation, since there was not yet an ozone layer to absorb the UV light. A secondary atmosphere was formed by **outgassing**; gas that had been trapped in the Earth's interior was released when the liquid Earth differentiated. The secondary atmosphere consisted largely of CO_2 and H_2O. The CO_2 dissolved in the early oceans, then reacted with other dissolved substances to form solid carbonates such as $CaCO_3$ (calcium carbonate), which then precipitated out. The loss of carbon dioxide meant that the atmosphere then consisted mostly of less reactive molecules such as N_2.

Something interesting happened on Earth some 3 billion years ago; the first life-forms appeared. Cyanobacteria (blue-green algae) began producing energy by photosynthesis, which creates molecular oxygen (O_2) as a byproduct. Oxygen is highly reactive; it combines with carbon and hydrogen in the process we know as "burning," and with iron in the process we know as "rusting." The only reason the Earth's atmosphere contains a large fraction of oxygen today is that the oxygen is being constantly replenished by green plants and other photosynthetic organisms. The current composition of the Earth's atmosphere is given in Table 9.1. The contribution of water vapor is omitted in the table, since it is highly variable. The fraction of atmospheric mass contributed by H_2O varies from less than 0.01% at the south pole during winter to more than 3% in the tropics during the rainy season.

The dense body of the Earth exerts a gravitational force on the Earth's atmosphere; the atmosphere is stabilized against gravitational collapse, however, by a pressure gradient. Consider, as shown in Figure 9.4, a small cylindrical portion of the Earth's atmosphere, oriented so that the faces of the cylinder are perpendicular to the gravitational force $\vec{\mathbf{F}}_{\text{grav}}$. Now let's examine the vector sum of the forces applied to the cylindrical volume element of area A and thickness Δr. Gas pressure pushes on both the top and bottom faces of the cylinder; the pressure pushing the bottom face upward is P and the pressure pushing

TABLE 9.1 Composition of the Earth's Present Atmosphere (Dry Air)

Fraction (by mass)	Fraction (by no. of molecules)	Species
75.5%	78.1%	N_2 (molecular nitrogen)
23.1%	20.9%	O_2 (molecular oxygen)
1.3%	0.93%	Ar (argon)
0.05%	0.04%	CO_2 (carbon dioxide)
trace	trace	Ne, He, CH_4, Kr (neon, helium, methane, krypton)

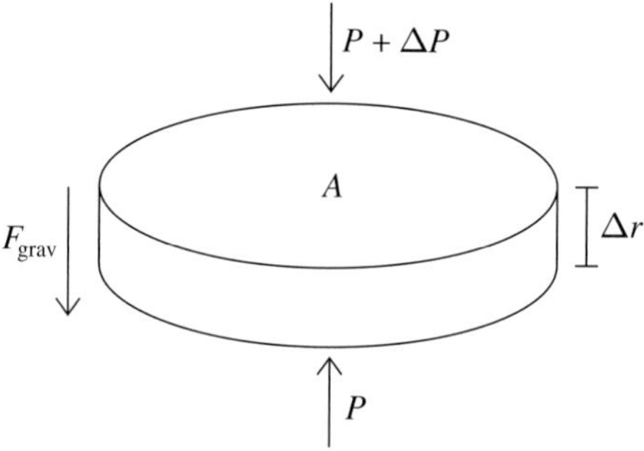

FIGURE 9.4 A small volume of gas is in hydrostatic equilibrium if the differential gas pressure balances the gravitational force.

the top face downward is $P + \Delta P$. The net force on the volume element in the vertical direction is then

$$F_{\text{pres}} = A[P - (P + \Delta P)]. \tag{9.4}$$

Note that we have chosen our sign convention so that a positive force is acting in the upward direction. The mass of the volume element is its mass density ρ times its volume $A\Delta r$; thus, the gravitational force acting on the volume element is

$$F_{\text{grav}} = -\frac{GM_r \rho A \Delta r}{r^2}, \tag{9.5}$$

where M_r is the mass enclosed within a sphere of radius r centered on the Earth's center. Since the mass of the atmosphere is negligible compared to the mass of the solid Earth, $M_r = M_\oplus$. In equilibrium, the atmosphere is neither expanding upward nor contracting downward. This requires that the sum of the differential pressure force (equation 9.4) and the gravitational force (equation 9.5) be zero:

$$A[P - (P + \Delta P)] - \frac{GM_r \rho A \Delta r}{r^2} = 0, \tag{9.6}$$

or

$$\Delta P = -\frac{GM_r\rho}{r^2}\Delta r. \tag{9.7}$$

In the limit that the cylinder becomes infinitesimally thin ($\Delta r \to 0$), this becomes the differential equation

$$\frac{dP}{dr} = -\frac{GM_r\rho}{r^2}. \tag{9.8}$$

Equation (9.8) is the **equation of hydrostatic equilibrium** for a spherically symmetric body. This equation holds true for any spherical object supported by its internal pressure. It is, for instance, one of the key equations that describe the structure of stars (as discussed in Sections 14.1 and 15.1).

The equation of hydrostatic equilibrium is particularly simple when the pressure is described by the ideal gas law, as it is in the atmospheres of planets and all throughout the Sun. The ideal gas law (equation 7.16) states that

$$P = \frac{\rho kT}{\mu m_p}, \tag{9.9}$$

where μ is the mean molecular mass of the gas particles. Inserting this into equation (9.8), we find that

$$\frac{dP}{dr} = -\frac{GM_r}{r^2}\rho = -\frac{GM_r}{r^2}\frac{\mu m_p}{kT}P, \tag{9.10}$$

or

$$\frac{dP}{P} = -\frac{GM_r\mu m_p}{kT}\frac{dr}{r^2}. \tag{9.11}$$

Since the gravitational acceleration at a radius r is $g = GM_r/r^2$, we can write the equation of hydrostatic equilibrium for an ideal gas as

$$\frac{dP}{P} = -\frac{g\mu m_p}{kT}dr. \tag{9.12}$$

In general, the acceleration g, the mean molecular mass μ, and the temperature T are all functions of r. However, the atmosphere of the Earth (and the other terrestrial planets) is thin compared to the radius of the planet. Thus, as a first approximation, we may assume that g, μ, and T are constant throughout the relatively thin atmosphere. With this approximation, equation (9.12) can be integrated to yield

$$\ln P = -\frac{g\mu m_p}{kT}r + C, \tag{9.13}$$

where C is a constant of integration. Using the average atmospheric pressure at sea level as a boundary condition, $P(R_\oplus) = P_\oplus$, we find the **barometric equation**,

$$P(r) = P_\oplus \exp\left(-\frac{r - R_\oplus}{H}\right), \tag{9.14}$$

where the **scale height** H of the atmosphere is

$$H = \frac{kT}{g\mu m_p}. \tag{9.15}$$

For the Earth's atmosphere, we can take $T_\oplus = 290$ K for the average temperature, and $g_\oplus = 9.8$ m s^{-2} for the gravitational acceleration. To evaluate the mean molecular mass, let's assume that the atmosphere is 78% N_2, 21% O_2, and 1% Ar. This yields

$$\mu \approx 0.78(2 \times 14) + 0.21(2 \times 16) + 0.01(40) \approx 29. \tag{9.16}$$

The scale height of the Earth's atmosphere is then

$$H = \frac{kT}{g\mu m_p} = \frac{(1.38 \times 10^{-23}\,\text{kg m}^2\,\text{s}^{-2}\,\text{K}^{-1})(290\,\text{K})}{(9.8\,\text{m s}^{-2})(29)(1.67 \times 10^{-27}\,\text{kg})}$$
$$\approx 8 \times 10^3\,\text{m} \approx 8\,\text{km}. \tag{9.17}$$

The atmospheric pressure falls exponentially with elevation, decreasing by a factor of $1/e \approx 0.37$ for every 8 km.[7] Although the air pressure $P_\oplus$ at sea level varies with location and time, the **standard atmospheric pressure** is conventionally taken to be $P_\oplus = 1.01325 \times 10^5\,\text{N m}^{-2} \equiv 1\,\text{atmosphere} \equiv 1\,\text{atm}$.[8]

Our calculation of the scale height H of the Earth's atmosphere assumed that the atmosphere was isothermal, that is, the same temperature throughout. This is not strictly true. The measured temperature profile of the real atmosphere is shown in Figure 9.5. Note that the temperature doesn't vary monotonically with elevation h above the Earth's surface. The points where the temperature gradient changes sign mark the boundaries between the different layers of the atmosphere.

The **troposphere** is the lowest layer of the atmosphere, in direct contact with the Earth's surface. It is heated primarily by absorption of the thermal infrared emission from the surface. Thus, the troposphere is warmest at the lowest elevations. Since warm air rises, this configuration is unstable, and results in circular convective motions of the air.[9] The convective motions are responsible for weather on the Earth's surface. About 80% of the mass of the atmosphere is contained in the troposphere.

The **stratosphere** starts ~ 11 km above the surface, where the convective motion of the troposphere ends.[10] The stratosphere is heated primarily by the absorption of direct solar ultraviolet light by ozone (O_3). Photons with $\lambda \lesssim 3000$ Å are capable of dissociating

[7] The elevation at the peak of Mount Everest is $h = 8.85\,\text{km} > H$. Thus, the air pressure at the summit is less than 37% of the pressure at sea level. This is why supplemental oxygen is a good idea when you are climbing Everest.

[8] At a latitude $\ell \approx 45°$, the average sea level air pressure is 1 atm. The lowest sea level air pressure ever recorded on Earth was 0.86 atm, in the eye of a typhoon.

[9] The name "troposphere" literally means "sphere of turning," in reference to the turning convective motions.

[10] The name "stratosphere" means "sphere of strata (or layers)," in reference to the smooth, stratified, laminar flow of the air in this layer. Large jet aircraft generally have cruising altitudes in the stratosphere, to avoid the turbulent motions in the troposphere.

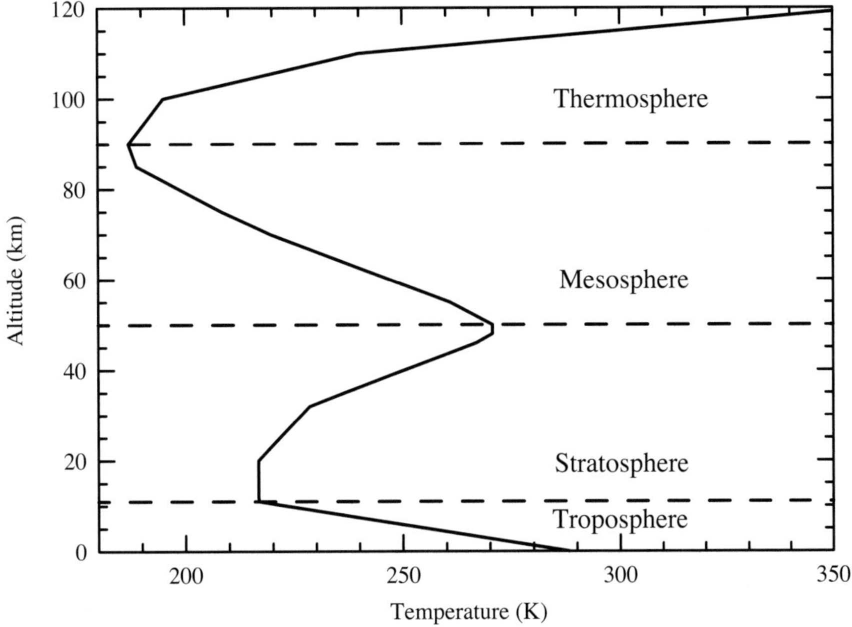

FIGURE 9.5 The lower layers of the Earth's atmosphere, defined by reversals in the temperature gradient.

ozone, so the "ozone layer" of the Earth's atmosphere, at $h \sim 25\,\text{km}$, is effective at absorbing UV light.

In the **mesosphere**, which extends from ~ 50 km to ~ 90 km above the surface, there is little O_3 or CO_2 to absorb UV or IR radiation. The mesosphere is the layer of the atmosphere in which most meteors burn up.

The **thermosphere** extends from $h \sim 90$ km to ~ 500 km. The thermosphere is sufficiently hot to be partially ionized. The regions within the atmosphere that have high concentrations of charged particles are known collectively as the **ionosphere**. The ionosphere is essentially a conducting sheet, reflecting long-wavelength $\lambda \gtrsim 20\,\text{m}$ radio waves. This was convenient for Guglielmo Marconi, the inventor of wireless communication, since it enabled him to send radio signals for long distances over the curved Earth by bouncing them off the ionosphere. It is inconvenient for radio astronomers, however, since it prevents them from doing long-wavelength radio astronomy from the Earth's surface.

Beyond the thermosphere is the **exosphere**, which begins at $h \sim 500$ km and gradually merges with the interplanetary medium. Within the exosphere, collisions between gas particles are sufficiently rare that they are on essentially ballistic trajectories, determined solely by gravitational forces. Since particles in the exosphere are noninteracting, the scale height for each component of the atmosphere is different, depending inversely on its molecular mass μ.

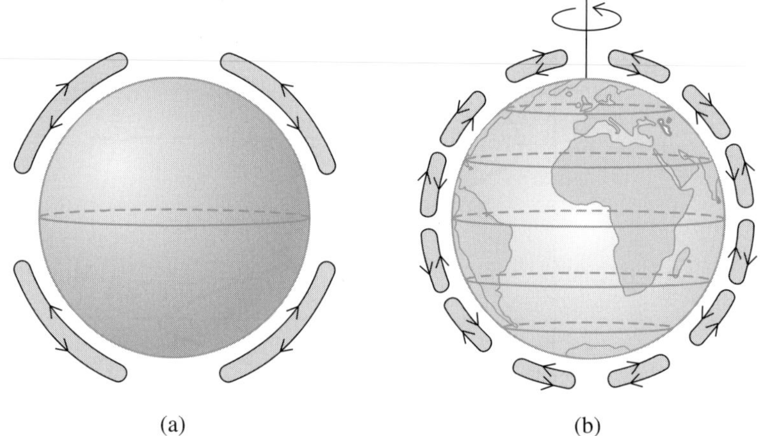

FIGURE 9.6 (a) Single cell Hadley circulation, as seen on a nonrotating planet. (b) The Coriolis effect modifies Hadley circulation by breaking the single cell pattern into several cells.

The air within the troposphere circulates in the north–south direction because it is heated differentially. The heating at the base of the troposphere is greatest at the subsolar point (the point on the Earth where the Sun is at the zenith), and lesser near the poles. The simplest type of atmospheric circulation is **Hadley circulation**, which is illustrated in Figure 9.6.[11] The warmest air at the subsolar point rises, and air from high latitudes flows toward the subsolar point to replace it, thus setting up the "single cell" pattern in Figure 9.6a. This simple model of circulation is actually a good description of atmospheric circulation for a slowly rotating planet, like Venus.

However, on the rapidly rotating Earth, the circulation is modified by the Coriolis effect, which breaks up the circulating Hadley cells into a more complex pattern, as shown in Figure 9.6b. In both hemispheres, the winds at midlatitudes ($\ell \approx 30° \rightarrow 60°$ are **prevailing westerlies**; that is, the winds blow primarily from west to east. By contrast, nearer the equator ($\ell \approx 0° \rightarrow 30°$), the **trade winds** blow primarily from east to west.[12]

Astronomers have a particular interest in how the changeable atmosphere of the Earth affects observations of celestial objects. The atmosphere absorbs, scatters, and refracts light from stars, and does so in a wavelength-dependent manner. As mentioned in Section 6.6, the Earth's atmosphere absorbs many wavelengths of radiation, with some transparent or semitransparent "windows" in the near UV, visible, near IR, and radio portions of the spectrum (see Figure 6.17). Even at wavelengths where the atmosphere

[11] George Hadley, an eighteenth-century English lawyer (and amateur meteorologist) first thought up the idea of Hadley circulation. He is not to be confused with his older brother John Hadley, who invented the sextant.

[12] The regions near $\ell \sim 30°$ are known as the "horse latitudes," supposedly because ships becalmed in this area of light, changeable winds had to toss horses and other livestock overboard when they ran short of fodder and fresh water.

is transparent, scattering of light can redirect photons, without necessarily changing their energy. The cross-section for scattering depends on the ratio L/λ, where L is the size of the scattering particle, and λ is the wavelength of light.

- $L \ll \lambda$. This case is known as **Rayleigh scattering**. When the scattering particles are much smaller than the wavelength of the light interacting with them, the effective scattering cross-section is $\sigma \propto \lambda^{-4}$. That is, shorter wavelengths are much more likely to be scattered than are longer wavelengths. Visible light, with $\lambda \sim 5000$ Å, undergoes Rayleigh scattering from individual oxygen and nitrogen molecules ($L \sim 3$ Å). It also is scattered by aerosols, the collective name for the tiny smoke and dust particles that are small enough to remain suspended in the Earth's atmosphere, rather than settling rapidly to the ground. The tiniest aerosol particles have $L \sim 100$ Å and thus can produce Rayleigh scattering of visible light.

  It's Rayleigh scattering by molecules and aerosols that makes the sky appear blue. When we look away from the Sun at the daytime sky, we are looking at scattered sunlight. Because of the strong wavelength dependence of Rayleigh scattering, we see much more scattered blue light than red light.[13] Conversely, when we look directly at the Sun through a large column of atmosphere, most of the blue light has been scattered away; thus, sunsets and sunrises appear red (particularly after large volcanic eruptions and forest fires, which inject lots of aerosols into the atmosphere).
- $L \sim \lambda$. This case occurs when dust particles of size $L \sim 1\,\mu$m scatter red and near infrared light. In this regime, the wavelength dependence of the scattering cross-section is $\sigma \propto \lambda^{-1}$.
- $L \gg \lambda$. This case occurs when water droplets with $L \sim 10\,\mu$m, of the sort found within clouds, scatter visible light. In this regime, the scattering is independent of wavelength. Thus, clouds seen by reflected sunlight appear white because they scatter all visible wavelengths of light equally.

The atmosphere also refracts light; for a celestial object observed close to the horizon (that is, at low altitude), the refraction can be significant and must be corrected for when determining the object's true position on the celestial sphere. Since the refraction is wavelength dependent, the atmosphere will also disperse the light from objects at low altitude.[14]

[13] Why isn't the sky violet, since that's the shortest wavelength of visible light? Two reasons: our eyes are not very sensitive to violet, and the Sun doesn't produce much violet light.

[14] The dispersion of light by atmospheric refraction explains the "green flash" that can sometimes be seen at sunset. As the Sun dips below the horizon, a brief flash of green light is seen. Why green? The short wavelengths of light are refracted through the greatest angle and thus can be seen for the longest time after the Sun dips below the horizon; however, the short wavelengths of light are also dimmed most by atmospheric Rayleigh scattering. Green wavelengths strike the optimum balance between maximum refraction and minimum dimming.

9.3 ▪ THE EARTH'S MAGNETOSPHERE

The Earth has a magnetic dipole field that is generated by convective motions within its molten outer core. Geological evidence tells us that the field reverses polarity at irregular intervals, with the average time between reversals being $\sim 10^5$ yr.[15] The magnetic axis of the Earth does not coincide with the rotation axis; the magnetic axis is tilted by $\sim 12°$ and tends to wobble erratically about. The magnetic north and south poles of the Earth are currently moving at speeds $\gtrsim 10\ \mathrm{km\ yr^{-1}}$.

The Earth's magnetic field is strong enough to deflect the charged particles of the solar wind. The condition for successfully deflecting the solar wind is that the energy density of the Earth's magnetic field B must be equal to the kinetic energy density of the solar wind. That is,

$$\frac{B^2}{2\mu_0} = \frac{\rho v^2}{2}, \tag{9.18}$$

where ρ and v are the mass density and speed of the solar wind at the Earth's location. Along the magnetic axis, the field strength of a magnetic dipole can be written in the form

$$B(r) = \frac{\mu_0}{2\pi}\frac{\mu}{r^3}, \tag{9.19}$$

where μ is the **magnetic dipole moment** of the field. We can thus write the radial dependence of the Earth's magnetic field in terms of its surface strength:

$$B(r) = \frac{B_\oplus R_\oplus^3}{r^3}, \tag{9.20}$$

where $B_\oplus = 3.1 \times 10^{-5}$ T at $r = R_\oplus$.[16] The average kinetic energy density of the solar wind at 1 AU from the Sun is

$$\rho_\mathrm{K} = \frac{\rho v^2}{2} \approx 10^{-9}\ \mathrm{J\ m^{-3}}. \tag{9.21}$$

Thus, equation (9.18) can be rewritten to find the distance r from the center of the Earth at which the solar wind is deflected:

$$\frac{r}{R_\oplus} = \left(\frac{B_\oplus^2}{2\mu_0\rho_K}\right)^{1/6} \approx 8.5. \tag{9.22}$$

This is the distance to the terrestrial **magnetopause** in the direction of the Sun (Figure 9.7). Some of the charged particles in the solar wind can partially penetrate the Earth's

[15] The most recent magnetic field reversal for the Earth was the Brunhes-Matuyama reversal, which occurred $\sim$ 780,000 years ago.

[16] This $B \propto r^{-3}$ radial dependence differs from the $B \propto r^{-2}$ dependence that we deduced for the Sun's magnetic field (equation 7.29); this is because the Sun's complex magnetic field is not well described by a magnetic dipole.

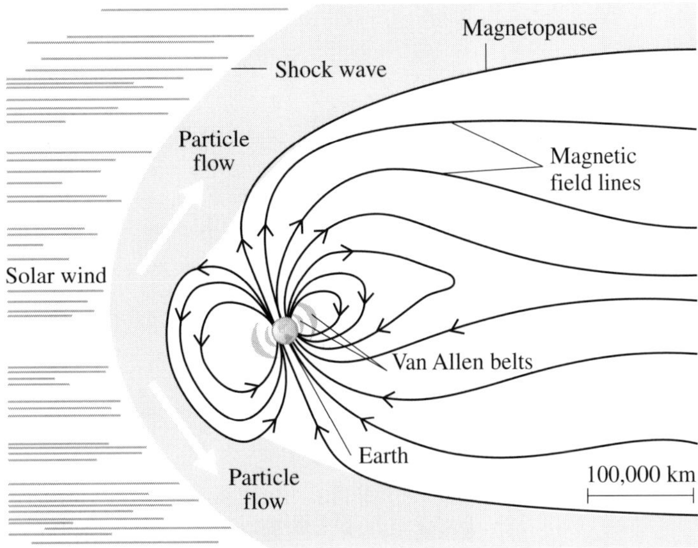

FIGURE 9.7 Interaction between the solar wind and the Earth's magnetic field.

magnetic field and be trapped in the **van Allen belts**. The inner belt, at $r \sim 1.1 \rightarrow 2.0 R_\oplus$, contains high-energy particles, with proton kinetic energies $K \gtrsim 50$ MeV and electron kinetic energies $K \gtrsim 30 \text{ MeV} \gg m_e c^2$. The outer belt, at $r \sim 3 \rightarrow 4 R_\oplus$, contains less energetic particles. Charged particles become trapped in these belts and bounce back and forth between the magnetic poles.

The Earth's magnetic field can also channel charged particles into the Earth's atmosphere at high latitudes, where the magnetic field lines converge near the magnetic poles (see Figure 9.7). These particles collisionally excite and ionize atoms in the upper atmosphere (typically at $h \sim 80 \rightarrow 160$ km, in the boundary region between the mesosphere and thermosphere). Here on the Earth's surface, we see the emission lines produced by the subsequent radiative de-excitation. The light is seen as the **aurorae** mentioned in Section 7.2. The colors of an aurora are usually due to a pair of forbidden transitions in atomic oxygen: one at $\lambda = 5577$ Å (green) and one at $\lambda = 6300$ Å (red). Color Figure 6 shows an aurora seen from the nonstandard viewpoint of an observer in orbit.

9.4 ▪ THE MOON'S INTERIOR AND SURFACE

The Earth's Moon has an average density of $\rho = 3370 \text{ kg m}^{-3}$, about the same as the density of the Earth's crust. Because of the Moon's small size, it has cooled more rapidly than the Earth, leaving it nearly dead, geologically speaking. Moonquakes are rare and weak. However, seismic instruments left at the Apollo landing sites were able to record artificial moonquakes, generated by discarded Apollo ascent stages and Saturn V third-stages crashing onto the surface, and by explosive charges set by astronauts. From

FIGURE 9.8 The lunar maria are seen as darker, smoother regions on the Moon's surface.

the analysis of the seismological data, it is known that the Moon has a solid core of density $\rho \approx 3500$ kg m^{-3}, a mantle of density $\rho \approx 3200$ kg m^{-3}, and a crust of density $\rho \approx 2900$ kg m^{-3}. Unlike the terrestrial planets, the Moon does not have an iron core (which would have a density $\rho \sim 8000$ kg m^{-3}), but it's at least somewhat differentiated. Because of the lack of a liquid iron core, the lunar magnetic field is weak ($B \sim 10^{-9}$ T). However, magnetization of surface rocks indicates that the magnetic field was stronger in the past ($B \sim 2 \times 10^{-6}$ T about 3.3 billion years ago).

Even to the naked eye, the Moon obviously has both darker and lighter regions, as shown in Figure 9.8. The lighter-colored regions are seen to be mountainous and heavily cratered when seen through a telescope; these regions are known as the lunar **highlands**. The darker-colored regions are smoother and lower-lying. Some early telescopic observers thought that the smooth dark regions were oceans, and thus called them **maria**, the Latin word for "seas."[17] Although the maria have long been shown to be solid rock, the inaccurate name has stuck. In fact, the maria aren't perfectly smooth—they do have a few scattered craters on their surface. (They aren't nearly as cratered as the heavily pocked highlands, however.) The maria are concentrated on the side of the Moon facing the Earth.

The lunar highlands are an old surface, largely unchanged since the time of heavy bombardment with planetesimals. The highland surface is saturated with craters; that is, there is no part of the surface that doesn't lie within a crater. The ages of igneous rocks

[17] The singular form of "maria" is "mare" (pronounced with two syllables: "mah′-ray").

(either of lunar or terrestrial origin) can be determined by the process of **radioactive dating**, described in the appendix to this chapter. Radioactive dating of rock samples brought back by the Apollo astronauts reveals that the typical age of a highland rock (that is, the time since it solidified from lava) is ~ 4 billion years. The typical age of a maria rock is only ~ 3.5 billion years. The ages of Moon rocks are in marked contrast to the measured ages of surface rocks on Earth, which show a wide range of ages, from ~ 0 yr for rocks solidifying now at volcanically active sites, to ~ 4 Gyr for the oldest rocks in geologically stable locations.[18]

The most obvious features on the surface of the Moon are **impact craters**: roughly circular depressions, shaped like shallow bowls with a distinct rim, that are the result of planetesimals and smaller debris hitting the Moon's surface. Crater formation occurs in four steps that occur in rapid succession over a matter of seconds:

1. **Impact.** An object collides with the lunar surface. This can occur at speeds as high as $73\ \text{km s}^{-1}$. The local speed limit for any object orbiting the Sun in the vicinity of the Earth–Moon system is the escape speed from the Sun at a distance $r = 1$ AU from the Sun's center. This is simply (from equation 3.62)

$$v_{\text{esc}} = \left(\frac{2GM_{\odot}}{r}\right)^{1/2} = 42\ \text{km s}^{-1}. \tag{9.23}$$

   If the Earth, orbiting the Sun at $v_{\oplus} \approx 30\ \text{km s}^{-1}$, has a head-on collision with such an object, the relative speed will be $v = v_{\text{esc}} + v_{\oplus} = 42 + 30\ \text{km s}^{-1} = 72\ \text{km s}^{-1}$. The Moon is orbiting the Earth at a speed of $1\ \text{km s}^{-1}$, so a similar head-on collision with the Moon can have a relative speed as high as $73\ \text{km s}^{-1}$.

2. **Deep penetration and vaporization.** The colliding object will penetrate the surface like a bullet; it is moving faster than the sound speed in rock ($\sim 8\ \text{km s}^{-1}$), which determines how fast rock can break up into fragments. As the colliding object buries itself in the lunar crust, it is rapidly decelerated, and its kinetic energy is converted into thermal energy. The released heat vaporizes the object and the surrounding lunar rock. As a specific example, consider a rocky object of radius $R = 1$ km striking the Moon at $v = 73\ \text{km s}^{-1}$. If the mass density of the rock is $\rho = 3000\ \text{kg m}^{-3}$, the kinetic energy of the object before it hits the Moon is

$$K = \frac{1}{2}mv^2 = \frac{2\pi R^3 \rho v^2}{3} \approx 3 \times 10^{22}\ \text{J}, \tag{9.24}$$

   equivalent to the energy released by 10 million megatons of TNT.[19]

3. **Formation of crater and ejecta.** The hot gas formed by vaporizing the colliding object and surrounding rock will expand explosively, forming a large circular crater where the explosion bursts through the Moon's surface. Impact craters are almost

[18] The time unit of 10^9 yr (one gigayear) is useful in geological, astronomical, and cosmological contexts. Thus, its abbreviation (Gyr) will show up repeatedly in this text.

[19] One megaton (Mt) is equal to 4.2×10^{15} J. The largest fusion bomb ever exploded released ~ 50 Mt of energy.

FIGURE 9.9 The crater Copernicus, imaged from the Apollo 17 spacecraft in low lunar orbit.

always circular, regardless of the angle of impact, because the crater is formed by the isotropic release of energy beneath the surface. Ejecta thrown outward from the explosion will cover the surrounding area, forming what is called an "ejecta blanket." The ejecta blanket may contain glass droplets formed when silicates are fused under intense heat; thus, young craters may be surrounded by bright rays containing these reflective bits of glass. The impact described above will produce a crater with diameter ~ 100 km, and depth ~ 5 km.[20]

4. **Formation of crater walls and central peak.** Rebound of the lunar crust produces high crater walls and a central peak in the crater itself. The lunar crater Copernicus (Figure 9.9) is a young crater with a diameter $D \approx 93$ km, a central peak, and terraced crater walls.

   It is estimated that Copernicus was formed by the impact of an object with diameter ~ 1 km approximately 0.8 Gyr ago.

Very large impacts can fracture the lunar crust to a great depth and allow subsequent flooding by molten rock from the mantle; this is how the maria were formed. Figure 9.10 shows Mare Orientale, a circular mare that is close to the Moon's limb as seen from Earth. The difference in cratering between the relatively smooth maria, with ages averaging 3.5 Gyr, and the crater-saturated highlands, with ages averaging 4.0 Gyr, shows that an era of heavy cratering took place more than 3.5 Gyr ago (Figure 9.11). This is about the time the solar system was being cleared of remaining planetesimals and debris after the formation of the planets. In addition to the Moon, all the terrestrial planets underwent

[20] As a rule of thumb, high-speed impacting bodies will blast out a crater ~ 50 times their own diameter.

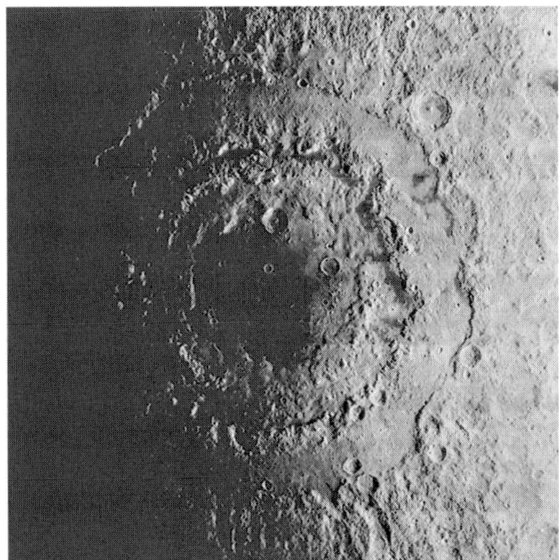

FIGURE 9.10 Mare Orientale, as imaged by the Lunar Orbiter 4 spacecraft. Like smaller impact craters, maria are nearly circular.

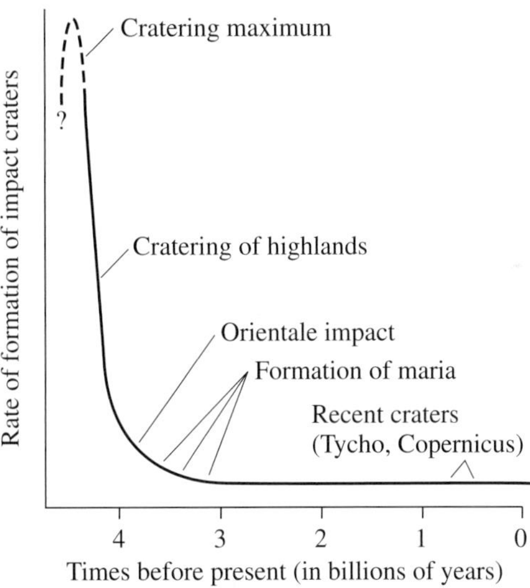

FIGURE 9.11 Cratering rate as a function of time. The period of heavy bombardment ended more than 3.5 Gyr ago.

this heavy bombardment, and all bear the scars to a greater or lesser degree. The Earth seems less affected than the other terrestrial bodies, but this is simply because erosion by wind and water is more efficient on the Earth and has obliterated craters that formed long ago, during the epoch of heavy bombardment.

The six Apollo missions that successfully landed on the surface of the Moon returned a total of ~ 400 kg of surface rocks from both maria and highlands. After study, lunar rocks were discovered to differ from terrestrial rocks in several fundamental respects:

- All lunar rocks are igneous (solidified lava). There are no sedimentary rocks (made of weathered bits of rock cemented together) or metamorphic rocks (made from igneous or sedimentary rocks subjected to high temperature and pressure).
- Lunar rocks do not contain detectable amounts of water. By contrast, Earth rocks can contain as much as 3% water, and nearly always contain detectable amounts of water.
- The iron in lunar rocks is not oxidized. Iron reacts quickly with free oxygen to form ferric oxide (Fe_2O_3), otherwise known as rust. This indicates that the lunar rocks have never been exposed to an oxygen atmosphere.
- Lunar rocks are depleted in volatile elements. This suggests that the rocks of the lunar crust were exposed at some point to higher temperatures than were the rocks of the Earth's crust. The higher temperatures would also explain the lack of water in the lunar crust; it was boiled away.

The Moon shows no evidence of large volcanos similar to those seen on the Earth, Venus, and Mars. However, features called **rilles** (Figure 9.12) provide evidence of past lava flows on the Moon. Rilles are sinuous valleys that are the remnants of lava rivers. Similar, although shorter, lava channels are seen on Hawaii, where hot lava flows between "levees" of igneous rock. If the lava drains from the channel before solidifying, an empty rille is left behind.

The entire lunar surface is covered with a layer of dust and rubble, most of it simply lunar crust that has been pulverized by impacts. This layer, the **regolith**, is $2 \rightarrow 10$ m thick.[21] The top of this layer is only loosely packed, and the Apollo astronauts left footprints a few centimeters deep in it.

We have said nothing about the Moon's atmosphere so far; this is because the Moon has virtually no atmosphere. Its slow rotation makes the daylight side much hotter than that of the Earth. Because of the Moon's low albedo ($A \approx 0.07$ for the dark maria), the daytime temperature of the lunar crust is close to the subsolar blackbody temperature $T \approx 395$ K. The combination of high temperature and low surface gravity doesn't permit the Moon to retain common atmospheric molecules.

[21] The word "regolith" comes from the Greek *rhegos*, meaning "blanket," and *lithos*, meaning "rock." Think of it as the sum of all the ejecta blankets of all the cratering events in the Moon's history.

FIGURE 9.12 Hadley rille, a sinuous rille with a total length of 120 km. (Hadley rille is named after John Hadley, the inventor of the sextant, not after his younger brother George, the describer of Hadley circulation.)

9.5 ▪ THE ORIGIN OF THE MOON

Astronomers have long wondered why the Earth has such an anomalously large satellite. Of the other terrestrial planets, Mercury and Venus have no natural satellites, and Mars has two tiny irregular satellites. The Moon, with a radius equal to $0.27R_\oplus$, is remarkably large in comparison with its parent planet. Over the course of time, many hypotheses for the Moon's formation have been proposed, then discarded.

- The "fission" hypothesis states that the Moon was flung away from the equator of a very rapidly rotating proto-Earth. One problem with this hypothesis is that the Moon is not in the Earth's equatorial plane, as you would expect in such a scenario. Another problem is the observed difference in chemical composition between the Moon's crust and the Earth's crust; in this scenario, they came from the same material.
- The "capture" hypothesis states that the Moon formed elsewhere in the solar system, ventured close to the Earth, and was gravitationally captured. The main problem with this hypothesis is that to go from a Sun-centered orbit to an Earth-centered orbit, the Moon would have to lose a great deal of kinetic energy. All the mechanisms proposed for braking the Moon (passage through a dust cloud? interaction with a third body?) are highly improbable.
- The "co-creation" hypothesis states that the Earth and the Moon formed side-by-side at their present distance from the Sun. The main problem with this hypothesis

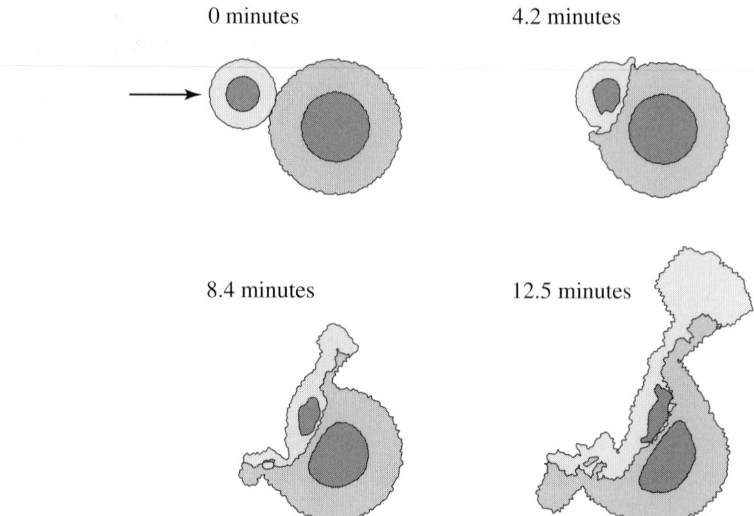

FIGURE 9.13 Four frames, at intervals of 4.2 min, of a computer simulation of the impact that ejected the Moon. In the upper left panel, the arrow indicates the initial velocity of the impacting protoplanet.

is the very different chemical compositions of the Earth and Moon. Why did the high-density Earth get so much more iron than the low-density Moon if they formed in the same place from the same population of planetesimals?

The currently favored explanation for the formation of the Moon is the **giant impact** theory. This scenario states that just over 4.5 billion years ago, a protoplanet roughly the size of Mars struck the proto-Earth. (Mars is nine times the mass of the present Moon and 1/9 the mass of the present Earth.) Both the proto-Earth and the colliding protoplanet were differentiated, containing rocky mantles over iron cores. The impact was not head-on, but oblique. Computer simulations (as shown in Figure 9.13) reveal that, during the impact, the rocky mantle of the colliding protoplanet was ejected into an Earth-centered orbit, eventually cooling to form the Moon. The iron core of the colliding protoplanet sank to the center of the Earth and merged with the Earth's core. Thus, the Moon ended up being made almost entirely of rock, while the iron content of the Earth was enhanced by its cannibalism of the colliding protoplanet's iron core.

▪ APPENDIX: RADIOACTIVE DATING

Many atomic nuclei, especially highly massive nuclei, are unstable against spontaneous disintegration. In this process of disintegration, known as **radioactive decay**, a "parent" isotope is converted into one or more "daughter" isotopes plus additional particles such as electrons, positrons, and neutrinos. A well-known example is the radioactive decay of the parent isotope uranium-235 into the daughter isotope thorium-231:

$$^{235}_{92}\text{U} \rightarrow {}^{231}_{90}\text{Th} + {}^{4}_{2}\text{He}. \quad (9.25)$$

This is the first of a series of decays that ultimately produces lead-207.

Radioactive decay is a random process. For an individual atom of a parent isotope, the moment of its decay can't be predicted; the only thing known is its decay constant λ, representing the probability per unit time that it will decay. Uranium-235, for instance, has the decay constant $\lambda = 9.85 \times 10^{-10}\,\text{yr}^{-1}$. That is, each year it has roughly a one-in-a-billion chance of decaying. If there are N atoms of a parent isotope initially present, then after some time period dt, the cumulative number dN of decays is given by the relation

$$dN = -\lambda N dt, \quad (9.26)$$

which can be rewritten as

$$\frac{dN}{N} = -\lambda dt. \quad (9.27)$$

This can be integrated with the initial conditions $N = N_0$ at $t = 0$ to yield

$$N(t) = N_0 e^{-\lambda t}. \quad (9.28)$$

It is common to express the decay constant λ in terms of the **half-life** of the parent isotope. The half-life τ_0 is the time it takes half the parent isotope to decay; that is, $N(\tau_0) = N_0/2$. It is left as an exercise for the reader to show that equation (9.28) can be written in the form

$$N(t) = N_0 \exp\left(-\ln 2 \frac{t}{\tau_0}\right). \quad (9.29)$$

Thus, the half-life and decay constant are related by the equation $\tau_0 = \ln 2/\lambda \approx 0.693/\lambda$, and the half-life of uranium-235, to revert to our example, is $\tau_0 \approx 0.693/9.85 \times 10^{-10}\,\text{yr}^{-1} \approx 700$ million years.

One radioactive decay used to determine the age of igneous rocks is the decay of uranium-238 to lead-206. The half-life for this decay is dominated by the slowest reaction, the initial decay

$$^{238}_{92}\text{U} \rightarrow {}^{234}_{90}\text{Th} + {}^{4}_{2}\text{He}, \quad (9.30)$$

which has a half-life $\tau_0 = 4.6 \times 10^9$ yr. In principle, the age t of a sample of material that contains uranium-238 is determined from equation (9.29). The current number of uranium-238 atoms is $N(t)$, and the current number of uranium-238 and lead-206 atoms combined is N_0, *if* there were no lead-206 atoms in the sample initially.

The tricky part, of course, is that we don't know how much lead-206 was in the sample initially. If there was already lead-206 in the sample when the uranium-238 started to decay, then assuming that all the lead-206 came from uranium-238 leads to an overestimate of the age of the rock. We can compute the initial quantity of lead-206, however, by determining the amount of the isotope lead-204 in the sample. Lead-204 is special among all isotopes of lead, in that it's nobody's daughter; that is, it's not the endpoint of any radioactive decay process. Technically, lead-204 is itself radioactively unstable; however, its half-life is more than 30 million times the age of the Earth and Moon. Thus, the amount of lead-204 you find in a rock is the amount the rock started with. After measuring the amount of lead-204 present, we assume that the initial ratio of lead-206 to lead-204 in our radioactive sample is the same as the ratio measured in nonradioactive rocks of similar history and composition.

An alternative method is to look at other radioactive decays, such as the decay of potassium-40 to argon-40:

$$ {}^{40}_{21}\mathrm{K} \rightarrow {}^{40}_{22}\mathrm{Ar} + e^{+}, \tag{9.31}$$

which has a half-life $\tau_0 = 1.3 \times 10^9$ yr. The particular appeal of this decay is that argon-40 is an inert gas that doesn't react with other elements. The presence of a gas trapped in a sample of solid rock means that the gas must have been produced in the rock after the rock solidified. If the potassium-40 had decayed while the rock was molten, the resulting argon-40 would have bubbled out of the rock. Thus, the relative amounts of potassium-40 and argon-40 in a rock provide a fairly unambiguous measurement of the time since the rock last solidified.

PROBLEMS

9.1 At what elevation does the Earth's atmospheric pressure fall to 50% of its sea-level value? At what elevation is it 10% of its sea-level value?

9.2 What is the Larmor radius r_c for electrons in the inner van Allen belt?

9.3 At what rate, in watts, is the Earth losing rotational kinetic energy due to tidal braking?

9.4 The continent of Europe (on the Eurasian plate) and the continent of North America (on the North American plate) are moving apart from each other at $v \sim 3\,\mathrm{cm\,yr^{-1}}$. Estimate how long it has taken them to attain their current separation of $d \sim 4500$ km.

9.5 Show explicitly that the half-life τ_0 and the decay constant λ of a radioactively unstable isotope are related by $\tau_0 = \ln 2/\lambda$.

9.6 At some point along the line between the Earth's center and the Moon's center, the gravitational force exerted by the Earth on a test mass exactly cancels the gravitational force exerted by the Moon. How far is this point from the center of the Earth?

9.7 (a) Show that if a particle is moving upward with a speed v at the Earth's exobase, it will reach a maximum height $h = v^2/(2g)$ above the exobase, where $g = GM_\oplus/R_{\mathrm{ex}}^2$ is the gravitational acceleration at the exobase.

(b) Show that the typical height reached by a molecule of mass m at a temperature T will be $h \sim kT/(mg)$. In the Earth's exosphere, what is the typical height for N_2, O_2, and H_2? (Assume $T_{ex} \approx 1000$ K.) Do you expect the oxygen-to-nitrogen ratio to increase or decrease with height in the exosphere?

9.8 We can approximate the Earth as consisting of a dense core 3500 km in radius, containing 31% of the Earth's mass, and a lower-density mantle 2900 km thick, containing 69% of the Earth's mass. With this approximation, what is the moment of inertia of the Earth?

9.9 If we make the approximation that the Earth's atmosphere is isothermal, with $T = 290$ K, what is the mass of the Earth's atmosphere? What is the ratio of the mass of the Earth's atmosphere to the total mass of the Earth?

9.10 Imagine an impacting body, traveling at $v = 72$ km s^{-1}, striking the Earth. How large would such an impactor have to be to physically destroy the Earth, that is, to gravitationally unbind it? Consider only the energy requirements.

10 The Planets

10.1 ▪ TERRESTRIAL PLANETS

The terrestrial planets, in order of increasing distance from the Sun, are Mercury ($a = 0.39$ AU), Venus (0.72 AU), Earth (1.00 AU), and Mars (1.52 AU). In order of decreasing mass, they are Earth ($M = 1.0M_\oplus$), Venus ($0.82M_\oplus$), Mars ($0.11M_\oplus$), and Mercury ($0.056M_\oplus$). The terrestrial planets are primarily rocky and metallic bodies; a group portrait (Color Figure 7) shows their relative sizes. We'll discuss each of the terrestrial planets (aside from the Earth, which we've already covered in the previous chapter), in order of increasing distance from the Sun.

10.1.1 Mercury

Mercury is difficult to observe from the Earth because it is never more than 30° away from the Sun. This difficulty led to early mistakes about its rotation. Prior to the mid-twentieth century, it was thought that Mercury was in synchronous rotation, with one side always facing the Sun, just as one side of the Moon always faces the Earth. The true rotation period was discovered using radar astronomy; the rotational Doppler broadening gave the rotation speed, which when combined with the known radius of Mercury gave the sidereal rotation period, $P_{\rm rot} = 58.65$ days. The sidereal orbital period of Mercury is $P_{\rm orb} = 87.97$ days, which tells us that

$$P_{\rm rot} = \frac{2}{3} P_{\rm orb}. \tag{10.1}$$

In other words, during the time it takes Mercury to go twice around the Sun, it completes three full rotations about its axis. The length of the solar day on Mercury is thus (compare to equation 1.3)

$$\begin{aligned} P_{\rm sol} &= \left(\frac{1}{P_{\rm rot}} - \frac{1}{P_{\rm orb}} \right)^{-1} = \left(\frac{3}{2} \frac{1}{P_{\rm orb}} - \frac{1}{P_{\rm orb}} \right)^{-1} \\ &= 2P_{\rm orb} = 175.9 \text{ days}. \end{aligned} \tag{10.2}$$

Because of the large eccentricity ($e = 0.206$) of Mercury's orbit, observers on Mercury would see a great variation in the angular speed of the Sun relative to the horizon. The rotational angular speed of Mercury is constant, with

$$\omega_{\rm rot} = \frac{2\pi}{P_{\rm rot}} = \frac{3}{2}\frac{2\pi}{P_{\rm orb}} = 1.5\omega_{\rm orb}, \tag{10.3}$$

where $\omega_{\rm orb} \equiv 2\pi/P_{\rm orb} = 4.1°/$ day is the *average* orbital angular speed of Mercury. However, the actual angular speed of Mercury varies with time. At aphelion, it is (equation 3.71)

$$\omega_{\rm ap} = \frac{2\pi}{P_{\rm orb}}\frac{(1-e)^{1/2}}{(1+e)^{3/2}} = \omega_{\rm orb}\frac{(0.794)^{1/2}}{(1.206)^{3/2}} = 0.67\omega_{\rm orb}. \tag{10.4}$$

At perihelion, the orbital angular speed of Mercury is (equation 3.70)

$$\omega_{\rm pe} = \frac{2\pi}{P_{\rm orb}}\frac{(1+e)^{1/2}}{(1-e)^{3/2}} = \omega_{\rm orb}\frac{(1.206)^{1/2}}{(0.794)^{3/2}} = 1.55\omega_{\rm orb}. \tag{10.5}$$

The angular speed with which an observer on Mercury sees the Sun move relative to the horizon is the same as the angular speed of Mercury's rotation as measured in a frame of reference co-rotating with a line drawn from Mercury to the Sun. (See Section 1.5 for a similar analysis in the case of the Earth and Sun.) When Mercury is at aphelion, an observer on Mercury will see the Sun move relative to the horizon with angular speed (compare to equation 1.2)

$$\omega_{\rm sol,ap} = \omega_{\rm rot} - \omega_{\rm ap} = (1.5 - 0.67)\omega_{\rm orb} = 0.83\omega_{\rm orb}, \tag{10.6}$$

or about 3.4 degrees per day. When Mercury is at perihelion, an observer on Mercury will see the Sun move relative to the horizon with angular speed

$$\omega_{\rm sol,pe} = \omega_{\rm rot} - \omega_{\rm pe} = (1.5 - 1.55)\omega_{\rm orb} = -0.05\omega_{\rm orb}, \tag{10.7}$$

or about 0.2 degrees per day in a *retrograde* direction. That is, at perihelion, an observer will see the Sun cease its usual east-to-west motion relative to the horizon and move slowly from west to east!

The rotation period $P_{\rm rot}$ and orbital period $P_{\rm orb}$ of Mercury are **commensurate periods**; that is, their ratio is equal to the ratio of two small integers. The commensurate periods of Mercury are a consequence of Mercury's eccentric orbit, combined with the fact that Mercury is slightly prolate, rather than perfectly spherical. The prolate shape of Mercury means that we can think of Mercury as having two permanent, fixed bulges, in addition to the changeable, moving tidal bulges raised by the Sun.

As shown in Figure 10.1, one or the other of the permanent bulges always points toward the Sun when Mercury is at perihelion. In addition, because the rotational angular speed and the orbital angular speed are nearly equal around the time of perihelion, the bulges remain pointing close to the Sun for the entire time that Mercury is nearest to the Sun. This makes the commensurate relationship between the orbital and rotational periods very stable. If Mercury's rotation rate were to slow down, then the bulges would be misaligned at perihelion. The gravitational forces exerted by the Sun on the permanent bulges would then torque up Mercury's rotation until the bulges were again aligned at

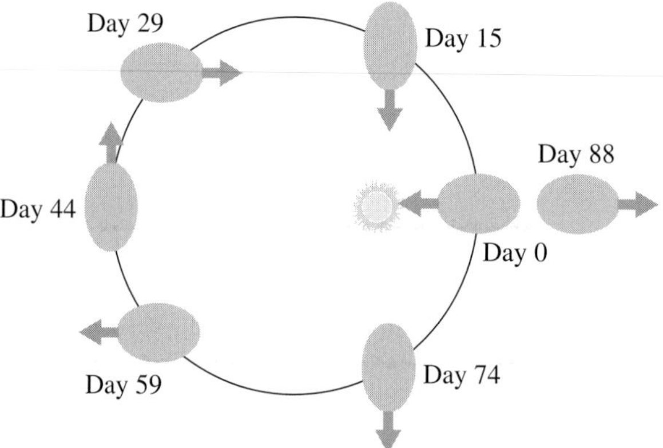

FIGURE 10.1 The orientation of the permanent bulges of Mercury during one sidereal orbital period (equal to 1.5 sidereal rotation periods, or 0.5 solar day).

perihelion. The misalignment of the permanent bulges near aphelion is not as physically significant, since the gravitational forces are smaller at aphelion than at perihelion.

The gravitational effects of the planet Mercury on the spacecraft *Mariner 10* and *MESSENGER* have allowed astronomers to compute the mass of Mercury accurately, despite the fact that it has no natural satellite. The mean (uncompressed) density of Mercury turns out to have the remarkably high value of $\rho = 5400 \text{ kg m}^{-3}$. This indicates that Mercury has an unusually large iron core. To give Mercury an uncompressed density of 5400 kg m^{-3}, a skimpy 600-kilometer-thick mantle of rock must lie atop an iron core with a radius of 1800 kilometers. The disproportionately large core may indicate that Mercury was once more massive but then lost most of its mantle in a large protoplanetary collision (similar to the collision in the giant impact theory for the formation of the Moon).

At first glance (see Color Figure 7), the surface of Mercury is similar to that of the Moon; the primary surface features are impact craters. The largest impact crater on Mercury, the Caloris Basin, is approximately 1300 kilometers in diameter. However, Mercury does have some unique features not found on the Moon:

- Mercury has linear features called **scarps** (Figure 10.2). These are cliffs that can be hundreds of kilometers long and as high as two kilometers. The height of the scarps, combined with the low surface gravity of Mercury ($g_{\text{Mer}} = 0.38 g_{\oplus} = 3.7 \text{ m s}^{-2}$), means that if you slipped from the top of a scarp with height $h = 2000$ m, you'd fall for a long time:

$$t = \left(\frac{2h}{g_{\text{Mer}}} \right)^{1/2} = \left(\frac{4000 \text{ m}}{3.7 \text{ m s}^{-2}} \right)^{1/2} \approx 33 \text{ s}. \tag{10.8}$$

By the time you reached the base of the scarp, you'd be traveling at a speed $v = g_{\text{Mer}} t \approx 120 \text{ m s}^{-1} \approx 440 \text{ km hr}^{-1}$, an unhealthy speed at which to slam into

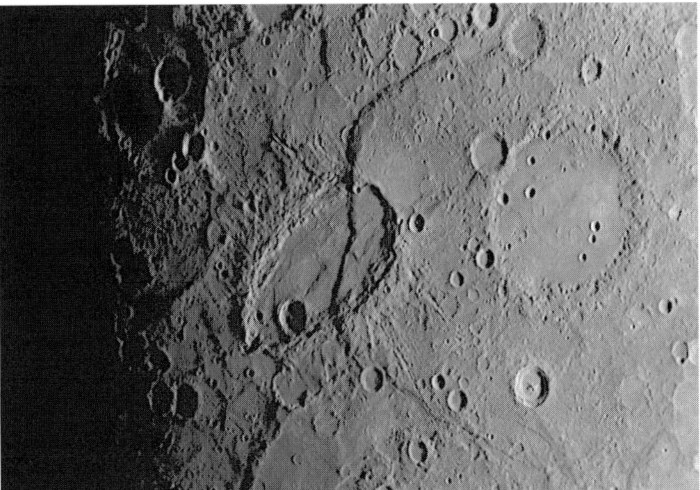

FIGURE 10.2 A scarp runs down the middle of this image of Mercury; since scarps cross craters, they were evidently formed after the era of heavy bombardment.

solid rock. The long tall scarps on Mercury were due to shrinkage of the planet's core by a few kilometers as it cooled.

- Unlike the highlands of the Moon, the surface of Mercury is not totally saturated with craters. This implies that the solid surface of Mercury formed near the tail end of the era of heavy bombardment.

Because of its high subsolar temperature and its low escape speed, Mercury has no permanent atmosphere. However, it does have a transient, tenuous atmosphere consisting primarily of sodium and oxygen released when micrometeorites vaporize bits of Mercury's crust. In addition, there are traces of hydrogen and helium from the solar wind, trapped temporarily by Mercury's magnetic field. Mercury's magnetic field strength, measured at the surface of the planet, is $B_{\rm mer} \approx 3 \times 10^{-7}\,{\rm T} \approx 0.01 B_\oplus$. This is a surprisingly strong magnetic field for a small, slowly rotating planet. Mercury's magnetic field, like that of the Earth, is strong enough to prevent solar wind particles from striking the planet's surface.

10.1.2 Venus

With the obvious exception of the Moon, Venus is the most brilliant object that can be seen in the Earth's night sky. Venus is easy to observe at its greatest elongation of $\theta = 47^\circ$. However, when we observe Venus at visible wavelengths, all we can see is a nearly featureless white cloud deck (Figure 10.3). Ground-based spectra, later supplemented by data from space probes to Venus, reveal that the clouds are made of sulfuric acid (H_2SO_4). The sidereal rotation period of Venus, like that of Mercury, can be determined by radar

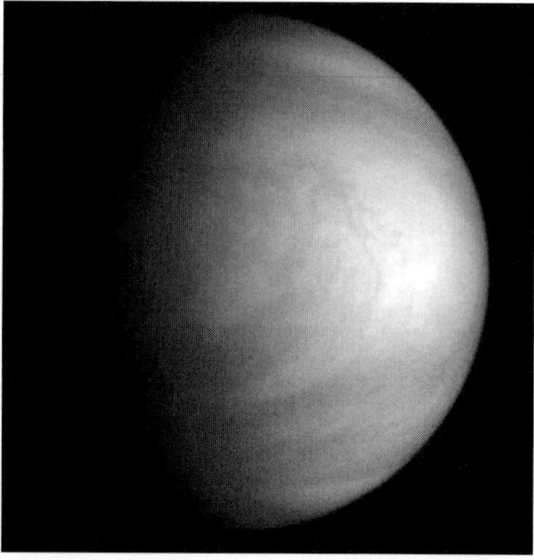

FIGURE 10.3 In this image, taken through a violet/blue filter, faint markings can be seen in the thick clouds of Venus.

astronomy. Venus turns out to be the slowest rotator of all the planets, with a sidereal rotation period of $P_{\rm rot} = 243$ days. The rotation is not only slow, it is also *retrograde*; that is, if we looked at the solar system from above the Earth's north pole, we would see Venus orbiting in a counterclockwise direction (the same as the other planets) but rotating in a *clockwise* direction. Probably as a consequence of its slow rotation, Venus does not have a detectable magnetic field.

Because the orbital angular momentum and the rotational angular momentum of Venus are in opposite directions, the sidereal orbital period $P_{\rm rot}$ and the solar day $P_{\rm sol}$ are related by the equation (compare to eq. 1.3)

$$\frac{1}{P_{\rm rot}} = \frac{1}{P_{\rm sol}} - \frac{1}{P_{\rm orb}}, \tag{10.9}$$

where $P_{\rm orb} = 225$ days is the orbital period of Venus. Thus,

$$P_{\rm sol} = \left(\frac{1}{P_{\rm rot}} + \frac{1}{P_{\rm orb}}\right)^{-1} = \left(\frac{1}{243} + \frac{1}{225}\right)^{-1} \text{ days} = 117 \text{ days}. \tag{10.10}$$

The surface temperature of Venus is high: $T \approx 740$ K, hot enough to melt tin, lead, and zinc. Despite the long solar day on Venus, the surface temperature is nearly uniform over the entire planet. This is because the circulation patterns in the thick atmosphere are highly effective at carrying hot air from the daytime side of Venus to the nighttime side, smoothing out the temperature differences. The high temperature at the surface of Venus is a result of the greenhouse effect, as discussed on page 199. The sulfuric acid clouds of Venus reflect about 75% of the light that strikes them; below the clouds, however, the atmosphere is nearly transparent at visible wavelengths.[1] However, the energy absorbed by the planet's surface is re-radiated as infrared light, to which the atmosphere is opaque, thanks to its high CO_2 content. The surface air pressure on Venus is $P_{\rm Ven} \approx 92$ atm, where 1 atm $\approx 10^5$ N m^{-2} is the average sea level air pressure on the Earth. The composition of the atmosphere on Venus is $\sim 96.5\%$ carbon dioxide and $\sim 3.5\%$ molecular nitrogen (by number of molecules present).

One intriguing question is why the atmospheres of Venus and the Earth are so different from each other. Venus and the Earth are nearly twins in radius ($R_{\rm Ven} = 0.95 R_\oplus$) and mass ($M_{\rm Ven} = 0.815 M_\oplus$). Why, then, does Venus have nearly 100 times as much atmospheric pressure as the Earth? The key difference between the two planets, in this context, is that the Earth has liquid water covering most of its surface. Carbon dioxide dissolves in water, where it forms, among other compounds, negatively charged bicarbonate ions:

$$\mathrm{CO_2 + H_2O \rightleftharpoons H^+ + HCO_3^-}. \tag{10.11}$$

The bicarbonate ions combine readily with any calcium ions dissolved in the water:

$$\mathrm{Ca^{2+} + 2\,HCO_3^- \rightleftharpoons CO_2 + H_2O + CaCO_3}. \tag{10.12}$$

Calcium carbonate ($CaCO_3$) is poorly soluble in water, so it precipitates out to the ocean floor. Limestone, which occurs in abundance near the Earth's surface, is a sedimentary rock that consists largely of calcium carbonate. Since Venus is too hot for liquid water to exist, it lacks this mechanism for removing carbon dioxide from the atmosphere and locking it up inside rocks. If you could somehow remove the CO_2 from the venusian atmosphere, the remaining atmosphere, consisting mostly of nitrogen, would resemble the atmosphere of Earth prior to the emergence of photosynthesis. Conversely, if you could unlock all the carbon dioxide presently contained in the Earth's limestone, the Earth's atmosphere would strongly resemble that of Venus.

The Earth is unique among the terrestrial planets in having large quantities of liquid water. To understand why this is so, consider Figure 10.4, a **phase diagram** for water. It shows which phase of water (solid ice, liquid water, or gaseous water vapor) is stable at a given temperature and pressure. There exists a unique "triple point" at $T_{\rm tp} = 273.2$ K $=$ 0.0° C and $P_{\rm tp} = 611.7$ N m$^{-2} = 0.0060$ atm at which all three phases can coexist. Note that ice is stable only at $T < T_{\rm tp}$, and liquid water is stable only at $P > P_{\rm tp}$. At low pressures ($P < P_{\rm tp}$), water goes straight from the solid state to the gaseous state at the temperature increases; in other words, it "sublimes." Figure 10.4 also shows average surface temperatures and pressures of the terrestrial planets with atmospheres. Venus is too hot for liquid water to exist. Mars, on average, is too cool for liquid water to exist; it was only in the warmest, highest pressure regions of Mars that liquid water flowed in the past (as discussed next in Section 10.1.3). However, the average temperature and pressure on Earth has been close to the triple point for much of the planet's history, meaning that gaseous, liquid, and solid water could all exist in significant quantities near the Earth's

[1] If you walked on the surface of Venus (in a highly heat-resistant spacesuit), the light level would be about the same as during an overcast day on Earth.

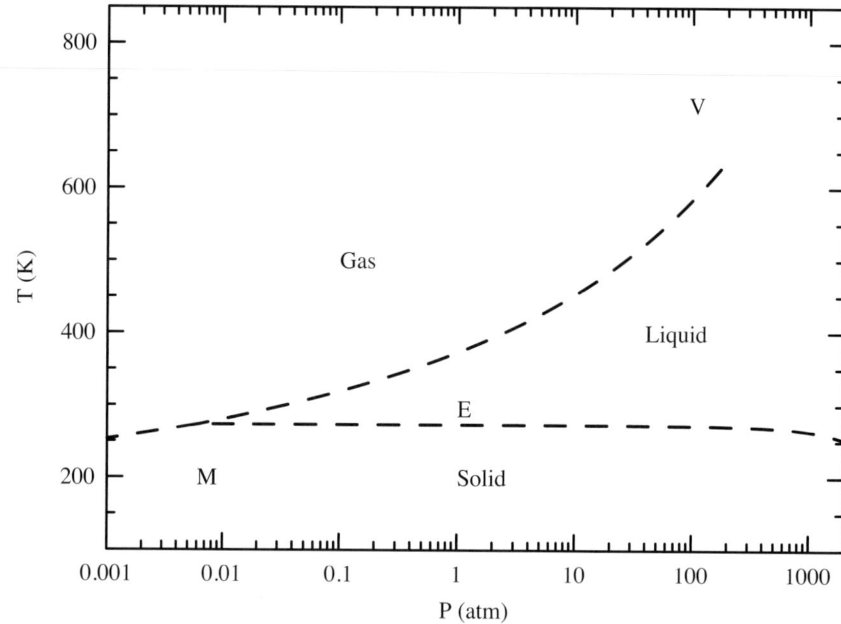

FIGURE 10.4 The phase diagram for water shows its phase (solid, liquid, or gas) as a function of temperature and pressure. Also shown are average surface temperatures and atmospheric pressures for Mars (M), Earth (E), and Venus (V).

surface. Astronomers, thinking back to the stories of their childhood, sometimes refer to the **Goldilocks effect**: Venus is too hot for liquid water, Mars is too cold, but the Earth is "just right."

The clouds of Venus are transparent to radio waves; thus, the *Magellan* spacecraft was able to map the surface of Venus using radar. The resulting radar image of Venus is shown in Color Figure 7. Most of Venus consists of low, rolling plains, but there are two prominent highland regions: Ishtar Terra (about the size of Australia) and Aphrodite Terra (about the size of Africa).

Venus has many volcanos, and extensive lava flows on its surface. The average age of the crust on Venus is roughly 0.5 Gyr, indicating that the surface has been extensively repaved by lava during the last billion years. Venus has relatively few impact craters (*Magellan* saw only a thousand of them more than 100 meters across). The scarcity of craters is partially due to the thick atmosphere of Venus (small objects break up in the atmosphere and are vaporized), and partly due to the relatively recent lava flows.

There is no evidence for plate tectonics on Venus; no rift zones or subduction zones, for instance. The volcanos of Venus are spread evenly across the planet's surface instead of being concentrated at plate boundaries, as they are on Earth. Because of the lack of plate tectonics, the lithosphere of Venus remains stationary relative to the planet's interior. This means that some of the volcanos on Venus are large shield volcanos; a single spot on the crust remains in place above a hot spot on the mantle and is supplied with lava for billions of years. The lack of plate tectonics on Venus is thought to be due to the high temperatures at the planet's surface; although $T = 740$ K isn't hot enough to melt rock, it does soften the rock a bit. Instead of being brittle and breakable, like the Earth's lithosphere, the surface layers of Venus are more pliable.

10.1.3 Mars

To the naked eye, Mars (shown at higher resolution in Color Figure 7) has a perceptibly reddish color. It was a natural series of associations that led ancient astronomers to associate the red planet Mars with the god of war. (Red $\rightarrow$ blood $\rightarrow$ war $\rightarrow$ war god $=$ Mars.) The ancient fascination with Mars has lasted to the present day.

Mars has an orbit with $a = 1.52$ AU, and hence an orbital period $P = 1.88$ yr. Mars is the terrestrial planet farthest from the Sun, and hence the terrestrial planet with the coldest surface. Mars has a number of superficial similarities to Earth. Its equator is tilted by 25.0° relative to its orbital plane; this is similar to the Earth's 23.5° tilt. Thus, the seasonal variations on Mars are similar to those on Earth (only with seasons 88% longer, thanks to the longer martian orbital period). The length of the solar day on Mars is 24 hours, 40 minutes; this is only 3% longer than the Earth's solar day.

Although Mars has an escape speed comparable to that of Mercury, its daytime temperatures are much lower; thus, Mars has been able to retain an atmosphere. However, the air on Mars is thin; the average air pressure at the surface of Mars is $P_{\rm Mars} \approx 0.006$ atm. The martian atmosphere is 95% CO_2, similar to the atmosphere of Venus. There is little water vapor in the atmosphere of Mars; this is because UV light dissociates it. (Mars, with no oxygen in its atmosphere, doesn't have an ozone layer to absorb UV light.) A water molecule (H_2O) is dissociated into a hydrogen molecule (H_2) and an oxygen atom; the low-mass hydrogen molecule escapes into space, and the oxygen combines with iron in the martian soil, making the ferric oxide that gives Mars its distinctive reddish color.[2] Martian clouds are thin and wispy, like cirrus clouds on Earth; they are made of solid ice crystals, not liquid droplets. The clouds of Mars are made of both frozen water and frozen carbon dioxide.

Geologically speaking, the two hemispheres of Mars (as illustrated in Figure 10.5) are very different from each other.

The *northern* hemisphere, which is relatively low in elevation, contains few craters and shows signs of recent volcanic activity (like the larger terrestrial planets, Earth and Venus). The *southern* hemisphere, which is relatively high in elevation, contains many craters and is geologically dead (like the smaller terrestrial planet Mercury, and like the Moon).

There are a pair of strikingly large geological features on Mars:

1. Olympus Mons (Figure 10.6) is an enormous shield volcano, about 600 km across and 25 km high. By contrast, Mauna Loa, the largest shield volcano on Earth, is

[2] The surface of Mars is about 44% silicon dioxide (sand) and 19% ferric oxide (rust). The remainder is a mix of various minerals.

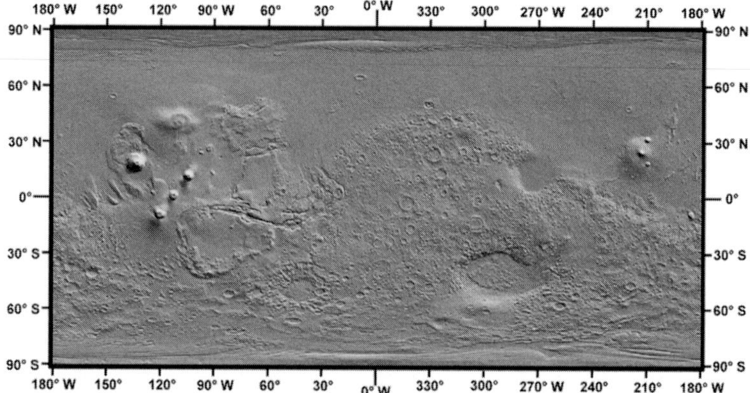

FIGURE 10.5 Relief map of Mars, using data from the Mars Orbiter Laser Altimeter (MOLA) on the Mars Global Surveyor spacecraft.

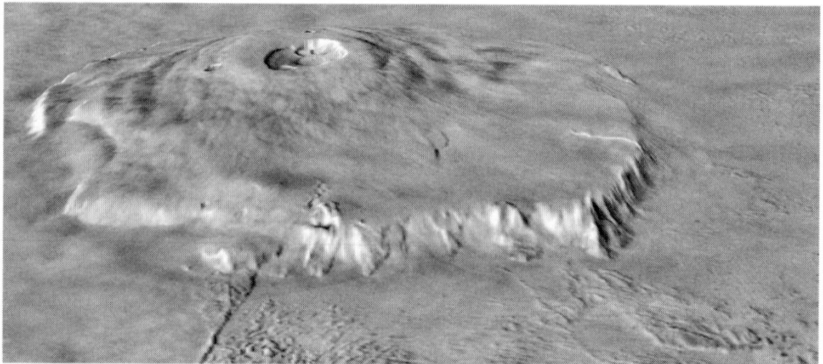

FIGURE 10.6 Olympus Mons is the largest shield volcano in the solar system.

about 200 km across and rises only 9 km above the ocean floor. The enormous size of Olympus Mons shows that there is little plate motion on Mars, and one location of the crust can stay above a "hot spot" on the mantle for a long time. There is little cratering on the slopes of Olympus Mons, indicating that it is geologically young.[3] The steep cliffs at the base of Olympus Mons are due to wind erosion.

2. Valles Marineris (Figure 10.7) is a tremendously long rift valley, stretching nearly 5000 km, or about one-fourth of the circumference of Mars. Valles Marineris is sometimes called the "Grand Canyon of Mars," but this is misleading on two counts. First, Valles Marineris is much larger than the Grand Canyon of the

[3] Judging from the number of impact craters on its slopes, Olympus Mons last erupted some 300 million years ago.

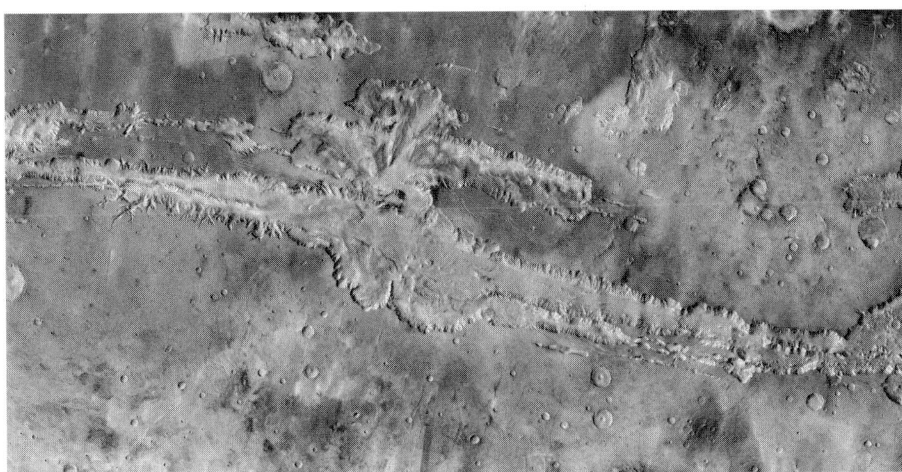

FIGURE 10.7 Valles Marineris, as imaged by the Thermal Emission Imaging System on NASA's *Mars Odyssey* orbiter.

Colorado River; Valles Marineris could stretch from Boston to Los Angeles, with length to spare. Second, Valles Marineris is not the result of erosion by water but is instead a *rift* valley, caused by convection currents in the mantle pulling the crust apart. The walls of Valles Marineris have subsequently been sculpted by erosion (the atmosphere is denser in the deep valleys than on the highlands).

As noted previously, Mars has similar seasons to the Earth because of its similar obliquity. However, the seasons on Mars are significantly affected by the relatively high eccentricity of the martian orbit ($e = 0.093$ for Mars, compared to $e = 0.017$ for Earth). When Mars is at perihelion, it is winter in the north and summer in the south; when it is at aphelion, it is summer in the north and winter in the south, as shown in Figure 10.8. As a result of the high orbital eccentricity, seasons are moderated in the northern hemisphere of Mars; winter occurs in the north when Mars is nearest to the Sun, and summer when Mars is farthest from the Sun. By contrast, seasons are enhanced in the southern hemisphere of Mars, since winter occurs there when Mars is farthest from the Sun, and summer when Mars is nearest to the Sun. The enhanced seasonal temperature swings in the south lead to strong surface winds. These winds produce major dust storms that can on occasion envelop the entire planet.

The polar caps of Mars also vary with the seasons, as shown in Figure 10.9. The area covered by the polar caps can change very rapidly, so they must be thin, like a layer of frost. During winter, the temperature is cold enough to freeze CO_2 out of the atmosphere. In the winter, the martian polar caps—particularly the southern cap—are large, since they are primarily CO_2, which freezes at $T \approx 150$ K at martian air pressure. In the summer, the CO_2 sublimes (that is, it goes directly from the frozen to gaseous state), but there is still a residual polar cap. The residual summer polar cap is probably frozen water, which sublimes at $T \approx 190$ K at martian air pressure.

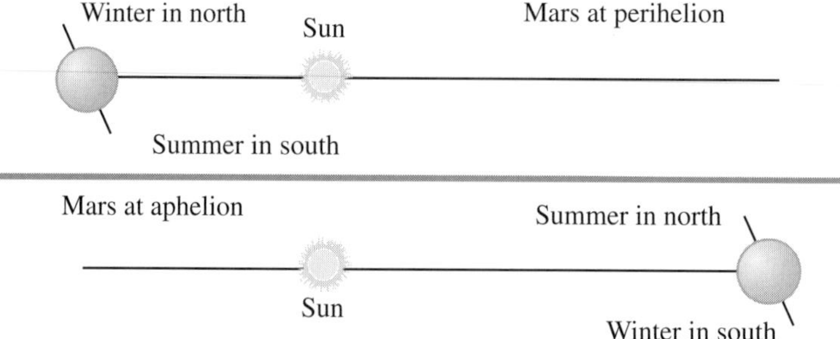

FIGURE 10.8 The orbit of Mars is seen edge-on at the times of the martian solstices. Because of the eccentricity of its orbit, martian seasons are moderated in the northern hemisphere and enhanced in the southern hemisphere.

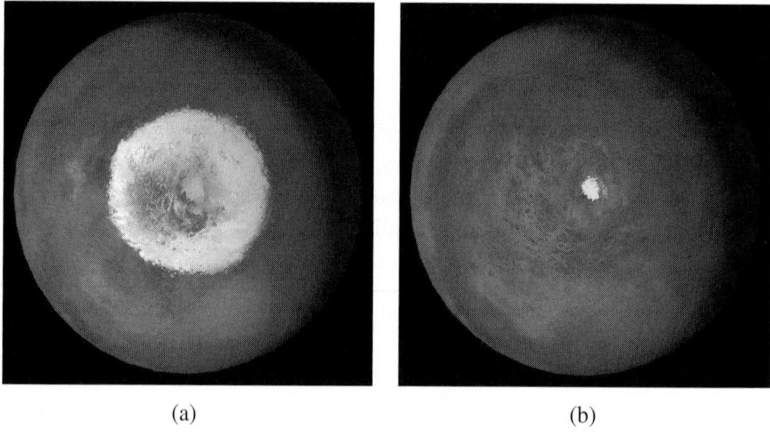

FIGURE 10.9 The martian south polar cap is large in the winter (left), but small in the summer (right).

Although at present, the water on Mars is primarily in its frozen state (in polar caps, clouds, or permafrost below the planet's surface), there is evidence that liquid water existed more abundantly on Mars in the past. The most obvious form of evidence consists of features on the surface of Mars that have been cut by running water. Some narrow dry riverbeds are found on Mars, meandering across the surface of the planet like rivers on Earth. The dried-up martian riverbeds are approximately 3 billion years old, judging from the density of impact craters atop them. In addition to the narrow, relatively orderly riverbeds, there are also broad flash-flood channels found on Mars, the result of catastrophic floods rushing across the surface. The outflow channels can be hundreds of kilometers wide, and contain "teardrop" islands, as shown in Figure 10.10. Thus,

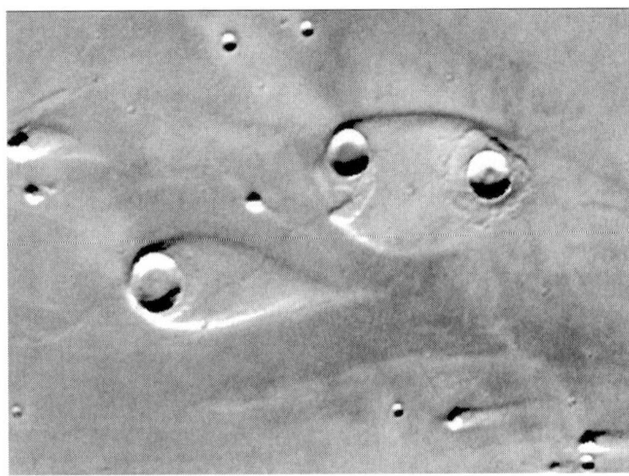

FIGURE 10.10 In the region shown in this image, water flowed from left to right; the crater that diverted the flow to form the teardrop island on the left is ~ 8 km in diameter.

liquid water existed abundantly on Mars in the past, although it does not at present. This indicates that Mars may once have had a thicker atmosphere and a higher temperature in the past than it does now.

Mars has two small, irregular satellites called Phobos and Deimos ("fear" and "panic"). Phobos and Deimos are both heavily pitted with craters, showing a history of collisions with other, smaller bodies. They are also undifferentiated, indicating they have never been hot enough to melt. Phobos and Deimos both resemble asteroids. (Asteroids are examined in Section 11.1). As mentioned in Section 4.3.1, Phobos is on an orbit with $a_{\rm Phobos} = 9400$ km $= 2.76 R_{\rm Mars}$. Not only does this put Phobos inside the Roche limit, it also means that its orbital period is 7.7 hours, less than the rotation period of Mars. Since Phobos orbits from west to east, the same direction as the rotation of Mars, this means that future inhabitants of Mars will be able to see Phobos rise in the west and set in the east, while Deimos rises in the east and sets in the west.

10.2 ▪ JOVIAN PLANETS

The Jovian (Jupiter-like) planets, in order of increasing distance from the Sun, are Jupiter ($a = 5.2$ AU), Saturn (9.6 AU), Uranus (19.2 AU), and Neptune (30.0 AU). In order of decreasing mass, they are Jupiter ($M = 318 M_\oplus$), Saturn ($95 M_\oplus$), Neptune ($17.1 M_\oplus$), and Uranus ($14.5 M_\oplus$). There are many properties that the Jovian planets have in common. They are all massive bodies that formed far from the young Sun, where the temperatures were low enough to allow condensation of volatile ices as well as refractory rocks and metals. The Jovian planets all have strong magnetic fields, since they are rapid rotators and contain large amounts of electrically conducting fluids in their interiors. They all have multiple satellites, the largest of which formed simultaneously with their parent planet, and they all have ring systems (discussed in Section 10.3), of which Saturn's is by far the most spectacular.

However, as a group portrait of the Jovian planet indicates (Color Figure 8), there are some significant differences between the inner Jovian planets, Jupiter and Saturn, and the outer Jovian planets, Uranus and Neptune. Jupiter and Saturn are large in both mass and radius ($M > 95M_\oplus$, $R \sim 10R_\oplus$) compared to the relatively petite Uranus and Neptune ($M < 18M_\oplus$, $R \sim 4R_\oplus$). Their colors at visible wavelengths are also quite different. As we have seen (in Section 8.2), the blue-green color of Uranus and Neptune results from methane. The brown and gold colors seen in Jupiter and Saturn are the result of trace amounts of complex compounds in their clouds. The readily visible differences between the Jupiter/Saturn pair and the Uranus/Neptune pair are accompanied by differences in their internal structure. As a consequence, we will find it useful to first investigate Jupiter and Saturn, in Section 10.2.1, and then Uranus and Neptune, in Section 10.2.3.

10.2.1 Jupiter and Saturn

The Jovian planets, like the terrestrial planets, have interiors that are in hydrostatic equilibrium. Consequently, as described in Section 9.2, the pressure gradient in their interiors must be given by the equation of hydrostatic equilibrium (eq. 9.8):

$$\frac{dP}{dr} = -\frac{GM_r\rho}{r^2}, \tag{10.13}$$

where M_r is the mass inside the radius r,

$$M_r = 4\pi \int_0^r \rho(r) r^2 dr. \tag{10.14}$$

(We have assumed, in the above equations, that planets are spherical. This is a good first approximation, although it is not strictly true, especially for the rapidly rotating Jovian planets.) In general, to solve equation (10.13) to yield $P(r)$, we need to know the density profile $\rho(r)$ of the planet in question. However, if we want a rough estimate of the central pressure of a planet, we can make the rough approximation that the planet's density is constant: $\rho(r) = \overline{\rho}$. With this assumption, the enclosed mass is

$$M_r = \frac{4\pi}{3}\overline{\rho} r^3, \tag{10.15}$$

and the equation of hydrostatic equilibrium can be written

$$dP = -\frac{4\pi}{3}\overline{\rho}^2 Gr\, dr. \tag{10.16}$$

Integrating this equation between $r = 0$, where $P = P_c$ is the central pressure, and the radius of the planet $r = R$, where $P \approx 0$, we find

$$\int_{P_c}^{0} dP = -\frac{4\pi}{3}\overline{\rho}^2 G \int_0^R r dr$$

$$P_c = \frac{2\pi}{3}\overline{\rho}^2 G R^2. \tag{10.17}$$

By scaling this result to the properties of the Earth, $\overline{\rho}_\oplus = 5500\,\text{kg m}^{-3}$ and $R_\oplus = 6.4 \times 10^6$ m, we can write the approximate central pressure of any planet as

$$P_c \approx 1.7 \times 10^{11}\,\text{N m}^{-2} \left(\frac{\overline{\rho}}{\overline{\rho}_\oplus}\right)^2 \left(\frac{R}{R_\oplus}\right)^2. \tag{10.18}$$

For Jupiter, which has $\overline{\rho} = 0.24\overline{\rho}_\oplus$ and $R = 11.2R_\oplus$, the estimated central pressure is $P_{c,\text{Jup}} \approx 1 \times 10^{12}\,\text{N m}^{-2} \approx 10^7$ atm. Compared with Jupiter, Saturn is lower in density, with $\overline{\rho} = 0.125\overline{\rho}_\oplus$, and smaller in radius, with $R = 9.45R_\oplus$. Thus, Saturn's central pressure is smaller: $P_{c,\text{Sat}} \approx 2 \times 10^{11}\,\text{N m}^{-2} \approx 2 \times 10^6$ atm.

At a pressure of millions of atmospheres, hydrogen becomes a metal, as shown in the phase diagram in Figure 10.11. A **metallic** substance is one in which a regular, latticelike structure of positive ions is surrounded by a cloud of delocalized electrons, not bound to any particular atomic nucleus. The delocalized electrons (called "conduction electrons") are what give metals their characteristic properties: metals typically are reflective, are

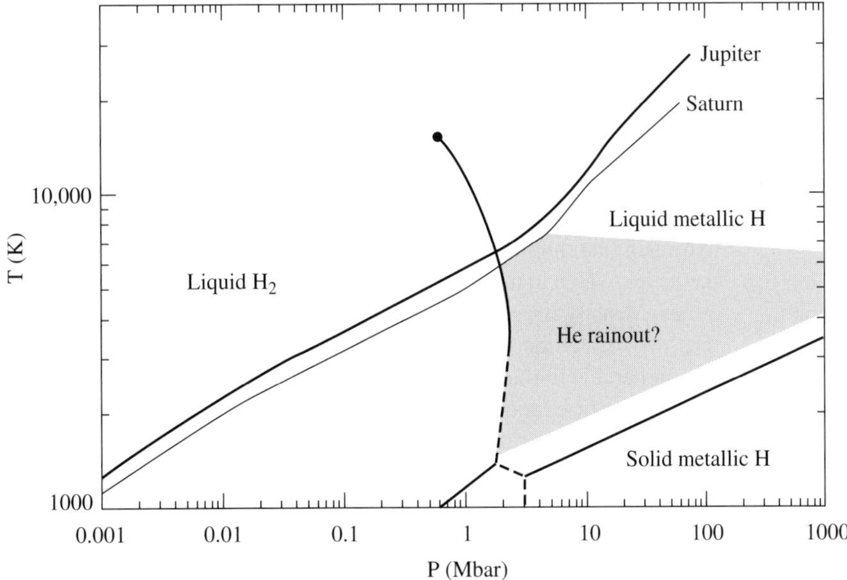

FIGURE 10.11 The phase of hydrogen as a function of pressure and temperature. The temperatures and pressures inside Jupiter and Saturn are shown. (Note that 1 bar = 1 atm.)

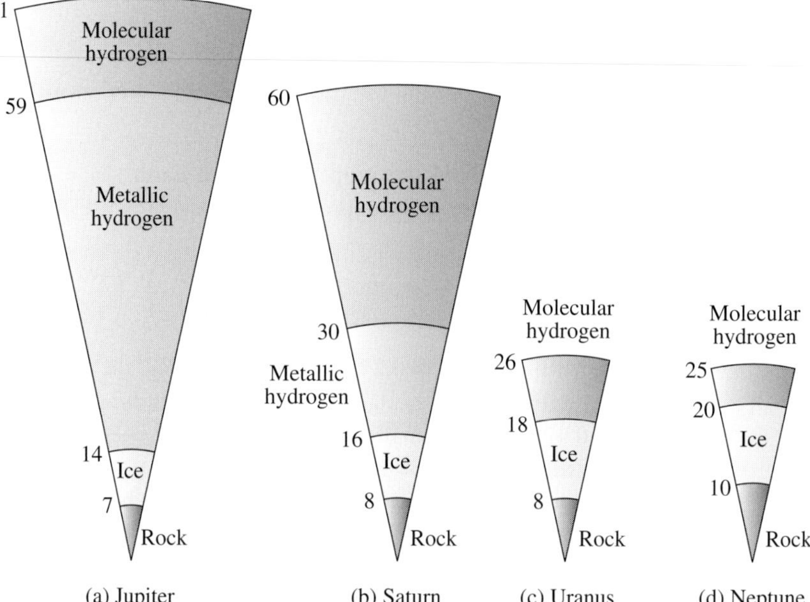

FIGURE 10.12 The radial structure of the Jovian planets: radii are indicated in units of 1000 km.

good conductors of heat, and are good conductors of electricity. Both Jupiter and Saturn have high enough interior pressures to have layers of liquid metallic hydrogen; however, Jupiter, with its higher central pressure, has a much greater amount.

Figure 10.12 shows the radial structure of all four Jovian planets, based on detailed solutions of the hydrostatic equilibrium equation, combined with a knowledge of their chemical composition. The Jovian planets are differentiated, with a low-density layer of hydrogen (and helium) on top of a medium-density layer of ices (water, ammonia, and methane), which in turn lies on top of a dense rocky core.[4] Uranus and Neptune each have a mean density roughly twice that of Saturn, and a radius roughly half that of Saturn; thus, we expect the central pressure of Uranus and Neptune, from equation (10.18), to be comparable to that of Saturn. However, as shown in Figure 10.12, the hydrogen-rich outer layers of Uranus and Neptune are relatively thin; even at their base, the pressure is not high enough for the hydrogen to take its metallic form.

Jupiter and Saturn radiate more energy than they absorb in the form of sunlight. For example, the rate at which Jupiter absorbs energy from sunlight is (equation 8.6)

$$W_{\rm Jup} = \frac{L_\odot}{4\pi a_{\rm Jup}^2}(1 - A_{\rm Jup})\pi R_{\rm Jup}^2 = 3.8 \times 10^{17}\ {\rm W}. \tag{10.19}$$

[4] Since each Jovian planet has a rocky core several thousand kilometers in radius, we might say that inside every Jovian planet is a terrestrial planet screaming to be let out.

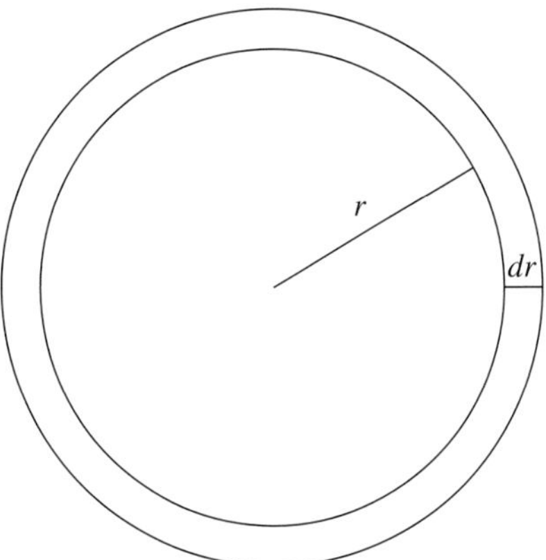

FIGURE 10.13 The gravitational potential energy of a shell of radius r and thickness dr is given by equation (10.23).

However, observations at infrared wavelengths reveal that Jupiter radiates energy at *twice* this rate. Obviously, Jupiter must have an additional source of energy. As we've seen in the case of Earth (page 212), radioactive decay of unstable elements is one source of internal heating. However, the rate at which the Earth is heated by radioactive decay is only $\sim 2 \times 10^{13}$ W, a factor of 5000 lower than the rate $W_\oplus \sim 10^{17}$ W at which it absorbs solar energy. Even scaling upward by Jupiter's mass ($M_{\rm Jup} = 318 M_\oplus$), and ignoring the fact that Jupiter is relatively low in heavy unstable nuclei, radioactivity falls short of explaining the total energy radiated by Jupiter.

In Jupiter, and Saturn as well, heat is generated by the global contraction of the planet. Again, let's make the simplifying assumption that a planet has constant mass density $\overline{\rho}$. Consider a spherical shell centered on the planet's center with radius r and thickness dr, as illustrated in Figure 10.13. The gravitational potential energy of the mass shell is

$$dU = -\frac{GM_r}{r} dm, \tag{10.20}$$

where

$$M_r = \frac{4\pi}{3} r^3 \overline{\rho} \tag{10.21}$$

is the mass enclosed within the shell and

$$dm = 4\pi r^2 \overline{\rho} dr \tag{10.22}$$

is the mass of the shell itself. Thus, the potential energy of the shell can be written in the form

$$dU = -3G \left(\frac{4\pi\overline{\rho}}{3} \right)^2 r^4 dr. \tag{10.23}$$

The potential energy of a planet with radius R is obtained by integrating over all shells from $r = 0$ to $r = R$:

$$U = -3G\left(\frac{4\pi\bar{\rho}}{3}\right)^2 \int_0^R r^4 dr = -3G\left(\frac{4\pi\bar{\rho}}{3}\right)^2 \frac{R^5}{5}$$

$$= -\frac{3}{5}G\left(\frac{4\pi R^3\bar{\rho}}{3}\right)^2 \frac{1}{R} = -\frac{3}{5}\frac{GM^2}{R}. \tag{10.24}$$

By decreasing the radius R of the planet while keeping its mass M constant, we can decrease the gravitational potential energy U:

$$\frac{dU}{dt} = \frac{3}{5}\frac{GM^2}{R^2}\frac{dR}{dt}. \tag{10.25}$$

Solving for the contraction rate of the planet, we find that

$$\frac{dR}{dt} = \frac{5}{3}\frac{dU}{dt}\frac{R^2}{GM^2}. \tag{10.26}$$

If Jupiter's excess radiation of $\sim 4 \times 10^{17}$ W comes from gravitational potential energy, then Jupiter must be contracting at the rate

$$\frac{dR}{dt} \approx \frac{5}{3}(-4 \times 10^{17}\,\mathrm{J\,s^{-1}})\frac{(6.96 \times 10^7\,\mathrm{m})^2}{(6.67 \times 10^{-11}\,\mathrm{J\,m\,kg^{-2}})(1.9 \times 10^{27}\,\mathrm{kg})^2}$$

$$\approx -1.6 \times 10^{-11}\,\mathrm{m\,s^{-1}} \approx -500\,\mathrm{km\,Gyr^{-1}}. \tag{10.27}$$

Thus, during the lifetime of the solar system, $t \sim 4.6$ Gyr, Jupiter need only have contracted by $\Delta R \sim 2300$ km, about 3% of its current radius, in order to maintain radiation at its current rate.

The appearance of Jupiter and Saturn depends on the temperature (and the resulting chemistry) at the highest levels in their atmospheres. The rapid rotation of Jupiter and Saturn stretches clouds into bands that run parallel to the equator. The banded structure is most clearly seen in the case of Jupiter (Figure 10.14). However, Saturn shows bands parallel to its equator as well; Figure 10.15 is an image of Saturn with its distracting rings removed. In the case of Jupiter, the visible clouds are primarily in three layers that are 75 km deep. The uppermost layer consists of ammonia (NH_3) crystals. In the middle layer, ammonia and hydrogen sulfide (H_2S) combine to form crystals of ammonium hydrosulfide (NH_4SH). In the lower level, the clouds are made of water crystals. All of these substances are white in their crystallized form; the varied colors visible in Figure 10.14 are the result of the complex chemistry of trace constituents in the clouds. In the case of Saturn, which is less than a third the mass of Jupiter, the lower gravity spreads the three cloud layers over a greater range in radius, about 300 km. The increased optical depth of the cloud layers makes the lower layers appear somewhat hazy, and accounts for the more muted colors of Saturn, when compared to Jupiter.

The upper atmosphere of Jupiter is divided into **zones** and **belts**. The zones are the lighter-colored bands in Figure 10.14; they are lower in temperature and higher in altitude than the belts and represent the tops of high-pressure clouds that are rising upward. The

FIGURE 10.14 Jupiter's clouds are stretched into bands known as belts (the darker bands) and zones (the lighter bands). Also visible in this image is the large circular storm known as the Great Red Spot.

FIGURE 10.15 Saturn with its rings removed by computer image processing.

belts are the darker-colored bands; they represent lower-pressure clouds that are sinking downward. Large cyclonic storms are sometimes seen in the atmospheres of the Jovian planets. The best known of these storms is the Great Red Spot on Jupiter (easily visible in Figure 10.14). The Great Red Spot varies in size and shape, but at its maximum size it is 40,000 kilometers across (about equal to the circumference of the Earth). The Great

Red Spot is a high-pressure system in Jupiter's southern hemisphere. As air flows away from the center of the Spot, it is deflected to the left by the Coriolis effect. As a result, the Great Red Spot rotates counterclockwise. The Great Red Spot was first recorded by astronomers on Earth in the mid-seventeenth century.

Careful observations of cloud motions reveal that the Jovian planets rotate differentially. For instance, Jupiter has a rotation period that is shorter at the equator than at the poles, by approximately 1%. The extremely rapid rotation of Jupiter ($P_{\rm rot} \approx 9.8$ hr) and Saturn ($P_{\rm rot} \approx 10.5$ hr), combined with their relatively low density, means that they are the most rotationally flattened of all the planets. The polar diameter of Jupiter is 6.5% shorter than its equatorial diameter; for Saturn, the polar diameter is a whopping 10% shorter than its equatorial diameter. [5] By comparison, the polar flattening of the stiffer, more slowly spinning Earth, as mentioned in Section 4.1, is only 0.3%.

Both Jupiter and Saturn have strong magnetic fields; the field of Jupiter, thanks to its rapid rotation and thick layer of liquid metallic hydrogen, is exceptionally strong. At the top of Jupiter's cloud deck, the magnetic field strength is $B_{\rm Jup} \sim 5 \times 10^{-4}\,{\rm T} \sim 10 B_\oplus$. The magnetosphere of Jupiter has a radius of $\sim 200 R_{\rm Jup}$; if it could be seen by human eyes, it would loom in the Earth's sky as large as the full Moon. Charged particles released by eruptions on Jupiter's volcanic satellite Io (discussed in Section 10.2.2) are trapped by Jupiter's magnetic field into a torus around Jupiter. This torus of charged particles is a strong source of radio emission.

As was noted in Section 7.2, electrons in a magnetic field of strength B will spiral in a helix with radius equal to the Larmor radius,

$$r_{\rm c} = \frac{m_e v_\perp}{eB}, \tag{10.28}$$

where $-e$ is the charge of the electron. The pitch angle α_c of the helix will be given by the relation

$$\tan \alpha_c = v_\perp / v_\parallel, \tag{10.29}$$

where $v_\perp$ is the component of the electron's velocity perpendicular to $\vec{\mathbf{B}}$, and $v_\parallel$ is the component parallel to $\vec{\mathbf{B}}$.[6] The angular frequency with which the electron spirals is the **cyclotron frequency**,[7]

$$\omega_{\rm c} = \frac{v_\perp}{r_{\rm c}} = \frac{eB}{m_e} = 1.76 \times 10^{11}\,{\rm s}^{-1} \left(\frac{B}{1\,{\rm T}} \right). \tag{10.30}$$

Notice that the cyclotron frequency, in radians per second, depends only on the magnetic field strength B and the charge-to-mass ratio of the electron, $-e/m_e$; it is independent

[5] If Saturn appears distinctly noncircular in Color Figure 8, it's not the result of a printer's error; it's because Saturn really is flattened!

[6] In other words, if $v_\perp = 0$, the electron moves parallel to $\vec{\mathbf{B}}$ ($\alpha_c = 0$), and if $v_\parallel = 0$, the electron moves in a circular orbit in a plane perpendicular to $\vec{\mathbf{B}}$ ($\alpha_c = \pi/2$).

[7] A cyclotron was an early type of particle accelerator, in which charged particles moved on circular orbits in a nearly uniform magnetic field.

of $v_\perp$ as well as $v_\parallel$. Because the electron is continuously accelerated in a direction perpendicular to the magnetic field $\vec{\mathbf{B}}$, it radiates energy according to the Larmor formula (equation 5.15):

$$P = \frac{2}{3}\frac{e^2}{4\pi\epsilon_0}\frac{a^2}{c^3}. \tag{10.31}$$

The acceleration of the electron is $a = v_\perp^2/r_c = v_\perp\omega_c$, and so the power radiated can be written in the form

$$P = \frac{2}{3}\hbar c\alpha\frac{v_\perp^2\omega_c^2}{c^3} = \frac{2}{3}\hbar\alpha\left(\frac{v_\perp^2}{c^2}\right)\omega_c^2, \tag{10.32}$$

where we have used the definition of the fine structure constant α (equation 5.11).

If the electrons are nonrelativistic ($v \ll c$), the radiation they produce is called **cyclotron radiation**. The spectrum of cyclotron radiation is narrow in frequency, and peaks at a frequency

$$\nu_{\rm max} \approx \frac{\omega_c}{2\pi} \approx 28.0\,{\rm GHz}\left(\frac{B}{1\,{\rm T}}\right). \tag{10.33}$$

As an example, the Earth's magnetic field, with $B_\oplus \sim 5\times10^{-5}\,{\rm T}$, produces radio cyclotron emission with $\nu_{\rm max} \sim 1\,{\rm MHz}$.[8]

If the electrons are relativistic, this will change the spectrum of radiation they produce. A relativistic electron will spiral around the magnetic field lines at the **electron gyro frequency**,

$$\omega_s = \frac{eB}{\gamma m_e} = \frac{\omega_c}{\gamma}, \tag{10.34}$$

where $\gamma \equiv (1 - v^2/c^2)^{-1/2}$ is the Lorentz factor. Note that $\omega_s < \omega_c$ for a fixed magnetic field strength B, and thus a relativistic particle will move on a helix with a radius larger than the Larmor radius:

$$r_s = \frac{v}{\omega_s} \approx \frac{c\gamma}{\omega_c} \approx \gamma r_c > r_c. \tag{10.35}$$

The radiation produced by relativistic electrons in a magnetic field is called **synchrotron radiation**.[9] Synchrotron radiation is strongly beamed in the direction of motion of the charged particle, emerging in a cone of angular width $\sim 1/\gamma$ radians. Thus, for a distant observer, the radiation from the charged particle appears to be pulsed as the cone of

[8] To avoid any confusion, note that the symbol ν represents a frequency in units of cycles per second (or Hertz); the symbol ω represents an angular frequency, or angular speed, in units of radians per second. Since there are 2π radians per cycle, that accounts for the factor of 2π in going from angular frequency ω to frequency ν.

[9] A synchrotron is a more sophisticated particle accelerator than an old-fashioned cyclotron and is able to accelerate charged particles to relativistic speeds.

radiation sweeps across the direction to the observer. The duration of each pulse in the rest frame of the observer is

$$\tau \approx \frac{1}{\omega_c \gamma^2 \sin \alpha_c}. \tag{10.36}$$

The time between pulses in the rest frame of the observer is

$$T \approx \frac{2\pi}{\omega_c} \gamma \sin^2 \alpha_c. \tag{10.37}$$

The spectrum of synchrotron radiation is much broader than the spectrum of cyclotron radiation, and peaks at a frequency

$$\begin{aligned} \nu_{\rm max} &\approx 0.070\, \omega_c \gamma^2 \sin \alpha_c \propto \frac{1}{\tau} \\ &\approx 12\ {\rm GHz}\ \gamma^2 \left(\frac{B}{1\,{\rm T}}\right) \sin \alpha_c. \end{aligned} \tag{10.38}$$

(Aside: the factor of 0.070 in equation 10.38 comes from finding the zeros of a modified Bessel function, so its origin is not intuitively obvious.) The magnetic field of Jupiter has $B_{\rm Jup} \sim 5 \times 10^{-4}$ T; if the trapped electrons were nonrelativistic, they would therefore emit radiation primarily at $\nu_{\rm max} \approx 0.01$ GHz. Instead, they emit a broad spectrum peaking at $\nu_{\rm max} \approx 1$ GHz. This tells us that the electrons are relativistic and are emitting synchrotron radiation. If we assume typical pitch angles of $\alpha_c \sim \pi/4$, then the Lorentz factor of the electrons must be $\gamma \sim 15$, corresponding to an energy $E = \gamma m_e c^2 \sim 8$ MeV.

10.2.2 Satellites of Jupiter and Saturn

Jupiter and Saturn have many natural satellites. At the time of writing, the number of known moons stood at 63 satellites for Jupiter and 60 satellites for Saturn, although the count may well be higher by the time you read this. Most of these satellites are small and irregular in shape and are on highly eccentric and highly inclined (sometimes even retrograde) orbits. The small irregular satellites are probably bodies that were captured after formation of the planet. Some of the satellites, however, are large spherical bodies, comparable in size to the Earth's Moon, that have nearly circular orbits close to the equatorial plane of their parent planet. These large satellites were probably formed at the same time as their parent planet.

It is found, in general, that small satellites are nonspherical, whereas the larger satellites are close to perfect spheres. Whether or not a satellite (or other celestial object) is spherical depends on whether its compressional strength is great enough to resist the force of gravity. The compressional strength of a material is the pressure required to significantly deform its shape. Engineers determine the compressional strength of different materials by putting a sample into a hydraulic press and increasing the pressure until the sample is either squashed flat or shattered, depending on how brittle the material is. For instance, iron has a compressional strength of $S \sim 4 \times 10^8$ N m^{-2}, typical igneous

TABLE 10.1 Minimum Radius for Spherical Satellites

Material	Strength (S) ($\mathrm{N\,m^{-2}}$)	Density (ρ) ($\mathrm{kg\,m^{-3}}$)	R_{sph} (km)
Iron	4×10^8	8000	210
Rock	2×10^8	3500	340
Ice	1×10^7	900	300

rocks have $S \sim 2 \times 10^8\,\mathrm{N\,m^{-2}}$ (similar to high-strength concrete), and solid ice has $S \sim 10^7\,\mathrm{N\,m^{-2}}$. For an approximately spherical body, we have found that the central pressure will be (equation 10.17)

$$P_c \approx \frac{2\pi}{3}\bar{\rho}^2 G R^2. \tag{10.39}$$

The critical radius R_{sph} at which the pressure is just great enough to overcome the compressional strength and squeeze the object into a sphere is determined by the relation $P_c \approx S$, or

$$\frac{2\pi}{3}\bar{\rho}^2 G R_{\mathrm{sph}}^2 \approx S$$

$$R_{\mathrm{sph}} \approx \left(\frac{3S}{2\pi G}\right)^{1/2} \frac{1}{\bar{\rho}}. \tag{10.40}$$

The values of R_{sph} for iron, rocky, and icy bodies are given in Table 10.1. The stronger of the materials of which natural satellites are made (iron, rock, and ice) are also the more dense. Thus, the value of R_{sph} is always ~ 300 km and doesn't depend strongly on whether the satellite is mostly metallic, mostly rocky, or mostly icy.[10]

Of the spherical satellites in the solar system, seven have radii $R > 1300\,\mathrm{km} \sim 0.2 R_\oplus$ and thus are larger than any known dwarf planets. These seven **giant satellites** are listed in Table 10.2. The giant satellites Ganymede and Titan are actually larger in radius than the planet Mercury ($R_{\mathrm{Mer}} = 0.38 R_\oplus$). The high mass density of the planet Mercury, however, means that its mass is greater than that of Ganymede and Titan combined ($M_{\mathrm{Mer}} = 0.0553 M_\oplus$).

The four giant satellites of Jupiter are known as the **Galilean satellites**, after their discoverer Galileo Galilei. As we discussed in Section 2.4, the discovery of the Galilean satellites was an important early support for the Copernican system. In order of increasing distance from Jupiter, the Galilean satellites are Io, Europa, Ganymede, and Callisto

[10] A few materials combine great compressive strength with relatively low density. For instance, a diamond has $\rho \approx 3500\,\mathrm{kg\,m^{-3}}$, comparable to an ordinary igneous rock, but a compressive strength of $S \approx 10^{10}\,\mathrm{N\,m^{-2}}$. Thus, a "diamond in the sky" could be as large as $R_{\mathrm{sph}} \approx 2400$ km in radius before being squeezed into a spherical shape.

TABLE 10.2 Giant Satellites in the Solar System

Planet	Satellite	Mass ($M_\oplus$)	Radius ($R_\oplus$)
Earth	Moon	0.0123	0.27
Jupiter	Io	0.0150	0.29
	Europa	0.0080	0.24
	Ganymede	0.0248	0.41
	Callisto	0.0180	0.38
Saturn	Titan	0.0225	0.40
Neptune	Triton	0.0036	0.21

(Color Figure 9).[11] The Galilean satellites probably formed as would a miniature solar system, with Jupiter at the center. That is, the gas close to Jupiter was hottest, allowing condensation of only the least volatile substance (metal and rock), while the gas farther away was cooler, allowing condensation of the more volatile ices. Within the Galilean satellite system, we see the familiar pattern of decreasing density with increasing distance from the parent body: Io has $\rho \approx 3600\ \text{kg m}^{-3}$; Europa has $\rho \approx 3000\ \text{kg m}^{-3}$; and the two outermost Galilean satellites, Ganymede and Callisto, have $\rho \approx 1900\ \text{kg m}^{-3}$. Such a low density indicates that Ganymede and Callisto cannot consist entirely of rock and metal but must contain substantial amounts of lower-density ice.

Io, the innermost Galilean satellite, is the most geologically active object in the solar system. When Io is viewed at infrared wavelengths (Color Figure 10), the "hot spots" associated with active volcanos are readily visible. The colors of Io in the visible range of the spectrum are due to the presence of sulfur and sulfur compounds. Because of Io's low surface gravity ($g = 0.18 g_\oplus$), the volcanic ejecta can rise high above the surface and spread over a large area. Volcanism on Io results in a layer of ejecta more than 100 m thick being laid down every million years. Thus, any impact craters that may be created on Io are rapidly paved over by the volcanic ejecta. The hottest volcanos on Io are spewing out lava that is ~ 2000 K in temperature. This is too high an internal temperature to be maintained solely by radioactive heating. Io's dominant source of heating is tidal flexing of its interior. Because Io is so close to the massive planet Jupiter, it experiences large tidal distortions. The orbital period of Io, $P_{\text{Io}} = 1.77$ days, is equal to half the orbital period of Europa, $P_{\text{Eur}} = 3.55$ days. The orbital resonance between Io and Europa causes the orbit of Io to change its shape continuously; thus, the tidal bulges of Io are always changing in amplitude. The constant flexing of Io's interior as it changes shape causes internal heating. Io has a very tenuous ($P \sim 10^{-10}$ atm) atmosphere of sulfur and sodium, which

[11] The Galilean satellites are named after four of the god Jupiter's innumerable lovers. The name of Juno, Jupiter's wife, was given to an asteroid that never comes within 100 million kilometers of Jupiter. This is probably appropriate, given that the Greek gods were not terribly keen on traditional family values.

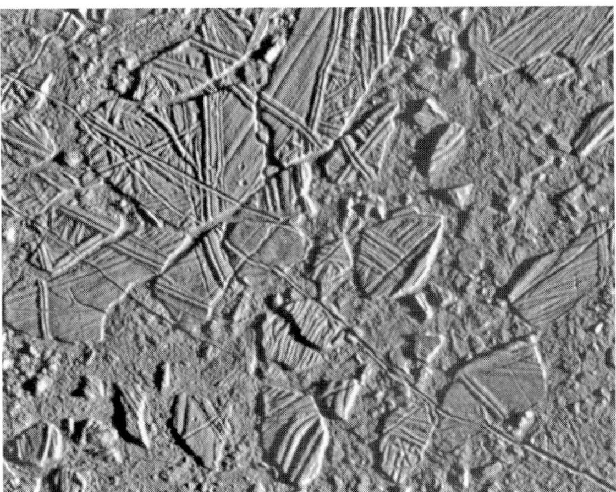

FIGURE 10.16 A portion of Europa's surface measuring 34 km $\times$ 42 km (about half the area of Rhode Island).

is continuously leaking away into the torus of charged particles around Jupiter and being replenished by fresh volcanic eruptions.

Europa, the next of the Galilean satellites, has a smooth, high-albedo surface ($A = 0.67$). The reflection spectrum of Europa reveals that its surface is made of frozen water. A high-resolution image of Europa (Figure 10.16) reveals that the icy surface is fractured into numerous ice rafts and ice floes measuring several kilometers across, similar to ice rafts seen on the Earth's Arctic Ocean. This may indicate that the outer ice layer of Europa lies atop an ocean of liquid water. The extreme scarcity of impact craters on Europa would then be explained by liquid water flowing up through cracks, then spreading out and freezing. The relatively high mean density of Europa ($\rho \approx 3000\,\text{kg m}^{-3}$) means that the icy surface and liquid ocean must be relatively thin layers atop a large central rocky core.

Ganymede, the third Galilean satellite, is the largest satellite in the solar system. Part of Ganymede's surface is dark and covered with impact craters. However, part of its surface is covered with enigmatic "grooved terrain," illustrated in Figure 10.17b. The grooves are about 10 km apart and 300 m deep; judging from the number of impact craters lying on top of them, the grooves formed ~ 1 Gyr ago. They may be stretch marks in the crust due to tectonic motions that have since ceased.

Callisto, the outermost Galilean satellite, has a surface that consists of dirty ice, that is, frozen water with dust and hydrocarbons embedded within it. The surface is heavily cratered, as shown in Figure 10.17c. The largest craters on Callisto have a higher albedo than the surrounding ice, since the impacting bodies that formed them broke through the superficial layer of dirty ice to reveal the pristine ice underneath.

The Galilean satellites, like the terrestrial planets, show different amounts of volcanic activity. Among the terrestrial planets, the amount of internal heat, and thus the amount of volcanic activity, is determined by the size of the planet. However, among the Galilean

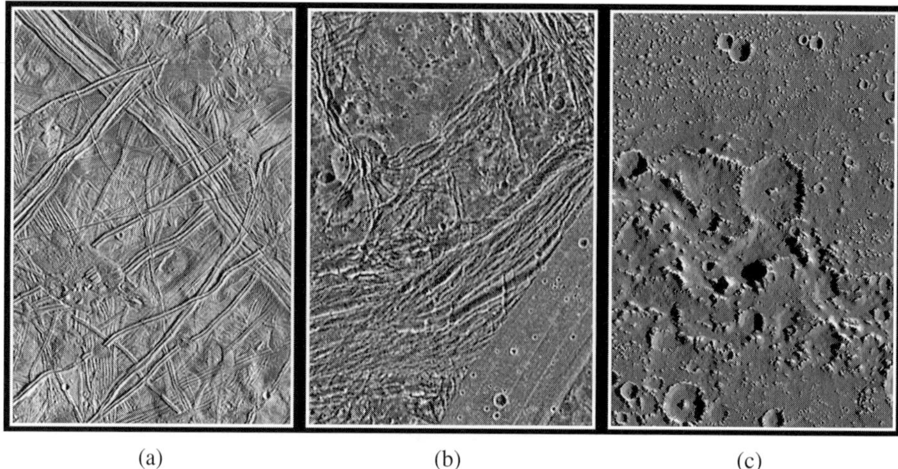

FIGURE 10.17 Regions on Europa (a), Ganymede (b), and Callisto (c). Each region is 220 km × 100 km (about the area of West Virginia).

satellites, the amount of internal heat is determined by the proximity to Jupiter. Io, the closest to Jupiter, has the most tidal heating; Io is volcanically hyperactive. Callisto, the farthest from Jupiter, has the least tidal heating; Callisto is volcanically dead and is covered with impact craters.

Titan, the giant satellite of Saturn, is observed to have a thick atmosphere, with surface pressure $P_T = 1.6$ atm; it consists primarily of N_2, with smaller amounts of argon and hydrocarbons such as methane (CH_4) and ethane (C_2H_6). The atmosphere of Titan is readily apparent in a view of Titan backlit by the Sun (Figure 10.18), since the refraction of sunlight by the atmosphere makes it look like a ring of light. Sunlight striking the atmosphere triggers the polymerization of hydrocarbons. The resulting long hydrocarbon chains constitute a "smog layer" in the upper atmosphere. The temperature and pressure in Titan's atmosphere is close to the triple point of methane. As a consequence, methane can exist on Titan in its gaseous, liquid, and solid forms. Infrared images of the surface of Titan, taken by the *Cassini* spacecraft, indicate that Titan may have lakes of liquid methane and ethane on its surface.

10.2.3 Uranus and Neptune

Uranus and Neptune are twin planets, similar in their properties. Uranus is 3% larger in radius than Neptune, but 15% smaller in mass. Their internal structure, as illustrated in Figure 10.12, is quite similar, with a layer of ordinary molecular hydrogen lying atop a layer of liquid water and ammonia ("ices," in the jargon of planetary scientists), in turn lying atop a solid rocky core. Although the pressure at the base of the hydrogen layer is insufficient to compress hydrogen into its metallic form, the layer of liquid water is electrically conducting; thus, the rapid rotation of Uranus and Neptune produces significant magnetic fields.

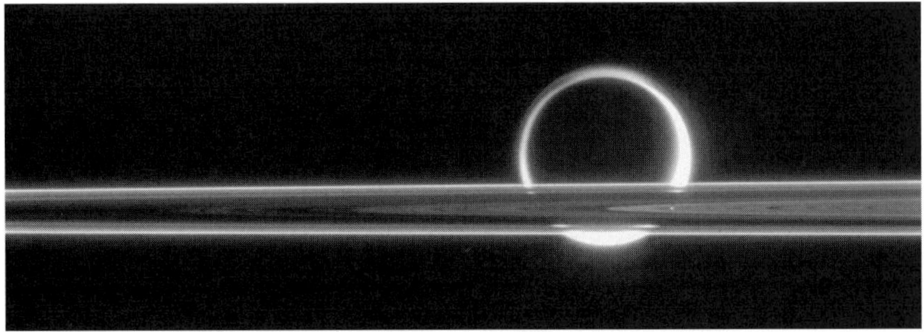

FIGURE 10.18 A view of Saturn's rings in front of Saturn's giant satellite Titan. The image was taken in visible light by the *Cassini* spacecraft; the Sun–Titan–spacecraft angle was 158° degrees.

Uranus and Neptune are difficult to observe from the Earth. Although Uranus can barely be detected by the unaided human eye under ideal conditions, it was not recognized as a planet until William Herschel resolved it with his telescope in 1781 March. The existence of the planet Neptune was first deduced from its gravitational influence on Uranus. In the 1840s, two astronomers, John Couch Adams and Urbain Le Verrier, computed where the unseen planet had to be located in order to produce the observed perturbations to the orbit of Uranus. The planet Neptune was discovered by Johann Gottfried Galle in 1846 September, only 1° away from the position where Le Verrier told him to look.

Since the maximum angular diameter of Uranus as seen from Earth is 4.1″ and that of Neptune is 2.4″, ground-based telescopes cannot obtain high-resolution images of these two planets. Much of what we know about Uranus and Neptune comes from the *Voyager 2* spacecraft, which made a flyby of Uranus in 1986 January, then a flyby of Neptune in 1989 August. In the *Voyager 2* images of Uranus, the planet looked very uniform, without the colored bands that characterize Jupiter and Saturn (the image of Uranus in Color Figure 5, for instance, was taken by *Voyager 2* and looks extraordinarily bland and boring). However, an image taken by the *Hubble Space Telescope* in 2006 August, more than 20 years after the *Voyager 2* flyby, shows a distinct banded pattern in the atmosphere of Uranus (Figure 10.19).

The change in the appearance of Uranus is linked to the planet's seasons. The axial tilt of Uranus is 98°, meaning that its rotation axis lies nearly in its orbital plane; this causes extreme seasonal variations in solar heating. The most recent solstice on Uranus occurred in 1985 October, just three months before the *Voyager 2* flyby. At this time, the northern hemisphere of Uranus experienced continuous sunlight and the southern experienced continuous darkness.[12] The resulting heat flow in the north–south direction disrupted the easterly and westerly flows that produce bands. Since the orbital period of Uranus is $P \approx 84$ years, the succeeding equinox occurred ~ 21 years later, in 2007

[12] We are defining the north pole of Uranus as the pole about which Uranus's rotation is counterclockwise as viewed from above.

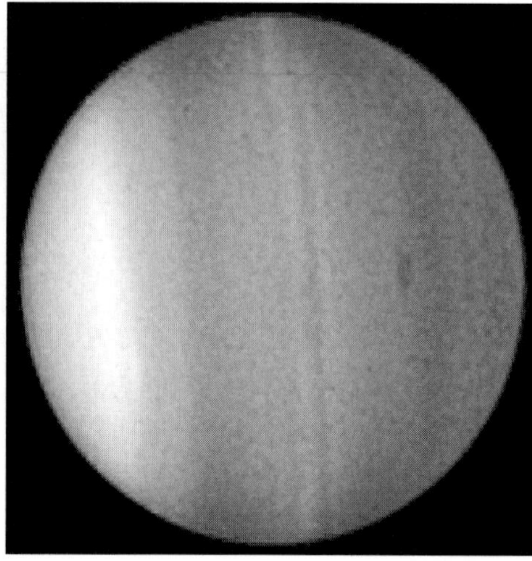

FIGURE 10.19 Banded cloud patterns on Uranus near the time of an equinox.

December, just a year after the image in Figure 10.19 was taken. Around the time of the equinox, the rotation axis of Uranus is nearly perpendicular to the Uranus–Sun line. Near equinox, a highly tilted planet like Uranus temporarily resembles a low-tilt planet like Jupiter ($\theta = 3°$) and has a Jupiter-like banded structure in its cloud layer.

Uranus and Neptune, as mentioned earlier, are similar in mass and radius; they are also similar in surface temperature. They both have a temperature, estimated from their infrared radiation, of $T \approx 59$ K. This is odd, given that Neptune's distance from the Sun is 1.56 times the distance of Uranus from the Sun; this implies that Neptune intercepts only $\sim 40\%$ as much solar power per square meter as Uranus. When the energy accounting is done in full, Uranus is found to radiate about as much energy as it absorbs in the form of sunlight. However, Neptune radiates roughly 2.5 times as much energy as it absorbs. This means that Neptune has an internal heat source, whereas Uranus does not. The reason why two very similar planets should differ so radically in their internal heating is not clear.

Neptune's satellite Triton (Figure 10.20) is unusual in that it is the only giant satellite that has a retrograde orbit. This suggests that Triton is a captured object, rather than one that formed close to Neptune.[13] The retrograde motion of Triton has important implications for the evolution of its orbit. As we saw in Section 4.2, the Earth's tidal bulges act to increase the Moon's orbital angular momentum; thus, the size of the Moon's orbit increases with time. However, for a satellite on a retrograde orbit, the pull of the parent planet's tidal bulges acts to decrease the satellite's angular momentum; the size of a retrograde satellite's orbit *decreases* with time. The current size of Triton's orbit,

[13] Neptune's second largest satellite, Nereid, is on a highly eccentric orbit, with $e = 0.75$, possibly a result of being gravitationally perturbed during the capture of Triton.

FIGURE 10.20 The *Voyager 2* spacecraft looks back at a crescent Neptune (above) and Triton (below).

in units of Neptune's radius, is $a \approx 5.9 R_{\rm Nep}$. However, as a result of the shrinking of its orbit, Triton will move inside the Roche limit a few billion years from now. Depending on its strength, Triton will either plunge intact into Neptune's atmosphere or be broken by the tidal forces into numerous fragments, forming a massive set of planetary rings.

10.3 ▪ PLANETARY RINGS

All of the Jovian planets have **planetary rings** in their equatorial planes. The ring system of Saturn (shown in part in Figure 10.18) is the most massive, but the other three Jovian planets also have rings. Planetary rings, for the most part, lie inside the Roche limit (Figure 10.21). Saturn's rings were first observed by Galileo in the year 1610. However, his telescope was too low in resolution to reveal the shape of the rings; he saw merely a small blob on either side of Saturn. By the end of the year 1612, the two blobs had disappeared, to Galileo's surprise. By the year 1616, to Galileo's further surprise, the extensions to Saturn had reappeared; with the superior telescope that Galileo was using, he was able to tell that the extension looked like *ansae*, or "handles."[14] It was not until the year 1655 that the Dutch astronomer Christiaan Huygens correctly deduced that Saturn was surrounded by a thin, flat ring. Huygens also pointed out that the rings were sufficiently thin as to be invisible when viewed edge-on; this accounted for the disappearance of the rings reported by Galileo in 1612. In the year 1675, Giovanni Cassini

[14] In the sketch made by Galileo, Saturn looked like this:

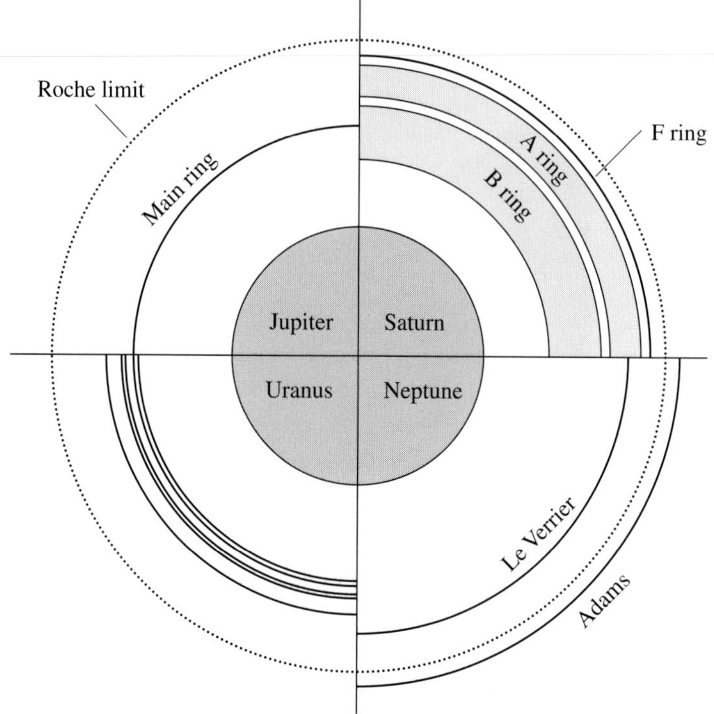

FIGURE 10.21 Each quadrant shows the main rings of the Jovian planets, plotted on a scale where the planet's radius equals 1. The dotted line shows the Roche limit for a satellite whose density equals that of the parent planet.

found a gap, now called the Cassini division, in the middle of the ring, showing that the ring was not a monolithic, solid disk.

The great theoretical leap in understanding Saturn's rings came in 1857, when the physicist James Clerk Maxwell demonstrated mathematically that a solid disk would be unstable; if the center of the ring were displaced infinitesimally from the center of Saturn, the displacement would grow until the ring slammed into Saturn. Maxwell thus deduced that the rings of Saturn were made of "an indefinite number of unconnected particles." In 1895, a classic observation by James Keeler of the Doppler shift of reflected sunlight from the rings revealed that the ring particles are in Keplerian orbits. That is, $v_{\rm orb} \propto r^{-1/2}$, where r is the distance from the center of Saturn.

The spectrum of reflected sunlight can also be used to determine the chemical composition of the ring particles. Spectra of Saturn's rings reveal that its ring particles are made of water ice, with an albedo $A \approx 0.6$. (Or, at least, they are coated with frost.) The size of ring particles can be determined from their interactions with light. Large particles—those with diameters $d > \lambda$, where λ is the wavelength at which you are observing—are best seen in reflected light. Small particles—those with $d \lesssim \lambda$—are best seen in forward scattering, that is, with the ring particles between the observer and the light source. Saturn's ring particles vary in size, but most are between $d \sim 1$ cm and $d \sim 3$ m.[15]

The rings of Saturn are stunningly thin, especially when we consider that the diameter of the entire ring system is $\sim 270{,}000$ km, about 70% of the distance from the Earth to the Moon. The rings have a typical thickness of only ~ 30 m. A scale model of Saturn's rings, made out of a sheet of paper 0.1 mm thick, would have to be almost a kilometer in diameter. It is no surprise, then, that the ultrathin rings disappear for observers on Earth whenever the Earth passes through Saturn's ring plane. All the ring particles orbiting Saturn, if collected in one place, would make an icy satellite with $d \sim 400$ km, and mass $M \sim 10^{-7} M_{\rm Sat}$.

The extreme brightness of Saturn's rings relative to those of the other Jovian planets can be deduced from a chronology of ring discoveries:

- Saturn: Rings discovered in the seventeenth century, using a small telescope.
- Uranus: Rings discovered in 1977, during a stellar occultation. (The light from a distant star was dimmed momentarily as it passed behind each of Uranus's narrow rings.)
- Jupiter: Rings discovered in 1979, in a picture taken by the *Voyager 2* spacecraft as it passed Jupiter.
- Neptune: Rings discovered in 1985, during a stellar occultation.

Uranus and Neptune have narrow, dark rings, separated by broad gaps, as shown in Figure 10.21. The rings of Neptune, in addition, are patchy, with clumps of dusty dark material, rather than being one continuous ring around the planet. Jupiter, the largest Jovian planet, has the least impressive ring. It has a single dark ring, made of fine dust particles, plus a tenuous "gossamer ring," which (as its name implies) is nearly transparent. If all the material in the rings of Jupiter, Uranus, and Neptune combined were swept into a single body, they would make a rocky satellite with $d \sim 10$ km.

Planetary rings might represent either primordial chunks of condensed matter or material from tidally disrupted satellites. Current evidence favors the latter explanation, since rings are not expected to be long-lived phenomena. For instance, large ring particles will undergo collisions that will gradually grind them into dust. The small dust particles will then be swept away by the solar wind and by radiation pressure from sunlight.[16] If planetary rings are made from tidally disrupted bodies, then the relatively bright, massive rings of Saturn are the result of a relatively recent tidal breakup. (As mentioned in Section 10.2.3, in a few billion years, the tidal disruption of Triton may give Neptune a counterrotating ring system that will put Saturn's current rings to shame.) The tidal-disruption hypothesis also helps to explain why the Jovian planets have rings and the less massive terrestrial planets, having smaller Roche limits, do not.

[15] The planet Uranus has ring particles of comparable size, but they are much harder to see, since they have an albedo $A \approx 0.05$, which is about as dark as coal.

[16] Radiation pressure is discussed in more detail in Section 11.3.

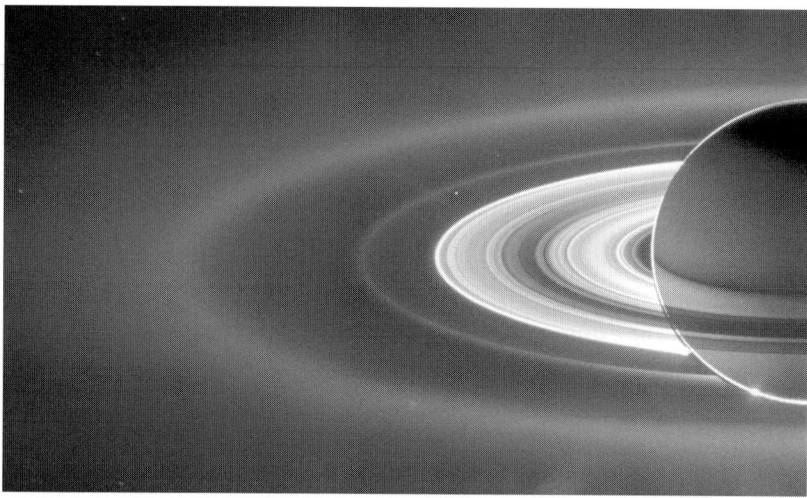

FIGURE 10.22 A panoramic backlit image of Saturn and its rings; a mosaic of images taken by the *Cassini* spacecraft.

The ring system of Saturn contains a great deal of complex structure, as seen in Figure 10.22. Much of the structure is due to orbital resonances with satellites of Saturn. As an example, consider the Cassini division within Saturn's ring system. A particle in the middle of the Cassini division has an orbital period $P = 11.3$ hr. When we compare this with the $P = 22.6$ hr orbital period of the satellite Mimas (the most massive of Saturn's inner satellites), we see that such a particle would be in a 2:1 orbital resonance with Mimas. As a consequence, Mimas would always be in opposition, and hence exerting its largest gravitational effect, at the same point in the ring particle's orbit. The repeated tugs by Mimas at opposition gradually make the particle's orbit more eccentric, stretching its major axis in the direction of Mimas at opposition. Ring particles in noncircular orbits will cross the orbits of other ring particles, increasing the likelihood of a collision that will suddenly alter their orbits significantly. By this process, particles are gradually removed from the Cassini division, and from the locations of other resonant orbits.

Gravitational interactions between ring particles and satellites also explain why narrow rings, such as the rings of Uranus, remain narrow. If a ring starts out narrow, it has a natural tendency to spread, thanks to the collisions between ring particles. However, narrow rings can be kept from spreading by the gravitational influence of **shepherd satellites**. Figure 10.23 shows a pair of shepherd satellites in action. The F ring of Saturn is a narrow ring that lies outside the main, bright rings of that planet. Just inside the F ring orbits a small irregular satellite called Prometheus; just outside the F ring orbits another small irregular satellite, this one called Pandora. The gravitational interaction among the inner shepherd, the outer shepherd, and the ring particle tends to add orbital angular momentum to particles that stray inward, thus causing them to return to the ring. Similarly, the gravitational interactions subtract angular momentum from particles that

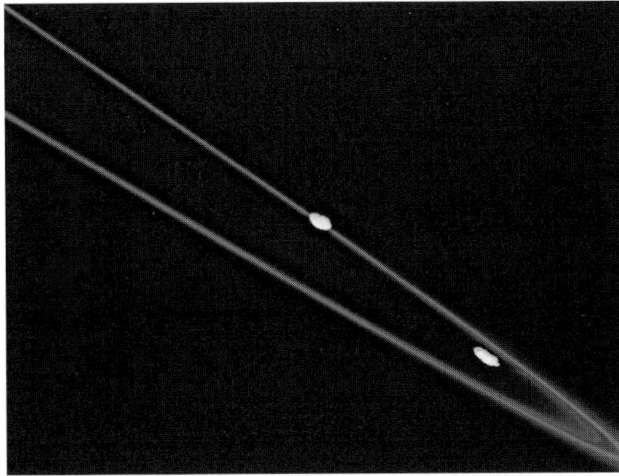

FIGURE 10.23 The shepherd satellites Pandora (left) outside the F ring of Saturn, and Prometheus (right) inside the F ring.

stray outward, causing them to return to the fold. The most prominent ring of Uranus is seen to have a pair of shepherd satellites, called Ophelia and Cordelia. Presumably other fainter rings have shepherds as well, but tiny satellites are difficult to spot.

Persistent features called "spokes" are sometimes seen in Saturn's rings. As their name implies, they rotate like rigid spokes on a wheel, not in a Keplerian fashion. These spokes are probably due to microscopic dust particles that are electrically charged and trapped by Saturn's magnetic field, with which they co-rotate. The origin of these tiny dust particles is not clear; it seems most likely that the dust is from impacts with meteoroids. This argues that Saturn's rings must be young, since prolonged accumulation of dust would make them dark. (Compare the old, dark surface of Callisto with the new, light surface of Europa, for instance.)

PROBLEMS

10.1 The Hill radius (discussed in Section 4.3.2) is the maximum orbital size for a satellite orbiting a planet.

(a) What is the Hill radius of the planet Mercury?
(b) What is the Hill radius of the planet Jupiter?

10.2 The mean mass density of Mercury is 3.9 times that of the Sun. What is the Roche limit for the Mercury–Sun system, expressed as a multiple of the Sun's photospheric radius? What is the Roche limit for the Mercury–Sun system, expressed as a fraction of the semimajor axis of Mercury's orbit?

10.3 At what frequency ν_{max} would you expect to detect the strongest cyclotron radiation from the Sun's K corona?

10.4 Compute the ratio of the differential tidal force on Io due to Jupiter to the differential tidal force on the Moon due to the Earth.

10.5 How often does an observer at the Sun's location see the rings of Saturn exactly edge-on?

10.6 Saturn's moon Titan has a mass $M = 1.3 \times 10^{23}$ kg and a radius $R = 2580$ km. The temperature at the surface of Titan is $T = 94$ K.

(a) What is the gravitational acceleration g at the surface of Titan?
(b) Would Titan be able to retain H_2 in its atmosphere? Would Titan be able to retain CO_2 in its atmosphere?
(c) If you approximate the atmosphere of Titan as consisting entirely of molecular nitrogen, what is the scale height of Titan's atmosphere?

10.7 Mars has an orbit with $a_{\text{Mars}} = 1.524$ AU and $e_{\text{Mars}} = 0.093$.

(a) If we assume that the orbit of the Earth is circular, and is coplanar with the orbit of Mars, what is the minimum possible distance between the Earth and Mars?
(b) The Moon orbits the Earth on an orbit with $a = 384{,}000$ km and $e = 0.055$. What is the maximum possible angular separation between the Earth and the Moon as seen from Mars?
(c) Suppose that the angular resolution of your eyes is $\theta = 1$ arcmin. If you go vacationing on Mars, will you be able to see the Earth and the Moon as separate points with your unaided eyes?

10.8 Infrared observations of the planet Saturn indicate that it emits radiation at the rate $L_{\text{Sat}} = 1.98 \times 10^{17}$ W. However, it absorbs sunlight at the rate $W_{\text{Sat}} = 1.11 \times 10^{17}$ W. If the excess radiated power comes from Saturn's gravitational potential energy, at what rate dR/dt is Saturn shrinking in radius? (Hint: you may make the approximation that Saturn is of uniform density.) At this rate of shrinkage, how long would it take for Saturn's radius to decrease by 1%?

10.9 Using the equation of hydrostatic equilibrium, compute the approximate central pressure of all four Jovian planets.

10.10 The rotation speed of Venus was first measured with ground-based radar.

(a) A radar signal with initial frequency ν_0 is bounced off a target that is receding at speed v. Show that the returned signal is shifted to a frequency

$$\nu = \nu_0 \left(1 - \frac{2v}{c}\right)$$

for $v \ll c$.

(b) Suppose that you are in the equatorial plane of Venus and you bounce a $\nu_0 = 1$ GHz radar signal off the entire planet. What is the spread in frequencies of the returned signal? (Neglect any Doppler shift due to the motion of the planet as a whole.)

10.11 The orbital planes of Phobos and Deimos are very close to the equatorial plane of Mars. Thus, when you are standing near the equator on Mars, you can see Phobos and Deimos pass through the zenith.

(a) You see Phobos at the zenith. How long will it take Phobos to reach the horizon? At what point on the horizon will Phobos set?

(b) You see Deimos at the zenith. How long will it take Deimos to reach the horizon? At what point on the horizon will Deimos set?

11 Small Bodies in the Solar System

Thus far, we have discussed the largest bodies in the solar system (those with $M > 0.003 M_\oplus$). These comprise the Sun, the four Jovian planets, the four terrestrial planets, and the seven giant satellites (the Moon, Io, Europa, Ganymede, Callisto, Titan, and Triton). However, as shown in Figure 8.1, there are a great many midsize bodies (with $10^{-10} M_\oplus \lesssim M \lesssim 0.003 M_\oplus$) orbiting the Sun. There are two main classes of these midsize objects, found in different regions of the solar system. **Asteroids** are rocky and metallic bodies found primarily in the asteroid belt between Mars and Jupiter ($1.5\ \text{AU} < a < 5.2\ \text{AU}$). **Trans-Neptunian objects** (TNOs) are icy and rocky bodies found primarily in the Kuiper belt beyond the orbit of Neptune ($a > 30$ AU).

There doesn't seem to be a low-mass cutoff in the distribution of icy or rocky objects in the solar system; it's just that the low mass objects are more difficult to detect. Low-mass rocky or metallic objects are called **meteoroids**; they are easily detected only when they enter the Earth's atmosphere and form a bright meteor. Low-mass icy objects are called **comets**; they are easily detected only when they come close to the Sun and develop a bright coma and tail.

In this chapter, we first discuss the midsize objects in the solar system: the dense, refractory asteroids and the low-density, volatile TNOs. Then we discuss the smaller bodies that clutter up the solar system: comets, meteoroids, and tiny interplanetary dust grains.

11.1 ▪ ASTEROIDS

Asteroids were originally discovered as a result of a search for a suspected "missing planet." The search was stimulated by what is known as the **Titius–Bode rule**. Ever since the time of Kepler, astronomers have been searching for numerical relationships among the orbital properties of planets. In the year 1766, an astronomer named Johann Titius pointed out such a relation. In modern notation, his rule was that the semimajor axis of each planet's orbit was given by the formula

$$a[\text{AU}] = 0.4 + 0.3(2^n), \tag{11.1}$$

where $n = -\infty$ for Mercury, and $n = 0, 1, 2, \ldots$ for the other planets. This rule was quoted, without proper attribution, by the more prominent astronomer Johann Bode, and

TABLE 11.1 Titius-Bode Rule

Titius-Bode Rule (AU)	Actual (AU)	Planet
$0.4 + 0.0 = 0.4$	0.39	Mercury
$0.4 + 0.3 = 0.7$	0.72	Venus
$0.4 + 0.6 = 1.0$	1.00	Earth
$0.4 + 1.2 = 1.6$	1.52	Mars
$0.4 + 2.4 = 2.8$	??	??
$0.4 + 4.8 = 5.2$	5.20	Jupiter
$0.4 + 9.6 = 10.0$	9.58	Saturn
$0.4 + 19.2 = 19.6$	??	??

thus it is frequently referred to as "Bode's law." As shown in Table 11.1, the Titius–Bode rule predicted the existence of planets at $a = 2.8$ AU, between Mars and Jupiter, and at $a = 19.6$ AU, beyond Saturn.

The Titius–Bode rule was regarded as a mildly intriguing bit of numerology until the year 1781, when the planet Uranus was discovered. The semimajor axis of Uranus's orbit turned out to be $a = 19.18$ AU; this is close enough to the predicted value, $a = 19.6$ AU, that astronomers started to take the Titius–Bode rule more seriously. In particular, they started an organized search for a planet with $a = 2.8$ AU, in the gap between Mars and Jupiter. A group of two dozen astronomers, who called themselves the "Celestial Police," divided the ecliptic into sectors 15° across; each astronomer would search his assigned region for the missing planet. They realized that the planet, since it had not been previously seen, was likely to be very small. However, even if it were not resolved by their telescopes, they could identify it by its motion relative to the background stars.

Despite their careful planning, the Celestial Police were "scooped" by an astronomer making a serendipitous discovery. Giuseppe Piazzi, director of the Palermo Observatory, was making a highly precise star chart. To reduce observational error, he observed each star multiple times. In 1801 January, he found that one of his "stars" had moved from one night to the next. The newly discovered object, which Piazzi named Ceres, after the Roman goddess of agriculture,[1] is on an orbit with $a = 2.77$ AU, reasonably close to the Titius–Bode prediction. Piazzi had no doubt that he had discovered a planet, albeit a small one. Ceres, it turns out, is an approximately spherical body with radius $R = 480$ km, about 20% of the radius of Mercury.[2]

Things became more interesting, however, when the Celestial Police went into action and discovered additional objects in the gap between Mars and Jupiter. During the course

[1] "Ceres" is pronounced somewhat like "series," with two syllables.

[2] As a historical footnote, the Titius–Bode rule fell out of favor when Neptune was discovered at $a = 30.1$ AU, whereas the value predicted by the Titius–Bode rule was $a = 38.8$ AU. By then, however, it had already played its historical role of stimulating the search for planets.

of seven years, astronomers discovered Pallas ($a = 2.77$ AU), Juno ($a = 2.67$ AU), and Vesta ($a = 2.36$ AU). This new class of relatively small objects on closely spaced orbits deserved, it was thought, a new name. William Herschel proposed the name **asteroid**, meaning "starlike," since he could barely resolve them even in his largest telescope; they have also been called **minor planets**, as a reference to their small size compared to the terrestrial and Jovian planets.

There are currently over 14,000 asteroids whose orbits have been accurately determined and which have therefore been assigned names. (The list of Roman deities has been pretty much exhausted, so the discoverer of an asteroid is permitted to select any name of 16 characters or fewer that has not yet been used—subject to the constraints of good taste, as interpreted by the Committee for Small Body Nomenclature of the International Astronomical Union.) The position of the larger asteroids at one arbitrary moment in time is shown in Figure 11.1. The great majority of asteroids have orbits with $1.8\text{ AU} < a < 3.3\text{ AU}$. However, as seen in the figure, some asteroids have orbits that take them inside the orbit of the Earth. Earth-crossing asteroids with $a > 1$ AU, but with a perihelion inside the Earth's orbit, are called **Apollo asteroids**; there are currently over 2800 Apollo asteroids known. Earth-crossing asteroids with $a < 1$ AU, but with an aphelion outside the Earth's orbit, are called **Aten asteroids**; there are fewer than 500 Aten asteroids known. Most of the known Apollo and Aten asteroids are tiny; the largest, named Sisyphus, is about 10 km across. However, even a small asteroid can gouge out a big hole if it strikes the Earth, which accounts for the careful way in which astronomers have been cataloging near-Earth asteroids. At present, no known asteroids are in an orbit that poses any danger to life on Earth.

Typical orbital eccentricities for asteroids lie in the range $0.05 < e < 0.3$, while typical inclinations are $0° < i < 30°$. The relatively high inclinations of asteroidal orbits result in an asteroid belt that is not a thin disk, like Saturn's rings, but is toroidal.

A plot of the distribution of orbital semimajor axis length (Figure 11.2) reveals that the distribution of a for asteroidal orbits is not a smooth function. Instead, there are gaps in the distribution at certain values of the semimajor axis a; these gaps are known as the **Kirkwood gaps**, after the astronomer who first noted their existence.[3] The main Kirkwood gaps are at values of a that correspond to orbital resonances with Jupiter. The planet Jupiter has $a = 5.20$ AU, and thus, from Kepler's third law, a sidereal orbital period of $P = 11.86$ yr. If an asteroid had $a = 3.28$ AU, it would have $P = 5.93$ yr, and thus would be in a 2:1 orbital resonance with Jupiter, just as a ring particle in the Cassini division (see Section 10.3) would be in a 2:1 orbital resonance with the satellite Mimas. Just as the resonance with Mimas clears particles out of the Cassini division, the resonance with Jupiter clears asteroids out of orbits with $a = 3.28$ AU. Additional Kirkwood gaps are seen at semimajor axes where the orbital period is commensurate with the orbital period of Jupiter (see Figure 11.2).

Some asteroids are in orbits that are nearly identical to that of Jupiter, except that they lead or follow Jupiter by 60° (see Figure 11.1). These asteroids are known as

[3] The Kirkwood gaps are not apparent in a plot of asteroid positions in space (see Figure 11.1), because the relatively large eccentricity of asteroidal orbits smears out the gaps.

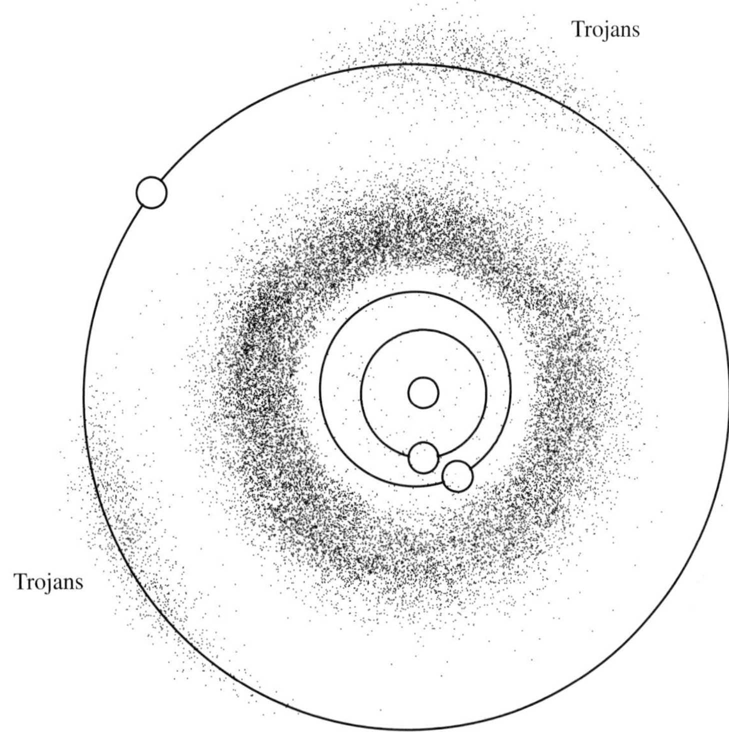

FIGURE 11.1 Location of asteroids with well-determined orbits, at a particular instant in time (which happens to be the summer solstice of the year 2003). The orbits and positions of Earth, Mars, and Jupiter are also shown. Note that the Trojan asteroids are in approximately the same orbit as Jupiter, but leading or following Jupiter by $\sim 60°$.

the **Trojan asteroids**.[4] The Trojan asteroids reside near two of the five **Lagrangian points** in the Sun–Jupiter system. In the eighteenth century, the mathematician Joseph Lagrange demonstrated that if two bodies of dissimilar mass (the Sun and Jupiter, for instance) are on circular orbits about their center of mass, there exist five points where the combined gravitational acceleration of the Sun and Jupiter is exactly that required to keep a small test body on a circular orbit co-rotating with the Sun–Jupiter line. As shown in Figure 11.3, three of these Lagrangian points, called L_1, L_2, and L_3, are colinear with the Sun and Jupiter. The L_1, L_2, and L_3 points are unstable equilibrium points. For instance, if we take a particle at the L_1 point (between the Sun and Jupiter) and move it slightly, it

[4] The first Trojan asteroid to be discovered was named Achilles; when further asteroids on similar orbits were discovered, they were given names taken from the *Iliad*. This explains the otherwise puzzling name "Trojan" asteroid.

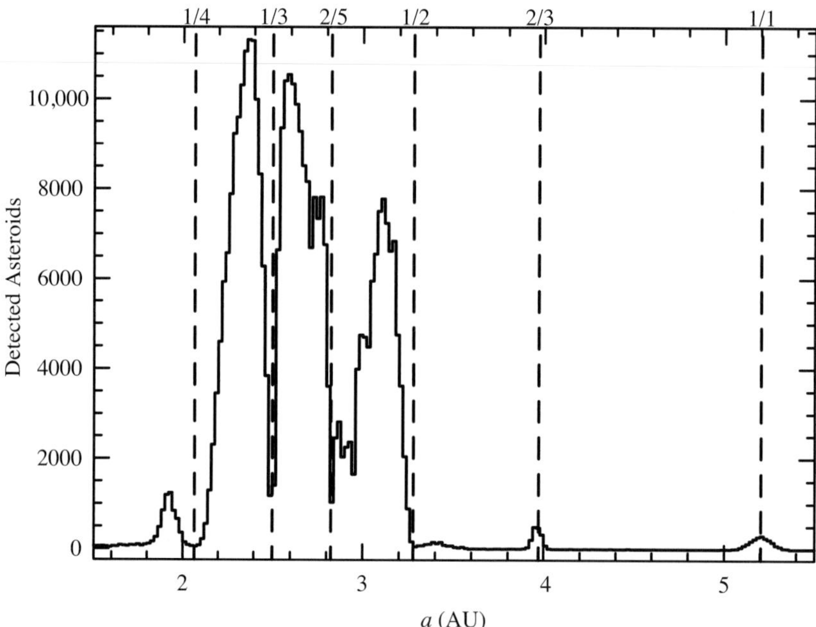

FIGURE 11.2 The number of detected asteroids as a function of semimajor axis. The vertical dashed lines are labeled with the orbital period, expressed as a fraction of Jupiter's orbital period.

will continue to drift steadily away from the L_1 point. The L_4 point, leading Jupiter on its orbit, and the L_5 point, trailing Jupiter on its orbit, can be shown to be *stable* equilibrium points. That is, if a particle is slightly displaced from the L_4 point or the L_5 point, it will oscillate around that point and not drift away.

Asteroids come in a variety of sizes. Ceres, with a diameter $d = 960$ km, is by far the largest asteroid. There are about 200 more asteroids that have $d > 100$ km, and an additional 500 that have 50 km $< d <$ 100 km. Most asteroids are much smaller than 50 km across, so they are faint and difficult to detect from Earth. It is estimated that there exist roughly 100,000 asteroids with $d > 1$ km. Given the large volume of the asteroid belt, this implies that the mean distance between asteroids is much larger than the size of an asteroid.[5] If you parked your spaceship next to an asteroid, the nearest neighboring asteroid more than 1 km across would be at a distance of a few million kilometers (roughly 10 times the Earth–Moon distance).

The brightness of asteroids commonly varies in a complex, periodic fashion, indicating that they have irregular shapes and are rotating (usually with periods of several hours). A few asteroids have been observed by spacecraft passing through the asteroid belt. For

[5] This is quite different from the situation within Saturn's rings, where the distance between ring particles is not vastly larger than the size of a ring particle.

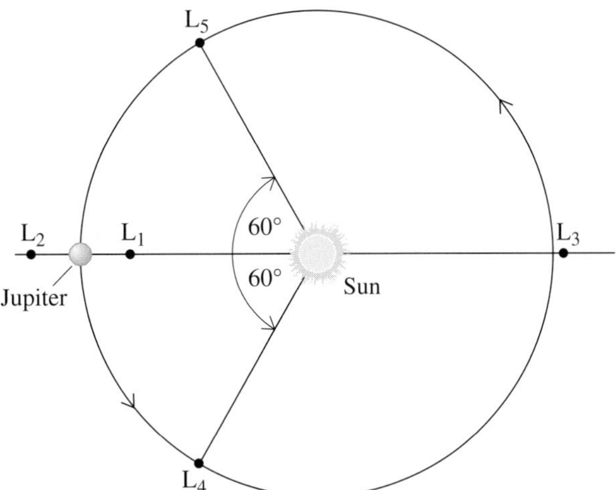

FIGURE 11.3 The Lagrangian points of the Sun–Jupiter system.

instance, the asteroid Gaspra (Color Figure 11) was studied by the *Galileo* spacecraft on its way to Jupiter. Given the sparse distribution of asteroids within the belt, they do not pose a major navigational hazard to spacecraft bound to the outer solar system. In fact, it was quite a feat to plot a trajectory that took *Galileo* within 5000 km of Gaspra and still got it to its rendezvous with Jupiter. The irregular, crater-pocked appearance of Gaspra is typical of those asteroids that have been imaged at high resolution.

In addition, the *Near Earth Asteroid Rendezvous* (*NEAR*) spacecraft was specifically designed, as its name implies, to study asteroids. First, *NEAR* made a flyby of the asteroid Mathilde; the surprisingly small gravitational force exerted by Mathilde led astronomers to deduce that its density was only $\rho \approx 1400$ km m^{-3}. Given the relatively dense material from which Mathilde is made, this means that the asteroid is a porous, loosely packed rubble pile. The *NEAR* spacecraft then flew on to Eros, going into orbit around that asteroid. The mean density of Eros is $\rho \approx 2400$ kg m^{-3}, indicating that it is much less porous than Mathilde. After a year in orbit, the *NEAR* spacecraft actually landed (more or less softly) on the surface of Eros. The closeup images sent back by the spacecraft revealed that the surface was covered with small rocks and loose regolith, much like the surface of the Moon.

The optical and infrared spectra of asteroids reveal that some are rocky (like Eros), while others have a metallic surface. None have icy coatings (if they did, they would have a much higher albedo than they do). The total mass of all the asteroids combined is estimated to be $\sim 0.001 M_\oplus$.

11.2 ▪ TRANS-NEPTUNIAN OBJECTS

A trans-Neptunian object, or TNO, can be broadly defined as any object orbiting the Sun on an orbit larger than that of Neptune ($a > 30.1$ AU). Ever since the discovery of

Neptune in 1846, astronomers have looked for TNOs. In the early twentieth century, the astronomer Percival Lowell instituted a search for "planet X," using the facilities of his private observatory in Flagstaff, Arizona. It was realized that a planet farther away than Neptune would be faint, and would not necessarily be resolved in angular size. The basic methodology used was to take photographs of areas near the ecliptic that were close to opposition. By comparing photographs of the same area taken several nights apart, objects in motion relative to the background stars can easily be detected. The Earth's average orbital speed is $v_\oplus \approx 2\pi$ AU yr^{-1}. Since $v \propto r^{-1/2}$ for Keplerian motion, an object on a nearly circular orbit with semimajor axis a will have an orbital speed

$$v \approx 2\pi \text{ AU yr}^{-1} (1 \text{ AU}/a)^{1/2} . \tag{11.2}$$

This means that when a superior object is seen at opposition from the Earth, its motion relative to the Earth will be retrograde, with a speed

$$\Delta v \approx 2\pi \text{ AU yr}^{-1} \left[1 - (1 \text{ AU}/a)^{1/2}\right] . \tag{11.3}$$

Since the superior object will be at a distance $a - 1$ AU from the Earth when it is at opposition, its retrograde motion will then translate into an angular speed on the celestial sphere of

$$\omega \approx \frac{\Delta v}{a - 1 \text{ AU}} \approx 2\pi \text{ rad yr}^{-1} \left(\frac{1 \text{ AU}}{a}\right) \frac{1 - (1 \text{ AU}/a)^{1/2}}{1 - (1 \text{ AU}/a)} . \tag{11.4}$$

This reduces to

$$\omega \approx 2\pi \text{ rad yr}^{-1} \frac{1 \text{ AU}/a}{1 + (1 \text{ AU}/a)^{1/2}}$$
$$\approx 59 \text{ arcsec day}^{-1} \frac{1 \text{ AU}/a}{1 + (1 \text{ AU}/a)^{1/2}} . \tag{11.5}$$

Thus, objects beyond Neptune, with $a > 30.1$ AU, have an apparent angular speed $\omega < 1.6$ arcmin day^{-1} at opposition, and can easily be distinguished from asteroids at $a \approx 3$ AU, which have the much faster retrograde motion $\omega \approx 12$ arcmin day^{-1}.

In the year 1930, a young astronomer named Clyde Tombaugh compared photographs of a small region of Taurus that had been taken on January 23 and January 29 of that year. He found one faint unresolved point of light that had moved by ~ 9 arcmin between one photograph and the next. Further observations revealed that the newly discovered object was on an orbit with $a = 39.5$ AU, eccentricity $e = 0.249$, and inclination $i = 17°$. The object was given the name Pluto, after the Roman god of the underworld, and was declared to be the "ninth planet." It was an unusual planet, given its high orbital eccentricity and inclination, but it really didn't fit into any other category in use at the time. It lay far outside the asteroid belt, and it was significantly larger than Ceres, the biggest of the asteroids. (Modern measurements of the radius of Pluto yield $R_{\rm Plu} =$ 1160 km, more than twice the radius of Ceres.)

For nearly half a century, Pluto was the only trans-Neptunian object known. Then, in 1979, Pluto was discovered to have a satellite; it was given the name **Charon**, after

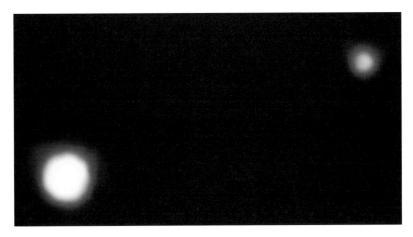

FIGURE 11.4 Pluto (left) and its satellite Charon (right), as imaged by *Hubble Space Telescope*.

the ferryman who transported dead souls into the realm of Pluto. Figure 11.4 is the best available image of the Pluto–Charon system, taken by the *Hubble Space Telescope*. The mass of Charon is 1/8 the mass of Pluto, making it the most massive satellite, relative to its parent body, in the Solar System. (For comparison, the Moon is 1/81 the mass of the Earth.) Pluto and Charon are separated by a distance of only 20,000 km, and orbit their center of mass with a period of $P = 6.4$ days. Pluto and Charon both show brightness variations with a period of $P = 6.4$ days, indicating that their rotation periods, as well as their orbital periods, are equal to 6.4 days. That is, both Pluto and Charon have been tidally braked until they are in synchronous rotation.

The discovery of two additional small satellites of Pluto in the year 2005 (named Nix and Hydra) enabled more accurate estimates of the mass, and thus the density, of Pluto and Charon. The mean density of Pluto is 2000 kg m^{-3}, consistent with its being 70% rock and 30% ice. The lower density of Charon, 1700 kg m^{-3}, indicates that it is probably half rock and half ice. Although the surfaces of Pluto and Charon have been observed with the *Hubble*, the limits on angular resolution mean that features smaller than a few hundred kilometers across cannot be seen.

Even at low resolution, however, it is apparent that Pluto has variations in its albedo. The darkest regions have $A < 0.5$, while the brightest regions have $A \approx 0.66$. This is surprisingly bright compared to other icy bodies in the outer solar system. Icy bodies darken over time because (among other causes) long exposure to solar UV radiation tends to darken methane ice. The explanation for the anomalous brightness of Pluto lies in its eccentric orbit. The surface temperature of Pluto at perihelion ($q = 29.6$ AU, $T = 51$ K) is considerably warmer than at aphelion ($Q = 49.2$ AU, $T = 40$ K). At perihelion, some of the atmosphere sublimes, giving Pluto a temporary methane atmosphere. As Pluto moves farther away from the Sun, however, the atmosphere freezes out again. This puts a fresh, highly reflective layer of methane snow on Pluto's surface once per orbit (every 248 years).

Pluto passed through perihelion in the year 1989. Because of the high eccentricity of Pluto's orbit, Pluto was actually closer to the Sun than Neptune was during the period 1979–1999. However, because of the high inclination of Pluto's orbit, the paths of Pluto and Neptune never intersect. When Pluto is at the same distance from the Sun as Neptune, it is several AU away from the plane of Neptune's orbit. In addition, Neptune, with orbital period $P_{\rm Nep} = 163.7$ yr, and Pluto, with orbital period $P_{\rm Plu} = 248.0$ yr, are nearly in a 3:2 orbital resonance. The synodic period of Pluto as seen from Neptune is (compare equation 2.10)

$$P_{\rm S} = \left(\frac{1}{163.7\ {\rm yr}} - \frac{1}{248.0\ {\rm yr}} \right)^{-1} = 481.7\ {\rm yr}, \tag{11.6}$$

which is close to three orbits of Neptune ($3 \times 163.7 = 491$ yr) and two orbits of Pluto ($2 \times 248.0 = 496$ yr). This means that as seen from Neptune, Pluto will be at opposition at approximately the same place in its orbit each time. At present, this happens to occur when Pluto is near aphelion; at aphelion, Pluto's distance from the Sun is $Q = 49.2$ AU, which means it is nearly 20 AU from Neptune.

As early as the mid-twentieth century, it was predicted that the region beyond Neptune should contain not just one or two ice-covered bodies, but a very large number. In 1951, Gerard Kuiper pointed out that large numbers of icy bodies condensed in the cold outer parts of the pre-solar nebula. Although many of those bodies were scattered by the gravitational influence of Neptune and the other Jovian planets, Kuiper predicted that some of the icy bodies should still remain in a circular ring beyond the orbits of Neptune and Pluto. As a consequence, the region close to the ecliptic plane at distances from ~ 30 AU to ~ 50 AU from the Sun is called the **Kuiper belt**.[6]

Small bodies beyond the orbit of Neptune are faint, and thus difficult to detect. The first trans-Neptunian object other than Pluto and Charon was detected in 1992. The object, named 1992 $\rm QB_1$, is on an orbit with $a = 43.7$ AU, $e = 0.065$, and $i = 2.2°$.[7] Thereafter, TNOs were discovered at an increasing rate; over a thousand TNOs were detected in the 15 years that followed the discovery of 1992 $\rm QB_1$.

The orbital parameters of the known trans-Neptunian objects are plotted in Figure 11.5. The upper panel shows orbital inclination i as a function of semimajor axis a; the lower panel shows orbital eccentricity e as a function of a. Uranus is the large dot at $a = 19.2$ AU and low inclination and eccentricity; Neptune is the large dot at $a = 30.1$ AU and low inclination and eccentricity. Figure 11.5 also includes a number of objects with orbits smaller than that of Neptune. These objects, which lie outside the orbit

[6] In fact, the presence of icy bodies in the belt beyond Neptune had previously been predicted by Kenneth Edgeworth as early as 1943. Thus, the alternate name "Edgeworth–Kuiper belt" is also used.

[7] Newly discovered minor bodies are given a name that consists of the year of discovery, a letter that tells the half-month during which it was discovered ("Q" indicates the period August 16–31), and another letter plus a subscripted number that tells the order of discovery in that half-month ("$\rm B_1$" indicates the 27th object discovered in a particular half-month).

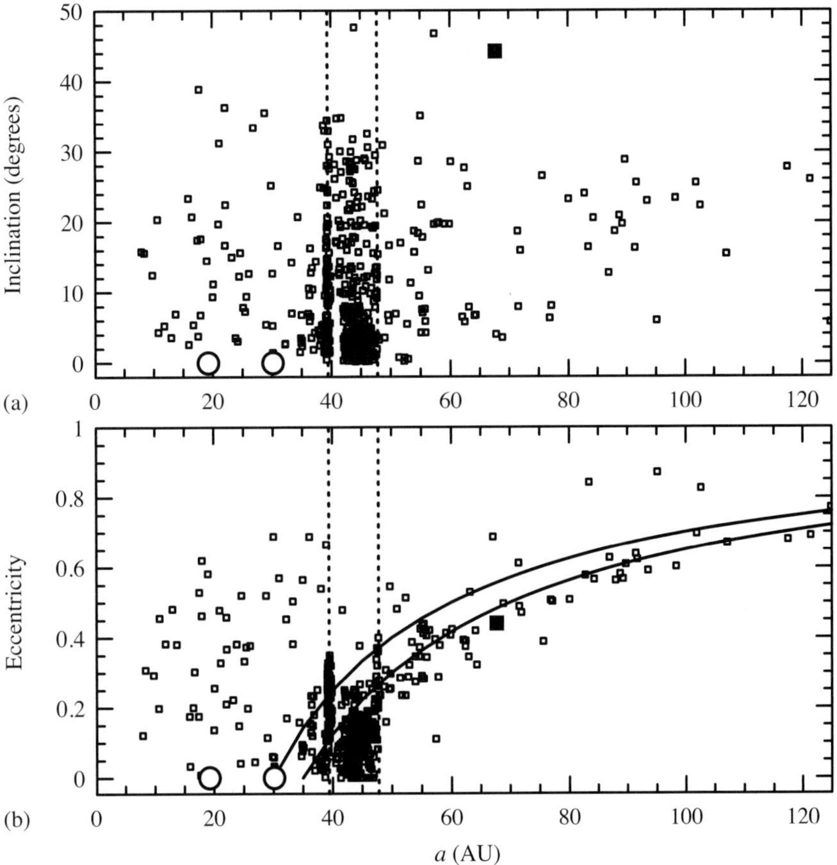

FIGURE 11.5 (a) Orbital inclination as a function of semimajor axis for all known TNOs. Vertical dotted lines are drawn at the 3:2 and 2:1 orbital resonance with Neptune. (b) Orbital eccentricity as a function of semimajor axis. Solid curved lines are drawn for orbits with perihelia of $q = 30$ AU and $q = 35$ AU.

of Jupiter but inside that of Neptune, are called **Centaurs**; this is because the first such object discovered was given the name Chiron, after the most famous of the mythological centaurs.

Notice, in Figure 11.5, that most of the trans-Neptunian objects have semimajor axes that place them in a narrow **classic Kuiper belt**, with $35 \text{ AU} \lesssim a \lesssim 50 \text{ AU}$. Most inclinations are relatively small, with the majority of TNOs having $i < 10°$, but this is partly a selection effect; the early searches for TNOs concentrated on the region near the ecliptic. Eccentricities have a fairly wide range; most TNOs have $0 < e < 0.3$, but a few have higher eccentricity. Notice that there is a large excess of TNOs with $a \approx 39.5$ AU, similar to the semimajor axis of Pluto's orbit. These objects, called **plutinos**, after their prototype, are all in a 3:2 orbital resonance with the planet Neptune. In addition, there is a smaller, but still significant, excess of TNOs with $a \approx 47.8$ AU. These bodies have sidereal orbital periods of $P \approx 330$ yr, and thus are in a 2:1 orbital resonance with Neptune. These objects have been given the name of **twotinos** (rhymes with "plutinos").

Interestingly, there are very few low-eccentricity TNOs beyond the twotinos. Most known TNOs with $a > 48$ AU have $e > 0.2$. In fact, as shown in Figure 11.5b, most of these distant TNOs have a combination of a and e that places their perihelion in the range 30 AU $\lesssim q \lesssim$ 40 AU, close to the orbit of Neptune. These distant, high-eccentricity TNOs are called **scattered disk objects**; the most plausible explanation for their orbits is that they were originally on low-eccentricity orbits within the classical Kuiper belt but were scattered onto larger, high-eccentricity orbits by their gravitational encounters with Neptune.

The TNOs have a range of sizes; at present, the largest known TNO is the scattered disk object Eris. The radius of Eris is $R = 1200$ km, which means it is a bit larger than Pluto. Eris was discovered in 2005 January, from images taken in 2003 October. In addition to being large, Eris also has a remarkably high albedo ($A = 0.86$). The reason why it was not discovered until 75 years after the discovery of Pluto is that it is very distant. In Figure 11.5, Eris is represented by the filled square at $a = 67.7$ AU, $i = 44.2°$, and $e = 0.44$. At the moment, Eris is close to its aphelion distance of $Q = a(1 + e) =$ 97.5 AU. This means that Eris is both very faint and very slowly moving as seen from Earth, which added to the difficulty of its detection.

The third largest known TNO, named Makemake, is significantly smaller than Eris or Pluto; its radius is $R \approx 750$ km.[8] However, its albedo ($A \approx 0.8$) is nearly as large as that of Eris, and its orbital size ($a = 45.8$ AU) places it significantly closer to us than Eris currently is. This means that the flux of Makemake is six times that of Eris. In fact, Clyde Tombaugh would have been able to detect it back in the 1930s, were it not for the unfortunate coincidence that Makemake was then passing close to the Milky Way, which provided a bright background against which Makemake could not be detected.

The discovery of hundreds of trans-Neptunian objects, one of them larger than Pluto, caused astronomers to ask both the specific question, Is Pluto a planet? and the more general question, What is a planet? In 2006, the International Astronomical Union (IAU) approved a definition of the word "planet" that seemed, to the majority of astronomers who voted on the issue, the most useful way of defining that word. According to the IAU definition, an object within the solar system is a planet if

1. It is in orbit around the Sun, and is not a satellite of another planet.
2. It has sufficient mass for its self-gravity to overcome its compressional strength, and thus assume a spherical, or spheroidal, shape in hydrostatic equilibrium.
3. It has cleared its orbital neighborhood (this criterion is sometimes referred to as "orbital dominance").

[8] "Makemake" is pronounced with four syllables: "Mah′-kay-mah′-kay". In the mythology of the Rapanui (the Polynesian inhabitants of Easter Island), Makemake is the creator of the human race.

COLOR PLATES

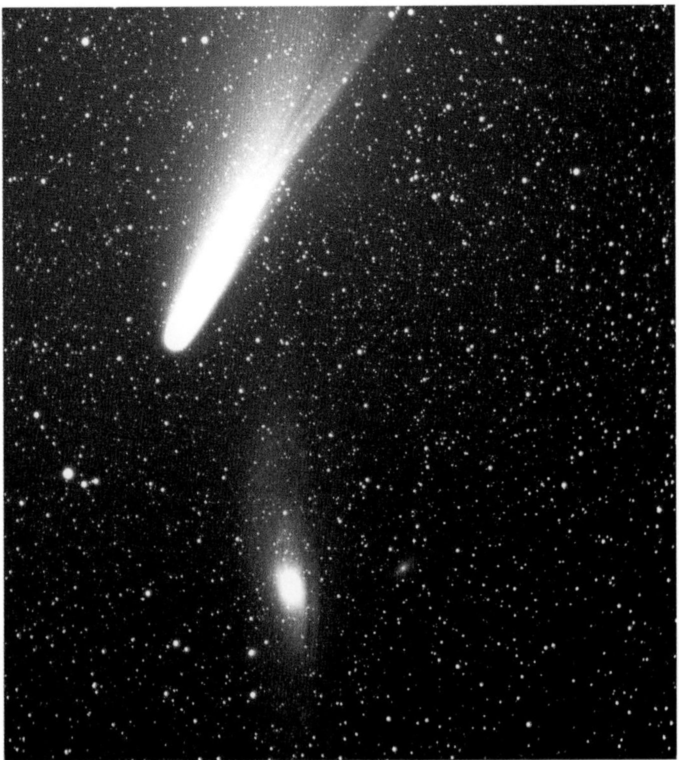

Color Figure 1: Comet Ikeya-Zhang (above) and the Andromeda Galaxy (below).

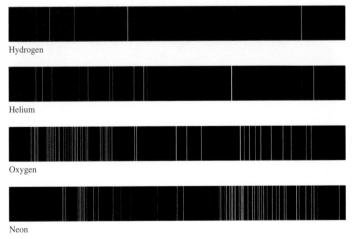

Color Figure 2: The emission spectrum of (top to bottom) hydrogen, helium, oxygen, and neon at visible wavelengths.

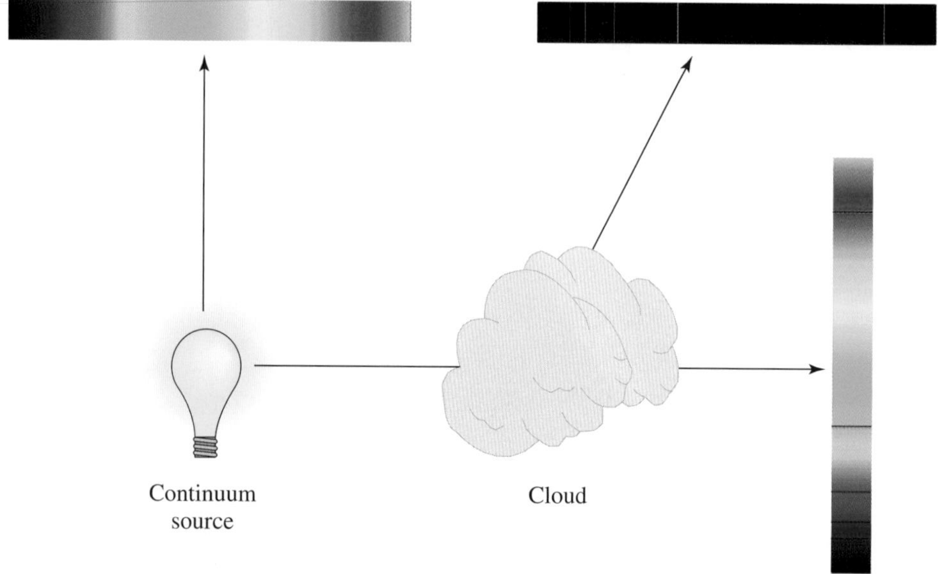

Color Figure 3: Kirchhoff's laws for the production of continuous spectra (upper left), emission spectra (upper right), and absorption spectra (lower right).

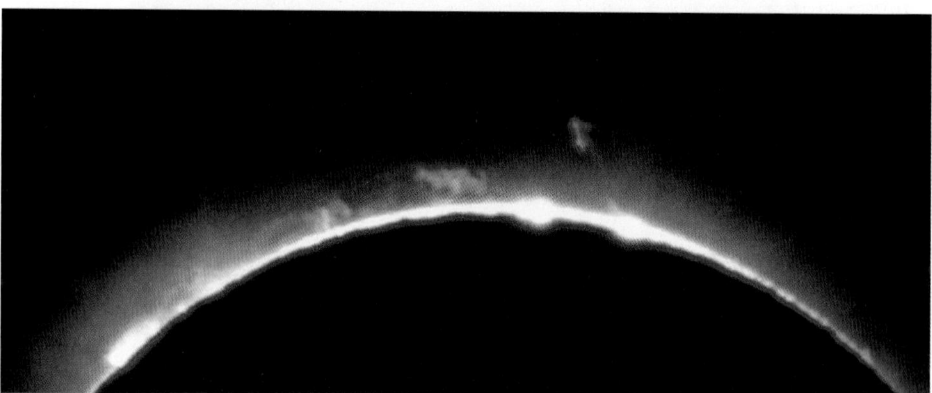

Color Figure 4: The brightly colored chromosphere of the Sun, seen during a total solar eclipse.

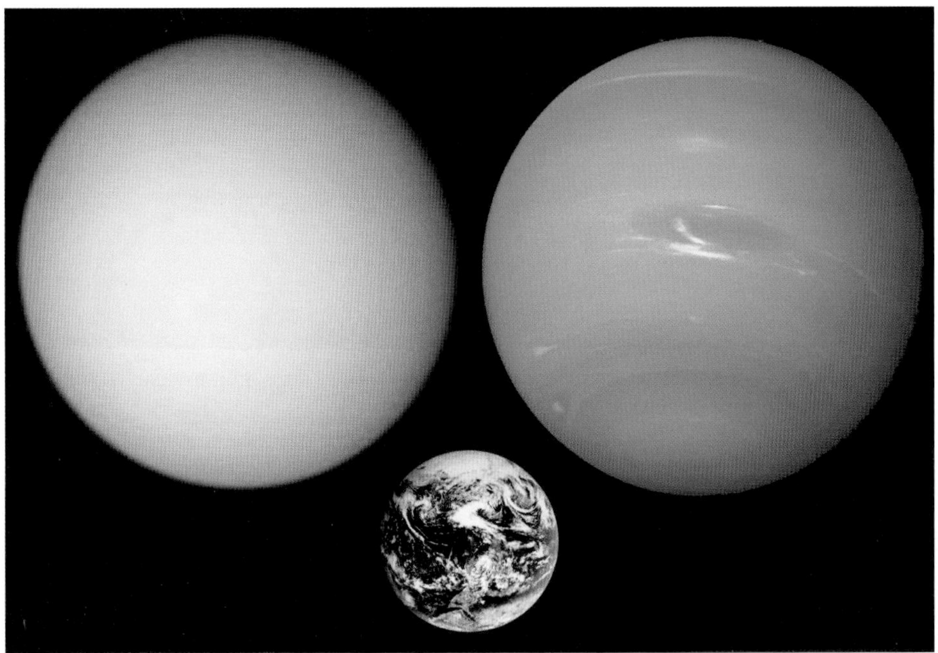

Color Figure 5: Uranus (left) and Neptune (right), with the Earth shown on the same scale. Notice the decidedly blue color of Uranus and Neptune.

Color Figure 6: Aurora australis seen from above, by an observer in the International Space Station ($h \approx 400$ km).

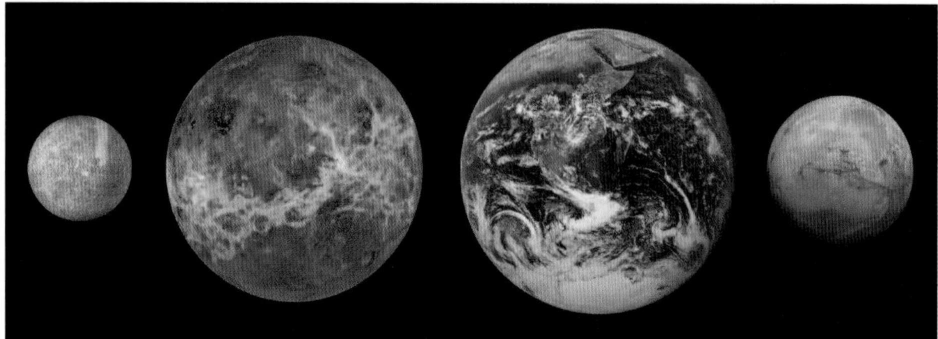

Color Figure 7: The terrestrial planets, shown at correct relative sizes. Left to right: Mercury, Venus (seen in a cloud-piercing radar image), Earth, and Mars.

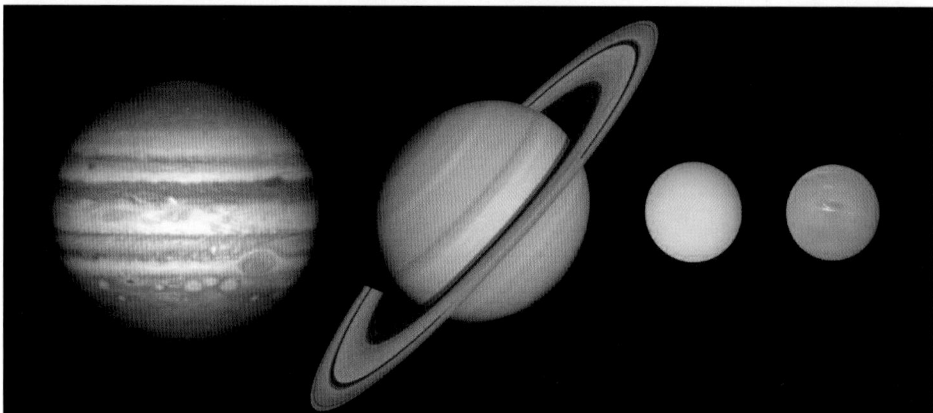

Color Figure 8: The Jovian planets, shown at correct relative sizes. Left to right: Jupiter, Saturn, Uranus, and Neptune.

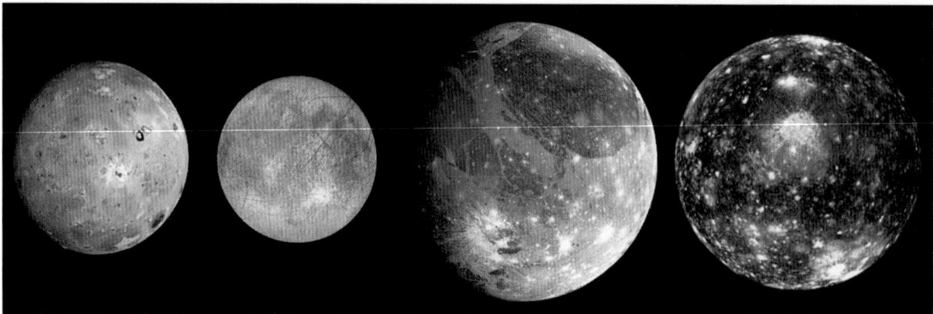

Color Figure 9: The Galilean satellites of Jupiter, shown at correct relative sizes. Left to right: Io, Europa, Ganymede, and Callisto.

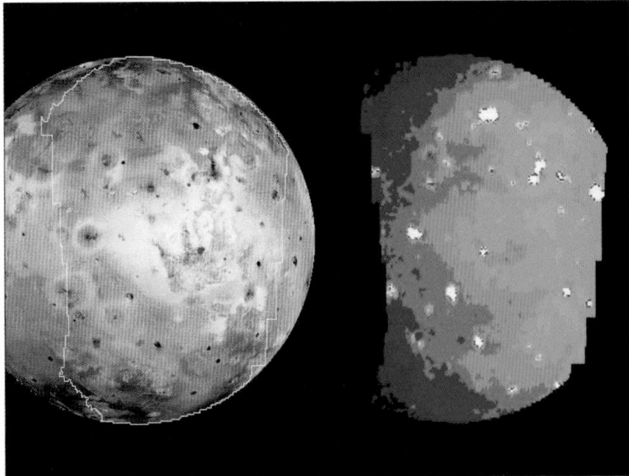

Color Figure 10: The satellite Io seen at visible wavelengths (left) and at the infrared wavelength $\lambda = 5\ \mu\text{m}$ (right). In the false-color infrared image, white indicates the highest temperature, and dark blue the lowest.

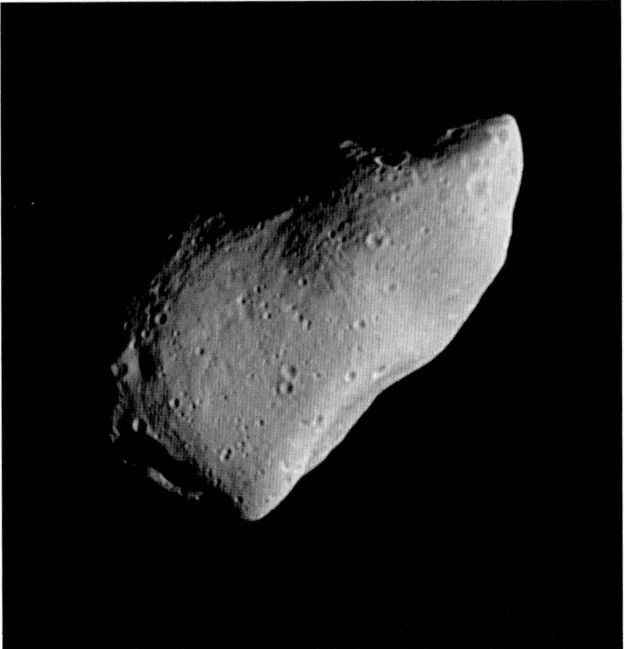

Color Figure 11: Gaspra, an irregular asteroid with axes of length $19 \times 12 \times 11$ km, as imaged by the spacecraft *Galileo* on its way to Jupiter. Colors are exaggerated to enhance small differences in surface composition.

Color Figure 12: The two tails of Comet Hale-Bopp. The ion tail (top) has an emission spectrum, and appears blue. The dust tail (bottom) reflects sunlight, and appears yellow.

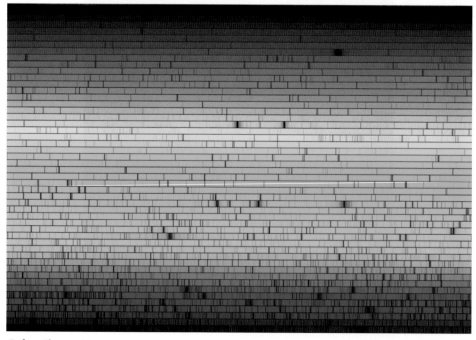

Color Figure 13: A high-resolution spectrum of the Sun at visible wavelengths.

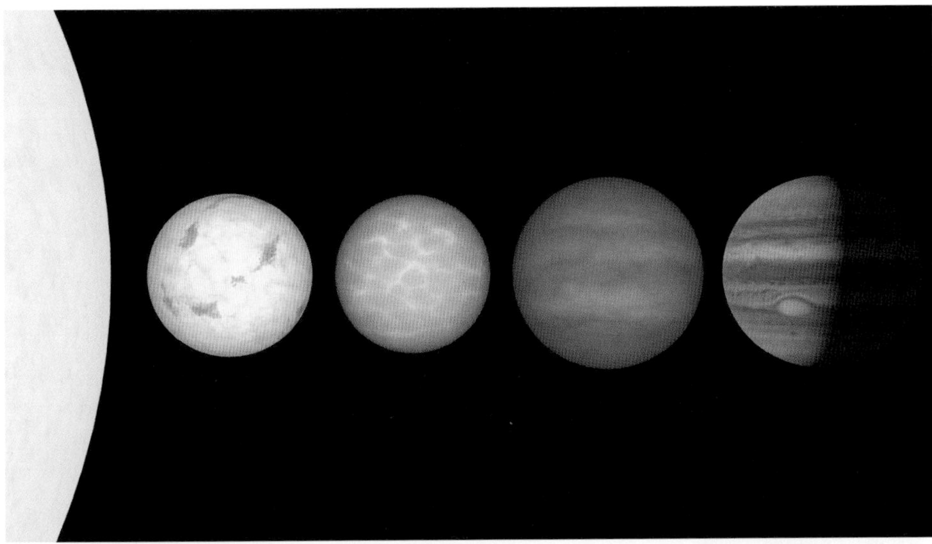

Color Figure 14: False-color image of (left to right) the Sun, an M dwarf, an L dwarf, a T dwarf, and Jupiter, as seen at near infrared wavelengths.

Color Figure 15: The Orion Nebula, an emission nebula at a distance $d \approx 400$ pc from the Sun. This *Hubble Space Telescope* image shows a region about 0.5 degrees across on the sky.

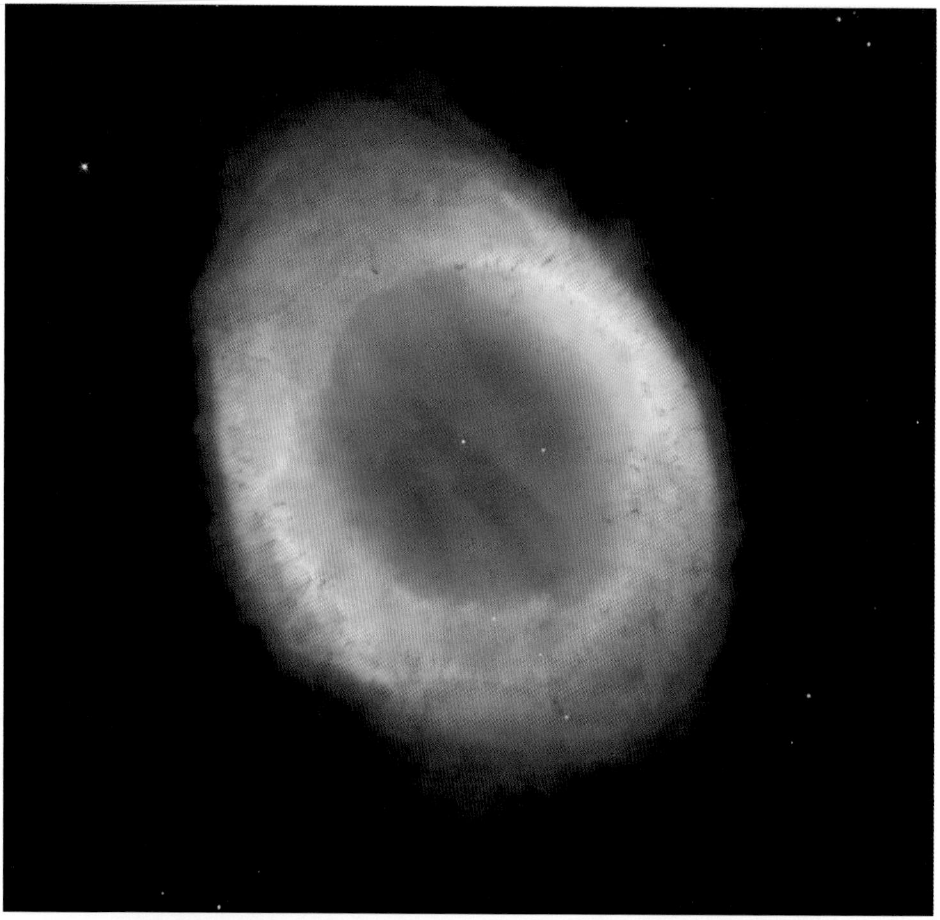

Color Figure 16: The Ring Nebula, a planetary nebula at a distance d $\approx$ 700 pc from the Sun in the constellation Lyra.

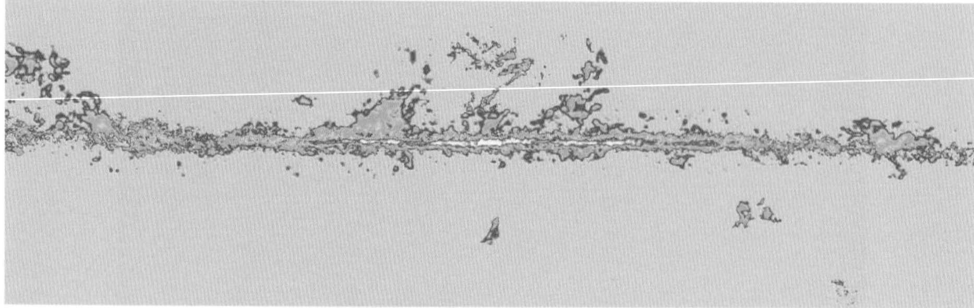

Color Figure 17: False-color map of the CO emission from the Milky Way. Red indicates the highest surface brightness, and blue the lowest.

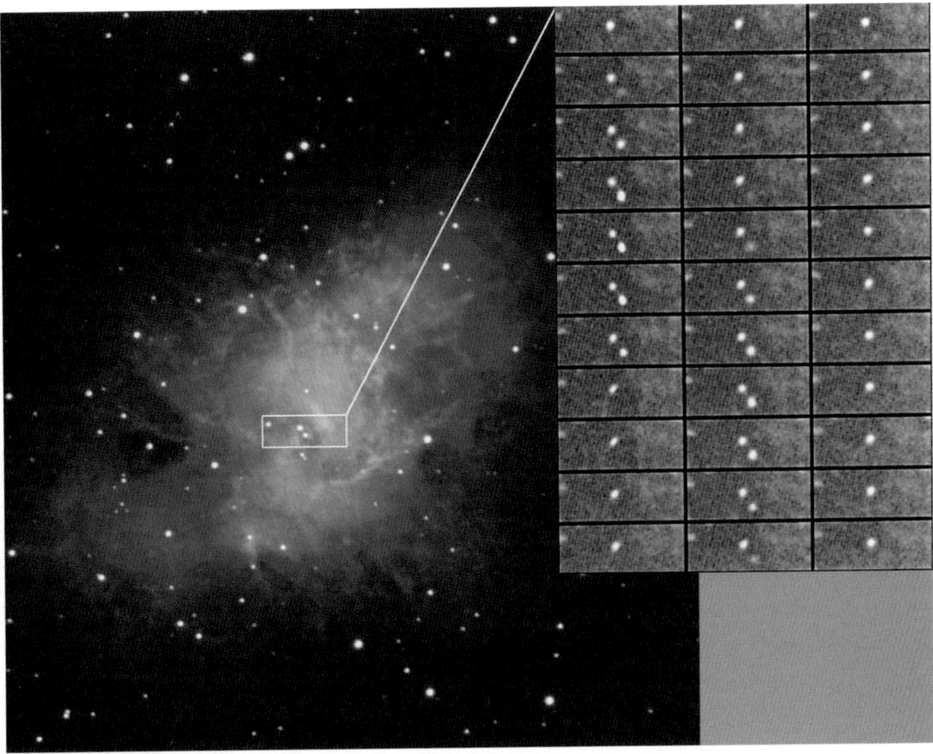

Color Figure 18: The Crab Nebula, a supernova remnant at a distance $d \approx 2$ kpc from the Sun. The panels at upper right show images of its pulsar taken 1 millisecond apart.

Color Figure 19: The Milky Way as seen at infrared wavelengths; notice the central bulge.

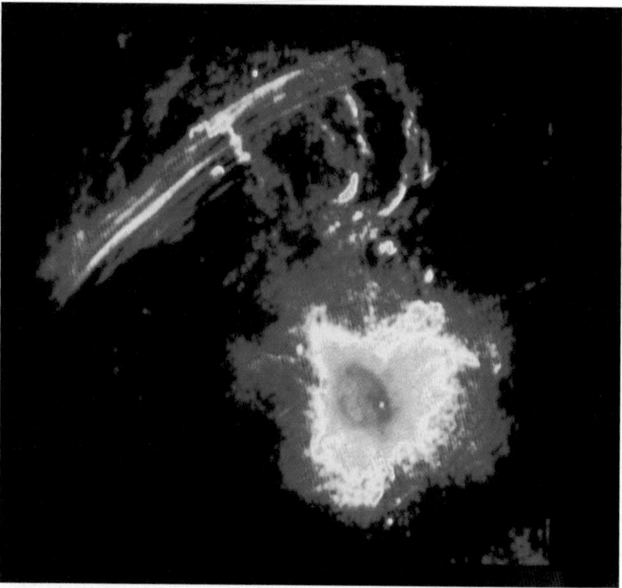

Color Figure 20: Sagittarius A, at a wavelength $\lambda = 20$ cm. This image shows a region 25 arcmin across, corresponding to a length scale $d \sim 60$ pc at the Galactic center. In this false-color image, red indicates the highest surface brightness.

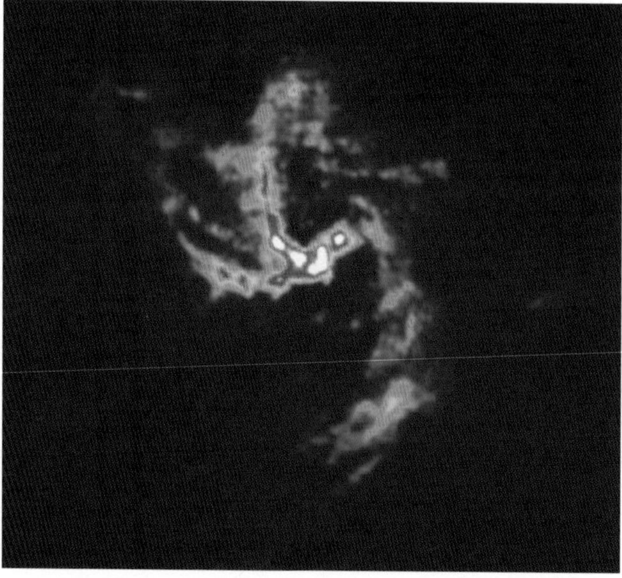

Color Figure 21: Sagittarius A West, at a wavelength $\lambda = 6$ cm. This image shows a region 3 arcmin across, corresponding to a length scale $d \sim 7$ pc at the Galactic center. In this false-color image, white indicates the highest surface brightness.

Color Figure 22: The *Hubble* Ultra Deep Field; the region of the sky shown here is approximately 3 arcmin on a side.

Color Figure 23: The galaxy M82 (at a distance $d \approx 3.5$ Mpc), seen at wavelengths ranging from X-ray (upper left), through visible (upper right) and infrared (lower right) to radio (lower left).

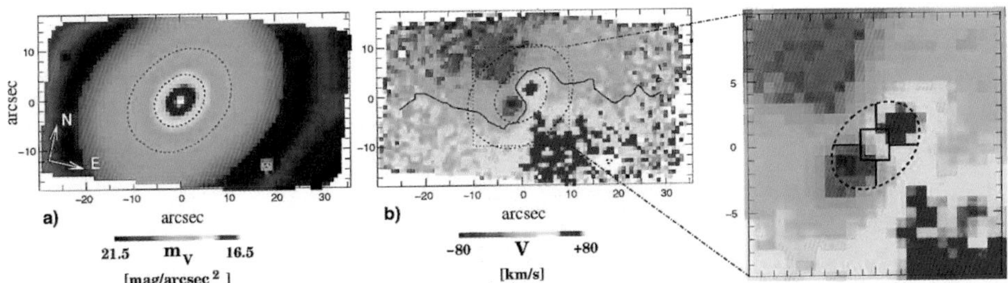

Color Figure 24: Surface brightness (left) and mean radial velocity (center) of the galaxy NGC 4365 (at a distance $d \approx 16$ Mpc). Note that the core (right inset) is rotating in a different direction from the rest of the galaxy.

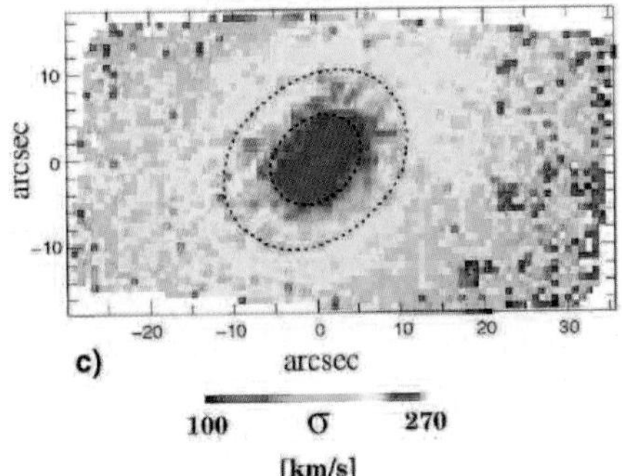

Color Figure 25: Line-of-sight velocity dispersion σ in the galaxy NGC 4365.

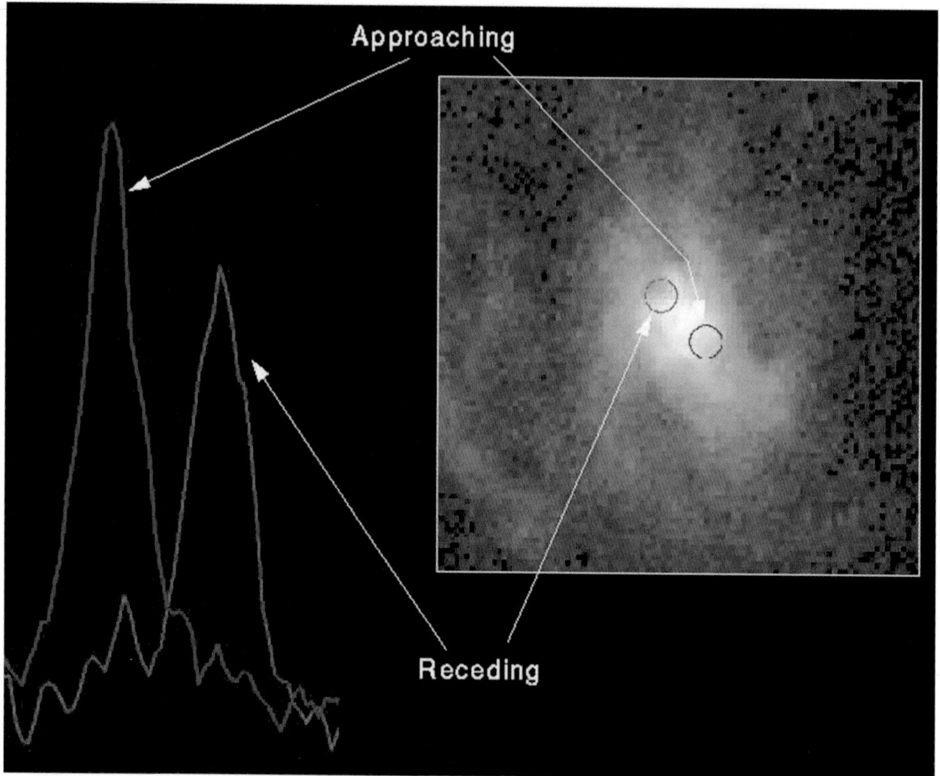

Color Figure 26: Gas disk at the center of the elliptical galaxy M87 (right), with Doppler shift information (left).

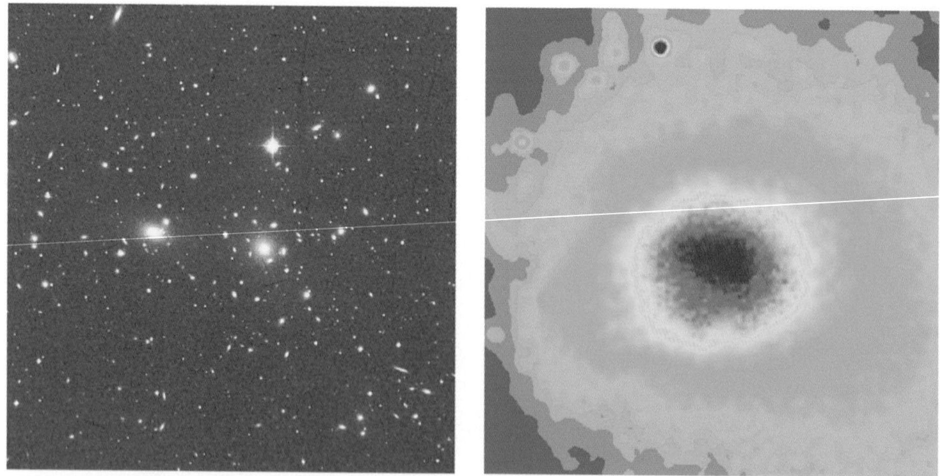

Color Figure 27: Left: The Coma Cluster of galaxies (at a distance $d \approx 100$ Mpc) seen at visible wavelengths. Right: The same cluster at X-ray wavelengths (false-color image, with red indicating the highest surface brightness).

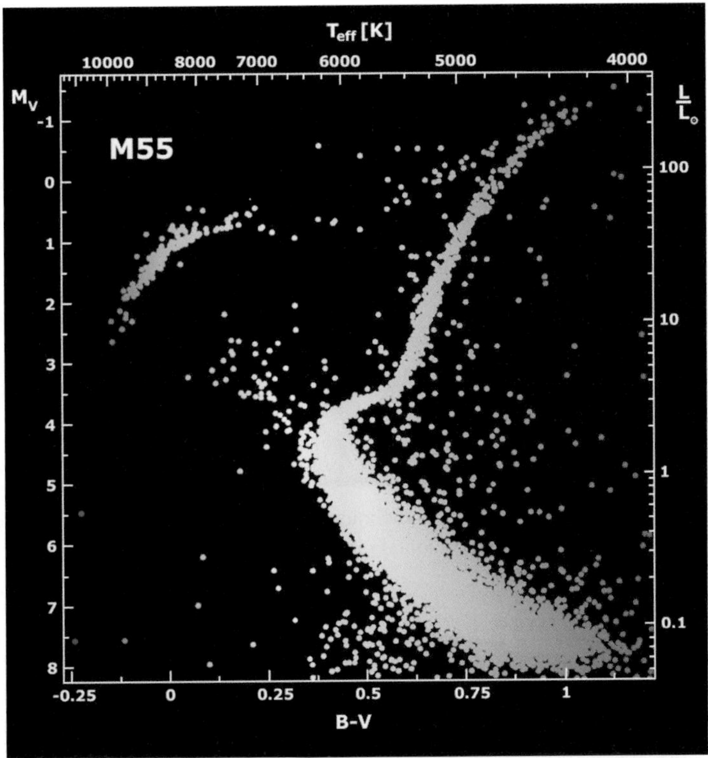

Color Figure 28: A Hertzsprung-Russell diagram of the globular cluster M55 (at a distance $d \approx 5$ kpc in the constellation Sagittarius). The blue straggler stars are circled.

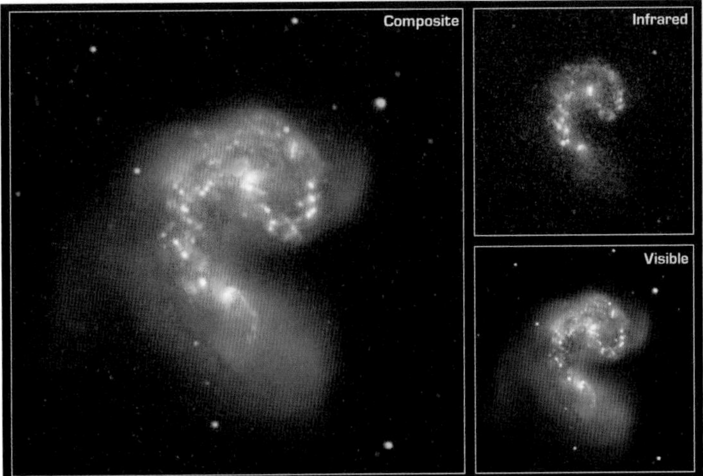

Color Figure 29: Infrared and visible images of the center of the Antennae, a pair of interacting galaxies at a distance $d \approx 14$ Mpc.

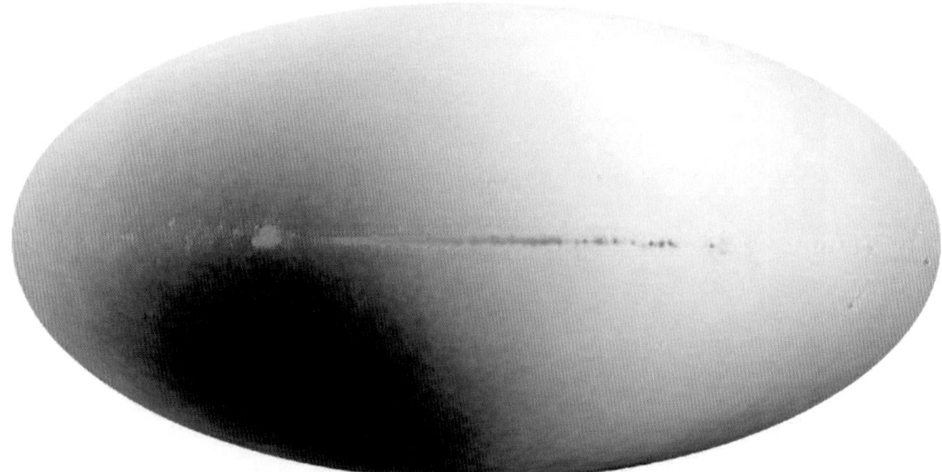

Color Figure 30: Temperature of the cosmic microwave background across the whole sky (black = cooler, white = hotter). The horizontal pink band is synchrotron (non-thermal) emission from the disk of our galaxy.

Color Figure 31: Temperature fluctuations of the cosmic microwave background, after removal of the dipole temperature distortion shown in Color Figure 30.

The four terrestrial planets and four Jovian planets satisfy all these criteria, and so they are labeled "planets." Objects that satisfy the first two criteria, but not the last, are called "dwarf planets." Eris, Pluto, Makemake, Haumea, and Ceres are dwarf planets. Ceres has an orbit similar to that of other asteroids but does not significantly affect their motions. Likewise, Pluto has an orbit similar to that of other plutinos but does not significantly affect their motions (in fact, all of the plutinos are dominated by Neptune). Eris has an orbit similar to that of other scattered disk TNOs but is not sufficiently massive to affect their motions.

11.3 ▪ COMETS

When we observe the asteroid belt, we can directly detect asteroids down to a size limit $d \sim 1$ km. We can deduce indirectly that smaller bodies exist in the asteroid belt because of the impact craters they create on larger asteroids (see, for instance, Color Figure 11). When we observe the Kuiper belt, because of its much greater distance, we can detect only TNOs larger than $d \sim 30$ km. Nevertheless, we have good reason to believe that smaller icy objects exist in trans-Neptunian space. Imagine an icy body ~ 1 km across in the Kuiper belt; if we moved such a body toward the inner solar system, then once it was within ~ 4 AU of the Sun, its ices would start to sublime. The resulting gases would form a large, tenuous atmosphere, or coma, that would be more easily visible than the tiny iceball at its center.

In fact, we have just described the behavior of a **comet**. The nucleus of a comet is an icy planetesimal left over from the formation of the solar system. Typical cometary nuclei are $1 \rightarrow 10$ km across, and have a mass of $10 \rightarrow 100$ billion tons. Cometary nuclei are often described as "dirty snowballs," since they consist of loosely packed ices, mixed with rocky material. Spacecraft have been sent to rendezvous with Comet Halley and with Comet Tempel 1 (Figure 11.6). Both these comets have nuclei that are cratered and irregular in shape. Their densities are quite low (as small as $\rho \sim 300 \text{ kg m}^{-3}$), indicating that the nuclei must be very loosely packed snowballs indeed. The surfaces have low albedos; the sublimation of ices has left a surface layer that consists of dark refractory rock.

If a cometary nucleus comes within ~ 4 AU of the Sun, the sublimation of ices produces a **coma** that may be as much as $\sim 10^5$ km in diameter. In addition, the sublimation releases chunks of rock and dust particles that were previously locked inside the ice. The gravitational force of the nucleus on the gas and dust of the coma is small. Thus, material from the coma is easily driven away by nongravitational forces. Ionized gas is swept away by the charged particles in the solar wind and forms a long, narrow **ion tail**. Tiny dust particles are pushed away from the coma by the gentler force of radiation pressure from sunlight; the dust grains form a broad, curved **dust tail**. Thus, comets generally have *two* tails (Color Figure 12), although from some viewing angles they are superimposed and appear as a single tail. In extreme cases, a comet's tails can be nearly 1 AU in length.

Comets can be divided into two classes, based on their orbital parameters. **Short-period comets** are those with $P < 200$ yr, and hence $a < 34$ AU. Short-period comets have other characteristics in common:

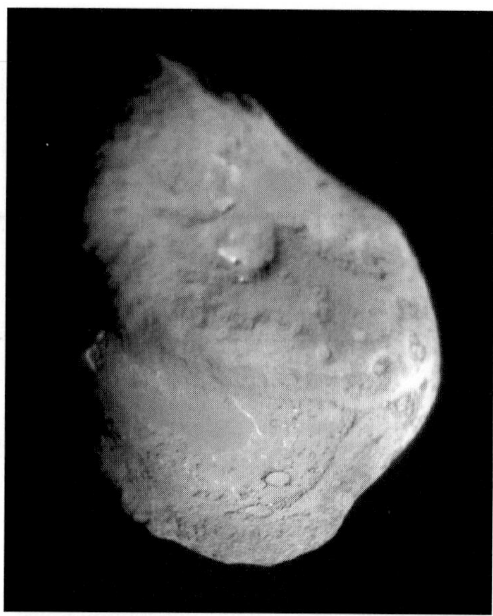

FIGURE 11.6 Composite image of the nucleus of Comet Tempel 1; the longest dimension of the nucleus is ~ 8 km.

- Most short-period comets have moderate inclination ($i < 30°$), though a few are retrograde ($i > 90°$).
- Most short-period comets have fairly large eccentricity ($e \sim 0.8$ is a typical value).

At present, there are roughly 200 short-period comets known. The most famous short-period comet is Comet Halley, with $P = 76$ yr; it is also one of the few retrograde short-period comets, with $i = 162°$. After a few hundred perihelion passes a comet tends to lose all its ices. Thus, after less than 10^5 years, a short-period comet becomes an inert pile of rubble. Given this brief time scale, there must exist a repository of cometary nuclei in the cold outer solar system, from which the supply of short-period comets is replenished. One obvious repository is the Kuiper belt, which is known to contain icy bodies. If short-period comets come from the Kuiper belt, this would explain why most of them have relatively small inclinations, which match the relatively small inclination of the majority of Kuiper belt objects (see Figure 11.5). But how can we explain the short-period comets that are on retrograde orbits? To explain the origin of retrograde short-period comets, it helps to consider the origin of long-period comets.

Long-period comets are defined as comets with $P > 200$ yr, and hence $a > 34$ AU. Some long-period comets have periods that are very long indeed, up to $P \sim 3 \times 10^7$ yr, implying $a \sim 100{,}000$ AU. Long-period comets have other characteristics in common, besides their long orbital periods:

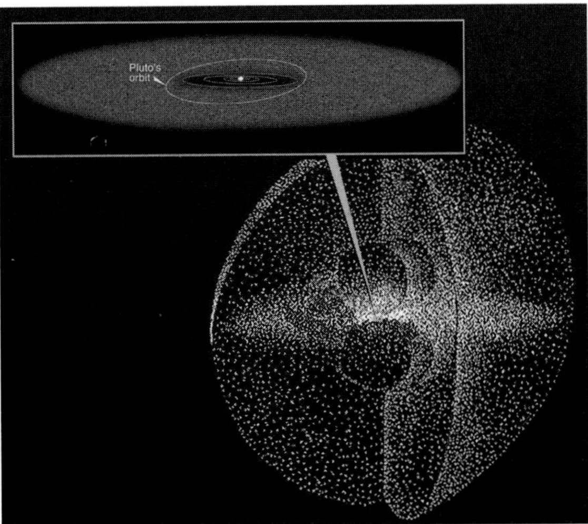

FIGURE 11.7 The Oort cloud is an extended distribution of cometary objects surrounding the Sun; the inner Oort cloud connects with the Kuiper belt.

- Long-period comets have random inclinations to the ecliptic, with as many being on retrograde orbits ($i > 90°$) as on prograde orbits ($i < 90°$).
- Long-period comets have extremely high orbital eccentricities ($e \approx 1$).

Over 800 long-period comets have been detected over the course of history. The Dutch astronomer Jan Oort proposed that long-period comets reside in a large, roughly spherical cloud of radius $\sim 50{,}000$ AU, centered on the Sun; this distribution of cometary bodies is called the **Oort cloud**. Figure 11.7 shows the relative sizes and shapes of the small, flattened Kuiper belt, and the large, quasi-spherical Oort cloud. The radius of the Oort cloud is not negligibly tiny compared to the distance of $\sim 270{,}000$ AU to the Sun's nearest neighboring star. The orbits of cometary bodies in the Oort cloud can be perturbed by passing stars; this alters their orbits and sometimes permits them to approach the inner solar system. Those few objects with perihelion distances $q \lesssim 4$ AU grow comas and tails and can be observed by us as comets. Retrograde short-period comets, such as Comet Halley, represent long-period comets that have had a close encounter with a Jovian planet that has flung them onto a smaller orbit.[9]

The chemical composition of cometary nuclei indicates that they were probably formed about $20 \rightarrow 30$ AU from the Sun, between the orbits of Uranus and Neptune, where the temperature of the pre-solar nebula was $T \sim 100$ K. If they had formed closer

[9] Close encounters of comets with Jovian planets are common enough to have been seen in modern times. Comet Shoemaker–Levy 9 was captured from its Sun-centered orbit onto a highly eccentric Jupiter-centered orbit. The fragile, low-density cometary nucleus was torn apart when it came within the Roche limit; on the next pericenter passage, in 1994 July, the fragments collided with the atmosphere of Jupiter.

to the Sun, they would not have been icy. If they had formed farther from the Sun, their water would have been locked up in solid hydrates of ammonia and methane. Since water vapor is detected in the spectra of comets, this clearly did not happen. The orbits of these small cometary bodies were easily altered by encounters with the massive Jovian planets. Thus, they were scattered onto a wide variety of different orbits, with a variety of ultimate fates:

- Comets on small orbits made frequent perihelion passages and sublimed early in the history of the solar system.
- Comets on large orbits constituted the Oort cloud.
- Comets on unbound (hyperbolic) orbits were able to escape the solar system altogether.

Those cometary bodies that suffered only minor perturbations to their initial orbit ended up in the Kuiper belt, a region of relative orbital stability outside the area dominated by the Jovian planets.

11.4 ▪ METEOROIDS AND DUST

In the standard scenario for the formation of the solar system, as outlined in Section 8.3, small, solid particles condensed from the pre-solar nebula. The small particles then accreted to form larger planetesimals, and the planetesimals coalesced to form planets. We know that the formation of planets from planetesimals was not completely efficient: leftover planetesimals are seen today, in the form of comets and asteroids. Similarly, we might expect that the formation of planetesimals from smaller particles was not completely efficient, leaving behind leftover dust grains and snowflakes in the solar system. However, as we shall see, nongravitational effects can remove small solid particles from the solar system.

The smallest solid particles can be removed by **radiation pressure**. The force of radiation acting on a particle is $F_{\rm rad} = P_{\rm rad}\sigma_{\rm pr}$, where $P_{\rm rad}$ is the pressure exerted by photons and $\sigma_{\rm pr}$ is the effective cross-section of the particle for interaction with photons. For a spherical dust grain of radius R, the cross-section can be written as $\sigma_{\rm pr} = Q_{\rm pr}\pi R^2$, where πR^2 is the geometrical cross-section, and $Q_{\rm pr} \leq 1$ is the radiation pressure coefficient. The value of $Q_{\rm pr}$ depends primarily on the size of the dust grain relative to the wavelength of the light acting on it. The spectrum of the Sun, as we have seen in Section 8.2, peaks at a wavelength $\lambda_{\rm p} \approx 5000$ Å. If a dust grain has $R \gg \lambda_{\rm p}$, then it has $Q_{\rm pr} = 1$; its cross-section for interactions with light is equal to its geometric cross-section. However, if the grain is tiny, with $R \ll \lambda_{\rm p}$, it will have $Q_{\rm pr} \ll 1$. If tiny grains interacted only by Rayleigh scattering, as described in Section 9.2, we would expect $Q_{\rm pr} \sim (R/\lambda_{\rm p})^4$ when $R \ll \lambda_{\rm p}$. Since dust grains can absorb as well as scatter light, the dependence of $Q_{\rm p}$ is different from that predicted by pure Rayleigh scattering. Empirically, it is found that dust grains made of rock or ice have $Q_{\rm pr} \propto R^2$ in the limit that $R \ll \lambda_{\rm p}$.

Since pressure is equivalent to momentum flux, and since the momentum of a photon is $p = E/c = h\nu/c$, the pressure of solar radiation is simply related to the energy flux:

$$P_{\rm rad} = \frac{1}{c}\frac{L_\odot}{4\pi r^2}, \tag{11.7}$$

where r is the distance of the particle from the Sun. Thus, radiation pressure will push the spherical particle away from the Sun, with a force

$$F_{\rm rad} = \frac{L_\odot}{4\pi r^2 c} Q_{\rm pr} \pi R^2 = \frac{L_\odot}{4c}\frac{R^2}{r^2} Q_{\rm pr}. \tag{11.8}$$

If the spherical particle has a density ρ, then its mass will be $m = 4\pi\rho R^3/3$, and the gravitational force exerted on the particle by the Sun will be

$$F_{\rm grav} = -\frac{GM_\odot m}{r^2} = -\frac{GM_\odot}{r^2}\frac{4\pi\rho R^3}{3}. \tag{11.9}$$

Since the radiative force and the gravitational force both follow inverse square laws, the ratio of radiative force to gravitational force will be independent of distance from the Sun:

$$\begin{aligned} \frac{F_{\rm rad}}{|F_{\rm grav}|} &= \frac{3}{16\pi}\frac{L_\odot}{GM_\odot c}\frac{Q_{\rm pr}}{\rho R} \\ &= 5.8 \times 10^{-7}\left(\frac{1000\,{\rm kg\,m^{-3}}}{\rho}\right)\left(\frac{1\,{\rm m}}{R}\right) Q_{\rm pr}. \end{aligned} \tag{11.10}$$

For small dust grains, with $R \ll \lambda_{\rm p}$, the ratio of radiative force to gravitational force increases linearly with R, since $Q_{\rm pr} \propto R^2$ in this range of sizes. For larger dust grains, with $R \gg \lambda_{\rm p}$, the same ratio decreases as $1/R$, since $Q_{\rm pr} = 1$ for these large grains. Thus, the ratio of radiative force to gravitational force has a maximum at a grain size $R \sim \lambda_{\rm p} \sim 5000$ Å.

Dust grains will be accelerated out of the solar system if the magnitude of the outward radiative force is greater than that of the inward gravitational force. This will occur if the grain radius R satisfies the criterion

$$R < 5800\,\text{Å}\left(\frac{1000\,{\rm kg\,m^{-3}}}{\rho}\right) Q_{\rm pr}. \tag{11.11}$$

Thus, unless dust grains are extremely low in density ($\rho \ll 1000\,{\rm kg\,m^{-3}}$), tiny dust grains, with $R \ll \lambda_{\rm p}$, will not be blown away; their values of $Q_{\rm pr}$ fall off too rapidly as R decreases. On the other hand, large dust grains, with $R \gg \lambda_{\rm p}$, will not be blown away either; their masses increase too rapidly as R increases. Thus, only those grains with $R \sim \lambda_{\rm p} \sim 5000$ Å will be accelerated out of the solar system by radiation pressure.

For dust grains with $R \gg 5000$ Å, where the Sun's gravity dominates over radiation pressure, a more subtle mechanism, the **Poynting–Robertson effect**, acts as a brake on orbiting particles, decreasing their angular momenta so that they slowly spiral into the Sun. In the frame of reference of an orbiting particle, the phenomenon of *aberration* (described in Section 2.6.2) causes a slight displacement in the direction of motion for photons striking the particle. Just as astronomers have to tilt their telescopes slightly in

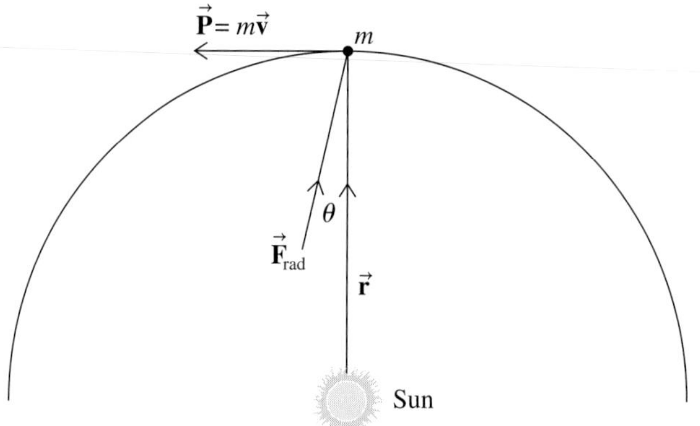

FIGURE 11.8 The Poynting–Robertson effect: an orbiting particle loses angular momentum since the aberrated sunlight exerts a force that is not exactly parallel to the particle–Sun line.

the Earth's direction of motion to capture photons from a distant star, a particle receives photons from the Sun in a direction that is slightly aberrated in the particle's direction of motion (Figure 11.8).

As the particle moves along its orbit about the Sun, its orbital angular momentum is $\vec{\mathbf{L}} = \vec{\mathbf{r}} \times \vec{\mathbf{p}}$, where the linear momentum of the particle is $\vec{\mathbf{p}} = m\vec{\mathbf{v}}$.[10] The torque exerted on the particle by radiation forces is

$$\vec{\boldsymbol{\tau}} \equiv \vec{\mathbf{r}} \times \vec{\mathbf{F}}_{\text{rad}} = \frac{d\vec{\mathbf{L}}}{dt}. \tag{11.12}$$

By referring to Figure 11.8, we see that

$$\tau = |\vec{\mathbf{r}} \times \vec{\mathbf{F}}_{\text{rad}}| = -rF_{\text{rad}} \sin\theta, \tag{11.13}$$

where θ is the angle of aberration. In Section 2.6.2, we learned that the angle of aberration is given by the relation $\tan\theta = v/c$, where v is the speed of the moving object. Since orbital motions in the solar system are nonrelativistic, the aberration angles are small, and we may write

$$\frac{v}{c} = \tan\theta \approx \sin\theta \approx \theta. \tag{11.14}$$

Therefore, the torque due to the radiation force of aberrated sunlight is

$$\tau \approx -rF_{\text{rad}}\theta \approx -rF_{\text{rad}}\frac{v}{c}. \tag{11.15}$$

[10] In this derivation, please distinguish carefully between L, the magnitude of the orbital angular momentum, and $L_\odot$, the luminosity of the Sun.

If we approximate our particle as a sphere of radius R, we can use the relation for radiation force that we found in equation (11.8) and write

$$\tau = -r\left(\frac{L_\odot}{4c}\frac{R^2}{r^2}Q_{\rm pr}\right)\frac{v}{c}, \tag{11.16}$$

or, since $\tau = dL/dt$,

$$\frac{dL}{dt} = -\frac{L_\odot}{4c^2}R^2 Q_{\rm pr}\frac{v}{r}. \tag{11.17}$$

This equation for angular momentum loss becomes particularly simple if we assume the particle is on a nearly circular orbit. In that case, $v = (GM_\odot/r)^{1/2}$, $L = mvr = m(GM_\odot r)^{1/2}$, and equation (11.17) becomes

$$m(GM_\odot)^{1/2}\frac{1}{2}r^{-1/2}\frac{dr}{dt} = -\frac{L_\odot}{4c^2}R^2 Q_{\rm pr}(GM_\odot)^{1/2}r^{-3/2}, \tag{11.18}$$

or

$$r dr = -\frac{L_\odot}{2c^2}\frac{R^2}{m}Q_{\rm pr}dt. \tag{11.19}$$

This equation can be integrated from a time $t = 0$, when the particle is at its initial distance from the Sun, $r = a$, to the time $t = t_{\rm pr}$ at which $r = 0$. The time $t_{\rm pr}$ is the **Poynting–Robinson timescale**, which tells us how long it takes for the particle to spiral in to the Sun's location. The integration of equation (11.19) yields

$$-\frac{a^2}{2} = -\frac{L_\odot}{2c^2}\frac{R^2}{m}Q_{\rm pr}t_{\rm pr}, \tag{11.20}$$

or

$$t_{\rm pr} = \frac{a^2c^2}{L_\odot}\frac{m}{R^2}\frac{1}{Q_{\rm pr}} = \frac{a^2c^2}{L_\odot}\frac{4\pi\rho R}{3}\frac{1}{Q_{\rm pr}}, \tag{11.21}$$

where we have used the expression $m = 4\pi\rho R^3/3$ for the mass of the spherical particle. Expressed numerically, the Poynting–Robertson timescale is

$$t_{\rm pr} \approx 0.7\,\text{Gyr}\left(\frac{a}{1\,\text{AU}}\right)^2\left(\frac{\rho}{1000\,\text{kg m}^{-3}}\right)\left(\frac{R}{1\,\text{m}}\right), \tag{11.22}$$

if we assume $R \gg \lambda_{\rm p} \sim 5000$ Å and hence $Q_{\rm pr} = 1$. If we assume a density $\rho \sim 3000\,\text{kg m}^{-3}$, comparable to that of rock, primordial chunks of matter less than 1 m across will be cleared away out to a distance $a \sim 1.5$ AU from the Sun. Primordial chunks of matter less than 1 cm across will be cleared away out to a distance $a \sim 15$ AU.

Despite the presence of radiation pressure and the Poynting–Robertson effect, small solid particles are known to exist in the inner solar system. At visible wavelengths, these particles manifest themselves as the **zodiacal light** (Figure 11.9), a faint glow of diffuse light centered on the ecliptic. The zodiacal light is sunlight that is backscattered from the small opaque particles that exist near the ecliptic plane. These particles are called

FIGURE 11.9 The diffuse glow of zodiacal light is seen near the horizon in this long exposure.

meteoroids.[11] The fact that meteoroids with $d \lesssim 10$ cm exist in the inner solar system implies that these objects must constantly be replenished. The sources of these small objects are comets and asteroids. Comets release small solid particles as their surfaces sublime. The smallest particles ($d \sim 0.5\,\mu$m) are blown away by radiation pressure from sunlight and form the comet's dust tail; larger particles travel on nearly Keplerian orbits about the Sun, gradually spiraling inward due to the Poynting–Robertson effect. Asteroids release small particles as the result of the collisions that create impact craters on their surfaces.

Occasionally, these meteoroids encounter the Earth and enter the Earth's atmosphere. In that case, friction with air molecules heats them to incandescence. The outer layer of the meteoroid vaporizes, leaving a trail of ionized atoms and molecules. These long, glowing streaks of ionized gas are called **meteors.**[12] At 1 AU from the Sun, a meteoroid can be traveling as fast as the solar escape speed, which is $v_{\rm esc} = 42\ {\rm km\ s^{-1}}$ at $r = 1$ AU. Given the Earth's orbital speed of $v_\oplus = 30\ {\rm km\ s^{-1}}$, this means that a meteoroid on a retrograde orbit can enter the Earth's atmosphere at a speed as high as $72\ {\rm km\ s^{-1}}$, but only after local midnight, when such head-on encounters are geometrically possible. The typical size of a meteoroid that gives rise to a naked-eye meteor is $\lesssim 1$ cm across. By observing the same meteor from different locations on the Earth, we find by triangulation that meteors typically occur at an elevation of $h \sim 90$ km, at the top of the mesosphere.

[11] The boundary between small asteroids and large meteoroids is a fuzzy one. A commonly used convention is that rocky and metallic bodies smaller than ~ 300 m across are called meteoroids and larger objects are called asteroids.

[12] The word "meteor" comes from the Greek word *meteoron*, meaning "a thing up high." Initially, the term "meteor" applied to any atmospheric phenomenon, which is why the study of weather is called "meteorology." (The study of meteors, meteoroids, and meteorites is called "meteoritics.")

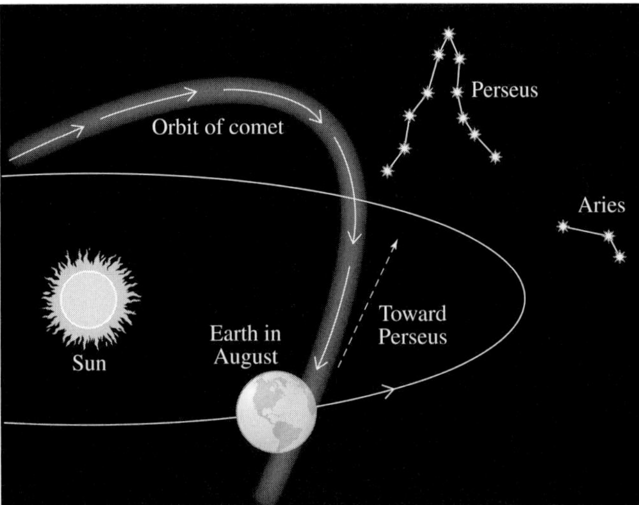

FIGURE 11.10 Origin of the Perseid meteor shower. At the point where the Earth's orbit and the comet's orbit intersect, the direction of the meteor stream appears to originate in the constellation Perseus. This gives the meteor shower its name.

Meteor showers happen when the Earth crosses the path of a comet, as illustrated in Figure 11.10. Many of the solid particles that were liberated from the subliming comet are in orbits similar to that of the comet from which they came. Thus, if the Earth's orbit passes close to a comet's orbit, the Earth is likely to pass through the particles shed by the comet. There are two types of meteor shower:

- **Annual showers** occur when meteoroids have been distributed along the entire orbital path of a comet or ex-comet. These showers occur with high predictability each year. For instance, the Perseid meteor shower occurs around August 12 each year, as the Earth passes through the solid particles released by Comet Swift–Tuttle.
- **Periodic showers** occur when meteoroids are clumped into certain locations along the comet's orbit. For a periodic meteor shower to take place, the Earth must cross the cometary orbit when the clump of meteoroids is near the intersection point. For instance, the Leonid meteor shower occurs around November 17 but is only a spectacular event once every 33 years, when the Earth passes through a clump of particles liberated from Comet Tempel–Tuttle. (The particles, like their parent comet, are on retrograde orbits with $P \approx 33$ yr.)

Generally, meteor showers are not very spectacular events. A meteor shower usually includes one or two visible meteors per minute. On rare occasions, however, larger, more spectacular displays have occurred. The greatest meteor shower in recorded history happened in 1833, when the Leonid meteor shower produced up to 100 visible meteors per *second*.

It is estimated that over 25 million potentially observable meteors plunge through the Earth's atmosphere every day. By accreting small particles in this way, the Earth's mass grows by ~ 100 tons per day. However, the vast majority of meteoroids that enter the Earth's atmosphere are completely vaporized. Only objects larger than $d \sim 3$ cm are able to survive the passage through the Earth's atmosphere. A meteoroid remnant that reaches the Earth's surface intact is called a **meteorite**. The relatively rare, large meteoroids that survive as meteorites are primarily chunks of asteroids; the much more common small meteoroids that completely vaporize are primarily material from comets. The general rule of thumb is Most meteorites originate in asteroids; most meteors originate in comets.

Meteorites come in two major categories:

- **Stony** meteorites are made, as their name implies, from silicate rock, with mass density $\sim 3000\,\mathrm{kg\,m^{-3}}$. About 95% of meteorites are stony. However, they are difficult to distinguish from surface rocks on the Earth, so the best place to search for them is on the thick icecaps of Antarctica.
- **Iron** meteorites are made of an iron–nickel alloy (about 85% iron and 15% nickel). Freshly fallen iron meteorites are easy to recognize, since the oxygen in the atmosphere ensures that the native iron is in the form of iron oxides.

Some meteorites fall into a special subcategory called "carbonaceous chondrites." These are primitive, undifferentiated bodies that are rich in carbon and complex carbon compounds (some even contain amino acids). The complex molecules in carbonaceous chondrites are fragile, and would have broken apart if the meteoroid had ever been at $T > 500$ K for a significant length of time.

Many meteorites, by contrast with the primitive carbonaceous chondrites, have been processed; they consist either of metallic material (presumably from the core of a differentiated body) or of volcanic rock (presumably from the mantle or crust of a differentiated body). By studying the crystal sizes in the processed iron meteorites, we can learn how rapidly the iron cooled as it solidified (Figure 11.11). As a general rule, the slower the cooling, the larger the crystals. By examining iron meteorites, we find that the iron was originally at a temperature of $T \sim 800$ K and cooled at a very slow rate, dropping only ~ 3 K per million years. The cooling time for these meteorites is consistent with an object having an iron–nickel core covered with an outer shell of rock hundreds of kilometers thick. Thus, we can deduce that iron meteorites probably originated in asteroids with sizes of 300 to 600 kilometers. It is hypothesized that some of these differentiated asteroids were destroyed by giant impacts in the early solar system.

Some of the processed meteorites have distinctive compositions that allow us to identify their origin. For instance, some meteorites prove to have originated on the Moon or on Mars.[13] Violent collisions with large meteoroids, of the sort that create big impact craters, can accelerate small bits of debris to speeds greater than the escape speed from the Moon ($v_{\mathrm{esc}} = 2.4\,\mathrm{km\,s^{-1}}$) or Mars ($v_{\mathrm{esc}} = 5.0\,\mathrm{km\,s^{-1}}$). These ejected bits of matter go into orbit around the Sun, and a few of them eventually collide with the Earth.

[13] The martian meteorites have gases trapped within them that are identical in composition to the martian atmosphere.

FIGURE 11.11 When an iron meteorite is sliced in half and etched with acid, it reveals its inner crystal pattern, indicative of the conditions under which its parent body formed and cooled.

Impacts of large meteoroids can have disastrous results for life on Earth. Some fairly large impacts of the recent past include

- **Tunguska, Siberia, 1908.** This event was probably the impact of a small cometary body that detonated before reaching the surface, since no impact crater was created. It flattened trees, however, for tens of kilometers in all directions, and shattered windows hundreds of kilometers away.
- **Barringer Crater, Arizona.** This well-known impact crater (seen in Figure 11.12) is approximately 1.5 km in diameter. It was created by the impact of an iron meteorite $\sim$ 50 m across, about 50,000 years ago.

There are reasons to believe that large impacts might be related to large-scale extinctions of species on Earth. The most famous mass extinction occurred at the end of the Cretaceous period, 65 million years ago. This extinction resulted in the disappearance of dinosaurs, as well as the death of 99% of all organisms and the extinction of 75% of all species. The origin of this Cretaceous–Tertiary extinction is believed to be the impact of an asteroid or comet with the Earth. If this hypothesis is correct, then the dinosaurs were not killed directly by the impact, but indirectly by the events in its aftermath. Tiny dust particles kicked up by the impact would have lingered in the atmosphere for years, severely inhibiting photosynthesis. This prolonged dimming of sunlight is referred to as **nuclear winter**, since it was first recognized as a possible side effect of global nuclear war. The extinction of the dinosaurs enabled mammals (which were mostly small, omnivorous, mouselike creatures at the time of the Cretaceous extinction) to flourish in the newly opened ecological niches.

FIGURE 11.12 Aerial view of the Barringer Crater, in northern Arizona.

Part of the evidence for a large impact at the time of the Cretaceous–Tertiary extinction is iridium-rich sediments in geological strata that are 65 million years old. These iridium-enhanced layers are found all over the world. While iridium is rare in the Earth's crust, it is much more common in meteorites. The total amount of iridium found would require an asteroid about 10 km in diameter. The most likely impact site for the asteroid is the Chicxulub Crater on the Yucatan peninsula. This crater is buried under thick layers of sediment and was only discovered by drilling during the course of oil exploration. The Chicxulub Crater has a diameter of 180 km (about right for an impacting body 10 km across), and radioactive dating of rocks melted by the impact yielded an age of 64.98 million years (about right for the time of the Cretaceous–Tertiary extinction).

PROBLEMS

11.1 The asteroid Eugenia has a small natural satellite orbiting it. The orbital period of the satellite is $P = 4.76$ days. The semimajor axis of its orbit is $a = 1180$ km. What is the mass of Eugenia? (Hint: it is safe to assume that the mass of the satellite is tiny compared to the mass of Eugenia.)

11.2 A cometary nucleus rotates rapidly and has an albedo $A = 0.05$. When its surface temperature reaches $T \approx 150$ K, the ice of which it's mostly made starts to sublime and forms a gaseous coma. How far is the cometary nucleus from the Sun when the coma starts to form?

11.3 The Oort cloud is thought to contain as many as 10^{12} cometary bodies. Estimate the total mass of the Oort cloud and compare it to the mass of the Earth. State the assumptions made in your calculation.

11.4 Imagine a regulation bowling ball (mass $m = 7.2$ kg, diameter $d = 21.6$ cm) orbiting the Sun at $a = 1$ AU. How long will it take the Poynting–Robertson effect to cause the bowling ball to spiral into the Sun?

11.5 In Section 4.3.2, we computed the Hill radius of the Earth, that is, the maximum stable radius of a satellite around the Earth, given the differential tidal force provided by the Sun. What is the maximum stable radius of a comet around the Sun, given the differential tidal force provided by the Alpha Centauri system, which has $M \approx 2M_\odot$? In view of this calculation, what do you expect the approximate radius of the Oort cloud to be?

11.6 An asteroid is 3 AU away from the Sun when it is observed from the Earth at opposition. How large does the asteroid have to be in order to occult a Sun-like star at a distance $d = 10$ pc?

11.7 Comet Halley has a perihelion distance $q = 0.586$ AU and orbital eccentricity $e = 0.967$.

(a) What is the semimajor axis of its orbit?
(b) What is its orbital speed at perihelion?
(c) What is its aphelion distance?
(d) What is its orbital speed at aphelion?

11.8 The *Wilkinson Microwave Anisotropy Probe*, which we encounter again in Chapter 24, is a satellite located near the L_2 point of the Sun–Earth system, 1.5 million km "down-Sun" of the Earth (see Figure 11.3). It is protected from solar radiation by a circular sunshield with diameter $d = 4.5$ m.

(a) Calculate the radiation force exerted on the sunshield. (For simplicity, assume that the shield is perpendicular to the Sun's rays, that it absorbs every photon that strikes it, and that it's outside the Earth's penumbra.)
(b) The satellite and sunshield together have a mass $m = 830$ kg. What is their acceleration as a result of the radiation force?
(c) Estimate the force exerted by the solar wind on the sunshield. Is this greater than or less than the radiation force?

12 The Solar System in Perspective

Our aim in studying planets in the solar system (aside from the intrinsic thrill of exploring new worlds) is to understand how the solar system was formed and how it has evolved with time. We have approached this problem through **comparative planetology**, comparing and contrasting the properties of objects within the solar system. In addition, recent advances have permitted astronomers to detect planets around stars other than the Sun. The properties of these extrasolar planets, or **exoplanets**, as they are called, can be compared and contrasted with the properties of planets within the solar system. This new information enables us to address the broader questions of how planetary systems in general are formed, and how they evolve with time.

12.1 ▪ COMPARATIVE PLANETOLOGY WITHIN THE SOLAR SYSTEM

Comparative planetology, as practiced in the preceding chapters, has a number of lessons to teach us.

Lesson 1: Surfaces of planets are shaped by competing internal mechanisms and external mechanisms. The primary internal mechanisms are volcanism and (in the case of the Earth) plate tectonics. The primary external mechanism is impact cratering due to bombardment. We expect that the internal mechanisms will be more important for the larger bodies, since these will retain their internal heat longer and thus remain geologically active for longer periods of time. Among the terrestrial planets, this is indeed correct. The larger planets, Venus and the Earth, have surfaces on which volcanism and erosion have all but obliterated the evidence of the early bombardment era. The smallest planet, Mercury, has surface features that are primarily the result of bombardment early in the history of the solar system, with little modification by volcanism and erosion. Mars is the intermediate case, with ample cratering in some locations and large volcanos in other locations; moreover, there is ample evidence for past erosion by water and current erosion by wind.

Lesson 2: More massive, colder planets are better able to retain atmospheres. Whether or not a body can retain an atmosphere, and the chemical composition of the atmosphere it does retain, depends on the competition between atmospheric temperature and gravity. More massive planets, all other things being equal, have higher escape speeds and are better able to retain an atmosphere. At a given temperature, the lightest atmospheric particles are moving most rapidly and are the most easily lost to space. In order of increasing mass, consider an array of objects within the solar system:

- Pluto is extremely cold ($T \approx 50$ K at perihelion), but it is too low in mass to retain anything other than a thin methane atmosphere.
- The Moon is more massive than Pluto, but it is too hot ($T \approx 350$ K at the subsolar point) to retain any atmosphere at all.
- Mercury is too low in mass and too hot to retain an atmosphere, aside from transient solar wind particles and atoms sputtered from its surface.
- Mars is cool ($T \approx 290$ K on a warm summer day), but its mass is too small for it to retain anything more than a weak CO_2 atmosphere.
- Venus is hot ($T \approx 740$ K), but it is massive enough to retain a thick CO_2 atmosphere.
- The Earth is warmer than Mars, but it is massive enough to have retained a dense CO_2 atmosphere initially. This evolved into a thinner N_2 and O_2 atmosphere because of the presence of liquid water, which removed the CO_2, and plant life, which added the O_2.
- The Jovian planets are sufficiently cold and massive to retain thick atmospheres of hydrogen and helium.

Lesson 3: Giant satellites of Jovian planets show patterns consistent with our ideas about formation of the planets. The satellite systems surrounding the Jovian planets formed in a process that mirrored that of the solar system as a whole, with a central, massive, warm object surrounded by an orbiting disk of material. The inner satellites of Jovian planets (Io and Europa, for instance) formed under relatively warmer conditions than the outer satellites (Ganymede and Callisto, for instance). The higher densities of the inner satellites imply a higher fraction of dense, refractory rock and a lower fraction of low-density, volatile ice. Inner satellites show evidence of volcanism, driven by tidal heating in this case, and resurfacing (Figure 12.1). In the solar system, the largest planets, Jupiter and Saturn, formed in the middle of the system, where there was a favorable trade-off between temperature (cool enough for volatiles to condense) and density (high enough to make massive bodies). Similarly, in the satellite systems of the Jovian planets, the largest satellites formed in the middle of the system: Jupiter's Ganymede, Saturn's Titan, and Uranus's Titania.[1]

Lesson 4: Many unusual features of the solar system can be attributed to giant impacts. It is only within the last few decades that astronomers have realized that some of the more unusual properties of solar system bodies can best be explained by "giant impacts" that occurred relatively late in the formation of the solar system. Giant impacts seem to be the best way to account for the relatively large satellites of small planets. In the case of the Earth–Moon system, the nonequatorial orbit of the Moon and the differences

[1] Neptune's giant satellite, Triton, is a special case, since it seems to be a captured TNO that didn't form in Neptune's immediate vicinity.

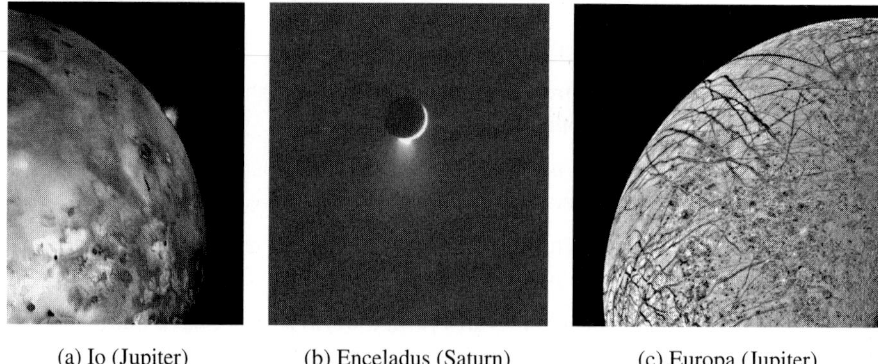

(a) Io (Jupiter) (b) Enceladus (Saturn) (c) Europa (Jupiter)

FIGURE 12.1 The Galilean satellites Io and Europa, and Saturn's midsize satellite Enceladus (not to scale), all show evidence of tidally driven volcanism that has renewed their surfaces.

in surface composition between the Earth and the Moon suggest that the histories of these objects are quite different.

A giant impact between the Earth and a Mars-sized planetoid seems the best way to account for the current Earth–Moon system. In the case of Neptune's satellite Triton, its retrograde motion strongly argues that it is a captured body; the most obvious way of slowing Triton sufficiently for it to be captured by Neptune is to have it collide with a satellite already orbiting that planet. In addition, the retrograde rotation of Venus and the "sideways" rotation of Uranus is best accounted for by impacts with large bodies that severely altered the orientation of their rotation axes. On bodies with the oldest surfaces, we see clear direct evidence for major collisions in the impact basins on the Moon, Mercury, and Callisto (Figure 12.2). Collisions with large objects can have a global impact on a planet's topography, as the Caloris Basin on Mercury attests: at the antipodes of the Caloris Basin, the surface has been shattered into chaotic terrain by the converging seismic waves set off by the impact.

12.2 ▪ ORIGIN OF THE SOLAR SYSTEM

Our understanding of the history and evolution of the solar system is still incomplete, but a comprehensive, self-consistent picture has begun to emerge. Our mental picture of the solar system forming from a rotating protoplanetary disk explains many of the basic features of the solar system:

1. Planetary orbits are all in nearly the same plane.
2. The Sun's equator lies close to this plane.
3. Planetary orbits are nearly circular.

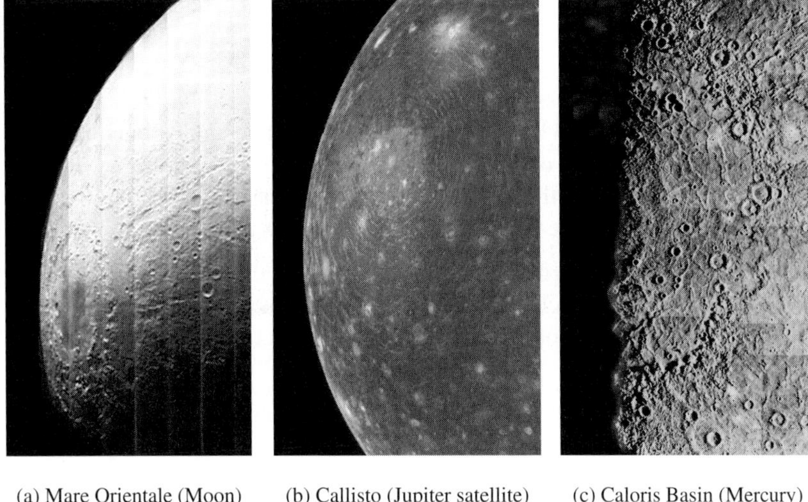

(a) Mare Orientale (Moon) (b) Callisto (Jupiter satellite) (c) Caloris Basin (Mercury)

FIGURE 12.2 The Moon, Callisto, and Mercury all show evidence of large impacts that altered their surfaces.

4. Planets all orbit in the same direction.
5. Most planets (and the Sun) rotate in the same direction as the planets' orbital motion.

We have also noted that although the Sun contains 99.8% of the mass in the solar system, it contains less than 5% of the angular momentum. The planets represent the material in the pre-solar nebula that had too much angular momentum to contract into the Sun as it was forming.

If we refer back to the list of questions at the end of Section 8.1, about the differences between terrestrial and Jovian planets, we can provide plausible answers to all of them. We have already noted that formation from a rotating gaseous disk explains why planetary orbits are circular, and we can address the other questions as well.

- **The nature of small bodies.** In the picture we have developed, comets are primitive bodies, planetesimals left over from the first phases of formation of the solar system. Asteroids are a mix of primitive planetesimals and fragments of larger differentiated bodies that were broken up in collisions. Some of these bodies have been captured as satellites by planets. Other satellites, particularly larger satellites in prograde, equatorial, nearly circular orbits, were probably formed at the same time as their central planet, mimicking on a smaller scale the formation of the solar system as a whole. Small debris in the solar system is replenished from comets and asteroids; otherwise, the Poynting–Robertson effect would clear out the solar system of these small bodies.

- **Differences in chemical composition.** The broad chemical composition differences within the solar system reflect the temperature gradient within the protoplanetary disk. The inner planets are primarily composed of the least volatile substances. The outer planets were able to grow to larger sizes because volatile ices could condense at the lower temperatures found in the outer disk. The combination of icy, rocky, and metallic planetesimals let the outer planets grow to a sufficiently high mass to retain hydrogen and helium (the most abundant elements in the pre-solar nebula) in their atmospheres.
- **Rings around Jovian planets.** All the Jovian planets have ring systems inside their Roche limits. The ring systems represent either bodies that failed to form because of tidal stresses or bodies that were destroyed by tidal forces—or perhaps both.
- **Chemical differentiation of the terrestrial planets** Radioactivity is the key to understanding why all but the very smallest rocky and metallic bodies in the solar system are chemically differentiated. In the early solar system, radioactivity provided enough heat to melt iron in planetary interiors, thus precipitating a complete liquefaction of the interior, and subsequent stratification by density.

12.3 ▪ DETECTING EXOPLANETS

Although astronomers have developed a broad outline of how the solar system formed, details are still lacking. One of the most active fields in astronomy is the study of **exoplanets**, or planets around stars other than the Sun. As more exoplanets are discovered, the statistical study of planetary systems will allow a better general understanding of how planets form and evolve, and whether our own solar system is ordinary or anomalous.

Detecting planets around other stars is far more difficult than detecting planets orbiting the Sun. Neptune was discovered when it was close to opposition, ~ 29 AU from the Earth. By comparison, the Sun's nearest neighbor among the stars, Proxima Centauri, is at a distance $d \sim 270{,}000$ AU ~ 1.3 parsecs from the Earth. To illustrate the difficulties involved in detecting exoplanets, let's imagine we are located at the position of Proxima Centauri and want to detect Jupiter as it orbits the Sun. (Since Jupiter is the largest planet in the solar system, we expect it to be the most easily detected.) Jupiter is on a nearly circular orbit, with semimajor axis $a = 5.2$ AU. Its maximum angular separation from the Sun, as seen from a distance $d = 270{,}000$ AU (and shown in Figure 12.3), will be

$$\theta = \frac{a}{d} = \frac{5.2\text{ AU}}{270{,}000\text{ AU}} = 1.9 \times 10^{-5}\text{ rad}. \tag{12.1}$$

Converting from radians to arcseconds, and scaling to an arbitrary orbit size a at an arbitrary distance d, we find

$$\theta = 4.0\text{ arcsec}\left(\frac{a}{5.2\text{ AU}}\right)\left(\frac{d}{270{,}000\text{ AU}}\right)^{-1}. \tag{12.2}$$

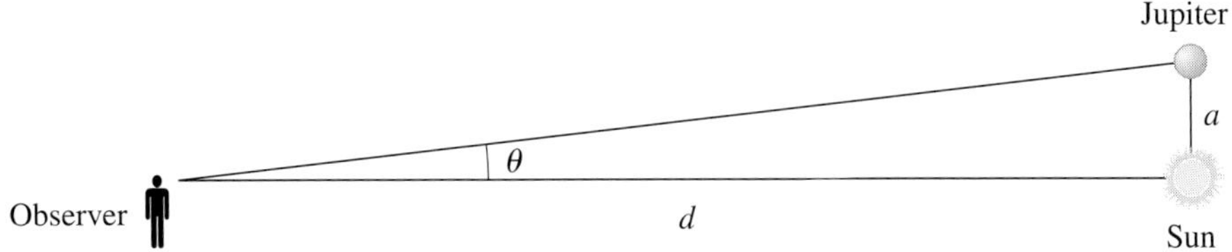

FIGURE 12.3 Finding the maximum angular separation θ between the Sun and Jupiter, as measured from a distance d.

Thus, in the diffraction-limited case (see equation 6.7), a telescope with $D \gtrsim 4$ cm would provide sufficient angular resolution to distinguish Jupiter from the Sun, as seen from Proxima Centauri. In reality, the difficulty in detecting Jupiter would come not from its proximity to the Sun but from its dimness relative to the Sun.

At visible wavelengths, Jupiter's luminosity is due to reflected sunlight. Jupiter's distance from the Sun is $a \approx 5.2$ AU $\approx 7.8 \times 10^8$ km, its radius is $R_{\rm Jup} \approx 11 R_\oplus \approx 7.2 \times 10^4$ km, and its albedo is $A \approx 0.51$. Its luminosity of reflected light is thus

$$L_{\rm Jup} = \left(\frac{L_\odot}{4\pi a^2}\right)(\pi R_{\rm Jup}^2)A = \frac{L_\odot}{4}\left(\frac{R_{\rm Jup}}{a}\right)^2 A. \tag{12.3}$$

(Compare this to equation 8.6, which gives the rate at which a planet *absorbs* sunlight.) The ratio of Jupiter's reflected luminosity to the Sun's intrinsic luminosity is

$$\frac{L_{\rm Jup}}{L_\odot} = \frac{A}{4}\left(\frac{R_{\rm Jup}}{a}\right)^2 \approx \frac{0.51}{4}\left(\frac{7.2 \times 10^4\ \rm km}{7.8 \times 10^8\ \rm km}\right)^2 \approx 4 \times 10^{-9}. \tag{12.4}$$

If an object with a luminosity $L \sim 4 \times 10^{-9} L_\odot$ were seen against a very dark background, it would be detectable from the distance of Proxima Centauri. However, the Sun's image would provide a bright background for the image of Jupiter, making the unambiguous detection of Jupiter's image far more difficult.

One way we can increase the apparent brightness of Jupiter relative to the Sun is to look for the thermal emission from Jupiter at photon energies $h\nu \lesssim kT_{\rm Jup}$. Given Jupiter's temperature of $T_{\rm Jup} \approx 150$ K, this implies $\nu \lesssim 3 \times 10^{12}$ Hz or $\lambda \gtrsim 100\ \mu$m. At these low frequencies, the emission from both Jupiter and the Sun is in the Rayleigh–Jeans limit (equation 5.87). Thus, the ratio of the specific flux of Jupiter to that of the Sun is

$$\frac{F_{\nu,\rm Jup}}{F_{\nu,\odot}} = \frac{\pi I_{\nu,\rm Jup}}{\pi I_{\nu,\odot}} \approx \frac{2kT_{\rm Jup}\nu^2}{c^2}\frac{c^2}{2kT_\odot \nu^2} \approx \frac{T_{\rm Jup}}{T_\odot}, \tag{12.5}$$

independent of ν. Of course, we must also take into account the fact that the surface area of Jupiter is smaller than that of the Sun. Approximating both Jupiter and the Sun as spherical blackbodies, the ratio of their specific luminosities (that is, their luminosities per unit frequency) is

$$\frac{L_{\nu,\mathrm{Jup}}}{L_{\nu,\odot}} = \frac{4\pi R_{\mathrm{Jup}}^2 F_{\nu,\mathrm{Jup}}}{4\pi R_\odot^2 F_{\nu,\odot}} \approx \frac{R_{\mathrm{Jup}}^2}{R_\odot^2} \frac{T_{\mathrm{Jup}}}{T_\odot}, \tag{12.6}$$

as long as we stay in the low-frequency Rayleigh–Jeans limit. Given $R_{\mathrm{Jup}} \approx 0.1 R_\odot$ and $T_{\mathrm{Jup}} \approx 0.026 T_\odot$, this implies that

$$\frac{L_{\nu,\mathrm{Jup}}}{L_{\nu,\odot}} \approx 3 \times 10^{-4}. \tag{12.7}$$

Thus, by switching from visible light ($\lambda \sim 0.5\ \mu\mathrm{m}$) to far infrared light ($\lambda \gtrsim 100\ \mu\mathrm{m}$), we increase the ratio of Jupiterlight to sunlight from a tiny 4×10^{-9} to a more respectable 3×10^{-4}.

Because direct imaging of exoplanets is difficult, planet searches have so far concentrated on indirect methods of finding exoplanets. For instance, the existence of a dim exoplanet can be deduced from its gravitational effect on its parent star. Newton's law of universal gravitation tells us that the force exerted by Jupiter on the Sun, for instance, is equal in magnitude to the force exerted by the Sun on Jupiter. Thus, instead of saying Jupiter orbits the Sun, we should strictly say that both Jupiter and the Sun orbit the center of mass of the Jupiter–Sun system. The location of the **center of mass** of a two-body system can be readily calculated. Suppose that two spherical objects, such as a star and a planet, are gravitationally bound to each other. The more massive object has a mass M_{A} and the less massive object has a mass M_{B}; the two objects are separated by a distance a, as shown in Figure 12.4. The center of mass lies on the line segment connecting the centers of the two objects. If a_{A} is the distance from M_{A} to the center of mass, and a_{B} is the distance from M_{B} to the center of mass, then

$$\frac{a_{\mathrm{B}}}{a_{\mathrm{A}}} = \frac{M_{\mathrm{A}}}{M_{\mathrm{B}}}. \tag{12.8}$$

The center of mass is always closer to the more massive object.

To revert to our example of the Jupiter–Sun system, the mass of Jupiter is $M_{\mathrm{B}} \approx 10^{-3} M_{\mathrm{A}}$, where $M_{\mathrm{A}} = 1 M_\odot$. Thus, as Jupiter travels on a large elliptical orbit with semimajor axis $a_{\mathrm{B}} = 5.2\ \mathrm{AU} = 7.8 \times 10^8\ \mathrm{km}$, the Sun travels on a small elliptical orbit with semimajor axis $a_{\mathrm{A}} = 5.2 \times 10^{-3}\ \mathrm{AU} = 780{,}000\ \mathrm{km} \approx 1.1 R_\odot$.

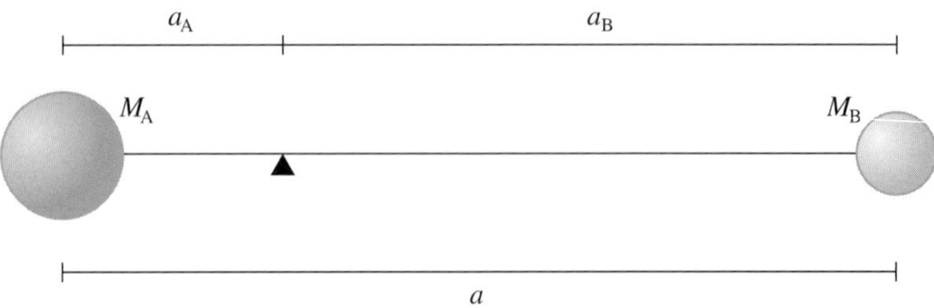

FIGURE 12.4 The center of mass of a two-body system.

As we computed above, the semimajor axis of Jupiter's orbit subtends an angle of 4 arcseconds as seen from the distance of Proxima Centauri; this implies that the semimajor axis of the Sun's orbit will subtend an angle of only 4 milliarcseconds. Generalizing, if a star of mass $M_{\rm A}$ has an exoplanet of mass $M_{\rm B} \ll M_{\rm A}$ at a distance a, the semimajor axis of the star's orbit will subtend an angle

$$\theta_{\rm A} = 0.004 \text{ arcsec} \left(\frac{M_{\rm B}/M_{\rm A}}{0.001}\right)\left(\frac{a}{5.2 \text{ AU}}\right)\left(\frac{d}{270{,}000 \text{ AU}}\right)^{-1}, \tag{12.9}$$

where d is the distance from the observer to the star. Thus, the angular size of the star's orbit is greatest for nearby stars with massive planets on large orbits.

If we were in the vicinity of Proxima Centauri, discovering Jupiter by **direct imaging** (that is, taking a "snapshot" of Jupiter) or by **astrometry** (that is, watching the Sun trace out a small ellipse) we would face different technical difficulties. The orbit of Jupiter would be relatively large in angular size—a point in favor of direct imaging—but detecting Jupiter would be difficult because of its tiny luminosity compared to the Sun. The Sun itself would be simple to detect—a point in favor of astrometry—but its orbit would be too small to resolve with a standard optical telescope.

Because of the difficulties involved in direct imaging and astrometry, most exoplanets discovered to date have been found using other techniques. One such technique is the **radial velocity** method. If the orbital velocity of a star, caused by an unseen exoplanet, has a component v_r along the line of sight to an observer, it will cause a Doppler shift

$$\frac{\Delta\lambda}{\lambda} = \frac{v_r}{c} \tag{12.10}$$

in the absorption lines of the star's spectrum. Consider the simple case of a star, with mass $M_{\rm A}$, and a planet, with mass $M_{\rm B}$, on circular orbits about their center of mass; this is illustrated in Figure 12.5a. The orbital periods of the star and planet about the center of mass are identical. Thus, the orbital speed $v_{\rm A}$ of the star is related to the orbital speed $v_{\rm B}$ of the planet by the equation

$$P = \frac{2\pi a_{\rm A}}{v_{\rm A}} = \frac{2\pi a_{\rm B}}{v_{\rm B}}. \tag{12.11}$$

With the assistance of equation (12.8), we deduce

$$\frac{v_{\rm A}}{v_{\rm B}} = \frac{a_{\rm A}}{a_{\rm B}} = \frac{M_{\rm B}}{M_{\rm A}}. \tag{12.12}$$

Let's use the Jupiter–Sun system as our example, once again. If we approximate the orbit of Jupiter as a circle, its orbital speed is

$$v_{\rm B} = \frac{2\pi a_{\rm B}}{P} = \frac{2\pi(5.2 \text{ AU})}{11.9 \text{ yr}}\left(\frac{1 \text{ yr}}{3.16\times10^7 \text{ s}}\right)\left(\frac{1.50\times10^8 \text{ km}}{1 \text{ AU}}\right) = 13 \text{ km s}^{-1}. \tag{12.13}$$

The orbital speed of the Sun will then be

$$v_{\rm A} = \frac{M_{\rm B}}{M_{\rm A}} v_{\rm B} = 0.001(13 \text{ km s}^{-1}) = 13 \text{ m s}^{-1}. \tag{12.14}$$

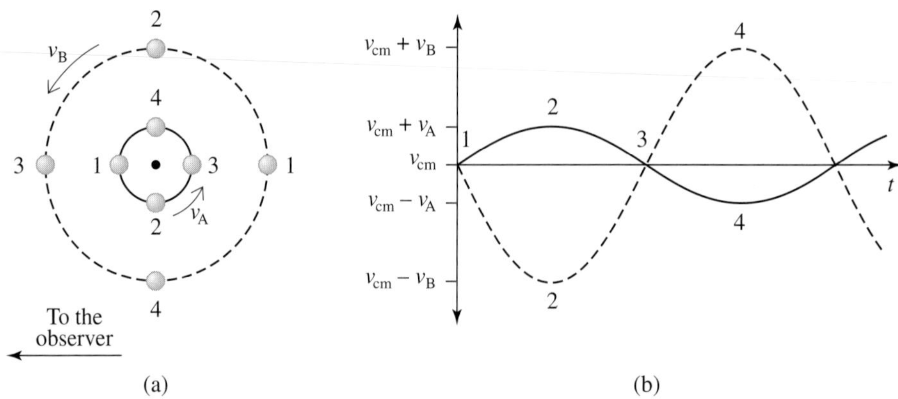

FIGURE 12.5 (a) A star (solid orbit) and a planet (dashed orbit) moving about their center of mass. The observer is in the orbital plane, far to the left. The numbers indicate the locations of the star and planet at specific times. (b) The radial velocity of the star (solid curve) and the planet (dashed curve) relative to the observer. The numbered positions correspond to the locations indicated in part (a).

Thus, the Sun's orbital speed is a modest $13\,\mathrm{m\,s^{-1}}$, or about 29 mph.[2] With current techniques, the radial velocity of relatively bright stars can be determined to within $\sim 1\,\mathrm{m\,s^{-1}}$; thus, it is technologically feasible to detect Jupiter-like exoplanets from their effect on the radial velocity of Sun-like stars.

Suppose that we are looking at a system that consists of a star and a single planet. The center of mass of the system will be moving with some radial velocity $v_{\rm cm}$. In addition, if the star and planet are on circular orbits, and we are in the orbital plane of the system (as shown in Figure 12.5a), the radial velocity of the star will show sinusoidal variations. The sinusoidal curve (as shown in Figure 12.5b) will have an amplitude $v_{\rm A}$ and a period P equal to the orbital period of the planet. In general, though, we will not be in the orbital plane of the system; we will be observing the orbital plane at an inclination i, as illustrated in Figure 12.6. Thus, the measured amplitude of a radial velocity curve is only equal to $v_{\rm A}$ if we are observing the orbital plane edge-on ($i = 90°$); at all other orbital inclinations, the amplitude is $v_{\rm A} \sin i < v_{\rm A}$.

From the velocity curve of a star with an unseen companion, we can determine the orbital period P and the value of $v_{\rm A} \sin i$. Let's see how we can use these measured quantities to deduce the properties of the perturbing exoplanet. Kepler's third law, as modified by Newton (equation 3.53), tells us that

$$(a_{\rm A} + a_{\rm B})^3 = \frac{G(M_{\rm A} + M_{\rm B})P^2}{4\pi^2}. \tag{12.15}$$

[2] For comparison, this also happens to be the average air speed of an unladen European swallow (Park et al., 2001, *Journal of Experimental Biology*, vol. 204, p. 2741).

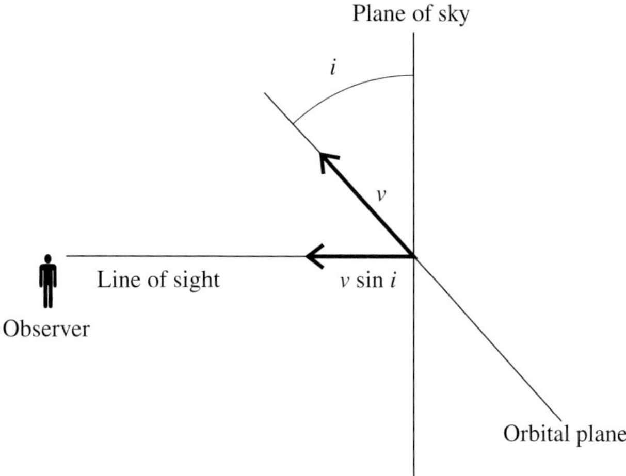

FIGURE 12.6 The maximum radial velocity v_r that we measure is the projection of the orbital speed v onto the line of sight: $v_r = v \sin i$.

In the case that $M_A \gg M_B$, and hence $a_A \ll a_B$, we can simplify this to

$$a_B^3 \approx \frac{GM_A P^2}{4\pi^2}. \tag{12.16}$$

The mass M_A of the star, as we see in Chapter 13, can usually be deduced from its spectrum. For instance, stars with spectra similar to the Sun's have masses close to a solar mass. Knowledge of the star's mass enables us to estimate the size of the planet's orbit:

$$a_B \approx 1\,\text{AU} \left(\frac{M_A}{1M_\odot}\right)^{1/3} \left(\frac{P}{1\,\text{yr}}\right)^{2/3}. \tag{12.17}$$

In addition, the amplitude of the velocity curve, $v_A \sin i$, gives us some knowledge of the mass of the planet. Using the relation $a_B = (M_A/M_B)a_A$, we start by rewriting equation (12.16) in the form

$$\frac{M_A^3}{M_B^3} a_A^3 \approx \frac{GM_A P^2}{4\pi^2}. \tag{12.18}$$

Solving for M_B, the planet's mass, we find

$$M_B \approx \left(\frac{4\pi^2 M_A^2}{GP^2} a_A^3\right)^{1/3}. \tag{12.19}$$

Although a_A, the size of the star's orbit, is generally too tiny to be measured by the astrometric method, we can make the substitution $2\pi a_A = P v_A$ into equation (12.19) to yield

$$M_B \approx \left(\frac{M_A^2 P}{2\pi G} v_A^3 \right)^{1/3} . \tag{12.20}$$

Unfortunately, we can't measure v_A; we can measure only $v_A \sin i$. Thus, we can't compute M_B; we can compute only

$$M_B \sin i \approx \left(\frac{M_A^2 P}{2\pi G} \right)^{1/3} v_A \sin i$$

$$\approx 11 M_\oplus \left(\frac{M_A}{1 M_\odot} \right)^{2/3} \left(\frac{P}{1\,\text{yr}} \right)^{1/3} \left(\frac{v_A \sin i}{1\,\text{m s}^{-1}} \right) . \tag{12.21}$$

In the above equation, $M_\oplus = 6.0 \times 10^{24}$ kg is the Earth's mass; it takes 318 Earths to equal the mass of one Jupiter.

If an exoplanet is to be detected by the radial velocity method, it must produce $v_A \sin i$ large enough to measure. Thus, exoplanets whose orbits are more nearly edge-on ($\sin i \sim 1$) are more readily detected. In addition, since $v_A \propto M_B M_A^{-2/3} P^{-1/3}$ (from equation 12.21), the radial velocity method favors the detection of massive, short-period planets around relatively low-mass stars. In addition, measuring P accurately requires observing the star for a time $\sim P$ or longer; this adds to the difficulty of detecting long-period exoplanets.

The detection of exoplanets by the **transit** method requires a different set of observations, and favors the detection of exoplanets with somewhat different properties than those found by the radial velocity method. In the language of astronomers, a "transit" is the passage of a planet between its parent star and the observer. Within the solar system, for instance, observers on Earth can occasionally see transits of Mercury and Venus across the Sun. During such a transit, the inferior planet appears as a small, dark circle against the bright circle of the Sun, as shown in Figure 7.5. When an exoplanet makes a transit, neither the star nor the exoplanet can be resolved in angle as seen from the Earth. However, the transiting planet can be detected indirectly since it blocks part of the star's flux. If the cross-section of the planet is πR_B^2 and the cross-section of the star is πR_A^2, then when the planet lies directly between the star and an observer, the star's measured flux F drops by a fractional amount

$$\frac{\delta F}{F} = \frac{\pi R_B^2}{\pi R_A^2} = \left(\frac{R_B}{R_A} \right)^2 . \tag{12.22}$$

If a distant observer saw Jupiter transit across the Sun, for instance, the Sun's flux would drop by a fraction

$$\frac{\delta F}{F} = \left(\frac{R_{\text{Jup}}}{R_\odot} \right)^2 = \left(\frac{69{,}900\,\text{km}}{696{,}000\,\text{km}} \right)^2 = 0.010, \tag{12.23}$$

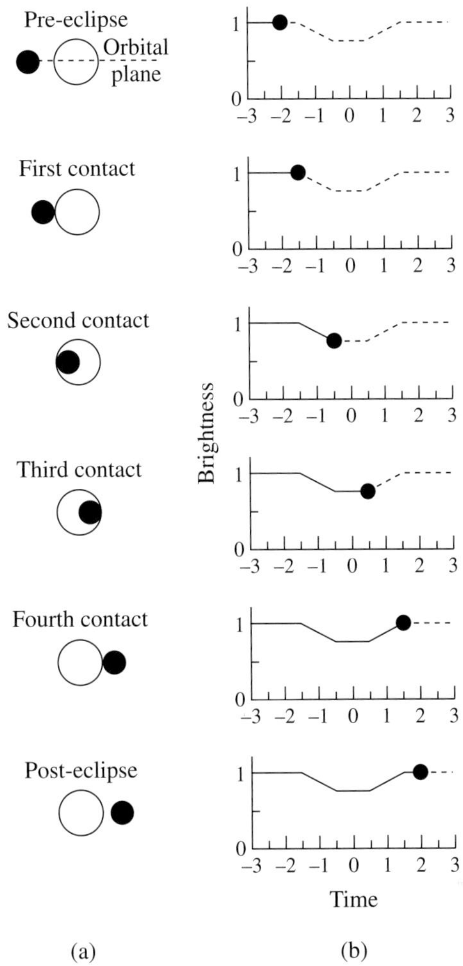

FIGURE 12.7 (a) Events when a planet transits a star. (b) The light curve (measured flux as a function of time) during the transit, with a marker indicating the flux during the corresponding event in part (a).

since Jupiter would cover 1% of the Sun's area.[3]

The light curve of a star during a transit is shown in more detail in Figure 12.7.

From the light curve, we can measure the drop in flux at mid-transit, $\delta F/F$; we can also measure the times of first, second, third, and fourth contact, as defined in the figure. If we wait patiently, we can also measure the time between successive transits; this is

[3] This simple calculation assumes that the Sun's surface brightness is uniform. The effects of limb darkening (described in Section 7.1) slightly complicate matters.

equal to the orbital period P of the exoplanet. The drop in flux tells us the relative sizes of the exoplanet and the star (equation 12.22):

$$\frac{R_B}{R_A} = \left(\frac{\delta F}{F}\right)^{1/2}. \tag{12.24}$$

If the planet is much less massive than the star, its orbital semimajor axis a_B is given by Kepler's third law (equation 12.16), and its speed (assuming a circular orbit) is

$$v_B \approx \frac{2\pi a_B}{P} \approx 30 \text{ km s}^{-1} \left(\frac{M_A}{1M_\odot}\right)^{1/3} \left(\frac{P}{1 \text{ yr}}\right)^{-1/3}. \tag{12.25}$$

If we are seeing the exoplanet's orbit exactly edge-on, then between the time of first contact (t_1) and the time of second contact (t_2), the planet moves a distance equal to its own diameter. In this case, the radius of the planet is

$$R_B = \frac{1}{2} v_B (t_2 - t_1) \approx 54{,}000 \text{ km} \left(\frac{M_A}{1M_\odot}\right)^{1/3} \left(\frac{P}{1 \text{ yr}}\right)^{-1/3} \frac{t_2 - t_1}{1 \text{ hr}}. \tag{12.26}$$

Similarly, between the time of first and third contact, if the exoplanet's orbit is edge-on, the planet moves a distance equal to the *star's* diameter, enabling the radius of the star to be calculated as well.

For an arbitrary exoplanet, our probability of seeing it in transit across its star is small. This is because a transit can be seen only if we are very close to the orbital plane of the planet. Figure 12.8 shows a star (A) and planet (B) separated by a distance a. The star is at a distance d from an observer and the planet is at a distance d'; however, when $d \gg a$, as we expect when observing exoplanets, we can make the approximation $d' \approx d$. The angular radius of the star, as seen by the observer, will be $\theta_A \approx R_A/d$. The angular radius of the planet will be $\theta_B \approx R_B/d' \approx R_B/d$. The observer will see a transit when the angle θ between the planet's center and the star's center satisfies the relation

$$\theta \leq \theta_A + \theta_B \approx \frac{R_A + R_B}{d}. \tag{12.27}$$

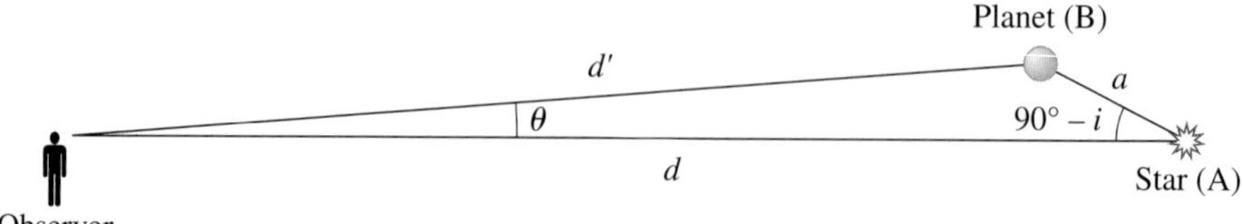

FIGURE 12.8 The geometry that determines whether an observer will see a planet (B) transit a star (A).

From the rule of sines applied to the triangle in Figure 12.8, we find

$$\frac{\sin\theta}{a} = \frac{\sin(90^\circ - i)}{d'} \approx \frac{\cos i}{d}, \tag{12.28}$$

where i is the inclination of the exoplanet's orbit around the star. Since θ must be small when $d \gg a$, we can write

$$\theta \approx \sin\theta \approx \frac{a}{d}\cos i. \tag{12.29}$$

By combining equations (12.27) and (12.29), we can find the range of inclination i for which a transit is observed:

$$\frac{a}{d}\cos i \lesssim \frac{R_A + R_B}{d}, \tag{12.30}$$

or

$$\begin{aligned} \cos i &\lesssim \frac{R_A + R_B}{a} \\ &\lesssim 0.0046 \left(\frac{R_A + R_B}{1R_\odot}\right)\left(\frac{1\,\text{AU}}{a}\right). \end{aligned} \tag{12.31}$$

Unless the planet is only a few stellar radii away from its parent star, $\cos i$ must be small for a transit to be observed, and thus i must be very close to 90°. A distant observer would see Mercury transit the Sun only if Mercury's orbit were at an inclination $i \gtrsim 89.3^\circ$. However, a distant observer would see Neptune transit the Sun only if the inclination were $i \gtrsim 89.99^\circ$.

The range of inclinations for which an exoplanet produces a transit is largest for planets close to their parent star. In addition, if an exoplanet is to be detected by the transit method, it must produce $\delta F/F$ large enough to be measured. Thus, exoplanets that are large in radius compared to their parent star are more readily detected. For a dedicated exoplanet hunter, the jackpot comes when a transiting exoplanet produces measurable radial velocity variations in the star that it orbits. The combination of transit data and radial velocity data provides a large payout of planetary parameters. The transit allows the radius of the planet to be known. Since the inclination of a transiting planet must be $i \approx 90^\circ$, this permits an exact calculation of the planetary mass from equation (12.21).

For instance, the star HD 209458 is a Sun-like star ($M_A \approx 1M_\odot$, $R_A \approx 1R_\odot$) that is orbited by a transiting planet. The planet's orbital period is $P = 3.525$ days, implying an orbital size $a_B \approx 0.045$ AU. The reduction of flux during the transit is $\delta F/F \approx 0.02$, implying that the radius of the planet is $R_B \approx 0.14R_\odot \approx 1.4R_{\text{Jup}}$. The orbital velocity of the star is $v_A = 85\ \text{m s}^{-1}$, implying a planetary mass (equation 12.21) $M_B \approx 200M_\oplus \approx 0.7M_{\text{Jup}}$. Thus, we are presented with a planet astonishingly close to its star; its volume is greater than that of Jupiter, but its mass is smaller, implying an average density of just $\rho \sim 400\ \text{kg m}^{-3}$. Obviously, exoplanets are not all carbon copies of planets within our solar system.

12.4 ▪ PROPERTIES OF EXOPLANETS

Recent searches for exoplanets have contributed greatly to our knowledge of planetary properties. First, and most fundamental, these searches have shown that exoplanets exist, and that the Sun is not unique in having an entourage of planets. By the middle of the year 2008, about 300 exoplanets had been discovered, primarily by the radial velocity and transit methods. With such large numbers of planets known, it becomes useful to do statistical analysis of planetary properties. When doing the statistics, however, we must keep in mind the selection biases involved in searches for exoplanets. The radial velocity method favors the discovery of high-mass planets on small orbits; the transit method favors the discovery of large-radius planets on small orbits. In general, big things are easier to detect than small things; it's easier to detect a hippopotamus in your living room than it is to detect an ant. Despite the ease of detection, it is still surprising, however, to find a hippo lounging on your sofa.

One of the most startling results of exoplanet searches is the astronomical equivalent of finding a living-room hippo; searches have revealed a significant population of **hot Jupiters**, where a hot Jupiter is a planet with a mass comparable to that of Jupiter, on an orbit with $a \lesssim 0.1$ AU.[4] It is estimated that $\sim 1\%$ of stars surveyed have planets with $M_B \sin i \geq 0.5 M_{Jup}$ on orbits with $a \leq 0.1$ AU. Going out to larger radii, $\sim 6\%$ of the stars have planets in the same mass range on orbits with $a \leq 5$ AU, comparable to the size of Jupiter's orbit.

Figure 12.9a shows $M_B \sin i$ as a function of semimajor axis length a for exoplanets detected with the radial velocity technique. Exoplanets lying below the dashed line in the figure would produce a radial velocity of $v_r \lesssim 3 \text{ m s}^{-1}$ in their parent star, and thus would be difficult to detect using current techniques. In addition, exoplanets with $a \gtrsim 5$ AU would have orbital periods of $P \gtrsim 11$ yr, assuming a $1 M_\odot$ parent star; few stars have been monitored for a long enough time at high enough spectral resolution to detect such long-period planets. Of the eight major planets in our solar system, indicated as the squares in Figure 12.9, only Jupiter falls within the region of parameter space where exoplanets can be reliably detected. Thus, current techniques for finding exoplanets cannot tell us about Earth-mass planets on Earth-like orbits. They are most effective at finding types of planets that are *not* seen in our own solar system: "hot Jupiters" and "hot Neptunes."

In the previous section, we discussed, for the sake of simplicity, the case of exoplanets on nearly circular orbits. If an exoplanet is on a circular orbit with constant orbital speed, then the radial velocity variations of its parent star will trace out a sinusoidal curve. If the star's radial velocity variations are not sinusoidal, that is a sign that the perturbing exoplanet is on an eccentric orbit with varying orbital speed. If the shape of the radial velocity curve is measured with sufficient accuracy, the eccentricity of the exoplanet's orbit can be calculated. Figure 12.9b shows the orbital eccentricity e as a function of semimajor axis length a for exoplanets detected with the radial velocity method. Exoplanets on extremely small orbits ($a \lesssim 0.05$ AU) tend to have nearly circular orbits; this is a result of the very strong tides acting on these planets. At larger radii, however,

[4] For comparison, Mercury has $a \approx 0.4$ AU and $M \approx 0.0002 M_{Jup}$.

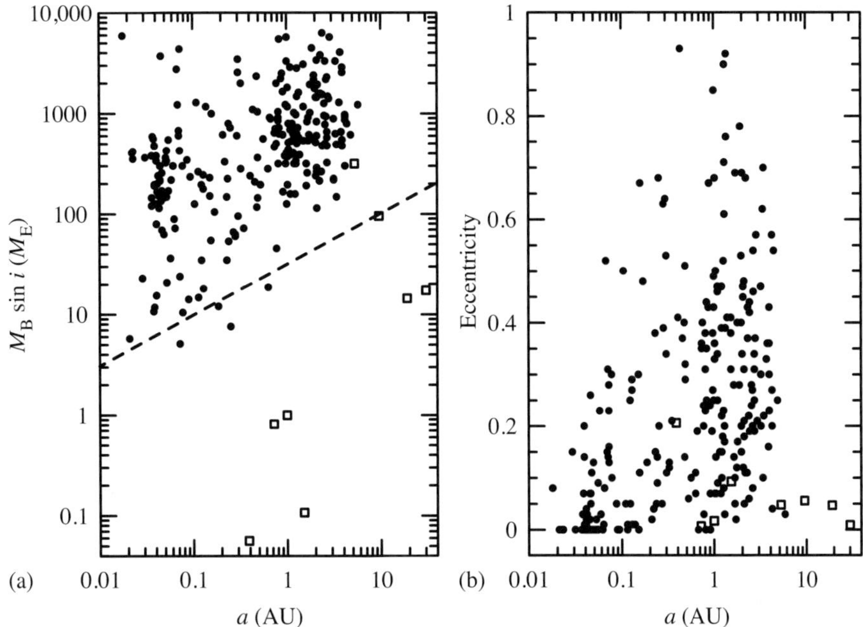

FIGURE 12.9 (a) $M_B \sin i$ as a function of orbit size a for planets detected by the radial velocity technique. The dashed line is the value of $M_B \sin i$ that would produce $v_r \approx 3\ \mathrm{m\ s^{-1}}$ in a Sun-like star. (b) Eccentricity e as a function of orbit size a. In both (a) and (b), solar system planets are indicated by squares.

exoplanets have a wide range of eccentricities. This is a strong contrast to the eccentricity of planetary orbits within the solar system (shown as the squares in Figure 12.9b).

The search for exoplanets has thus provided some surprising results. The discovery of hot Jupiters posed a challenge to existing models for the formation and evolution of planetary systems. Jovian planets should not be able to form within 0.1 AU of a Sun-like star; the temperature there is too hot for even the most refractory substances to condense into solids. Thus, hot Jupiters must be formed far from their parent stars, then migrate inward. To move from a large orbit to a smaller one, a planet must lose orbital angular momentum. There are various mechanisms by which a young planet can lose angular momentum. Planets initially form in gaseous protoplanetary disks (see Figure 8.3). The viscous gas of the disk can exert a torque on protoplanets as they form, decreasing their orbital angular momentum and driving them to smaller orbits. At a later stage, close encounters with planetesimals can transfer angular momentum from the exoplanet to the planetesimals. Finally, if two planets have a close encounter (as opposed to a collision, like the one that formed the Moon), orbital angular momentum will be transferred from one planet to the other. In this case, one planet will be driven inward to a smaller orbit, while the other will be driven outward (and may even attain escape speed). It is not yet known which of these mechanisms is the primary one that drives hot Jupiters inward.

Future exoplanet surveys will increase our knowledge about planets on larger orbits, and about planets of lower mass. Even our limited knowledge so far, however, has opened our eyes to new types of planetary systems. The existence of hot Jupiters, and of high-mass planets on eccentric orbits, has taught us that not all planetary systems are like the solar system.

PROBLEMS

12.1 What would be the minimum Earth–Moon distance at which the center of mass of the Earth–Moon system would lie outside the Earth's surface?

12.2 The star 51 Pegasi has a mass $M_{\rm A} = 1.06 M_\odot$. Its radial velocity varies sinusoidally with a period $P = 4.23$ days and an amplitude $v_{\rm A} \sin i = 56\ {\rm m\ s^{-1}}$. What is $M_{\rm B} \sin i$ of the exoplanet causing these velocity variations?

12.3 What fractional decrease in flux, $\delta F/F$, would be caused by an Earth-like planet transiting a Sun-like star? If a transit of the Earth across the Sun were viewed by a very distant observer in the Earth's orbital plane, how long would the transit last?

12.4 Compute the ratio of Jupiter's luminosity to the Sun's luminosity, $L_{\nu,{\rm Jup}}/L_{\nu,\odot}$, at a wavelength $\lambda = 20\ \mu{\rm m}$.

12.5 What is the rotational angular momentum of Jupiter? What is the sum of the orbital angular momenta of the four Galilean satellites? Explain how the analogy of a "solar system in miniature" breaks down here.

12.6 While doing a transit study, you find an exoplanet around a nearby Sun-like star. The time between transits is $P = 32$ days. During a transit, the time from first to second contact is $t_2 - t_1 = 30$ minutes and the time from first to third contact is $t_3 - t_1 = 5$ hours. The depth of the transit is $\delta F/F = 0.01$. Doing follow-up radial velocity measurements of the star, you find that its peak radial velocity is $v_r = 65\ {\rm m\ s^{-1}}$.

(a) What is the radius of the planet?
(b) What is the mass of the planet?
(c) What is the semimajor axis of the planet's orbit?

13 Properties of Stars

Although astronomers study a wide range of objects, from dust grains to superclusters of galaxies, the study of *stars*, and the laws dictating their behavior, is a key part of astronomy. A star can be concisely defined as a luminous ball of gas powered by nuclear fusion in its interior. This definition distinguishes stars from smaller objects like planets and brown dwarfs that are too cool inside for fusion to take place. And it also distinguishes stars from stellar remnants like white dwarfs and neutron stars that were stars once but no longer host a fusion reactor in their interiors.[1]

13.1 ▪ HOW FAR IS A STAR?

Astronomers have developed many methods of estimating distances to stars and other celestial objects. We review only a few of the more useful techniques. Within the solar system, the distances between planets can be accurately measured using **radar**. A brief, powerful burst of radio waves is sent toward a planet, using the dish of a large radio telescope to collimate the radiation. After a time δt, a radio "echo" is detected; the time δt, which can be measured with great accuracy, is the round-trip travel time for a photon. The one-way distance from the radio telescope to the planet is then $d = c(\delta t)/2$.

Thanks to careful radar measurements, distances within the solar system are known with great accuracy. For instance, the length of the astronomical unit is known to better than one part per billion ($1\ \text{AU} = 149{,}597{,}870.7\ \text{km}$). The radar technique for measuring distances is useful only within the solar system. Even if you had the patience to wait more than 8 years for a reflected radio signal from Proxima Centauri (the nearest star other than the Sun), the signal would be far too faint to detect with current technology.

A useful tool in the astronomer's kit for measuring the distance to nearby stars is **stellar parallax**. Stellar parallax was mentioned in Chapter 2 (see Section 2.6) as a proof of the Earth's motion around the Sun. The Earth's motion around the Sun creates an apparent motion of nearby stars in a tiny ellipse. The semimajor axis of the ellipse

[1] Despite their name, "neutron stars" are not gaseous and not fusion-powered; thus, they are not stars by the strict definition we have adopted.

has an angular size π'', given by the formula (equation 2.33)

$$\pi'' = \frac{206{,}265 \text{ arcsec}}{d[\text{AU}]}. \tag{13.1}$$

Thus, if the parallax π'' of a star is large enough to be measured with the equipment at hand, the distance d to the star can be calculated:

$$d = \frac{206{,}265 \text{ AU}}{\pi''[\text{arcsec}]}. \tag{13.2}$$

Since we know the length of the astronomical unit very well, the accuracy with which we know the distances to nearby stars is determined by the accuracy with which we can measure π''. The astronomical unit is an inconveniently small unit for measuring stellar distances. Thus, astronomers tend to measure stellar distances in **parsecs**, where

$$d = \frac{1 \text{ parsec}}{\pi''[\text{arcsec}]}. \tag{13.3}$$

Thus, the parsec (abbreviated pc) is the distance at which a star has a parallax angle π'' equal to one arcsecond.[2] In metric units, $1 \text{ pc} = 3.086 \times 10^{16}$ m.

The angle π'', even for the Sun's nearest neighbors among the stars, is less than one arcsecond. When Friedrich Wilhelm Bessel first measured stellar parallax in 1838, he found that the star 61 Cygni had $\pi'' = 0.3$ arcsec; that's the size of a 1-cent coin (U.S. or Canadian) as seen 14 km away. Currently, the best measurements of parallax angles at visible wavelengths are those made by the *Hipparcos* satellite, which measured π'' for over 100,000 stars in our galaxy, with a typical accuracy of a milliarcsecond (0.001 arcsec).

The Sun's nearest neighbors are the three stars of the Alpha Centauri (α Cen) system. The nearest of the three is Alpha Centauri C, also known, from its proximity to us, as **Proxima Centauri**. From the *Hipparcos* measurements, we know the parallax, and thus the distance, to each star in the system:

- Proxima Centauri:
  - $\pi'' = 0.7723 \pm 0.0024$ arcsec
  - $d = 1 \text{ pc}/\pi'' = 1.295 \pm 0.004 \text{ pc} = 267{,}100 \pm 800 \text{ AU}$
- Alpha Centauri A and B:
  - $\pi'' = 0.7421 \pm 0.0014$ arcsec
  - $d = 1.348 \pm 0.003 \text{ pc} = 277{,}900 \pm 500 \text{ AU}$

Alpha Centauri A and B form a closely bound binary pair, with an average separation of $a = 24$ AU (slightly larger than the distance between the Sun and Uranus). Proxima Centauri, however, is $0.053 \text{ pc} = 10{,}800$ AU closer to us than Alpha Centauri A and B are. In addition, Proxima Centauri is separated from Alpha Centauri A and B by

[2] Because of this definition, there are 206,265 astronomical units in 1 parsec, just as there are 206,265 arcseconds in 1 radian.

7849 arcseconds (more than 2 degrees) as seen from Earth. This angular separation corresponds to a physical distance

$$D = \left(\frac{7849 \text{ arcsec}}{206{,}265 \text{ arcsec rad}^{-1}} \right) 267{,}100 \text{ AU} = 10{,}200 \text{ AU} \quad (13.4)$$

at the distance of Proxima Centauri. Thus, the three-dimensional distance between Proxima Centauri and the Alpha Centauri A and B pair is

$$D = \left[(10{,}800 \text{ AU})^2 + (10{,}200 \text{ AU})^2 \right]^{1/2} = 14{,}900 \text{ AU}, \quad (13.5)$$

or about 0.07 pc.

Even with the accurate angular measurements provided by *Hipparcos*, stellar parallax is useful only for stars within 200 parsecs of us. That's less than 3% of the distance from here to the center of our galaxy. *Hipparcos* measured the parallax of $\sim 42{,}000$ stars with an error of less than 20%; compared to the estimated 200 billion stars in our galaxy, the number with distances estimated by stellar parallax is small. For the overwhelming majority of stars in our galaxy, we must develop techniques other than stellar parallax to find their distances. We will return to the problem of distance determination later in the text. For the moment, let us divert our attention to the problem of determining the **brightness** of a star.

13.2 ▪ HOW BRIGHT IS A STAR?

In casual conversation, the word "brightness" is often used loosely. In astronomy, it is useful to distinguish between **intrinsic brightness** and **apparent brightness**. The intrinsic brightness of a star is a measure of how much light the star emits in a given time. The apparent brightness of a star is a measure of how much starlight per unit area enters our pupils (or the aperture of our telescope) in a given time. The apparent brightness of a star, or any other luminous object, depends on both its intrinsic brightness and its distance; the farther away a star is, the lower its apparent brightness.

The **intrinsic brightness** of a star is also known as its **luminosity** L. The luminosity of a star is the rate at which it emits energy in the form of electromagnetic radiation. Luminosity is commonly measured in watts (W). For example, the Sun has a luminosity of

$$L_\odot = 3.86 \times 10^{26} \text{ W}. \quad (13.6)$$

This luminosity includes all electromagnetic radiation emitted by the Sun, from radio waves to gamma rays. It is also an average over time, since the Sun's luminosity is slightly variable, changing by about 0.1% over the course of a sunspot cycle.[3]

The **apparent brightness** of a star is also known as its **flux** F. The flux of a star is the rate per unit area at which its energy strikes a surface held perpendicular to the star's

[3] Strange though it may seem, the Sun's luminosity is greatest when the number of sunspots is largest. This is because increased numbers of dark sunspots are correlated with increased numbers of bright plages. The increase in light from the plages more than compensates for the decrease in light from the sunspots.

rays. Flux is commonly measured in watts per square meter. The light emitted by stars is usually isotropic; that is, it's the same in all directions. Consider a transparent sphere of radius d centered on a star of luminosity L. The flux of light energy through the sphere is the luminosity of the star divided by the sphere's area:

$$F = \frac{L}{4\pi d^2}. \tag{13.7}$$

The observed flux of a star falls off as the inverse square of its distance d. For example, the Sun's flux at the Earth's location is

$$F = \frac{L_\odot}{4\pi (1\,\mathrm{AU})^2} = \frac{3.86 \times 10^{26}\,\mathrm{W}}{4\pi (1.496 \times 10^{11}\,\mathrm{m})^2} = 1370\,\mathrm{W\,m^{-2}}. \tag{13.8}$$

Sunlight is a potentially potent power source on Earth; unfortunately, solar panels are inefficient, our atmosphere is not transparent, clouds are frequent, and half the Earth is in shadow at any given time.

In equation (13.8), we computed the Sun's flux, given its luminosity and distance. In practice, astronomers more commonly work in the other direction: after measuring the flux and distance of a star, they compute its luminosity. Consider, for example, the star Sirius, also known as Alpha Canis Majoris (α CMa), the apparently brightest star in our night sky. From the Earth's northern hemisphere, Sirius can be seen in the winter sky, "dogging the heels" of the constellation Orion. The flux of Sirius is

$$F_\mathrm{S} = 1.2 \times 10^{-7}\,\mathrm{W\,m^{-2}}. \tag{13.9}$$

To intercept 1370 watts of sunlight, you'd need a panel one meter on a side; to intercept 1370 watts of Siriuslight, you'd need a panel roughly the size of Connecticut. The distance to Sirius, computed from its parallax, is

$$d_\mathrm{S} = 2.637\,\mathrm{pc} = 8.14 \times 10^{16}\,\mathrm{m}. \tag{13.10}$$

Thus, we can compute the luminosity of Sirius to be

$$L_\mathrm{S} = 4\pi d_\mathrm{S}^2 F_\mathrm{S} = 1.0 \times 10^{28}\,\mathrm{W} = 26 L_\odot. \tag{13.11}$$

This tells us that stars don't all have the same luminosity.

Measuring the total flux of a star, integrated over all wavelengths of light, is a difficult task. The history of stellar flux measurement began when a prehistoric human looked up at Sirius and said the prehistoric equivalent of "Gosh, that's a bright star!" As a flux measurement, this has two drawbacks. First, the human eye can only detect light in the wavelength range 4000–7000 Å. Second, the exclamation "Gosh, that's a bright star!" is not quantitative.

The first recorded attempt to quantify stellar flux at visible wavelengths was made by the Greek astronomer Hipparchus in the second century BC. After noting that stars differed in their apparent brightness, he classified them in six categories. The stars with the greatest flux were stars of the 1st magnitude. The stars with the next highest flux were stars of the 2nd magnitude, and so on down to stars of the 6th magnitude, which are the faintest stars visible to the human eye. After the invention of the telescope, the **apparent magnitude** scheme of Hipparchus was extended to fainter stars (7th magnitude, 8th magnitude, and so forth). A 6-inch amateur telescope at a dark site can reach to 13th magnitude or so; the faintest objects in the *Hubble* Ultra Deep Field, shown as Color Figure 22, are about 29th magnitude. Improvements in measuring flux led to the introduction of fractional magnitudes; careful photometry can routinely measure a star's flux to within 0.01 magnitudes, and in high-precision applications, errors as small as 0.001 magnitudes can be achieved.

The apparent magnitude system was placed on a firm mathematical basis by Norman Pogson in the mid-nineteenth century, when he realized that a difference of 5 magnitudes represents a multiplicative factor of 100 in flux. To illustrate this point, consider two stars; star #1 has an apparent magnitude m_1 and star #2 has an apparent magnitude m_2. If $m_2 - m_1 = 5$, we say that star #1 is 5 magnitudes brighter than star #2.[4] With $m_2 - m_1 = 5$, the ratio of the stars' fluxes is

$$\frac{F_1}{F_2} = 100. \tag{13.12}$$

If $m_2 - m_1 = 1$ (that is, if star #1 is 1 magnitude brighter than star #2), the ratio of fluxes is

$$\frac{F_1}{F_2} = 100^{1/5} = 10^{0.4} \approx 2.512. \tag{13.13}$$

In general, the relation between apparent magnitude and flux is

$$\frac{F_1}{F_2} = 100^{(m_2 - m_1)/5} = 10^{0.4(m_2 - m_1)}, \tag{13.14}$$

or

$$m_2 - m_1 = 2.5 \log(F_1/F_2), \tag{13.15}$$

where "log" represents a common logarithm, with base 10.

The apparent magnitude m can be thought of as a logarithmic measure of the flux, with

$$m = C - 2.5 \log F. \tag{13.16}$$

The constant C has historically been chosen so that the star Vega (Alpha Lyra, or α Lyr) has an apparent magnitude of zero.[5] Thus, $C = 2.5 \log F_{\rm Vega}$. Here are a few other apparent magnitudes of stars, taking into account only the flux at visible wavelengths:

[4] Please note: the apparent magnitude system is "bass-ackwards," in that *smaller* values of m correspond to *larger* flux.

[5] Why Vega? It's an apparently bright star, which makes its flux easier to measure. It doesn't have a binary companion providing contaminating light. It is a fairly hot star, so it emits a significant amount of light at visible wavelengths, where cooler stars are comparatively faint. Finally, its luminosity doesn't vary significantly.

Star	Apparent Magnitude, m
Sirius	-1.5
Alpha Centauri A	0.0
Alpha Centauri B	1.4
Proxima Centauri	10.7
Sun	-26.75

This tells us that we receive the same flux of visible light from Alpha Centauri A as we do from Vega. However, the flux from Vega is greater than that from Proxima Centauri by a factor

$$\frac{F_{\rm Vega}}{F_{\rm Prox}} = 10^{0.4(10.7-0.0)} \approx 19{,}000. \tag{13.17}$$

Since a star's flux depends on both luminosity and distance, so does the star's apparent magnitude. If we want a logarithmic measure of the luminosity alone, we use the **absolute magnitude** of a star, designated by the symbol M. The absolute magnitude M of a star is defined as the apparent magnitude it would have if it were at a distance $d = 10$ pc.[6] Since the apparent magnitude of a star is

$$m = C - 2.5 \log F \tag{13.18}$$

$$= C - 2.5 \log L + 2.5 \log(4\pi) + 5 \log d, \tag{13.19}$$

the absolute magnitude of the star is

$$M = C - 2.5 \log L + 2.5 \log(4\pi) + 5 \log(10\,\text{pc}) \tag{13.20}$$

$$= C - 2.5 \log L + 2.5 \log(4\pi) + 5. \tag{13.21}$$

If we measure the apparent magnitude of a star (by comparing its flux to that of Vega or some other standard star) and then measure the distance to the star (by finding its parallax), the absolute magnitude can then be computed:

$$M = m - 5 \log d + 5, \tag{13.22}$$

where d is measured in parsecs, or

$$M = m - 5 \log(d/10\,\text{pc}). \tag{13.23}$$

For instance, consider Proxima Centauri. In Section 13.1, we found that its distance from us is $d = 1.295$ pc; its apparent magnitude at visible wavelengths, as tabulated above, is $m = 10.7$. The absolute magnitude of Proxima Centauri is then

$$M = 10.7 - 5 \log(1.295/10) = 15.1. \tag{13.24}$$

[6] Why 10 parsecs? When the absolute magnitude was first defined in the early twentieth century, the only stars whose distances were accurately known were those within 10 parsecs of the Sun (and hence with parallaxes of more than 0.1 arcsec).

If Proxima Centauri were 10 parsecs away from us, it would be 4.4 magnitudes fainter than it actually is.

The difference between a star's apparent magnitude m and its absolute magnitude M is known as its **distance modulus**. From equation (13.23), the distance modulus is

$$m - M = 5\log(d/10\,\text{pc}). \tag{13.25}$$

Thus, the distance modulus is a logarithmic measure of the distance to a star. If you hear an astronomer say "That star has a distance modulus of 10," you know that the star in question is 1000 parsecs away. Let's look at the apparent magnitude, absolute magnitude and distance modulus of our example stars:

Star Name	m	M	$m - M$
Sirius	−1.5	1.4	−2.9
Alpha Centauri A	0.0	4.4	−4.4
Alpha Centauri B	1.4	5.8	−4.4
Proxima Centauri	10.7	15.1	−4.4
Sun	−26.75	4.83	−31.58

Note the wide range of absolute magnitudes for these stars. The most luminous (Sirius) is 13.7 magnitudes brighter than the least luminous. That's a factor of $10^{0.4\times 13.7} \approx 300{,}000$ in luminosity.

13.3 ▪ HOW HOT IS A STAR?

Stars are not monochromatic; they emit light with a wide range of wavelengths. Let F_λ be the specific flux of a star, defined so that $F_\lambda d\lambda$ is the star's flux at wavelengths in the range $\lambda \to \lambda + d\lambda$. Figure 13.1 shows the specific flux F_λ of the star Vega at visible and near infrared wavelengths. The spectrum of Vega, like most stellar spectra, consists of a continuum (approximately a blackbody) with absorption lines superimposed. Since our eyes can only detect radiation from wavelengths in the range 4000–7000 Å, they can give a misleading impression of the total flux of a star. This point was driven home to astronomers in the nineteenth century, when it became feasible to photograph stars. Since the photographic emulsions they used were insensitive to wavelengths longer than $\sim$ 5400 Å, stellar fluxes measured photographically differed from those measured by the human eye. Thus, astronomers began to distinguish between the photographic apparent magnitude of a star, m_{pg}, and its visual apparent magnitude, m_{vis}. Both systems were calibrated so that Vega had $m_{\text{pg}} = m_{\text{vis}} = 0$; this meant that red stars, to which photographic plates were insensitive, had $m_{\text{pg}} > m_{\text{vis}}$.

The total flux of a star, integrated over all wavelengths,

$$F_{\text{bol}} = \int_0^\infty F_\lambda d\lambda, \tag{13.26}$$

is also known as the **bolometric flux**. (A bolometer is a very sensitive thermometer that absorbs all the photons that strike it; it was invented by Samuel Langley in the nineteenth

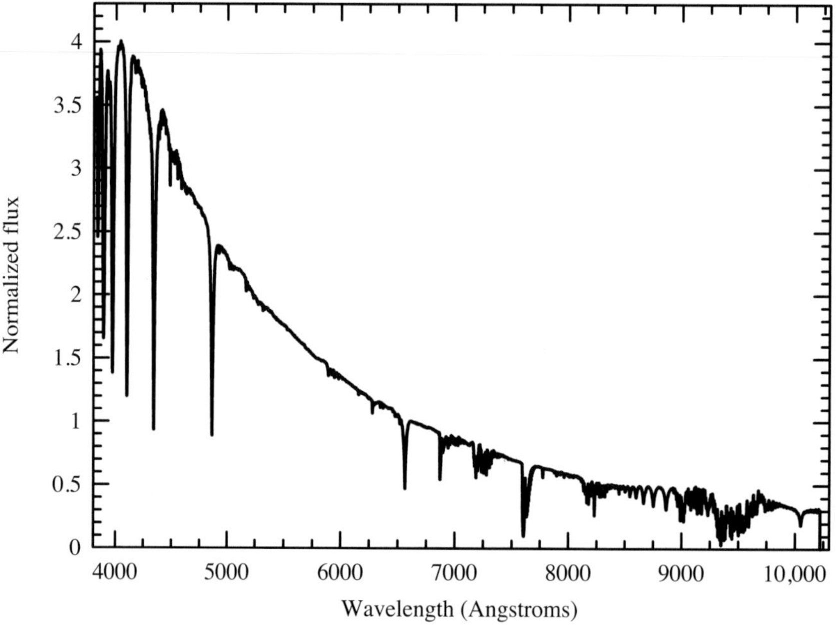

FIGURE 13.1 Spectrum (flux per unit wavelength) of the star Vega. The flux is in arbitrary units.

century, when he was attempting to measure the Sun's total flux.) Unfortunately, measuring the bolometric flux of apparently faint stars is difficult. For one thing, the Earth's atmosphere is opaque at many wavelengths, requiring you to place your detectors above the atmosphere.

If you manage to measure the bolometric flux $F_{\rm bol}$ of a star, the apparent bolometric magnitude of the star is

$$m_{\rm bol} = C_{\rm bol} - 2.5 \log F_{\rm bol}. \tag{13.27}$$

Because of the difficulty of measuring bolometric fluxes, there was long debate about the appropriate value of the constant $C_{\rm bol}$. Finally, in 1997, Commission 25 of the International Astronomical Union (which deals with stellar photometry) requested that everyone use a scale on which the Sun's absolute bolometric magnitude is $M_{\rm bol,\odot} = 4.74$. The Sun, rather than Vega, was chosen for the honor of normalizing the bolometric magnitude scale because the Sun is the only star whose bolometric flux has been measured with extremely high accuracy. With the IAU-approved normalization, the relation between absolute bolometric magnitude and luminosity is

$$M_{\rm bol} = 4.74 - 2.5 \log(L/L_\odot), \tag{13.28}$$

or

$$L/L_\odot = 10^{0.4(4.74 - M_{\rm bol})}. \tag{13.29}$$

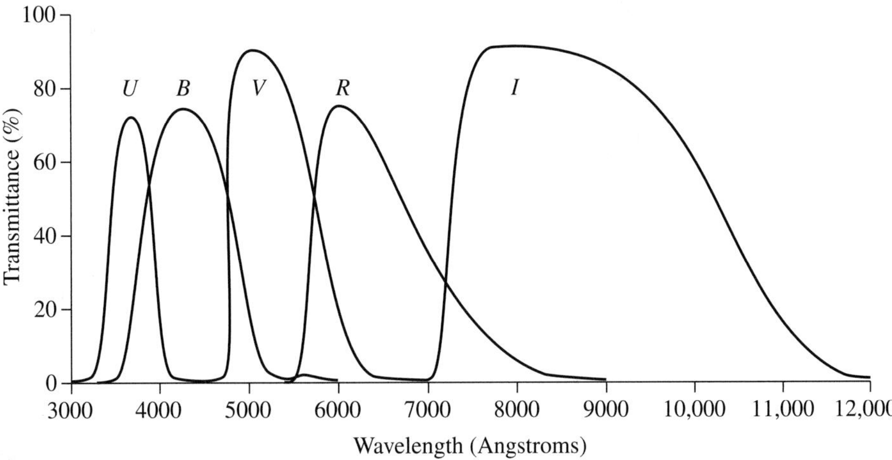

FIGURE 13.2 Spectral sensitivity $S(\lambda)$ of the Johnson–Cousins filters.

Given the difficulty of measuring bolometric fluxes, astronomers usually content themselves with measuring the flux over a strictly defined range of wavelengths. This is done, in practice, by putting a colored filter in the light path of a telescope, thus eliminating the unwanted wavelengths of light. Many different filter systems are in use. One of the most durably popular is the **Johnson–Cousins system** devised by Harold Johnson, A. W. J. Cousins, and their collaborators. There are five main filters in the Johnson–Cousins system: U (ultraviolet), B (blue), V (visual), R (red), and I (infrared). The spectral sensitivity, $S(\lambda)$, of the five basic Johnson–Cousins filters is displayed in Figure 13.2. The spectral sensitivity S gives the fraction of the light at wavelength λ that can pass through a filter. The Johnson-Cousins filters were chosen to highlight different spectral regions. The V filter, in particular, was chosen to approximate what our eyes see (although the V filter admits a narrower range of wavelengths than the human eye can detect).[7]

In practice, the flux of starlight detected through a particular filter depends on the specific flux F_λ of the star and the combined spectral sensitivity $S(\lambda)$ of the filter and the detector used. A star's flux seen through the V filter, for instance, would be

$$F_V = \int_0^\infty F_\lambda S_V(\lambda) d\lambda, \tag{13.30}$$

where F_λ is the star's differential flux and $S_V(\lambda)$ is the combined spectral sensitivity of the V filter and the detector. The apparent magnitude of the star in the V band is then

$$m_V = C_V - 2.5 \log F_V. \tag{13.31}$$

[7] The Johnson–Cousins filters are popular in part because they can be constructed from relatively inexpensive colored glass, available "off the shelf" from specialty glass manufacturers like Schott Glaswerke and Corning Incorporated.

Similarly, the U band apparent magnitude is

$$m_U = C_U - 2.5 \log F_U, \tag{13.32}$$

the B band apparent magnitude is

$$m_B = C_B - 2.5 \log F_B, \tag{13.33}$$

and so forth, through all the filters. In keeping with astronomical tradition, the *UBVRI* magnitudes are normalized so that Vega has an apparent magnitude close to zero. More precisely, it turns out that Vega has $m_U = m_B = m_V = m_R = m_I = +0.03$.

A filter can be approximately described by its **bandwidth** and **effective wavelength**. The bandwidth $\Delta\lambda$ can be defined in different ways; one common definition is the width of the spectral response $S(\lambda)$ at half its maximum value. The V band, for instance (see Figure 13.2), has a bandwidth $\Delta\lambda_V = 840$ Å. The effective wavelength of the V band is

$$\lambda_{\text{eff},V} = \frac{\int_0^\infty \lambda S_V(\lambda) F_\lambda d\lambda}{\int_0^\infty S_V(\lambda) F_\lambda d\lambda}, \tag{13.34}$$

and similarly for the other bands in the Johnson–Cousins system. Obviously, the effective wavelength depends on the shape of the spectrum; the effective wavelength is shorter for a bluer star. For the Sun, the effective wavelength of the V band is $\lambda_{\text{eff},V} = 5502$ Å; for Vega, a bluer star, the effective wavelength is only $\lambda_{\text{eff},V} = 5448$ Å. When effective wavelengths are given without reference to a particular star, it is usually the case that equation (13.34) has been computed assuming a blackbody in the Rayleigh–Jeans limit where $F_\nu \propto \nu^2$, or equivalently, $F_\lambda \propto \lambda^{-4}$.

Multicolor photometry (that is, the practice of measuring a star's flux through multiple filters) is useful for two main reasons. First, the more filters you look through, the more accurate your reconstruction of the star's spectrum and bolometric flux. Second, multicolor photometry yields information about the color, and hence the **temperature**, of a star. The **color index** of a star is the difference of its apparent magnitude seen through two different filters. For instance, one popular color index is

$$B - V = m_B - m_V. \tag{13.35}$$

Other frequently used color indices are $U - B = m_U - m_B$, and $V - R = m_V - m_R$.[8] Color indices are useful because they depend on the temperature of the star's photosphere (usually called the star's "surface temperature," even though stars don't have a sharply defined surface).

By definition, the star Vega has $B - V = 0$. The surface temperature of Vega is $T \approx 10{,}000$ K. If a star has $T > 10{,}000$ K, it will be bluer than Vega (its B flux will be enhanced relative to its V flux), and thus it will have $B - V < 0$, thanks to the backward nature of the magnitude scale. By contrast, if a star has $T < 10{,}000$ K, it will be redder than Vega (its V flux will be enhanced relative to its B flux), and thus it will have

[8] The universal convention is that color indices are m(shorter effective wavelength) $-$ m(longer effective wavelength).

$B - V > 0$. The relation between the surface temperature T and the color index $B - V$ can be computed for a blackbody. However, since stars aren't exactly blackbodies, it's more useful to use a purely empirical relation such as

$$T \approx \frac{9000 \text{ K}}{(B - V) + 0.93}, \tag{13.36}$$

which provides a good fit for stars with color index $-0.1 \lesssim B - V \lesssim 1.4$, implying surface temperatures $4000 \text{ K} \lesssim T \lesssim 11{,}000 \text{ K}$. Measuring a color index such as $B - V$ thus gives you a quick and cheap way of estimating a star's surface temperature. It's usually much easier than taking a star's spectrum and doing a detailed spectral analysis.

Because bolometric fluxes are so difficult to measure, what astronomers generally do is measure the V band apparent magnitude and then add a **bolometric correction**:

$$BC = m_{\text{bol}} - m_V = M_{\text{bol}} - M_V. \tag{13.37}$$

For instance, the Sun has $M_{\text{bol}} = 4.74$ and $M_V = 4.83$, so its bolometric correction is $BC = 4.74 - 4.83 = -0.09$. The calibration of the bolometric magnitude scale ($M_{\text{bol},\odot} = 4.74$) was chosen so that normal stars have $BC \leq 0$. The bolometric correction is equal to zero for stars with $T \approx 6700$ K, since these are the stars whose emission peaks in the V band; for these stars, estimating the bolometric flux from the V band flux yields a pretty good approximation. For hotter stars, most of the energy escapes at shorter wavelengths; for cooler stars, most of the energy escapes at longer wavelengths. The bolometric correction can be calculated by using model stellar atmospheres and computing how much of the total flux is emitted in the V band. In practice, the bolometric correction is something that you can look up, because someone else has done the dirty work of calculating it for you.[9]

To review how magnitudes, color indices, and bolometric corrections work, let's do an example. The star Epsilon Eridani (ϵ Eri) has $m_V = 3.73$ and $m_B = 4.61$; it's visible to the naked eye but is not eye-catchingly bright. Its color index is $B - V = 4.61 - 3.73 = 0.88$ (redder than Vega). The empirical relation given in equation (13.36) tells us that the surface temperature of Epsilon Eridani is $T \approx 5000$ K (cooler than Vega). The bolometric correction for a normal star of this temperature is $BC = -0.40$. The apparent bolometric magnitude of Epsilon Eridani is $m_{\text{bol}} = 3.73 - 0.40 = 3.33$. The distance to Epsilon Eridani, found by stellar parallax, is $d = 3.218$ pc, so the absolute bolometric magnitude is

$$M_{\text{bol}} = m_{\text{bol}} - 5 \log(d/10 \text{ pc}) = 5.80 \tag{13.38}$$

and its luminosity is

$$L/L_\odot = 10^{0.4(4.74 - M_{\text{bol}})} = 10^{-0.42} = 0.38. \tag{13.39}$$

[9] Exactly how this dirty work is done, in the relatively simple case of a blackbody, is outlined in the appendix to this chapter.

Thus, measuring the flux of a star through two filters and measuring its parallax is sufficient for us to compute its surface temperature and luminosity.

Warning: we have been making the assumption that the space between the star and our telescope is completely transparent. In the real universe, we must take the effect of **extinction** into account. **Atmospheric extinction**, due to the Earth's atmosphere, causes about 0.2 magnitudes of dimming in the V band when a telescope is pointing straight up (toward the zenith). The exact amount of extinction depends on your location and is also variable with time. Let Z be the angle between the zenith direction and the direction in which the telescope is pointing; $Z = 0°$ when the telescope is pointing straight up, and $Z = 90°$ when it's pointing toward the horizon. When $Z < 60°$, a useful approximation to the amount of dimming is

$$m_V(\text{above atmosphere}) \approx m_V(\text{observed}) - 0.2 \sec Z. \tag{13.40}$$

(When $Z > 60°$, you should think twice about observing something so close to the horizon.) When you look up the apparent magnitude of a star in a reference work, it will already be corrected for atmospheric extinction. **Interstellar extinction**, due to scattering by interstellar dust, can also be significant, especially within the disk of our galaxy, where most of the dust lies. We discuss interstellar extinction in more detail in Chapter 17, which deals with the interstellar medium.

You might, at this point, be asking yourself why astronomers persist in use of the magnitude scale when it is possible to measure fluxes in real physical units (energy per unit time per unit area). There is a good, but somewhat subtle, reason for retaining the magnitude system; namely, that astronomers can measure *relative* brightnesses to greater accuracy than *absolute* brightnesses. The magnitude system is a relative scale: the magnitude of a given star tells us precisely how bright that star is relative to Vega. The next step, that of calibrating the magnitude system in physical units (as in equation 13.31), is thus a separate, and in fact a more difficult, process. Use of the magnitude system allows us to keep the observations separate from the "zero-point" calibration (the constant C_V in equation 13.31), which can change as observations improve.

13.4 ▪ HOW BIG IS A STAR?

If you know the distance d to a star, and its angular diameter α, then its radius R can be determined by a simple bit of trigonometry:

$$\frac{R}{d} = \tan\left(\frac{\alpha}{2}\right). \tag{13.41}$$

Therefore,

$$R = d \tan\left(\frac{\alpha}{2}\right) \approx \frac{d\alpha}{2}, \tag{13.42}$$

in the small angle limit, where $\alpha \ll 1$ rad. Computing the Sun's radius is easy. The average distance to the Sun is $d = 1\text{ AU} = 1.496 \times 10^8$ km. The average angular diameter

FIGURE 13.3 *Hubble Space Telescope* image of Betelgeuse, with finding chart.

of the Sun is $\alpha = 1919$ arcsec $= 9.30 \times 10^{-3}$ rad, so

$$R_\odot = \frac{(1.496 \times 10^8\,\text{km})(9.30 \times 10^{-3})}{2} = 696{,}000\,\text{km}. \tag{13.43}$$

Stars other than the Sun have angular diameters that are very small, and hence difficult to measure. If we were to view the Sun from the location of Proxima Centauri ($d = 1.295$ pc $= 267{,}000$ AU), its angular size would be

$$\alpha = 1919\,\text{arcsec}\left(\frac{1\,\text{AU}}{267{,}000\,\text{AU}}\right) = 7.2 \times 10^{-3}\,\text{arcsec}. \tag{13.44}$$

Measuring an angular size of 7 milliarcseconds is difficult.

Only one star other than the Sun has had its angular diameter resolved by direct imaging. The star Betelgeuse (Alpha Orionis, or α Ori) has been resolved by the *Hubble Space Telescope* (Figure 13.3). The distance to Betelgeuse, determined from its parallax, is

$$d_\text{B} = 131\,\text{pc} = 2.70 \times 10^7\,\text{AU}. \tag{13.45}$$

The angular diameter of Betelgeuse, measured at ultraviolet wavelengths using the *Hubble Space Telescope*, is

$$\alpha_\text{B} = 0.125\,\text{arcsec} = 6.06 \times 10^{-7}\,\text{rad}. \tag{13.46}$$

Thus, its radius is

$$R_\text{B} = d_\text{B}\alpha_\text{B}/2 = 8.2\,\text{AU} = 1.2 \times 10^9\,\text{km} = 1800R_\odot. \tag{13.47}$$

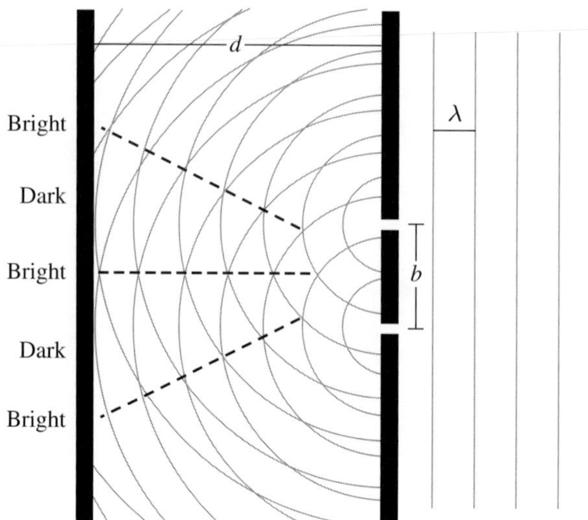

FIGURE 13.4 Two-slit interference; light coming in from the right passes through two slits and falls on a screen to the left.

Betelgeuse is a supergiant star, swollen to an immense radius. If you placed its center at the Sun's location, it would extend far beyond Jupiter's orbit. Supergiant stars are rare; Betelgeuse is the nearest supergiant to the Sun. All the stars closer than Betelgeuse are smaller in radius, and all the stars larger than Betelgeuse are farther away.

Stars less bloated than Betelgeuse can have their radii measured using **interferometry**. To see how the principle of interference can tell you the angular size of a star, consider the classic "two-slit" interference experiment illustrated in Figure 13.4. Light of wavelength λ comes from a point source far to the right of the image and strikes a wall in which two narrow slits have been cut. The light passing through one slit interferes with the light passing through the other slit, causing **constructive interference** where the wavecrests add together, and **destructive interference** where wavecrests from one slit encounter wave troughs from the other slit. If a screen is placed to the left of the slits, parallel to the wall, a pattern of bright and dark bands is seen (bright = constructive interference, dark = destructive interference). The distance between the bright bands on the screen is $x = \lambda d/b$, where d is the distance between the wall and the screen, and b is the distance between the slits.

If stars were perfect point sources, they would produce interference bands in exactly this way. However, stars have a finite (although small) angular size α. Thus, light from the upper limb of the star approaches the slits from a slightly different angle than light from the lower limb of the star. This means that when light from the *upper* limb of the star falls on the screen, its bright and dark bands are slightly displaced from the bright and dark bands produced by light from the *lower* limb. The interference pattern produced by a star of angular diameter α will be smeared out when

$$\alpha[\text{radians}] > \lambda/b. \tag{13.48}$$

Thus, if you observed a star of angular diameter α at a wavelength λ through a pair of movable slits, the interference pattern would be smeared out once the baseline between the slits increased to

$$b > \lambda/\alpha. \tag{13.49}$$

Scaling to a plausible angular diameter for a nearby star,

$$b > 10\,\mathrm{m}\left(\frac{\lambda}{5000\,\text{Å}}\right)\left(\frac{\alpha}{0.01\,\mathrm{arcsec}}\right)^{-1}. \tag{13.50}$$

In practice, the technique of stellar interferometry doesn't use two slits in a wall; it uses two telescopes separated by a distance b.

The angular diameter of Betelgeuse was measured using interferometry as early as the year 1920.[10] Currently, the interferometer at the Very Large Telescope (VLT) in Chile uses telescopes separated by a distance as large as 140 meters; thus, it can measure angles smaller than a milliarcsecond. The VLT interferometer has been used, for instance, to find the angular diameters of the stars in the Alpha Centauri system:

- Alpha Centauri A: $\alpha = 8.5 \times 10^{-3}$ arcsec, $R = 1.23 R_\odot$.
- Alpha Centauri B: $\alpha = 6.0 \times 10^{-3}$ arcsec, $R = 0.87 R_\odot$.
- Proxima Centauri: $\alpha = 1.0 \times 10^{-3}$ arcsec, $R = 0.14 R_\odot$.

Note that the most luminous star in the system (α Cen A) is also the largest in size. It is particularly interesting to discover that Proxima Centauri isn't much larger than Jupiter, which has $r = 0.10 R_\odot$. We tend to think of stars as huge balls of gas, but the smallest stars aren't much larger than the biggest planets. At the moment, there are roughly a thousand stars whose radii have been measured using interferometric techniques. They span the range from supergiants like Betelgeuse to dwarfs like Proxima Centauri.

When a star's radius and luminosity are both known, we have a new way of estimating its temperature. For a spherical blackbody,

$$L = 4\pi R^2 \sigma_{\mathrm{SB}} T^4, \tag{13.51}$$

where σ_{SB} is the Stefan–Boltzmann constant. Whether a star is really a blackbody or not, we can assign it an **effective temperature**, defined as

$$T_{\mathrm{eff}} = \left(\frac{L}{4\pi R^2 \sigma_{\mathrm{SB}}}\right)^{1/4}. \tag{13.52}$$

If this temperature doesn't agree closely with the temperature as estimated from the star's spectrum or color index, then the star must be far from being a blackbody.

[10] The angular diameter of Betelgeuse at visible wavelengths is found to be $\alpha = 0.058$ arcsec, less than half its size measured at ultraviolet wavelengths. This indicates that Betelgeuse has an extended atmosphere that is far more opaque to ultraviolet light than to visible light.

13.5 ▪ HOW MASSIVE IS A STAR?

The mass of a star can be determined using Kepler's third law, as modified by Isaac Newton (Section 3.1.3):

$$M_A + M_B = \frac{4\pi^2}{G} \frac{a^3}{P^2}, \tag{13.53}$$

where M_A and M_B are the masses of two objects orbiting their mutual center of mass, P is their orbital period, and a is the semimajor axis of their relative orbit.[11] Kepler's law restricts us to measuring the total mass ($M_A + M_B$) of a binary system. Finding the mass of an isolated star is like recording the sound of one hand clapping. Fortunately, the majority of stars in the solar neighborhood have companions: a star, a substellar object (a brown dwarf or planet), or a formerly stellar object (a white dwarf).

Binary stellar systems are usually classified by the way in which they are detected. There are three main classes.

- **Visual binary.** The two stars in the binary system are individually resolved in your telescope. You can tell they are not a chance superposition of stars at different distances, because one star moves on an elliptical orbit relative to the other. Visual binaries tend to have large separations, and hence long periods.
- **Spectroscopic binary.** The two stars are unresolved, appearing as a single blob. However, the spectrum of the binary system shows absorption lines that oscillate in wavelength as the stars' radial velocities change. Spectroscopic binaries tend to have high orbital speeds, and hence small orbits and short periods.
- **Eclipsing binary.** The two stars are unresolved. However, the orbital plane of the system is seen nearly edge-on, so the stars periodically eclipse each other, causing dips in the flux. Eclipsing binaries tend to have small separations, and hence short periods.

We discuss each of these in turn.

13.5.1 Visual Binaries

An example of a visual binary is the Sirius system. It was realized in the nineteenth century that the star we know as Sirius has a much fainter companion. The more luminous component of the binary system is now known as Sirius A, while its dim companion is called Sirius B. At visible wavelengths, Sirius A is roughly 8000 times brighter than Sirius B; at X-ray wavelengths, however, Sirius B is the brighter of the two, as shown in Figure 13.5. Sirius A has a surface temperature $T = 9900\,\mathrm{K}$ (estimated from its spectrum) and a luminosity $L = 26L_\odot$. By contrast, Sirius B has

[11] Unfortunately, the letter "M" does double duty, indicating both a star's mass and its absolute magnitude. The context should make it clear which meaning is appropriate.

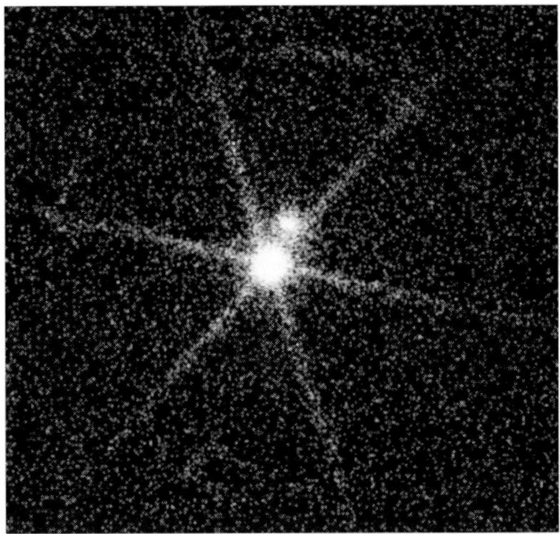

FIGURE 13.5 Sirius B (with diffraction spikes) and Sirius A (above and to the right) as seen in X-rays.

a surface temperature $T = 24{,}800$ K but a luminosity of only $L = 0.024L_\odot$, most of which emerges at wavelengths too short for the human eye to detect.

Sirius B is much less luminous than Sirius A, despite having a higher surface temperature. This means that Sirius B must have a much smaller surface area. From the relation among radius, temperature, and luminosity for a spherical blackbody,

$$\frac{R_\mathrm{B}}{R_\mathrm{A}} = \left(\frac{L_\mathrm{B}}{L_\mathrm{A}}\right)^{1/2} \left(\frac{T_\mathrm{A}}{T_\mathrm{B}}\right)^2 = (0.00092)^{1/2}(0.40)^2 = 0.0048. \tag{13.54}$$

Since the radius of Sirius A is known to be $r_\mathrm{A} = 1.71R_\odot$, from stellar interferometry, the radius of Sirius B must be

$$R_\mathrm{B} = 0.0048R_\mathrm{A} = 0.0084R_\odot = 0.92R_\oplus. \tag{13.55}$$

When Sirius B and similar objects were determined to have high temperatures but low luminosities, they were termed "white dwarfs," since they are small in size (comparable in size to the Earth) but high enough in temperature to be white hot.

The motion of Sirius B relative to Sirius A has been traced for well over a century; thus, the relative orbit of the two objects, as projected onto the plane of the sky, is well known (Figure 13.6). The orbital period of the Sirius system is $P = 50.05$ yr. To find the semimajor axis a of the relative orbit, we must *deproject* the observed ellipse. That is, we must ask the question What ellipse, with Sirius A at one focus, appears (when viewed at an angle), like the ellipse in Figure 13.6? This question has a unique answer.

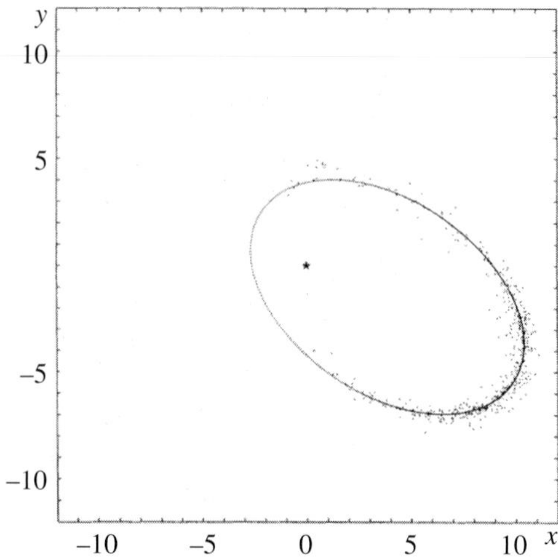

FIGURE 13.6 The projected orbit of Sirius B relative to Sirius A. The five-pointed star is the position of Sirius A, and the small points are observed positions of Sirius B, from the year 1862 to 1979; coordinates are in arcseconds. The best-fitting ellipse for the orbit of Sirius B is shown.

We won't go into the geometrical details,[12] but will simply quote the answer. The true three-dimensional orbit of Sirius B relative to Sirius A has an eccentricity of $e = 0.59$ and a semimajor axis of angular length

$$a'' = 7.50 \text{ arcsec} = 3.64 \times 10^{-5} \text{ rad} \tag{13.56}$$

and is viewed at an inclination $i = 43.4°$.[13] Since the distance to the Sirius system is $d = 2.637$ pc $= 544{,}000$ AU, the semimajor axis of the orbit of Sirius B relative to Sirius A is

$$a = a''d = 19.78 \text{ AU}, \tag{13.57}$$

comparable to the average distance from Uranus to the Sun, and slightly smaller than the average distance between Alpha Centauri A and B.

We now have the necessary information to compute the total mass of the Sirius system. If the masses M_A and M_B are in solar masses, a is in AU, and P is in years, Kepler's third law can be written in the form

$$M_A + M_B = \frac{a^3}{P^2}. \tag{13.58}$$

[12] A good textbook on celestial mechanics will give you all the details you might want.

[13] An inclination $i = 0°$ means we are looking at the orbit face-on; an inclination $i = 90°$ means we are looking at the orbit edge-on.

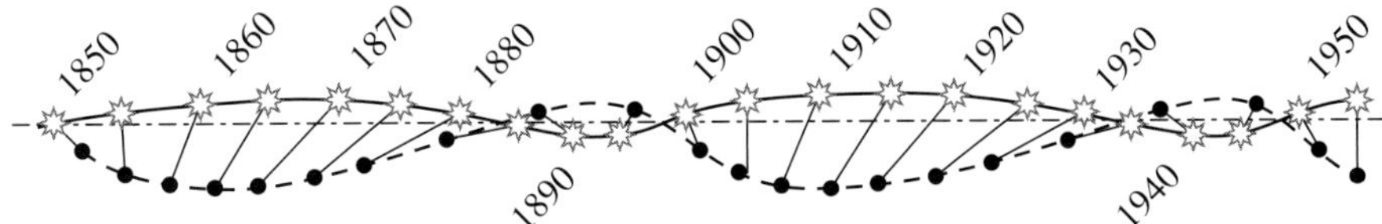

FIGURE 13.7 The long-term motion of the Sirius system. The asterisk represents Sirius A, and the dot represents Sirius B.

For the Sirius system,

$$M_A + M_B = \frac{(19.78)^3}{(50.05)^2} = 3.09 M_\odot. \tag{13.59}$$

We know that Sirius A and Sirius B, taken together, have a mass more than three times that of the Sun; but how is the mass allocated between the two objects?

To determine how the total mass is divided between Sirius A and Sirius B, we must locate the **center of mass** of the binary system. When the positions of Sirius B and Sirius A are plotted relative to background objects (such as distant quasars), they show the "wobbly" motion displayed in Figure 13.7. This motion is the combination of the linear motion of the center of mass of the Sirius system plus the elliptical motions, with period $P = 50.05$ yr, of Sirius A and Sirius B relative to the center of mass. (The small annual wiggles due to stellar parallax are not shown in this plot.)

If Sirius B had a negligibly small mass—as you might expect from its tiny volume—then the motion of Sirius A would be a straight line. Since Sirius A does show wobbles in its motion, the mass of Sirius B must be significant.[14] The motion of the center of mass is a straight line if we assume $a_B/a_A = 2.2$. Thus, from equation (12.8), we may conclude

$$\frac{M_A}{M_B} = \frac{a_B}{a_A} = 2.2. \tag{13.60}$$

That is, Sirius A is just over twice the mass of Sirius B. When we combine our two bits of information, $M_A + M_B = 3.09 M_\odot$ and $M_A = 2.2 M_B$, we find the solution

$$M_B = 0.97 M_\odot, \qquad M_A = 2.12 M_\odot. \tag{13.61}$$

When astronomers first learned that white dwarfs like Sirius B had masses comparable to the Sun, but volumes comparable to the Earth, they were flabbergasted. The average density of Sirius B is more than 2 tons per cubic centimeter.

[14] In fact, the wobbles in the motion of Sirius A were first noticed by Friedrich Bessel (of parallax fame) in 1844, nearly 20 years before Sirius B was seen through a telescope.

13.5.2 Spectroscopic Binaries

Spectroscopic binaries are detected by variability in the Doppler shift of one or both stars in the binary system. In a **double-lined spectroscopic binary**, the absorption lines of *both* stars can be seen in the spectrum of the binary system. In a **single-lined spectroscopic binary**, the absorption lines of only one star in the system can be seen; the existence of the dimmer star can only be inferred from the Doppler shift it gives to its brighter companion. An exoplanet and its parent star, as discussed in Section 12.3, can be thought of as a variety of single-lined spectroscopic binary, in which the Doppler shift of one object (the star) can be detected, but the shift of the other (the exoplanet) cannot be.

Suppose we start with the case of a double-lined spectroscopic binary. This is the case in which we have more information and can learn more about the properties of the binary system. For a double-lined binary, the radial velocity curve of both stars can be plotted; the amplitudes of the two radial velocity curves (see Figure 12.5b for an example) yield $v_{\rm A} \sin i$ and $v_{\rm B} \sin i$. By convention, $v_{\rm A}$ is the orbital speed of the more massive star, and $v_{\rm B} > v_{\rm A}$ is the orbital speed of the less massive star; the parameter i is the inclination of their orbital plane (as illustrated geometrically in Figure 12.6).

For spectroscopic binaries, it is generally safe to assume that the orbits are approximately circular. This would be a bad assumption for visual binaries, since many visual binaries, such as the Sirius system, have large eccentricity. However, spectroscopic binaries tend to have larger orbital speeds, and hence smaller orbits: small orbits, as we have seen when looking at hot Jupiters, tend to become circularized with time by tidal friction. The radial velocity curves also tell us the orbital period P of the two stars.

What properties of the double-lined spectroscopic binary can we compute, given the measurable quantities $v_{\rm A} \sin i$, $v_{\rm B} \sin i$, and P? Starting with equation (12.12), which applies to all binary systems on circular orbits, we can write

$$\frac{M_{\rm B}}{M_{\rm A}} = \frac{v_{\rm A}}{v_{\rm B}} = \frac{v_{\rm A} \sin i}{v_{\rm B} \sin i} \tag{13.62}$$

and thus find the ratio of the stellar masses in the binary system. In addition, we know that

$$P = \frac{2\pi a_{\rm A}}{v_{\rm A}} = \frac{2\pi a_{\rm B}}{v_{\rm B}}, \tag{13.63}$$

and thus we can compute

$$a_{\rm A} \sin i = \frac{P v_{\rm A} \sin i}{2\pi} \tag{13.64}$$

and

$$a_{\rm B} \sin i = \frac{P v_{\rm B} \sin i}{2\pi}. \tag{13.65}$$

Note that we cannot determine the size of the orbits without some independent method of determining the inclination i. This also implies that we cannot determine the total mass, $M_A + M_B$, of the spectroscopic binary system. To see why this is so, let's start with Kepler's third law in the form (compare to equation 12.15):

$$M_A + M_B = \frac{4\pi^2}{GP^2}(a_A + a_B)^3. \tag{13.66}$$

However, as we have just seen, we cannot determine a_A and a_B using the spectroscopic information alone; we know only $a_A \sin i$ and $a_B \sin i$. Thus, we can compute only the quantity

$$\begin{aligned}(M_A + M_B)\sin^3 i &= \frac{4\pi^2}{GP^2}(a_A \sin i + a_B \sin i)^3 \\ &= \frac{P}{2\pi G}(v_A \sin i + v_B \sin i)^3, \end{aligned} \tag{13.67}$$

where we have made use of equations (13.64) and (13.65). Thus, although we can compute the mass ratio M_B/M_A without knowing the inclination, we can only place a lower limit on the total mass, since $M_A + M_B \geq (M_A + M_B)\sin^3 i$.

There are two ways in which we can deal with our ignorance of the inclination i. First, we can look solely at spectroscopic binaries that are also *eclipsing* binaries. A binary will undergo eclipses, with one star passing in front of the other, only in the case $i \approx 90°$. Thus, for an eclipsing spectroscopic binary, we can make the substitution $\sin i \approx 1$ in equation (13.67). Second, we can look at a large sample of spectroscopic binaries, which will allow us to make statements about the mass in a statistical fashion. If we average the observed properties of many double-lined spectroscopic binaries, we find, from equation (13.67),

$$\langle M_A + M_B\rangle\langle\sin^3 i\rangle = \frac{1}{2\pi G}\langle P(v_A \sin i + v_B \sin i)^3\rangle. \tag{13.68}$$

To find the average mass, $\langle M_A + M_B\rangle$, for the sample of binaries in question, we must obtain an average value $\langle\sin^3 i\rangle$. In general, to obtain the average value of a function $f(i)$ of the inclination angle, as shown in Figure 13.8, we must compute

$$\langle f(i)\rangle = \frac{\int_0^{\pi/2} f(i)\sin i\, di}{\int_0^{\pi/2}\sin i\, di}. \tag{13.69}$$

Since the denominator on the right-hand side is unity, we can compute

$$\langle\sin^3 i\rangle = \int_0^{\pi/2}\sin^4 i\, di = \left.\frac{3i}{8} - \frac{\sin 2i}{4} + \frac{\sin 4i}{32}\right|_0^{\pi/2} = \frac{3\pi}{16} = 0.59. \tag{13.70}$$

This tells us that inclination effects cause us to underestimate masses typically by about 41%.

There is, however, an important subtlety: binary systems are decreasingly likely to be discovered as the inclination of the system decreases. Indeed, a binary at $i = 0°$ cannot be detected as a spectroscopic binary, since the radial component of its stars' velocity is zero. We should thus adjust our calculation to account for this selection effect. A simple

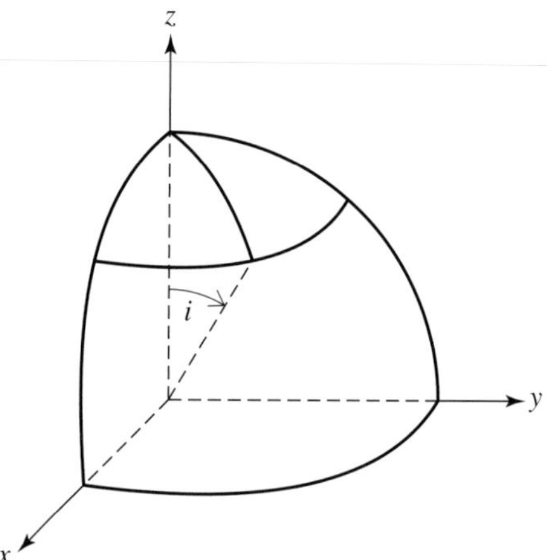

FIGURE 13.8 The polar angle i can in this case take values between 0 and $\pi/2$. A circumscribed circle obtained by rotating a line around the z axis has a differential area $2\pi \sin i \, di$.

approximation might be that all binaries with $i \geq i_0$ are detected, and no binaries with $i < i_0$ are detected. In this case, the average value of $\sin^3 i$ for the detected spectroscopic binaries will be

$$\langle \sin^3 i \rangle = \frac{\int_{i_0}^{\pi/2} \sin^4 i \, di}{\int_{i_0}^{\pi/2} \sin i \, di}. \tag{13.71}$$

If $i_0 = 45°$, for instance, then $\langle \sin^3 i \rangle \approx 0.77$ for the detected binaries, and the underestimate in mass will be 23%. If we set a lower detection limit of $i_0 = 60°$, then $\langle \sin^3 i \rangle \approx 0.88$, and the underestimate will be only 12%. By eliminating from our observation the nearly face-on systems for which the underestimates are most severe, the selection effects produce estimates that are closer to the true value of $M_A + M_B$.

Although we have been discussing double-lined spectroscopic binaries, the more limited information provided by a single-lined binary is not totally useless. If we know P and $v_A \sin i$, but not $v_B \sin i$, it is useful to write equation (13.67) in the form

$$(M_A + M_B) \sin^3 i = \frac{P}{2\pi G} (v_A \sin i)^3 \left(1 + \frac{v_B \sin i}{v_A \sin i}\right)^3. \tag{13.72}$$

Substituting from equation (13.62), we can eliminate the unknown value of $v_B \sin i$:

$$(M_A + M_B) \sin^3 i = \frac{P}{2\pi G} (v_A \sin i)^3 (1 + M_A/M_B)^3. \tag{13.73}$$

Rearranging, we find

$$\frac{M_B}{(1 + M_A/M_B)^2} \sin^3 i = \frac{P}{2\pi G} (v_A \sin i)^3. \tag{13.74}$$

Thus, from P and $v_A \sin i$, observable properties of a single-lined spectroscopic binary, we can compute the **mass function**

$$f(M_A, M_B) = \frac{M_B}{(1 + M_A/M_B)^2} \sin^3 i = \frac{M_B^3}{(M_A + M_B)^2} \sin^3 i. \qquad (13.75)$$

Since $f(M_A, M_B) \leq M_B$, the mass function gives us a lower limit on the mass of the unseen component of the single-lined spectroscopic binary. Note that in the limit $M_B \ll M_A$, $f(M_A, M_B) \approx (M_B^3/M_A^2) \sin^3 i$. In that case, equation (13.75) reduces to

$$M_B^3 \sin^3 i \approx \frac{M_A^2 P}{2\pi G} (v_A \sin i)^3, \qquad (13.76)$$

a result that we have already derived (equation 12.21) for the case of a star–exoplanet system, which can be thought of as a single-lined spectroscopic binary with M_B (the exoplanet's mass) much smaller than M_A (the star's mass).

13.5.3 Eclipsing Binaries

Eclipsing binaries are detected as **variable stars**, whose brightness changes with time. In Section 17.3, we discuss **pulsating variable stars** whose luminosities actually do change on timescales that can by detected by humans; however, eclipsing binaries vary in brightness only because each star completely or partially eclipses the other star once each orbit. Obviously, this requires the observer to be quite close to the orbital plane of the system, so $i \approx 90°$ for eclipsing systems. Many eclipsing binaries are in close pairs and have periods of a few days or less.

The light curve for a typical eclipsing binary is shown in Figure 13.9. Unlike the light curve of a star transited by a dark exoplanet, the light curve of an eclipsing binary shows

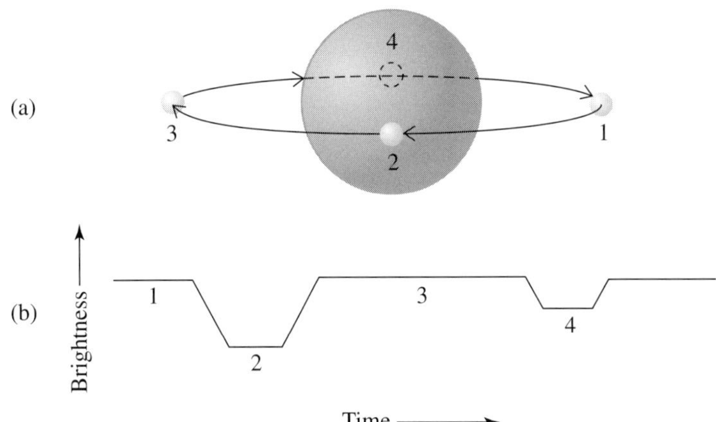

FIGURE 13.9 (a) The orbit of the smaller component of an eclipsing binary relative to the larger component. (b) The light curve of the eclipsing binary. The numbers on the light curve correspond to the corresponding positions in part (a).

two dips in flux per orbit. This reflects the fact that there are two eclipses during one orbit: one when star A is the closer of the two stars to the observer, and the other when star B is closer. The depth of each eclipse depends on how much radiating area is blocked from the observer's view. Unless the two stars have the same surface temperature, the two eclipses will have different values of $\delta F/F$. It is left as an exercise for the reader to demonstrate that the **primary eclipse**, the deeper of the two, occurs when the *hotter* star is eclipsed.

13.6 ▪ HOW ARE MASS, RADIUS, AND LUMINOSITY RELATED?

If we plot radius R versus mass M for stars whose mass we know (Figure 13.10), we find that most stars lie along a fairly well-defined mass–radius relation. The stars that obey the mass–radius relation are called "main sequence" stars. The Sun, Sirius A, Epsilon Eridani, and the stars of the Alpha Centauri system are all main sequence stars. As you can see from the figure, more massive main sequence stars are larger in radius, but not with the $R \propto M^{1/3}$ relation that you would expect if all stars had the same density. Stars that are larger in radius are lower in density. Although a single power law doesn't give a good fit over the full range of the mass–radius relation, a reasonable fit[15] is given by a pair of power laws:

$$\begin{aligned} R/R_\odot &= 1.06(M/M_\odot)^{0.945} \quad M < 1.66M_\odot \\ R/R_\odot &= 1.33(M/M_\odot)^{0.555} \quad M > 1.66M_\odot. \end{aligned} \tag{13.77}$$

Among the stars that don't fall on the usual mass–radius relation are Betelgeuse (which has an overly large radius for its mass) and Sirius B (which has an overly small radius for its mass). Thus, supergiants like Betelgeuse and white dwarfs like Sirius B are special cases that don't follow the same relations as ordinary "main sequence" stars.

If we plot stellar luminosity versus mass (Figure 13.11), we find that most stars lie along a well-defined mass–luminosity relation. More massive stars are higher in luminosity. A reasonably good empirical fit to the mass–luminosity relation[16] is

$$\begin{aligned} L/L_\odot &= 0.35(M/M_\odot)^{2.62} \quad M < 0.7M_\odot \\ L/L_\odot &= 1.02(M/M_\odot)^{3.92} \quad M > 0.7M_\odot. \end{aligned} \tag{13.78}$$

Note the steep dependence of luminosity upon mass, particularly for high-mass stars; this has important implications for stellar evolution. A star is its own fuel tank; that is, it powers itself by fusion of the material that it contains. Thus, the total fuel supply of a star is proportional to its mass M. The rate at which it uses fuel is proportional to its luminosity. The lifetime τ of a star before it exhausts its fuel is then $\tau \propto M/L$. For the observed mass–luminosity relation,

[15] From Demircan and Kahraman 1991, *Astrophysics and Space Science*, pp. 181, 313.

[16] Also from Demircan and Kahraman 1991, *Astrophysics and Space Science*, pp. 181, 313.

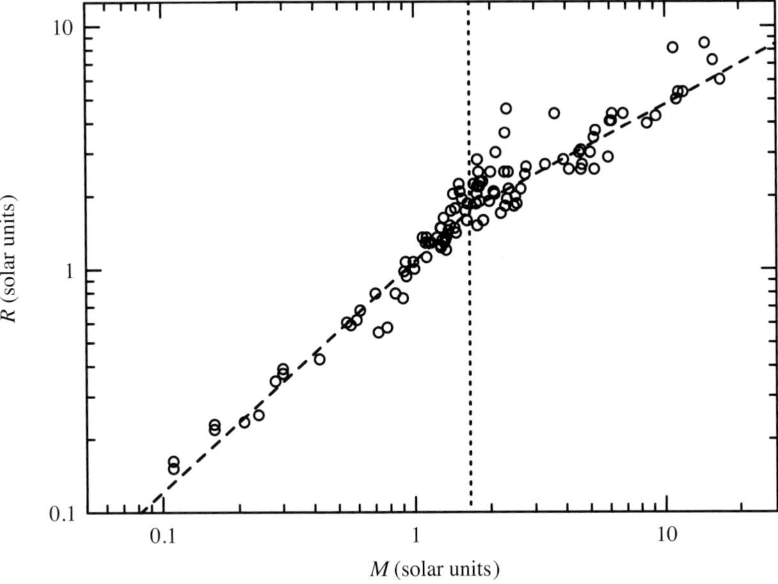

FIGURE 13.10 Stellar radius versus mass (logarithmic scale). The dashed line is the fit given in equation (13.77).

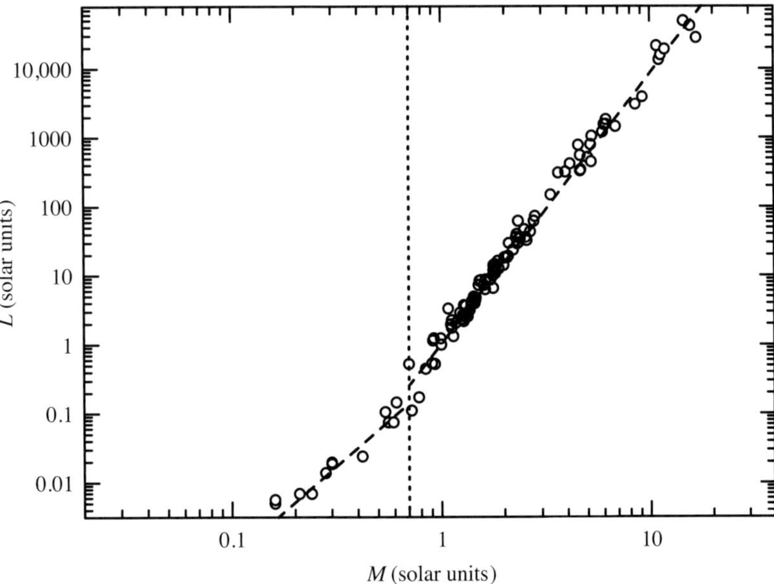

FIGURE 13.11 Stellar luminosity versus mass (logarithmic scale). The dashed line is the empirical fit given in equation (13.78).

$$\tau \propto M/L \propto M^{-1.62} \quad M < 0.7M_\odot \tag{13.79}$$

$$\tau \propto M/L \propto M^{-2.92} \quad M > 0.7M_\odot. \tag{13.80}$$

For example, Sirius A, which has a mass more than twice that of the Sun, will have a lifetime less than 1/8 as long, before it runs out of fuel for its fusion "engine."

APPENDIX: DETERMINATION OF BOLOMETRIC CORRECTIONS

A star's bolometric correction takes account of the fact that much of a star's flux is emitted at wavelengths outside the band in which we are observing. In the case of a blackbody, the bolometric correction is a relatively simple function of temperature. For illustrative purposes, we will derive the bolometric correction of a blackbody, and then compare it to the numerically computed bolometric corrections found for model stellar atmospheres.

From equation (5.98), we recall that the luminosity of a spherical blackbody is

$$L = 4\pi R^2 \sigma_{\rm SB} T^4, \tag{13.81}$$

where R is the radius of the blackbody and T is its surface temperature. The bolometric absolute magnitude is then (equation 13.28)

$$\begin{aligned} M_{\rm bol} &= 4.74 - 2.5 \log(L/L_\odot) \\ &= 4.74 + 2.5 \log L_\odot - 2.5 \log(4\pi R^2 \sigma_{\rm SB} T^4) \\ &= K' - 10 \log T, \end{aligned} \tag{13.82}$$

where the constant K' contains all terms that are not temperature dependent.

The specific luminosity of the blackbody—that is, the luminosity per unit wavelength—is proportional to the Planck function I_λ (equation 5.90):

$$\begin{aligned} L_\lambda &= 4\pi R^2 \pi I_\lambda \\ &= 4\pi R^2 \frac{2\pi hc^2}{\lambda^5} \frac{1}{e^{hc/\lambda kT} - 1}. \end{aligned} \tag{13.83}$$

The absolute V magnitude of the blackbody can then be written as

$$M_V = K_V - 2.5 \log \left(\int L_\lambda S_\lambda(V) d\lambda \right), \tag{13.84}$$

where $S_\lambda(V)$ is the spectral sensitivity of the V filter. To compute M_V with extreme accuracy, we would need to determine the exact shape of $S_\lambda(V)$. However, since the V band is not extremely wide in wavelength, we can approximate the integral in equation (13.84) as the bandwidth $\Delta\lambda_V$ times the value of L_λ evaluated at the effective wavelength $\lambda_{{\rm eff},V}$. Thus,

$$M_V = K_V - 2.5 \log \left(\Delta\lambda_V \cdot 4\pi R^2 \frac{2\pi hc^2}{\lambda_{{\rm eff},V}^5} \frac{1}{e^{hc/kT\lambda_{{\rm eff},V}} - 1} \right). \tag{13.85}$$

This can be rewritten as

$$M_V = K'_V + 2.5 \log(e^{hc/kT\lambda_{\mathrm{eff},V}} - 1), \tag{13.86}$$

where the constant K'_V contains all terms that are not temperature dependent.

The effective wavelength $\lambda_{\mathrm{eff},V}$ depends to some extent on the temperature of the observed blackbody; however, for temperatures similar to the Sun's surface temperature, it is adequate to use the approximation $\lambda_{\mathrm{eff},V} = 5500$ Å. This characteristic wavelength can be used to define a characteristic temperature

$$T_V \equiv \frac{hc}{k\lambda_{\mathrm{eff},V}} = 26{,}160 \text{ K}. \tag{13.87}$$

Using this parameter, we can write the absolute V magnitude as

$$M_V = K'_V + 2.5 \log(e^{T_V/T} - 1) \tag{13.88}$$

and the bolometric correction (using equation 13.82) as

$$\begin{aligned} BC &= M_{\mathrm{bol}} - M_V \\ &= K'' - 10 \log(T/T_V) - 2.5 \log(e^{T_V/T} - 1), \end{aligned} \tag{13.89}$$

where the constant K'' contains all terms that are not dependent on the dimensionless temperature T/T_V.

The bolometric correction BC for a blackbody (equation 13.89) is plotted as the solid line in Figure 13.12. The extremum in BC as a function of T can be found from inspection

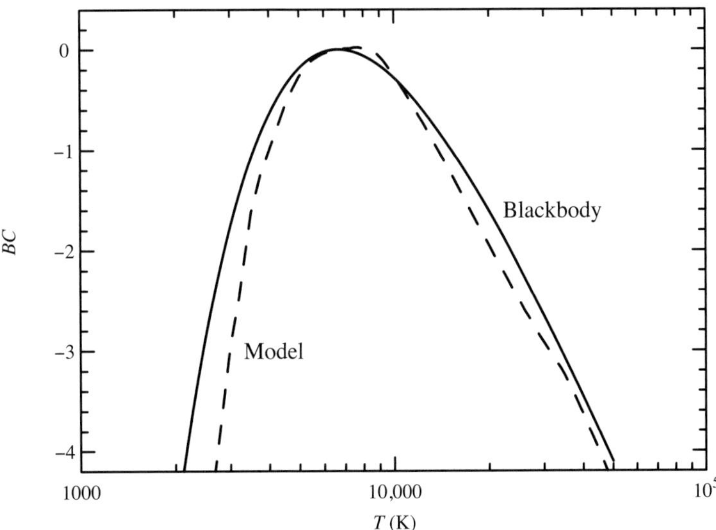

FIGURE 13.12 Bolometric correction as a function of temperature. The solid line shows the bolometric correction of a blackbody; the dotted line shows the bolometric correction for a more realistic stellar atmosphere model.

of the figure, or by differentiating BC with respect to T and finding the temperature at which $d(BC)/dT = 0$. This temperature turns out to be $T = 0.255T_V = 6670$ K, representing the temperature at which a blackbody has the largest fraction of its flux in the V band. The usual convention is to have $BC = 0$ mag at this temperature, which requires setting $K'' = -1.70$ in equation (13.89).

The dotted line in Figure 13.12, for comparison, indicates the bolometric correction calculated for a realistic model stellar atmosphere. The blackbody approximation is reasonably good for stars with small bolometric corrections but underestimates the required bolometric correction for very hot ($T \gtrsim 10{,}000$ K) and very cool ($T \lesssim 3000$ K) stars.

PROBLEMS

13.1 What is the apparent magnitude of the Sun as seen from Mercury at perihelion? What is the apparent magnitude of the Sun as seen from Eris at aphelion?

13.2 Considering absolute magnitude M, apparent magnitude m, and distance d or parallax π'', compute the unknown for each of these stars:

(a) $m = -1.6$ mag, $d = 4.3$ pc. What is M?
(b) $M = 14.3$ mag, $m = 10.9$ mag. What is d?
(c) $m = 5.6$ mag, $d = 88$ pc. What is M?
(d) $M = -0.9$ mag, $d = 220$ pc. What is m?
(e) $m = 0.2$ mag, $M = -9.0$ mag. What is d?
(f) $m = 7.4$ mag, $\pi'' = 0.0043''$. What is M?

13.3 What are the *angular diameters* of the following, as seen from the Earth?

(a) The Sun, with radius $R = R_\odot = 7 \times 10^5$ km
(b) Betelgeuse, with $M_V = -5.5$ mag, $m_V = 0.8$ mag, and $R = 650R_\odot$
(c) The galaxy M31, with $R \approx 30$ kpc at a distance $D \approx 0.7$ Mpc
(d) The Coma cluster of galaxies, with $R \approx 3$ Mpc at a distance $D \approx 100$ Mpc

13.4 The Luyten 726-8 star system contains two stars, one with apparent magnitude $m = 12.5$ and the other with $m = 12.9$. What is the combined apparent magnitude of the two stars?

13.5 A cluster of stars contains 100 stars with absolute magnitude $M = 0.0$, 1000 stars with $M = 3.0$, and 10,000 stars with $M = 6.0$. What is the absolute magnitude of the cluster taken as a whole?

13.6 A visual binary has a parallax $\pi'' = 0.4$ arcsec, a maximum separation $a'' = 6.0$ arcsec, and an orbital period $P = 80$ yr. What is the total mass of the binary system? Assume a circular orbit.

13.7 The stars β Aurigae A and β Aurigae B constitute a double-lined spectroscopic binary with an orbital period $P = 3.96$ days. The radial velocity curves of the two stars have amplitudes $v_A \sin i = 108\ \mathrm{km\ s^{-1}}$ and $v_B \sin i = 111\ \mathrm{km\ s^{-1}}$. If $i = 90°$, what are the masses of the two stars?

13.8 The star Procyon A has an effective temperature $T_A = 6530$ K and a radius $R_A = 2.06 R_\odot$. Its companion Procyon B has a radius $R_B = 0.0096 R_\odot$ and an absolute bolometric magnitude $M_{bol,B} = 12.9$.

(a) What is the ratio of the two objects' luminosities?
(b) What is the ratio of their surface temperatures?

13.9 Astronomers often use the approximation that a 1% change in brightness of a star corresponds to a change of 0.01 magnitudes. Justify this approximation.

13.10 At visible wavelengths, what is the ratio of the flux of the full Earth as seen from the Moon to the flux of the full Moon as seen from the Earth? (Hint: the albedos of the Earth and the Moon are given in Section 8.2.)

13.11 Prove that in an eclipsing binary system, the primary (deeper) eclipse always occurs when the *hotter* (not necessarily the larger or more luminous) star is eclipsed.

13.12 Show that for a blackbody at low temperatures, the relation between the effective temperature and the $B - V$ color is

$$T_{\mathrm{eff}} \approx \frac{7090\ \mathrm{K}}{(B - V) + 0.71}.$$

(Hint: use the Wien approximation, and assume $\lambda_{\mathrm{eff},B} = 4450$ Å and $\lambda_{\mathrm{eff},V} = 5500$ Å.) This relation is a really lousy approximation at high temperatures. Why?

14 Stellar Atmospheres

Most of what we know about stars other than the Sun comes from gathering photons that the stars emit. The problem with photons is that they tell us only what is happening in the *photosphere*, the relatively thin layer of a star from which the photons escape. When we compute the "radius of a star," for instance, we are really computing the radius of the star's photosphere. When astronomers talk about the "temperature of a star," they mean the temperature of the star's photosphere, unless they explicitly state otherwise.

14.1 ▪ HYDROSTATIC EQUILIBRIUM

To understand how a star's spectrum is produced, we must understand the basic physics of stellar atmospheres. In some ways, the atmosphere of a star is like the Earth's atmosphere; despite winds and storms, both types of atmosphere are in **hydrostatic equilibrium**, as described in Section 9.2. In other ways, a star's atmosphere is unlike the Earth's. For one thing, the Earth's atmosphere rests upon a solid or liquid surface; since stars are completely gaseous, you can think of them as being nothing but atmosphere. Another difference between stellar atmospheres and the Earth's atmosphere is that the atmospheres of stars are relatively hot, and ionization becomes important.

For a spherical star in hydrostatic equilibrium (equation 9.8),

$$\frac{dP}{dr} = -\frac{GM(r)\rho}{r^2}, \tag{14.1}$$

where r is the distance from the star's center, P is the local pressure, $M(r)$ is the mass contained within a sphere of radius r, ρ is the local mass density, and G is the Newtonian gravitational constant. In other words, the upward force due to the pressure gradient (the left-hand side of equation 14.1) is exactly balanced by the downward force due to gravity (the right-hand side of equation 14.1).

The pressure at any point inside the star is well approximated by the ideal gas law (equation 7.16):

$$P = nkT = \frac{\rho kT}{\mu m_p}, \tag{14.2}$$

where the mean molecular mass μ depends on the mix of elements present as well as the degree of ionization. If we characterize the elements present as hydrogen (atomic number = 1), helium (atomic number = 2), and "metals" (atomic number > 2), the total mass density ρ can be broken down into three components:

$$\rho = \rho_{\rm H} + \rho_{\rm He} + \rho_{\rm metal}. \tag{14.3}$$

The chemical composition of the atmosphere can then be expressed in terms of the hydrogen mass fraction,

$$X \equiv \rho_{\rm H}/\rho, \tag{14.4}$$

the helium mass fraction,

$$Y \equiv \rho_{\rm He}/\rho, \tag{14.5}$$

and the "metal" mass fraction,

$$Z \equiv \rho_{\rm metal}/\rho = 1 - X - Y. \tag{14.6}$$

In Section 7.1, for instance, you learned that the Sun's photosphere has $X_\odot = 0.734$, $Y_\odot = 0.250$, and $Z_\odot = 0.016$. Most of the mass of "metals" is contributed by oxygen and carbon, with neon, iron, and nitrogen rounding out the list of the top five metals.

The mean molecular mass μ of a gas depends both on its mass fractions (X, Y, and Z) and on its ionization state. Consider, for instance, a gas consisting of pure atomic hydrogen ($X = 1$). If it is neutral, it will have a mean molecular mass $\mu = 1$ and a number density $n = \rho/m_p$.[1] If the hydrogen is fully ionized, the number of particles is doubled, since one electron is freed from each atom. Thus, the number density of a hydrogen gas doubles to $n = 2\rho/m_p$ if it is fully ionized, and its mean molecular mass drops to $\mu = 1/2$.

Now consider a gas made of pure helium ($Y = 1$). If it is neutral, it will have a mean molecular mass $\mu = 4$ and a number density $n = \rho/(4m_p)$, since each helium atom contains four nucleons—two protons and two neutrons.[2] If the helium is fully ionized, the number of particles is tripled, since two electrons are freed from each atom. Thus, the number density of a helium gas triples to $n = 3\rho/(4m_p)$ if it is fully ionized, and its mean molecular mass drops to $\mu = 4/3$.

Finally, consider a gas made of "metals" ($Z = 1$). If the average number of nucleons per atom is A, the number density of atoms for the gas in its neutral state is $n = \rho/(Am_p)$, and the mean molecular mass is $\mu = A$. If the number of protons and neutrons in each nucleus is roughly equal, then $\sim A/2$ electrons will be liberated from each atom when the gas is fully ionized. The number density of the gas will then be $n \approx \rho/(2m_p)$, assuming $A/2 \gg 1$, and its mean molecular mass will be $\mu \approx 2$.

[1] Our calculations will be sufficiently accurate if we assume that the mass of an electron is negligibly small compared to the mass of a proton.

[2] Our calculations will be sufficiently accurate if we assume that the mass of a neutron and the mass of a proton are the same.

Thus, if a gas consists of a mix of hydrogen, helium, and metals, its number density of particles, in a fully ionized state, will be

$$n \approx X\left(\frac{2\rho}{m_p}\right) + Y\left(\frac{3\rho}{4m_p}\right) + Z\left(\frac{\rho}{2m_p}\right) \tag{14.7}$$

$$\approx \left(2X + \frac{3}{4}Y + \frac{1}{2}Z\right)\frac{\rho}{m_p}. \tag{14.8}$$

For a fully ionized gas, the mean molecular mass will then be

$$\mu(\text{ionized}) = \frac{\rho}{nm_p} = \left(2X + \frac{3}{4}Y + \frac{1}{2}Z\right)^{-1}. \tag{14.9}$$

For the Sun's photosphere, the mean molecular mass, assuming total ionization, is

$$\mu_\odot = \frac{1}{2(0.734) + 0.75(0.250) + 0.5(0.016)} = \frac{1}{1.664} = 0.60. \tag{14.10}$$

By contrast, for a gas of neutral atoms, the mean molecular mass will be

$$\mu(\text{neutral}) = \left(X + \frac{Y}{4} + \frac{Z}{A}\right)^{-1}. \tag{14.11}$$

If the Sun's photosphere were not ionized, its mean molecular mass would be $\mu_\odot \approx 1.25$; since the rare metals contribute so little to the computation of μ, it doesn't really matter what their value of A is. We conclude that no matter how highly ionized the Sun's atmosphere is, its mean molecular mass will be roughly 1.

Ionized gas has a higher pressure than a gas of neutral atoms with the same mass density for two reasons:

- Maintaining a gas in an ionized state requires a high temperature T; higher temperature implies a higher thermal velocity, and hence more pressure per particle.
- Ionization frees large quantities of electrons, thus increasing the number density n of particles at a given mass density.

Ionizing a gas thus increases both T and n in the relation $P = nkT$.

Stars are usually stable. The Sun has been shining away for 4.6 billion years without exploding or imploding. Some stars pulsate in and out perceptibly, and now and then a supernova explodes, but for the most part, stars are in sedate hydrostatic equilibrium:

$$\frac{dP}{dr} = -\frac{GM(r)}{r^2}\rho = -g\rho, \tag{14.12}$$

where

$$g \equiv \frac{GM(r)}{r^2} \tag{14.13}$$

is the gravitational acceleration (directed inward) at a distance r from the star's center. Since g and ρ are nonnegative, the pressure gradient is $dP/dr \le 0$. That is, the pressure becomes greater as you dive inward to the center of a star.

At the Sun's photosphere, $M(r) \approx M_\odot = 1.99 \times 10^{30}$ kg; the mass of the chromosphere and corona above the photosphere are negligible. The radius of the photosphere is $R_\odot = 6.96 \times 10^8$ m. The gravitational acceleration at the Sun's photosphere (often called the "surface gravity" of the Sun) is then

$$g_\odot = \frac{GM_\odot}{R_\odot^2} = 274 \text{ m s}^{-2}, \tag{14.14}$$

about 28 times the gravitational acceleration at the Earth's surface. For an ideal gas, the mass density and pressure are related by the law

$$\rho = \frac{\mu m_p}{kT} P, \tag{14.15}$$

so the equation of hydrostatic equilibrium (eq. 14.12) can be written in the form

$$\frac{dP}{dr} = -\frac{g\mu m_p}{kT} P. \tag{14.16}$$

If we are in a region of the star where g, μ, and T are roughly constant with radius, equation (14.16) has a solution of the form

$$P \propto \exp\left(-\frac{r}{H}\right), \tag{14.17}$$

where the scale height H is

$$H = \frac{kT}{g\mu m_p}. \tag{14.18}$$

For the Sun's photosphere, with $T_\odot \approx 5800$ K and $\mu_\odot \approx 0.60$, the scale height is

$$H_\odot = \frac{kT_\odot}{g_\odot \mu_\odot m_p} \approx 300 \text{ km}. \tag{14.19}$$

This is much longer than the scale height $H \approx 8$ km of the Earth's atmosphere, which is cooler and has a higher mean molecular mass. However, it is much shorter than the Sun's radius of $R_\odot \approx 700{,}000$ km. Thus, our assumption of constant $g_\odot$, $\mu_\odot$, and $T_\odot$ within the photosphere is a reasonable first approximation.

14.2 ▪ SPECTRAL CLASSIFICATION

A star's spectrum contains information about the photosphere's chemical composition and temperature. Every absorption line that you see in a star's spectrum represents the transition of an electron from a lower energy level to a higher level, in a particular element in a particular ionization state.[3] The solar spectrum (Color Figure 13) contains absorption lines from most of the elements found in nature.[4]

The strength of absorption lines, as measured by their equivalent width W, depends on the temperature of the photosphere. For example, the absence of helium absorption lines in the Sun's photospheric spectrum doesn't mean that all the helium has run off to the chromosphere; rather, it indicates that the photosphere isn't hot enough to excite helium atoms above their ground state. As a relatively simple example of how line strength depends on temperature, let's consider the **Balmer lines** of hydrogen, created when electrons in the $n = 2$ energy level absorb photons of the correct energy to lift them to a higher ($n > 2$) energy level. As we computed in Section 5.6, when the temperature T of gaseous hydrogen increases, the fraction of neutral atoms that have their electron in the $n = 2$ level *increases*. However, as T increases, the fraction of atoms that are neutral *decreases*. These two competing effects maximize the total number of electrons in the $n = 2$ level at $T \approx 10{,}000$ K, as shown in Figure 5.13. This is roughly the temperature of the photospheres of Vega and Sirius A.

The Balmer lines can be used to estimate the temperature of a star; the stars with the strongest Balmer lines are those with $T \approx 10{,}000$ K. What about stars with weak Balmer lines? There are three possible reasons why the Balmer lines might be weak or nonexistent:

- There's little or no hydrogen present in the photosphere. (Given the ubiquity of hydrogen in the universe, this is an implausible explanation.)
- The photospheric temperature is $T \gg 10{,}000$ K.
- The photospheric temperature is $T \ll 10{,}000$ K.

We can distinguish between the high-temperature case and the low-temperature case by looking at the absorption lines of elements other than hydrogen.

Consider a neutral helium atom. Its ionization potential is $\chi = 24.5$ eV, roughly twice the ionization potential of hydrogen. The energy required to lift an electron from the ground level to the first excited level is $E_2 - E_1 = 20.9$ eV, roughly twice the equivalent energy in a hydrogen atom. Given the relative energy scales, we expect absorption by helium to occur at energies (and hence temperatures) roughly twice that of hydrogen. If a star has weak Balmer lines and strong helium lines, we conclude that it has $T > 10{,}000$ K.

Now consider an atom with a single electron in its outer shell (lithium, sodium, and so forth). These atoms have low ionization potentials (for sodium, $\chi = 5.1$ eV). Thus, lines of neutral sodium are seen only in relatively cool stars; in hotter stars, the sodium is nearly all ionized. If a star has weak Balmer lines and strong neutral sodium lines, we conclude that it has $T < 10{,}000$ K. We can also conclude that a star is cool if it has molecular absorption bands. The dissociation energy for molecules tends to be small compared to the ionization energy of hydrogen. Titanium oxide is a relatively tough

[3] In some cooler stars, you also see absorption bands due to molecules.

[4] As we have seen, however, the element helium (the second most abundant element in the universe) reveals its presence in the Sun only by emission lines from the chromosphere.

molecule; its dissociation energy is about 6.9 eV. Strong TiO absorption bands are seen in stars with $T \approx 3000$ K.

Detailed study of the strength of different absorption lines provides a good estimate of the photospheric temperature. Given the vagaries of history, however, it shouldn't astonish you greatly that the spectral classification of stars long predates the realization that the *spectral sequence* of stars is a *temperature sequence*. The spectral classification scheme we use today had its origin around the year 1890, when Edward Pickering, at the Harvard College Observatory, undertook the task of sorting out thousands of stellar spectra. Needing a collaborator who was experienced at creating order out of chaos, he hired his housekeeper, Williamina Fleming, as his assistant.

Pickering and Fleming proposed a scheme in which each spectrum was assigned a letter, from "A" through "Q." The letter "J" was not used (apparently because it looks like "I" when scribbled quickly). This meant that Pickering and Fleming had 16 types in all. The letter "P" was assigned to planetary nebulae (a type of small gaseous nebula) and the "Q" was their wastebasket type: if a spectrum was too bizarre to fit any other type, it went there. The other letters in their scheme were assigned in order of decreasing strength of the Balmer lines, with A stars having the strongest Balmer lines, and O stars the weakest.

The classification of Pickering and Fleming started out as a purely empirical scheme, like sorting buttons by their color. At the beginning of the twentieth century, however, astronomers began to have a clearer idea of how the strength of the Balmer lines depended on temperature. Another Harvard astronomer, Annie J. Cannon, tossed out the redundant types in the Pickering–Fleming scheme and reordered the remaining types according to temperature. The order, from hot to cold, was **OBAFGKM**.[5] With higher resolution spectra, Cannon was able to refine the classification further. A single letter was subdivided into numbered subtypes. For instance, G stars are subdivided into G0 stars (the hottest), G1, G2, G3, G4, G5, G6, G7, G8, and G9 stars (the coolest). The Sun, for instance, is a G2 star.

The coolest stars in the standard classification scheme were M9 stars, with $T \approx 2400$ K. For many decades, astronomers toiled to discover starlike objects cooler than 2400 K. Such relatively cool bodies are extremely difficult to detect at visible wavelengths, since most of their luminosity emerges in the infrared. A blackbody with $T \approx 1500$ K, for instance, has a spectrum peaking at $\lambda_{\rm max} \approx 2\ \mu$m.

Not coincidentally, an all-sky infrared survey called the 2 Micron All-Sky Survey (2MASS, for short) examined the sky at wavelengths of $\sim 2\ \mu$m, with one of its major goals being the discovery of objects cooler than M9 stars. This goal was attained. The 2MASS survey found a number of cool dwarfs, comparable in size to an ordinary "main sequence" M9 star but with temperatures $T < 2400$ K. Cool dwarfs with a temperature of $T \sim 2000$ K have been given the name L dwarfs.[6] L dwarfs have distinctive spectral characteristics:

[5] The traditional mnemonic for this sequence is "Oh Be A Fine Girl; Kiss Me." However, if you prefer kissing guys (or goats or gorillas), feel free to make the appropriate substitution.

[6] Why "L"? It was available, since it was one of the types discarded by Cannon.

- Disappearance of TiO (and other oxide) absorption bands common in M dwarfs.
- Appearance of metal hydride absorption bands (FeH, CrH, and so forth).
- Greater flux in the infrared than in the visible.

The nearest known L dwarf is at a distance $d = 5.0$ pc; the estimated number density of L dwarfs in our vicinity is $n \sim 0.01\,\text{pc}^{-3}$, about 1/10 the number density of M stars.

Encouraged by their success, astronomers looked for still cooler dwarfs. With considerable effort, they found a few dim cool dwarfs that display methane absorption bands, similar to those seen in the spectrum of Jupiter. Methane is a relatively fragile molecule, compared to metal oxides and metal hydrides; it dissociates at $T \geq 1300$ K. The ultracool dwarfs that have $T < 1300$ K have been given the name T dwarfs.[7] The distinctive spectral characteristic of T dwarfs is their methane absorption bands. Because of their lower luminosity, T dwarfs are harder to detect than L dwarfs. It is estimated, though, that their number density is $n \sim 0.01\,\text{pc}^{-3}$, comparable to the number density of L dwarfs. The nearest known T dwarfs are a pair of companions to the star ϵ Indi, at a distance $d = 3.6$ pc from the Sun.

The extended spectral classification scheme is now **OBAFGKMLT**. An approximate (but handy) translation of spectral type into temperature is

Spectral Type	Temperature
O	40,000 K
B	20,000 K
A	9000 K
F	7000 K
G	5500 K
K	4500 K
M	3000 K
L	2000 K
T	$<$ 1300 K

You may have noticed that we've been using the names L dwarfs and T dwarfs rather than L stars and T stars. This is because L and T dwarfs are not hot and dense enough to fuse hydrogen into helium in their cores. Thus, they are not stars by the strict definition of the term. Rather, L and T dwarfs are examples of **brown dwarfs**, objects that fall into the gap between stars and planets. Brown dwarfs are balls of gas, but they are balls of gas without central fusion reactors. Thus, they cool down with time. A brown dwarf that starts as an L dwarf will end as a T dwarf. Main sequence stars, by contrast, maintain a roughly constant surface temperature as long as they have a supply of hydrogen to fuse into helium at their centers.

[7] Why "T"? It was available, since it wasn't part of the original Pickering–Fleming classification scheme.

As seen in Color Figure 14, M, L, and T dwarfs are all comparable in size to Jupiter. Because of the difference in temperatures, $L_M > L_L > L_T > L_{Jup}$. Color Figure 14 is astronomically correct in that it shows "clouds" and "weather" on the T dwarf, similar to the clouds and weather on Jupiter.

14.3 ▪ LUMINOSITY CLASSES

The spectral types O through T are a temperature sequence. Although the photospheric temperature is the most important parameter determining a star's spectrum, it is not the only one. In the 1930s, W. W. Morgan and Philip Keenan added an extension to the old OBAFGKM scheme by introducing the concept of **luminosity classes**. Empirically, the six luminosity classes, I, II, III, IV, V, and VI, correspond to different absorption line widths, with luminosity class I having the narrowest lines at a given temperature and luminosity class VI having the broadest. As an example, Figure 14.1 shows the spectra of three stars, each of spectral type A0. Spectrum (a), which has luminosity class I, has perceptibly narrower Balmer absorption lines than spectrum (c), which has luminosity class V. (In fact, the lowest spectrum is that of Vega.)

In practice, it is found that the six luminosity classes correspond to stars of different radii:

Luminosity Class	Star Size
I	supergiant
II	bright giant
III	giant
IV	subgiant
V	dwarf (main sequence)
VI	subdwarf

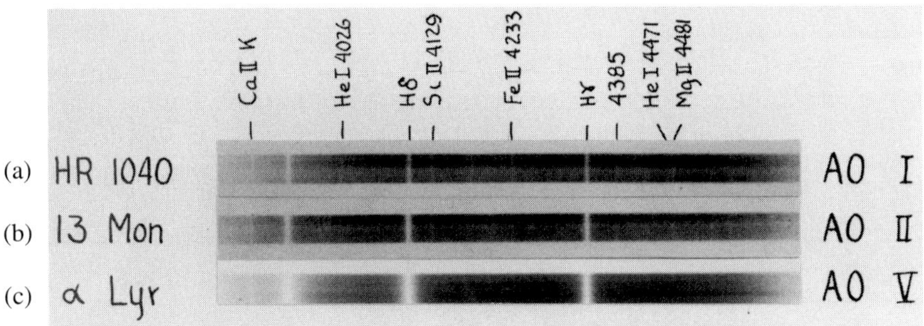

FIGURE 14.1 Spectra (negative photographic images) of three A0 stars. (a) Luminosity class I. (b) Luminosity class II. (c) Luminosity class V.

The majority of stars (like the Sun, Sirius A, Alpha Centauri A, Alpha Centauri B, Proxima Centauri, and Vega) are of luminosity class V. Betelgeuse is a supergiant, of luminosity class I. Arcturus and Capella are examples of giants, with luminosity class III.[8]

Why should supergiants (luminosity class I) have narrower absorption lines than ordinary main sequence stars (luminosity class V) of the same surface temperature? To understand, let's first think about what it means to look at the photosphere of a star. If you look at a star from a distance, the optical depth τ increases as you look farther into the star's interior. For a thin spherical shell of radius r and thickness dr, the relation between the shell's optical depth $d\tau$ and physical thickness dr is (equation 5.60):

$$d\tau = -n(r)\sigma(r)dr, \tag{14.20}$$

where n is the number density of absorbers (or scatterers) and σ is the average cross-section. Alternatively, we can write

$$d\tau = -\rho(r)\kappa(r)dr, \tag{14.21}$$

where ρ is the mass density of the shell and κ is the **opacity**.[9] The opacity κ is found by adding together the cross-sections of all the absorbers and scatterers in the shell, and dividing by the total mass of the shell. Thus, the units of κ are $\text{m}^2\,\text{kg}^{-1}$, and from equations (14.20) and (14.21), we find

$$\kappa = \frac{n\sigma}{\rho}. \tag{14.22}$$

In general, the opacity κ is a function of temperature, density, and chemical composition; in the Sun's photosphere, the opacity is $\kappa \sim 3\,\text{m}^2\,\text{kg}^{-1}$. The negative sign in equation (14.21) is a reminder that τ increases as r decreases. For a thin shell, $d\tau$ is just the probability that a photon is absorbed or scattered as it passes through the shell. As you dive inward into a star, the photosphere is where the optical depth reaches a value $\tau \approx 1$.

Since the equation of hydrostatic equilibrium tells us that

$$\frac{dP}{dr} = -g\rho, \tag{14.23}$$

we can write the dependence of pressure on optical depth as

$$\frac{dP}{d\tau} = \frac{dP}{dr}\frac{dr}{d\tau} = \frac{g}{\kappa}. \tag{14.24}$$

If we assume that g/κ is roughly constant in the star's atmosphere,

$$P \approx \frac{g}{\kappa}\tau. \tag{14.25}$$

[8] White dwarfs are sometimes classified as luminosity class VII, and supergiants are often subdivided into class Ia ("hypergiants") and class Ib ("ordinary" supergiants).

[9] In general, the cross-section σ_λ is dependent on the wavelength of light; thus, κ_λ will be wavelength-dependent, too.

As you dive into a star's atmosphere, the pressure and optical depth both exponentially increase with the same scale height. Since $\tau \approx 1$ in the photosphere, we can compute the pressure at the location of the photosphere as

$$P_{\text{phot}} \approx \frac{g_{\text{phot}}}{\kappa_{\text{phot}}}. \tag{14.26}$$

For the Sun, for instance, the pressure in the photosphere is $P_\odot \approx 100\,\text{N m}^{-2} \approx 10^{-3}$ atm. In combination with the photospheric temperature of $T_\odot = 5800$ K and a mean molecular mass of order unity, this implies a density in the photosphere of $\rho_\odot \approx 10^{-5}\,\text{kg m}^{-3}$.

In general, stars with a higher gravitational acceleration g will have higher pressures P in their photosphere, since the photospheric opacity κ_{phot} doesn't vary wildly from one star to another. This means that stars with high acceleration will have more pressure broadening of their absorption lines.[10] As a specific example, let's compare Betelgeuse with Proxima Centauri. Betelgeuse is an M2 I star, and has an absolute magnitude $M_V \approx -5.5$ mag. Proxima Centauri is an M5 V star, and has an absolute magnitude $M_V \approx 15$ mag. Betelgeuse and Proxima Centauri are similar in surface temperature ($T \approx 3000$ K) but differ by over 20 magnitudes in M_V; that's a difference of $\sim 10^8$ in luminosity. Betelgeuse is more luminous than Proxima Centauri by a factor of $\sim 10^8$ because it's larger in radius[11] by a factor $\sim 10^4$. Because Betelgeuse is much larger than Proxima Centauri, the gravitational acceleration in its photosphere is much smaller:

$$\frac{g_{\text{Betel}}}{g_{\text{Prox}}} = \frac{M_{\text{Betel}}}{M_{\text{Prox}}} \left(\frac{R_{\text{Betel}}}{R_{\text{Prox}}}\right)^{-2} \approx (200)(13{,}000)^{-2} \approx 10^{-6}. \tag{14.27}$$

Thus, the pressure at Proxima Centauri's photosphere will be greater by a factor $\sim 10^6$, leading to a greater amount of pressure broadening of its absorption lines.

The Sun's complete spectral classification is **G2 V**. This grouping of three symbols contains a wealth of information. Empirically, the spectral type G2 indicates that the Sun's spectrum has weak Balmer lines and very strong Ca II lines. (Ca II is calcium with one electron stripped away; given the low ionization potential of calcium, this does not imply a high temperature.) By deduction from the relative strength of Balmer lines and Ca II lines, the spectral type G2 corresponds to a surface temperature $T = 5800$ K. Empirically, the luminosity class V means that the Sun's absorption lines are broad. By deduction, the luminosity class V corresponds to an ordinary main sequence star, with relatively high pressure in its photosphere.

14.4 ▪ HERTZSPRUNG–RUSSELL DIAGRAMS

In the early twentieth century, when the OBAFGKM spectral classification system was being sorted out, it occurred independently to a pair of astronomers that it might be

[10] Pressure broadening is described in Section 5.3.

[11] Recall from Section 13.4 that $R_{\text{Bet}} \approx 1800 R_\odot$, whereas $R_{\text{Prox}} \approx 0.14 R_\odot$.

interesting to plot the absolute visual magnitude of stars versus their spectral type. (This is the approximate equivalent of plotting luminosity versus surface temperature.) The two scientists who had this idea were the Danish astronomer Ejnar Hertzsprung and the American astronomer Henry Norris Russell. Their joint invention—the plot of absolute magnitude versus spectral type—is called the Hertzsprung–Russell diagram, or **H–R diagram** for short.

Russell's first published H–R diagram is shown in Figure 14.2a. To generate his plot, Russell first computed the absolute magnitude of stars with known distances. He then plotted the absolute magnitude of each star versus its spectral type, determined from its spectrum. Russell found that the stars were *not* splattered randomly around the plot. Instead, they fell along a broad band from the upper left (hot, luminous stars) to the lower right (cool, dim stars). The general results of Russell are confirmed by the more data-rich plot seen in Figure 14.2b. Part (b) plots M_V versus $B - V$ color index for a sample of more than 16,000 stars whose parallaxes were measured with an error $< 10\%$ by the *Hipparcos* satellite. Since the $B - V$ color index and the OBAFGKM spectral type are both related to surface temperature, an H–R diagram can use either quantity along its x axis.

The diagonal band from upper left to lower right in an H–R diagram is called the **main sequence** and is the origin of the term "main sequence stars." All the stars on the main sequence have luminosity class V and are relatively dense, small dwarf stars. The small number of stars above and to the right of the main sequence (cool but luminous stars) are of luminosity class III. The scattered points below and to the left of the main sequence are white dwarfs, those dense stellar remnants of which Sirius B is an example.

It is instructive to look at the H–R diagrams for different populations of stars. As an example, let's start by looking at the 25 stars that are closest to Earth. These are stars whose distances, and hence absolute magnitudes, are well known (the most distant is at $d \approx 4$ pc). These nearby stars should also constitute a "fair sample" of the stars in our galaxy, since we don't think there's anything particularly special about our location. What do we find when we look at the H–R diagram for local stars, as shown in Figure 14.3a? We find a strong temptation to paraphrase Abraham Lincoln, and say "M main sequence stars are the best in the universe: that is the reason the Lord makes so many of them."[12] There are no giants or supergiants in the Sun's immediate neighborhood. On the main sequence, there is nothing hotter or more luminous than Sirius A (spectral type A1; $T \approx 10{,}000$ K). Over half the stars in our neighborhood are type MV: small, cool, main sequence stars.

As another example, let's look at the apparently brightest stars (that is, the stars with the highest flux as seen from Earth). The H–R diagram of the 25 highest-flux stars is shown in Figure 14.3b. The apparently brightest stars are an unrepresentative sample of stars. Most of them are hot, luminous main sequence stars, giants, and supergiants. Although supergiants are extremely rare, their high luminosity makes them visible over

[12] What Lincoln actually said, according to his secretary John Hay, was "Common looking people are the best in the world: that is the reason the Lord makes so many of them."

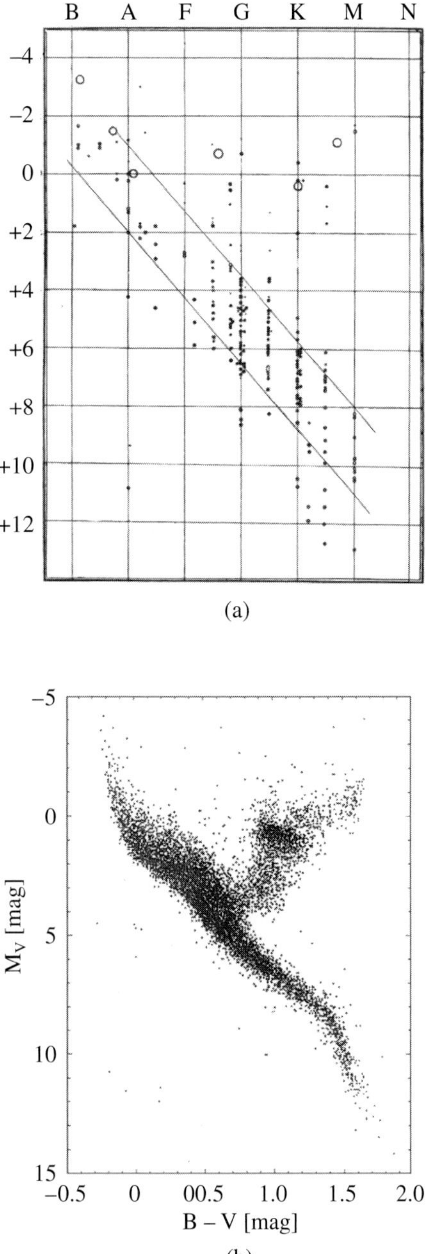

FIGURE 14.2 (a) Henry Russell's original H–R diagram. (b) H–R diagram for stars observed by the *Hipparcos* satellite.

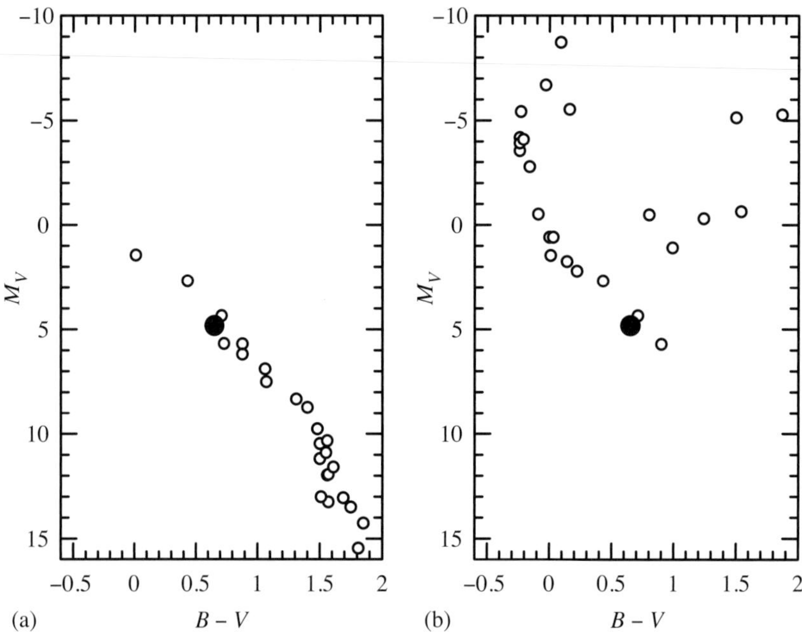

FIGURE 14.3 (a) H–R diagram for the 25 stars closest to Earth. (b) H–R diagram for the 25 apparently brightest stars as seen from Earth. In both (a) and (b), the filled dot represents the Sun.

long distances.[13] The M dwarfs that make up more than half the population of stars are all invisible to the naked eye. Even Proxima Centauri, the closest M dwarf, would have to be moved to 1/10 its present distance to be visible.

The fact that the main sequence is a relatively narrow band on the H–R diagram (see Figure 14.2) gives us a new way of estimating the distances to stars; this new method is called **spectroscopic parallax**. To see how spectroscopic parallax works, suppose that you take the spectrum of a star and find that it has luminosity class V; that is, it's on the main sequence. For any main sequence star, its absolute magnitude (or equivalently, its luminosity) is determined by its spectral type. For instance, an O5 main sequence star has $M_V = -6.0$ mag; an M5 main sequence star has $M_V = 12.3$ mag.[14] If you measure the apparent magnitude m_V of a star, you can find its distance from the relation

$$\log(d/10\ \text{pc}) = (m_V - M_V)/5. \qquad (14.28)$$

[13] Supergiants are the "celebrities" of the galaxy; although there aren't many of them, they have a talent for publicity, so everyone knows about them.

[14] The absolute magnitudes of main sequence stars can be read from an H–R diagram or taken from a table.

There will be some error in this calculation. Not all stars of the same spectral classification have *exactly* the same absolute magnitude; the main sequence is a band, not an infinitesimally thin line. In addition, we have ignored any extinction by **interstellar dust** along the line of sight to the star; we return to this topic in Section 16.1.2.

PROBLEMS

14.1 The star 9 Sagittarii is a main sequence star with spectral type O5. Its apparent magnitude is $m_V = 6.0$. What is the distance to 9 Sagittarii (ignoring any extinction by dust)?

14.2 At the center of the Sun, the mass density is $\rho = 1.52 \times 10^5$ kg m^{-3} and the mean opacity is $\kappa = 0.12$ m^2 kg^{-1}. What is the mean free path for a photon at the Sun's center?

14.3 Use the equation of hydrostatic equilibrium and the assumption of constant density to compute approximate central pressures for each of the following:

(a) a K0 V star ($M = 0.8M_\odot$, $R = 0.85R_\odot$)
(b) a K0 III star ($M = 4M_\odot$, $R = 16R_\odot$)
(c) a K0 I star ($M = 13M_\odot$, $R = 200R_\odot$)

14.4 Show explicitly that if the Sun's photosphere were not ionized, its mean molecular mass would be $\mu \approx 1.25$.

14.5 When a Hertzsprung–Russell diagram is constructed from observed data (as in Figure 14.2b), part of the width of the main sequence is due to errors in distance measurements. If a typical uncertainty in parallax is 10%, in which direction and by how much (in magnitudes) will stars typically be displaced from their true positions on the H–R diagram?

14.6 How does surface gravity vary as a function of luminosity along the main sequence?

14.7 Consider the two stars whose properties are described below:

Star	V	$B - V$	M_V	$T_{\rm eff}(K)$	Spectral Class	BC
Betelgeuse	0.45	1.50	−0.60	3370	M2 Ib	−1.62
Gliese 887	7.35	1.48	9.76	3520	M2 V	−1.89

How much larger in radius is Betelgeuse than Gliese 887?

15 Stellar Interiors

Our observations of stars are only skin-deep. The mass of the Sun's photosphere, chromosphere, and corona (the portions of the Sun we can see directly) is only 10^{-10} of the Sun's total mass. We are not entirely ignorant of the 99.99999999% of the Sun that is opaque, however. Because the structure of the Sun, and other stars, is dictated by well-understood laws of physics, we can make mathematical models of stellar interiors, using the observed surface properties of stars as our boundary conditions.

15.1 ▪ EQUATIONS OF STELLAR STRUCTURE

The internal structure of a spherical star in equilibrium is dictated by a few basic **equations of stellar structure**. The first equation of stellar structure is the familiar **equation of hydrostatic equilibrium** (eq. 9.8):

$$\frac{dP}{dr} = -\frac{GM(r)\rho(r)}{r^2}. \tag{15.1}$$

Make note of the assumptions that have gone into this equation: the star is spherical and nonrotating; the star is neither expanding nor contracting; and gravity and pressure gradients provide the only forces. Equation (15.1) is a single equation with three unknowns—$P(r)$, $M(r)$, and $\rho(r)$—so even with known boundary conditions, we can't solve it to find a unique solution for the pressure and density inside the star. However, we can still extract interesting information from the equation of hydrostatic equilibrium. For instance, we can make a very crude estimate of the central pressure of the Sun.

A rough approximation to the equation of hydrostatic equilibrium is

$$\frac{\Delta P}{\Delta r} \approx -\frac{G\langle M\rangle\langle\rho\rangle}{\langle r\rangle^2}, \tag{15.2}$$

where ΔP is the difference in pressure between the Sun's photosphere and its center; Δr is the difference in radius between the Sun's photosphere and its center; and $\langle M\rangle$, $\langle\rho\rangle$, and $\langle r\rangle$ are typical values of mass, density, and radius in the Sun's interior. As a rough guess, we can set $\langle\rho\rangle \approx \rho_\odot \approx 1400\ \text{kg m}^{-3}$, the average density of the Sun. We can also guess that $\langle M\rangle \approx M_\odot/2 \approx 1.0 \times 10^{30}$ kg and $\langle r\rangle \approx R_\odot/2 \approx 3.5 \times 10^8$ m. The pressure at the photosphere will be much less than the central pressure, so we can rewrite equation (15.2) as

$$\frac{0 - P_c}{R_\odot - 0} \approx -\frac{G(M_\odot/2)\rho_\odot}{(R_\odot/2)^2} \approx -\frac{2GM_\odot\rho_\odot}{R_\odot^2}, \tag{15.3}$$

implying a central pressure

$$P_c \approx 2\frac{GM_\odot\rho_\odot}{R_\odot} \approx \frac{8\pi}{3}G\rho_\odot^2 R_\odot^2. \tag{15.4}$$

For comparison, in Section 10.2.1, we computed the central pressure for a planet of uniform density. With the uniform-density assumption, we computed a central pressure (given in equation 10.18) that was smaller by a factor of 4 than that given in equation (15.4). From this discrepancy, we learn that although the relation $P_c \sim GM\rho/R$ should hold true for any sphere in hydrostatic equilibrium, we should regard our computed values of P_c as order-of-magnitude approximations, unless we know the exact density profile $\rho(r)$.

With the numerical values of $M_\odot$, $\rho_\odot$, and $R_\odot$ inserted into equation (15.4), we find that

$$P_c \approx 2\frac{GM_\odot\rho_\odot}{R_\odot} \approx 5 \times 10^{14}\,\mathrm{N\,m^{-2}} \approx 5 \times 10^9\,\mathrm{atm}. \tag{15.5}$$

When we compare this to the pressure $P_{\rm phot} \approx 10^{-3}$ atm in the Sun's photosphere, as computed in Section 14.3, we see that the center of a star is a high-pressure place.

The second equation of stellar structure is the **equation of mass continuity**:

$$\frac{dM}{dr} = 4\pi r^2 \rho(r). \tag{15.6}$$

This simply tells us that the total mass of a spherical star is the sum of the masses of the infinitesimally thin spherical shells of which it is made. It tells us the relation between $M(r)$, the mass enclosed within a radius r, and $\rho(r)$, the local mass density at r. Combining equations (15.1) and (15.6) gives us two equations in the three unknowns, $P(r)$, $M(r)$, and $\rho(r)$. We need further information before we can compute unique solutions for P, M, and ρ.

Of course, we do have another equation that relates P to ρ: the **equation of state**, which tells the relation among density, temperature, and pressure for a gas. For most stars, the appropriate equation of state is the ideal gas law:

$$P(r) = \frac{\rho(r)kT(r)}{\mu m_p}. \tag{15.7}$$

Strictly speaking, we should also include the radiation pressure exerted by the photons,

$$P_{\rm rad}(r) = \frac{a}{3}T(r)^4, \tag{15.8}$$

where $a = 4\sigma_{SB}/c = 7.56 \times 10^{-16}\,\text{J m}^{-3}\,\text{K}^{-4}$ is the radiation constant. However, for all but the hottest stars, the radiation pressure is negligibly tiny compared to the gas pressure. Generally, the mean molecular mass μ in equation (15.7) is a function of r, since the chemical composition and ionization state change with radius inside a star.

Most of the Sun is almost completely ionized, and the chemical composition is nearly constant outside the central regions where hydrogen is fused to helium; thus, for most of the Sun's radius, the mean molecular mass is $\mu \approx 0.6$ (see Section 14.1). If we approximate the mean molecular mass as being constant in a star, we now have three equations in four unknowns, T, P, M, and ρ. Although this is insufficient for a complete solution within the solar interior, we can make a crude estimate of the central temperature of the Sun, using the ideal gas law as the equation of state:

$$T_c \approx P_c \frac{\mu_\odot m_p}{\rho_\odot k} \approx \frac{2GM_\odot \mu_\odot m_p}{R_\odot k}. \tag{15.9}$$

With $\mu_\odot = 0.60$, this yields

$$T_c \approx 3 \times 10^7\,\text{K}. \tag{15.10}$$

Careful computer models of the Sun's interior yield a central temperature $T_c = 1.47 \times 10^7$, so our guesstimate is off by a factor of 2. Note that we are able to guess the central temperature of the Sun without knowing *anything* about how energy is generated in the Sun. The central temperature of a star of mass M and radius R is dictated by the fact that it is a sphere made of ideal gas in hydrostatic equilibrium. We have used the central temperature of the Sun as our example, but note that

$$T_c \propto \frac{M\mu}{R} \tag{15.11}$$

for any sphere of ideal gas in hydrostatic equilibrium. In Section 13.6, we found that main sequence stars with $M < 1.66M_\odot$ have $R \propto M$, approximately. Since all main sequence stars have similar mean molecular masses ($\mu \sim 1$), this implies that low-mass main sequence stars have

$$T_c \propto \frac{M}{R} \approx \text{constant}. \tag{15.12}$$

Thus, all main sequence stars with $M < 1.66M_\odot$ should have central temperatures close to that of the Sun.

15.1.1 Energy Transport in Stars

One of the defining characteristics of stars (and one that's been ignored so far in this chapter) is that they glow in the dark. A basic question about stars—one so simple that a child might ask it—is Why do stars shine? The basic answer to that question is Stars shine because they are hot. If you place a hot, bright object in the middle of cool, dark space, then energy, in the form of photons, will flow away from the hot object. Not only do stars shine because they are hot, but within the star, energy flows from the very hot center ($T_c \approx 14{,}700{,}000$ K for the Sun) to the not-so-hot photosphere ($T_{\text{phot}} \approx 5800$ K for the Sun). The rate at which thermal energy flows outward is dictated by the next equation of stellar structure, the **equation of energy transport**.

Thermal energy can be transported by one of three methods: radiation, convection, and conduction. **Radiative** energy transport occurs when energy is carried from one location to another by photons. Radiation tends to be the dominant form of energy transport in transparent media. **Convective** energy transport occurs when thermal energy is carried by the bulk motion of hot fluids. Convection is thus the dominant form of energy transport in relatively opaque liquids and gases. **Conductive** energy transport occurs when the random thermal motion of atoms or molecules causes them to collide with adjacent atoms or molecules, with a resulting transfer of kinetic energy. Conduction is the dominant form of energy transport in relatively opaque solids and thus can be disregarded in gaseous stars. Within a star, energy can be transported from the center to the photosphere either by convection (hot blobs of gas move upward, while cooler blobs sink downward to take their place) or by radiation. In the Sun, it happens that over most of the distance from the core to the photosphere, the energy is transported by photons. Thus, we'll look first at radiative energy transport.[1]

15.1.2 Radiative Transport

Consider a thin spherical shell centered on a star's center. The inner radius of the shell is r; the outer radius is $r + dr$, with $dr \ll r$. The temperature at the inner surface of the shell is T; the temperature at the outer surface is $T + dT$, where $|dT| \ll T$. Typically, stars have $dT < 0$, meaning that the temperature drops as you move away from the center. The radiation pressure at the inner surface of the shell is

$$P_{\text{rad}}(r) = \frac{a}{3}T^4, \tag{15.13}$$

while the radiation pressure at the outer surface is

$$P_{\text{rad}}(r + dr) = \frac{a}{3}[T + dT]^4 = \frac{a}{3}T^4\left[1 + \frac{dT}{T}\right]^4 \approx \frac{a}{3}T^4\left[1 + 4\frac{dT}{T}\right], \tag{15.14}$$

where in the last step we have used a first-order expansion, assuming that $|dT/T| \ll 1$. The net radiation force acting on the shell will be the pressure difference between the inside and outside, multiplied by the shell's area:

$$F_{\text{rad}} = \left[P_{\text{rad}}(r) - P_{\text{rad}}(r + dr)\right] 4\pi r^2 \tag{15.15}$$

$$\approx -\frac{a}{3}4T^4\frac{dT}{T}4\pi r^2 = -\frac{16\pi}{3}ar^2T^3dT. \tag{15.16}$$

Thus, a temperature gradient across a thin shell is accompanied by a net radiation force, caused by photons shoving on the material inside the shell. We've already seen that the optical depth of a thin spherical shell is (equation 14.21)

$$d\tau = -\rho(r)\kappa(r)dr, \tag{15.17}$$

[1] To avoid accusations of helio-chauvinism, we'll also look at convective energy transport a little later on.

where κ is the opacity, in units of $\text{m}^2\,\text{kg}^{-1}$. If $d\tau \ll 1$, the probability that a photon will be absorbed while crossing the shell is $dP \approx d\tau$. The total rate at which photons carry energy through the shell is just the luminosity, $L(r)$. Since a photon has a momentum $p = E/c$, where E is the photon energy, the rate at which photons carry momentum through the shell is $L(r)/c$. Thus, the rate at which photon momentum is transferred to the shell (in other words, the *force* on the shell) is

$$F_{\text{rad}}(r) = \frac{L(r)}{c}d\tau = -\frac{L(r)}{c}\rho(r)\kappa(r)dr. \tag{15.18}$$

Setting equations (15.16) and (15.18) equal to each other, we have an equation that relates the temperature, luminosity, and opacity of a star:

$$-\frac{16\pi}{3}ar^2T(r)^3dT = -\frac{\rho(r)\kappa(r)L(r)dr}{c}. \tag{15.19}$$

With a bit of rearrangement, this becomes the **equation of radiative energy transport**:

$$\frac{dT}{dr} = -\frac{3\rho(r)\kappa(r)L(r)}{16\pi acT(r)^3r^2}. \tag{15.20}$$

This equation can be written in alternate forms. For instance, we can take advantage of the equality $a = 4\sigma_{\text{SB}}/c$, where σ_{SB} is the Stefan–Boltzmann constant, to write the equation of radiative energy transport in the form

$$\frac{dT}{dr} = -\frac{3\rho(r)\kappa(r)L(r)}{64\pi\sigma_{\text{SB}}T(r)^3r^2}. \tag{15.21}$$

Equation (15.21) links the luminosity to the temperature gradient of the star, in much the same way that the equation of hydrostatic equilibrium (eq. 15.1) links the mass to the pressure gradient of the star. A perfectly transparent star ($\kappa = 0$) would have no temperature gradient, since it wouldn't absorb any of the gamma rays generated by fusion reactions in the star's central core. The actual temperature gradient in the Sun, between the center and the photosphere, averages to

$$\frac{\Delta T}{\Delta r} \approx \frac{T_{\text{phot}} - T_c}{R_\odot - 0} \approx \frac{5800\text{ K} - 1.47 \times 10^7\text{ K}}{6.96 \times 10^5\text{ km}} \approx -20\text{ K km}^{-1}. \tag{15.22}$$

For every kilometer you move outward in the Sun, on average, you cool down by 20 degrees.

Since the Sun's temperature is not uniform, we may ask whether the assumption of local thermodynamic equilibrium applies. We recall, from Section 5.6, that a necessary condition for LTE is that the mean free path of the photons is small compared to the distance over which the temperature T varies significantly. The mean free path of a photon (that is, the average distance it travels between interactions with massive particles), is

$$\langle \ell \rangle = \frac{1}{n\sigma} = \frac{1}{\kappa\rho}. \tag{15.23}$$

In a fully ionized gas, the source of opacity is **Thomson (or electron) scattering**, for which the cross-section is

$$\sigma_e = \frac{8\pi}{3}\left(\frac{e^2}{4\pi\epsilon_0 m_e c^2}\right)^2 = 6.65 \times 10^{-29}\,\mathrm{m}^2. \tag{15.24}$$

To total up the electron density, we consider (as we did in Section 14.1) how many particles are contributed by various elements. The number density of hydrogen atoms is $X\rho/m_p$, and each of these contributes one electron. The number density of helium atoms is $Y\rho/4m_p$, and each of these contributes two electrons. Metals have number density $Z\rho/Am_p$, and each of these contributes $\sim A/2$ electrons. Thus, the total electron density can be written as

$$\begin{aligned} n_e &= \frac{X\rho}{m_p} + 2\frac{Y\rho}{4m_p} + \frac{A}{2}\frac{Z\rho}{Am_p} \\ &= \frac{\rho}{m_p}\left[X + \frac{1}{2}Y + \frac{1}{2}Z\right] \\ &= \frac{\rho}{2m_p}(1+X), \end{aligned} \tag{15.25}$$

where in the last step we used $(Y+Z)/2 = (1-X)/2$. The mean free path of a photon in the Sun is thus

$$\langle \ell \rangle = \frac{1}{n_e\sigma_e} = \frac{2m_p}{\sigma_e\rho(1+X)} \approx 0.02\,\mathrm{m}, \tag{15.26}$$

assuming $\rho = \rho_\odot = 1400\,\mathrm{kg\,m^{-3}}$ and $X = 0.73$. From equation (15.22), we see that over this small distance, the temperature change will typically amount to only $\Delta T \sim 4 \times 10^{-4}$ K. So while the temperature gradient in the Sun causes energy to flow outward, it is so small that local thermodynamic equilibrium (see Section 5.6) is a very good approximation in the solar interior.

The equation of radiative energy transport (eq. 15.21) can be used to make a crude estimate of the Sun's luminosity:

$$\frac{\Delta T}{\Delta r} \approx \frac{-T_c}{R_\odot} \approx -\frac{3\langle\kappa\rangle\rho_\odot L_\odot}{64\pi\sigma_{\mathrm{SB}}(T_c/2)^3(R_\odot/2)^2}, \tag{15.27}$$

and so

$$L_\odot \approx \frac{2\pi\sigma_{\mathrm{SB}}T_c^4 R_\odot}{3\langle\kappa\rangle\rho_\odot}, \tag{15.28}$$

where $\langle\kappa\rangle$ is the typical opacity inside the Sun. With $T_c = 1.47 \times 10^7$ K, $R_\odot = 6.96 \times 10^8$ m, and $\rho_\odot = 1400\,\mathrm{kg\,m^{-3}}$, we find

$$L_\odot \approx 3 \times 10^{27}\,\mathrm{W}\left(\frac{\langle\kappa\rangle}{1\,\mathrm{m^2\,kg^{-1}}}\right)^{-1}. \tag{15.29}$$

If we assumed that the opacity throughout the Sun had the same value as in the photosphere, $\kappa \approx 3\,\mathrm{m^2\,kg^{-1}}$, we would compute a solar luminosity of $L_\odot \sim 10^{27}$ W, about 2.5 times the actual value of $L_\odot = 3.90 \times 10^{26}$ W. This implies that the average opacity inside the Sun is actually

$$\langle\kappa\rangle \approx 8\,\mathrm{m^2\,kg^{-1}}. \tag{15.30}$$

Since opacity depends on temperature and density, its value varies with radius in the Sun; however, $\kappa \approx 8\,\mathrm{m^2\,kg^{-1}}$ is not absurd as an average value for the Sun as a whole.

It is also useful to ask at this point how long it takes photons to diffuse outward from the center of the Sun to the bottom of the convective zone at $\sim 0.7R_\odot$ (discussed in Section 15.1.3) keeping in mind, of course, that radiative transport involves repeated absorption and re-emission processes rather than scattering of photons whose identity is preserved as they wander through the solar interior. We have already seen that the mean free path for a photon in the completely ionized solar interior is $\ell \approx 2$ cm (equation 15.26). So a typical photon travels about 2 cm though the solar interior until it is absorbed and another photon is emitted to replace it; since the second photon retains no "memory" of the direction in which the original photon was headed, photons essentially "random walk" their way out of the solar interior by diffusion. A three-dimensional random walk (see the appendix to this chapter) requires $N = 3(0.7R_\odot/\ell)^2$ individual steps of size ℓ to diffuse out to a distance of $0.7R_\odot$ from the starting point, so

$$N = \frac{3(0.7R_\odot)^2}{\ell^2} = \frac{3(0.7 \times 6.96 \times 10^8\,\mathrm{m})^2}{(2 \times 10^{-2}\,\mathrm{m})^2} \approx 1.8 \times 10^{21} \tag{15.31}$$

individual absorptions and re-emissions. The total distance traveled is $d = N\ell \approx 3.6 \times 10^{19}$ m, and with photons traveling at the speed of light, the time for photons to escape is the **diffusion time**

$$t_{\mathrm{dif}} = \frac{d}{c} \approx 1.2 \times 10^{11}\,\mathrm{s} \approx 3.8 \times 10^3\,\mathrm{yr}. \tag{15.32}$$

Thus, even if there were changes in the energy-generation rates in the center of the Sun on timescales of 1000 years or less, the photon diffusion would smooth over these variations.

15.1.3 Convective Transport

Suppose, as a thought experiment, you inserted a layer into the Sun that was perfectly opaque, with $\kappa = \infty$. Equation (15.21) seems to imply that an infinite temperature gradient would occur across that layer, with the photons absorbed at the bottom of the layer driving the temperature at the bottom arbitrarily high. In reality, this would not happen; as the bottom of the layer absorbed energy, the hot gas would become buoyant and rise upward. Energy would then be transported across the opaque layer by hot, rising blobs of gas, rather than by photons. Generally, in stars with high opacity, energy is transported outward by convection, with hot gas rising and cooler gas sinking to take its place.

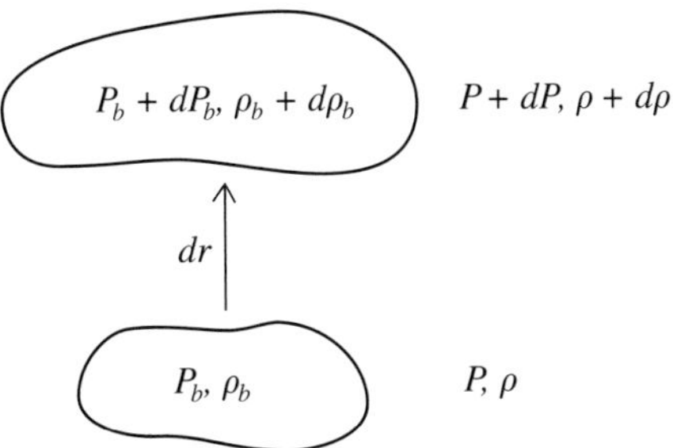

FIGURE 15.1 A blob of gas is moved upward by a small distance dr.

Convection is a chaotic, turbulent process, as you can verify by watching a pot of simmering soup on a stove, so the detailed physics tends to be messy. Nevertheless, we can derive, in a fairly straightforward calculation, the physical conditions under which a gas becomes unstable against the onset of convection. We will do this by perturbing a small blob of gas in the upward direction and then determining whether this perturbed gas element will sink back down to its initial position (that is, the gas is *stable* against convection) or if the element will continue to rise; convection will occur if a small displacement of a gas element triggers a runaway process of vertical motion.

Consider a small blob of gas in a star, as shown in Figure 15.1. The pressure and density within the blob are P_b and ρ_b. The pressure and density in the gas immediately outside the blob are P and ρ. Initially, the blob is in hydrostatic equilibrium with its surroundings, requiring $P_b = P$. Now, let's consider what happens when we move the gas blob upward by a short distance dr, to a region of the star where the pressure is $P + dP$ (with $dP < 0$, since we're moving upward) and the density is $\rho + d\rho$. As the blob is moved upward, it expands until it is again in hydrostatic equilibrium with its surroundings; at its new position, the internal pressure in the blob will be $P_b + dP_b = P + dP$. If, in its new position, the blob has a density $\rho_b + d\rho_b > \rho + d\rho$ (that is, if it's denser than its surroundings), it will sink back down, and the gas is stable to convection. On the other hand, if $\rho_b + d\rho_b < \rho + d\rho$, the blob is buoyant and will continue to rise; this marks the onset of convection. Thus, the condition for stability against convection is $\rho_b + d\rho_b > \rho + d\rho$, or since $\rho_b = \rho$ initially,

$$d\rho_b > d\rho. \tag{15.33}$$

Thus, to find whether a star of known pressure $P(r)$ and density $\rho(r)$ is stable against convection, we need to compute the value of $d\rho_b$ when a blob of gas is moved upward through a distance dr.

If the blob is moved upward rapidly, it will not have time to exchange heat with its surroundings. Such a process, in which heat is not gained or lost, is called an **adiabatic process**; in an adiabatic process, entropy is conserved. For a blob of gas undergoing an adiabatic process,

$$PV^\gamma = \text{constant}, \tag{15.34}$$

where V is the volume of the blob and γ is the **adiabatic index**.[2] Since for a blob of fixed mass, $V \propto 1/\rho$, we may write equation (15.34) as $P_b \rho_b^{-\gamma} =$ constant, or

$$\ln \rho_b = \frac{1}{\gamma} \ln P_b + \text{constant}. \tag{15.35}$$

Taking the derivative of equation (15.35), we find that an adiabatic process that changes the pressure of a gas blob by an amount dP_b will change its density by an amount $d\rho_b$, given by the relation

$$\frac{d\rho_b}{\rho_b} = \frac{1}{\gamma} \frac{dP_b}{P_b}. \tag{15.36}$$

Given the initial conditions for the blob ($\rho_b = \rho$, $P_b = P$) and the requirement that it end in hydrostatic equilibrium with its surroundings ($dP_b = dP$), this means that the change in the blob's density as it moves upward adiabatically will be

$$d\rho_b = \frac{\rho_b}{\gamma} \frac{dP_b}{P_b} = \frac{\rho}{\gamma} \frac{dP}{P} = \frac{\rho}{\gamma P} \frac{dP}{dr} dr. \tag{15.37}$$

The change in the external density can be written as

$$d\rho = \frac{d\rho}{dr} dr. \tag{15.38}$$

Given equations (15.37) and (15.38), the criterion for stability against convection, $d\rho_b > d\rho$, can be rewritten in the form

$$\frac{\rho}{\gamma P} \frac{dP}{dr} > \frac{d\rho}{dr}, \tag{15.39}$$

or

$$\frac{1}{\gamma P} \frac{dP}{dr} > \frac{1}{\rho} \frac{d\rho}{dr}. \tag{15.40}$$

It is often more useful to state the stability criterion in terms of the temperature gradient dT/dr of a star, rather than its density gradient $d\rho/dr$. For an ideal gas, $P = \rho kT/\mu m_p$, and thus, if the mean molecular mass is constant,

$$\begin{aligned} \frac{dP}{dr} &= \frac{\rho k}{\mu m_p} \frac{dT}{dr} + \frac{kT}{\mu m_p} \frac{d\rho}{dr} \\ &= \frac{P}{T} \frac{dT}{dr} + \frac{P}{\rho} \frac{d\rho}{dr}, \end{aligned} \tag{15.41}$$

[2] The adiabatic index has a value $\gamma = 5/3$ for simple atomic gases and fully ionized gases. For a gas of diatomic molecules, such as H_2, N_2, and O_2, the adiabatic index is $\gamma = 7/5$.

which can be rearranged to

$$\frac{1}{\rho}\frac{d\rho}{dr} = \frac{1}{P}\frac{dP}{dr} - \frac{1}{T}\frac{dT}{dr}. \tag{15.42}$$

The stability equation then becomes

$$\frac{1}{\gamma P}\frac{dP}{dr} > \frac{1}{P}\frac{dP}{dr} - \frac{1}{T}\frac{dT}{dr}. \tag{15.43}$$

Collecting terms and multiplying by T yields the final form of the stability equation:

$$-\left(1 - \frac{1}{\gamma}\right)\frac{T}{P}\frac{dP}{dr} > -\frac{dT}{dr}. \tag{15.44}$$

Here we leave in the minus sign since both the pressure and the temperature decrease with radius in a star. The left-hand side of this equation is the **adiabatic temperature gradient**. The right-hand side of the equation is the *actual* temperature gradient in the star. Both of these quantities are evaluated at every point in the stellar structure calculation; as long as equation (15.44) is satisfied, energy transport will be radiative. However, if equation (15.44) is *not* satisfied, then convection ensues. In this condition, the temperature gradient then approaches the adiabatic temperature gradient, which is the largest realizable value. Thus, the **equation of convective energy transport** is

$$\frac{dT}{dr} = \left(1 - \frac{1}{\gamma}\right)\frac{T(r)}{P(r)}\frac{dP}{dr}. \tag{15.45}$$

When energy is transported by convection, the temperature gradient is proportional to the pressure gradient.

In general, energy is carried outward in a star either by radiation or by convection, whichever is more efficient at shuttling joules toward the photosphere. The more efficient process is the one that leads to the smaller temperature gradient (equations 15.21 and 15.45). Within a single star, energy can be carried by radiation in one region and by convection in another. In the Sun, for example, radiation is the more efficient process out to $r = 0.7R_\odot$ (Figure 15.2). In the outer 30% of the Sun (by radius), convection is the dominant means of energy transport.

15.2 ▪ ENERGY GENERATION IN STARS

So far, we've talked about how energy is carried to the photosphere of the star, but not about how it is generated in the star's interior. The energy that a star tosses away into space must come from some source inside the star. The generation of energy within a star is described by the last of the equations of stellar structure, the **equation of energy generation**. Consider the usual thin spherical shell of inner radius r and outer radius $r + dr$, centered on the star's center. A luminosity L flows outward through the inner surface, and a luminosity $L + dL$ flows outward through the outer surface. Where does the extra bit of power dL come from? Even if we don't know from a physics standpoint,

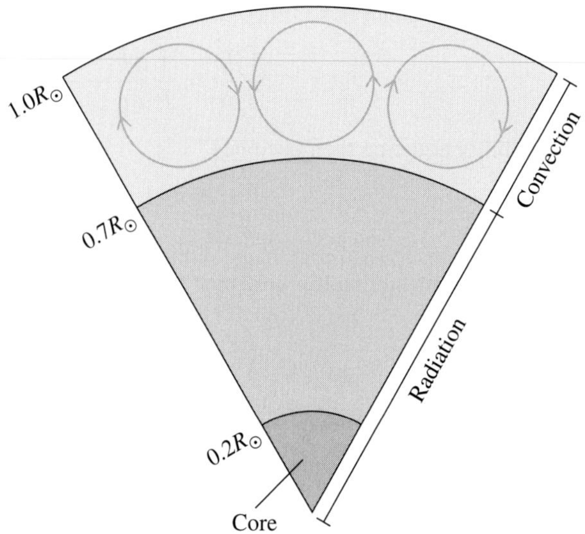

FIGURE 15.2 Radiative and convective energy transport within the Sun.

we can still express it mathematically in terms of the rate of energy production ϵ (the units of ϵ are watts per kilogram). The equation of energy generation can be written as

$$dL = (4\pi r^2 dr)\rho\epsilon, \tag{15.46}$$

or

$$\frac{dL}{dr} = 4\pi r^2 \rho(r)\epsilon(r). \tag{15.47}$$

All we need to do now is find the physical process by which energy is generated, and determine how ϵ depends on the temperature, density, and chemical composition within a star.

The answer to the question Why do stars shine? is Stars shine because they are hot. The obvious follow-up question is Why don't they cool down? There are several possible answers to this question. One possible source of energy for stars is **gravitational potential energy**. The current gravitational potential energy of the Sun is

$$U_\odot = -q\frac{GM_\odot^2}{R_\odot}, \tag{15.48}$$

where q is a factor of order unity. For a sphere of uniform density, as we computed in equation (10.24), the factor is $q = 3/5$. Stars, however, are centrally concentrated and have $q \approx 1.5$. For the Sun, then,

$$U_\odot = -1.5\frac{GM_\odot^2}{R_\odot} \approx -5.7 \times 10^{41}\,\text{J}. \tag{15.49}$$

Since the Sun started as a gas cloud with $r \gg R_\odot$, in collapsing to its present size, it lost 5.7×10^{41} J of gravitational potential energy. If all this energy were converted to photons, it would keep the Sun at its present luminosity for a time equal to the **Kelvin–Helmholtz time**:

$$t_{\mathrm{KH}} \equiv \frac{|U_\odot|}{L_\odot} = \frac{5.7 \times 10^{41}\,\mathrm{J}}{3.9 \times 10^{26}\,\mathrm{J\,s^{-1}}} \approx 1.5 \times 10^{15}\,\mathrm{s} \approx 50\,\mathrm{Myr}. \tag{15.50}$$

In the 1850s, Helmholtz proposed that the Sun was powered by gravitational potential energy, and computed its age as being $t_{\mathrm{KH}} \sim 20$ million years. (With no knowledge of the interior structure of the Sun, Helmholtz assumed it was of uniform density, thus underestimating $|U_\odot|$ and t_{KH}.) Slightly later, Kelvin, taking into account the nonuniform density of the Sun, calculated a value of $t_{\mathrm{KH}} \sim 60$ million years. Nineteenth-century geologists were dubious of the results of Helmholtz and Kelvin. They pointed out, quite rightly, that the fossil record implies that the Sun has been shining at a roughly constant luminosity for a time much longer than 60 million years.[3]

It was not until the 1930s that astrophysicists had the grand realization that **nuclear fusion** provides the necessary energy to keep the stars hot. The Sun, like other main sequence stars, fuses hydrogen into helium:

- Mass of 4 hydrogen nuclei = 6.6905×10^{-27} kg
- Mass of 1 helium nucleus = 6.6447×10^{-27} kg
- Mass difference = 0.0458×10^{-27} kg

When four hydrogen atoms fuse to form one helium atom, the lost mass, $\Delta m = 4.58 \times 10^{-29}$ kg, is converted to energy. The conversion rate is given by Einstein's formula:

$$\Delta E = (\Delta m)c^2 = 4.1 \times 10^{-12}\,\mathrm{J}. \tag{15.51}$$

Fusing together four hydrogen atoms doesn't create a lot of energy: 4.1×10^{-12} J is about enough to lift a nickel through a height of 1 Å against the Earth's gravity at sea level.[4] However, there are a whole lot of hydrogen atoms inside a star. If the Sun had started out made entirely of hydrogen, it would have contained N_{H} hydrogen atoms, where

$$N_{\mathrm{H}} = \frac{M_\odot}{m_p} \approx \frac{1.99 \times 10^{30}\,\mathrm{kg}}{1.67 \times 10^{-27}\,\mathrm{kg}} \approx 1.2 \times 10^{57}. \tag{15.52}$$

Fusing all the hydrogen atoms into $N_{\mathrm{H}}/4$ helium atoms would release an amount of energy

$$E_{\mathrm{fus}} = \frac{N_{\mathrm{H}}}{4}\Delta E = \frac{1.2 \times 10^{57}}{4}(4.1 \times 10^{-12}\,\mathrm{J}) = 1.2 \times 10^{45}\,\mathrm{J}. \tag{15.53}$$

[3] The Kelvin–Helmholtz time for brown dwarfs and Jovian planets is hundreds of times longer than the Sun's Kelvin–Helmholtz time. Thus, the dim light from these objects is powered by gravitational potential energy.

[4] A nickel has the same mass as a British 20-pence coin and is slightly less massive than a euro 20-cent piece.

This is about 2000 times the magnitude of the Sun's gravitational potential energy. Thus, fusion can keep the Sun shining at a constant rate for a time

$$t_{\text{fus}} = \frac{E_{\text{fus}}}{L_\odot} \approx 3.3 \times 10^{18}\,\text{s} \approx 100\,\text{Gyr}. \tag{15.54}$$

In truth, the Sun wasn't pure hydrogen when it started out, and the conversion of hydrogen to helium in the Sun isn't total; only the central $\sim 10\%$ of the Sun's mass is at temperatures high enough for fusion to take place. The lifetime of the Sun, as a consequence, is only ~ 10 gigayears instead of ~ 100 gigayears. It is still comfortably longer than the Kelvin–Helmholtz time, though.

All main sequence stars are powered by the fusion of hydrogen into helium in their central regions. The Sun's main sequence lifetime is $\tau_\odot \approx 10\,\text{Gyr}$. Since the main sequence lifetime is $\tau \propto M/L$ and since the approximate mass-luminosity relation is $L \propto M^4$ for stars with $M > 0.7M_\odot$ (Section 13.6), the main sequence lifetime of a massive star is

$$\tau \approx 10\,\text{Gyr}\left(\frac{M}{1M_\odot}\right)^{-3}. \tag{15.55}$$

The lifetime of a $20M_\odot$ star (with spectral type 09) will be only 1 Myr. The lifetime of a $0.5M_\odot$ star (spectral type M0) will be 80 Gyr, longer than the age of our galaxy. Every M dwarf ever made is still fusing hydrogen into helium; they aren't going to run out of fuel any time soon.

15.3 ▪ NUCLEAR FUSION REACTIONS

The fundamental source of energy in stars over most of their lifetime is fusion of light elements into heavier elements. The first of these reactions is the fusion of two protons into a deuteron. This can happen only if the protons can approach close enough that the very short-range but powerful **strong nuclear force** can overcome the long-range electronic Coulomb repulsion between the two protons, as illustrated in Figure 15.3. The potential energy of two protons separated by a distance r is

$$U = \frac{e^2}{4\pi\epsilon_0}\frac{1}{r}. \tag{15.56}$$

The strong nuclear force has a range of only 10^{-15} m, and the electrostatic potential energy of two protons at this separation is

$$\begin{aligned} U &\approx \frac{(1.60 \times 10^{-19}\,\text{C})^2}{4\pi(8.8 \times 10^{-12}\,\text{C}^2\,\text{J}^{-1}\,\text{m}^{-1})(1.60 \times 10^{-19}\,\text{J}\,\text{eV}^{-1})(10^{-15}\,\text{m})} \\ &\approx 1.4 \times 10^6\,\text{eV} \approx 1.4\,\text{MeV}. \end{aligned} \tag{15.57}$$

To overcome the repulsion, the kinetic energy of a proton must exceed this potential energy; this is possible in a high-temperature gas, so the fusion processes in such

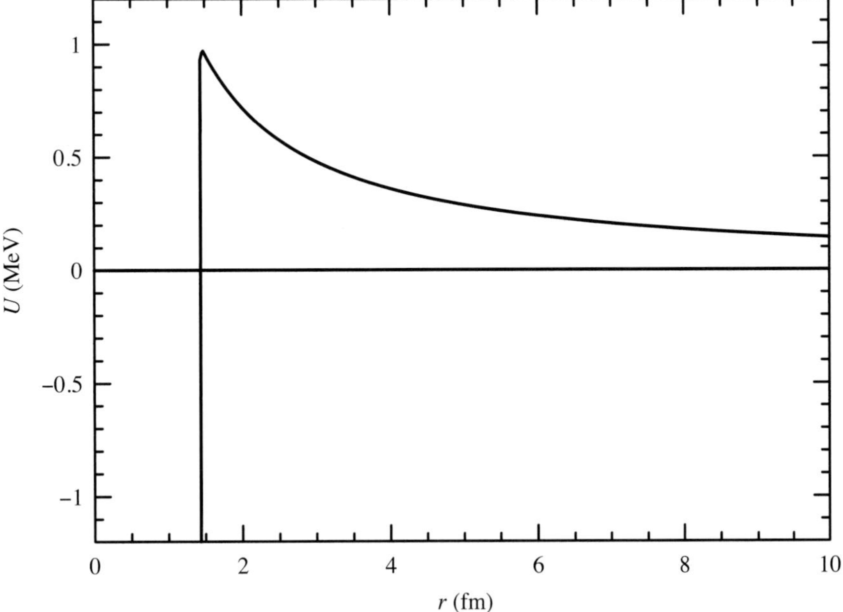

FIGURE 15.3 Potential energy of two protons as a function of their separation (in units of 10^{-15} m = 1 fm). At large separations, the Coulomb repulsion dominates; at small separations, the strong nuclear force overcomes the electronic repulsion between the protons.

environments are often called **thermonuclear reactions**. The typical kinetic energy of a proton at the center of the Sun is (see equation 5.46)

$$\langle E \rangle = \frac{3kT_c}{2} \approx \frac{3(1.38 \times 10^{-23}\,\mathrm{J\,K^{-1}})(1.47 \times 10^{7}\,\mathrm{K})}{2(1.60 \times 10^{-19}\,\mathrm{J\,eV^{-1}})}$$
$$\approx 2 \times 10^{3}\,\mathrm{eV} \approx 2\,\mathrm{keV}. \tag{15.58}$$

It thus seems that even at the Sun's center, protons have insufficient energy to overcome their mutual Coulomb repulsion.[5] The nuclear reactions can occur nevertheless because of **quantum mechanical tunneling**. An elementary particle like a proton can't be treated like a macroscopic object such as a tennis ball, which must be on one side or the other of a barrier such as a net. A typical proton near the Sun's center has a kinetic energy $E \approx 2$ keV and a speed $v = (2E/m_p)^{1/2} \approx 0.002c$. The proton is thus nonrelativistic and has a momentum $p = m_p v \approx 1 \times 10^{-21}$ kg m s^{-1}. This leads to a de Broglie wavelength

[5] The astronomer Arthur Eddington was among the first scientists to propose that the Sun is powered by nuclear fusion. When critics pointed to the relatively low kinetic energy of protons in the Sun, Eddington's response was, "The critics lay themselves open to an obvious retort; we tell them to go and find *a hotter place*."

for the proton of $\lambda_{\rm dB} \equiv h/p \approx 7 \times 10^{-13}$ m, much larger than the range of the strong nuclear force. Thus, when discussing fusion reactions in the Sun, we can't treat protons like classical tennis balls; their wavelike quantum nature must be taken into account.

Tunneling is a quantum mechanical process; for a proton to have any significant chance of tunneling through the Coulomb barrier around another proton, two criteria must be satisfied. First, the proton–proton separation must be comparable to the de Broglie wavelength, $\lambda_{\rm dB} = h/p = h(2m_pE)^{-1/2}$. Second, the kinetic energy E of the proton must be comparable to the electrostatic potential energy U at that separation. These two criteria require that

$$U = \frac{e^2}{4\pi\epsilon_0}\frac{1}{\lambda_{\rm dB}} = \frac{e^2}{4\pi\epsilon_0}\frac{(2m_pE)^{1/2}}{h} \approx E. \tag{15.59}$$

Solving for E, we find that the minimum kinetic energy at which tunneling has a significant probability is

$$E \approx \left(\frac{e^2}{4\pi\epsilon_0}\right)^2 \frac{2m_p}{h^2} \approx \frac{1}{2\pi^2}\alpha^2 m_p c^2 \approx 3\,\text{keV}. \tag{15.60}$$

This energy is comparable to the average kinetic energy, $\langle E\rangle \sim 2$ keV, in the Sun's core. Thus, we expect tunneling to be possible in the Sun.

To do a slightly more sophisticated calculation of fusion rates in the Sun's central regions, start by considering a single proton with kinetic energy E and de Broglie wavelength $\lambda_{\rm dB} = h(2m_pE)^{-1/2}$. If the number density of protons at the Sun's center is n_p, then the mean free path of our proton before it undergoes fusion with another proton is

$$\ell_{\rm pp} = \frac{1}{n_p\sigma_{\rm pp}}, \tag{15.61}$$

where $\sigma_{\rm pp}$ is the cross-section for the fusion of two protons.[6] Since a proton must come within a de Broglie wavelength of another proton before fusing, we might expect a cross-section $\sigma_{\rm pp} \sim \pi\lambda_{\rm dB}^2 \propto 1/E$. However, coming within a de Broglie wavelength doesn't guarantee a successful quantum tunneling event. The probability of tunneling, given a separation of $\lambda_{\rm dB}$ or less, is given by the **Gamow factor**

$$P_G \sim \exp\left(-\sqrt{\frac{E_G}{E}}\right), \tag{15.62}$$

where the **Gamow energy** for proton–proton encounters is $E_G = \pi^2\alpha^2 m_p c^2 = 490$ keV.[7] Second, quantum tunneling doesn't guarantee a successful fusion event.

[6] Compare equation (15.61) with equation (5.62), which gives the mean free path of a photon through a medium filled with absorbers of number density n and cross-section σ.

[7] The Gamow factor and Gamow energy are named after the Ukrainian American physicist George Gamow, who first investigated quantum mechanical tunneling.

Proton–proton fusion is mediated by the weak nuclear force and has a tiny cross-section. The probability P_{fus} that fusion occurs, given that quantum tunneling through the Coulomb barrier has already taken place, is a small number—but one that is only weakly dependent on the proton speed. The cross-section for proton–proton fusion can thus be written in the form

$$\sigma_{\text{pp}} = \pi \lambda_{\text{dB}}^2 P_G P_{\text{fus}} \approx \frac{\pi \hbar^2}{2m_p E} \exp\left(-\sqrt{\frac{E_G}{E}}\right) P_{\text{fus}}. \quad (15.63)$$

Since our proton is traveling along with a speed $v = (2E/m_p)^{1/2}$, its average time spent before fusing with another proton is

$$t_{\text{pp}} = \frac{\ell_{\text{pp}}}{v} = \frac{1}{n_p \sigma_{\text{pp}} v}. \quad (15.64)$$

If all protons in the Sun had the same speed v, then the total number of proton–proton fusions per unit volume per unit time would be

$$N_{\text{pp}} = \frac{1}{2}\frac{n_p}{t_{\text{pp}}} = \frac{1}{2} n_p^2 \sigma_{\text{pp}} v. \quad (15.65)$$

(The factor of 1/2 enters because it takes two protons to perform one fusion.) Expressed as a function of proton kinetic energy E, the proton–proton fusion rate is

$$N_{\text{pp}}(E) = \frac{n_p^2}{2} \sigma_{\text{pp}} \sqrt{2E/m_p} \approx \frac{\pi}{2\sqrt{2}} \frac{\hbar^2 n_p^2 P_{\text{fus}}}{m_p^{3/2}} \frac{1}{E^{1/2}} \exp\left(-\sqrt{\frac{E_G}{E}}\right). \quad (15.66)$$

Of course, real protons at the Sun's center don't all have the same kinetic energy E; instead, they have a range of kinetic energies, given by a Maxwell–Boltzmann distribution (equation 5.46):

$$F(E) \propto E^{1/2} \exp(-E/kT). \quad (15.67)$$

The proton–proton fusion rate, averaged over all proton kinetic energies at a given temperature T, is

$$\langle N_{\text{pp}} \rangle = \int_0^\infty N_{\text{pp}}(E) F(E) dE. \quad (15.68)$$

Using equation (15.66) for the fusion rate at energy E, and equation (15.67) for the distribution of E, we find, after a bit of algebra,

$$\langle N_{\text{pp}} \rangle \propto \frac{n_p^2 P_{\text{fus}}}{(kT)^{3/2}} \int_0^\infty \exp\left[-\sqrt{\frac{E_G}{E}} - \frac{E}{kT}\right] dE. \quad (15.69)$$

Here we have taken the liberty of ignoring the mild dependence of the fusion probability P_{fus} on proton kinetic energy and have taken it outside the integral.

The integrand in equation (15.69) is the product of the Gamow factor (which goes to zero at energies much smaller than the Gamow energy E_G) and a Boltzmann exponential

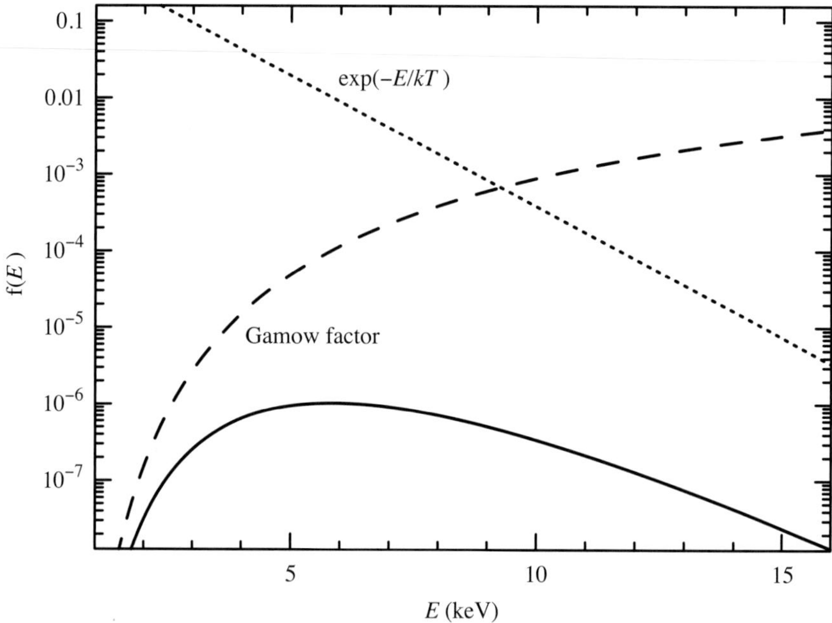

FIGURE 15.4 The Gamow factor (dashed line), which gives the probability of tunneling as a function of E, and the Boltzmann exponential $\exp(-E/kT)$ (dotted line). The solid line shows the product of the two functions.

(which goes to zero at energies much greater than the thermal energy kT). The Gamow factor, and also the Boltzmann exponential for the Sun's central temperature $T_c = 1.47 \times 10^7$ K, are shown in Figure 15.4. The solid line in the figure represents the integrand in equation (15.69) and graphically shows us the proton energy at which fusion is most likely to occur. At low energies, the protons are more numerous, but the tunneling probability is smaller. At high energies, the tunneling probability is higher, but the number of protons is dropping steeply. The maximum reaction rate is expected at $E \sim 6$ keV, roughly three times the average proton energy. The relatively narrow energy range between ~ 3 keV and ~ 11 keV, where the reaction rate is highest, is called the **Gamow window**.

Fusion of hydrogen into helium occurs by a series of two-body collisions, instead of a single, grand four-body collision. For stars with central temperatures less than 18 million Kelvin (this includes the Sun), helium is created from hydrogen via the **PP chain**, illustrated in Figure 15.5. In the first step of the chain, two protons (p) fuse together to form a deuteron (^{2}H). A deuteron is the nucleus of a deuterium (or "heavy hydrogen") atom, and consists of a proton and neutron held together by the strong nuclear force. When one of the protons is converted to a neutron, a positron (e^+) is emitted to conserve charge, and an electron neutrino (ν_e) is emitted to conserve electron quantum number. Because P_{fus} is so small for proton–proton fusion, the first step of the PP chain is the

$$\boxed{p} + \boxed{p} \longrightarrow {}^{2}H + e^{+} + \nu_e \text{ (twice)}$$
$${}^{2}H + p \longrightarrow {}^{3}He + \gamma \text{ (twice)}$$
$${}^{3}He + {}^{3}He \longrightarrow \boxed{{}^{4}He} + p + p$$

FIGURE 15.5 The PP chain, the dominant form of hydrogen fusion at $T_c < 1.8 \times 10^7$ K.

slowest one. During the past 4.6 billion years, only half the protons in the Sun's core have undergone fusion. In the second step of the PP chain, the deuteron (^{2}H) fuses with a proton to form light helium (^{3}He), which contains two protons and only one neutron. The excess energy from the fusion is carried away by a gamma-ray photon (γ). In the final step of the PP chain, two light helium nuclei fuse together to form ordinary helium (^{4}He), which contains two protons and two neutrons. The excess protons are spat out, ready to begin a new PP chain.

The net result of the PP chain is

$$4\text{p} \rightarrow {}^{4}\text{He} + 2\text{e}^{+} + 2\nu_{\text{e}} + 2\gamma. \tag{15.70}$$

The positrons quickly annihilate with electrons to form additional gamma rays. The neutrinos carry away only 2% of the energy released in the PP chain; gamma rays take away the rest. The neutrinos, because of their extremely tiny cross-sections for interactions, stream freely through the Sun. In other words, although the Sun is opaque to photons, it is transparent to neutrinos. The Sun emits about 2×10^{38} neutrinos per second (of which roughly 10^{15} are destined to pass through your body, which is also transparent to neutrinos).

In main sequence stars with central temperatures greater than 18 million Kelvin (this includes O, B, A, and F stars), hydrogen is fused into helium via the **CNO cycle**, illustrated in Figure 15.6. In the CNO cycle, carbon (C), nitrogen (N), and oxygen (O) act as catalysts to speed the fusion of hydrogen. The net result of the CNO cycle is

$$4p \rightarrow {}^{4}\text{He} + 2\text{e}^{+} + 2\nu_{\text{e}} + 3\gamma. \tag{15.71}$$

Again, the positrons annihilate with electrons to form additional gamma rays.

Fusion of hydrogen into helium is a reasonably efficient form of energy; about 0.7% of the hydrogen's mass is converted into energy in the process. However, still more energy can be squeezed out of a star if the helium is fused into heavier and heavier elements, until iron is reached. Iron has the lowest mass per nucleon of any element, so it is the end of the line as far as fusion is concerned.[8] The process by which stars convert helium

[8] Elements more massive than iron can release energy by fission, splitting into lower-mass nuclei.

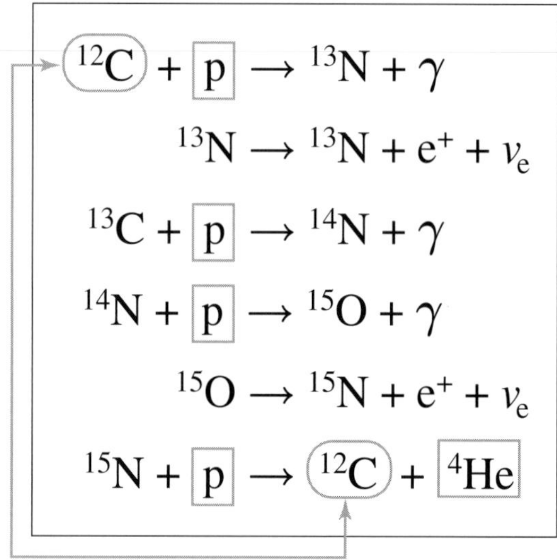

FIGURE 15.6 The CNO cycle, the dominant form of hydrogen fusion at $T_c > 1.8 \times 10^7$ K.

to carbon is the **triple alpha process**. In the first step of the triple alpha process, two helium nuclei fuse to form a beryllium nucleus:

$$^4\text{He} + {}^4\text{He} \rightarrow {}^8\text{Be} + \gamma. \tag{15.72}$$

The ^{8}Be nucleus is extremely unstable; it decays back into a pair of helium nuclei with a half-life of only $t_{1/2} \sim 2 \times 10^{-16}$ s.[9] However, if the ^{8}Be nucleus encounters a ^{4}He nucleus during the brief period before it decays, the two nuclei can fuse to form a stable ^{12}C nucleus:

$$^4\text{He} + {}^8\text{Be} \rightarrow {}^{12}\text{C} + \gamma. \tag{15.73}$$

Thus, the net result of the triple alpha process is the conversion of three ^{4}He nuclei into a single ^{12}C nucleus.[10] Because the beryllium nucleus has such a brief life, it will encounter a helium nucleus and fuse only if the surroundings are very dense (which increases the number density of helium nuclei) and very hot (which increases the average speed of the helium nuclei). In practice, the triple alpha process occurs at a significant rate only when $T_c > 10^8$ K. The Sun isn't currently fusing helium into carbon, because it's not hot enough.

[9] The stable isotope of beryllium—the kind found in emeralds—is ^{9}Be.

[10] In nuclear physics, a ^{4}He nucleus is also called an "alpha particle," which helps to explain the odd terminology "triple alpha process." Electrons are "beta particles," and high-energy photons are "gamma particles," or "gamma rays."

15.4 ▪ MODELING STELLAR INTERIORS

We now have the basic equations that govern the structure of stellar interiors. In particular, there are four differential equations that astronomers refer to collectively as **the equations of stellar structure**. First, the equation of hydrostatic equilibrium is

$$\frac{dP}{dr} = -\frac{GM(r)\rho(r)}{r^2}. \tag{15.74}$$

Second, the equation of mass continuity is

$$\frac{dM}{dr} = 4\pi r^2 \rho(r). \tag{15.75}$$

Third, the equation of energy transport is

$$\frac{dT}{dr} = -\frac{3\kappa(r)\rho(r)L(r)}{64\pi\sigma_{\rm SB} r^2 T(r)^3} \quad \text{[for radiative transport]} \tag{15.76}$$

or

$$\frac{dT}{dr} = \left(1 - \frac{1}{\gamma}\right)\frac{T(r)}{P(r)}\frac{dP}{dr} \quad \text{[for convective transport]} \tag{15.77}$$

with the energy transport mechanism, radiative or convective, determined by which gives the smaller temperature gradient. Fourth, the equation of energy generation is

$$\frac{dL}{dr} = 4\pi r^2 \rho(r)\epsilon(r). \tag{15.78}$$

To solve this set of equations, we need boundary conditions at the photosphere. We also need to know the equation of state; the ideal gas law usually works just fine within stars:

$$P(r) = \frac{k\rho(r)T(r)}{\mu(r)m_p}. \tag{15.79}$$

In adddition, however, we need to know how the mean molecular mass $\mu(\rho, T)$, opacity $\kappa(\rho, T)$, and energy generation rate $\epsilon(\rho, T)$ depend on density and temperature within the star. The energy generation rate ϵ, in particular, is extremely sensitive to temperature. For the PP chain, $\epsilon \propto T^4$, and for the CNO cycle, $\epsilon \propto T^{20}$.

Given all this information, models of stellar interiors can be built up by numerically solving the five equations of stellar structure. In the mid-twentieth century, back in the time of slide rules and mechanical calculating machines, you could earn a PhD by modeling a single star. Nowadays, computers can crank out stellar models on an assembly line. The result of a model of the Sun's interior is shown in Figure 15.7. Note in particular that most of the Sun's luminosity comes from $r < 0.2R_\odot$. Because of the strong dependence of ϵ on temperature, as the temperature T gradually drops with r, the energy generation rate ϵ plummets. It is also interesting to note that the Sun's central density is roughly 150 times the density of water. The high temperatures in the core keep the material in ionized gaseous form, despite its high density.

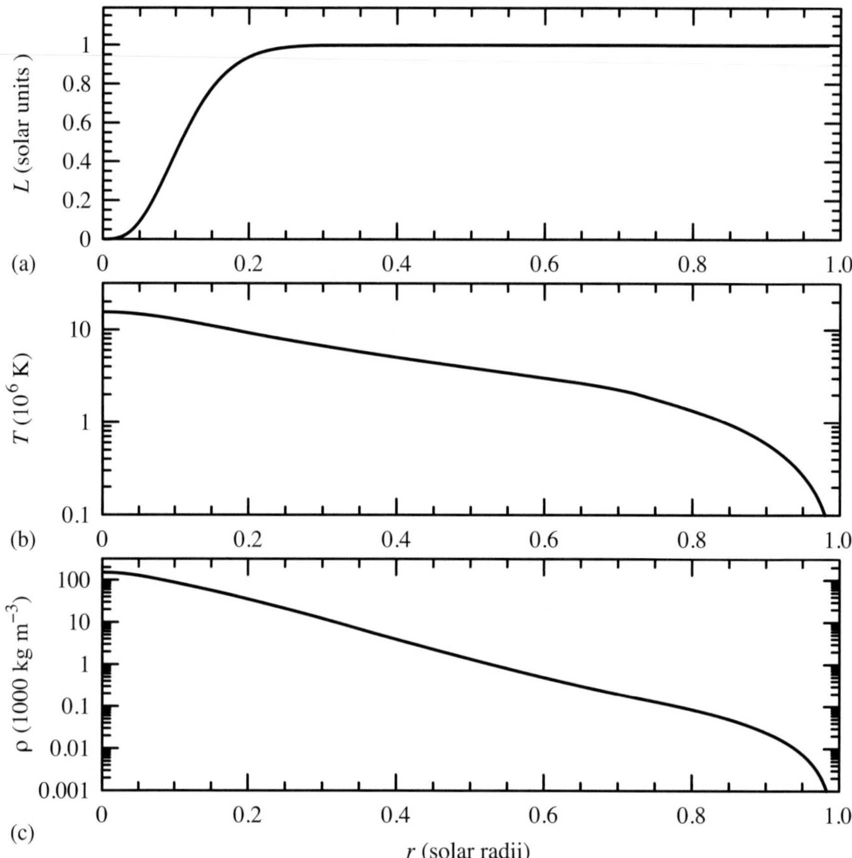

FIGURE 15.7 (a)–(c) Enclosed luminosity, temperature, and mass density as a function of radius within the Sun.

How do we know that our models of the solar interior are correct? The boundary conditions are determined by the well-observed properties of the photosphere. The equations of stellar structure (eqs. 15.74–15.78) are based on well-understood physics. Nevertheless, it is a good thing to verify our models by comparison with observations. Although we cannot see directly into the Sun's interior, there are indirect methods by which we can deduce the Sun's interior structure. For instance, **helioseismology** (the study of seismic waves in the Sun's interior) can tell us the sound speed inside the Sun. In the interior of the Earth, as mentioned in Section 9.1, both S-waves (shear waves) and P-waves (pressure waves) can propagate. In the interior of the Sun, S-waves, which can only propagate through solids, are not found. However, P-waves, which are sound waves, are free to move throughout the Sun's interior. Because the sound speed in a gas is $c_s \propto T^{1/2}$, it increases as you go farther into the Sun's interior. This causes P-waves (sound waves) to be refracted upward.

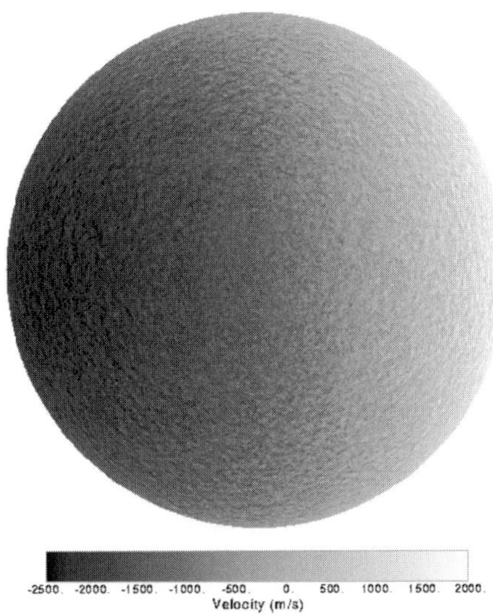

FIGURE 15.8 Dopplergram of the Sun's surface.

Although it is impractical to put a seismometer in the Sun's photosphere, we can see the vertical motions of the photosphere in a "dopplergram" (Figure 15.8), which shows the Doppler shift as a function of position on the visible hemisphere of the Sun. The rotation of the Sun can be seen in Figure 15.8, as well as the upward and downward motions due to P-waves reaching the photosphere. The sound oscillations in the Sun can be decomposed into different modes using spherical harmonics (in much the same way that the sound from a piano can be decomposed into different frequencies using Fourier transforms). The observed modes of oscillation can be used to determine the sound speed as a function of radius within the Sun. The sound speeds measured in this way agree with those predicted by the best solar models with errors of $< 0.1\%$.

Another source of information about the Sun's interior is **solar neutrinos**. The PP chain that provides most of the fusion energy in the Sun's core produces two electron neutrinos for every helium nucleus created. These electron neutrinos fly straight through the Sun with a very tiny chance of interaction. Thus, if we could manage to capture a few of the neutrinos, we would have a direct window on the fusion reactions at the Sun's center. Although neutrinos have small cross-sections for interaction with "ordinary" matter, they are capable of undergoing reactions such as

$$\nu_e + {}^{37}\text{Cl} \to {}^{37}\text{Ar} + e^- \tag{15.80}$$

and

$$\nu_e + {}^{71}\text{Ga} \to {}^{71}\text{Ge} + e^-. \tag{15.81}$$

Over the past three or four decades, typical solar neutrino experiments have involved filling large tanks with carbon tetrachloride or gallium and waiting for the infrequent neutrino reactions. Numerous hunts for solar neutrinos all found that only a third the expected number of electron neutrinos were detected. Astronomers and physicists scrambled to solve the "solar neutrino problem," as it was called. One suggested solution was that the solar interior was slightly cooler than the standard solar models suggested; this would drive down the rate of energy generation, since $\epsilon \propto T^4$. However, helioseismology confirmed that the temperatures predicted by the standard solar model were correct.

The ultimate resolution of the solar neutrino problem came from particle physics. There are three types, or flavors, of neutrinos. In addition to the electron neutrinos, ν_e, there are also muon neutrinos, ν_μ, and tau neutrinos, ν_τ. Although nuclear fusion produces only electron neutrinos, if neutrinos have mass and if the masses of the three types differ, then electron neutrinos can spontaneously convert into muon or tau neutrinos. Recent neutrino experiments indicate that the three types of neutrino really do have (small) masses. In addition, the Sudbury Neutrino Observatory, located in a former nickel mine in Ontario, has searched for neutrinos of all three types. The heart of the Sudbury detector is a large tank filled with heavy water (D_2O). The deuterium (or "heavy hydrogen") has a small probability of being split by a neutrino:

$$\nu + \mathrm{D} \rightarrow \mathrm{p} + \mathrm{n} + \nu. \tag{15.82}$$

This reaction occurs for any kind of neutrino and thus doesn't distinguish among ν_e, ν_μ, and ν_τ. The total number of all neutrinos detected by Sudbury is consistent with the number predicted by the standard solar model. This indicates that some of the electron neutrinos have been converted to muon and tau neutrinos during their flight from the Sun to Ontario.

APPENDIX: RANDOM WALK PROCESSES

We suppose that a test particle moves away from an initial point in a series of "steps"; in our particular case of interest, the particles are photons and a step is the mean free path that a photon travels before it is absorbed. The premise of random walk processes is that the direction of each step is completely uncorrelated with the direction of the previous step. Again, in our particular case, it means that the direction of each re-emitted photon is independent of the direction in which its absorbed predecessor was traveling.

Consider a large ensemble of photons generated at the origin and constrained to move in either positive or negative x direction in steps of length ℓ. Their average position after zero steps (that is, their starting position) is obviously $\langle x_0 \rangle = 0$. In the first step, about half of the photons will move a distance ℓ in the $+x$ direction, and the other half will move a distance ℓ in the $-x$ direction. The average position of all the photons after one step will be $\langle x_1 \rangle \approx 0$.

Things get more interesting in the second step: of the half of the original photons at position $+\ell$, about half of these will move another step in the same direction to position $+2\ell$, and the remainder will move back to the origin. Similarly, about half of those photons at position $-\ell$ will move to position -2ℓ, and the rest will return to the origin. After only two steps, one-quarter are at position $+2\ell$, one-quarter are at position -2ℓ, and half are back at the origin. Again, their average position remains $\langle x_2\rangle \approx 0$, but what is clear is that the distribution in x is becoming broader. Their average position after N steps will remain $\langle x_N\rangle \approx 0$, but it is clear that their root-mean-square displacement $\langle x_N^2\rangle^{1/2}$ after N steps is a good characterization of the increasing width of the distribution. We can compute this by induction.

For any photon, its position at step N will be one step removed from its position at the previous step,

$$x_N = x_{N-1} \pm \ell. \tag{15.83}$$

Thus,

$$x_N^2 = (x_{N-1} \pm \ell)^2 = x_{N-1}^2 + \ell^2 \pm 2\ell x_{N-1}. \tag{15.84}$$

Averaging over a large number of photons, the mean square position becomes

$$\langle x_N^2\rangle = \langle x_{N-1}^2\rangle + \ell^2 \pm 2\ell\langle x_{N-1}\rangle, \tag{15.85}$$

and the last term is zero because $\langle x_{N-1}\rangle = 0$. Similarly,

$$\langle x_{N-1}^2\rangle = \langle x_{N-2}^2\rangle + \ell^2. \tag{15.86}$$

Putting this in equation (15.85), we have

$$\langle x_N^2\rangle = \langle x_{N-2}^2\rangle + 2\ell^2. \tag{15.87}$$

By induction, we conclude that

$$\langle x_N^2\rangle = \langle x_{N-N}^2\rangle + N\ell^2, \tag{15.88}$$

and since $N - N = 0$ and $x_0 = 0$, we have the general result that

$$\langle x_N^2\rangle = N\ell^2, \tag{15.89}$$

or

$$\langle x_N^2\rangle^{1/2} = \sqrt{N}\ell. \tag{15.90}$$

The root-mean-square (rms) displacement of photons from the origin after N steps is simply $\sqrt{N}\ell$. Although, for simplicity, we have computed the result for a one-dimensional random walk, the same result holds true in two or three dimensions: rms displacement equals $\sqrt{N}\ell$, where ℓ is the length of a single step.

If we took a direct route from the center of the Sun to the bottom of the convective layer at $0.7R_\odot$, the number of steps it would take would be $n = 0.7R_\odot/\ell$. Comparing this to equation (15.90) and taking $\langle x_N^2\rangle^{1/2} = 0.7R_\odot$, we see that $\sqrt{N} = n$, so $N = n^2$. This is the easiest way to remember the random walk: if the direct path requires n steps, a random walk of n^2 steps is typically required to reach the same distance from the origin.

PROBLEMS

15.1 What is the rate (in kilograms per second) at which the Sun is currently converting hydrogen into helium?

15.2 How much energy, in MeV, is produced per proton in the PP chain?

15.3 Approximately half the original hydrogen in the Sun's core has now been converted to helium. Compute the mean molecular mass μ (a) at the surface of the Sun, given standard abundances ($X_\odot = 0.734$, $Y_\odot = 0.250$, $Z_\odot = 0.016$), and (b) at the center of the Sun.

15.4 If a star has $M = 100M_\odot$ and $L = 10^6 L_\odot$, how long can it shine at that luminosity if it started as pure hydrogen and is able to convert all its H to He? If a star has $M = 0.5M_\odot$ and $L = 0.1L_\odot$, how long can it shine under the same conditions?

15.5 Consider the Sun to be a sphere of uniform density that derives its luminosity from steady contraction. What fractional decrease in the Sun's radius, $\delta R/R$, would be required over historical times (say, the last 6000 years) to account for the Sun's constant luminosity over that period of time?

15.6 Suppose that the Sun is 100% carbon (coal, for instance) and that burning this can extract 3 eV per carbon nucleus. How long, assuming an inexhaustible supply of oxygen from outside, could burning carbon maintain the Sun's current luminosity?

15.7 On a clear day, Mount Fuji can be seen from central Tokyo, 100 km away. Under these conditions, what is the maximum possible opacity κ of the atmosphere, in $\mathrm{m^2\,kg^{-1}}$? (Assume that the density of air along the line of sight is $\rho \approx 1\,\mathrm{kg\,m^{-3}}$.)

15.8 Under ideal conditions, scuba divers in clear tropical waters can see objects as far as 50 m away. What is the opacity of the water?

15.9 In low-mass main sequence stars, the opacity is due primarily to photoionization of heavy elements. For this case, the opacity can be approximated by Kramers' law, which is written as

$$\kappa \propto \rho T^{-3.5}.$$

Use the equations of stellar structure to show that this implies

$$L \propto M^{5.5} R^{-0.5}.$$

15.10 Suppose the mass density of a star as a function of radius is

$$\rho(r) = \rho_0 \left[1 - \left(\frac{r}{R}\right)^2\right],$$

where R is the radius of the star.

(a) Find the mass M of the star in terms of ρ_0 and R.
(b) Find the mean density of the star in terms of ρ_0.
(c) Show that the central pressure of the star is

$$P_c = \frac{15}{16\pi} \frac{GM^2}{R^4}.$$

16 The Interstellar Medium

Stars are formed by gravitational compression of the **interstellar medium**, that is, the low-density mix of dust and gas that lies in the space between stars. In this chapter, we introduce the basic characteristics and physical properties of the interstellar medium as a prelude to a discussion of star formation and evolution.

16.1 ▪ INTERSTELLAR DUST

16.1.1 Evidence for Interstellar Dust

We first noted in Section 13.3 that interstellar dust affects the light of distant stars by **extinction**,[1] or the diminution of light, and **reddening**, as the shorter-wavelength light is more highly extinguished than longer-wavelength light. Observations of the effects of dust yield clues that help us to determine the size, shape, and composition of dust grains.

Clue one. Dust causes reddening at both visible and ultraviolet wavelengths. Figure 16.1 shows the extinction A_λ plotted as a function of λ. At all wavelengths $\lambda > 2.2\,\mu\text{m} = 2200$ Å, the amount of extinction decreases with increasing wavelength. This differential extinction can happen only if the individual dust grains are smaller than the wavelength of light that they are scattering ($d \leq \lambda$, where d is the length of a dust grain). If interstellar matter were made of pebbles or boulders rather than dust grains, it would absorb all wavelengths of visible and near ultraviolet light equally. Detailed studies of the extinction as a function of wavelength give an estimate of $d \approx 50 \to 2000$ Å for the size of the dust particles. These minuscule grains are much smaller than the dust particles that you sweep from under the bed; they are more like the particles in cigarette smoke.

Clue two. Starlight is polarized by dust grains. A Polaroid filter polarizes light because it contains long polymer molecules that are aligned preferentially in one direction. Dust grains polarize light because they are nonspherical and are aligned

[1] When "extinction" occurs, light is "extinguished," not "extincted." Beware of committing this particular gaffe in the presence of professional astronomers.

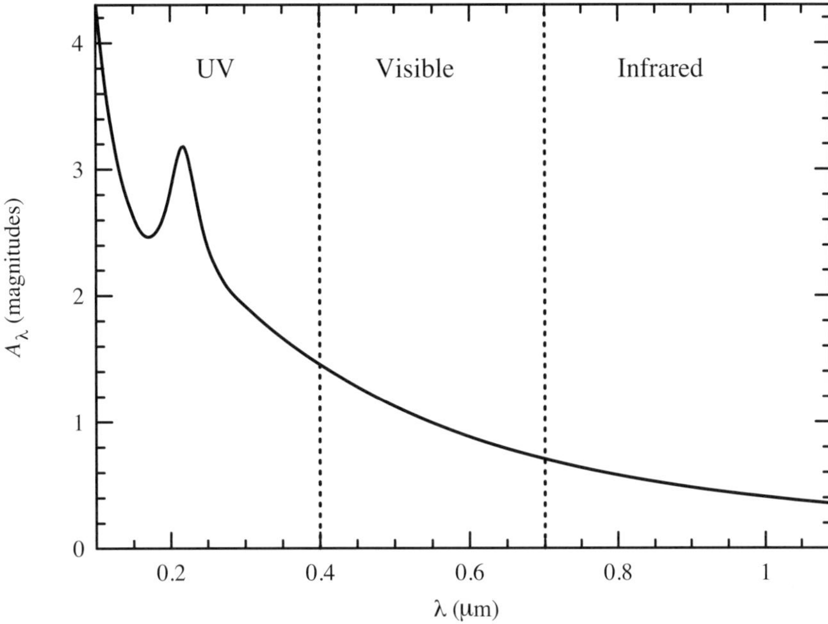

FIGURE 16.1 Dust extinction as a function of wavelength, normalized so that $A_V = 1$ magnitudes. The dotted vertical lines indicate the range of visible wavelengths.

preferentially in one direction.[2] If dust grains were perfect spheres, they wouldn't cause polarization of light.

Clue three. The plot of extinction versus wavelength (see Figure 16.1) has a "bump" at $\lambda \sim 0.22\ \mu$m. What causes the excess extinction at this wavelength? It is known experimentally that in graphite the bonds between carbon atoms absorb and emit light at wavelengths $\lambda \sim 0.22\ \mu$m. In the laboratory, an excellent fit to the extinction bump is given by graphite grains with $d \sim 0.02\ \mu\text{m} \sim 200$ Å. (On Earth, you find such tiny graphite particles in soot.)

Clue four. If you look at the infrared spectrum from isolated dusty clouds, you find that at long wavelengths ($\lambda > 100\ \mu$m), it's a blackbody spectrum with a typical temperature of $T \sim 20$ K. At shorter wavelengths (Figure 16.2), there are absorption bands at $\lambda \sim 4 \rightarrow 7\ \mu$m due to ices (frozen water, carbon dioxide, carbon monoxide, and so forth), and at $\lambda \sim 10\ \mu$m due to silicates (a.k.a. "rock"). Thus, dust seems to contain a mix of volatile material, like ices, and refractory material, like silicates and graphite.

[2] Dust grains tend to line up perpendicular to the interstellar magnetic field.

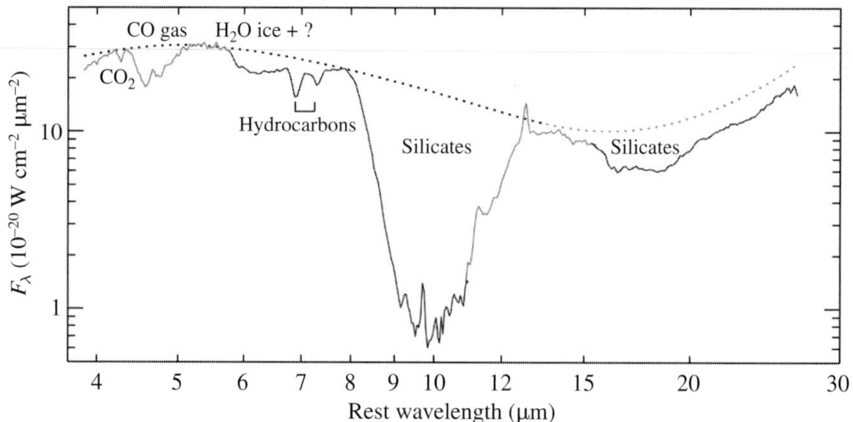

FIGURE 16.2 Infrared emission from a luminous, dust-rich galaxy.

Models of dust featuring ice-covered flecks of silicate, mixed with small graphite grains, provide a good match with observations. Where does the dust come from? Cool supergiant stars, like Betelgeuse, have very strong stellar winds, with mass loss rates a million times greater than that of the solar wind. As the wind expands, it cools, and refractory material like graphite and silicates can condense out into tiny grains. A thin layer of frost can later form on these grains in the coolest, densest regions of the interstellar medium.

16.1.2 Observable Effects of Dust on Starlight

Here we consider from an observational point of view how light from distant stars is attenuated by interstellar dust. Absorption and scattering of starlight in the interstellar medium is a radiative transfer problem that we addressed for the general case in Section 5.4. Specifically, the flux received by an observer is

$$F = F_0 e^{-\tau} = F_0 e^{-n\sigma r}, \tag{16.1}$$

where F_0 is the flux in the absence of absorption; τ is the optical depth; n is the number density of absorbers, σ is the cross-section of an absorbing particle; and r is the path length through the absorbing medium. Converting this equation to magnitudes, we have

$$\begin{aligned} m_{\text{obs}} &= C - 2.5 \log F = C - 2.5 \log F_0 - 2.5 \log(e^{-\tau}) \\ &= m_0 + 2.5\tau \log e \\ &= m_0 + 1.086\tau, \end{aligned} \tag{16.2}$$

where C is an arbitrary constant (see equation 13.16), and m_0 is the apparent magnitude in the absence of extinction. We thus see that extinction appears as an *additive* term in magnitudes and is linear with τ.

Interstellar extinction is a function of wavelength,

$$m_{\rm obs}(\lambda) = m_0(\lambda) + A(\lambda), \tag{16.3}$$

where the extinction term is

$$A(\lambda) = 1.086\tau_\lambda = 1.086 n\sigma_\lambda r. \tag{16.4}$$

The extinction-corrected distance modulus is then

$$m_{\rm obs} - M = 5\log r - 5 + A. \tag{16.5}$$

Extinction has the effect of increasing the apparent magnitude of a star and apparently increasing the distance modulus, since extinguished objects appear to be farther away than they actually are.

Consider now a color index such as $B - V$; since the observed magnitudes are $V = V_0 + A_V$ and $B = B_0 + A_B$, the observed color will be

$$(B - V) = (B - V)_0 + (A_B - A_V) = (B - V)_0 + E(B - V), \tag{16.6}$$

where the final term, $E(B - V)$, is known as the **color excess**. We can explicitly calculate the color excess in terms of the optical depth as

$$E(B - V) = A_B - A_V = 1.086(\tau_B - \tau_V) = 1.086\tau_V\left(\frac{\tau_B}{\tau_V} - 1\right). \tag{16.7}$$

The wavelength dependence of the extinction curve is often characterized by the **ratio of total to selective extinction** R defined by

$$R \equiv \frac{A_V}{E(B - V)} = \frac{1}{(\tau_B/\tau_V) - 1}. \tag{16.8}$$

In general, $\tau_\lambda \propto \lambda^{-n}$, where $n > 0$: in other words, extinction is larger at shorter wavelengths. Not only is light extinguished, it is also **reddened**. As noted in Section 9.2, scattering by particles of sizes comparable to the wavelengths of light results in wavelength-dependence of the optical depth of the form $\tau \propto \lambda^{-1}$. In this case, the ratio of total to selective absorption is

$$R = \frac{1}{(\tau_B/\tau_V) - 1} = \frac{1}{(\lambda_{{\rm eff},B}/\lambda_{{\rm eff},V})^{-1} - 1} \approx 4.2, \tag{16.9}$$

whereas in the last step we have understood the effective wavelengths to be $\lambda_{{\rm eff},V} =$ 5500 Å and $\lambda_{{\rm eff},B} = 4450$ Å.

By comparing the observed colors of stars of the same spectral class (say, A0V stars), one can determine observationally that in the solar neighborhood, $R \approx 3.1$.

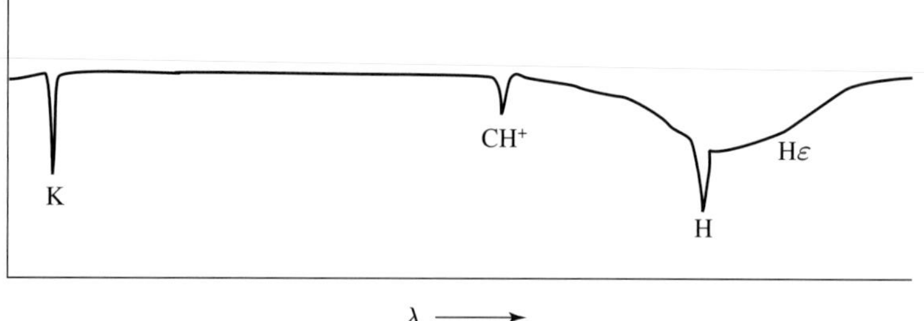

FIGURE 16.3 Narrow absorption lines of Ca II (labeled K and H) and of CH^+, due to interstellar gas between us and ζ Ophiuchi.

16.2 ▪ INTERSTELLAR GAS

At visible wavelengths, dust is a highly prominent component of the interstellar medium. However, dust contributes only a minority of the material between stars. By mass, the interstellar medium is just 1% dust; the remaining 99% is contributed by **interstellar gas.**[3] Although the bulk of the gas is low-density hydrogen and helium, which is not necessarily easy to detect, there are many ways in which interstellar gas makes its presence known.

Method 1 of gas detection. Interstellar gas can be detected when it creates **absorption lines** in a star's spectrum. Kirchhoff's laws (Section 5.3) tell us that a relatively cool gas cloud seen against a hotter source of continuum emission (such as a star) will produce an absorption line spectrum. Figure 16.3 shows a small portion of spectrum of the star Zeta Ophiuchi, an O9 main sequence star that has $A_V \sim 1.5$ mag from intervening dust. The narrow absorption lines are due to cool interstellar gas between us and the star. How do we know the absorption lines are from interstellar gas and not from the outer layers of Zeta Ophiuchi itself?

- The lines are narrow, with little thermal or pressure broadening. (Note how broad the Balmer ϵ line is in Figure 16.3; this line comes from the star's photosphere, which has both high temperature and high pressure, since Zeta Ophiuchi has a spectral type O9V.)
- The lines are either of atoms in a low-ionization state or of molecules, which indicates a lower temperature than you would find in the star's atmosphere.
- The lines show a different Doppler shift from the star's absorption lines.
- In general, interstellar absorption lines can be multiple if there happen to be many individual gas clouds, with slightly differing radial velocities, along the line of sight to the star.

The study of absorption lines from gas clouds indicates that interstellar clouds, like stellar photospheres, are mostly hydrogen and helium.

[3] Parenthetic aside: the processes that cause gas to curdle up into stars are not ultraefficient; in our galaxy, the total mass of interstellar gas is equal to 10–20% of the total stellar mass.

Method 2 of gas detection. Interstellar gas can be detected when it produces an **emission spectrum**. Kirchhoff's laws tell us that a relatively warm gas cloud seen against a dark background will produce an emission line spectrum. An interstellar gas cloud that is hot enough to emit light at visible wavelengths is called an **emission nebula**. The Orion Nebula, shown in Color Figure 15, is a famous example of an emission nebula. To produce emission lines, electrons have to be lifted above the ground state (the $n = 1$ orbit) of their atom. Since interstellar gas is low in density, excitation of electrons is unlikely to occur by collisions between atoms. Instead, it occurs by the absorption of photons. In general, nebular gas absorbs ultraviolet light from nearby hot stars, which ionizes some of its atoms. Eventually, the ions and electrons recombine, but when they do, the recombination of the electron is frequently to a high energy level. As the electron cascades down to the lowest vacant energy level, it emits a photon with each downward quantum leap it makes. Thus, emission nebulae are an example of fluorescence, in which high-energy ultraviolet photons are converted to lower-energy visible photons. The source of ultraviolet photons varies between different types of emission nebulae:

- **H II** region: an emission nebula in which the source of ultraviolet light is one or more hot stars (spectral type O or B). The Orion Nebula and the Trifid Nebula are examples of H II regions.
- **Planetary nebula:** an emission nebula in which the source of ultraviolet light is the hot, exposed core of an AGB star, which is rapidly evolving into a white dwarf. The Ring Nebula (Color Figure 16) and the Spirograph Nebula are examples of planetary nebulae.[4]

The physics of these hot, ionized components of the interstellar medium is discussed further in Section 16.3.

Method 3 of gas detection. Interstellar gas can be detected when it produces continuum radio emission. When a hot gas is largely ionized, it can produce photons by the process known as **bremsstrahlung**, a German term whose literal English translation is "braking radiation." Bremsstrahlung, also known as **free–free emission**, is produced when a free electron is accelerated by the electrostatic attraction of a positively charged ion. The electron emits a photon as it is accelerated. The initial kinetic energy of the electron, before it encounters the ion, will be comparable to the thermal energy, kT. Since the electron can't radiate away more energy than it has originally, the energy of the photon that it emits must be $hc/\lambda < kT$, implying

$$\lambda > \frac{hc}{kT} \sim 0.1\,\text{mm}\left(\frac{T}{10^4\,\text{K}}\right)^{-1} \tag{16.10}$$

for bremsstrahlung emission from a hot, ionized gas. Bremsstrahlung is thus of interest mainly to radio astronomers. If the ionized gas is in the presence of a

[4] "Planetary nebulae" received their odd and inappropriate name from William Herschel in the eighteenth century. Seen through a small telescope, they look like a fuzzy disk, just as the planet Uranus did when Herschel discovered it.

magnetic field, it can also produce continuum radio emission via **synchrotron radiation**, as the electrons are accelerated along helical paths following the magnetic field lines. Synchrotron emission is produced copiously by young supernova remnants, in which ionized gas surrounds a highly magnetized neutron star.

Method 4 of gas detection. Interstellar gas can be detected when hydrogen atoms produce **21 centimeter** line emission. When a neutral hydrogen atom has its electron in the $n = 1$ energy level, the spin of the electron can be either parallel or antiparallel to the spin of the central proton. When the spins are antiparallel, the energy of the atom is slightly smaller than when the spins are parallel. Because the antiparallel state is lower in energy, an electron in a parallel state can undergo a spontaneous "spin-flip" transition that inverts its spin. Since this transition violates a quantum mechanical selection rule, it is a forbidden transition. (Remember from Section 5.1 that a "forbidden" transition is not absolutely forbidden; it is merely far less probable than a "permitted" transition that doesn't violate any selection rules.) The half-life of the higher energy parallel state is roughly 10 million years. Thus, if you have a population of $\sim 3 \times 10^{14}$ hydrogen atoms in the parallel state, 1 atom per second will undergo the spontaneous spin-flip transition. As it does so, the small amount of energy lost is carried away by a photon with $\lambda = 21$ cm. An individual hydrogen atom will undergo a spin-flip transition only rarely, and the density of hydrogen atoms in interstellar space is low. However, space is big. Along a sufficiently long line of sight, the spin-flip transitions of hydrogen will produce a detectable signal at $\lambda = 21$ cm.

Method 5 of gas detection. Interstellar gas can be detected when molecules produce **radio line emission**. In the denser regions of interstellar space, individual atoms can join together to form molecules. When a small molecule such as CH or CO spins about one of its axes, it can undergo quantum transitions from one spin state to another. If the transition is from a rapidly rotating state to a less rapidly rotating state, the lost kinetic energy of rotation is carried away by a photon. For instance, when a carbon monoxide (CO) molecule makes a transition from the $J = 1$ rotational state to the $J = 0$ ground rotational state, it emits a photon with $\lambda = 2.6$ mm, a wavelength at which the Earth's atmosphere is conveniently transparent. A map of the Milky Way's emission at 2.6 mm (Color Figure 17) reveals that the CO—and by inference, the other molecular gas—in our galaxy is confined to a thin layer near the midplane of the galaxy. At higher resolution, it is found that most of the molecules are in relatively small, dense, and cold molecular clouds.

The interstellar medium is by no means homogeneous. Some consists of very hot ionized gas, while some consists of much cooler gas containing H_2 and other molecules. Astronomers have labeled different components of the interstellar gas:

- Cold Molecular Clouds: $T \sim 10$ K, $n > 10^9\ \mathrm{m^{-3}}$
- Cold Neutral Medium (H I regions): $T \sim 100$ K, $n \sim 10^8\ \mathrm{m^{-3}}$
- Warm Neutral Medium (intercloud medium): $T \sim 7000$ K, $n \sim 4 \times 10^5\ \mathrm{m^{-3}}$

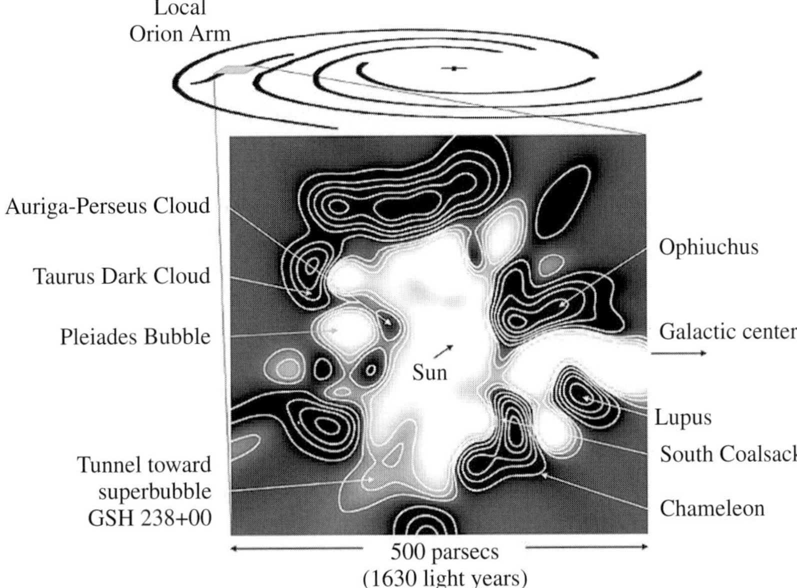

FIGURE 16.4 Density of gas in the vicinity of the Sun. White areas represent low density "Local Bubble"; black areas represent higher-density gas.

- Warm Ionized Medium (H II regions): $T \sim 10^4$ K, $n \sim 10^6$ m^{-3}
- Hot Ionized Medium (coronal gas): $T \sim 10^6$ K, $n < 10^4$ m^{-3}

Which component is the most common? That depends on how you do the accounting: coronal gas comprises $\gtrsim 50\%$ of the *volume* of the interstellar medium but only a tiny fraction of the *mass*. The coronal gas takes the form of "bubbles" blown by supernovae.[5] The Sun is within such a bubble of hot, low-density gas, about 100 parsecs across, shown in Figure 16.4. When most of the gas was cleared from this "Local Bubble," most of the dust was cleared as well. Thus, stars less than ~ 50 pc from us suffer little extinction from dust. Although a coronal bubble contains gas at a very high temperature, it contains very little thermal energy, thanks to its low density. If you were tossed out the airlock of your spaceship without a spacesuit, you wouldn't have to worry about roasting or freezing to death. Heat transfer by convection and conduction would be negligible, and it would take about 5 minutes for your temperature to drop by 1° Celsius due to radiative heat losses. By that time, you would have asphyxiated.[6]

[5] Supernovae are exploding stars, discussed in Section 18.4.

[6] Other unpleasant things happen when you undergo explosive decompression. You rapidly dehydrate as your water content evaporates into the near vacuum of space. Any gas trapped within your body expands rapidly, so you're likely to burst your eardrums and experience excruciating sinus pains (not to mention the fact that your intestinal tract contains about a liter of gas at any given time). However, it will be the lack of oxygen that kills you.

16.3 ▪ THE PHYSICS OF NON-LTE GASES

Luminous **H II** regions around stars of spectral type O and B constitute a conspicuous component of the interstellar medium. Main sequence stars of type O and B are the hottest, most luminous, and shortest-lived stars on the main sequence. Because of their short main sequence lifetimes, O and B stars never stray far from the site where they formed, and hence they are surrounded by the relatively high-density gas ($n \gtrsim 10^7\ \mathrm{m}^{-3}$) characteristic of star formation regions. The copious ultraviolet radiation emitted by these hot stars ionizes the gas surrounding them, creating a bright H II region.

16.3.1 Ionization Balance

The ionized gas of an H II region is optically thin at most wavelengths; thus the photons in the H II region are not in local thermodynamic equilibrium with the electrons and ions (Section 5.6). The fact that H II regions are non-LTE systems should not intimidate us. We can still characterize the light emitted by the central star as having a temperature T_* equal to the effective surface temperature of the star ($T_* \gtrsim 20{,}000$ K for O and B stars), and we can still characterize the free electrons in the ionized gas as having a temperature T_e, even though we don't expect T_e to equal T_*.

Let us assume, for simplicity, that the gas around a hot star consists of pure hydrogen. In this case, provided that the column density of hydrogen is large enough, the gas will be optically thick at wavelengths $\lambda < 912$ Å; this corresponds to photon energies $h\nu >$ 13.6 eV, sufficient to ionize hydrogen. Each photon with $\lambda < 912$ Å will ionize a single atom of hydrogen. Photons with $\lambda > 912$ Å will be absorbed only if they correspond to a Lyman transition, arising from the ground state of hydrogen (see page 116). At all wavelengths longer than 1216 Å, corresponding to the Lyα transition from $n = 1$ to $n = 2$, the nebula is optically thin. The odds of a Balmer transition occurring, for instance, are negligible, since a hydrogen atom with an electron in the $n = 2$ level will undergo spontaneous radiative decay to the $n = 1$ level long before it is likely to absorb a Balmer-series photon. It is a good approximation to assume that nearly all the hydrogen in the H II region is ionized; of the small fraction of hydrogen that is neutral, nearly all is in the ground state.

In a gas of ionized hydrogen, emission lines are produced when free electrons recombine with protons to form neutral hydrogen. Often, the electron recombines to an excited state; this results in the emission of one or more photons with $h\nu < 13.6$ eV, as the electron cascades down through the lower energy levels. For instance, Figure 16.5 shows the possible downward transitions of an electron that has recombined to the $n = 3$ state. The decay from the $n = 3$ state can produce either a Lyβ photon, as shown in Figure 16.5a, or an Hα photon followed by a Lyα photon, as shown in Figure 16.5b. If the H II region is optically thick in the ionizing continuum ($\lambda < 912$ Å) and in the Lyman lines, but optically thin elsewhere, then the Hα photon will escape freely.[7] However, the

[7] In fact, it is the $\lambda = 6563$ Å photons of Hα that give H II regions their characteristic reddish color, as seen in Color Figure 15.

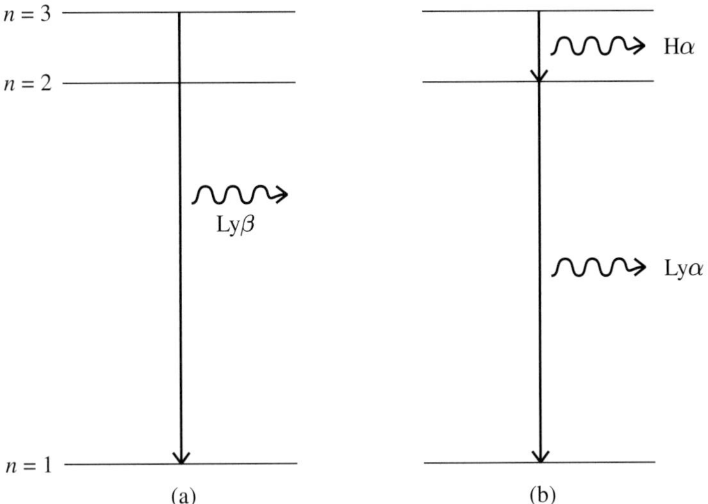

FIGURE 16.5 Possible outcomes of a recombination to the $n = 3$ level in hydrogen. (a) Spontaneous decay to the $n = 1$ level, yielding a Lyβ photon. (b) Decay to the $n = 2$ level, producing an Hα photon, followed by a decay to the $n = 1$ level, producing a Lyα photon.

emitted Lyman photons will be rapidly reabsorbed by neutral hydrogen atoms in their ground state. Eventually (sometimes only after multiple absorptions and re-emissions), each Lyβ photon ($h\nu = 12.1$ eV) will be converted to an Hα photon (1.9 eV) plus a Lyα photon (10.2 eV). Thus, each electron that recombines to the $n = 3$ level will end by producing an Hα photon plus a Lyα photon. Similarly, an electron that recombines to the $n = 4$ level will produce a combination of Balmer and/or Paschen photons plus a Lyα photon. An electron that recombines to the $n = 5$ level will produce a combination of Balmer, Paschen, and/or Brackett photons plus a Lyα photon, and so forth for higher values of n. In short, every recombination into an excited state eventually results in a single Lyα photon that random-walks its way out of the H II region.[8]

If the hot star is embedded in a region of uniform density gas, then it will be surrounded by a sphere of almost completely ionized hydrogen. This sphere is known as the **Strömgren sphere**, named after the Danish astronomer Bengt Strömgren, who did pioneering theoretical work on ionized nebulae. At the surface of the sphere, within a layer whose thickness is comparable to one photon mean free path, the ionized fraction of hydrogen drops from nearly 1 to nearly zero. The transition between the inner ionized zone, or H II region, and outer neutral zone, or H I region, occurs at the **Strömgren**

[8] To be technically correct, the more likely path out of an emission nebula for a Lyα photon is by diffusion in velocity. As the photon is absorbed and re-emitted by atoms with different random velocities, it is randomly Doppler shifted to shorter or longer wavelengths; eventually, it is shifted far enough into the line wings that the optical depth is less than 1, and the photon can escape.

radius. The Strömgren radius can be calculated by noting that if the ionized and neutral populations are in statistical equilibrium, the number of ionizations per second within the Strömgren sphere must equal the number of recombinations per second. The total number of ionizations per second equals the rate Q_* at which the central star is producing hydrogen-ionizing photons with $h\nu > 13.6\,\text{eV}$. If the specific luminosity of a star at frequency ν is L_ν, the number of photons produced per second at that frequency is given by dividing the specific luminosity by the energy per photon $h\nu$. Thus, the rate at which ionizing photons are produced by the star is

$$Q_* = \int_{\nu_0}^{\infty} \frac{L_\nu}{h\nu} d\nu, \tag{16.11}$$

where $h\nu_0 = 13.6\,\text{eV}$ is the minimum photon energy required to ionize hydrogen.

To calculate the number of recombinations, we start with the realization that a free electron can recombine to form a neutral hydrogen atom only if it encounters a proton. Thus, the mean free path for an electron (that is, the average distance it travels before recombining) can be written as

$$\ell_{\text{rec}} = \frac{1}{n_p \sigma_{\text{rec}}}, \tag{16.12}$$

where n_p is the number of protons per unit volume and σ_{rec} is the cross-section for recombination of an electron with a proton. The average time an electron spends in its free state before recombining is simply

$$t_{\text{rec}} = \frac{\ell_{\text{rec}}}{v_e} = \frac{1}{n_p \sigma_{\text{rec}} v_e}, \tag{16.13}$$

where v_e is the electron's speed. If there are n_e free electrons per unit volume, and their average lifetime before recombining is t_{rec}, then the total number of recombinations per unit volume per unit time will be

$$N_{\text{rec}} = \frac{n_e}{t_{\text{rec}}} = n_e n_p \sigma_{\text{rec}} v_e. \tag{16.14}$$

The cross-section for recombination, σ_{rec}, is a function of electron speed: low-speed electrons recombine much more readily than do high-speed electrons, in part because they spend more time loitering near each proton they pass. The mathematical dependence of the cross-section on electron speed is $\sigma_{\text{rec}} \propto v_e^{-2}$. At a temperature T_e, electron speeds are distributed according to a Maxwell–Boltzmann distribution (equation 5.40). Thus, to find the recombination rate at a given temperature T_e, we must take into account both the dependence of the cross-section on electron speed and the dependence of electron speed on temperature. It is customary to express the average recombination rate in terms of a temperature-dependent **recombination coefficient** α, defined as

$$\alpha(T_e) \equiv \langle \sigma_{\text{rec}} v_e \rangle = \int_0^{\infty} \sigma_{\text{rec}}(v_e) v_e F(v_e)\, dv_e, \tag{16.15}$$

where $F(v_e)$ is the Maxwell–Boltzmann distribution. Since $\sigma_{\rm rec} \propto v_e^{-2}$, and since typical electron speeds are $v_e \propto T_e^{1/2}$, we see that

$$\alpha(T_e) \propto v_e^{-2} v_e \propto v_e^{-1} \propto T_e^{-1/2}. \tag{16.16}$$

The recombination rate per unit volume will then be (from equation 16.14)

$$N_{\rm rec} = n_e n_p \alpha(T_e) \propto T_e^{-1/2}. \tag{16.17}$$

We can now compute the total number of recombinations per second in the nebula by multiplying the recombination rate per unit volume (equation 16.17) by the volume of the nebula, taken to be a sphere of radius $R_{\rm S}$. We set the recombination rate equal to the ionization rate (equation 16.11) to find

$$Q_* = n_e n_p \alpha(T_e) \frac{4\pi}{3} R_{\rm S}^3, \tag{16.18}$$

where $R_{\rm S}$ is the Strömgren radius. Thus, noting that $n_p = n_e$ for ionized hydrogen,

$$R_{\rm S} = \left[\frac{3}{4\pi} \frac{Q_*}{\alpha(T_e) n_e^2} \right]^{1/3}. \tag{16.19}$$

A typical nebular temperature is $T_e \approx 10^4$ K; at that temperature, the recombination coefficient[9] has a value $\alpha \approx 2.6 \times 10^{-19}\ {\rm m^3\,s^{-1}}$. An O6V star produces ionizing photons at the rate $Q_* \approx 5 \times 10^{48}\ {\rm s^{-1}}$. Taking as a typical nebular density $n_e \approx 10^7\ {\rm m^{-3}}$, the size of the Strömgren sphere around an O6V star is

$$\begin{aligned} R_{\rm S} &= \left(\frac{3}{4\pi} \frac{5 \times 10^{48}\ {\rm s^{-1}}}{(2.6 \times 10^{-19}\ {\rm m^3\,s^{-1}})(10^7\ {\rm m^{-3}})^2} \right)^{1/3} \\ &\approx 3.4 \times 10^{17}\ {\rm m} \approx 10\ {\rm pc}. \end{aligned} \tag{16.20}$$

16.3.2 Thermal Balance

The temperature of the free electrons in an H II region, as mentioned above, is generally $T_e \approx 10{,}000$ K, corresponding to a mean kinetic energy $E = (3/2)kT_e \approx 1.3$ eV. This electron temperature is remarkably independent of the temperature of the central star in the H II region. The favored temperature of 10,000 K must be the result of a balance between heating processes, which drive the mean kinetic energy of electrons upward, and cooling processes, which drive the energy downward.

[9] A subtlety to note is that this recombination coefficient counts recombinations to every state of hydrogen *except the ground state*: ground-state recombinations produce photons with $h\nu > 13.6$ eV, which almost immediately reionize another hydrogen atom. In nebular physics, this is called the "on-the-spot approximation" as it assumes that ionizing photons produced in the nebula itself are immediately reabsorbed.

When free electrons are freshly liberated from hydrogen atoms by photoionization, their average kinetic energy will be

$$\langle E \rangle = \langle h\nu \rangle - \chi, \tag{16.21}$$

where $\langle h\nu \rangle$ is the mean energy of an ionizing photon, and $\chi = 13.6\,\text{eV}$ is the ionization potential of hydrogen. In computing the mean energy $\langle h\nu \rangle$ of ionizing photons, we must weight the photon energy $h\nu$ by the number of photons at frequency ν above the ionizing threshold $\nu_0 = \chi/h$. This weighting factor is proportional to the mean intensity of light divided by the energy per photon; for blackbody radiation (equation 5.86), this is

$$\frac{J_\nu}{h\nu} \propto \frac{\nu^3}{h\nu} \frac{1}{e^{h\nu/kT_*} - 1}. \tag{16.22}$$

For a central star with $T_* < \chi/k \approx 160{,}000$ K, the photoionizing photons are all on the high-energy Wien tail of the Planck function, which permits the approximation

$$\frac{J_\nu}{h\nu} \propto \nu^2 e^{-h\nu/kT_*}. \tag{16.23}$$

We must also weight the photon energy by the cross-section for photoionization by a photon with frequency ν. For hydrogen above the ionizing threshold ν_0, the cross-section $\sigma_{\rm ion}$ has the frequency dependence $\sigma_{\rm ion} = \sigma_0(\nu/\nu_0)^{-3}$.

Thus, the average energy per ionizing photon can be computed as

$$\langle h\nu \rangle = \left[\int_{\nu_0}^{\infty} \left(\frac{J_\nu}{h\nu} \right) (h\nu)\, \sigma_{\rm ion}(\nu)\, d\nu \right] \left[\int_{\nu_0}^{\infty} \left(\frac{J_\nu}{h\nu} \right) \sigma_{\rm ion}(\nu)\, d\nu \right]^{-1}. \tag{16.24}$$

Canceling constants and inserting frequency dependence explicitly, we have

$$\langle h\nu \rangle = \left[h \int_{\nu_0}^{\infty} \left(\nu^2 e^{-h\nu/kT_*} \right) \nu(\nu^{-3})\, d\nu \right] \times \left[\int_{\nu_0}^{\infty} \left(\nu^2 e^{-h\nu/kT_*} \right) \nu^{-3}\, d\nu \right]^{-1}. \tag{16.25}$$

We make the useful substitution $x = h\nu/kT_*$ (and define $x_0 \equiv h\nu_0/kT_*$), so this simplifies to

$$\langle h\nu \rangle = kT_* \left[\int_{x_0}^{\infty} e^{-x}\, dx \right] \left[\int_{x_0}^{\infty} e^{-x} x^{-1}\, dx \right]^{-1}. \tag{16.26}$$

The integral in the numerator is trivial and has value e^{-x_0}. The integral in the denominator is known as the first exponential integral $E_1(x_0)$ and for $x_0 \gg 1$ (or $h\nu \gg kT_*$) can be expanded as

$$E_1(x_0) \equiv \int_{x_0}^{\infty} \frac{e^{-x}}{x}\, dx \approx \frac{e^{-x_0}}{x_0} \left(1 - \frac{1}{x_0} + \frac{2}{x_0^2} + \cdots \right). \tag{16.27}$$

If we keep only the first two terms, a little algebra then gives

$$\langle h\nu \rangle \approx h\nu_0 + kT_*, \tag{16.28}$$

where the first term on the right-hand side is the ionization potential χ and the second term kT_* is thus the mean kinetic energy per liberated electron:

$$\langle E \rangle = \langle h\nu \rangle - \chi \approx kT_*. \tag{16.29}$$

In other words, if the central star in an H II region is an O star with $T_* \approx 50{,}000$ K, the newly liberated electrons will have an average kinetic energy of $\langle E \rangle \approx kT_* \approx 4.3\,\text{eV}$. There must therefore be a cooling mechanism that reduces the average kinetic energy of the free electrons to its observed value of $E \approx 1.3\,\text{eV}$.

In H II regions, the most effective cooling mechanism is the collisional excitation of atoms and ions in the gas, followed by spontaneous de-excitation. To see how collisional excitation can reduce the average kinetic energy of free electrons, let's consider the simple two-level atom illustrated in Figure 5.3, which has a ground state ($n = 1$) and an excited state ($n = 2$), separated by an energy ΔE. If an electron with kinetic energy $E > \Delta E$ collides with the atom, it can lift a bound electron from the ground state to the excited state. However, the energy gained by the bound electron is lost by the free electron, whose kinetic energy drops from E to $E - \Delta E$. Thus, if we wanted to cool an H II region, we could sprinkle it with two-level atoms that have $\Delta E \lesssim kT_*$. (If the atoms had a level spacing $\Delta E \gg kT_*$, then very few of the electrons produced by photoionization would have enough energy to collisionally excite the atoms.)

If the H II region is heated by photoionization and cooled by collisional excitation, then the equilibrium temperature is the temperature at which the heating rate equals the cooling rate. The heating rate (or "gain") G, in units of watts per cubic meter, is found by multiplying the number of photoionizations per second per cubic meter by the average energy, in joules, of the free electron produced. In equilibrium, the photoionization rate equals the recombination rate N_{rec}, and we have already determined that the average energy of a newly freed electron is $\langle E \rangle = kT_*$. Therefore,

$$G = N_{\text{rec}}\langle E \rangle = n_e n_p \alpha(T_e) kT_*, \tag{16.30}$$

where we have used equation (16.17) for the recombination rate. Since $\alpha(T_e) \propto T_e^{-1/2}$, and the other quantities entering equation (16.30) are independent of the electron temperature, we find that $G \propto T_e^{-1/2}$; the higher the electron temperature, the less effective the heating by photoionization.

To compute the collisional cooling rate (or "loss") L, we start by exploiting some of the similarities between collisional excitation and radiative recombination. For recombination to occur, a free electron must interact with a proton; the cross-section for the interaction leading to recombination is σ_{rec}. As we saw in equation (16.17), the recombination rate is

$$N_{\text{rec}} = n_e n_p \langle \sigma_{\text{rec}} v_e \rangle = n_e n_p \alpha(T_e). \tag{16.31}$$

For collisional excitation of our two-level atom to occur, a free electron must interact with the atom in its ground state; the cross-section for this collisional interaction is σ_{12}, which in general is a function of the electron speed. By analogy with the recombination rate (equation 16.31), we can write the collisional excitation rate as

$$N_{12} = n_e n_1 \langle \sigma_{12} v_e \rangle, \tag{16.32}$$

where n_1 is the number density of two-level atoms in the ground state. Mathematically, the recombination rate (equation 16.31) and the collisional excitation rate (equation 16.32) look similar. There is one important physical difference between them, however. A free electron with an arbitrarily small speed v_e can undergo recombination; by contrast, a free electron with $m_e v_e^2/2 < \Delta E$ will be unable to trigger a collisional excitation. The velocity-weighted cross-section for collisional excitation will therefore have a form

$$\langle \sigma_{12} v_e \rangle \propto T_e^{-1/2} \exp\left(-\frac{\Delta E}{kT_e}\right). \tag{16.33}$$

The Boltzmann factor, $\exp(-\Delta E/kT_e)$, results from the fact that at temperatures $T_e < \Delta E/k$, very few of the electrons will be energetic enough to collisionally excite the upper level.

Since the energy lost by the free electron during a collisional excitation is ΔE, the collisional cooling rate can be written as

$$L = n_e n_1 \langle \sigma_{12} v_e \rangle \Delta E \propto T_e^{-1/2} \left(-\frac{\Delta E}{kT_e}\right). \tag{16.34}$$

For fixed n_e and n_1, the cooling rate L has a maximum at a temperature $kT_e = 2\Delta E$. Thus, for effective cooling in the temperature range $T_e = 10^4 \to 5 \times 10^4$ K, we need to have collisional excitations with energies in the range $\Delta E = 0.4 \to 2$ eV. This implies that collisional excitation of *hydrogen* will be relatively unimportant for cooling H II regions. Within an H II region, most of the neutral hydrogen is in the ground state ($n = 1$); collisional excitation to the next higher state requires $\Delta E = 10.2$ eV, far more energy than is possessed by a typical free electron in an H II region.

Most of the cooling of electrons in H II regions is done by collisional excitation of "metals" such as oxygen and nitrogen, which have excited states with ΔE equal to a few electron volts. In Figure 16.6, we show the heating rate and the sum of all cooling rates for realistic nebular conditions, with a typical mix of metals. The electron temperature is determined by where the heating and cooling rate lines cross; it is important to notice that the cooling curve is very steep around 10^4 K, so fairly large differences in heating rates (due to hot stars with different temperatures) will have little effect on the electron temperatures, which are thus typically $\sim 10{,}000$ K.

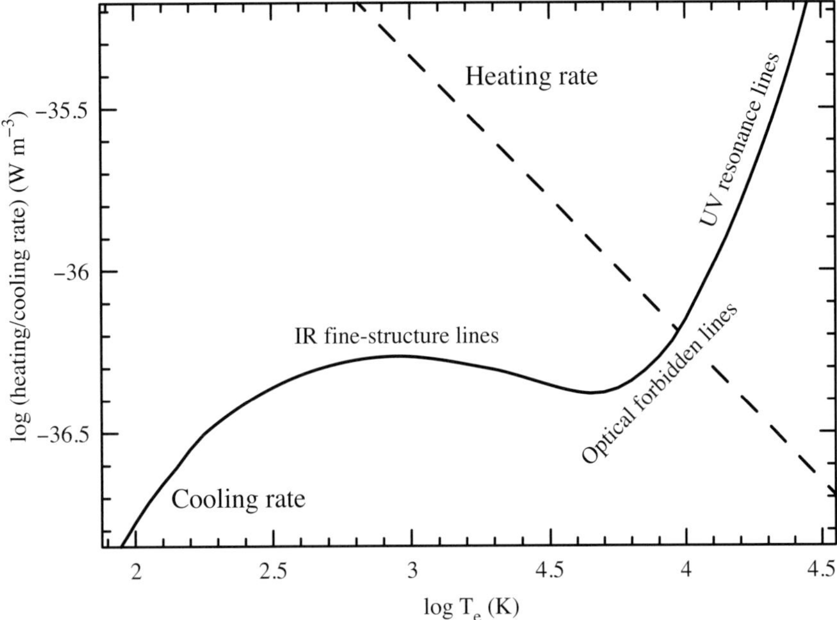

FIGURE 16.6 Heating and cooling functions for an O star with $T_* = 40{,}000$ K. The principal cooling mechanisms in various regimes are indicated. The locus where the heating and cooling functions cross is the equilibrium temperature for the nebula.

PROBLEMS

16.1 Compute $R = A_V/E(B-V)$, the ratio of total to selective absorption, for the case of Rayleigh scattering, $\tau \propto \lambda^{-4}$. (This would be appropriate if interstellar dust particles were very small compared to the wavelength of visible light, like aerosols in the Earth's atmosphere.)

16.2 The Sun emits 5×10^{23} photons per second with $h\nu > 13.6$ eV. If the density of hydrogen atoms in interplanetary space is $n = 10^9$ m^{-3}, what is the size of the Sun's Strömgren sphere? Assume a recombination coefficient $\alpha = 2.6 \times 10^{-19}$ m^3 s^{-1}.

16.3 An A0 V star is observed to have $m_V = 14.0$ and $B - V = 1.5$. What is the distance to the star?

16.4 Dust grains made of graphite will sublime (that is, turn from solid to gas) at a temperature $T \approx 1500$ K. The albedo of graphite is $A \approx 0.04$.

(a) How close to an O5 V star ($T_{\text{eff}} = 42{,}000$ K, $R = 12R_\odot$) can graphite grains survive?

(b) How close to an M2 III star ($T_{\text{eff}} = 3540$ K, $R = 0.5R_\odot$) can graphite grains survive?

16.5 Demonstrate that equation (16.28) follows from equations (16.26) and (16.27).

16.6 Suppose that two cold ($T = 100$ K) interstellar clouds of $1M_\odot$ each collide with a relative velocity $v = 10\text{ km s}^{-1}$, with *all* the kinetic energy of the collision being converted into heat. What is the temperature of the merged cloud after the collision? You may assume the clouds consist of 100% hydrogen.

16.7 In general, an F0 main sequence star has absolute magnitude $M_V = 2.7$ and intrinsic color $(B - V)_0 = 0.30$. A specific F0 main sequence star is observed to have $m_V = 12.00$ and $m_B = 12.56$.

(a) What is the color excess $E(B - V)$ for this star?

(b) What is the extinction A_V for this star? (Assume $R = 3.1$).

(c) What is the distance to this star?

(d) What distance would you have computed if you had ignored extinction?

16.8 Consider the H II region surrounding an O6 V star, as described at the end of Section 16.3.1.

(a) What is the recombination time in this H II region?

(b) What is the light travel time across the H II region?

16.9 We observe an interstellar cloud, with temperature $T = 80$ K and neutral hydrogen density $n_H = 10^8\text{ m}^{-3}$, at a distance $d = 100$ pc. Suppose that the cloud is spherical and that the column density of neutral hydrogen atoms through its middle is $N_H = 1.5 \times 10^{24}\text{ m}^{-2}$.

(a) What is the diameter of the cloud?

(b) How many neutral hydrogen atoms are in the cloud?

(c) What is the mass of the cloud (in units of $M_\odot$)?

(d) If 75% of the atoms are in the higher-energy parallel state, how many 21 cm photons are emitted per second by the cloud?

(e) What is the luminosity of the cloud in 21 cm photons (in units of $L_\odot$)?

(f) What is the flux in 21 cm photons as seen from Earth?

16.10 When observing a star behind the interstellar cloud described in Problem (16.9), we detect absorption in the Na I $\lambda = 5889.973$ Å line. What is the thermal broadening, $\Delta\lambda/\lambda_0$, of this absorption line? What is the thermal broadening of the same line in the solar spectrum?

17 Formation and Evolution of Stars

The equations of stellar structure (eqs. 15.74–15.78) do not contain any explicit time dependence. However, though the properties of stars usually change slowly, they must necessarily change. Stars have a nonzero luminosity and a finite fuel supply. This implies that they began nuclear fusion at some time in the past and will exhaust their fuel supply at some time in the future. In this chapter, we will address some of the issues of stellar evolution, starting with the interstellar medium that supplies the raw material of star formation, and ending when a star wrings the last possible joule from nuclear fusion. In Chapter 18, we discuss stellar remnants—the dense "corpses" left over when a star no longer is powered by fusion.

17.1 ▪ STAR FORMATION

Stars form by the gravitational collapse of the densest, coolest regions of the interstellar medium. These regions are the relatively dense cores of molecular clouds, which have densities as high as

$$n_{\rm mc} \sim 10^{12} \text{ molecules/ m}^3. \tag{17.1}$$

To put things in perspective, the Earth's atmosphere at sea level has $n \sim 10^{25}$ molecules per cubic meter; even the densest part of the interstellar medium would qualify as a high-grade vacuum in a terrestrial laboratory. If the gas is pure molecular hydrogen, its mass density will be

$$\rho_{\rm mc} \approx 2m_p n_{\rm mc} \approx 3 \times 10^{-15} \text{ kg m}^{-3} \approx 5 \times 10^{-12} M_\odot \text{ AU}^{-3}. \tag{17.2}$$

Mixing in heavier particles, like He atoms and CO molecules, will raise the mass density, but this is an adequate order-of-magnitude estimate.

In order to become a star, even the densest regions of the interstellar gas must obviously be greatly compressed. The average density of the Sun is $\rho_\odot \approx 1400 \text{ kg m}^{-3}$. For a spherical molecular cloud core of radius $R_{\rm mc}$ to become as dense as the Sun, the ratio of its final to initial radius must be

$$\frac{R_\odot}{R_{\rm mc}} = \left(\frac{\rho_{\rm mc}}{\rho_\odot}\right)^{1/3} \approx \left(\frac{3 \times 10^{-15} \text{ kg m}^{-3}}{1400 \text{ kg m}^{-3}}\right)^{1/3} \approx 10^{-6}. \tag{17.3}$$

If a star ends with a radius of $1R_\odot$, it must have started with a radius $R \sim 10^6 R_\odot \sim 4000$ AU ~ 0.02 pc.

The compression required to create a star from interstellar gas is provided by gravity, the universal trash compactor. Consider a spherical gas cloud with an initial radius $r_0 \approx 4000$ AU and mass $M \approx 1M_\odot$. If the cloud is not rotating, and is very cold, the molecules of which it is made will fall inward on radial orbits, with eccentricity $e \approx 1$. An orbiting molecule must satisfy Kepler's third law:

$$P^2 = \frac{4\pi^2}{G}\frac{a^3}{M}, \tag{17.4}$$

where P is the orbital period of the molecule, a is the semimajor axis of its highly eccentric orbit, and M is the mass contained in a sphere whose radius is equal to the molecule's distance from the cloud's center. As the molecule plummets inward, M should be constant, since the molecules closer to the cloud's center will fall inward with an equal or shorter period. If the molecule starts at a radius r_0 and falls on a radial orbit, the semimajor axis will be $a = r_0/2$. The freefall time $t_{\rm ff}$ is defined as the time it takes the molecule to reach the center; thus, $t_{\rm ff} = P/2$. From Kepler's third law,

$$4t_{\rm ff}^2 = \frac{4\pi^2}{G}\frac{r_0^3}{8M} = \frac{4\pi^2}{G}\frac{r_0^3}{8}\frac{3}{4\pi r_0^3 \rho_0}, \tag{17.5}$$

where ρ_0 is the initial average density of the cloud. Thus,

$$t_{\rm ff} = \left(\frac{3\pi}{32G\rho_0}\right)^{1/2} \approx 4 \times 10^4 \text{ yr} \left(\frac{3 \times 10^{-15} \text{ kg m}^{-3}}{\rho_0}\right)^{1/2}, \tag{17.6}$$

scaling to the typical density of a molecular cloud core. Note that the freefall collapse time depends only on the average density ρ_0 of a cloud, and not on its initial radius r_0.

It takes a mere 40 millennia for a dense molecular cloud core to collapse under its own gravity. Nevertheless, our galaxy, which has been around for 10 million millennia, is still cluttered with molecular cloud cores that have not collapsed. This is because the clouds are in hydrostatic equilibrium; the inward force due to gravity is balanced by the outward force provided by a pressure gradient. However, not every equilibrium is a *stable* equilibrium.[1] Consider a spherical gas cloud of radius r_0 and mass M, initially in hydrostatic equilibrium. We give it a slight squeeze, so that its radius goes from r_0 to $r_0(1 - \epsilon)$. The gravitational force at its surface increases, and the pressure gradient between its center and surface must therefore increase if hydrostatic equilibrium is to be maintained. Consider what happens at the molecular level as the cloud is squeezed:

- Molecules near the surface are shoved closer together, increasing their density and pressure.
- These molecules jostle against molecules a littler farther into the cloud, increasing their density and pressure.

[1] Think of a pencil balanced on its point; it's in equilibrium, but a tiny disturbance will topple it.

- These in turn jostle against molecules still farther in, and so on, until the center of the cloud is reached.

In short, the pressure within the sphere is altered by a pressure wave—otherwise known as a **sound wave**—traveling through the cloud. The change in pressure required to restore hydrostatic equilibrium thus travels at the speed of sound. If the inward-traveling sound wave reaches the center in a time less than the freefall time, the cloud is saved. If not, the cloud collapses.

If you're a fan of science fiction, you may be familiar with the tag line of the movie *Alien*: "In space, no one can hear you scream." If this is true, the concept of sound in outer space may seem absurd. However, sound waves can travel through space if their wavelength is long enough. The propagation of sound depends on molecules and atoms bumping into each other. In the low density of an interstellar cloud, a molecule will travel $\sim 10^4$ km before bumping into another molecule. As a consequence, only sound waves with $\lambda \gtrsim 10^4$ km can travel through a molecular cloud. If you had vocal cords $\sim 10^4$ km long, someone could hear you scream in space—if their eardrums were sensitive to frequencies of $\sim 10^{-5}$ Hz.

The collapse time for a cloud is (equation 17.6)

$$t_{\text{ff}} = \left(\frac{3\pi}{32 G \rho_0} \right)^{1/2} . \tag{17.7}$$

The time required to build up a pressure gradient within a cloud is

$$t_{\text{press}} = \frac{r_0}{c_s}, \tag{17.8}$$

where c_s is the sound speed within the cloud:

$$c_s = \left(\frac{\gamma k T}{\mu m_p} \right)^{1/2} . \tag{17.9}$$

In equation (17.9), γ is the adiabatic index, T is the gas temperature, and μ is the mean molecular mass. The cloud is unstable if

$$t_{\text{ff}} < t_{\text{press}}, \tag{17.10}$$

or

$$\left(\frac{3\pi}{32 G \rho_0} \right)^{1/2} < r_0 \left(\frac{\mu m_p}{\gamma k T} \right)^{1/2} . \tag{17.11}$$

The freefall collapse time t_{ff} is independent of the cloud's initial radius r_0 and the sound travel time is linearly proportional to r_0. This implies that, for a given density ρ_0 and temperature T, there is a *maximum* radius r_{J} for which a cloud is stable against collapse. The critical radius r_{J} is known as the **Jeans length**, after the astronomer James Jeans, who first realized its importance.

From equation (17.11), the Jeans length is

$$r_J = \left(\frac{3\pi \gamma k T}{32 G \rho_0 \mu m_p} \right)^{1/2} . \tag{17.12}$$

For pure molecular hydrogen, with $\mu = 2$ and $\gamma = 7/5$, the numerical value of the Jeans length is

$$r_J \approx 2000 \text{ AU} \left(\frac{T}{10 \text{ K}} \right)^{1/2} \left(\frac{\rho_0}{3 \times 10^{-15} \text{ kg m}^{-3}} \right)^{-1/2} . \tag{17.13}$$

The mass within a sphere of radius r_J is called the Jeans mass:

$$M_J \approx 0.2 M_\odot \left(\frac{T}{10 \text{ K}} \right)^{3/2} \left(\frac{\rho_0}{3 \times 10^{-15} \text{ kg m}^{-3}} \right)^{-1/2} . \tag{17.14}$$

This hints at why stars form only in the coolest, densest regions of the interstellar gas. In hotter, lower-density regions, the Jeans mass is much bigger than a star's mass.

If a dense core inside a molecular cloud is bigger than its Jeans length, then squeezing it (with a supernova shock wave, for instance) will trigger a gravitational collapse. What stops the collapse? Usually, the collapse is stopped by conservation of angular momentum. When we look at dense interstellar clouds, they are usually rotating slowly. Consider, for instance, the Horsehead Nebula, a dusty, dense cloud, ~ 500 pc away from us in the constellation Orion (Figure 17.1). From the Doppler shift of radio emission on different sides of the nebula, it's known that the Horsehead Nebula is rotating with a speed $\sim 1 \text{ km s}^{-1}$. The Horsehead Nebula itself is too large to collapse into a single star: it's ~ 1 pc from one end of its "mane" to the other, and the total mass of the nebula is $\sim 1000 \text{ M}_\odot$. However, smaller dense cores—small enough to form a single star—are also found to be rotating, with a typical rotation speed $\sim 0.1 \text{ km s}^{-1}$. If a cloud starts with

FIGURE 17.1 The Horsehead Nebula, an opaque dusty cloud seen in projection against a bright emission nebula.

rotation speed v_0 and radius r_0, then when it collapses to a final radius r_f, its rotation speed will be given by the law of conservation of angular momentum: $v_0 r_0 = v_f r_f$, or

$$v_f = \left(\frac{r_0}{r_f}\right) v_0. \tag{17.15}$$

If the radius of a molecular cloud core decreases by a factor $\sim 10^{-6}$, as suggested earlier, its final rotation velocity will be $v_f \sim 10^6 v_0$. For an initial rotation speed of $v_0 \sim 0.1\,\text{km s}^{-1}$, the final rotation speed will be $v_f \sim 100{,}000\,\text{km s}^{-1}$. However, no young stars have been seen to rotate with these relativistic speeds.

In fact, if angular momentum is conserved, the cloud will stop falling inward when it forms a rotationally supported disk, in which the gravitational acceleration is just sufficient to keep material on a circular orbit:

$$\frac{GM}{r_f^2} = \frac{v_f^2}{r_f}. \tag{17.16}$$

Combining equations (17.15) and (17.16), we find that the disk's radius will be

$$r_f = \frac{v_0^2 r_0^2}{GM} \approx 200\,\text{AU} \left(\frac{v_0}{0.1\,\text{km s}^{-1}}\right)^2 \left(\frac{r_0}{4000\,\text{AU}}\right)^2 \left(\frac{M}{1 M_\odot}\right)^{-1}. \tag{17.17}$$

Even when the initial rotation speed is small, the cloud will collapse to a disk much larger than a star. Such protoplanetary disks can be seen, for instance, in the denser regions of the Orion Nebula (see Figure 8.3).

A protoplanetary disk contains far more angular momentum than a star does. In order to create a star from a rotationally supported planetary disk, some of the material must lose angular momentum and fall to the center, where it forms a **protostar**. (A protostar is the slowly rotating ball of gas that will eventually become a star but which has not yet started fusion in its center.) One hint of where a protostar dumps its angular momentum is provided by the present solar system. The orbital angular momentum of Jupiter is

$$L_{\text{Jup}} \approx 2 \times 10^{43}\,\text{kg m}^2\,\text{s}^{-1}. \tag{17.18}$$

The rotational angular momentum of the Sun is only

$$L_\odot \approx 1 \times 10^{42}\,\text{kg m}^2\,\text{s}^{-1}. \tag{17.19}$$

Although the Sun contains more than 99.8% of the mass of the solar system, it contains less than 5% of the angular momentum. A relatively small amount of mass on a very large orbit can act as a "scapegoat" for angular momentum, carrying most of the initial angular momentum of the collapsing cloud. Part of the proto-Sun's angular momentum was carried by viscous torques to the outer part of the protoplanetary disk.[2] Another part of the proto-Sun's angular momentum is carried away by a magnetized protosolar wind; this is the same mechanism described in Section 7.3.

The main steps in the formation of a star can be summarized as follows:

- The dense core of a molecular cloud is perturbed by a shock wave and starts to collapse.
- The core collapses to form a rotationally supported disk. The gas in the disk is threaded by magnetic field lines. The central dense region of the disk is the protostar.
- Hot gas from the disk moves out along the magnetic field lines, forming a strong stellar wind. Since the magnetic field lines rotate along with the disk, the gas traveling along the field lines carries much angular momentum.
- Within the disk, dust clumps to form planetesimals; planetesimals collide to form planets. The remaining gas and dust in the disk is blown away by the stellar wind.

The protostar contracts on the Kelvin–Helmholtz timescale, $t_{\rm KH} \propto M^2/(RL)$, as it radiates away its gravitational potential energy. When its central regions become hot and dense enough for hydrogen fusion to ignite, the protostar becomes a star.

17.2 ▪ EVOLUTION OF SUN-LIKE STARS

Once the Sun started fusing hydrogen in its core, its time evolution became very gradual. Solving the time-dependent equations of stellar evolution indicates that the Sun's fusion-powered life began 4.6 billion years ago. Since the Sun's main sequence lifetime is 10 billion years, the Sun has exhausted nearly half the hydrogen in its core. During the Sun's 10 billion years on the main sequence, there have been few changes in its global structure. The most significant differences are due to the changing composition of the Sun's core. Fusing hydrogen into helium increases the mean molecular mass μ of the gas. (As shown in Section 14.1, fully ionized hydrogen has $\mu = 1/2$, while fully ionized helium has $\mu = 4/3$.) In order to maintain the central pressure P_c required for the Sun to remain in hydrostatic equilibrium, either the central temperature T_c or the central mass density ρ_c, or both, will have to increase. This in turn increases the energy generation rate ϵ and drives up the luminosity of the Sun. When the Sun began fusion, 4.6 billion years ago, its luminosity was $\sim 0.7 L_\odot$; about 6 billion years from now, right before it runs out of hydrogen in its core, the Sun's luminosity will be $\sim 2.2 L_\odot$.[3] A minor side effect of the Sun's luminosity increase is that ~ 3.5 Gyr from now, the increased flux at the Earth's location will trigger a runaway greenhouse effect; the Earth's oceans will evaporate, and the Earth's climate will resemble that of Venus today.

[2] To visualize viscous torques, think of a disk made of extremely sticky, syrupy material. If the inner regions are moving more rapidly than the outer regions, they will tend to drag the outer regions along with them, speeding up the outer regions and increasing their angular momentum.

[3] Tripling the Sun's luminosity sounds impressive, but when it occurs over the course of 10 billion years, it requires an increase of only 0.01% every million years.

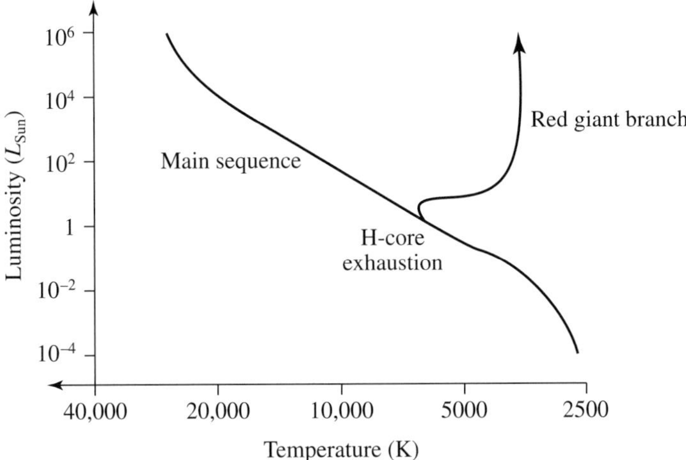

FIGURE 17.2 Climbing the red giant branch.

What happens when the Sun (or a Sun-like main sequence star) runs out of hydrogen in its core? When the central hydrogen is depleted, the Sun is still losing energy from its surface. The photosphere doesn't have a "gas gauge" to monitor and thus is unaware that the core has run out of fuel; it just keeps on radiating because it keeps on being hotter than its surroundings. The energy lost to interstellar space has to come from somewhere; the source on which the star falls back is the old standby, gravitational potential energy. The helium core of the Sun slowly contracts inward, converting its gravitational potential energy into thermal energy.

The layer above the core is heated to the point at which it starts fusing its hydrogen into helium. During this phase of the Sun's life, the main energy source is much closer to the surface than it was before. The outer layers of the Sun absorb the energy emitted by the hydrogen-fusing shell; their temperature and pressure increase, and they expand outward. The radius of the photosphere increases from $\sim 1.6R_\odot$ to $\sim 170R_\odot$.[4] The swollen photosphere drops in temperature from 6000 K to 3000 K, but the large increase in surface area means that the Sun's luminosity climbs from $2.2L_\odot$ to $2300L_\odot$. During this stage of the Sun's evolution, its location on a Hertzsprung–Russell diagram (Figure 17.2) moves upward and to the right. When the Sun fuses hydrogen into helium in a shell outside its core, it is very big and very red. Hence it is called a **red giant**, and its evolution away from the main sequence is referred to as "climbing the red giant branch."

As the hydrogen-fusing shell eats its way outward through the Sun, a larger and larger sphere of helium is left behind in the center. As the core of ionized helium becomes more massive, it is squeezed together by its own gravity until the free electrons become **degenerate**. (The term "degenerate" is not a value judgment about the electrons' lifestyle; to a physicist, electrons are degenerate when they are sufficiently close together

[4] Since $170R_\odot \approx 0.8$ AU, at this point, we can kiss Mercury and Venus goodbye.

that the Pauli exclusion principle comes into play and prevents any two electrons from occupying the same state.) The degenerate electrons produce a pressure that is entirely independent of the temperature of the helium nuclei. We defer an in-depth discussion of degeneracy to Section 18.1.1.

As the helium core continues to grow, it is supported by degenerate electron pressure even as the temperature of the helium nuclei increases. When the temperature of the helium nuclei reaches 10^8 K, fusion of helium into carbon by the triple alpha process begins (as detailed in Section 15.3). As the energy released by fusion is dumped into the core, the temperature of the helium nuclei, and hence the rate of the triple alpha process, shoots up rapidly. Since the core pressure is provided by degenerate electrons, the pressure initially remains constant despite the rapid temperature rise. This means that the mass density of the core remains constant while the temperature increases—just the conditions needed for runaway fusion. The initiation of helium fusion in a degenerate core is called the **helium flash**. The fusion runaway continues until the temperature rises to 3.5×10^8 K. At this temperature, the electrons finally become nondegenerate, and the dominant source of pressure is ordinary thermal motions.

After the helium flash, the Sun settles into a steady state; helium is fused into carbon (and a little oxygen) in the core, while hydrogen is fused into helium in a shell outside the core. At this stage of the evolution, the Sun is "on the horizontal branch" (Figure 17.3). This odd name is given because all stars have nearly the same luminosity—about $100L_\odot$—when they are at this stage of their evolution. Thus, they fall along a horizontal line in the H–R diagram. As a horizontal branch star, the Sun is smaller ($R \sim 10R_\odot$) and hotter ($T \sim 5000$ K) than during its red giant phase.

There's enough helium in the core to fuel the star by the triple alpha process for 100 million years. When helium is depleted, the core contracts, the released gravitational potential energy heats the shell above, and fusion of helium into carbon begins in the

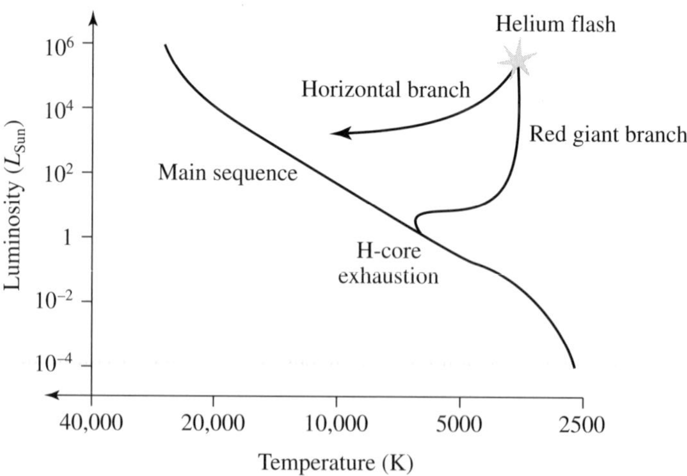

FIGURE 17.3 On the horizontal branch.

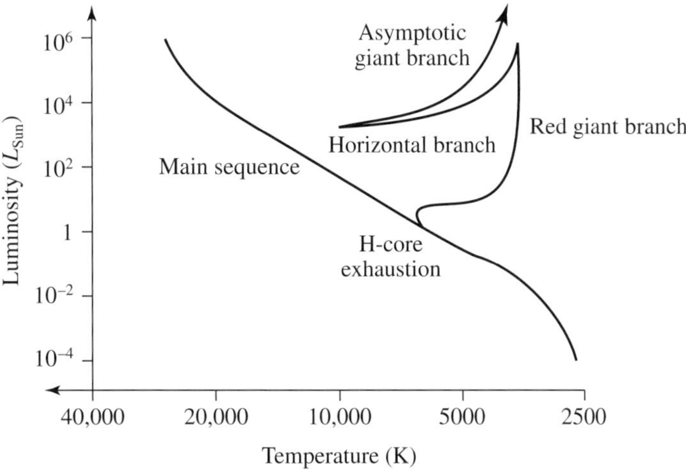

FIGURE 17.4 Climbing the asymptotic giant branch.

shell above the carbon core. During this phase, the Sun is a red giant again, only with *two* fusion shells; the inner converts helium to carbon (and a little oxygen), and the outer converts hydrogen to helium. At this stage, the Sun is an "asymptotic giant branch" star, or AGB star.

This name is given because the star's path on the H–R diagram (Figure 17.4) is asymptotic to the path it followed on its first climb up the giant branch.

The "layered look" of an AGB star is shown schematically in Figure 17.5. The relative sizes of the different layers are not shown to scale; the outer cool envelope is actually much larger than the central fusion shells and the carbon/oxygen core. The life of the Sun as an AGB star is relatively brief; this is the most luminous stage of the Sun's existence, with a predicted maximum luminosity of $\sim 3000 L_\odot$. An AGB star is unstable and pulsates in and out. The outward pulsations eject the outer layers of the star in huge gusts of stellar wind. The final pulses blow away the hydrogen-fusing and helium-fusing shells. Only the inert carbon/oxygen core remains.

The naked core is extremely hot, and emits copious ultraviolet light. It ionizes the surrounding gas and produces a planetary nebula (Color Figure 16). The carbon/oxygen core is small, dense, and supported by degenerate electron pressure; it is the type of stellar remnant that we call a "white dwarf." It is expected that the Sun will be able to blow away $\sim 40\%$ of its mass, leaving behind a white dwarf with mass $M_{wd} \sim 0.6 M_\odot$. The white dwarf is too cool to fuse its carbon and oxygen to heavier elements, and too stiff to collapse to a smaller volume and decrease its gravitational potential. With no energy source available other than thermal energy, the white dwarf gradually cools down, until it becomes a "black dwarf" in the distant future.[5]

[5] This may be disappointing to you, but the Sun will never become a supernova or a black hole. Sorry.

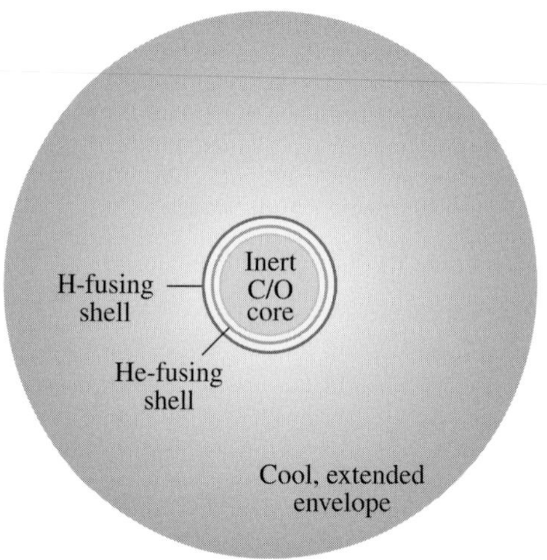

FIGURE 17.5 Layers of an asymptotic giant branch star.

The life of the Sun is a play in five acts, plus a prologue and epilogue:

Prologue: Protostar ($t \sim 50$ Myr)
No fusion; powered by gravity.

Act I: Main sequence ($t \sim 10$ Gyr)
Fusion of H to He in core.

Act II: Red giant branch ($t \sim 1$ Gyr)
Fusion of H to He in shell.

Act III: Horizontal branch ($t \sim 100$ Myr)
Fusion of H to He in shell, He to C in core.

Act IV: Asymptotic giant branch ($t \sim 20$ Myr)
Fusion of H to He in outer shell, He to C in inner shell.

Act V: Planetary nebula ($t \sim 50$ kyr)
No fusion; hot core emits UV radiation, and gas shell fluoresces.

Epilogue: White dwarf ($t \to \infty$)
No fusion; white dwarf cools down.

17.3 ▪ PULSATING VARIABLE STARS

During the Sun's main sequence lifetime, its properties change slowly. However, there do exist stars whose luminosities, and other properties, vary periodically on timescales of less than a year. These stars are known as **pulsating variable stars**. We'll focus on the two most celebrated types: **Cepheid stars** (sometimes called "classical Cepheids") and **RR Lyrae** stars. Each of these classes is named after its archetype, that is, the first star of that class to be recognized. This was δ Cephei in the case of the Cepheids, and RR Lyrae itself in the case of the RR Lyrae stars.

TABLE 17.1 Pulsating Variable Stars Compared

Property	Cepheid	RR Lyrae
M_V (average)	-0.5–-6	0.5–1
Spectral Type	F, G, K	A, F
Pulsation Period	1–50 days	1.5–24 hours
Mass	3–18 $M_\odot$	0.5–0.7 $M_\odot$
Evolutionary Stage	supergiant	horizontal branch
Metallicity	high	low

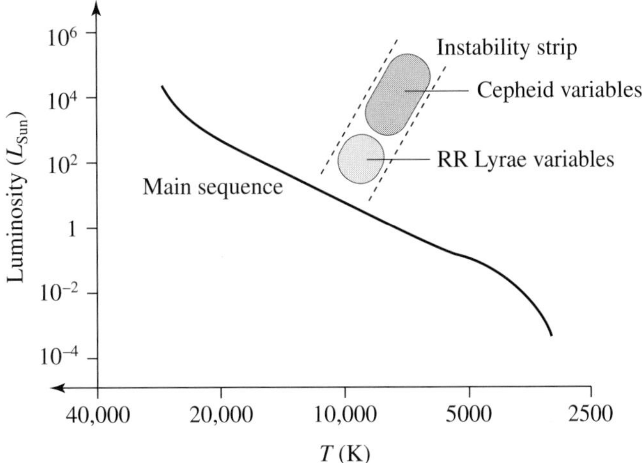

FIGURE 17.6 Location of Cepheids and RR Lyrae stars on a Hertzsprung–Russell diagram.

Table 17.1 shows the difference between Cepheids and RR Lyrae stars by comparing some of their properties.

Cepheid and RR Lyrae variables differ in many properties. However, if you plot their locations on an H–R diagram (Figure 17.6), you find they are adjacent to each other, on a diagonal stripe called the **instability strip**. Unlike the Sun and other stable main sequence stars, Cepheid and RR Lyrae stars pulsate in and out: they actually grow substantially larger and smaller in radius.

As an example of a pulsating variable star, consider δ Cephei. Over a period of $P =$ 5.366 days, its apparent magnitude varies by $\Delta m_V \approx 1$; this corresponds to $F_{max}/F_{min} \approx 10^{0.4} \approx 2.5$. A steep rise in flux (Figure 17.7) is followed by a slow decline. The radial velocity of the star δ Cephei, relative to the Sun, is approximately $-15\,\mathrm{km\,s^{-1}}$. Because

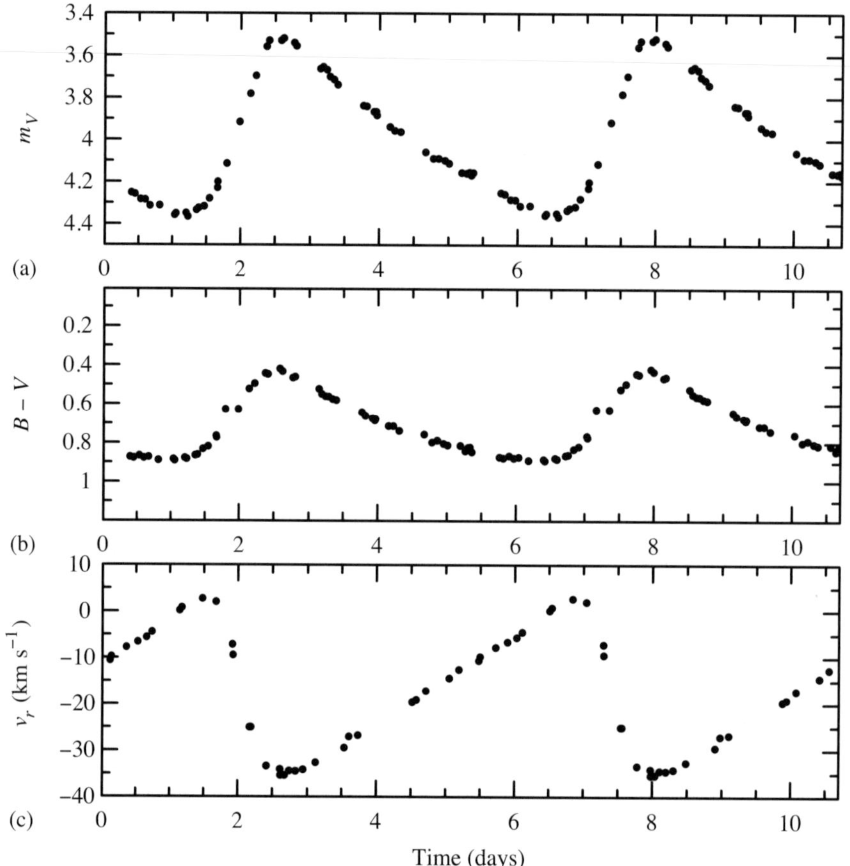

FIGURE 17.7 Time-variable properties of δ Cephei. (a) V-band apparent magnitude. (b) $B-V$ color index. (c) Radial velocity of photosphere.

of the expansion and contraction of the photosphere, the *observed* radial velocity varies from $v_r \approx -35\text{ km s}^{-1}$ when the photosphere is expanding to $v_r \approx +5\text{ km s}^{-1}$ when the photosphere is contracting. The effective temperature changes from $T_{\text{eff}} \approx 5600$ K when the star is near its minimum luminosity to $T_{\text{eff}} \approx 6600$ K when the star is near its maximum luminosity. The variations in radius can be deduced from the star's changes in luminosity and effective temperature: $R \propto L^{1/2}/T_{\text{eff}}^2$. The radius of δ Cephei varies by nearly 15% over the course of one cycle; this means that the maximum volume is 50% greater than the minimum volume.

To see why pulsating variable stars pulsate, while the Sun is content to be stable, start by taking a star of radius R and squeezing slightly, so that its radius decreases to $R - dR$. As we saw in Section 17.1, squeezing a gas sphere makes a sound wave travel toward its center. Since a star in hydrostatic equilibrium is smaller than its Jeans length, the sound wave will reach the center in a time less than the collapse time. When the sound wave reaches the center, it expands back to the surface; when it reaches the surface, it reflects back to the center. Thus, squeezing a star in a spherically symmetric manner results in a standing sound wave, or **acoustic oscillation**.

The period P of the acoustic oscillation is the time required for a sound wave to make a round trip from photosphere to center back to photosphere. This time is

$$P = 2t_{\text{press}} = \frac{2R}{c_s} = 2R\left(\frac{\mu m_p}{\gamma k\langle T\rangle}\right)^{1/2}, \tag{17.20}$$

where c_s is the average sound speed in the star; the relation between c_s and the average temperature $\langle T\rangle$ is taken from equation (17.9). If we take the average temperature to be roughly half the central temperature (as approximated in equation 15.9), we find a period of

$$P \approx 2R\left(\frac{2\mu m_p}{\gamma k T_c}\right)^{1/2} \approx 2R\left(\frac{R}{\gamma GM}\right)^{1/2}. \tag{17.21}$$

Note that the period for acoustic oscillations is

$$P \approx \left(\frac{4R^3}{\gamma GM}\right)^{1/2} \approx \left(\frac{3}{\pi\gamma G\langle\rho\rangle}\right)^{1/2}, \tag{17.22}$$

so stars of greater density have shorter periods for acoustic oscillations. This tells us that the short-period RR Lyrae stars are higher in density than the longer-period Cepheids. Notice also that the sound travel time from surface to center, $P/2$, and the freefall time t_{ff} (equation 17.6) are both proportional to $\langle\rho\rangle^{-1/2}$. However, the ratio of the two timescales is

$$\frac{P/2}{t_{\text{ff}}} \approx \frac{2}{\pi}\left(\frac{2}{\gamma}\right)^{1/2} \approx 0.7, \tag{17.23}$$

assuming an adiabatic index of $\gamma = 5/3$, appropriate for an ionized gas. Thus, a star in hydrostatic equilibrium is stable, but not extravagantly so; the laws of stellar structure ensure that the sound travel time is only 30% shorter than the freefall time.

Any star will undergo acoustic oscillations when it is squeezed or otherwise perturbed. In the Sun, helioseismology reveals that the oscillations are of small amplitude; this is true in most other stars as well. The acoustic oscillations are damped by the viscosity of the gas. Cepheids and RR Lyraes have high-amplitude pulsations because their acoustic oscillations are *driven*. Like a little girl on a swing who gets repeated pushes from Mom at the same point of her swing, a pulsating variable star is driven by a periodic force that has the same period P as the acoustic oscillations.

The driving force behind the pulsations of Cepheids and RR Lyrae stars is related to changes in opacity. To see how opacity changes can drive oscillations, let's consider the process step by step.

1. A layer near the surface is heated to a temperature of $T \sim 40{,}000$ K. At this temperature, He^{+} is ionized to He^{++}.
2. The opacity shoots upward, due to scattering by the newly freed electrons.

3. Light is absorbed at the bottom of the opaque layer, increasing the temperature and pressure gradient across the layer.
4. Driven by the increased pressure gradient, the opaque layer expands outward and cools.
5. He^{++} recombines with free electrons to form He^{+}.
6. The opacity plummets, and previously trapped photons rush outward through the newly transparent layer.
7. The layer contracts and reheats to 40,000 K, where the whole cycle begins again.

A layer with $T \sim 40{,}000$ K is near the photosphere of the star. The central regions, where the fusion reactions occur at much higher temperatures, are unaffected by the opacity changes far above them. The fusion generates energy at a steady rate; however, the changes in opacity cause the energy to be released in periodic outbursts.[6]

Stars in the instability strip (see Figure 17.6) have a helium ionization layer in which the natural period for the transition from opaque to transparent to opaque just matches the period P for acoustic oscillations in the star. The Sun is too low in mass ever to evolve into a Cepheid variable, but it might lose enough mass as a red giant to become an RR Lyrae star when it's on the horizontal branch. Within the instability strip, the most luminous stars are large, low-density supergiants. Since their density ρ is low, their acoustic oscillation period, $P \propto \rho^{-1/2}$, is long. As a result, there is a **period–luminosity relation** for Cepheid stars, with the more luminous stars having longer periods.

In fact, the period–luminosity relation for Cepheids was discovered empirically long before anyone knew the precise mechanism driving their luminosity fluctuations. In 1912, Henrietta Swan Leavitt was studying variable stars in the Large and Small Magellanic Clouds. When she plotted the apparent magnitude of each Cepheid as a function of its oscillation period P, she found $m \propto \log P$. Since the Clouds are small compared to their distance from us, the stars in each Cloud are all at roughly the same distance from us, implying

$$M \propto \log P. \tag{17.24}$$

To calibrate the period–luminosity relation, and find exactly which absolute magnitude M corresponds to a given period P, we need to know the distance d to at least some of the Cepheids out there. Unfortunately, Cepheid variables, like all highly luminous stars, are rare; only the closest Cepheids have had their parallax measured accurately.[7] A widely used form of the Cepheid period–luminosity relationship is

$$\overline{M}_V = -2.76 \log(P/10 \text{ days}) - 4.16, \tag{17.25}$$

[6] Think of a pot of boiling water with a heavy lid. The water evaporates at a steady rate; however, the water vapor escapes only in periodic "puffs" when the pressure grows high enough to lift the lid.

[7] The nearest Cepheid is Polaris, at $d = 130 \pm 10$ pc. The next nearest is δ Cephei, at $d = 270 \pm 10$ pc.

where $\overline{M}_V$ is the absolute magnitude in the V band, averaged over a complete period. The scatter in absolute magnitude at a given period is about 0.3 magnitudes.

If you measure the period P of a Cepheid, you can compute its absolute magnitude from equation (17.25). After measuring its average apparent magnitude $\overline{m}_V$, you can then compute its distance using the relation

$$\log(d/10\,\text{pc}) = 0.2(\overline{m}_V - \overline{M}_V), \tag{17.26}$$

assuming there's no extinction. Thus, Cepheids make excellent **standard candles**; this is the term used by astronomers for an object whose luminosity is known, and whose distance can thus be calculated once its flux is measured. Cepheids are bright enough to be seen in the Virgo Cluster of galaxies, whose distance (as measured using Cepheids) is approximately 18 million parsecs. RR Lyrae stars can also be used as standard candles, but since they are less luminous, they can't be seen as far away as Cepheids and be seen.

PROBLEMS

17.1 A protostellar cloud starts as a sphere of radius $R = 4000$ AU and temperature $T = 15$ K. If it emits blackbody radiation, what is its total luminosity? What is the wavelength λ_p at which it emits the most radiation?

17.2 A Cepheid star in the Large Magellanic Cloud is observed to have an average apparent magnitude $\overline{m}_V = 11.80$ and a period $P = 95$ days. Compute the distance to the Large Magellanic Cloud, ignoring any effects due to dust.

17.3 Consider two clouds in the interstellar medium. A molecular (H_2) cloud has $T = 10$ K and $n = 10^{12}\,\text{m}^{-3}$; a neutral atomic (H) cloud has $T = 120$ K and $n = 10^7\,\text{m}^{-3}$.

(a) What is the Jeans mass for each of the two clouds?
(b) What is the minimum radius each cloud must have to collapse?
(c) What is the timescale for the gravitational collapse of each cloud?

17.4 In this problem, you will estimate the duration of the horizontal branch phase in a $1M_\odot$ star.

(a) Compute the energy released in the net triple alpha reaction $3\,{}^4\text{He} \to {}^{12}\text{C}$. The masses of ^{4}He and ^{12}C are 4.0026 amu and 12.0000 amu, respectively, where 1 amu (atomic mass unit) = 1.6606×10^{-27} kg.
(b) Assume that at the beginning of the horizontal branch phase, 10% of the original mass of the star is in the form of ^{4}He in the stellar core. Estimate the total energy released by fusing this amount of helium into carbon via the triple alpha process.

(c) Assume that during the horizontal branch phase, $L = 100L_\odot$. If all this luminosity is provided by fusion of helium to carbon in the core, how long will the horizontal branch phase last?

17.5 Make an order-of-magnitude estimate of the length of the protostar phase for the Sun.

17.6 What is the thermal energy density, in joules per cubic meter, of a typical giant molecular cloud? What is the thermal energy density of the Earth's atmosphere at sea level?

18 Stellar Remnants

The life history of a star is determined primarily by its **mass**; if you know a star's initial mass, you know (at least in broad outline) how it will evolve. Other factors, like the star's initial chemical composition and initial angular momentum, have a smaller effect.[1] For instance, the fusion history of a star depends on its initial mass (all mass limits below are approximate):

- $M < 0.08M_\odot$: No fusion. The center of a low-mass ball of gas never becomes hot enough for hydrogen fusion. This mass range represents not stars but brown dwarfs of spectral type L and T.
- $0.08M_\odot < M < 0.5M_\odot$: Fusion of ^{1}H to ^{4}He. The center of the star never becomes hot enough to fuse ^{4}He to ^{12}C and ^{16}O. Stars in this mass range are M stars on the main sequence and will eventually end up as white dwarfs made of helium.
- $0.5M_\odot < M < 5M_\odot$: Fusion of ^{1}H to ^{4}He and ^{4}He to ^{12}C and ^{16}O. The center of the star never becomes hot enough to fuse ^{12}C and ^{16}O to heavier elements. Stars in this mass range are A, F, G, and K stars on the main sequence and end up as white dwarfs made of carbon and oxygen.
- $5M_\odot < M < 7M_\odot$: Fusion of ^{1}H to ^{4}He, ^{4}He to ^{12}C and ^{16}O, and ^{12}C and ^{16}O to ^{20}Ne and ^{24}Mg. Stars in this mass range are B stars on the main sequence.
- $M > 7M_\odot$: Fusion of ^{1}H to ^{4}He, ^{4}He to ^{12}C and ^{16}O, ^{12}C and ^{16}O to ^{20}Ne and ^{24}Mg, ^{20}Ne and ^{24}Mg to ^{28}Si, and ^{28}Si to ^{52}Fe and ^{56}Ni. Stars in this mass range are O stars on the main sequence. (Note that the elements produced have atomic masses that are multiples of 4; this is because the heavier elements are built up by fusing on additional ^{4}He nuclei.)

The initial mass of a star also determines what its "corpse," or stellar remnant, will be. The lowest-mass stars become dense white dwarfs, supported by electron degeneracy pressure. Higher-mass stars leave behind even denser neutron stars, and the most massive

[1] The statement "A star's properties are determined primarily by its mass" is known as the Russell–Vogt theorem, after the ubiquitous Henry Norris Russell and the German astronomer Heinrich Vogt, who were the first to realize its truth.

stars of all end up as black holes, representing the ultimate in density. We will examine in turn the evolutionary roads that lead to white dwarfs, neutron stars, and black holes.

18.1 ▪ WHITE DWARFS

A white dwarf is a stellar remnant supported by electron degeneracy pressure. The name "white dwarf," as it turns out, is something of a misnomer. Nobody objects to the "dwarf" part; white dwarfs really are small compared to main sequence stars of comparable mass. As we computed in Section 13.5, the radius of the white dwarf Sirius B is $R_{\rm wd} = 0.0084 R_\odot$, and its mass is $M_{\rm wd} = 0.96 M_\odot$. However, not all white dwarfs are white-hot. Although the first discovered white dwarfs, such as Sirius B, have high surface temperatures (Sirius B has $T \approx 25{,}000$ K), some white dwarfs have surface temperatures as low as $T \approx 4000$ K.

White dwarfs are high in density compared to main sequence stars. The average density of Sirius B is

$$\rho_{\rm wd} = \frac{M_{\rm wd}}{M_\odot}\left(\frac{R_\odot}{R_{\rm wd}}\right)^3 \rho_\odot = 0.96(0.0084)^{-3}\rho_\odot \tag{18.1}$$

$$\approx 2 \times 10^6 \rho_\odot \approx 2 \times 10^9 \text{ kg m}^{-3}, \tag{18.2}$$

about 200,000 times the density of lead.[2] Supporting such a dense object in hydrostatic equilibrium requires a high internal pressure. In Section 15.1, we computed the approximate central pressure for a sphere in hydrostatic equilibrium:

$$P_c \sim \frac{2GM\langle\rho\rangle}{R} \sim \frac{3GM^2}{2\pi R^4}. \tag{18.3}$$

The central pressure in a white dwarf must then be

$$P_c \sim \left(\frac{M_{\rm wd}}{M_\odot}\right)^2 \left(\frac{R_\odot}{R_{\rm wd}}\right)^4 P_{c,\odot} \sim (0.96)^2 (0.0084)^{-4} P_{c,\odot} \tag{18.4}$$

$$\sim 2 \times 10^8 P_{c,\odot} \sim 10^{18} \text{ atm}. \tag{18.5}$$

If this pressure were provided by ordinary thermal motions, the temperature at the center of Sirius B would have to be $T_c \sim 6 \times 10^9$ K. However, the pressure inside a white dwarf is *not* due primarily to thermal motions. Instead, it is provided by the degenerate electrons.

18.1.1 Degeneracy Pressure

As noted in Section 17.2, electrons become **degenerate** when they are packed closely enough that the Pauli exclusion principle produces an additional form of pressure to keep

[2] This might seem more impressive if we point out that 2×10^9 kg m^{-3} is equivalent to 10 tons per teaspoon.

them apart.[3] The **electron degeneracy pressure** is a consequence of the Heisenberg uncertainty principle, which states that you can't simultaneously specify the position x and momentum p of a particle to arbitrary accuracy. There is always an uncertainty in each such that

$$\Delta x \Delta p \geq \hbar, \tag{18.6}$$

where $\hbar$ is the reduced Planck constant introduced in Section 5.1. Suppose that the degenerate electrons have a number density n_e. In their cramped conditions, each electron is confined to a volume $V \sim n_e^{-1}$. Thus, the location of each electron is determined with an uncertainty $\Delta x \sim V^{1/3} \sim n_e^{-1/3}$. From the uncertainty principle, the minimum uncertainty in the electron momentum is

$$\Delta p \sim \frac{\hbar}{\Delta x} \sim \hbar n_e^{1/3}. \tag{18.7}$$

If the electrons are nonrelativistic,

$$\Delta v = \frac{\Delta p}{m_e} \sim \frac{\hbar n_e^{1/3}}{m_e}, \tag{18.8}$$

where m_e is the mass of the electron.

Thanks to the uncertainty principle, degenerate electrons are zipping around with a speed $v_e \propto n_e^{1/3}$ regardless of how low the temperature drops. These "Heisenberg speeds" contribute to the pressure, just as the thermal speeds do. For ordinary thermal motions, the electron speeds are

$$v_{\rm th} \sim \left(\frac{kT}{m_e}\right)^{1/2}, \tag{18.9}$$

and the pressure contributed by thermal motions of electrons is

$$P_{\rm th} = n_e kT \sim n_e m_e v_{\rm th}^2. \tag{18.10}$$

By analogy, the "Heisenberg speeds" contribute a pressure

$$P_{\rm degen} \sim n_e m_e (\Delta v)^2 \sim n_e m_e \left(\frac{\hbar n_e^{1/3}}{m_e}\right)^2 \sim \hbar^2 \frac{n_e^{5/3}}{m_e}. \tag{18.11}$$

We label a population of electrons as "degenerate" when $P_{\rm degen} > P_{\rm th}$.

[3] Electrons, neutrons, and protons are all fermions, particles with half-integral spin, to which the Pauli exclusion principle applies. Photons are examples of bosons, particles with integral spin, to which the exclusion principle does not apply.

18.1.2 Mass–Radius Relationship

Because white dwarfs are supported by electron degeneracy pressure, we can derive a simple mass–radius relation for white dwarfs. (Warning for the faint-hearted: since we want only the correct proportionality between radius and mass, and are not concerned with exact numbers, we'll be omitting numerical factors like π and 2 from the following analysis.) Since a white dwarf is in hydrostatic equilibrium (equation 18.3), its central pressure must be

$$P_c \sim \frac{GM^2}{R^4}. \tag{18.12}$$

This is true for *all* spheres in hydrostatic equilibrium, regardless of the pressure source. If the pressure is provided by degenerate electrons, then from equation (18.11),

$$P_c \sim \hbar^2 \frac{n_e^{5/3}}{m_e} \sim \hbar^2 \frac{\rho^{5/3}}{m_p^{5/3} m_e} \sim \frac{\hbar^2}{m_p^{5/3} m_e} \frac{M^{5/3}}{R^5}. \tag{18.13}$$

(Since a carbon/oxygen white dwarf is made of ionized "metals," its mean molecular mass is $\mu \approx 2$. In keeping with our policy of ignoring small numerical factors, we have set $2 \approx 1$.) Equating the pressure required for hydrostatic equilibrium (equation 18.12) with the pressure provided by degenerate electrons (equation 18.13), we find that

$$G\frac{M^2}{R^4} \sim \frac{\hbar^2}{m_p^{5/3} m_e} \frac{M^{5/3}}{R^5}, \tag{18.14}$$

or

$$R \sim \frac{\hbar^2}{G m_e m_p^2} \left(\frac{M}{m_p} \right)^{-1/3}. \tag{18.15}$$

Notice the counterintuitive result that more massive white dwarfs have a smaller radius. We don't expect 40 pounds of cow manure to fit in a smaller bag than 20 pounds of cow manure, but we do expect a $1 M_\odot$ white dwarf to fit in a smaller volume than a $0.5 M_\odot$ white dwarf. Since larger masses correspond to smaller radii, we expect the average density to increase rapidly with mass:

$$\langle \rho \rangle \sim \frac{M}{R^3} \sim \frac{G^3 m_e^3 m_p^5}{\hbar^6} M^2. \tag{18.16}$$

The exact value of the radius for a white dwarf of a given mass depends slightly on the chemical composition of the dwarf. However, a good empirical fit is found to be

$$R \approx 0.01 R_\odot \left(\frac{M}{0.7 M_\odot} \right)^{-1/3}. \tag{18.17}$$

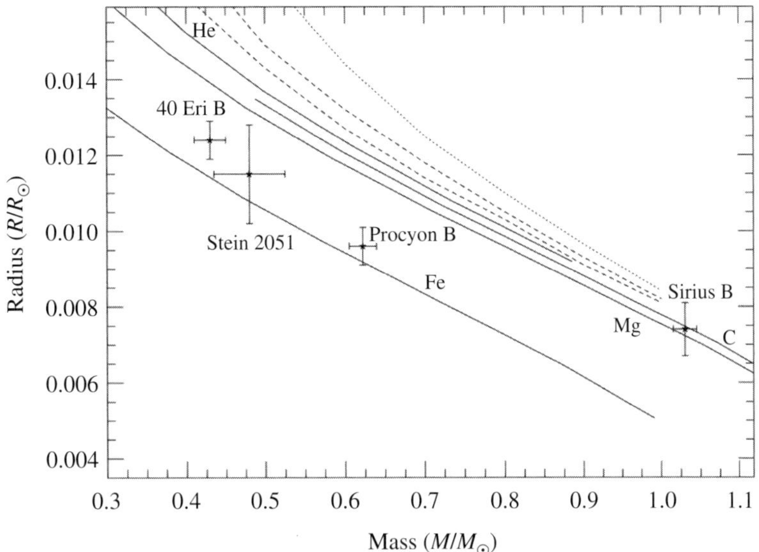

FIGURE 18.1 Observed masses and radii for nearby white dwarfs. The solid and dashed lines represent models for different chemical compositions.

The measured masses and radii of some nearby white dwarfs are plotted in Figure 18.1. The best-determined masses and radii are for Sirius B, Procyon B, and 40 Eri B, all part of visual binary systems.

As the pressure P drops with distance from the center of the white dwarf, so does the density $\rho \propto P^{3/5}$ (equation 18.13). The outermost layers of a white dwarf are thus low enough in density for ordinary thermal pressure to be the dominant pressure source. The nondegenerate atmosphere of the white dwarf is different in some ways from the atmospheres of ordinary stars. The gravitational acceleration in the white dwarf's atmosphere is large:

$$g_{\rm wd} = \frac{GM_{\rm wd}}{R_{\rm wd}^2} \approx 7000 g_\odot \left(\frac{M}{0.7M_\odot}\right)^{5/3} \approx 2 \times 10^6 \text{ m s}^{-2} \left(\frac{M}{0.7M_\odot}\right)^{5/3} . \quad (18.18)$$

Thus, a white dwarf's atmosphere will have a small scale height, despite the high photospheric temperature ($T \sim 10^5$ K $\sim 20T_\odot$) of a newly unveiled white dwarf:

$$H_{\rm wd} = \frac{kT_{\rm wd}}{g_{\rm wd}\mu_{\rm wd}m_p} \sim 0.4 \text{ km} \left(\frac{T_{\rm wd}}{10^5 \text{ K}}\right) \left(\frac{M_{\rm wd}}{0.7M_\odot}\right)^{-5/3} . \quad (18.19)$$

Because of the high gravitational acceleration in the photosphere, the spectra of white dwarfs typically show extreme pressure broadening of absorption lines (Figure 18.2). Thus, white dwarfs can be distinguished from hot main sequence stars by their spectra alone.

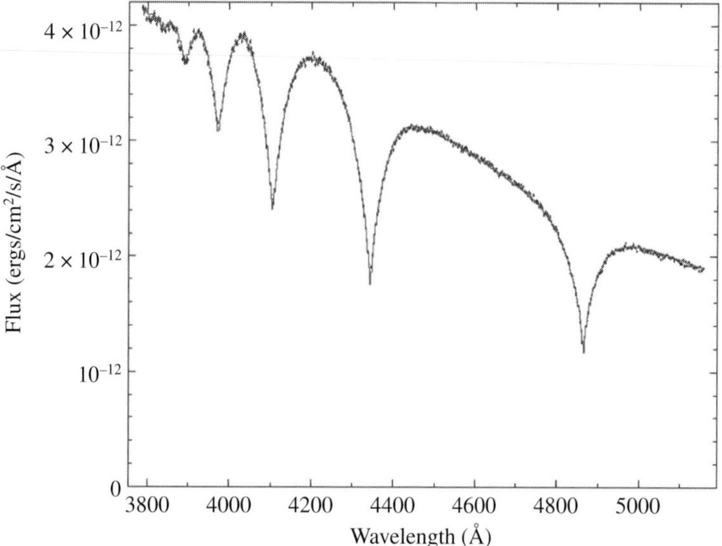

FIGURE 18.2 Spectrum of the white dwarf Sirius B; note the pressure-broadened Balmer lines.

Sirius B, with a mass equal to that of the Sun, is the most massive white dwarf in our immediate neighborhood (see Figure 18.1). White dwarfs cannot have an arbitrarily high mass. As additional mass is piled on to a white dwarf, the density increases, and the "Heisenberg speed" of the electrons,

$$\Delta v \sim \hbar \frac{n_e^{1/3}}{m_e}, \tag{18.20}$$

approaches the speed of light. When $\Delta v \sim c$, rearranging equation (18.20) gives

$$n_e \sim \left(\frac{c m_e}{\hbar} \right)^3 \sim 2 \times 10^{37}\ \mathrm{m}^{-3}, \tag{18.21}$$

so the degenerate electrons become relativistic at this density. Since a typical white dwarf has one proton and one neutron for each electron, this corresponds to a mass density

$$\rho \sim 2 m_p n_e \sim \frac{2 c^3 m_e^3 m_p}{\hbar^3} \sim 6 \times 10^{10}\ \mathrm{kg\ m}^{-3}. \tag{18.22}$$

When we compare this critical density to the mass–density relation for white dwarfs (equation 18.16), we find that the degenerate electrons in a white dwarf become relativistic when

$$\frac{G^3 m_e^3 m_p^5}{\hbar^6} M^2 \sim \frac{c^3 m_e^3 m_p}{\hbar^3}, \tag{18.23}$$

or when the white dwarf's mass is

$$M \sim \left(\frac{\hbar^3 c^3}{G^3 m_p^4} \right)^{1/2} \sim 4 \times 10^{30} \text{ kg}, \tag{18.24}$$

or about twice the mass of the Sun.

The transition of the degenerate electrons from nonrelativistic to relativistic has grave consequences for the structure of the white dwarf. Highly relativistic electrons have an energy much greater than their rest energy: $E_{\text{rel}} \gg m_e c^2$. In practice, we can treat them like massless particles (photons, for instance). For instance, photons have an energy related to their momentum by the equation $E = pc$. Similarly, highly relativistic degenerate electrons will each have an energy

$$E_{\text{rel}} \sim (\Delta p)c \sim \hbar n_e^{1/3} c, \tag{18.25}$$

providing a total electron energy density

$$u_{\text{rel}} \equiv E_{\text{rel}} n_e \sim \hbar c n_e^{4/3}. \tag{18.26}$$

By analogy, once again, with photons, which have a radiation pressure $P = u/3$, we can compute the pressure contributed by the highly relativistic degenerate electrons:

$$P_{\text{rel}} = \frac{1}{3} u_{\text{rel}} \sim \frac{\hbar c}{3} n_e^{4/3}. \tag{18.27}$$

Note the different dependence on n_e than that which held true for *nonrelativistic* electrons, which had $P \propto n_e^{5/3}$. The pressure in the relativistic case is less strongly dependent on density.[4]

If the relativistic white dwarf is to remain in hydrostatic equilibrium (equation 18.3), it must have

$$P_c \sim G \frac{M^2}{R^4}. \tag{18.28}$$

The pressure provided by the relativistic degenerate electrons is

$$P_{c,\text{rel}} \sim \hbar c \left(\frac{\rho}{m_p} \right)^{4/3} \sim \frac{\hbar c}{m_p^{4/3}} \frac{M^{4/3}}{R^4}. \tag{18.29}$$

Setting these two pressures equal, we find that a white dwarf supported by relativistic degenerate electrons will be in equilibrium when

$$G \frac{M^2}{R^4} \sim \frac{\hbar c}{m_p^{4/3}} \frac{M^{4/3}}{R^4}. \tag{18.30}$$

[4] In other words, the relativistic white dwarf isn't as stiff; a stiff material is one in which a small change in density produces a large change in pressure.

Note that the factors of R^{-4} cancel on either side. This means that a relativistic white dwarf is only in equilibrium for a specific mass,

$$M \sim \left(\frac{\hbar^3 c^3}{G^3 m_p^4}\right)^{1/2}. \tag{18.31}$$

But this is just the mass that we computed in equation (18.24) as the minimum mass required for the electrons to be relativistic! For any mass greater than this value, $M \sim 2M_\odot$, the central pressure required for hydrostatic equilibrium is *greater* than the pressure that relativistic degenerate electrons can supply, and the white dwarf collapses.

The maximum possible mass for a white dwarf, $M_{\rm Ch} \sim 2M_\odot$, is called the **Chandrasekhar mass**, after the astrophysicist Subramanyan Chandrasekhar, who was the first to calculate it. A more careful calculation of the Chandrasekhar mass for a carbon/oxygen white dwarf yields

$$M_{\rm Ch} = 1.4M_\odot. \tag{18.32}$$

Any star that can reduce its mass below the Chandrasekhar mass (generally by strong stellar winds during a giant phase) will end as a white dwarf. It is estimated that stars with $M \leq 7M_\odot$ will be able to slim themselves down below the Chandrasekhar mass. This calculation is uncertain, though, since mass loss during the giant phase is irregular by nature and difficult to model. Stars with initial masses less than $0.5M_\odot$ will eventually become helium white dwarfs; stars with $0.5M_\odot < M < 5M_\odot$ initially will leave behind carbon/oxygen white dwarfs; stars with $5M_\odot < M < 7M_\odot$ will leave behind neon/magnesium white dwarfs.

18.2 ▪ NEUTRON STARS AND PULSARS

What happens to stars whose initial mass is greater than $7M_\odot$? There aren't many stars that massive, but they do exist. Very massive stars have short lives and die spectacularly, leaving behind a badly crushed corpse. The main sequence lifetime of a massive star is short:

$$\tau \approx 30\ \text{Myr} \left(\frac{M}{7M_\odot}\right)^{-3}. \tag{18.33}$$

Life after the main sequence, when the star becomes a supergiant, is even shorter. The ultraluminous supergiant is relying on less efficient energy sources. Fusing hydrogen to helium releases about $6.4 \times 10^{14}\ \text{J kg}^{-1}$; thereafter, fusing helium all the way to iron releases only 24% as much energy per kilogram. In its last moments as a star, a massive supergiant has many concentric fusion layers over an iron core. A schematic diagram of such a star is shown in Figure 18.3.

As we saw in Section 13.4, a supergiant star such as Betelgeuse can have a radius $> 1000R_\odot$; however, models of stellar evolution indicate that the dense central fusion region of the supergiant has a radius $< 1R_\odot$. The central iron core is very dense (thou-

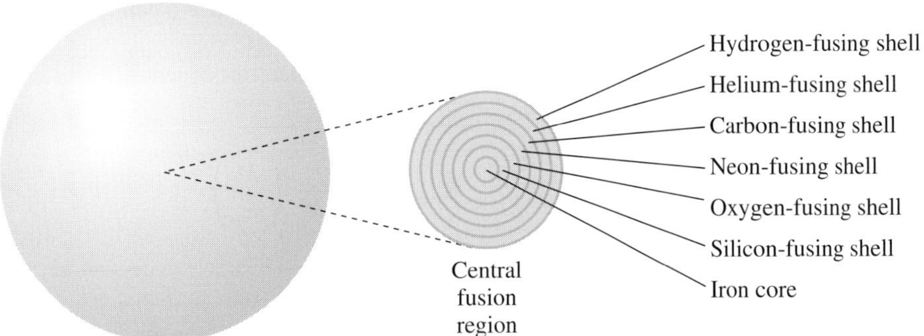

FIGURE 18.3 Fusion layers in a supergiant near the end of its life as a star.

sands of tons per cubic centimeter) and is supported by degenerate electron pressure. The iron core continues to grow in size as the silicon-fusing shell eats its way outward in the supergiant. When the iron core reaches the Chandrasekhar mass, it is no longer adequately supported by the relativistic degenerate electrons, and it starts to collapse—very rapidly. The collapse time $t_{\rm ff}$ at the high density of a degenerate core is less than 1/10 second.

As the density of the collapsing iron core rises, protons and free electrons start to combine to form neutrons:

$$p + e^- \to n + \nu_e. \tag{18.34}$$

One side effect of the collapse is thus a huge burst of electron neutrinos (ν_e). A Chandrasekhar mass of iron contains $\sim 10^{57}$ protons, so $\sim 10^{57}$ neutrinos are created during the collapse of the iron core. This burst of neutrinos escapes from the star, carrying away a large amount of energy, about 10^{46} J. (For comparison, the Sun will radiate $\sim 10^{44}$ J of energy during its main sequence lifetime.) The core is now a sphere of $\sim 2 \times 10^{57}$ neutrons that are essentially in free fall. Can anything stop the headlong collapse of the neutrons?

Yes! Core collapse can be stopped by **neutron degeneracy pressure**. Degenerate nonrelativistic electrons have a pressure

$$P_e \sim \hbar^2 \frac{n_e^{5/3}}{m_e}. \tag{18.35}$$

The Fermi exclusion principle applies to neutrons as well as to electrons. Degenerate nonrelativistic neutrons thus have a pressure

$$P_n \sim \hbar^2 \frac{n_n^{5/3}}{m_n}, \tag{18.36}$$

where n_n is the number density of neutrons and $m_n = 1839 m_e$ is the mass of a neutron. A sphere of neutrons supported by degenerate neutron pressure is called a **neutron star**.

(It doesn't satisfy our strict definition of a "star," since a neutron star isn't gaseous and isn't powered by fusion, but the name has stuck.)

A white dwarf—a sphere supported by electron degeneracy pressure—has a radius (equation 18.15)

$$R_{\rm wd} \sim \frac{\hbar^2}{G m_e m_p^2} \left(\frac{M}{m_p} \right)^{-1/3} . \tag{18.37}$$

By analogy, a neutron star—a sphere supported by neutron degeneracy pressure—should have a radius

$$R_{\rm ns} \sim \frac{\hbar^2}{G m_n m_p^2} \left(\frac{M}{m_p} \right)^{-1/3} . \tag{18.38}$$

A neutron star will thus be smaller than a white dwarf of comparable mass, by a ratio

$$\frac{R_{\rm ns}}{R_{\rm wd}} \sim \frac{m_e}{m_n} \left(\frac{M_{\rm wd}}{M_{\rm ns}} \right)^{1/3} \sim \frac{1}{1839} \left(\frac{M_{\rm wd}}{M_{\rm ns}} \right)^{1/3} . \tag{18.39}$$

If a white dwarf with $M_{\rm wd} = 0.7 M_\odot$ has a radius $R_{\rm wd} = 0.01 R_\odot$, then equation (18.39) leads us to expect

$$R_{\rm ns} \sim 3\,{\rm km} \left(\frac{M_{\rm ns}}{1.4 M_\odot} \right)^{-1/3} . \tag{18.40}$$

In truth, things are a little more complicated than we've been letting on. The density within a neutron star is comparable to the density within an atomic nucleus. Thus, the strong nuclear force between neutrons must be taken into account. Although the strong nuclear force is attractive at distances $d > 5 \times 10^{-16}$ m (this attraction is what keeps nuclei from flying apart), it is repulsive at shorter distances. Thus, the short-range repulsion between neutrons stiffens the neutron star and keeps the neutrons from being shoved arbitrarily close together.

For one plausible model of neutron star interiors, which takes the strong nuclear force into account, the mass–radius relation is

$$R_{\rm ns} \approx 11\,{\rm km} \left(\frac{M_{\rm ns}}{1.4 M_\odot} \right)^{-1/3} . \tag{18.41}$$

White dwarfs contain the mass of a star squashed into the volume of the Earth; neutron stars contain the mass of a star squashed into the volume of a small asteroid. It is a cliché for an astronomy text to show a neutron star juxtaposed with a city (Figure 18.4). Nevertheless, there's a reason why so many books include such a figure; it's an effective way of showing how tiny neutron stars are for their mass.

Understanding the structure of neutron stars is difficult for two reasons. Not only must the strong nuclear force be taken into account, but also gravity must be treated using general relativity rather than Newtonian gravity. The escape speed from a neutron

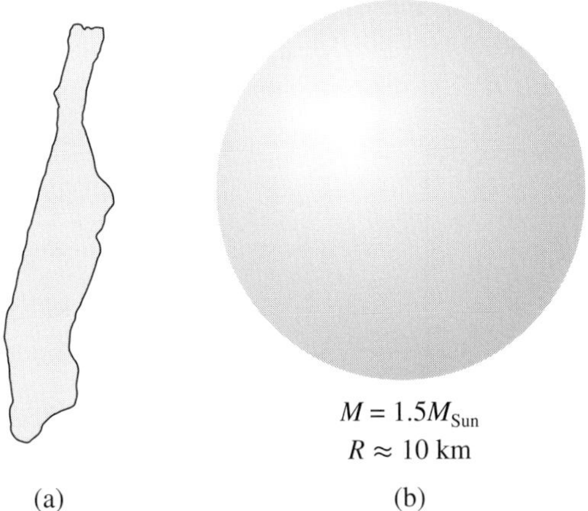

FIGURE 18.4 Manhattan (a) compared with a neutron star (b).

star, using the Newtonian formula (equation 3.62), is

$$v_{\text{esc}} = \left(\frac{2GM_{\text{ns}}}{R_{\text{ns}}}\right)^{1/2} \approx 2 \times 10^8 \text{ m s}^{-1} \left(\frac{M_{\text{ns}}}{1.4M_\odot}\right)^{2/3}. \tag{18.42}$$

The escape speed from a neutron star is a significant fraction of the speed of light:

$$v_{\text{esc}}/c \approx 0.6 \left(\frac{M_{\text{ns}}}{1.4M_\odot}\right)^{2/3}. \tag{18.43}$$

If we crave accuracy, we really should be using general relativity in the vicinity of a neutron star, rather than classical Newtonian dynamics.

Since neutron stars are tiny, you might think they'd be difficult to detect in the vast darkness of interstellar space. In fact, there's more than one way to detect a neutron star. When neutron stars are first formed, they have hot surfaces, with $T_{\text{ns}} \approx 10^6 \text{ K} \approx 170T_\odot$. The radius of the neutron star is small, with $R_{\text{ns}} \approx 11 \text{ km} \approx 1.6 \times 10^{-5} R_\odot$. The blackbody luminosity of the neutron star is then

$$L_{\text{ns}} = \left(\frac{R_{\text{ns}}}{R_\odot}\right)^2 \left(\frac{T_{\text{ns}}}{T_\odot}\right)^4 L_\odot \approx (1.6 \times 10^{-5})^2 (170)^4 L_\odot \approx 0.2L_\odot. \tag{18.44}$$

A neutron star thus has a respectable luminosity. Its wavelength of maximum emission, given $T_{\text{ns}} \approx 10^6$ K, is $\lambda_{\text{p}} \approx 30$ Å, corresponding to a photon energy of $E \sim 400$ eV in the X-ray range of the spectrum. Thus, isolated neutron stars can be detected by X-ray satellites. Figure 18.5 shows the spectrum of a relatively nearby neutron star ($d \sim 120$ pc) observed by the *Chandra X-ray Observatory*.

Neutron stars are capable of creating photons in ways other than simple blackbody emission. Because angular momentum is conserved during core collapse, neutron stars

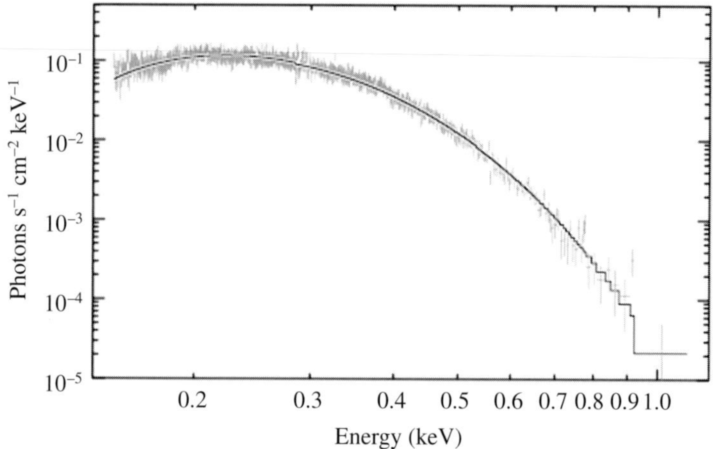

FIGURE 18.5 X-ray spectrum of the neutron star RX J1856.5-3754. The superimposed solid line represents the spectrum of a blackbody with $kT = 63$ eV.

rotate rapidly. Rotation speeds as great as $v \sim 0.1c$ are possible, with a corresponding rotation period of

$$P_{\rm ns} \sim \frac{2\pi R_{\rm ns}}{0.1c} \sim 2 \times 10^{-3}\ {\rm s}. \tag{18.45}$$

Because the magnetic flux threading through the core is also conserved during collapse, neutron stars have strong magnetic fields. Compared to the magnetic field strength of $B_\odot \approx 10^{-4}$ Tesla at the Sun's surface, a neutron star can have $B_{\rm ns} \approx 10^6$ Tesla. Rapidly rotating, highly magnetized neutron stars are **pulsars**. The name "pulsar" was given not because they pulsate in and out like Cepheids (they don't!) but because they produce pulses of electromagnetic radiation as seen from Earth. Pulsars were first detected in 1967, during a radio survey of the sky. No one knew what they were at first.[5] It was soon realized, however, that they must be neutron stars. Some pulsars have periods as short as a millisecond, but others have gradually spun down to periods of more than a second.

A neutron star has a surface gravitational acceleration of

$$g_{\rm ns} = \frac{GM_{\rm ns}}{R_{\rm ns}^2} \sim 6 \times 10^9 g_\odot \sim 1.5 \times 10^{12}\ {\rm m\ s^{-2}}. \tag{18.46}$$

The atmosphere of the neutron star consists of ionized gas supported by ordinary thermal pressure, with a scale height

$$H_{\rm ns} = \frac{kT_{\rm ns}}{g_{\rm ns}\mu m_p} \sim 0.3\ {\rm cm}. \tag{18.47}$$

[5] The first pulsar discovered was given the half-joking name "LGM-1," standing for **L**ittle **G**reen **M**en.

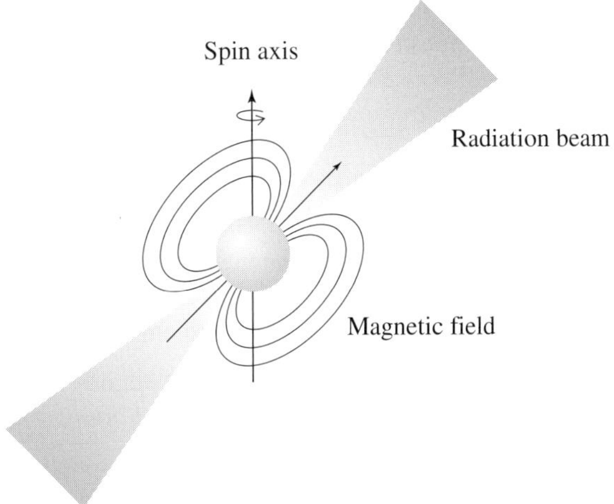

FIGURE 18.6 "Lighthouse model" of a pulsar.

The rotating magnetic field of the neutron star creates a strong electric field. This field is capable of ripping electrons and ions out of the neutron star's atmosphere and accelerating them along magnetic field lines, where they produce copious synchrotron radiation. The synchrotron emission tends to be beamed along the two magnetic axes of the neutron star (Figure 18.6). Since the magnetic axis isn't aligned perfectly with the rotation axis, the beams of synchrotron emission sweep around and around as the neutron star rotates. Neutron stars whose beams of synchrotron emission happen to sweep across the Earth's location are labeled "pulsars" by Earthlings, since we detect periodic pulses of synchrotron radiation from them. This is similar to the way in which we see pulses of light from a lighthouse whenever its rotating beams of light sweep across our location. There are about 500 pulsars known to us. However, since the synchrotron beams of pulsars are fairly narrow, we expect the number of pulsars (as seen from Earth) to be significantly smaller than the number of neutron stars.

Why are we sure that pulsars are rotating magnetized neutron stars? For one thing, we don't know of anything else that would produce strong pulses of light at such short periods. If a pulsar were actually undergoing radial pulsations with a period $P = 10^{-3}$ s, its density would have to be (see Section 17.3)

$$\langle \rho \rangle \sim \frac{1}{GP^2} \sim 10^{16}\,\text{kg m}^{-3}, \tag{18.48}$$

denser than any known white dwarf. If a pulsar were a rotating white dwarf, it would have to rotate with a speed

$$v = \frac{2\pi r}{P} \sim 100c \left(\frac{M}{0.7 M_\odot} \right)^{-1/3}. \tag{18.49}$$

A clinching piece of circumstantial evidence is that some pulsars are found within **supernova remnants**. Both neutron stars and supernova remnants are created as the consequence of core collapse in a massive star.

Let's briefly review events during core collapse:

- The iron core of a massive star grows larger than the Chandrasekhar mass. No longer supported by degenerate electron pressure, the core starts to collapse.
- Protons and electrons combine to form neutrons, and 10^{57} neutrinos are produced. The collapsing neutron sphere is actually dense enough to be opaque to neutrinos, so it takes them a while to work their way out in a random walk.
- The neutron sphere is compressed slightly beyond the point where degenerate neutron pressure balances gravity, and bounces back.
- The "core bounce" sends a shock wave through the outer layers of the star. After 30 milliseconds or so, the amount of matter swept up by the shock wave is sufficient to temporarily stall the shock's outward motion.
- The neutrinos making their way out of the neutron star interact with the dense gas in the shock wave, heating the gas and causing the shock wave to start moving outward again.
- The outer layers are then ejected at high speed (up to $\sim 15{,}000\ \mathrm{km\ s^{-1}}$, or $\sim 0.05c$).

One interesting side effect of the shock wave is that it produces small amounts of elements heavier than iron in the shock wave. (It takes energy to create these ultramassive elements, but the supernova has energy to spare.) Another interesting side effect is that the expanding gas becomes hot and emits lots of light.

The light emitted by the shock-heated gas is what we usually think of when we talk about a **supernova** (discussed further in Section 18.4). However, photons carry only a small amount of the energy associated with core collapse. Here's a more complete accounting:

- Neutrino energy $\sim 10^{46}$ J
- Kinetic energy of ejected gas $\sim 10^{44}$ J
- Photon energy $\sim 10^{42}$ J

This energy comes from the gravitational potential energy of the pre-collapse core. A neutron star has gravitational potential energy

$$U_{\rm ns} \approx -\frac{3}{5}\frac{GM_{\rm ns}^2}{R_{\rm ns}} \approx -3 \times 10^{46}\ \mathrm{J}. \tag{18.50}$$

This is about 10% of the rest energy of a neutron star, $M_{\rm ns}c^2 \approx 3 \times 10^{47}$ J; thus, compressing matter into a neutron star is a fairly efficient way of converting its mass into energy.

At maximum luminosity, the supernova emits photons with a luminosity $L_{\rm sn} \sim 10^9 L_\odot$. However, the supernova decreases in luminosity by a factor of 100 over a

FIGURE 18.7 Wall painting in Chaco Canyon, New Mexico, thought to depict the 1054 supernova next to the crescent Moon.

few months; supernovae are bright but brief-lived phenomena. An interesting tidbit of information is that the nearest star with $M > 7M_\odot$ is Betelgeuse, at $d \approx 130$ pc. When it becomes a supernova (any millennium, now!), its flux as seen from Earth will be comparable to that of the full Moon. There haven't been any Betelgeuse-scale supernovae during recorded history, but there have been some comparable in flux to Venus. On AD 1054 July 5 (in the Julian calendar), a supernova appeared in the constellation Taurus. Chinese astronomical records report that the "guest star" had a flux corresponding to $m \sim -4$ in the system of Hipparchus. The supernova was also commemorated in an Anasazi wall painting (Figure 18.7).

Today, nearly 1000 years later, when we turn our telescopes to the position recorded by the Chinese astronomers, we see the Crab Nebula, a supernova remnant (Color Figure 18). The Crab Nebula is 1.5 pc in radius, and its expansion speed, measured from Doppler shifts, is $1500\,\text{km s}^{-1}$. A little calculation soon shows that $1.5\,\text{pc} \approx (1500\,\text{km s}^{-1})(1000\,\text{yr})$. The Crab Nebula shows all the signs of being the expanding cloud of debris from an explosion about 1000 years earlier. In the center of the Crab Nebula is a pulsar. The Crab pulsar has a period of 33 milliseconds and is seen to pulsate at radio, visible, X-ray, and gamma-ray wavelengths.[6] The pulsar is in the Crab Nebula, just where you'd expect the neutron star; it's a classic "smoking gun" piece of evidence.

18.3 ▪ BLACK HOLES

Neutron stars are not the ultimate in compression for stellar remnants. There exists an upper limit to neutron star masses, the **Oppenheimer–Volkov limit**, that is analogous to the Chandrasekhar mass for white dwarfs. The upper mass limit for neutron stars is

[6] Curiously, astronomers had been observing the pulsar for years at visible wavelengths without noticing the strobe effect—the pulses were too fast.

difficult to calculate, because the strong nuclear force must be taken into account and gravity must be treated using general relativity. Although there is still some debate among neutron star mavens, an upper mass limit of

$$M_{\rm max} \approx 3M_\odot \tag{18.51}$$

for neutron stars is generally accepted. Stars with initial mass $M > 18M_\odot$ or so will leave behind remnants with $M > M_{\rm max}$. These ultramassive, ultraluminous, short-lived stars will leave behind **black holes** as their remnants.

A black hole can be defined, quite simply, as an object whose escape speed is greater than the speed of light. For a spherical body,

$$v_{\rm esc} = \left(\frac{2GM}{r}\right)^{1/2}, \tag{18.52}$$

so $v_{\rm esc} = c$ when a body of mass M has a radius

$$r = \frac{2GM}{c^2} \equiv r_{\rm Sch}. \tag{18.53}$$

The critical radius at which a mass M has an escape speed equal to the speed of light is called the **Schwarzschild radius**, $r_{\rm Sch}$, after the physicist Karl Schwarzschild, who first calculated it in a relativistically correct manner. (In general, you don't expect Newtonian calculations to give the correct results in the highly relativistic regime. In the case of computing the Schwarzschild radius, however, it works out correctly.)

Any object will become a black hole if you squeeze it until it is smaller than its Schwarzschild radius. An astronomer with $M \approx 70$ kg will become a black hole if he/she is squeezed to a radius $r_{\rm Sch} \approx 10^{-25}$ m. While it's not practical (nor in most cases desirable) to squeeze an astronomer to this submicroscopic size, it is practical to squeeze an extremely massive star down to its Schwarzschild radius. Just let gravity do the work. For a massive stellar remnant,

$$r_{\rm Sch} = \frac{2GM}{c^2} = 3\ \text{km}\left(\frac{M}{M_\odot}\right). \tag{18.54}$$

Every black hole is surrounded by an **event horizon**: a spherical surface whose circumference is equal to 2π times the Schwarzschild radius. It is possible to enter the event horizon, but it is not possible to emerge again. Nothing, not even light, travels fast enough to escape from inside the event horizon.

General relativity predicts the existence of singularities within event horizons. A singularity is a point of infinite density and infinite spacetime curvature. However, to test the predictions of general relativity, and see whether singularities really exist, you'd have to enter the event horizon of a black hole. Presumably, once there, you'd be able to discover the answer, but you wouldn't be able to communicate your results to the outside world. It would be the ultimate scientific tragedy: having a great result but being unable to publish it.

Lurid sci-fi movies sometimes regard black holes as dangerous "vacuum cleaners," sucking up everything within reach. In fact, when you are far outside the Schwarzschild radius of a black hole, its gravitational pull is just the same as you'd feel from any other object of the same mass. What makes black holes potentially dangerous is their extremely compact nature; as you approach the central singularity of a black hole, you feel stronger and stronger tidal forces, until you are eventually ripped apart. If you want to spare your friends and relatives the gory sight of your demise, you should take care to dive toward a high-mass black hole rather than a low-mass black hole. For a low-mass black hole, you will be ripped apart *before* entering the event horizon; for a high-mass black hole, you will be ripped apart *after* entering the event horizon. To see why this is so, let's review what happens as you drop (feet first) toward a black hole.

The tidal force pulling you apart will be, from Section 4.2,

$$\Delta F \approx \frac{GMm}{r^3}\ell. \tag{18.55}$$

Here m is your mass and ℓ is your height; if you're an average sort of person, $m \sim 70$ kg and $\ell \sim 1.8$ m, respectively. The black hole's mass is M, and your distance from the black hole is r. You will be torn apart when the tidal force reaches a critical value $F_{\rm rip}$. The classic studies of M. Python reveal that the force exerted by a 16-ton weight is adequate to crush a human being.[7] Since the human body is about as strong in extension as in compression, let's take $F_{\rm rip} = 16$ tons $= 32{,}000$ lb $= 1.4 \times 10^5$ N. Thus, the radius at which you'll be ripped apart is

$$r_{\rm rip} = \left(\frac{GMm\ell}{F_{\rm rip}}\right)^{1/3} \approx 480\,\text{km}\left(\frac{M}{1M_\odot}\right)^{1/3}. \tag{18.56}$$

Since the Schwarzschild radius of a black hole is

$$r_{\rm Sch} = 3\,\text{km}\left(\frac{M}{1M_\odot}\right), \tag{18.57}$$

the ratio of the "ripping radius" to the Schwarzschild radius is

$$\frac{r_{\rm rip}}{r_{\rm Sch}} \approx 160\left(\frac{M}{1M_\odot}\right)^{-2/3}. \tag{18.58}$$

You will be ripped apart exactly at the Schwarzschild radius when $M \approx 2000M_\odot$. For lower-mass black holes, $r_{\rm rip} > r_{\rm Sch}$, so you will be torn apart before having a chance to reach the event horizon. If you want to see what life is like inside an event horizon, be sure to choose a black hole with $M \gg 2000M_\odot$. (You can curl yourself into a sphere to decrease ℓ, but this delays your tidal disruption by only a small amount.)

[7] See, for example, *Monty Python's Flying Circus,* Season 1, Episode 4, "Self-defence Against Men Armed with Fruit" sketch. If more recent experiments on the compressional strength of the human body have been performed, the authors would greatly appreciate *not* hearing about them.

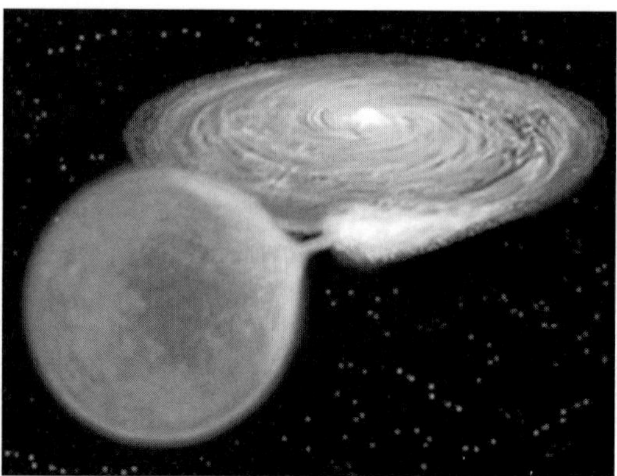

FIGURE 18.8 Artist's impression of a black hole accreting gas from a stellar companion.

If you want to dive into a black hole, you must first find a black hole. This can be difficult, since (after all) a black hole is black.[8] We can, however, detect black holes indirectly, by their gravitational effects on nearby matter. Consider, for instance, a black hole that is in a binary system with a normal star (Figure 18.8). If the normal star, as it evolves, swells until it is larger than its Roche limit (see Section 4.3.1), the tidal force exerted by the black hole will be great enough to strip away the outer layers of the star. The stripped gas, as it falls toward the black hole, will be compressed and heated to $\sim 10^6$ K or so. Since the falling gas will have some angular momentum, it will generally form an **accretion disk** around the black hole, as shown in Figure 18.8. The disk of hot gas will be detectable from the X-rays it emits.

One strong candidate for a binary system including a black hole is the X-ray source V404 Cygni. At visible wavelengths, V404 Cygni is a spectroscopic binary, with a relatively cool subgiant (spectral type K0 IV) orbiting a dark massive companion with a period of $P = 6.47$ days. If we are looking at the orbit of the subgiant edge-on, the orbital speed we deduce from its periodically varying Doppler shift is $v_c = 209\ \text{km s}^{-1}$. The minimum possible mass for the dark massive companion, given these observed parameters, is $M_{\text{bh}} = 6.1 M_\odot$, far higher than the maximum permissible mass for a neutron star. A more detailed model of the V404 Cygni system assigns a mass of $M_{\text{bh}} = 10 M_\odot$ to the black hole and of $M_{\text{star}} = 0.6 M_\odot$ to the subgiant star.

18.4 ▪ NOVAE AND SUPERNOVAE

On the evening of 1572 November 11, Tycho Brahe was contemplating the night sky when he saw what he thought was a new star in the constellation Cassiopeia. Since this

[8] The quantum effect known as Hawking radiation, which causes particles and antiparticles to be emitted from the region near the event horizon, is significant only for small black hole masses. A black hole with $M = 3M_\odot$ has a luminosity in Hawking radiation of only $L = 3 \times 10^{-22}$ W.

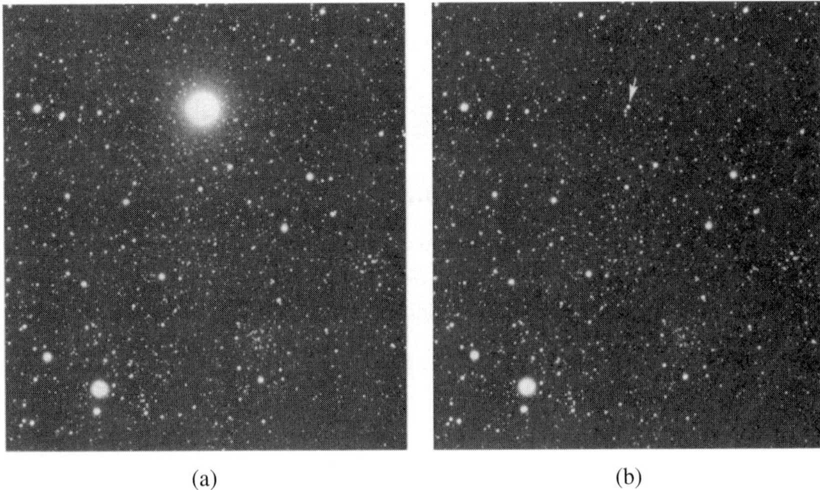

FIGURE 18.9 Nova V1500 Cygni, during its outburst (a) and after its outburst (b), when it had dimmed to $m_v \sim 15$.

was a blow to the Aristotelian dogma that no new stars can appear in the heavens, Tycho rushed into print with his new work *De Nova Stella* ("On the New Star").[9] Following the lead of Tycho, astronomers applied the term "nova" to unresolved celestial objects whose flux increases by a large amount over a short period of time, giving the impression of a new star appearing in the sky.

When a nova produces an outburst of light, its luminosity can briefly increase by a very large factor. For example, consider the nova V1500 Cygni, seen in the year 1975 (Figure 18.9). Before its outburst, V1500 Cygni was an inconspicuous object, with $m_V \approx 20$. At the peak of its outburst, it had $m_V \approx 2$, making it the second brightest source in Cygnus, after the star Deneb. At maximum, therefore, the flux of V1500 Cygni had increased by a factor $10^{0.4(20-2)} \approx 2 \times 10^6$. The sharp rise to maximum flux was followed by a gradual decline over the course of months.

The modern definition of **nova** states that a nova is a type of cataclysmic variable (as opposed to a pulsating variable like a Cepheid) in which a normal star dumps gas onto a white dwarf. Thus, a nova is actually a close binary system in which a star and a white dwarf orbit their mutual center of mass. Tossing material onto a compact object such as a white dwarf or neutron star is an excellent way to convert gravitational potential energy into thermal energy, and then into photons. However, we still need to explain why pouring gas onto a white dwarf produces sudden cataclysmic outbursts of light, rather than a steady glow.

Let's look at what happens when a star expands past its Roche limit and pours gaseous hydrogen onto its close companion, a white dwarf. (The outer layers of a star are actually

[9] More fully, *De Nova et Nullius Aevi Memoria Prius Visa Stella*, or "On the New and Never Previously Seen Star."

a mix of hydrogen and helium, with a smattering of heavier elements. For simplicity, though, let's assume it's pure hydrogen). Because the surface gravity of the white dwarf is high, the hydrogen is flattened into a layer with a short scale height (equation 18.19):

$$H = \frac{kT}{g\mu m_p} \sim 0.4\,\text{km}\left(\frac{T}{10^5\,\text{K}}\right)\left(\frac{M_{\text{wd}}}{0.7M_\odot}\right)^{-5/3}. \tag{18.59}$$

As hydrogen is poured on, as long as the atmospheric temperature T is roughly constant, the scale height H remains the same. Thus, as more and more hydrogen is poured into a layer of constant scale height, the density at the base increases steadily, until it reaches the point where the electrons become degenerate.

As the hydrogen relentlessly continues to pile on, the pressure and density at the base of the degenerate hydrogen layer must continue to increase to maintain hydrostatic equilibrium. When the density is sufficiently high, fusion of hydrogen to helium (via the proton–proton chain) begins with a "hydrogen flash," analogous to the helium flash with which solar-mass stars begin helium fusion. The material on the white dwarf's surface becomes, in effect, a hydrogen fusion bomb. A bright nova is observed to release as much as $E_{\text{nova}} \approx 10^{38}$ J of energy.

The energy released by fusing hydrogen to helium is

$$\epsilon = 6.4 \times 10^{14}\,\text{J kg}^{-1}. \tag{18.60}$$

Thus, a bright nova requires a mass of hydrogen equal to

$$M_H = \frac{E_{\text{nova}}}{\epsilon} \approx \frac{10^{38}\,\text{J}}{6.4 \times 10^{14}\,\text{J kg}^{-1}} \approx 10^{23}\,\text{kg}. \tag{18.61}$$

This is roughly the mass of the Moon, to give a comparison. The energy released by the fusion of degenerate hydrogen is sufficient to blow the outer layers of nondegenerate hydrogen into space. The nova GK Persei, for instance, had a major outburst in 1901 February; a century later, it was surrounded by an expanding gaseous nebula (Figure 18.10). Measurement of Doppler shifts tells us that the gas is expanding with a speed $v \approx 1200\,\text{km s}^{-1}$.

Since the white dwarf itself is unharmed by the nova explosion, novae tend to be recurrent phenomena; the more energetic the outburst, the longer the time until the next outburst. **Dwarf novae** have an interval between outbursts of 3 weeks to 2 years. **Recurrent novae** have an outburst every few decades. Bright novae, like V1500 Cygni, are probably recurrent phenomena as well, with outbursts every thousand to million years.

Novae are energetic events; the brightest nova outburst can produce as much light in a week as the Sun does in 10,000 years. However, novae pale in comparison with **supernovae**, which have absolute magnitudes that are at least 7 magnitudes brighter than the brightest novae. The term "super-nova" (the hyphen has since vanished) was first proposed by Fritz Zwicky and Walter Baade in 1934, when they suggested that outbursts of light much brighter than ordinary novae would be produced when the core of

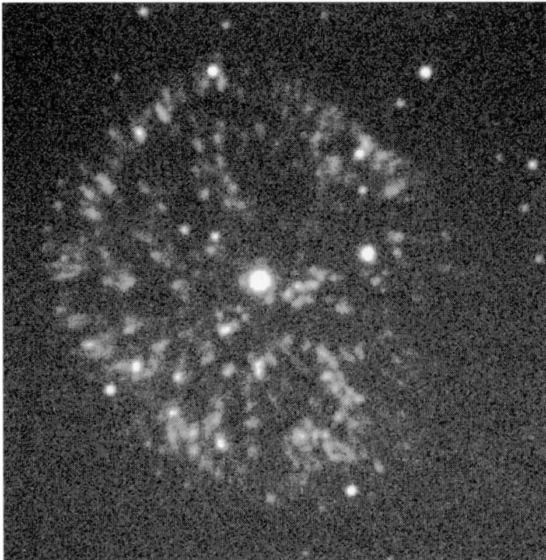

FIGURE 18.10 Expanding nebula around the nova GK Persei.

a massive star collapsed to form a neutron star.[10] Astronomers, once they acknowledged the existence of these extraordinarily luminous supernovae, classified them according to their spectra. Simply enough, they distinguished between "type I" and "type II" supernovae:

- **Type I supernovae** do not show hydrogen absorption lines, such as the Balmer lines, in their spectra.
- **Type II supernovae** do show hydrogen absorption lines.

It turned out that type II supernovae are the "core-collapse" supernovae we discussed in Section 18.2 when talking about the creation of neutron stars. They have hydrogen absorption lines because the outer layers of a massive star are typically rich in hydrogen; fusion destroys hydrogen only in the deeper, hotter layers.

Astronomers had difficulty understanding type I supernovae until they made a further subdivision into "type Ia" and "type Ib":

- **Type Ia supernovae** have neither hydrogen lines nor helium lines. After they decline from their maximum luminosity, the strongest absorption lines are of iron.
- **Type Ib supernovae** have strong helium lines and no hydrogen lines.

Type Ib supernovae (aside from their lack of hydrogen absorption lines) are very similar to type II supernovae. It is thought that type Ib supernovae are massive stars that

[10] Zwicky and Baade also coined the term "neutron star."

completely lost their hydrogen-rich outer layers in a strong stellar wind and *then* underwent core collapse. The difference between type Ib and type II supernovae is (literally) superficial; they have the same physics at heart.

A type Ia supernova is something completely different. They produce $\sim 10^{44}$ J of energy without detectable hydrogen or helium present and end up as a cloud of hot gaseous iron. To see how a type Ia supernova could be created, consider a white dwarf made of carbon that is just barely below the Chandrasekhar mass. Additional material from a companion star is then poured onto the white dwarf. As the last straw is placed on the camel's back (or less figuratively, as the last hydrogen atom is placed on the white dwarf), the white dwarf is nudged over the Chandrasekhar mass and starts to collapse. The density increases until runaway fusion of carbon begins. (Think of it as a "carbon flash.") The entire white dwarf—all $1.4M_{\odot}$ of it—becomes a fusion bomb. Suppose the initial mass of carbon is

$$M_{\rm C} = 1.4M_{\odot} = 2.8 \times 10^{30}\ {\rm kg}. \tag{18.62}$$

The energy released by fusing carbon into iron is

$$\epsilon = 1.0 \times 10^{14}\ {\rm J\ kg^{-1}}. \tag{18.63}$$

This is only 16% as much energy per kilogram as you get by fusing hydrogen into helium, but it's still a substantial energy release. The total energy released by fusing the white dwarf into iron is

$$E = \epsilon M_{\rm C} = 2.8 \times 10^{44}\ {\rm J}. \tag{18.64}$$

This is larger than the magnitude of the white dwarf's gravitational potential energy before the collapse. Thus, the entire white dwarf is blown to smithereens, and $1.4M_{\odot}$ of iron is mixed into the interstellar medium. Type Ia supernovae are a major source of iron in our galaxy. They don't produce detectable amounts of elements more massive than iron; these all come from core-collapse supernovae (types Ib and II).[11]

Tycho's "nova stella" of 1572 turns out to have been a supernova, according to our current scheme of nomenclature. When we turn our telescopes toward Cassiopeia, we see a supernova remnant at the position recorded by Tycho. The spectrum of Tycho's Supernova Remnant is rich in iron; there is no evidence for a pulsar or a central dense object of any kind. The evidence indicates that Tycho's "nova stella" was a type Ia supernova.

A brief summary of novae and supernovae:

- **Nova:** Energy source = fusion of H to He on the surface of a white dwarf. Energy $\leq 10^{38}$ J.
- **Type Ia supernova:** Energy source = fusion of C (and O) to Fe. Energy $\approx 10^{44}$ J.
- **Type II (or Ib) supernova:** Energy source = gravitational potential energy. Energy $\approx 10^{46}$ J, including neutrinos.

[11] There are alternate schemes proposed for inching a white dwarf over its Chandrasekhar mass, including the method of merging two white dwarfs.

The photon luminosity is actually slightly greater for type Ia supernovae than for type II supernovae. For type Ia supernovae, $L_{max} \approx 4 \times 10^9 L_\odot$ in the V band. Supernovae are rare events: about one per century in our galaxy, with type II supernovae slightly more common than type Ia. Unfortunately, supernovae within our galaxy are often hidden from us by dust. The most recent "naked eye" supernova was Supernova 1987a, in the Large Magellanic Cloud.

PROBLEMS

18.1 What would be the rotation period of the Sun if it collapsed to a radius $R = 6000$ km without losing angular momentum?

18.2 What is the radius of a $1.5M_\odot$ neutron star, expressed as a fraction of its Schwarzschild radius?

18.3 A star reaches "break-up" speed when it rotates so rapidly that the centrifugal acceleration at its equator equals the surface gravity g. Estimate the rotation period of a white dwarf rotating at break-up speed. (You may ignore any deformations of the white dwarf from a spherical shape.)

18.4 What is the mean density of a $1.5M_\odot$ neutron star? A carbon nucleus has a radius $r \approx 3 \times 10^{-15}$ m; what is its density? What is the ratio of the two densities?

18.5 Suppose that a supernova explosion results in the outer $4M_\odot$ of the dying star being ejected at a speed $v = 5000 \text{ km s}^{-1}$.

(a) What is the kinetic energy of the expanding ejecta?

(b) The ejecta are slowed by sweeping up the local interstellar gas. Assuming the density of the interstellar gas is $\rho = 2 \times 10^{-19} \text{ kg m}^{-3}$, how large a volume will be swept up by the time the outflow velocity has decreased to 10 km s^{-1}? (Hint: you may assume that the kinetic energy of expansion is conserved.)

18.6 Photons leaving the surface of a compact stellar remnant are gravitationally redshifted by an amount

$$\frac{\Delta\nu}{\nu_0} \approx -\frac{r_{\text{Sch}}}{2r},$$

where r_{Sch} is the Schwarzschild radius of the remnant and r is its actual radius. Calculate the gravitational redshift, in angstroms, for the Hβ λ4861 line of hydrogen from a $1M_\odot$ white dwarf. How can we distinguish this gravitational redshift from a possible Doppler shift due to the motion of the white dwarf?

18.7 On a plot of $\log T$ versus $\log \rho$, where T is temperature and ρ is mass density, plot the line along which the thermal pressure of an ideal gas is equal to the pressure provided by nonrelativistic degenerate electrons. (You may assume that all factors of order unity, including the mean molecular mass μ, are exactly equal to one.) Label the "degenerate" and "nondegenerate" regimes. Which regime does the center of the planet Jupiter ($\rho \approx 3000$ kg m^{-3}, $T \approx 40{,}000$ K) fall into?

18.8 The speed of sound can generally be written as

$$c_s = \left(\gamma \frac{P}{\rho}\right)^{1/2},$$

where P is the pressure, ρ is the mass density, and γ is the adiabatic index (a number of order unity). Show that within a white dwarf, the typical sound speed is

$$c_s \sim \left(\frac{G^2 m_e m_p^3}{\hbar^2}\right)^{1/2} \left(\frac{M}{m_p}\right)^{2/3}$$

and the sound crossing time is

$$t \sim \frac{R}{c_s} \sim \frac{\hbar^3}{G^2 m_e^2 m_p^2} \left(\frac{m_e}{m_p}\right)^{1/2} \frac{1}{M}.$$

What is the shortest possible sound crossing time for a white dwarf? How does this timescale compare to the shortest known pulsar periods?

19 Our Galaxy

Our study of stars began with a definition of the word "star." A sense of symmetry compels us to begin our study of galaxies with a definition of the word "galaxy." A galaxy is a collection of stars (between a million and a trillion of them, in round numbers), plus gas, dust, and dark matter, held together by gravity. A galaxy is bigger than a star cluster, such as the Pleiades, and smaller than a cluster of galaxies, such as the Virgo Cluster. Although there are hundreds of billions of galaxies within the volume accessible to our telescopes, we will start by looking in depth at the galaxy in which we live: the Milky Way Galaxy.

19.1 ▪ OVERVIEW: MORPHOLOGY OF OUR GALAXY

On a dark night, far from city lights, you can see a luminous band of light across the sky, forming a great circle[1] on the celestial sphere (Figure 19.1). In English, this band of light is called the **Milky Way** because it looks, to the naked eye, like a luminous white fluid. In ancient Greece, it was called the *galaktikos kuklos,* which translates literally as "milky circle." The Greek word *galaktikos* is the origin of the English word "galaxy."

Although the Milky Way looks as if someone spilled glow-in-the-dark milk across the celestial sphere, when Galileo examined it with his telescope, he found that it is actually composed of a very large number of stars, each individually very faint (see Section 2.4). A hypothesis that explains the existence of the Milky Way is that the Sun is embedded in a relatively thin disk of stars. When we look perpendicular to the disk, we see few stars, and the sky is dark. When we look in the plane of the disk, we see the many stars that make up the Milky Way. This disk of stars is a major component of the galaxy in which we live, which is therefore called the **Milky Way Galaxy**. It is also called the Galaxy (with a capital "G"), or if we're feeling particularly possessive, *our* galaxy.

The first method used to determine the size and shape of our galaxy was the method of **star counts**. To demonstrate how star counts work, let's start with some simplifying assumptions:

[1] The Milky Way is tilted by 60.2° relative to the ecliptic, and by 62.6° relative to the celestial equator.

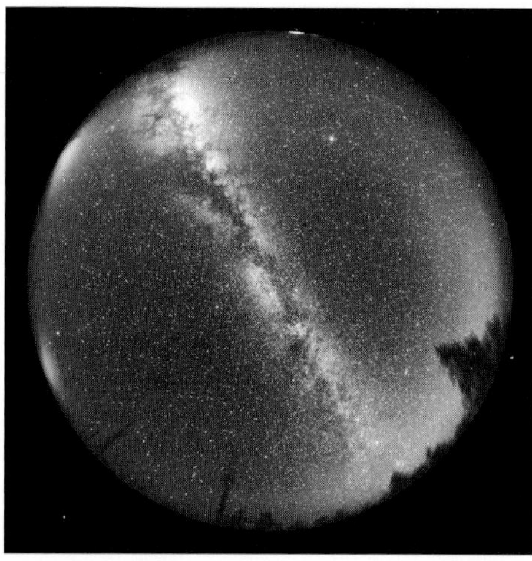

FIGURE 19.1 The Milky Way seen from Mount Graham, Arizona.

- All stars have the same absolute magnitude M. This is not true in general, but we can choose to look only at main sequence stars of a particular spectral type.
- The number density of stars, n, is constant within our galaxy.
- There is no absorption due to dust. (This is perhaps our most dubious assumption.)

A star of absolute magnitude M will have an apparent magnitude m when it is at a distance

$$d = 10^{0.2(m-M+5)}\ \text{pc}. \tag{19.1}$$

Every star closer than a distance d will be brighter than m. Thus, the total number of stars brighter than m will be

$$N(< m) = \frac{4\pi}{3} d^3 n = \frac{4\pi}{3} 10^{0.6(m-M+5)} n, \tag{19.2}$$

or, taking the logarithm,

$$\log_{10} N = 0.6m + \text{ constant}. \tag{19.3}$$

By going 1 magnitude fainter, you should increase the number of stars you see in a given patch of sky by a factor $10^{0.6} \approx 4$.

If the Galaxy were infinitely large, then as we counted stars as a function of their apparent magnitude, we would find that $\log N$ just kept increasing to arbitrarily large values of m. However, if there are no stars beyond a distance d_{max}, then there will be no stars fainter than m_{max}, where

$$m_{\text{max}} = M + 5 \log d_{\text{max}} - 5. \tag{19.4}$$

Thus, if we find m_{max} for a particular patch of sky, we can find

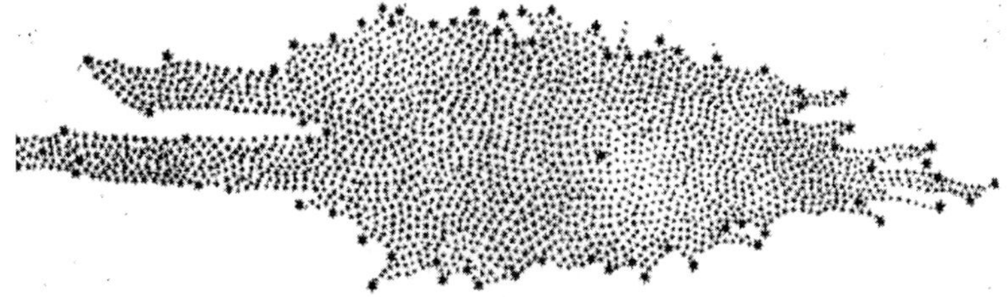

FIGURE 19.2 A cross-section through the Herschel's "grindstone" model of our galaxy. The large star slightly to right of center marks the Sun's position.

$$d_{\max} = 10^{0.2(m_{\max}-M+5)} \text{ pc} \tag{19.5}$$

in that direction. For example, suppose you are counting G0V stars in a particular small patch of sky. G0 main sequence stars have $M_v = 4.4$. The faintest G0V stars that you can find have $m_{v,\max} = 16.4$. Thus, you compute that the most distant stars in the patch are at a distance

$$d_{\max} = 10^{0.2(16.4-4.4+5)} \text{ pc} = 10^{3.4} \text{ pc} = 2500 \text{ pc}, \tag{19.6}$$

if you ignore the effects of dust.

The pioneers in using star counts to determine the shape of our galaxy were William and Caroline Herschel, the great sibling act of astronomy. In the late eighteenth century, they did star counts in various regions of the sky and came to the conclusion that our galaxy is shaped like a grindstone (Figure 19.2). (For those of you who haven't recently sharpened any knives the old-fashioned way, a grindstone is a thick disk made of coarse, abrasive stone.) Unfortunately, the Herschels were unaware of the existence of interstellar dust and didn't take into account extinction by dust. As a result, they came to the erroneous conclusion that we are near the center of a relatively small galaxy. The "notch" in the grindstone, shown on the left side of Figure 19.2, is actually the result of a dust lane down the center of the Milky Way.

A more accurate determination of our place in the Milky Way Galaxy was provided by Harlow Shapley in the early twentieth century. Shapley did it by looking away from the dust-laden Milky Way and examining the distribution of **globular clusters**. A globular cluster is a compact cluster of stars, containing as many as $\sim 10^6$ stars within a spherical region ~ 30 pc in diameter. The Hertzsprung–Russel (H–R) diagrams of globular clusters show an absence of hot main sequence stars, indicating that there are no young stars present. Typical ages estimated for globular clusters are ~ 12 Gyr. Our galaxy has about 150 globular clusters associated with it. The nearest globular cluster is M4, about ~ 2 kpc away from us (Figure 19.3); the most distant of our galaxy's globular clusters is ~ 100 kpc away. Shapley noted that the globular clusters are not uniformly distributed across the sky. Instead, they are concentrated in one-half of the celestial sphere, centered on the constellation Sagittarius.

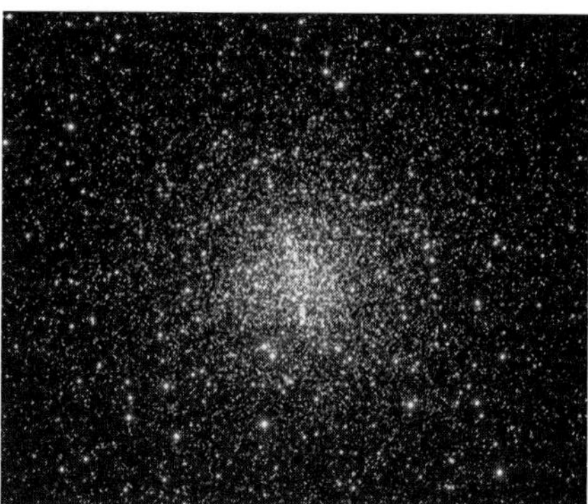

FIGURE 19.3 The globular cluster M4.

Shapley concluded that the globular clusters are all orbiting the center of our galaxy, which lies in the direction of Sagittarius. In addition, he measured the distances to globular clusters, using RR Lyrae variable stars (which all have $M_V \approx 0.5$) as standard candles. Shapley's distances were, as it turns out, too large; he thought that RR Lyrae stars were more luminous than they actually are. However, he had the right order of magnitude. Modern distance measurements tell us that the distance to the Galactic center is $R_0 = 8 \pm 1$ kpc, or about 26,000 light years.

Shapley's discovery that we are not at the center of our galaxy has been called an extension of the Copernican Revolution. Copernicus said that Earth is not at the center of the solar system. Shapley said that the solar system is not at the center of the Galaxy. (And before you ask, the Galaxy is not at the center of the universe. In fact, as far as we can tell, the universe doesn't have a center.)

In describing the shape of our galaxy, it is useful to break it down into three different components. The most luminous component of our galaxy is the **disk**. Defining the size of the disk is a bit tricky, since it doesn't have sharp edges. However, stars can be seen out to $R \sim 15$ kpc from the Galactic center; this places the Sun roughly halfway from the Galactic center to the edge of the visible disk. However, the disk of the Galaxy, when viewed in the 21 cm emission of atomic hydrogen, stretches farther, to $R \sim 25$ kpc from the Galactic center. The thickness of the disk is small compared to its radius. The vast majority of disk stars are less than 0.5 kpc from the midplane of the disk. A study of stellar properties as a function of z, the distance from the midplane, reveals that the disk can be divided into two components: a **thin disk**, containing stars of all ages (including stars that are just now forming), and a **thick disk**, made of stars older than $t \sim 5$ Gyr. In the vicinity of the Sun, the distribution of stars in both the thin disk and the thick disk falls exponentially with distance z from the midplane. For the thin disk, $n(z) = n_{\text{thin}} \exp(-|z|/h_{\text{thin}})$, with a scale height $h_{\text{thin}} \approx 0.3$ kpc. For the thick disk, $n(z) = n_{\text{thick}} \exp(-|z|/h_{\text{thick}})$, with $h_{\text{thick}} \approx 1$ kpc. In the midplane, $n_{\text{thin}} \approx 10 n_{\text{thick}}$.

The second component of our galaxy is the **bulge**. If you look at the Milky Way at infrared wavelengths, to minimize the effects of dust, you see a bulge in the direction of

Sagittarius, near the Galactic center (Color Figure 19). The central bulge of our galaxy is about 1 kpc in radius, so you can see it stick out "above" and "below" the disk. At the very center of the bulge, there is a tiny nucleus, bright at radio wavelengths.

The third component of our galaxy is the **halo**. The halo is a roughly spherical distribution of stars, about 100 kpc in radius; thus, it extends far beyond the disk. The halo has the same luminosity as the bulge, but its stars are spread over a volume $\sim (100)^3 \sim 10^6$ times larger. The three stellar components of our galaxy—disk, bulge, and halo—differ in their stellar populations as well as in their size, shape, and kinematic properties. The thin disk contains relatively young stars, for the most part, and is where most star formation is occurring today. Thin disk stars tend to be rich in "metals," with a mass fraction in metals of $Z \geq 0.01$. In the jargon of astronomers, stars that are young and metal-rich are called **population I** stars. By contrast, the halo contains stars that are relatively old and low in metals ($Z \leq 0.001$). The astronomical term for old, metal-poor stars is **population II** stars. The thick disk contains stars that are intermediate in their properties between population I and population II; thus, it is described as having an "intermediate population" of stars. The bulge contains a mixture of old and young stars, so it is a place where population I and population II coexist.[2]

Making a complete census of the stars in our galaxy is difficult because of all the dust. However, an estimate of the luminosity of different components can be made:

- Disk: $L_B = 19 \times 10^9 L_\odot$
- Bulge: $L_B = 2 \times 10^9 L_\odot$
- Halo: $L_B = 2 \times 10^9 L_\odot$
- Grand Total: $L_B = 23 \times 10^9 L_\odot$

Our galaxy's total luminosity of 23 billion solar luminosities is fairly bright, as galaxies go. If all the stars in the Milky Way Galaxy were identical to the Sun, then we'd conclude that our galaxy contains 23 billion stars. However, most stars are dim little M dwarfs that contribute little to the total luminosity (particularly in the B band). The best current estimate is that our galaxy contains 200 billion stars.

If you could view the Milky Way Galaxy from outside, oriented so that the disk were edge-on, it would probably look like the galaxy NGC 891 (Figure 19.4). Notice the prominent dustlane running down the middle of the galaxy. It is generally true, for disk-dominated galaxies like the Milky Way and NGC 891, that the gas and dust is confined to a much thinner disk than the stars.

If you could view the Milky Way Galaxy oriented so that the disk were face-on, it would probably look like the galaxy M83 (Figure 19.5). In this orientation, you can clearly see the most striking feature of galaxies similar to the Milky Way Galaxy; their **spiral arms**. Disk-dominated galaxies like M83 and the Milky Way Galaxy are thus

[2] Exasperating fact of the day: core collapse supernovae occur in massive, short-lived (and thus, of necessity, young) stars. This means that type II supernovae occur among population I stars. This is the sort of jargon confusion that prompts astronomers to clutch at their hair in despair.

FIGURE 19.4 NGC 891 ($d = 9.8$ Mpc), shown in the Large Binocular Telescope "first light" image.

FIGURE 19.5 The spiral galaxy M83 ($d = 4.5$ Mpc).

referred to as **spiral galaxies**.[3] Images taken at radio wavelengths indicate that the atomic and molecular gas tend to be concentrated along spiral arms, with lower-density ionized gas filling in the regions between arms. Images taken at ultraviolet and blue wavelengths indicate that the luminous, hot, and short-lived O and B stars also tend to lie along spiral arms. Thus, spiral arms are star-forming factories; they are where dense molecular clouds are converted into stars. The Sun is located in a short stub of a spiral arm, usually called

[3] The edge-on galaxy NGC 891 is part of the "spiral galaxy" class as well; we just can't see its spiral arms because it is edge-on with respect to us.

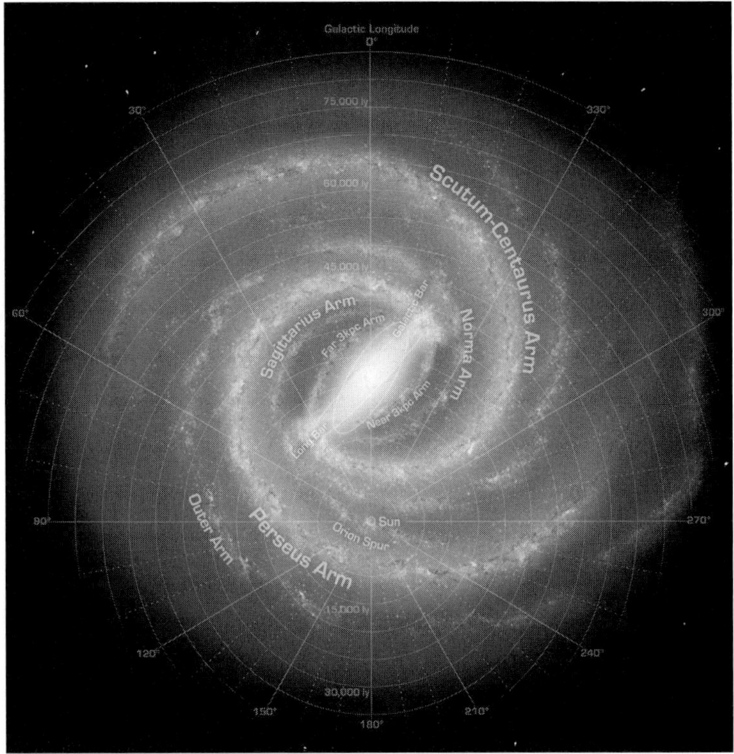

FIGURE 19.6 Artist's impression of spiral arms in our galaxy, along with its central bar.

the Orion Spur, nestled inside the Perseus Arm (Figure 19.6). The Orion Nebula, the local hotbed of star formation, is situated within the Orion Spur.

19.2 ▪ OVERVIEW: KINEMATICS AND DYNAMICS OF OUR GALAXY

An image of a spiral galaxy, like that of Figure 19.5, looks very dynamic, like a snapshot of a hurricane. In fact, if we were to see the Galaxy from the viewpoint of Figure 19.6, we would see the stars of the disk moving in a clockwise direction. That is, the disk is rotating such that the spiral arms are *trailing*. Stars in the disk are on nearly circular orbits, close to the midplane of the disk, all orbiting in the same direction around the center of the Galaxy. By contrast, stars in the halo are on elongated orbits, at random orientations relative to the disk, with some moving in the same sense as the disk stars and others moving in the opposite sense. (On much different length scales, the disk of our galaxy can be compared to the planets in the solar system, while the halo of our galaxy can be compared to the Oort Cloud.)

The Sun's orbital speed[4] around the Galactic center is estimated to be

$$v_0 = 220\ \text{km s}^{-1} = 225\ \text{kpc Gyr}^{-1}. \tag{19.7}$$

Later on, we'll discuss how the orbital speeds of the Sun and other stars are actually determined. For now, let's just consider the implications of this orbital motion. The distance of the Sun from the Galactic center is

$$R_0 = 8\ \text{kpc}. \tag{19.8}$$

If we make the approximation that the Sun is on a perfectly circular orbit, we find that its orbital period is

$$P_0 = \frac{2\pi R_0}{v_0} = \frac{50.3\ \text{kpc}}{225\ \text{kpc Gyr}^{-1}} = 0.22\ \text{Gyr}. \tag{19.9}$$

During the Sun's 4.6 Gyr lifetime, it has gone around the Galactic center just over 20 times.[5]

To find out how much mass is inside the Sun's orbit, we may use Kepler's third law:

$$M_\odot + M_G = \frac{a^3}{P^2}, \tag{19.10}$$

where M_G is the mass (measured in solar masses) inside a sphere of radius $R_0 = 8$ kpc centered on the Galactic center; $a = 8\ \text{kpc} = 1.65 \times 10^9$ AU; and $P = 2.2 \times 10^8$ yr. Strictly speaking, it must be confessed, we can only use Kepler's third law if the two bodies involved are spherical. Since the matter distribution of our galaxy is flattened, the mass M_G computed from equation (19.10) will be slightly inaccurate. Still, it will be good enough for a first estimate. Since the mass of the Galaxy is much greater than the mass of just one star, we may assume $M_G \gg M_\odot$ and

$$M_G \approx \frac{a^3}{P^2} \approx \frac{(1.65 \times 10^9)^3}{(2.2 \times 10^8)^2} M_\odot \approx 9.3 \times 10^{10} M_\odot. \tag{19.11}$$

Ninety-three billion solar masses is a lot of matter, especially when you consider that the total luminosity of our galaxy is merely 23 billion solar luminosities. The material of which our galaxy is made must have, on average, a higher mass-to-light ratio than the Sun. Moreover, the mass $M_G = 9.3 \times 10^{10} M_\odot$ contains only the mass *inside* the Sun's orbit. The mass outside the Sun's orbit has no net effect on the Sun's orbital motion (assuming that it's spherically distributed).

For stars and gas in the disk of our galaxy to be on stable circular orbits, we require

$$\frac{v(R)^2}{R} = \frac{GM(R)}{R^2}, \tag{19.12}$$

[4] Note the useful coincidence that $1\ \text{km s}^{-1} \approx 1\ \text{pc Myr}^{-1} \approx 1\ \text{kpc Gyr}^{-1}$.

[5] In some sense, the Sun is no longer a teenager.

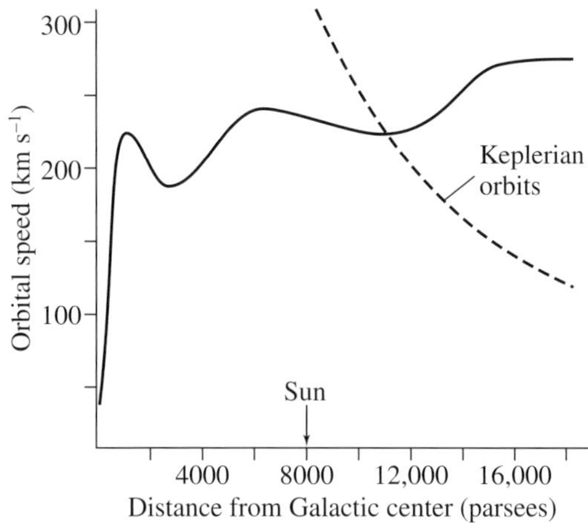

FIGURE 19.7 Rotation of our galaxy compared to a Keplerian system.

where $v(R)$ is the orbital speed of a star on an orbit of radius R, and $M(R)$ is the mass inside a sphere of radius R centered on the Galactic center. We can thus determine the mass M inside a star's orbit from its orbital speed v:

$$M(R) = \frac{v(R)^2 R}{G}. \tag{19.13}$$

Most of the Galaxy's luminosity is provided by the disk, whose brightness falls off exponentially with a scale length $R_s \approx 3\,\text{kpc}$. Thus, most of the Galaxy's luminosity lies inside the Sun's orbit. If most of the *mass* lies inside the Sun's orbit as well, we would expect $M \approx$ constant for $R > R_0$, and hence[6] $v \propto R^{-1/2}$.

Observations of stars and gas clouds in our galaxy reveal that the orbital speed v in the disk does not fall off with distance from the center (Figure 19.7). Instead, all the way to the outer fringes of the disk, the orbital speed is constant, or even slowly rising with radius. Thus, there is more mass outside the Sun's radius than inside. If we estimate from Figure 19.7 that $v = 270\,\text{km s}^{-1}$ at $R = 2R_0 = 16\,\text{kpc}$, we find that the mass inside that radius is

$$M(2R_0) \approx \frac{(2.7 \times 10^5\,\text{m s}^{-1})^2 (4.94 \times 10^{20}\,\text{m})}{6.67 \times 10^{-11}\,\text{m}^3\,\text{s}^{-2}\,\text{kg}^{-1}} \tag{19.14}$$

$$\approx 5.4 \times 10^{41}\,\text{kg} \approx 2.7 \times 10^{11} M_\odot. \tag{19.15}$$

[6] A system that has $v \propto R^{-1/2}$ is called a **Keplerian** system, because it obeys Kepler's third law. The solar system is a Keplerian system since nearly all its mass is concentrated in the Sun.

Since v is roughly constant outside the Sun's orbital radius, the mass must increase as $M \propto R$, implying an average mass density $\rho \propto R^{-2}$ in a region where the luminosity density is plummeting exponentially.

The leading explanation is that there must be **dark matter** in the outer regions of our galaxy. The phrase "dark matter" is the term used by astronomers to refer to matter that is too dim to be detected using current technology.[7] The obvious question to pose is What's the (dark) matter? It's hard to determine what something is made of when you can't see it. In recent years, there have been three major candidates to play the role of dark matter.

The first candidate is the **neutrino**. A neutrino can interact with other particles only via gravity or the weak nuclear force. The weak nuclear force is weak indeed; the overwhelming majority of solar neutrinos zip through the Sun as if it weren't there. Since neutrinos snub photons just as they snub other particles, they are a possible candidate for the dark matter, if they have enough mass. As mentioned in Section 15.4, recent experiments indicate that the three flavors of neutrino—electron, muon, and tau—have masses that differ from each other. Although the exact mass of each flavor has not been determined, there are fairly strict upper limits on their mass from various experiments. Even though neutrinos are very common particles, their low masses mean that they contribute at most a few percent of the dark matter present in the universe.

The second candidate is the **WIMP**. The term WIMP is an acronym for **w**eakly **i**nteracting **m**assive **p**article. Supersymmetric extensions to the Standard Model of particle physics predict massive particles that interact only through the weak nuclear force (and through gravity, of course). Think of them as the obese cousins of neutrinos. Particle physicists give these hypothetical particles names like photinos, gravitinos, axinos, sneutrinos, and gluinos. However, since they are massive and weakly interacting, astronomers lump them all together under the generic label of weakly interacting massive particles.

Although WIMPs have been searched for in particle accelerator experiments, they have not yet been found. Of course, since their predicted rest mass is quite large (typical numbers are ~ 100 times the mass of the proton), you wouldn't expect to produce them in the current generation of accelerators. However, the Large Hadron Collider at CERN is expected to search for supersymmetric particles with thousands of times the mass of the proton. Other experiments are searching for WIMPs in the same way you detect neutrinos, by building a really big detector and waiting for those very rare interactions mediated by the weak nuclear force. So far, there exist only upper limits on the WIMP interaction rate. But the search goes on . . .

The third candidate is something completely different: the **MACHO**. The term MACHO is a slightly strained acronym for **ma**ssive **c**ompact **h**alo **o**bject.[8] MACHOs are dim, dense objects with masses comparable to, or somewhat less than, the mass of the Sun. Brown dwarfs, old cold white dwarfs, neutron stars, and black holes can all be MACHOS if they are located in the halo of our galaxy.

[7] Dark matter might also be called "invisible matter," or maybe "transparent matter," but dark matter is the name that has stuck.

[8] The term MACHO was first devised as a spoof of the term WIMP.

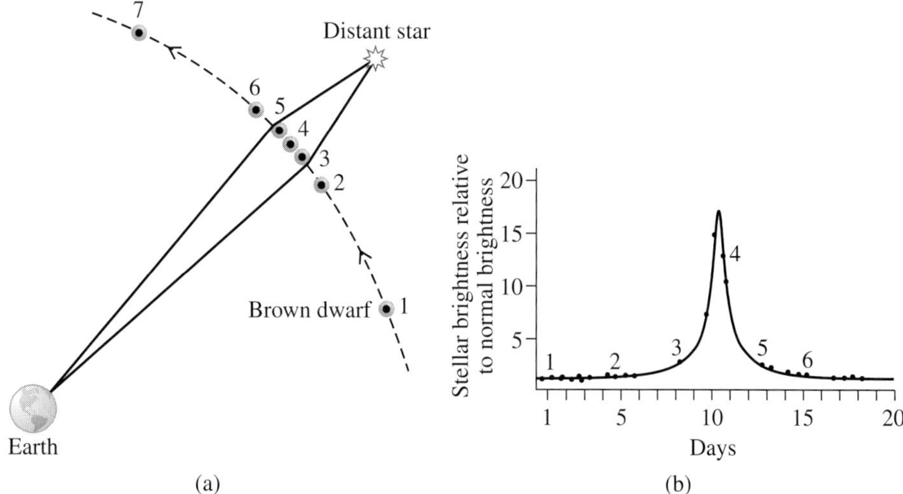

FIGURE 19.8 (a) Gravitational lensing of a distant star by a MACHO. (b) The resulting light curve of the star.

MACHOs, like any massive compact object, can act as **gravitational lenses**. One of the predictions of the Theory of General Relativity is that massive objects can bend the path of light. An early experimental support for General Relativity came in the year 1919, when observations of stellar positions during a solar eclipse revealed that the Sun had bent the path of the starlight, by an angle consistent with the predictions of Einstein.

Because massive compact objects bend light, they can act as lenses, making distant stars appear higher in flux than they ordinarily would. Suppose that a compact object (the **lens**) moves directly between us and a distant star (the **source**), as shown in Figure 19.8a. As the lens moves toward the source, as seen from our viewpoint, the flux of light we receive from the source grows larger. As the lens moves away again, the flux decreases to its original value. Thus, the light curve of the source, as shown in Figure 19.8b, shows a characteristic rise and fall. If the lens is a stellar-mass MACHO, and the source is a star in the Magellanic Clouds, for example, then the rise and fall typically occur over the course of a few weeks.[9] Research groups have carefully monitored the brightness of stars in the Magellanic Clouds, hoping to catch MACHOs in the act of lensing. Although lensing events have been seen, there turn out to be fewer MACHOs than are needed to contribute all the dark matter. Only $\sim 20\%$ of the dark matter in the halo can consist of MACHOs. There's still plenty of room for WIMPs in the Galaxy.

[9] One scientifically exciting application of gravitational lensing is the search for exoplanets around lensing objects. If a planet is orbiting the lensing object, then it's possible, if the geometry is right, for the gravitational influence of the planet to perceptibly change the shape of the light curve of the lensed source. Using this technique, exoplanets less massive than Uranus and Neptune have been detected.

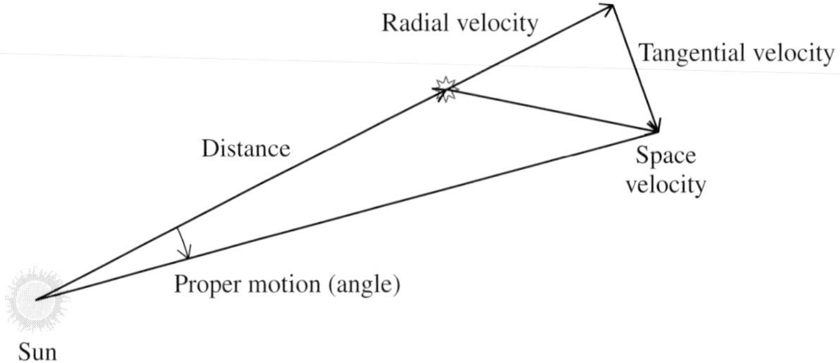

FIGURE 19.9 Components of a star's velocity relative to the Sun.

19.3 ▪ LOCAL STELLAR MOTIONS

Thus far, we have examined the global picture of our galaxy, outlining its general size, shape, mass, and composition. In this section, however, we will be thinking locally rather than globally, focusing on the motion of stars in the solar neighborhood, within 5 parsecs of the Sun. The location of these stars, relative to the Sun, is fairly easy to determine, since their parallaxes can be measured accurately. It is also relatively easy to determine their *velocity* relative to the Sun. Consider, for instance, the radial velocity v_r of a star relative to the Sun—that's just the rate at which the distance between the star and the Sun is changing (Figure 19.9). The radial velocity relative to your telescope can be found from the Doppler shift of the star's absorption lines:

$$v_r = \frac{\Delta\lambda}{\lambda} c. \tag{19.16}$$

It is important to correct the measured radial velocity for the Earth's orbital motion around the Sun ($v_{\rm orb} \approx 30\ {\rm km\ s^{-1}}$). If you are striving for high accuracy, you must also correct for the Earth's rotation speed at the location of your telescope ($v_{\rm rot} \leq 0.5\ {\rm km\ s^{-1}}$). If the star you are observing is part of a spectroscopic binary system, you can separate the radial velocity of the star relative to the center of mass and the radial velocity of the center of mass itself. This can be done by averaging the radial velocity of the star over an entire orbital period.

The radial velocity of 40 stellar systems within 5 parsecs of the Sun is plotted in Figure 19.10. Notice a few interesting results:

- The radial velocities show approximately equal numbers of blueshifts ($v_r < 0$) and redshifts ($v_r > 0$).
- There is one notable outlier on the plot: Kapteyn's star, at a distance $d \approx 3.9$ pc, which is moving away from the Sun with $v_r \approx 250\ {\rm km\ s^{-1}}$.
- If Kapteyn's star is left out of the sample, the root mean square radial velocity of nearby stars (relative to the Sun) is $v_r \sim 35\ {\rm km\ s^{-1}}$.

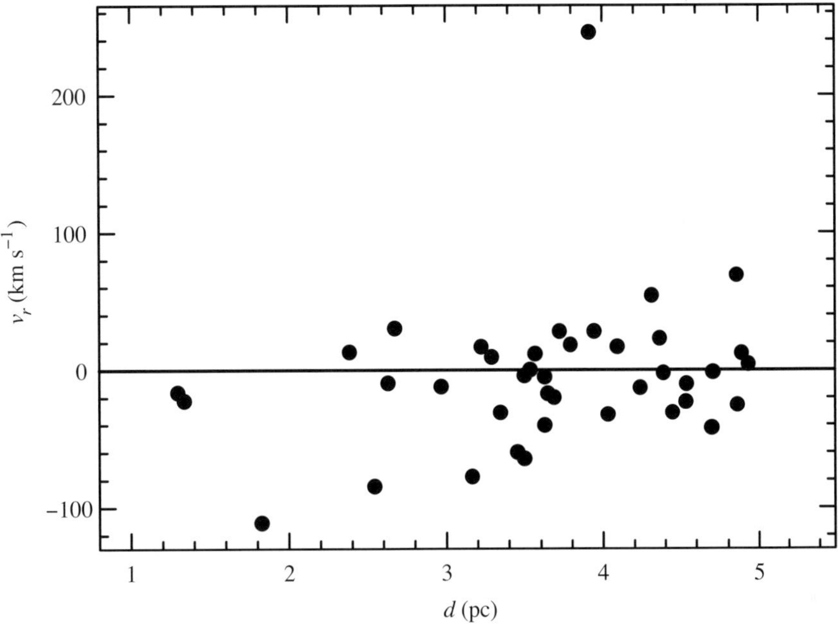

FIGURE 19.10 Radial velocity of stars within 5 pc of the Sun.

Kapteyn's star is distinctly different in its kinematic properties from the other neighborhood stars because it belongs to the *halo*, not the disk. The Sun, and its neighbors in the disk, are orbiting the Galactic center at $\sim 220\ \text{km s}^{-1}$, passing by Kapteyn's star, which is on a nearly radial orbit. Halo stars can be recognized by their high velocity relative to the Sun and by their low metallicity.[10]

The radial velocity v_r gives you just one component of the star's three-dimensional velocity. To know completely the star's velocity through space, you must also determine the star's tangential velocity v_t, the component of the velocity perpendicular to the Sun—star line (see Figure 19.9). In the nonrelativistic limit, the tangential velocity doesn't produce a Doppler shift. However, the tangential velocity can be determined indirectly because it produces a **proper motion** μ, which is the rate of change of the star's angular position on the celestial sphere. In the small angle limit,

$$\mu = \frac{v_t}{d}, \tag{19.17}$$

where μ is in radians per year, v_t is in parsecs per year, and d is in parsecs. In measuring μ, you must correct for the elliptical motion due to parallax, and for the orbital motion

[10] Kapteyn's star is an M subdwarf with a metallicity 1/7 that of the Sun. Halo stars in the solar neighborhood are often referred to as **high-velocity stars** because of their high speeds relative to most of the nearby stars, which are participating in the rotation of the Galactic disk. In a reference frame that is *not* corotating with our line-of-sight to the Galactic center, the halo stars are in fact low-velocity stars.

of the star, if it's in a binary system. For instance, Barnard's star, an M dwarf with $m_V = 9.6$ mag, has the largest proper motion of any star in the night sky. Barnard's star has a parallax of $\pi'' = 0.547$ arcsec, implying a distance[11] of only $d = 1.83$ pc. The proper motion of Barnard's star is $\mu = 10.358$ arcsec yr^{-1}. This is a huge proper motion by stellar standards; it would take Barnard's star less than two centuries to cross an angular distance equal to the width of the full Moon.

Knowing the distance d and proper motion μ, we can compute the tangential velocity from equation (19.17):

$$\frac{v_t}{\text{pc yr}^{-1}} = \left(\frac{d}{\text{pc}}\right)\left(\frac{\mu}{\text{rad yr}^{-1}}\right). \tag{19.18}$$

Among astronomers, however, the preferred unit of speed is not the parsec per year but the kilometer per second. The preferred unit of proper motion is not the radian per year but the arcsecond per year. We can make the translation to astronomer-approved units by noting that

$$1 \text{ rad yr}^{-1} = 206{,}265 \text{ arcsec yr}^{-1} \tag{19.19}$$

and that

$$1 \text{ pc yr}^{-1} = \frac{3.086 \times 10^{13} \text{ km}}{3.16 \times 10^{7} \text{ s}} = 9.77 \times 10^{5} \text{ km s}^{-1}. \tag{19.20}$$

Thus,

$$\begin{aligned} &\left(\frac{1 \text{ pc yr}^{-1}}{9.77 \times 10^{5} \text{ km s}^{-1}}\right)\left(\frac{v_t}{\text{pc yr}^{-1}}\right) \\ &= \left(\frac{1 \text{ rad yr}^{-1}}{2.063 \times 10^{5} \text{ arcsec yr}^{-1}}\right)\left(\frac{\mu}{\text{rad yr}^{-1}}\right)\left(\frac{d}{\text{pc}}\right). \end{aligned} \tag{19.21}$$

This means that in the preferred units,

$$\frac{v_t}{\text{km s}^{-1}} = 4.74 \left(\frac{d}{\text{pc}}\right)\left(\frac{\mu}{\text{arcsec yr}^{-1}}\right). \tag{19.22}$$

If the parallax π'' is measured in arcseconds, and the proper motion μ'' is measured in arcseconds per year, the tangential velocity of a star is

$$v_t = 4.74 \left(\frac{\mu''}{\pi''}\right) \text{ km s}^{-1}. \tag{19.23}$$

For Barnard's star, as an example,

$$v_t = 4.74 \left(\frac{10.358}{0.547}\right) \text{ km s}^{-1} = 89.8 \text{ km s}^{-1}. \tag{19.24}$$

[11] Barnard's star is the fifth closest star to the Earth, after the Sun, Proxima Centauri, and Alpha Centauri A and B.

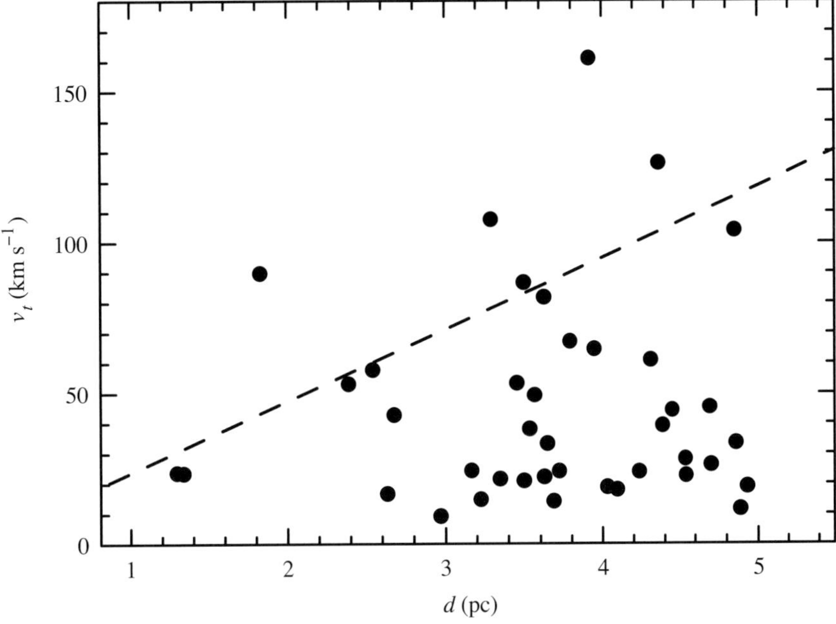

FIGURE 19.11 Tangential velocity of stars within 5 pc of the Sun. Stars above the dashed line have $\mu'' > 5$ arcsec yr^{-1}.

The tangential velocity of stellar systems within 5 parsecs of the Sun is shown in Figure 19.11. The high proper motion of Barnard's star is due both to the fact that it is exceptionally close to us and to the fact that its tangential velocity is higher than the average for nearby stars.

If the average tangential velocity v_t doesn't vary with distance from the Sun, then *on average*, nearby stars will have a higher proper motion than more distant stars. One way to search for nearby stars, if you don't have the time or patience to measure parallaxes for every star in the sky, is to start by looking at stars with high proper motion. The history books state that in the year 1838, Friedrich Wilhelm Bessel measured the parallax of 61 Cygni. What they sometimes don't tell you is why Bessel chose that star to observe: it is, after all, a humble 5th magnitude star. Bessel, in fact, chose 61 Cygni because it has the highest proper motion of any star visible to the naked eye,[12] with $\mu'' = 5.2$ arcsec yr^{-1}.

Of course, any star, no matter its distance, will have zero proper motion if it's moving straight toward us or straight away from us. An example of a nearby star with small proper motion is Gliese 710, which has $v_r = -13.9$ km s^{-1} and $\mu'' = 0.014$ arcsec yr^{-1}.

[12] The unusual nature of Kapteyn's star was first recognized in the year 1897, when this otherwise unobtrusive M dwarf was found to have a proper motion of nearly 9 arcseconds per year. At the time, this was the largest stellar proper motion known, surpassed only when E. E. Barnard measured the proper motion of Barnard's star in 1916.

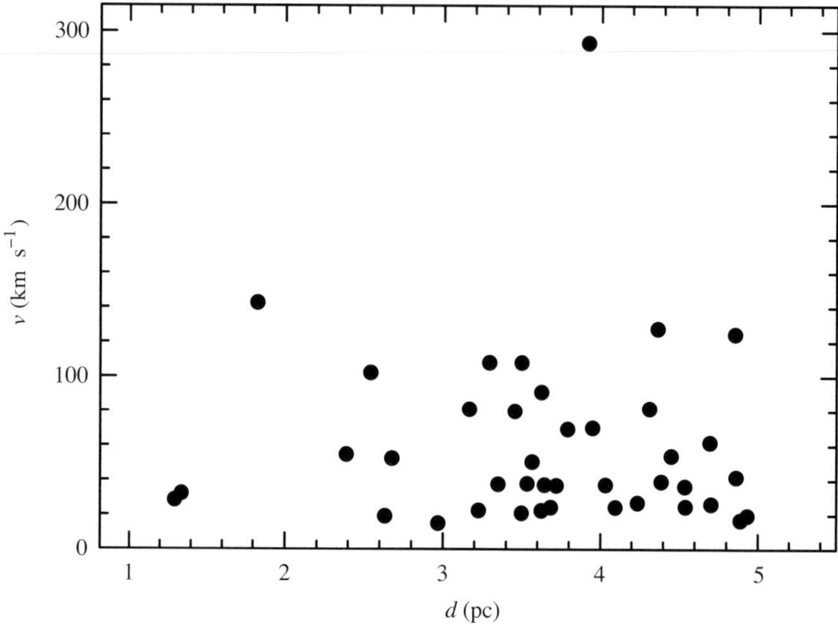

FIGURE 19.12 Space motion of stars within 5 pc of the Sun. The outlier at $d \approx 3.9$ pc, $v \approx 300$ km s^{-1} is Kapteyn's star.

At the moment, Gliese 710 is over 19 pc away from us, and has $m_V = 9.7$ mag. In 1.4 Myr, however, it will be a mere 0.34 pc away (about a fourth the present distance to Proxima Centauri) and have $m_V = 0.9$.

Putting it all together, the total speed relative to the Sun,

$$v = (v_r^2 + v_t^2)^{1/2}, \tag{19.25}$$

is called the **space motion**, or **space velocity**. A plot of the space motion for nearby stars (Figure 19.12) reveals that Kapteyn's star again stands out like the proverbial sore thumb. Its space motion of $v \approx 300$ km s^{-1} is twice that of Barnard's star, the next speediest star in the solar neighborhood. The average space motion of stars within 5 parsecs of the Sun is $v \sim 50$ km s$^{-1} \sim 50$ pc Myr^{-1}. This indicates that the list of stars within 5 parsecs of us will be thoroughly revised on timescales $t \sim 5\,\text{pc}/50\,\text{pc Myr}^{-1} \sim 0.1$ Myr.

19.4 ▪ THE LOCAL STANDARD OF REST

If the disk of our galaxy were perfectly orderly, with all the stars on exactly circular orbits in the same plane, it would be simple to compute the expected velocity of stars relative to the Sun. However, the Sun and the other disk stars are not on perfectly circular orbits. A typical orbit is neither circular nor elliptical, but forms a complicated "rosette" pattern,

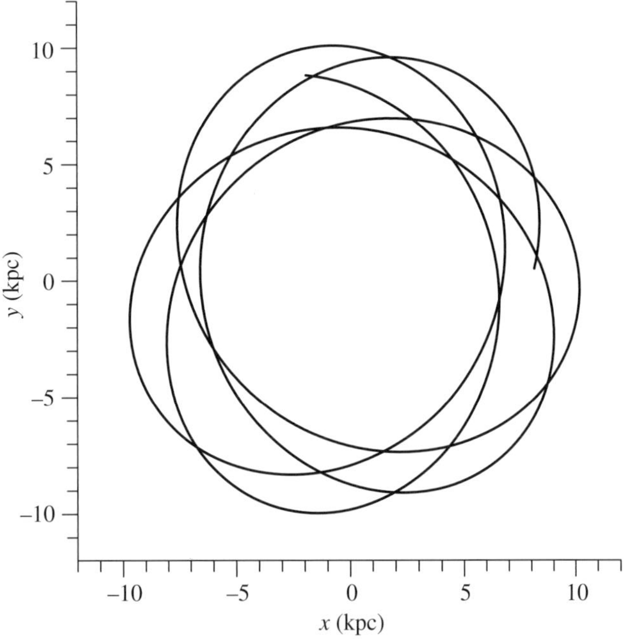

FIGURE 19.13 Orbit for a disk star in the solar neighborhood; the orbit has been integrated for a time $t = 2$ Gyr.

as shown in Figure 19.13. Since the Sun is on a complicated noncircular orbit, using it as the origin for our reference frame makes the mathematics unnecessarily complicated.[13]

It is mathematically convenient to use an idealized reference frame for our study of motion within the Galaxy. This reference frame is called the **Local Standard of Rest**, or LSR. The LSR has its origin at the Sun's location ($R_0 = 8$ kpc) and is moving in a circular orbit with $v_0 = 220\ \text{km s}^{-1}$. In other words, the LSR is doing what the Sun would be doing *if* it were on a perfectly circular orbit.

The Sun is moving with respect to the LSR at a speed of $\sim 20\ \text{km s}^{-1}$. How can we tell? In the solar neighborhood, the circular speed

$$v_c(R) \equiv \left(\frac{GM(R)}{R}\right)^{1/2} \approx 220\ \text{km s}^{-1} \qquad (19.26)$$

doesn't depend strongly on R, as shown in Figure 19.7. Thus, if the Sun were moving at the same velocity as the LSR, the average radial velocity of nearby stars ($d \ll R_0$) would be zero in all directions.

[13] Poor Ptolemy . . . since the Earth is on a noncircular orbit around the Sun, using it as the origin for his reference frame made the Ptolemaic model for the solar system unnecessarily complicated. Think of all those equants, epicycles, and deferents!

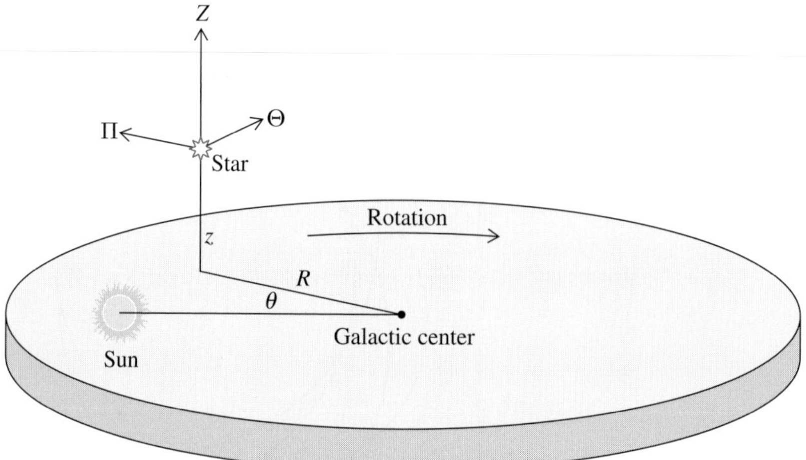

FIGURE 19.14 Cylindrical coordinates for position and velocity.

Now suppose that the Sun is moving at a velocity $\Delta\vec{v}$ relative to the LSR. Stars in the direction of the Sun's motion will be blueshifted, with $v_r = -\Delta\vec{v}$ on average (the Sun will be overtaking them). Stars opposite the direction of the Sun's motion with be redshifted, with $v_r = +\Delta\vec{v}$ on average (the Sun will be pulling away from them). The point toward which the Sun is moving, *relative to the LSR*, is called the **apex**; the opposite point on the celestial sphere is called the **antapex**. Statistical analysis of the radial velocity of nearby stars reveals that the apex of the Sun's motion is in the constellation Hercules, and the antapex is in the constellation Columba.

In the disk of our galaxy, it's convenient to use cylindrical coordinates, rather than Cartesian or spherical (Figure 19.14). The cylindrical coordinates (R, θ, z) are chosen such that $R = 0$ and $z = 0$ at the Galactic center. The azimuthal coordinate is $\theta = 0$ at the Sun's location and increases in the direction of the LSR's direction of motion. The z coordinate increases as you go north.[14] The location of the Sun in this cylindrical coordinate system is $(R_0, \theta_0, z_0) = (8\text{ kpc}, 0, 0\text{ kpc})$. The velocities in the three directions are as follows:

- Π = speed in the R direction (positive away from the Galactic center)
- Θ = speed in the θ direction (positive in the direction of motion of the LSR)
- Z = speed in the z direction (positive toward the north galactic pole)

The velocity of the LSR is then

$$(\Pi_0, \Theta_0, Z_0) = (0, 220\text{ km s}^{-1}, 0). \tag{19.27}$$

[14] The Milky Way divides the celestial sphere into two hemispheres. The hemisphere that happens to contain the north celestial pole is called the northern galactic hemisphere; its central point is the north galactic pole.

The velocity of the Sun relative to the LSR is

$$(\Pi_\odot - \Pi_0, \Theta_\odot - \Theta_0, Z_\odot - Z_0) \\ = (-10.4\,\mathrm{km\,s^{-1}}, 14.8\,\mathrm{km\,s^{-1}}, 7.3\,\mathrm{km\,s^{-1}}). \tag{19.28}$$

The Sun is currently moving *inward* (toward the Galactic center). The Sun is moving *forward* (faster than the LSR in the azimuthal direction). The Sun is moving *northward* (toward the north galactic pole). The net motion is $\Delta v = 19.5\,\mathrm{km\,s^{-1}}$ in the direction of Hercules, about 23° north of the Milky Way.

The motions of any arbitrary star relative to the LSR is referred to as the star's **peculiar velocity**,[15] which has components

$$u_* = \Pi_* - \Pi_0 \\ v_* = \Theta_* - \Theta_0 \\ w_* = Z_* - Z_0. \tag{19.29}$$

The LSR is itself moving at $\Theta_0 = 220\,\mathrm{km\,s^{-1}}$ in the direction of Cygnus, 90° away from the Galactic center in Sagittarius. Sometimes the Sun's extra velocity Δv, which is $< 9\%$ of the speed of the LSR, is small enough to be ignored, and we can pretend the Sun is on a circular orbit. At other times, it must be taken into account.[16]

19.5 ▪ DIFFERENTIAL ROTATION OF OUR GALAXY

Since we know the Sun's velocity relative to the Local Standard of Rest, if we measure the velocity of a star relative to the Sun, it is a straightforward piece of vector algebra to convert it into a velocity relative to the LSR. This is useful because the analysis of stellar velocities relative to the LSR tells us about the rotation of our galaxy. The orbital speed of a star on a circular orbit is

$$\Theta(R) = \left(\frac{GM(R)}{R}\right)^{1/2}. \tag{19.30}$$

The orbital speed Θ can be converted to an **angular velocity** ω:

$$\omega(R) \equiv \frac{\Theta(R)}{R}. \tag{19.31}$$

At the Sun's location, the angular velocity is

$$\omega_0 = \frac{\Theta_0}{R_0} = \frac{220\,\mathrm{km\,s^{-1}}}{8\,\mathrm{kpc}} = 27.5\,\mathrm{km\,s^{-1}\,kpc^{-1}}. \tag{19.32}$$

In other units, this becomes $0.028\,\mathrm{rad\,Myr^{-1}}$, or $5.8 \times 10^{-3}\,\mathrm{arcsec\,yr^{-1}}$.

[15] The word "peculiar" is used in the archaic sense of the velocity belonging to that particular star.

[16] Similarly, sometimes we can pretend the Earth's orbit is circular; sometimes we must take its eccentricity into account. It's all a matter of how accurate you need to be for a given problem.

There are different types of rotation, some of which we have already encountered:

- **Keplerian rotation**, in which all the mass is concentrated at the center of a system. $M =$ constant, $\Theta \propto R^{-1/2}$, $\omega \propto R^{-3/2}$.
- **Constant orbital speed**, a fair approximation for most of our galaxy. $\Theta =$ constant, $M \propto R$, $\omega \propto R^{-1}$.
- **Rigid-body rotation**, seen, for instance, in a rotating wheel. $\omega =$ constant, $\Theta \propto R$, $M \propto R^3$.

If the disk of our galaxy were in rigid-body rotation, then stars would have the same orbital period, $P = 2\pi/\omega$, regardless of distance from the Galactic center. However, the disk is actually in **differential rotation**, with ω decreasing with radius. We expect stars closer to the Galactic center to be passing us, while stars farther from the Galactic center fall behind.

Let's now reconstruct the analysis of the Galaxy's differential rotation performed by Jan Oort in the 1920s.[17] This analysis is an exercise in trigonometry, so we should start by examining the diagram in Figure 19.15. This diagram—let's call it the Oort diagram—is central to an understanding of Galactic rotation. The Sun's location, which defines the origin of the Local Standard of Rest, is at a distance $R_0 = 8$ kpc from the Galactic center. The LSR is moving on a circular orbit with speed $\Theta_0 = 220 \text{ km s}^{-1}$. We observe a star in the disk of the Galaxy at a Galactic longitude ℓ; the Galactic longitude is just the angle between the star and the Galactic center, assuming that the star is at $z = 0$, in the midplane of the disk.[18] The star is at a distance d from us. The quantities ℓ and d are things that we can measure. The star is at a distance R from the Galactic center and is moving on a circular orbit with speed Θ.[19]

How can we determine the orbital speed $\Theta(R)$ from observations of a star? One thing we can determine from observations of the star is v_r, its radial velocity relative to the LSR. From the Oort diagram (see Figure 19.15), we see that

$$v_r = \Theta \cos\alpha - \Theta_0 \cos(90° - \ell) = \Theta \cos\alpha - \Theta_0 \sin\ell, \tag{19.33}$$

where α is the angle between the star's velocity vector and the line from the star to the Sun. We cannot measure α directly, so we must eliminate α from equation (19.33) by using trigonometry. The lines from the Sun to the star to the Galactic center define a triangle whose vertex angles equal ℓ at the Sun, $90° + \alpha$ at the star, and thus $90° - \alpha - \ell$ at the Galactic center. From the Law of Sines,

$$\frac{\sin\ell}{R} = \frac{\sin(90° + \alpha)}{R_0}, \tag{19.34}$$

[17] This is the same Oort after whom the Oort Cloud is named. The dynamics of stellar systems, such as the Galaxy, aren't that different from the dynamics of a swarm of comets, such as the Oort Cloud.

[18] The value of ℓ runs, by convention, from 0° to 360°, with $\ell = 90°$ in the direction of motion of the LSR.

[19] In general, of course, stars in the disk aren't on *perfectly* circular orbits. However, unless we goof up and accidentally look at a halo star, the approximation of a circular orbit is close enough to give useful results.

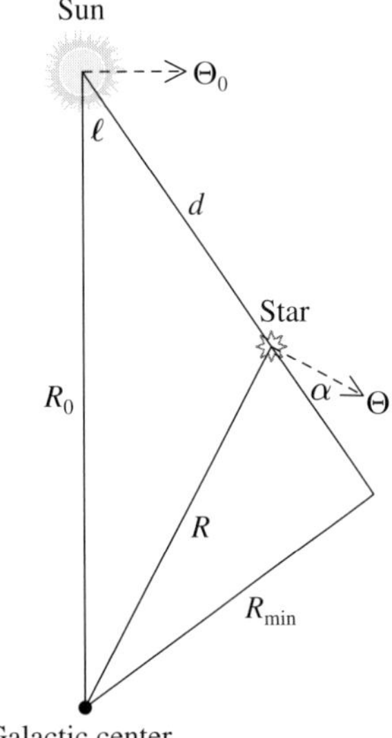

FIGURE 19.15 Oort diagram, giving the geometry of Galactic rotation.

and thus

$$\frac{\sin \ell}{R} = \frac{\cos \alpha}{R_0}. \tag{19.35}$$

By substituting equation (19.35) into equation (19.33), we find

$$v_r = \Theta \frac{R_0}{R} \sin \ell - \Theta_0 \sin \ell = \left(\frac{\Theta}{R} - \frac{\Theta_0}{R_0} \right) R_0 \sin \ell, \tag{19.36}$$

or

$$v_r = (\omega - \omega_0) R_0 \sin \ell. \tag{19.37}$$

Equation (19.37) is the **first Oort equation**, which permits you to compute ω in terms of the observables v_r and ℓ and the known values of ω_0 and R_0.[20]

Another thing we can determine from observations of the star is v_t, its tangential velocity relative to the LSR. From the Oort diagram (see Figure 19.15), we see that

$$v_t = \Theta \sin \alpha - \Theta_0 \cos \ell. \tag{19.38}$$

Once again, we need to eliminate the unmeasurable angle α. From inspection of the Oort diagram, we can deduce that

$$R_0 \cos \ell = d + R \sin \alpha \tag{19.39}$$

[20] Sanity check: for rigid-body rotation, $\omega = \omega_0$ and thus $v_r = 0$. In other words, the distance between points on a rigid wheel remains constant as the wheel spins.

and thus

$$\sin\alpha = \frac{1}{R}(R_0 \cos\ell - d). \tag{19.40}$$

By substituting equation (19.40) into equation (19.38), we find

$$v_t = \frac{\Theta}{R}R_0 \cos\ell - \frac{\Theta}{R}d - \Theta_0 \cos\ell = \left(\frac{\Theta}{R} - \frac{\Theta_0}{R_0}\right) R_0 \cos\ell - \frac{\Theta}{R}d, \tag{19.41}$$

or

$$v_t = (\omega - \omega_0)R_0 \cos\ell - \omega d. \tag{19.42}$$

Equation (19.42) is the **second Oort equation**, which permits you to compute ω in terms of the observables v_t, ℓ, and d, and the known values of ω_0 and R_0.[21]

The Oort equations in their full glory (eqs. 19.37 and 19.42) can be applied to a disk star at any distance. However, we can simplify them by considering only nearby stars, with $d \ll R_0$ (stars within 100 or 200 parsecs of us, for instance). For these nearby stars, we can expand the angular velocity $\omega(R)$ in a Taylor series around $R = R_0$:

$$\omega(R) \approx \omega(R_0) + \left.\frac{d\omega}{dR}\right|_{R=R_0}(R - R_0). \tag{19.43}$$

Thus,

$$\omega - \omega_0 \approx \left.\frac{d\omega}{dR}\right|_{R=R_0}(R - R_0), \tag{19.44}$$

and the first Oort equation becomes

$$v_r \approx R_0 \left(\frac{d\omega}{dR}\right)_{R=R_0}(R - R_0)\sin\ell. \tag{19.45}$$

For stars with $d \ll R_0$, $R - R_0 \approx -d\cos\ell$, meaning that we can write

$$v_r \approx -R_0 \left(\frac{d\omega}{dR}\right)_{R=R_0} d\cos\ell\sin\ell. \tag{19.46}$$

Using the trigonometric identity $2\cos\ell\sin\ell = \sin 2\ell$, the first Oort equation can be written in the simplified form

$$v_r \approx Ad\sin 2\ell, \tag{19.47}$$

[21] Sanity check: for rigid-body rotation, $\omega = \omega_0$ and thus $v_t = -\omega_0 d$. Imagine you are on a merry-go-round rotating counterclockwise as seen from above; if you look at someone else standing on the merry-go-round with you, he will appear to move right-to-left relative to distant background objects, with a proper motion $\mu = \omega_0$ and hence $v_t = d\mu = d\omega_0$.

where

$$A \equiv -\frac{R_0}{2}\left(\frac{d\omega}{dR}\right)_{R=R_0}. \tag{19.48}$$

The **Oort constant** A is a measurement of the shear, that is, the degree to which the disk of our galaxy does not rotate like a rigid body. Equation (19.47) implies that the radial velocity of nearby stars on circular orbits will be zero when $\ell = 0°$ and $180°$ (toward and away from the Galactic center, respectively) but also when $\ell = 90°$ and $270°$ (toward and away from the LSR's direction of motion, respectively).

A similar analysis (with details left to the reader) yields a similar simplification of the second Oort equation when $d \ll R_0$:

$$v_t \approx d(A \cos 2\ell + B), \tag{19.49}$$

where

$$B \equiv A - \omega_0 \tag{19.50}$$

is the second Oort constant.

A plot of v_r versus Galactic longitude ℓ for stars all at the same distance d from the Sun shows a sinusoidal pattern with amplitude Ad; see, for instance, the results of Figure 19.16. One recent fit to the radial velocities of Cepheids yielded $A = 14.8 \pm$

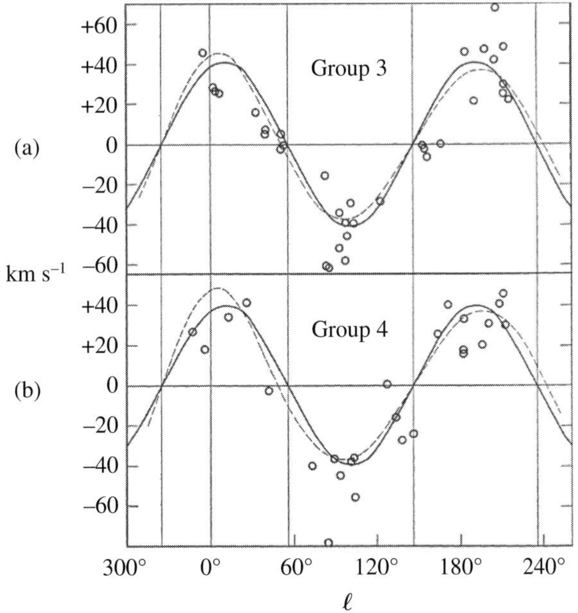

FIGURE 19.16 Radial velocity (corrected to LSR) versus ℓ for Cepheid stars. (a) Cepheids with $d \sim 1.5$ kpc. (b) Cepheids with $d \sim 3$ kpc.

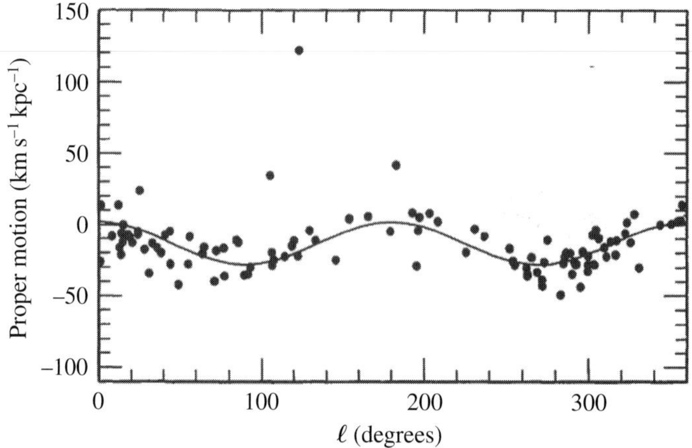

FIGURE 19.17 Proper motion of Cepheid stars within 2 kpc of the Sun.

0.8 km s^{-1} kpc^{-1}. A plot of proper motion ($\mu = v_t/d$) versus ℓ also shows a sinusoidal pattern, offset in the vertical direction by a value B (Figure 19.17). The best fit to the proper motion of Cepheids yields $B = -12.4 \pm 0.6$ km s^{-1} kpc^{-1}. Incidentally, Oort's original value for the Oort constants, found in 1927, were $A = 19 \pm 3$ km s^{-1} kpc^{-1} and $B = -24 \pm 5$ km s^{-1} kpc^{-1}. His nonzero value for A was the first definitive evidence that our galaxy is in differential rotation. Note that our best current values for the Oort constants yield a local angular velocity

$$\omega_0 = A - B = 27.2 \pm 1.0 \text{ km s}^{-1}\text{ kpc}^{-1} \tag{19.51}$$

in agreement with the value of $\omega_0 = 27.5$ km s^{-1} kpc^{-1} that we have been assuming.

19.6 ▪ DETERMINING THE ROTATION CURVE

The Oort constants A and B tell us the value of the angular velocity ω and its radial derivative $d\omega/dR$ at the Sun's location in the disk. This enables us to recreate the rotation curve of the disk in the Sun's immediate vicinity. To determine the full rotation curve for the Galaxy, we must use the Oort equations in their complete form. Suppose we observe a star at Galactic longitude ℓ and measure its distance d and its radial velocity v_r relative to the Local Standard of Rest (see Figure 19.15). From the Law of Cosines, we can determine the distance R of the star from the Galactic center:

$$R = \left(R_0^2 + d^2 - 2dR_0 \cos \ell\right)^{1/2} . \tag{19.52}$$

From the first Oort equation (eq. 19.37),

$$v_r = (\omega - \omega_0) R_0 \sin \ell , \tag{19.53}$$

we can determine the angular velocity at R:

$$\omega(R) = \omega_0 + \frac{v_r}{R_0 \sin \ell}. \tag{19.54}$$

We can then compute the orbital velocity $\Theta(R) = R\omega(R)$.

The problem with this method of determining the Galactic rotation curve (Θ as a function of R) is that if you want to find Θ for a wide range of R, you must observe stars several kiloparsecs away, near the midplane of the disk where the dust is thickest. Thus, the extinction corrections will be large, and the resulting errors in distance will be sizable. One way to pierce through the dust is to look at radio emission from gas clouds instead of visible light from stars. The 21 cm emission from atomic hydrogen and the 2.6 mm emission from carbon monoxide are largely unaffected by dust. The problem with using gas clouds as your source of emission is that determining the distance to a gas cloud is difficult. Measuring their parallax is impractical, since they are distant fuzzy blobs instead of nearby unresolved sources.[22]

Despite the difficulty in determining the distance to gas clouds, you can still derive some information from a gas cloud of known Galactic longitude ℓ and radial velocity v_r (but unknown distance d). We can start by rewriting the Law of Cosines (equation 19.52) to give the distance d in terms of R, R_0, and ℓ.

$$d = R_0 \cos \ell \pm \sqrt{R^2 - R_0^2 \sin^2 \ell}. \tag{19.55}$$

Along a line of sight with fixed Galactic longitude ℓ, every gas cloud will have $R \geq R_{\rm min}$, where $R_{\rm min} \equiv R_0 \sin \ell$ (see Figure 19.15); this assumes, implicitly, that $\sin \ell \geq 0$, which is true in the quadrants $0° < \ell < 90°$ and $270° < \ell < 360°$. The point along the line where $R = R_{\rm min}$ is called the **tangent point.**[23] For values of R in the range $R_0 > R > R_{\rm min}$, a single value of R corresponds to *two* values of d, one on the near side of the tangent point, as seen from the Sun, and the other on the far side. In general, even if we knew R for a particular gas cloud, its distance d from the Sun is still ambiguous. Is it on the near side of the tangent point, or the far side?

Cool atomic gas clouds and cold molecular clouds are sufficiently common in our galaxy that a line of sight through the disk usually passes through several of them. Since ω decreases with R in the disk, the gas cloud closest to the tangent point will have the largest angular velocity ω, and hence the largest radial velocity,

$$v_r = (\omega - \omega_0) R_{\rm min}. \tag{19.56}$$

This leads to the **tangent point method** for determining the rotation curve of the Galaxy. Start by pointing your radio telescope at a given Galactic longitude ℓ along the Milky Way. Along the chosen line of sight, the tangent point lies at a distance $R_{\rm min} = R_0 \sin \ell$

[22] If you knew the gas clouds' physical size in AU, you could measure the cloud's angular size in arcseconds, and thus compute their distance. Unfortunately, gas clouds aren't of uniform size.

[23] The tangent point is where the line of sight is *tangent* to a circle of radius $R_{\rm min}$ centered on the Galactic center.

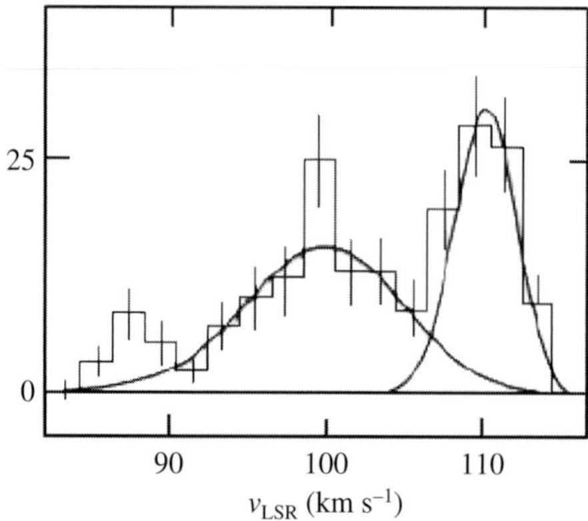

FIGURE 19.18 Carbon monoxide emission at Galactic longitude $\ell = 31.5°$.

from the Galactic center.[24] In the molecular or atomic emission lines that you observe, you will typically find several peaks; see, for instance, the CO emission line displayed in Figure 19.18. Each peak in the emission line corresponds to a different gas cloud with a different radial velocity v_r. The peak with the highest radial velocity tells you $v_{r,\max}$, the radial velocity of the most rapidly receding cloud along the line of sight. From the first Oort equation (eq. 19.37), this radial velocity corresponds to an angular velocity for the cloud of

$$\omega_{\max} = \omega_0 + \frac{v_{r,\max}}{R_{\min}}. \tag{19.57}$$

If we assume that the cloud is located *exactly* at the tangent point, then the angular velocity at the tangent point is

$$\omega(R_{\min}) = \omega_{\max} = \omega_0 + \frac{v_{r,\max}}{R_{\min}}, \tag{19.58}$$

and the orbital velocity at the tangent point is

$$\Theta(R_{\min}) = R_{\min}\omega(R_{\min}) = \omega_0 R_{\min} + v_{r,\max}. \tag{19.59}$$

Note the error built into this method; if the cloud isn't exactly at the tangent point, both R and Θ for the cloud will be *larger* than we have computed.

As an example, let's use the data presented in Figure 19.18, the result of observing CO emission at the Galactic longitude $\ell = 31.5°$, in the constellation Aquila. Along this line of sight, the tangent point is at a distance $R_{\min} = (8 \text{ kpc}) \sin 31.5° = 4.18$ kpc from the Galactic center. The emission peak with the highest radial velocity is at $v_r = +109.9 \text{ km s}^{-1}$, corresponding to an angular velocity (equation 19.57)

[24] Again, we are assuming that we are looking in the quadrants closest to the Galactic center, with $0° < \ell < 90°$, and $270° < \ell < 360°$.

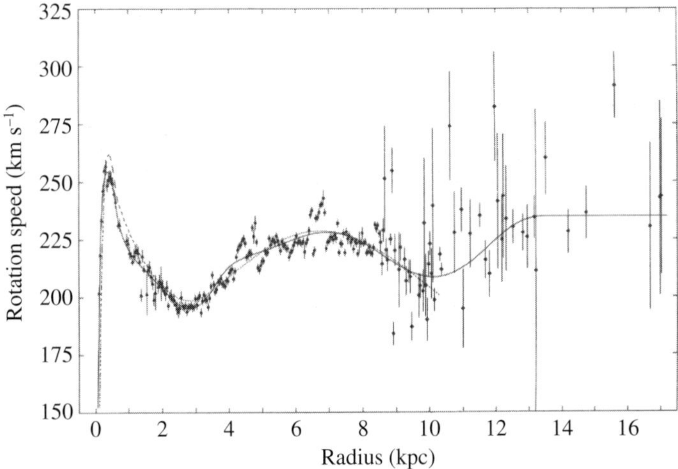

FIGURE 19.19 Rotation curve of our galaxy.

$$\omega_{\text{max}} = 27.5 \text{ km s}^{-1} + \frac{109.9 \text{ km s}^{-1}}{4.18 \text{ kpc}} = 53.8 \text{ km s}^{-1} \text{ kpc}^{-1}. \tag{19.60}$$

This angular velocity corresponds to an orbital velocity

$$\Theta(R_{\text{min}}) = \omega_{\text{max}} R_{\text{min}} = 225 \text{ km s}^{-1}, \tag{19.61}$$

if the molecular cloud is exactly at the tangent point.

By looking along many lines of sight at different Galactic longitudes, we can build up a plot of $\Theta(R)$ versus R; an example is shown in Figure 19.19. Note that the tangent point method works only for gas clouds with $R < R_0$. To find the rotation curve outside R_0, you really do need to observe objects whose distance is known unambiguously. Once you have a more or less accurate idea of $\omega(R)$, you can compute ω for each gas cloud along a line of sight and use the plot of ω versus R to determine the distance R of each gas cloud from the Galactic center. Finding d, the distance from the Sun, is still fraught with ambiguity. Often you can use the angular size of the gas cloud to guess whether it's at the nearer value of d or the farther value of d. Translating from ω to d makes it possible to make maps of the gas distribution in our galaxy. For instance, Figure 19.20 shows the surface density of atomic hydrogen (H I) in the Galaxy, in units of hydrogen atoms per square meter, projected onto the midplane of the Galaxy. Note the features in the map that seem to point directly away from the Sun. This is because errors in d tend to stretch out compact, nearly spherical structures into long smears along the line of sight. In addition, structures near the Galactic center, in the middle of the figure, are muddled and noisy. This is because there are strongly noncircular motions near the center of our galaxy.

Plots of the surface density of gas as a function of R are less noisy, since they don't require dealing with the ambiguity in d for a particular gas cloud. Figure 19.21 shows the surface density of molecular, atomic, and ionized gas within our galaxy. Outside

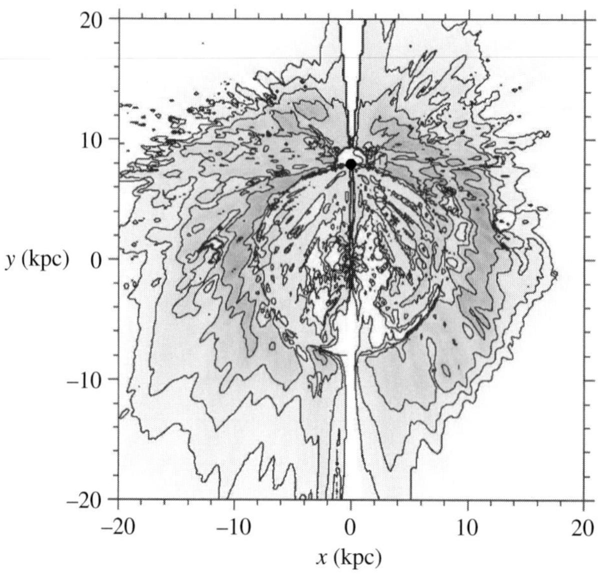

FIGURE 19.20 Surface density of atomic hydrogen in our galaxy. The small circle in the upper center marks the location of the Sun.

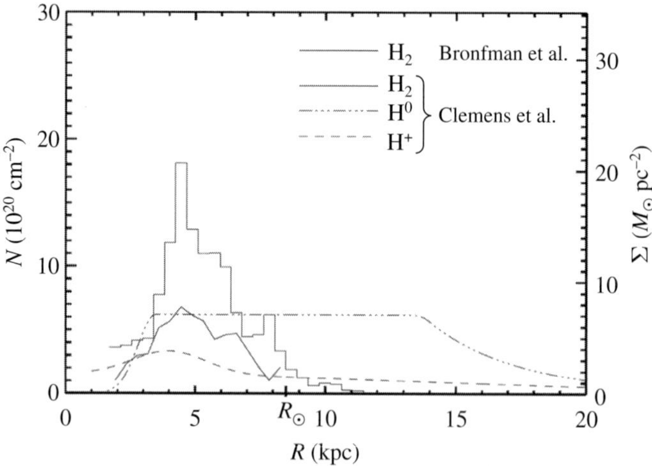

FIGURE 19.21 Surface density of molecular gas (solid lines), atomic gas (dot-dash line), and ionized gas (dashed line) as a function of galactocentric distance R.

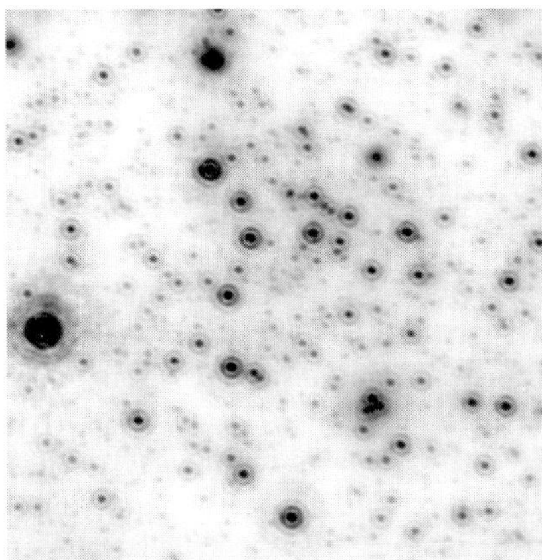

FIGURE 19.22 A negative image of stars near the Galactic center. The region shown is 13 arcsec on a side, corresponding to a physical length ~ 0.5 pc. Note that in this image, the effects of seeing are largely eliminated, and stars appear as Airy disks surrounded by diffraction rings.

$R \sim R_0 \sim 8$ kpc, most gas in the Galaxy is atomic; inside $R \sim R_0$, a large fraction of the gas is molecular. The key difference is density. Molecules can form only in relatively dense regions of interstellar space, where dust grains act as a catalyst for the formation of molecules.[25]

19.7 ▪ THE NUCLEUS OF OUR GALAXY

At the distance of the Galactic center, an angle of 1 arcsecond corresponds to a length $d = 8000$ AU $= 0.039$ pc. In the V band, there are $A_V = 28$ magnitudes of extinction between us and the Galactic center, which pretty well rules out observations at visible wavelengths. However, at infrared wavelengths of $\lambda \sim 2\ \mu$m, there are just 2 magnitudes of extinction. Adaptive optics (mentioned in Section 6.7) permits viewing the Galactic center at infrared wavelengths with a resolution of ~ 0.1 arcsec, as shown in Figure 19.22. This permits us to resolve structures as small as $d \sim 800$ AU at the Galactic center.

The stars in the infrared image of Figure 19.22 are mostly cool giants. If we assume that the ratio of giants to main sequence stars is the same at the Galactic center as in our neighborhood, we deduce that the number density of stars within a parsec of the center is $n_\star \sim 10^7\ \text{pc}^{-3}$. For comparison, the number density of stars in the solar neighborhood is $n \sim 0.1\ \text{pc}^{-3}$. If the Sun were half a parsec from the Galactic center:

[25] Note, in Figure 19.21, that the molecular gas density deduced by Bronfman et al. differs by a factor of ~ 2 from those of Clemens et al. This is an indication of the extreme difficulty of measuring the amount of molecular hydrogen present in space.

- The nearest star would be $\sim$ 1000 AU away.
- The night sky would contain $\sim 10^6$ stars brighter than Sirius.
- The total starlight would be $\sim$ 200 times brighter than the full Moon.
- The probability of stars colliding would not be negligible.

The central regions of our galaxy would be a good place to study stars but a bad place to study external galaxies, because of the high sky brightness.

At the center of our galaxy is a strong radio source called Sagittarius A. The total region of radio emission, shown in Color Figure 20, is about 50 parsecs across. The spectrum of the radio emission indicates that it is synchrotron emission, produced by relativistic electrons accelerated by a magnetic field. The long prominences stretching away from Sagittarius A resemble solar prominences (see Section 7.2) scaled up by a factor of 1 billion.

The early radio observations that detected Sagittarius A were of low angular resolution and merely revealed the presence of an unresolved blob of radio emission. More recent observations have found detailed substructure in Sagittarius A. If we zoom in on the highest surface brightness region of Sagittarius A, we find an interesting radio source called Sagittarius A West, depicted in Color Figure 21. Sagittarius A West is a rotating minispiral of partially ionized gas, about 5 parsecs across. Its radio spectrum looks much like that of an H II region, in which gas is excited by a central source of ultraviolet light. In the case of Sagittarius A West, the central source must be very luminous; but what is it?

If we zoom in on the center of Sagittarius A West, we find a highly compact radio source called **Sagittarius A***. (Don't look for a footnote; the asterisk is part of the name, which is pronounced "Sagittarius A Star.") The angular size of Sagittarius A* has been measured using radio interferometry. At a wavelength $\lambda = 7$ mm, the measured diameter of Sagittarius A* is $d'' \sim 0.8$ milliarcsec, corresponding to $d \sim 6$ AU in physical units.[26] The high angular resolution provided by radio interferometry also enables a measurement of the proper motion of Sagittarius A*. Recent measurements reveal $\mu'' = (6.38 \pm 0.02) \times 10^{-3}$ arcsec yr^{-1}, directed almost entirely along the plane of the Milky Way. If Sagittarius A* were perfectly stationary at the Galactic center, we'd expect the Sun's motion about the Galactic center to produce a proper motion of $\mu \approx \omega_0 \approx 5.8 \times 10^{-3}$ arcsec yr^{-1}, as outlined in Section 19.5.

Observations by the Chandra X-ray Observatory reveal that Sagittarius A* is an X-ray source as well as a radio source. The X-ray emission from Sagittarius A* varies significantly on timescales of less than 1 hour, revealing that the majority of its X-ray emission must come from a region less than 1 light-hour ($\sim$ 7 AU) across. Let's review what we know about Sagittarius A*: it is a fairly luminous, but highly compact, source of radio and X-ray emission located at the Galactic center. (The bolometric luminosity of Sagittarius A* is not exactly known, due to the high extinction at many wavelengths, but is estimated to be $L \sim 1000 L_\odot$.) The leading hypothesis is that Sagittarius A* is a

[26] Thus, Sagittarius A* would fit inside the orbit of Jupiter and is smaller than the star Betelgeuse.

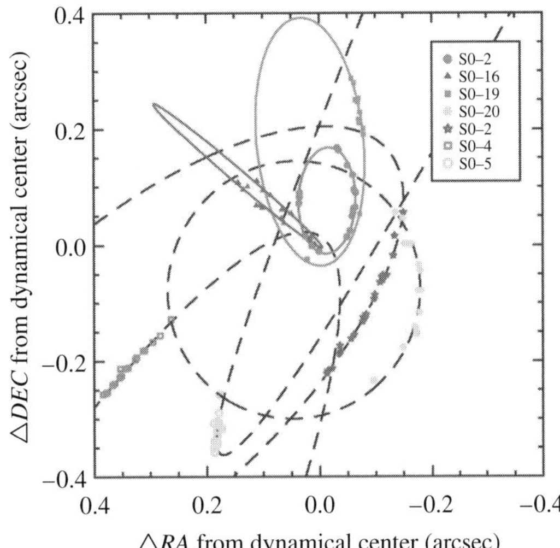

FIGURE 19.23 Observed orbits of bright stars near the Galactic center.

supermassive black hole that is accreting gas. A "supermassive" black hole is a black hole larger than would result from a massive star collapse alone.

The black hole hypothesis is testable by looking at the motion of stars in the vicinity of Sagittarius A*. If there's a black hole present, with mass $M_{\rm bh}$ much greater than a stellar mass, then stars on elliptical orbits around the black hole will obey Kepler's third law:

$$M_\star + M_{\rm bh} = \frac{a^3}{P^2}, \tag{19.62}$$

where a is the semimajor axis of the star's orbit (in AU) and P is its orbital period (in years). Adaptive optics imaging at $\lambda \approx 2\,\mu$ m has enabled astronomers to track a few bright stars near Sagittarius A* for more than a decade. One star in particular, called S0-2, has been observed to have a very small orbit (Figure 19.23), with semimajor axis $a = 920$ AU (assuming $R_0 = 8$ kpc) and orbital period $P = 14.5$ yr. The mass of the black hole is then given by the relation

$$M_\star + M_{\rm bh} = \frac{(920)^3}{(14.5)^2} M_\odot = 3.7 \times 10^6 M_\odot. \tag{19.63}$$

Since the star's mass is insignificant compared to that of the black hole, we can simply state $M_{\rm bh} = 3.7 \times 10^6 M_\odot$.

Within about 0.2 parsecs (or 40,000 AU) of Sagittarius A*, all the stars are on Keplerian orbits, indicating that their dynamics are dictated by a single massive object at the center. Combining all the orbits of stars near Sagittarius A* yields a mass

$$M_{\rm bh} = (3.7 \pm 0.2) \times 10^6 M_\odot \left(\frac{R_0}{8\ \rm kpc}\right)^3, \tag{19.64}$$

indicating that the biggest uncertainty in the mass of the central supermassive black hole is provided by the uncertainty in R_0 (which in turn translates into an uncertainty in a for stars near the center). One of the stars near Sagittarius A*, called S0-16, has an extremely eccentric ($e = 0.976$) elliptical orbit (see Figure 19.23). At pericenter, S0-16 comes within 45 AU of Sagittarius A*. Since the star's orbit is neatly elliptical, with no detectable deviations from a Keplerian orbit, the massive object it is orbiting must be spherically symmetric, with a radius less than 45 AU. The only way to cram 3.7 million solar masses into a volume that small would be to make it into a black hole.[27]

The Schwarzschild radius of a black hole with $M_{\rm bh} = 3.7 \times 10^6 M_\odot$ is $R_{\rm Sch} = 1.1 \times 10^7\,{\rm km} = 0.07\,{\rm AU}$, which subtends an angle of 9 microarcseconds as seen from Earth. The radio and X-ray emission we see from Sagittarius A* comes from outside $R_{\rm Sch}$, more or less by definition. It is emitted by gas that is compressed and heated as it falls toward the black hole. In order to grow to a mass of $M_{\rm bh} = 3.7 \times 10^6 M_\odot$ during the Galaxy's lifetime of $t_{\rm mw} \sim 10$ Gyr, the black hole would have to accrete mass at an average rate

$$\frac{dM}{dt} = \frac{M_{\rm bh}}{t_{\rm mw}} \approx \frac{3.7 \times 10^6 M_\odot}{1.0 \times 10^{10}\,{\rm yr}} \approx \frac{1 M_\odot}{3000\,{\rm yr}}. \tag{19.65}$$

By gobbling a solar mass of material every few millennia, the central black hole of our galaxy could have grown to its present mass.

In Section 18.3, we learned that black holes less massive than $M_{\rm bh} \sim 2000 M_\odot$ would tidally rip you apart before you could reach the event horizon. Thus, if you wanted to dive into our galaxy's supermassive black hole, you wouldn't be tidally destroyed until after passing the event horizon. But what would happen to a star approaching the black hole? Stars are held together by gravitational forces. Thus, a star of mass $M_\star$ and radius $r_\star$ would be tidally ripped apart at a distance $r_{\rm rip}$ from the black hole's singularity. The distance $r_{\rm rip}$ is where the differential tidal force across the star,

$$\Delta F \approx \frac{G M_{\rm bh} M_\star}{r_{\rm rip}^3} r_\star, \tag{19.66}$$

is equal to the gravitational force holding the two halves of the star together,

$$F_{\rm grav} \approx \frac{G M_\star^2}{r_\star^2}. \tag{19.67}$$

By combining equations (19.66) and (19.67), we find that the star would be tidally disrupted at a distance

$$r_{\rm rip} \approx \left(\frac{M_{\rm bh}}{M_\star} \right)^{1/3} r_\star. \tag{19.68}$$

[27] If you tried to make a cluster of neutron stars, for instance, the neutron stars would collide and merge on a timescale that was short compared to the age of our galaxy.

(This is the equivalent to the Roche limit for a moon orbiting a planet; see Section 4.3.1.) A star similar to our Sun, for instance, would be ripped apart at a distance

$$r_{\rm rip} \approx 10^8\,{\rm km}\left(\frac{M_{\rm bh}}{3.7\times 10^6 M_\odot}\right)^{1/3} \tag{19.69}$$

from a supermassive black hole. When we compare this to the Schwarzschild radius of the black hole,

$$r_{\rm Sch} = \frac{2GM_{\rm bh}}{c^2} \approx 10^7\,{\rm km}\left(\frac{M_{\rm bh}}{3.7\times 10^6 M_\odot}\right), \tag{19.70}$$

we find that Sun-like stars would be ripped apart before entering the event horizon so long as

$$M_{\rm bh} < \left(\frac{c^6 r_\odot^3}{8G^3 M_\odot}\right)^{1/2} \approx 2\times 10^{38}\,{\rm kg} \approx 10^8 M_\odot. \tag{19.71}$$

If a star is swallowed whole by a black hole, it doesn't produce a major outburst of radiation before it enters the event horizon. However, if the star is tidally shredded first, its gas forms a hot accretion disk around the star and produces copious emission as it spirals in toward the event horizon.

PROBLEMS

19.1 A star at rest with respect to the LSR is 60° away from the solar apex. The star's parallax is $\pi'' = 15$ milliarcseconds (mas). What are its radial velocity and proper motion? In what direction is the proper motion, relative to the solar apex?

19.2 Suppose the Milky Way consisted of 2.7×10^{11} stars, each of solar luminosity $M_{\rm B} = 4.7$. What would be the absolute magnitude of the whole Galaxy?

19.3 Show that in the case of Keplerian orbits with a centrally concentrated mass (that is, $M(r) = $ constant),

$$\frac{A-B}{A+B} = 2.$$

Does this agree with the observationally determined values of A and B for the Galaxy?

19.4 Derive equation (19.49), starting from equation (19.42).

19.5 Determine the proper motion relative to the LSR for a star in a circular orbit about the Galactic center, at a distance $d = 5$ kpc from the Sun and at galactic longitude $\ell = 45°$. Hint: the Galaxy's rotation curve is given in Figure 19.19.

19.6 The star Rigel has a radial velocity $v_r = 20.7\ \text{km s}^{-1}$, parallax $\pi'' = 4.22$ milliarcseconds (mas), and proper motion components $\mu_\alpha = 1.67\ \text{mas yr}^{-1}$ in right ascension and $\mu_\delta = 0.56\ \text{mas yr}^{-1}$ in declination. What are its total proper motion, tangential velocity, and space motion?

19.7 Derive the relation

$$\frac{A+B}{A-B} = -\frac{d\Theta}{dR} \Big/ \frac{\Theta}{R},$$

starting with the definitions of the Oort constants A and B.

19.8 Assume that a galaxy is spherical. What radial dependence of the mass density $\rho(R)$ gives a flat rotation curve (that is, $\Theta(R) =$ constant)? In this case, how does the enclosed mass $M(R)$ vary with radius R?

19.9 The Perseus spiral arm of the Galaxy can be traced from $R = 4$ kpc from the Galactic center to $R = 12$ kpc. Using the rotation curve in Figure 19.19, determine how long it takes for the stars at the inner end of the Perseus arm to gain one full orbit on stars at the outer end.

19.10 The star S0-2, of spectral class B1 V, orbits the central black hole at the Galactic center on an orbit with semimajor axis $a = 920$ AU and eccentricity $e = 0.867$.

(a) What is the star's distance from the black hole at pericenter?
(b) How close does the star get to the Roche limit?

20 Galaxies

The *Hubble* Ultra Deep Field (Color Figure 22) is the result of 800 exposures of a single field in the constellation Fornax with the *Hubble Space Telescope*, summing to a total exposure of over 11 days. The limiting magnitude in the V band is $m_V = 29$ mag. Within the 3 arcmin $\times$ 3 arcmin field of view of the Ultra Deep Field, there are $\sim 10{,}000$ galaxies. If you multiply the ~ 1100 galaxies per square arcminute within the *Hubble* Ultra Deep Field by the 150 million square arcminutes on the celestial sphere, it implies that there are ~ 170 billion galaxies potentially observable by our telescopes.

The universe is as full of galaxies as a pomegranate is of pips. However, as late as the year 1920, there was still considerable uncertainty among astronomers about whether large galaxies other than the Milky Way Galaxy actually existed. This uncertainty was encapsulated in a pair of talks, known to posterity as the "Great Debate," given by the astronomers Harlow Shapley and Heber Curtis. Shapley maintained that the Milky Way was by far the largest collection of stars in the universe; Curtis maintained that the Milky Way was only one of numerous large stellar systems, or "island universes," as they were sometimes called. The Great Debate involved the nature of small, spiral-shaped, nebulous objects known at the time as "spiral nebulae." Shapley identified them as either gas clouds within our galaxy (probably sites of star formation) or small satellite galaxies orbiting our own galaxy. Curtis, by contrast, identified them as distant, large "island universes" comparable in size and shape to the Milky Way Galaxy.

That Curtis was essentially correct, and that the Milky Way is just one of many galaxies, started to become clear when Edwin Hubble used the Mount Wilson 100-inch telescope to observe the Andromeda Nebula, the largest of the spiral nebulae. Hubble detected Cepheid variable stars within the Andromeda Nebula, and showed that it is actually the ***Andromeda Galaxy***. In 1929, Hubble published the result of his calculations; the Andromeda Galaxy, he believed, is at a distance $d = 275$ kpc. This is not the end of the story, however. In the year 1949, the 200-inch telescope at Mount Palomar was put into operation. Walter Baade, one of the first astronomers to use the new telescope, calculated that if the Andromeda Galaxy were at a distance $d = 275$ kpc, then he would be able to detect its RR Lyrae variable stars. However, when he actually observed the Andromeda Nebula with the 200-inch, he found no RR Lyrae stars at all. Consequently, he concluded that Edwin Hubble had badly underestimated the distance to the Andromeda Galaxy, which had to be at least twice as far as Hubble's distance estimate for its RR Lyrae stars to be undetected. (In fact, the best current estimate of the distance to the Andromeda Galaxy is $d = 780$ kpc.)

Hubble's fundamental mistake was that he was comparing two different types of variable stars. The Cepheids he observed in the Andromeda Galaxy were luminous blue Population I stars. He was comparing them, however, to variable stars in our galaxy that were Population II objects, within the galactic bulge. These Population II variable stars, called **W Virginis stars** after their prototype, are only 1/4 as luminous as a Cepheid star with the same pulsation period. Thus, by applying the W Virginis period-luminosity relation to the intrinsically more luminous Cepheid stars, Hubble was underestimating the distance to the Andromeda Galaxy by a factor of 2.

20.1 ▪ GALAXY CLASSIFICATION

Edwin Hubble, in addition to determining the true nature of the Andromeda Galaxy, also devised the classification scheme for galaxies that we use today. Classification is an important first step in understanding: the purely empirical classification of stellar spectra, for instance, led to the physical understanding that the OBAFGKM sequence of spectral types is a temperature sequence. Galaxies, unlike stars, are not customarily classified by their spectra. In practice, Hubble found it was most useful to classify galaxies by their shapes. The Hubble classification scheme for galaxies is thus a **morphological** classification.[1]

The Hubble scheme divides galaxies into three main classes: **elliptical**, **spiral**, and **irregular** galaxies. Our galaxy is an example of a spiral galaxy. As a useful mnemonic device, the different types of galaxies are laid out in what's generally called a "tuning fork" diagram, as shown in Figure 20.1. In the tuning fork diagram, elliptical galaxies are on the fork's handle; the two types of spiral galaxies (with and without central bars) provide the two tines of the fork; and irregular galaxies are dumped off to one side. Hubble erroneously thought that the sequence shown in the tuning fork diagram was an evolutionary sequence, with galaxies moving from left to right on the diagram as they evolved. We now know that Hubble was wrong on this point: elliptical galaxies do *not* evolve into spiral galaxies. Nevertheless, the tuning fork still appears in astronomy textbooks as a convenient visual aid to remembering the different classes of galaxies.

Elliptical galaxies derive their name from the fact that they look like smooth, glowing, elliptical blobs, with no dark dust lanes, no spiral arms, and no bright patches of star formation. If you approximate the shape of an elliptical galaxy as a perfect ellipse, the size and shape of the ellipse are given by the semimajor axis a and the semiminor axis b, where $b \leq a$. The shape of the ellipse can be described by a single number. It might be the axis ratio $q \equiv b/a$, or the ellipticity $\varepsilon = 1 - q$, or the eccentricity $e = (1 - q^2)^{1/2}$.

The Hubble classification scheme assigns to each elliptical galaxy a label "**En**," where **n** is equal to 10 times the ellipticity, rounded to the nearest integer. Thus, an E0 galaxy is nearly circular, while the flattest elliptical galaxies seen are around E6 or E7

[1] The term "morphological" comes from the Greek root *morphos*, meaning "shape."

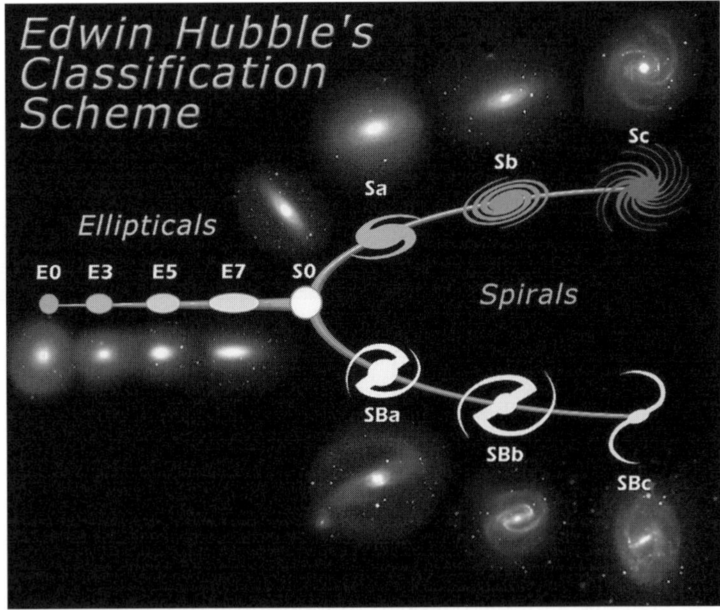

FIGURE 20.1 The tuning fork diagram of galaxies.

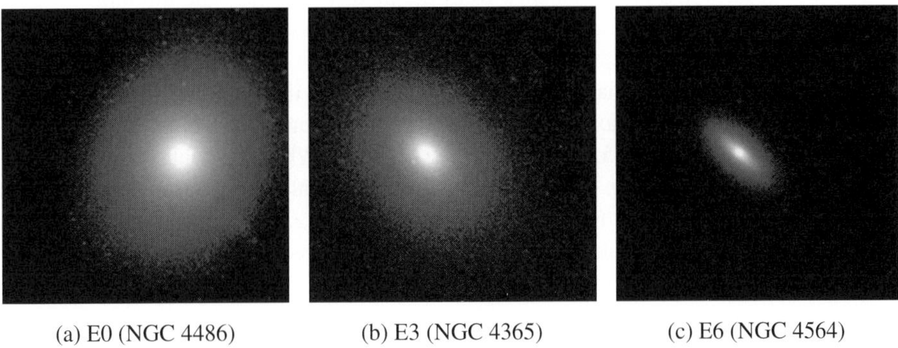

(a) E0 (NGC 4486) (b) E3 (NGC 4365) (c) E6 (NGC 4564)

FIGURE 20.2 Elliptical galaxies, from circular (E0) to flattened (E6).

(Figure 20.2). One unavoidable drawback to Hubble's method for classifying elliptical galaxies is that it relies on the projected, two-dimensional shape of ellipticals, and not on their intrinsic, three-dimensional shape. Unfortunately, we can't scoot around to the side of the galaxy for an alternate view, so we don't know whether an E0 galaxy is a sphere, or an oblate spheroid seen face-on, or a prolate spheroid seen end-on.

(a) Sa (NGC 1302) (b) Sb (NGC 4450) (c) Sc (NGC 4303)

FIGURE 20.3 Spiral galaxies, from Sa to Sc.

Although the shape of a single elliptical galaxy can't be unambiguously determined from its two-dimensional image, statistical statements can be made about the intrinsic shapes of elliptical galaxies, after looking at large data sets.[2] The typical elliptical galaxy must be a triaxial ellipsoid, with principal axes of three different lengths. The surface brightness of bright elliptical galaxies, $I(r)$, usually follows the law

$$\log I \propto -r^{1/4}, \tag{20.1}$$

where r is the distance from the galaxy's center. The luminosities of ellipticals cover a very wide range. The most luminous giant ellipticals have $M_V \sim -23$ mag, or $L_V \sim 10^{11} L_{V,\odot}$.[3]

Spiral galaxies derive their name from their spiral arms, seen most easily when we view the galaxy face-on. The spiral structure of these galaxies was first noted by Lord Rosse in 1845, when he viewed M51 (the Whirlpool Galaxy) with the 72-inch telescope, the *Leviathan of Parsonstown*.[4] Every spiral galaxy has a central bulge, a rotating disk, and spiral arms within the disk, containing gas, dust, and star-forming regions. There are three main subdivisions of spiral galaxies, as illustrated in Figure 20.3. The main classes are as follows:

- **Sa**: big bulge, tightly wound spiral arms, little gas and dust
- **Sb**: medium bulge, moderately wound spiral arms, middling amounts of gas and dust
- **Sc**: small bulge, loosely wound spiral arms, lots of gas and dust

[2] As a simple example, globular clusters must all be nearly spherical. Why? Because they all look nearly circular in projection. The only shape that always looks circular, from any angle, is a sphere.

[3] Cautionary note: all absolute magnitudes and V-band luminosities in this section are approximate.

[4] He wasn't certain, however, whether he was looking at a galaxy, a smaller cluster of stars in our own galaxy, or perhaps a nearby planetary system in the process of formation.

(a) SBa (NGC 4314) (b) SBb (NGC 4548) (c) SBc (NGC 613)

FIGURE 20.4 Barred spiral galaxies, from SBa to SBc.

Many spiral galaxies, perhaps even the majority of them, have an elongated central bar of stars. Barred spirals, like "ordinary" spirals, can be further subdivided into SBa, SBb, and SBc, with the capital "B" standing for Barred. Examples of barred spiral galaxies are shown in Figure 20.4. Our own galaxy has a bar, though its degree of "barrishness" is hard to tell from our location inside the disk. At a guess, the Hubble classification of our galaxy would be SBb. The surface brightness of the disks of spiral galaxies falls off exponentially with distance from the galaxy's center:

$$\log I \propto -r. \tag{20.2}$$

The bulges can frequently be fit with the $\log I \propto -r^{1/4}$ law that applies to elliptical galaxies. Our galaxy and M31 (the Andromeda Galaxy) are both bright, spiral galaxies, with $M_V \sim -21$, or $L_V \sim 2 \times 10^{10} L_{V,\odot}$.

The Hubble classification for spiral galaxies has been extended to type Sd, which represents spirals with minuscule bulges and huge amounts of gas and dust. A final type of spiral galaxy is the **Magellanic spiral,** also referred to as type Sm. Magellanic spirals are systems similar to the Large Magellanic Cloud, which has a prominent bar, rudimentary spiral arms, and lots of active star formation. Since the Large Magellanic Cloud has many young stars, some of them massive, it is host to the occasional core-collapse (type II) supernova, like Supernova 1987a.

There exist galaxies intermediate between spiral and elliptical galaxies; these are called **S0** galaxies.[5] S0 galaxies have flat, rotating disks, like spiral galaxies. However, like elliptical galaxies, they have very little gas and dust, and no spiral arms at all. In tribute to their hybrid nature, they are placed at the Y-junction of the "tuning fork," between the ellipticals and spirals (see Figure 20.1). S0 galaxies are sometimes also referred to as lenticular galaxies, since they have big, central bulges, which give them a shape like a convex lens (Figure 20.5).

[5] That's "S zero," not "S oh."

FIGURE 20.5 NGC 3115, an edge-on S0 galaxy ($d \approx 10$ Mpc).

FIGURE 20.6 Large Magellanic Cloud ($d \approx 55$ kpc) at left, and Small Magellanic Cloud ($d \approx 65$ kpc) at right.

Irregular galaxies are the last of Hubble's main classes. As their name implies, they are amorphous, lacking any regular shape. Irregular galaxies are rich in gas and dust and have copious star formation. A nearby example of an irregular galaxy is the Small Magellanic Cloud, in which there is not even a hint of spiral structure. The Small Magellanic Cloud looks like an egg-shaped smear of stars punctuated with emission nebulae. Figure 20.6 offers a comparison of the Large and Small Magellanic Clouds.

Although Hubble's classification scheme is useful, it has some restrictions. First of all, due to the technical limitations of Hubble's day, it applies only to luminous galaxies with high surface brightness. The lowest-luminosity galaxies are called **dwarf galaxies**. These dwarfs tend to be low in surface brightness as well as low in total luminosity;

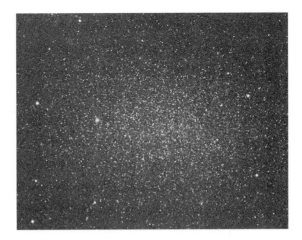

FIGURE 20.7 The dwarf spheroidal Leo I ($d \approx 250$ kpc).

that is, the stars they contain are spread over a relatively wide area across the sky. Thus, dwarf galaxies are hard to detect. Some dwarf galaxies are elliptical in shape and contain little gas and dust; these are called **dwarf ellipticals**. A dwarf elliptical generally has $M_V > -18$ mag, or $L_V < 10^9 L_{V,\odot}$. The dimmest dwarf ellipticals, with $M_V > -14$ mag, or $L_V < 3 \times 10^7 L_{V,\odot}$, are often called **dwarf spheroidals**. The dwarf spheroidal galaxy Leo I is shown in Figure 20.7. Note that individual stars can be resolved in this dwarf spheroidal, which is only 250 kpc away. Some dwarf galaxies do contain lots of gas and dust; these are called **dwarf irregulars**.[6] Just as inconspicuous M dwarfs are the most common type of star, inconspicuous dwarf galaxies are the most common type of galaxy. An estimated 90% of the galaxies in our immediate neighborhood (less than a megaparsec away) are dwarfs.

Finally, some galaxies don't fit into Hubble's classification scheme because they are just plain weird. The galaxy Centaurus A (shown in Figure 21.7) has been called "a pathological object." It resembles an elliptical galaxy, but it has a prominent dustlane slashing across its middle—something that ordinary ellipticals just don't have. Centaurus A is also a strong radio source; its name indicates that it is the brightest radio source in the constellation Centaurus. It is probable that Centaurus A has recently cannibalized a dust-rich companion galaxy. The galaxy NGC 7252 (Figure 20.8) has been called "a train wreck." Although NGC 7252 bears a single catalog number in the New General Catalog, it is actually a pair of galaxies that have not yet finished the process of merging together. The long tails extending away from the train wreck have been stretched out by tidal forces. Roughly 0.5% of nearby bright galaxies are estimated to be members of merging pairs.

[6] There are exceedingly few dwarf spiral galaxies; spiral structure apparently is an attribute of massive, luminous galaxies.

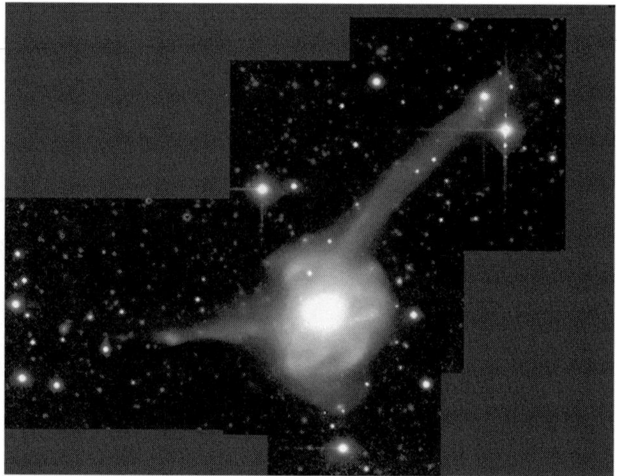

FIGURE 20.8 NGC 7252, the remnant of a galaxy merger ($d \approx 60$ Mpc).

20.2 ▪ GALAXY SPECTRA

Although a galaxy's shape contains interesting information, it doesn't tell the whole story. Useful information can also be derived from the spectrum of a galaxy. **Visible** light emitted by a galaxy is primarily produced by stars (which have an absorption line spectrum) and by hot gas (which has an emission line spectrum). Most of the starlight comes from a small number of very luminous stars, not from the huge number of dim M dwarfs. A single main sequence O star, with $M_V \approx -5$ mag, produces as much visible light as 100 million main sequence M stars, with $M_V \approx +15$ mag. In spiral and irregular galaxies, where stars are currently forming, the brightest stars are young, hot main sequence stars of spectral type O and B. In elliptical galaxies, where star formation has usually ceased long ago, the brightest stars are red giants. Thus, elliptical galaxies, whose bright stars are relatively cool, are redder in color than spiral and irregular galaxies, whose bright stars are hot.

The spectra of different types of galaxies are similar in appearance. Elliptical galaxies have strong absorption lines and no emission lines; the integrated light of all the stars in an elliptical galaxy produces a spectrum similar to a star of spectral type K. Spiral galaxies typically have both strong absorption lines and moderately strong emission lines; the absorption lines are similar to those of a star of spectral type F or G. (Irregular galaxies have spectra similar to those of spiral galaxies.) About 1 or 2% of bright galaxies are **active galaxies**, in which a large fraction of the light is nonstellar in origin.[7] The nonstellar light in an active galaxy comes from a small but luminous central nucleus; thus, active galaxies are also referred to as **active galactic nuclei**, or **AGNs** for short (see

[7] The fraction given here refers to the most luminous types of activity found in galactic nuclei.

Chapter 21). The spectrum of an AGN has extremely strong emission lines, indicating the presence of large quantities of hot gas.

Radio emission is usually stronger in spirals than in ellipticals, although there are many exceptions to the rule. In addition, many radio-loud galaxies, like Centaurus A (Figure 21.7), are peculiar in their morphology. Active galactic nuclei are strong sources of synchrotron emission. Gas clouds in spiral galaxies are strong sources of line emission. Seen at $\lambda = 21$ cm, the disk of a spiral galaxy appears larger than at visible wavelengths. This is what tells us that the gaseous disk is larger than the stellar disk.

Infrared light at 10–100 microns comes primarily from warm dust, with temperature $T \sim 100$ K. Infrared emission is stronger from spiral and irregular galaxies than from elliptical galaxies. Within spiral galaxies, the infrared emission is greatest from the spiral arms, where the dust is concentrated.

Ultraviolet light comes primarily from hot, short-lived stars. Thus, ultraviolet light traces the arms of spiral galaxies, where most of the star formation occurs. The small amount of ultraviolet light from elliptical galaxies comes from relatively hot, helium-fusing stars.

X-rays from galaxies come primarily from a relatively small number of X-ray binaries. In an X-ray binary, a black hole or neutron star accretes gas from a stellar companion.[8] In addition, some X-rays come from the hot coronal gas in a galaxy. In a class of galaxies known as active galaxies, described in Chapter 21, a small central nucleus is a luminous source of X-ray and ultraviolet light.

Because photons of different energy are created by different physical phenomena, a galaxy can change its appearance dramatically when viewed at different wavelengths. Consider, for example, the irregular galaxy M82, shown in Color Figure 23 at four different wavelengths. At infrared wavelengths, M82 has the highest flux of any galaxy in the sky. Although it's at a distance $d \approx 3.5$ Mpc, about five times the distance to M31, it is more than 25 times as luminous in the infrared. The excess infrared emission from M82 is caused by recent star formation inside dusty clouds. The dust absorbs the light from luminous young stars and re-radiates it at longer wavelengths. Note also that the hot, X-ray emitting gas stretches out beyond the star-inhabited region. It's thought that the gas in M82 has been heated by supernova explosions to the point where it is expanding outward in a "galactic wind." M82 is an example of a **starburst galaxy**, a galaxy that has recently experienced a major episode of star formation.[9]

The visible spectra of galaxies generally contain absorption or emission lines that can be used to compute a radial velocity for the galaxy. We measure a redshift, $z \equiv \Delta\lambda/\lambda$, and compute a radial velocity $v_r = cz$, assuming we are in the nonrelativistic limit, where $z \ll 1$ and $v_r \ll c$. Because galaxies are resolved, extended objects, we can measure the radial velocity as a function of position on the galaxy's image. From this measurement, we can deduce how fast the stars in the galaxy are orbiting, and hence how massive the galaxy is.

[8] The X-ray source V404 Cygni, discussed in Section 18.3, is an example of an X-ray binary consisting of a black hole and a star.

[9] In the case of M82, the star formation may have been triggered by a tidal encounter with its neighboring galaxy, M81.

As an example, suppose you are looking at a spiral galaxy in which the stars in the disk are on perfectly circular orbits about the galaxy's center. You see the disk at an inclination i, where $i = 0°$ if the disk is face-on and $i = 90°$ if the disk is edge-on. You see the intrinsically circular disk thanks to the effects of perspective, as an ellipse of axis ratio $q = b/a$, where a is the semimajor axis and b is the semiminor axis. If the disk is infinitesimally thin and perfectly circular, then the apparent axis ratio of the ellipse will be

$$q = \cos i. \tag{20.3}$$

Thus, a face-on disk ($i = 0°$) appears circular ($q = 1$), while an edge-on disk ($i = 90°$) appears as a line segment ($q = 0$). After you measure the apparent axis ratio q of a spiral galaxy, you can compute the inclination $i = \cos^{-1} q$. (There will be an error in your calculation, of course, since disks are neither infinitesimally thin nor perfectly circular, but in most circumstances, the error will be acceptably small.) For instance, the nearby spiral galaxy M31, as seen in Color Figure 1, has an apparent axis ratio $q = 0.3$. This implies that we are viewing M31 at a relatively high inclination of $i = 73°$.[10]

Now suppose we measure the radial velocity $v_r = cz$ along the apparent long axis of the disk of M31. The observed radial velocity v_r will be related to the orbital speed v_c by the relation

$$v_r(R) = v_c(R) \sin i + v_{r,0}, \tag{20.4}$$

where $v_{r,0}$ is the radial velocity of the center of M31, and R is the distance measured from the center of M31. If we want to know the orbital speed as a function of radial distance, we must compute

$$v_c(R) = \frac{v_r(R) - v_{r,0}}{\sin i} = \frac{v_r(R) - v_{r,0}}{\sqrt{1 - \cos^2 i}} = \frac{v_r(R) - v_{r,0}}{\sqrt{1 - q^2}}. \tag{20.5}$$

Although this equation applies to any rotationally supported disk, let's continue to use M31 as our example. M31 is at a distance $d = 700$ kpc from us; at this distance, 1 arcsec corresponds to a length $r = 700{,}000$ AU $= 3.4$ pc. The center of M31 is moving toward us, with a radial velocity $v_{r,0} = -270 \text{ km s}^{-1}$ relative to the Sun. At an angular distance $R'' = 600$ arcsec from the center of M31, along its apparent long axis, we measure a Doppler shift $z = -0.00010$, corresponding to a radial velocity $v_r = cz = -30 \text{ km s}^{-1}$. We can compute that

$$R = (600 \text{ arcsec})(3.4 \text{ pc arcsec}^{-1}) = 2040 \text{ pc} = 2.04 \text{ kpc}. \tag{20.6}$$

The orbital speed at this distance from M31 is (from equation 20.5)

$$v_c(R) = \frac{v_r(R) - v_{r,0}}{\sqrt{1 - q^2}} = \frac{-30 \text{ km s}^{-1} + 270 \text{ km s}^{-1}}{\sqrt{1 - (0.3)^2}} = 250 \text{ km s}^{-1}. \tag{20.7}$$

[10] Since M31 is at a low galactic latitude ($b \approx -22°$), inhabitants of M31 see the Milky Way at a high inclination, as well.

In fact, the rotation curve of M31 is observed to be flat out to nearly 3 degrees ($R'' \approx$ 10,800 arcsec) from the center of the galaxy, corresponding to a physical distance $R \approx 36$ kpc from the center. At $R \approx 36\text{ kpc} \approx 1.1 \times 10^{21}$ m, the calculated orbital speed is $v_c \approx 230\text{ km s}^{-1} \approx 2.3 \times 10^5\text{ m s}^{-1}$. The deduced mass of M31 is then (compare to equation 19.13):

$$M(R) \approx \frac{v_c^2 R}{G} \approx \frac{(2.3 \times 10^5\text{ m s}^{-1})^2 (1.1 \times 10^{21}\text{ m})}{6.67 \times 10^{-11}\text{ m}^3\text{ s}^{-2}\text{ kg}^{-1}}$$

$$\approx 9 \times 10^{41}\text{ kg} \approx 4 \times 10^{11} M_\odot, \tag{20.8}$$

or nearly half a trillion solar masses, with no sign of a Keplerian falloff, which would indicate the edge of a massive dark halo.

It is practical to reconstruct the rotation curve from measuring radial velocities v_r, but not from measuring proper motions μ. If we viewed a spiral galaxy face-on ($i = 0°$), a star with orbital speed v_c would have a proper motion μ'' given by the relation (equation 19.22):

$$\mu'' = \frac{v_c}{4.74d}\text{ arcsec yr}^{-1}, \tag{20.9}$$

where v_c is in kilometers per second and d is in parsecs. If we saw M31 face-on, we'd expect proper motions of approximately

$$\mu'' \approx \frac{250}{4.74(700{,}000)}\text{ arcsec yr}^{-1} \approx 8 \times 10^{-5}\text{ arcsec yr}^{-1}. \tag{20.10}$$

Future space-based interferometry missions may be able to measure proper motions of this magnitude, but at the moment, it's too small to measure.

The spectra of elliptical galaxies reveal different kinematics from spiral galaxies. In ellipticals, the mean rotation speed v_c is found to be small compared to that of comparably sized spiral galaxies. However, the width of the absorption lines in ellipticals is much greater than you would expect from the temperature of their stars. The added width is due to the velocity dispersion σ of the stars. In an elliptical galaxy, stars are not on orderly, near-circular orbits, like stars in the disk of a spiral galaxy. Instead, they are on eccentric, randomly oriented orbits, like the stars in the halo of a spiral galaxy. Thus, along any line of sight through an elliptical, you will see stars with a wide range of radial velocities, and hence a wide range of Doppler shifts.

An interesting nearby elliptical galaxy is NGC 4365, a bright galaxy in the Virgo Cluster, about 16 Mpc away from us. In the Hubble classification scheme, it is labeled an E3 galaxy; its apparent diameter is about 6 arcminutes, so it is well resolved from Earth. Its surface brightness, shown in false color in the left panel of Color Figure 24, is smooth and featureless, with no bright patches of star formation and no dark dustlanes. Along the semimajor axis, the surface brightness follows the usual law for bright elliptical galaxies: $\log I \propto -r^{1/4}$. The average radial velocity, $v_r - v_{r,0}$, is shown in the central panel of Color Figure 24. The main body of the galaxy rotates about the apparent major axis, with a maximum velocity of $v \sim 50\text{ km s}^{-1}$, much lower than the rotation speed in spiral galaxies of similar luminosity. Note also that the central core in NGC 4365 is rotating in a different direction from the rest of the galaxy! This is actually fairly common in elliptical galaxies; the "kinematically decoupled core," as the jargon goes, may be the remnants of a small but dense galaxy that has been cannibalized by the larger, fluffier galaxy.

The dispersion in radial velocity, σ, is quite large in NGC 4365, particularly when compared to the average radial velocity. The dispersion is as large as $\sigma \approx 275\ \text{km s}^{-1}$ in the central regions of NGC 4365 (Color Figure 25), and remains as high as $\sigma \approx 200\ \text{km s}^{-1}$ farther from the center. If we think of the individual stars in NGC 4365 as point masses in a gas, we can think of NGC 4365 as a system that is *pressure supported* rather than *rotationally supported*. The mean square velocity of the stars is then a measure of the "temperature" of the gas of stars.

The observed velocity dispersion σ along the line of sight can be used to estimate the mass of an elliptical galaxy, or any other system dominated by random stellar motions rather than ordered orbital motions. The mass estimate involves the use of the **virial theorem**, which states that if a self-gravitating system of stars, such as a galaxy or star cluster, is in equilibrium (neither expanding nor contracting), there is a simple relation between the total kinetic energy K of all the stars and the gravitational potential energy U of the system:

$$2K = -U. \tag{20.11}$$

A derivation of the virial theorem is given in Section 3.4.

The total kinetic energy of a system of N stars is

$$K = \sum_{i=1}^{N} \frac{1}{2} m_i v_i^2, \tag{20.12}$$

where m_i is the mass of the ith star, and $\vec{v}_i$ is its velocity with respect to the center of mass of the system. The kinetic energy can also be written in the form

$$K = \frac{1}{2} M \langle v^2 \rangle, \tag{20.13}$$

where M is the total mass of the stars and $\langle v^2 \rangle$ is the mass-weighted mean square velocity of the stars.

The potential energy of a system of stars will be

$$U \sim -\frac{GM^2}{r}, \tag{20.14}$$

where r is an appropriately defined radius for the system. Finding an "appropriate" radius for a galaxy might be difficult; remember that galaxies don't have sharp, clearly defined edges. For an elliptical galaxy, it's found that a good approximation is

$$U \approx -0.4 \frac{GM^2}{r_h}, \tag{20.15}$$

where r_h is the half-mass radius of the system, that is, the radius of a sphere (centered on the galaxy's center) large enough to contain half the mass of the galaxy. With this approximation, the virial theorem (equation 20.11) becomes

$$M\langle v^2\rangle = 0.4\frac{GM^2}{r_h}, \tag{20.16}$$

or

$$M \approx 2.5\frac{\langle v^2\rangle r_h}{G}. \tag{20.17}$$

Note the similarity of equation (20.17) to the equation we used to determine the mass of a spiral galaxy:

$$M = \frac{v_c^2 R}{G}. \tag{20.18}$$

In each case, we square a velocity, multiply it by a radius, and divide by Newton's gravitational constant G.

Unfortunately, we can't measure the mass-weighted mean square velocity $\langle v^2\rangle$ for an elliptical galaxy. The practical difficulties of measuring proper motions for stars in external galaxies means that we have information only about the velocity along the line of sight. Moreover, the line-of-sight dispersion σ that we measure is luminosity-weighted, not mass-weighted. If we assume that the red giants that provide the bulk of an elliptical galaxy's luminosity have the same dispersion as the rest of the galaxy's stars, we can ignore the difference between mass-weighting and luminosity-weighting. If, in addition, we assume that the velocity dispersion is *isotropic* (the same in all three dimensions), we may write

$$\langle v^2\rangle = 3\sigma^2 \tag{20.19}$$

if the galaxy's net rotation speed is small compared to its line-of-sight velocity dispersion. This leads to a mass estimate

$$M \approx 7.5\frac{\sigma^2 r_h}{G}. \tag{20.20}$$

Unfortunately, we can't measure the half-mass radius r_h, only the half-light radius (and in projection, at that!). In estimating the total mass of a galaxy, we sometimes have to sigh with resignation and make the additional assumption that the half-mass radius equals the half-light radius.

As an example of the virial theorem in action, consider the dwarf spheroidal galaxy Leo I (Figure 20.7), a member of the Local Group of galaxies, at a distance $d \approx 250$ kpc from us. The half-light radius of Leo I is $r_h'' \approx 4.0$ arcmin ≈ 240 arcsec; at a distance of $d \approx 250$ kpc, this corresponds to a physical distance

$$r_h \approx 290\ \text{pc} \approx 8.9 \times 10^{18}\ \text{m}. \tag{20.21}$$

Because the stars in Leo I are individually resolved, a radial velocity can be measured for each bright star in the galaxy. The dispersion in the radial velocities is

$$\sigma = 8.8\,\mathrm{km\,s^{-1}} \approx 8.8 \times 10^3\,\mathrm{m\,s^{-1}}. \tag{20.22}$$

The estimated mass of Leo I is then

$$\begin{aligned} M &\approx 7.5\frac{(8.8 \times 10^3\,\mathrm{m\,s^{-1}})^2(8.9 \times 10^{18}\,\mathrm{m})}{6.67 \times 10^{-11}\,\mathrm{m^3\,s^{-1}\,kg^{-1}}} \\ &\approx 8 \times 10^{37}\,\mathrm{kg} \approx 4 \times 10^7 M_\odot. \end{aligned} \tag{20.23}$$

Since the V-band luminosity of Leo I is $L_V = 4.9 \times 10^6 L_{V,\odot}$, this implies a mass-to-light ratio for Leo I of $M/L_V \approx 8M_\odot/L_{V,\odot}$. Such a high mass-to-light ratio suggests that Leo I may contain significant amounts of dark matter.

20.3 ▪ SUPERMASSIVE BLACK HOLES IN GALAXIES

The spectra of gas and stars in the central regions of galaxies reveals that most, if not all, bright galaxies harbor a supermassive black hole at their center. For instance, M31 (the Andromeda Galaxy) has a black hole with $M_{\rm bh} \approx 5 \times 10^7 M_\odot$, over 10 times the mass of our own galaxy's supermassive black hole. Even though the masses of these black holes are large, they are difficult to measure; the observable effects of the black hole dominate the dynamics of gas and dust only on the very smallest scales, thus requiring high angular resolution for accurate measurements. The gravitational potential of the black hole dominates over the gravitational potential of the surrounding stars only within the black hole **radius of influence**, defined by

$$r_{\rm bh} = \frac{GM_{\rm bh}}{\sigma_*^2} \approx 11\,\mathrm{pc}\left(\frac{M_{\rm bh}}{10^8\,M_\odot}\right)\left(\frac{\sigma_*}{200\,\mathrm{km\,s^{-1}}}\right)^{-2}, \tag{20.24}$$

where $M_{\rm bh}$ is the black hole mass and σ_* is the velocity dispersion of the stars in the bulge of the galaxy.

Black hole masses have been measured by modeling the dynamics of stars or gas disks (Figure 20.9) in galaxies in which the radius of influence is resolvable, which in practice means resolvable with the *Hubble Space Telescope*. Generally speaking, elliptical galaxies have a central black hole whose mass is proportional to the galaxy's luminosity L; for spiral galaxies, the black hole mass is proportional to the luminosity of the galaxy's *bulge* alone (Figure 20.10). Thus, the black hole mass depends only on the "bulge" component of a galaxy (we can think of an elliptical galaxy as being all bulge and no disk). Since $L \propto \sigma^4$ for elliptical galaxies and bulges of spiral galaxies, there is also a correlation between central black hole mass and the velocity dispersion σ of the "bulge" component.

It's not really surprising that big galaxies have big black holes. Astronomers are surprised, however, at the tight correlation between $M_{\rm bh}$ and σ; when the velocity

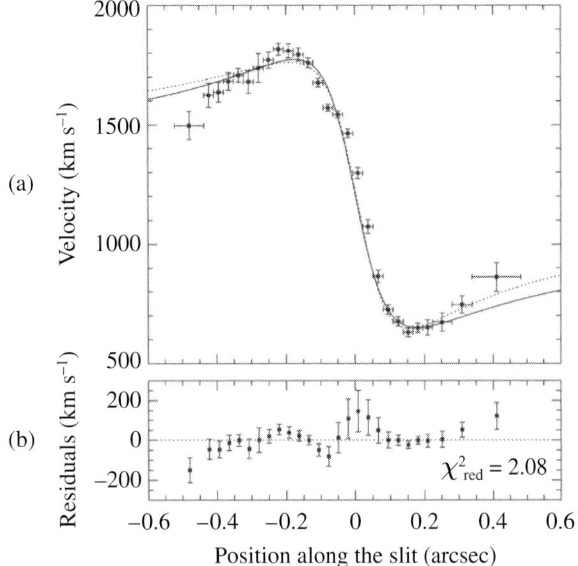

FIGURE 20.9 Rotation curve of the central gas disk surrounding the black hole in M87. Part (b) shows the residuals to the best fit in part (a).

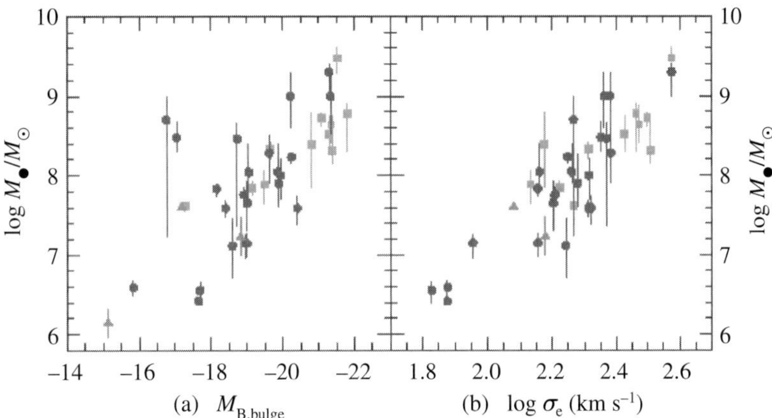

FIGURE 20.10 (a) Central black hole mass versus absolute magnitude of the bulge component of the host galaxy. (b) Central black hole mass versus the velocity dispersion of the bulge component.

dispersion σ is measured, it is dominated by stars far enough out that the black hole has a negligible effect on their velocity. Why a black hole's mass should be so tightly correlated with the velocity of stars that are ignorant of its existence is an unanswered question.

20.4 ▪ DISTANCES TO GALAXIES

Knowing the distance to galaxies is of great interest to astronomers, just as it was of great interest to know the distance to stars within our galaxy. In practice, we work our way outward to larger distances by using a **distance ladder**, with each "rung" in distance depending on a lower rung. We have already encountered the lowest rungs of the distance ladder:

- **Radar** determines distances out to ~ 10 AU. Radar distances depend on knowing the speed of light.
- **Stellar parallax** determines distances out to ~ 200 pc. Stellar parallax distances depend on knowing the length of the astronomical unit (determined using radar).
- **Spectroscopic parallax** determines distances out to ~ 10 kpc. Spectroscopic parallax distances depend on knowing the distance to nearby main sequence stars (determined using stellar parallax).

The technique of spectroscopic parallax is an example of a **standard candle** technique. To astronomers, a standard candle is an object whose luminosity L you know and whose flux F you can measure. The distance d can then be computed from the formula

$$d = \left(\frac{L}{4\pi F} \right)^{1/2}, \tag{20.25}$$

or expressed in terms of magnitudes,

$$d = 10^{0.2(m-M+5)}\ \text{pc}, \tag{20.26}$$

assuming no dust extinction.

To measure the distance to external galaxies (as opposed to the distance to stars within our own galaxy), a very luminous standard candle is required. Cepheid pulsating stars are a favorite standard candle for nearby galaxies. The distances to nearby Cepheids can be determined by stellar parallax, for the very nearest Cepheids, and by spectroscopic parallax, in the case of Cepheids in clusters with main sequence stars. The period–luminosity relation for Cepheids depends on the filter being used. In the V band, we can use equation (17.25). Using ground-based telescopes, the apparent magnitudes of Cepheids can be measured to $d \sim 4$ Mpc. Using the *Hubble Space Telescope*, the apparent magnitudes can be measured to $d \sim 25$ Mpc.

Most of the hundreds of billions of galaxies in the visible universe are at distances greater than 25 Mpc. In order to measure their distances, we need a brighter standard candle. Such a standard candle is provided by type Ia supernovae. Type II (core-collapse) supernovae, although very bright, are not very standardized; they have a variety of progenitor masses, and hence a variety of luminosities. Type Ia supernovae, however, are more standardized, since they all result from white dwarfs pushed over the Chandrasekhar limit.

In any given galaxy, type Ia supernova are rare. In galaxies similar to the Milky Way Galaxy, there are roughly one per century, on average. However, galaxies are sufficiently numerous that we expect an average of three type Ia supernovae to go off each year within 25 Mpc of our galaxy. Since these supernovae occur in galaxies that contain observable Cepheid stars, we can tie the supernova length scale to the Cepheid length scale, and add another rung to the distance ladder.

Careful study of nearby type Ia supernovae reveals that they don't all have exactly the same luminosity. This is bad news for a standard candle. However, the peak luminosity is found to be correlated with the rate of decline of the luminosity after the peak; brighter supernovae have a slower decline. This means that we can predict a supernova's peak luminosity from its rate of decline, just as we can predict a Cepheid's average luminosity from its pulsation rate. This makes type Ia supernovae a much more attractive standard candle. The average peak luminosity of nearby type Ia supernovae is $M_V = -19.2$ mag. This is about one-third the luminosity of the Milky Way Galaxy or M31, and about 3 times the luminosity of the Large Magellanic Cloud. Thus, a type Ia supernova is briefly equal in luminosity to a midsized galaxy. Basically, if a galaxy is bright enough for you to detect, a supernova in that galaxy will be bright enough to detect as well (unless it is deeply buried in dust).

Type Ia supernovae have the drawback of being rare. If you want to find the distance to any particular galaxy, you might have to wait decades—or even centuries—for a type Ia supernova to occur. If you are impatient, and want to estimate a galaxy's distance *right now*, there is an alternative. You can use the entire galaxy as a standard candle. For nearby ellipticals whose distance is known, the velocity dispersion σ is discovered to be correlated with the total V band luminosity, L_V. The relation between σ and L_V, known as the Faber–Jackson relation after its discoverers, is

$$\frac{L_V}{2 \times 10^{10} L_{V,\odot}} = \left(\frac{\sigma}{200 \text{ km s}^{-1}} \right)^4, \tag{20.27}$$

or in terms of absolute magnitudes,

$$M_V = -21.4 - 10 \log \left(\frac{\sigma}{200 \text{ km s}^{-1}} \right). \tag{20.28}$$

For spiral galaxies, there is a similar relation, called the Tully–Fisher relation, which relates the peak rotation speed v_c to the absolute magnitude M of the galaxy. In the B band,

$$M_B = -20.8 - 10.2 \log \left(\frac{v_c}{200 \text{ km s}^{-1}} \right). \tag{20.29}$$

There is plenty of scatter in the Faber–Jackson and Tully–Fisher relations, but sometimes it's the best you can do. The scatter can be reduced in the Faber–Jackson relation by including information about the galaxies' central surface brightness. The scatter can be reduced in the Tully–Fisher relation by making observations in the infrared, where the dimming effects of dust aren't as strong.

20.5 ▪ THE HUBBLE LAW

In addition to determining the distance to a galaxy from a standard candle, we can determine its radial velocity from its redshift. In the nonrelativistic limit, $v_r = cz = c(\Delta\lambda/\lambda)$. As noted in Section 19.3, nearby stars within our own galaxy show a mixture of redshifts ($z > 0$) and blueshifts ($z < 0$). This is no surprise; our galaxy is neither expanding nor contracting. Early in the twentieth century, Vesto Slipher began what was then the nontrivial task of measuring the wavelength shifts of nearby galaxies. By the year 1917, he had measured z for a sample of 25 nearby spiral galaxies (or "spiral nebulae," as Slipher called them back then). He was surprised to find that 21 out of the 25 galaxies were *redshifted*, and only four were blueshifted. This proportion of redshifts was unlikely to have occurred by chance.[11] Slipher was also surprised by the size of the deduced radial velocities; some of the spiral galaxies had $cz > 1000\text{ km s}^{-1}$.

Another surprise came in 1929, when Edwin Hubble examined how the wavelength shift z depended on d, the measured distance to a galaxy. Hubble had z for about 50 galaxies but had distance estimates for only 25 of them. When he plotted cz versus d, he found an approximate linear relation, as shown in Figure 20.11.

The linear relation between redshift and distance can be written in the form

$$cz = H_0 d, \tag{20.30}$$

where the constant H_0, pronounced "H naught," is now called the **Hubble constant.**[12] As a further tribute to Hubble, the plot of cz versus d is called a Hubble diagram, and the equation $cz = H_0 d$ is called the Hubble law.

As it turned out, Hubble severely underestimated the distance to nearby galaxies, since his standard candles were actually much more luminous than he thought.[13] The farthest galaxies in Figure 20.11 are in the Virgo Cluster of galaxies. Hubble thought they were ~ 2 Mpc away; current distance estimates put the Virgo Cluster at a distance $d \sim 16$ Mpc. Thus, although Hubble thought the Hubble constant was $H_0 \sim 1000\text{ km s}^{-1}/2\text{ Mpc} \sim 500\text{ km s}^{-1}\text{ Mpc}$, the current best estimate of the Hubble constant is much smaller:

$$H_0 = 70 \pm 5\text{ km s}^{-1}\text{ Mpc}^{-1}. \tag{20.31}$$

After Hubble's erroneously high value of H_0 was discredited, there was a decades-long debate over whether the true value of the Hubble constant was $H_0 \approx 50\text{ km s}^{-1}\text{ Mpc}^{-1}$

[11] If you flipped a coin 25 times, the probability of having 21 or more heads is less than 1 in 2000.

[12] In a rare display of modesty, Hubble called the constant "K," not H.

[13] He was using the brightest star in each galaxy as his standard candle. However, in more distant galaxies, what he thought was the brightest star was actually a compact, ultraluminous H II region, which can be 50 times more luminous than even the brightest stars.

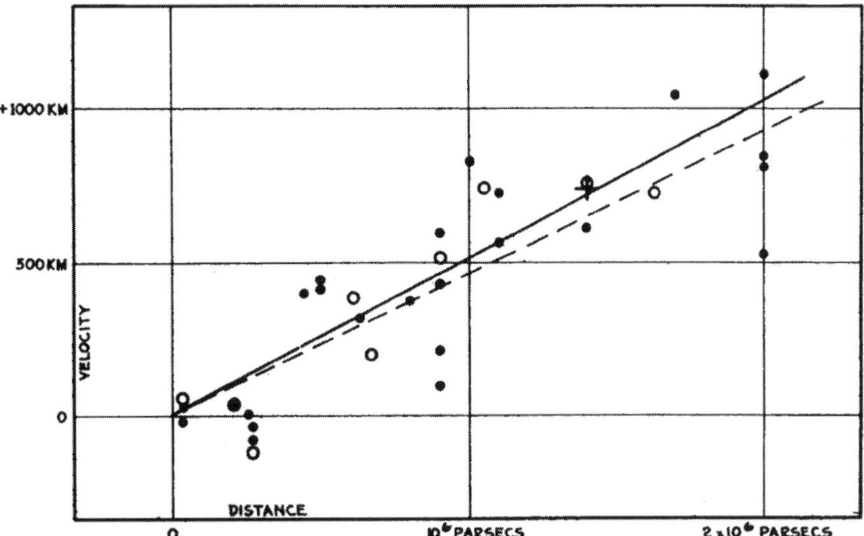

FIGURE 20.11 The Hubble diagram, according to Hubble.

or 100 km s^{-1} Mpc^{-1}. As a relic of that long debate, you sometimes see the Hubble constant written in the form

$$H_0 = 100h \text{ km s}^{-1}\text{ Mpc}^{-1}, \tag{20.32}$$

where $0.5 < h < 1$. Since the value of the Hubble constant is pinned down more accurately now, feel free to substitute $h = 0.7$.

When cosmologically naïve individuals first encounter the Hubble law, they are likely to say, "Why are all the galaxies moving away from *us*? Was it something we said? Do we have the galactic equivalent of bad breath?" In fact, there is nothing particularly special about us. We are *not* the center of the expansion; in fact, there *is* no center of expansion. The Hubble law is the natural result of the homogeneous, isotropic expansion of the universe.

Saying that the expansion is **homogeneous** means that it is the same at all locations. Saying that it is **isotropic** means that it is the same in all directions. For homogeneous expansion, H_0 is the same at all positions; for isotropic expansion, H_0 is the same in all directions at a given location. The usual analogy compares the universe to a loaf of raisin bread expanding homogeneously and isotropically, so that its shape remains the same as the loaf expands. In such an expansion, each raisin sees every other raisin moving away with a velocity proportional to its distance. Note that the raisins themselves don't expand, since they are held firmly together by intermolecular forces. Similarly, in the globally expanding universe, galaxies themselves don't expand, since they are held firmly together by gravity. The Hubble expansion is not the result of some mysterious force that is prying apart the entire universe down to the tiniest scales; it's simply an empirical statement about kinematics on large scales. Widely separated galaxies are moving away from each other, with a speed proportional to the distance between them.

If galaxies are moving apart from each other, they must have been closer together in the past. Thus, the discovery by Hubble of the expansion of the universe led naturally to the **Big Bang** model for the universe. A Big Bang model can be defined as a universe that starts in an extremely dense, compressed state, and then expands to increasingly low densities. Consider a pair of galaxies that are currently separated by a distance d and thus have a relative speed $v_r = H_0 d$. If the relative speed of the galaxies has been constant, then the time it took to reach their current separation is

$$t = \frac{d}{v_r} = \frac{d}{H_0 d} = \frac{1}{H_0}, \tag{20.33}$$

assuming that the galaxies started out very close to each other ($d_{\text{init}} \ll d$). Notice that t is independent of the present separation d. This means that at a time t before the present, all pairs of galaxies were very close to each other.

The Hubble law thus gives us a natural time scale for the expansion of the universe: the Hubble time, which is simply

$$H_0^{-1} = 14 \pm 1\,\text{Gyr} = (4.4 \pm 0.3) \times 10^{17}\,\text{s}. \tag{20.34}$$

In calculating that the universe has been expanding for 1 Hubble time, we assumed that the relative speed of any pair of galaxies has been constant; this is not necessarily true. The gravitational attraction between galaxies tends to decelerate the expansion; in this case, the expansion was faster in the past than it is now, and the universe is younger than H_0^{-1}. On the other hand, it has been theorized that the universe contains "dark energy," which causes the expansion to accelerate outward; in such a case, the expansion was slower in the past than it is now, and the universe is older than H_0^{-1}. It is reassuring to note, in any case, that the estimated age of the oldest globular clusters is close to 14 Gyrs. Hubble's initial (erroneously large) value for the Hubble constant implied a Hubble time of only 2 Gyr, which was embarrassingly short compared to the known age of the Earth.[14]

The expansion of the universe has profound cosmological and philosophical implications. At the moment, let's postpone the profound philosophy, and be practical and down-to-earth. The Hubble law, from a practical viewpoint, gives us another way of estimating the distance to a galaxy. We simply measure the redshift z and then compute

$$d = \frac{c}{H_0} z = (4300 \pm 300\,\text{Mpc}) z. \tag{20.35}$$

Notice that just as the Hubble time, $1/H_0$, gives a natural timescale in an expanding universe, the Hubble distance, c/H_0, gives a natural length scale. If a galaxy is moving away from us at 1% the speed of light, its distance from us is 1% of the Hubble distance, or 43 Mpc.

[14] The universe can't be younger than the objects it contains. In the words of a lady who was asked her age, "I am older than my teeth, and the same age as my tongue."

We should keep in mind, however, the sources of error involved when we use equation (20.35) to estimate distances. First of all, we don't know the Hubble distance exactly. Second, there is a significant amount of scatter in the Hubble diagram (see Figure 20.11). This is because in addition to the perfectly homogeneous Hubble expansion, galaxies have *peculiar velocities* of order $\sim 10^{-3}c$ caused by the gravitational attraction of their neighboring galaxies. These peculiar velocities cause errors of order $\sim 10^{-3}(c/H_0) \sim 4$ Mpc in the distance estimates.[15] Finally, we must keep in mind that the linear relation between d and z holds true only in the limit $z \ll 1$. At higher redshifts, nonlinear relativistic corrections must be taken into account.

PROBLEMS

20.1 At what distance (and at what redshift) does an object 1 kpc across subtend an angle of 1 arcsecond?

20.2 The Ca II H and K lines have rest wavelengths of $\lambda_0 = 3968.5$ Å and 3933.6 Å, respectively. In the spectrum of a galaxy in the cluster Abell 2065 (a.k.a. the Corona Borealis Cluster), the observed wavelengths of the two lines are $\lambda = 4255.0$ Å and 4217.6 Å.

(a) What is the redshift z of the galaxy?
(b) What is the distance to the galaxy?
(c) What is the distance modulus of the galaxy?

20.3 Rewrite the relation for the distance modulus (equation 13.25) in terms of the redshift z rather than the distance d.

20.4 A spiral galaxy, when seen face-on, appears circular; the flux you observe per square arcsecond of the galaxy is given by the relation

$$\Sigma(r) = \Sigma_0 e^{-r/r_0},$$

where r is the distance, in arcseconds, from the center of the galaxy. Show that the total flux you observe from the spiral galaxy is $F = 2\pi \Sigma_0 r_0^2$.

[15] For example, the radial velocity of M31 relative to the Milky Way Galaxy is negative, since the two galaxies form a gravitationally bound system in which they are on strongly radial orbits about their center of mass. If you stared straight at M31 while computing its distance using equation (20.35), you would conclude it was behind you!

20.5 You observe an E0 elliptical galaxy; the flux you observe per square arcsecond of the galaxy is given by the relation

$$\Sigma(r) = \Sigma_0 \exp\left[-(r/r_0)^{1/4}\right],$$

where r is the distance, in arcseconds, from the center of the galaxy. Show that the total flux you observe from the elliptical galaxy is $F = 8!\pi\,\Sigma_0 r_0^2$.

20.6 Consider a black hole of mass $M = 10^8 M_\odot$. What is the maximum distance at which its radius of influence could be resolved using the *Hubble Space Telescope* at a wavelength $\lambda \approx 1\,\mu$m?

20.7 Figure 20.9 shows the rotation curve for the galaxy M87, which is at a distance $d =$ 16 Mpc. Just outside 0.1 arcsec from the center, the rotation curve is approximately Keplerian. Estimate the mass inside this region.

20.8 At what redshift (and at what distance) do peculiar velocities of galaxies contribute less than 1% error to the distance measurement?

21 Active Galaxies

Active galaxies have many distinctive attributes that distinguish them from the "normal" galaxies discussed in the previous chapter.

- As briefly mentioned in Section 20.2, active galaxies produce large amounts of **nonstellar emission**, some of it **nonthermal** in origin, arising from energetic and violent processes.[1] Active galaxies produce more radio and X-ray emission than you'd expect if all their light came from the photospheres of stars.
- Active galaxies have much of their light concentrated in a small, central region known as an **active galactic nucleus**, or **AGN**.
- Light from AGNs is **variable** on short timescales at virtually all wavelengths. The timescale for significant variability is dependent on both luminosity and wavelength, with the most rapid variability seen at shorter wavelengths in lower-luminosity AGNs. In low-luminosity AGNs, X-ray emission can vary on time-scales of minutes.
- Some active galactic nuclei have **jets** that are detectable at X-ray, visible, and radio wavelengths. The jets contain ionized gas flowing outward at relativistic speeds.
- The ultraviolet, visible, and infrared spectra of AGNs are dominated by strong **emission lines**.

In order to be labeled as an active galaxy, a galaxy need not have every attribute listed above; it's more like a "choose any 3 out of 5" proposition. Thus, there are many classes of active galaxies, depending on which attributes a particular object displays.

The accumulated evidence indicates that the activity in galactic nuclei results from the accretion of matter onto supermassive black holes. As we saw in Chapter 20, most if not all bright galaxies harbor a supermassive black hole at their centers. Although bright galaxies contain supermassive black holes, not every bright galaxy is an active galaxy. A supermassive black hole is a necessary but insufficient condition for activity in galaxies: to be classified as an active galaxy, the central black hole must be accreting

[1] These energetic and violent processes are the "activity" that gives active galaxies their name.

gas sufficiently rapidly to have a luminosity high enough to compete with or outshine the stars in the host galaxy.

21.1 ▪ TYPES OF ACTIVE GALAXIES

We can identify three major classes of active galaxies: Seyfert galaxies, quasars, and radio galaxies. Seyfert galaxies and quasars are actually quite similar to each other; Seyfert galaxies can be thought of as low-luminosity, relatively nearby quasars. Radio galaxies are different in that their nuclear emission often is not prominent in wavelengths other than radio. We consider each of these three main classes below.

21.1.1 Seyfert Galaxies

Seyfert galaxies, whose unusual properties were first identified by Carl Seyfert in the 1940s, are spiral galaxies with luminous, variable nuclei that have strong emission lines in their spectra. Seyfert galaxies have nuclei with $L \sim 10^8 L_\odot \rightarrow 10^{12} L_\odot$. There are over 10,000 Seyfert galaxies known; it's estimated that about 5% of bright spiral galaxies are Seyferts, but this fraction might be as high as $\sim 40\%$ if very-low-luminosity Seyfert-like objects are included in the census. The spiral galaxy NGC 3516 (Figure 21.1) is a relatively nearby Seyfert galaxy.

Seyfert galaxies can be divided into two main subclasses, depending on the properties of the emission lines in their spectra. The spectrum of a **Seyfert 1** galaxy is shown in Figure 21.2a. In the spectrum of a Seyfert 1 galaxy, the emission lines from forbidden transitions (such as the O III lines at $\lambda = 4959$ Å and $\lambda = 5007$ Å) are relatively narrow, with Doppler widths $\sim 300 \text{ km s}^{-1}$. By contrast, the emission lines from permitted transitions (such as the Hα line at $\lambda = 6563$ Å) have two components. One component

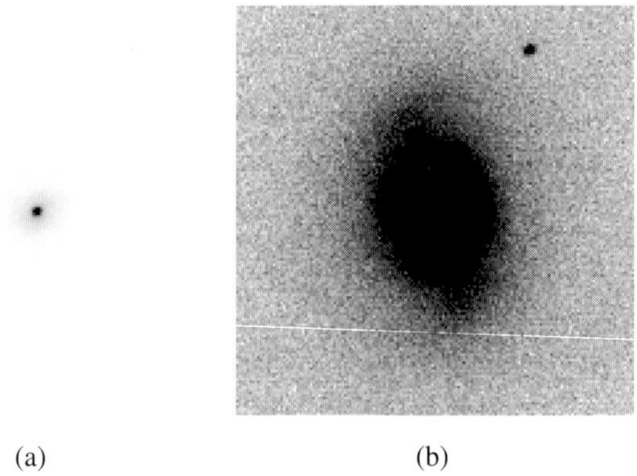

(a) (b)

FIGURE 21.1 Two images of the Seyfert galaxy NGC 3516 ($d \sim 36$ Mpc). (a) A brief exposure reveals the bright nucleus. (b) A longer exposure reveals the extended starlight surrounding the nucleus.

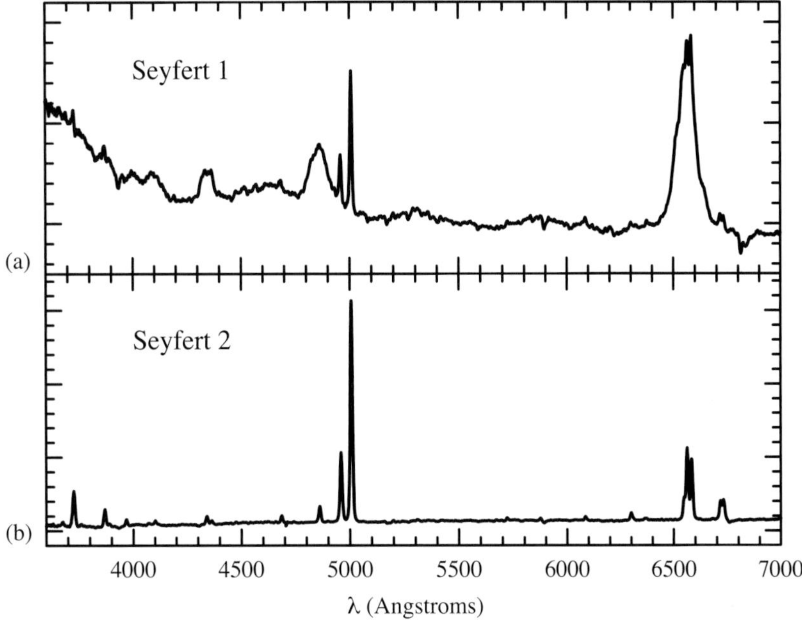

FIGURE 21.2 (a) Spectrum of a Seyfert 1 galaxy (both broad and narrow lines). (b) Spectrum of a Seyfert 2 galaxy (narrow lines only).

of a Seyfert 1 permitted line is narrow, but the other is very broad, with a Doppler width $\sim 5000\ \mathrm{km\ s^{-1}}$ to $10{,}000\ \mathrm{km\ s^{-1}}$.[2] The broad and narrow emission lines arise in two different regions of the AGN. The **broad-line region** contains relatively dense gas, in which the forbidden lines are collisionally suppressed. The width of the broad line is due to the proximity of the broad-line region to the central black hole, where gas motions are very rapid. The **narrow-line region** contains lower-density gas and is farther from the central black hole.

In the spectrum of a **Seyfert 2** galaxy, shown in Figure 21.2b, all the emission lines, both forbidden and permitted, are narrow. In at least some cases, a broad-line region is present in Seyfert 2 galaxies but is hidden from our direct view by dust.

21.1.2 Quasars

Radio telescopes first surveyed the sky in the mid-twentieth century. These first surveys showed that the optical counterparts of most radio sources were galaxies (the radio galaxies discussed in the next section). However, some radio sources did not have

[2] In extreme cases, this combination of narrow and broad components makes a permitted line from a Seyfert 1 galaxy look like a pickelhaube, the old-fashioned spiked helmet of the Prussian infantry.

FIGURE 21.3 The quasar 3C 273, and two galaxies, NGC 4527 and NGC 4536.

obvious counterparts; ambiguity arose because of the poor angular resolution of single-dish radio telescopes. Once better radio positions were obtained, it was found that the radio sources were associated with blue starlike objects. This was unusual; stars are not generally strong radio sources. It seemed clear that they weren't stars, so their starlike appearance led to the name **quasi-stellar radio source**, which was later contracted to **quasar**. Figure 21.3 shows the quasar 3C 273, one of the first quasars to be identified.[3]

Note how the quasar more strongly resembles the stars in the field of view than the two fuzzy, extended galaxies at the left of the figure.

When the spectra of these quasi-stellar radio sources were obtained, things became more unusual still. The spectra contained strong, broad emission lines that didn't correspond to any known element. The breakthrough to understanding came in 1963, when Maarten Schmidt had a "Eureka!" moment while looking at the spectrum of 3C 273 (Figure 21.4). He realized that the emission lines in the spectrum were Balmer lines with a redshift $z = 0.158$. No one had been expecting such a high redshift. It corresponds to a radial velocity $v_r \approx cz \approx 47{,}000 \text{ km s}^{-1}$ and a distance $d \approx (c/H_0)z \approx 680$ Mpc. The large distance was itself impressive at the time but hardly unprecedented: clusters of galaxies at redshifts as large as $z \approx 0.2$ were already known.

However, what was extraordinary was the luminosity implied by the large distance. A quasar is an AGN that has a very high luminosity: $L > 10^{38}$ W, or so. The most luminous known quasars have $L \sim 10^{41} \text{ W} \sim 3 \times 10^{14} L_\odot \sim 10^4 L_{\text{MW}}$. The reason why they look "quasi-stellar" is that quasars are much more luminous than the sum of all the stars in the galaxies that contain them; thus, the quasars usually appear like unresolved points of light, just like stars (Figure 21.5). It's initially difficult to believe that the highest and lowest luminosity AGNs in Figure 21.5 represent the same phenomenon. Indeed, as

[3] Many well-known quasars have names that start with "3C." This is their catalog number in the third Cambridge Catalog of radio sources.

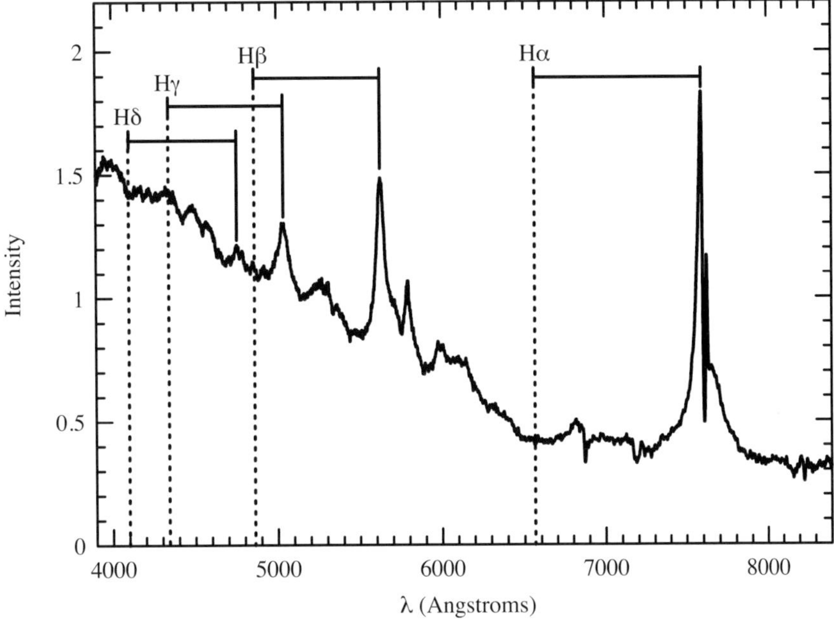

FIGURE 21.4 Spectrum of the quasar 3C 273, showing redshifted Balmer lines.

we'll describe next, the energy requirements needed to explain quasars were so radical in the 1960s that a 20-year controversy over the nature of quasars ensued: are they high-luminosity versions of Seyfert galaxies at the large distances implied by their redshifts, or are they low-luminosity, nearby objects of some sort, with their large redshifts due to some other effect than the expansion of the universe? For most astronomers, the matter was settled definitively in favor of the former hypothesis in the 1980s, with the detection of stellar absorption features in the faint "fuzz" surrounding quasars.

The remarkable properties of quasars soon led to more efficient ways of discovering them. For example, quasars are very blue compared to stars, so searches for objects with "ultraviolet excess" (relative to starlight, of course) proved to be especially effective. Moreover, it quickly became apparent that there was an even larger population of objects that had the optical properties of quasars but were weak or even undetectable emitters at radio wavelengths. These **radio-quiet** sources became known as **quasi-stellar objects**, or simply **QSOs**. The original **radio-loud** quasars make up only $\sim 5\%$ of the total population of QSOs.[4]

Intensive searches for quasars turned up yet another variant of the AGN theme, objects that are known as **BL Lac** objects. Unlike Seyfert galaxies, which are named after their discoverer, BL Lac objects are named after their archetype, BL Lacertae. The name "BL

[4] In practice, the terms QSO and quasar are now used interchangeably by most astronomers.

FIGURE 21.5 These three AGNs have comparable apparent brightness, but each step from (a) to (c) represents an increase in distance of about a factor of 10 and an increase in luminosity of almost a factor of 100.

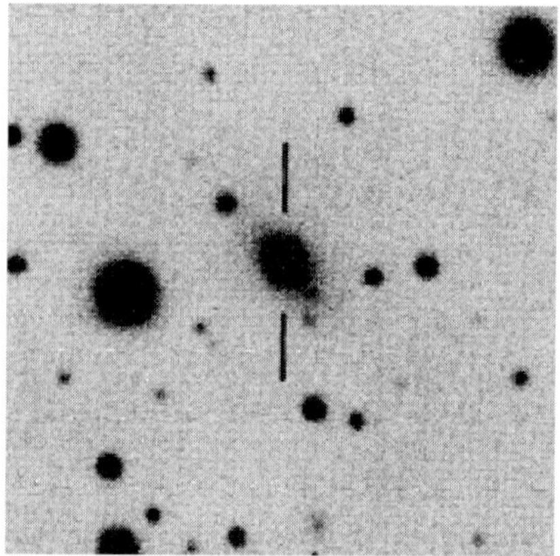

FIGURE 21.6 A negative image of the object BL Lac (between the two vertical lines); the area imaged is $\sim 1'$ on a side.

Lacertae" is the type of name (two letters plus the Latin genitive form of the constellation name) that is given to a variable star. In fact, BL Lacertae was originally thought to be a variable star within our own galaxy. However, deep exposures (Figure 21.6) reveal that BL Lacertae is not a star but a distant elliptical galaxy; its redshift of $z \approx 0.07$ puts it at a distance $d \approx (c/H_0)z \approx 300$ Mpc.

The reason why BL Lacertae was originally mistaken for a star was that its unresolved nucleus is brighter than the diffuse fuzz of starlight surrounding it. The nuclei of BL Lac objects are rapidly variable, and have a nonthermal spectrum, but do not show any emission lines. This makes it difficult to determine the distance to a BL Lac object, unless you can manage to detect the absorption lines in the fuzz of starlight. Doing so is very difficult for many BL Lac objects, since their nuclei are so extraordinarily luminous compared to their stars. (It also makes it difficult to determine their morphology; it's conjectured that all BL Lac objects are ellipticals, but no one is totally sure.) The flux from a BL Lac object can change significantly from one night to the next; this indicates that the bulk of the luminosity comes from a region less than one light-day (~ 200 AU) across.

Seyfert galaxies were first noticed because someone looked at images of spiral galaxies and said, "Wow! Those nuclei are really bright." BL Lac objects were first noticed because someone looked at the radio emission from BL Lacertae and said, "Wow! That's way too much radio emission for BL Lacertae to be a star."

21.1.3 Radio Galaxies

Radio galaxies are defined, simply enough, as galaxies that have strong radio emission. Sometimes it is stated that radio galaxies must have $L_{\text{radio}} > 10^{33}\,\text{W} \sim 3 \times 10^{6} L_{\odot}$ at radio wavelengths, but that's a fairly arbitrary cutoff, chosen because a bright but inactive spiral galaxy—like our own galaxy—has $L_{\text{radio}} \sim 10^{33}$ W from its interstellar gas. The strongest radio galaxies have $L_{\text{radio}} \sim 10^{38}\,\text{W} \sim 3 \times 10^{11} L_{\odot}$. Strong radio sources are most frequently associated with elliptical galaxies, but sometimes radio galaxies are classified as "peculiar," as in the case of Centaurus A (Figure 21.7).

In a radio galaxy, the radio emission is not necessarily confined to a central nucleus. **Extended** radio galaxies have long jets that can be much larger than the image of the galaxy at visible wavelengths. For instance, Centaurus A is an extended radio galaxy;

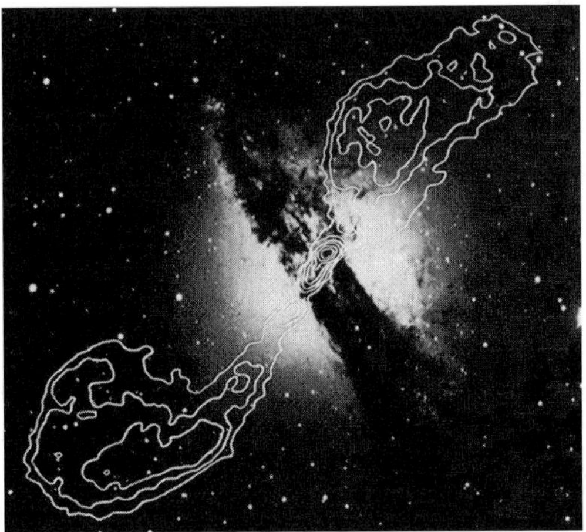

FIGURE 21.7 A contour map of $\lambda = 6$ cm radio emission from the extended radio galaxy Centaurus A, superimposed on a B-band image of the galaxy.

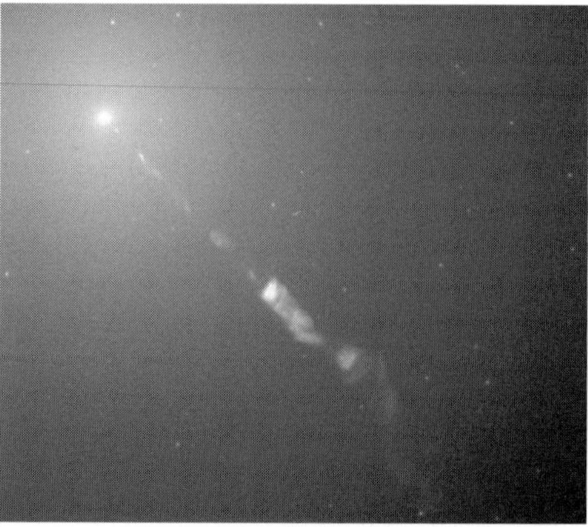

FIGURE 21.8 The nucleus and jet of the elliptical galaxy M87.

Figure 21.7 shows the long radio jets of Centaurus A superimposed on its visible-light image. The jets of an extended radio galaxy can be over a megaparsec long.

Compact radio galaxies have smaller jets, frequently accompanied by an unresolved radio source at the galaxy's nucleus. Very-long-baseline interferometry reveals that the material in jets is moving outward at relativistic speeds; indeed, the apparent motion of jets away from the nucleus often appears to be faster than the speed of light (or **superluminal**), although this is simply a projection effect (as described in the Appendix to this chapter). To find a compact radio galaxy, we need go no farther than the nearby Virgo Cluster of galaxies. The galaxy M87, a bright elliptical galaxy near the cluster's center, goes under the alias "Virgo A," indicating that it's the highest-flux radio source in the constellation Virgo.

In the central regions of M87, there's a 2 kpc long jet seen at radio, visible, and X-ray wavelengths (Figure 21.8). *Hubble Space Telescope* imaging of the central regions reveals a gas disk perpendicular to the jet. The disk (Color Figure 26) looks like the minispiral of hot gas in the center of our own galaxy, only an order of magnitude bigger. The radius of the central disk in M87 is $r = 16$ pc $= 4.9 \times 10^{17}$ m; we are seeing it at an inclination $i = 42°$. The observed spectrum of the disk reveals that it is rotating with

$$v_c \sin i = 460 \text{ km s}^{-1}, \tag{21.1}$$

or $v_c = 690 \text{ km s}^{-1}$. We can then calculate that the mass within 16 pc of the center is

$$M = \frac{v_c^2 r}{G} \approx 4 \times 10^{39} \text{ kg} \approx 2 \times 10^9 M_\odot. \tag{21.2}$$

This mass is primarily due to the supermassive black hole at the center of M87.[5]

[5] It is tempting to call it a "hypermassive" black hole, since it is more than 500 times the mass of the relatively dainty black hole at the center of our galaxy.

21.2 ▪ ACCRETION BY SUPERMASSIVE BLACK HOLES

21.2.1 Energetics

As matter falls toward a black hole, a large amount of energy can, in principle, be extracted from it. Suppose a mass m falls from a large distance $r \gg r_{\rm Sch}$ toward the Schwarzschild radius (equation 18.54), $r_{\rm Sch} = 2GM_{\rm bh}/c^2$, of a black hole. Its loss of gravitational potential energy will be

$$\Delta E = -\frac{GM_{\rm bh}m}{r} + \frac{GM_{\rm bh}m}{r_{\rm Sch}} \approx \frac{GM_{\rm bh}m}{r_{\rm Sch}} \approx \frac{1}{2}mc^2. \tag{21.3}$$

If the mass isn't halted before reaching the Schwarzschild radius, it will pass the event horizon with a speed $v \sim c/\sqrt{2}$, and its kinetic energy will go to swell the mass of the black hole. If the mass is decelerated by slamming into an accretion disk (as described in Section 18.3), its kinetic energy will be converted into thermal energy, and then into photon energy. The conversion of gravitational potential energy (equation 21.3) into photon energy is not perfectly efficient. (Some of the energy, for instance, goes into the kinetic energy of the outflowing jets.) It is customary to write the energy carried away by photons as

$$\Delta E_{\rm phot} = \eta mc^2, \tag{21.4}$$

where η is a dimensionless number, sometimes called the "efficiency" of the black hole. From equation (21.3), we expect $\eta \leq 1/2$. In practice, it is thought that a typical active galactic nucleus has

$$\eta \approx 0.1, \tag{21.5}$$

which means that a single gram of matter falling toward the central black hole yields 9 trillion joules of radiation energy.

As gas feeds into the black hole at a rate $\dot{M}$, the luminosity of the AGN is

$$L = \eta \dot{M} c^2. \tag{21.6}$$

If we know an AGN's luminosity, we can then deduce its infall rate $\dot{M}$:

$$\dot{M} = \frac{L}{\eta c^2} = 0.018 M_\odot\,{\rm yr}^{-1} \left(\frac{L}{10^{37}\,{\rm W}}\right)\left(\frac{\eta}{0.1}\right)^{-1}. \tag{21.7}$$

In general, we don't expect $\dot{M}$ to be constant with time, perhaps accounting for some of the variability detected in the luminosity of AGNs. The low luminosity of the Galaxy's central black hole, $L \sim 1000 L_\odot \sim 4 \times 10^{29}$ W, implies either a low infall rate at the present moment ($\dot{M} \sim 10^{-9} M_\odot\,{\rm yr}^{-1}$), or a low efficiency, or both.

21.2.2 The Eddington Limit

You might think that by shoveling in matter at higher and higher rates, you can make an AGN have arbitrarily high luminosity. In fact, there exists a maximum permissible luminosity for a given black hole mass $M_{\rm bh}$; crank up the luminosity too high, and the gas surrounding the black hole will be blown away by radiation pressure. Consider an accreting black hole of mass $M_{\rm bh}$ and luminosity $L = \eta \dot{M} c^2$. Let's assume that the black hole is surrounded by ionized hydrogen.[6]

At a distance r from the active nucleus, the photons have an energy flux

$$F = \frac{L}{4\pi r^2}. \quad (21.8)$$

In addition to an energy E, each photon has a momentum $p = E/c$. Thus, the outward flow of photons creates a *momentum flux*

$$F_p = \frac{F}{c} = \frac{1}{c}\frac{L}{4\pi r^2}. \quad (21.9)$$

Because the photons carry momentum, they can exert a force on the free electrons and protons in the ionized hydrogen. The force exerted on each particle is the rate at which momentum is transfered to it. The rate of momentum transfer depends in turn on the particle's cross-section for interaction with photons. Electrons have a much larger cross-section for photon interactions than protons do.[7] For electrons, the relevant cross-section is the Thomson cross-section (equation 15.24) $\sigma_e = 6.65 \times 10^{-29}$ m^2. The electron experiences an outward force due to radiation pressure. The amplitude of the force is equal to the momentum flux times the electron's cross-section:

$$F_{\rm rad} = \sigma_e F_p = \frac{\sigma_e L}{4\pi c r^2}. \quad (21.10)$$

As the electron is accelerated, it drags the nearest proton along with it, which maintains charge neutrality. Thus, every electron is burdened with a massive proton that it drags along like a ball and chain.

The inward force on the electron–proton pair is provided by gravity:

$$F_{\rm grav} = -\frac{G M_{\rm bh}(m_p + m_e)}{r^2} \approx -\frac{G M_{\rm bh} m_p}{r^2}, \quad (21.11)$$

since the electron mass m_e is insignificant compared to the proton mass m_p. The maximum possible luminosity for the accreting black hole is called the **Eddington luminosity**, or **Eddington limit**, after the astronomer Arthur Eddington. The Eddington luminosity L_E is the luminosity at which the outward radiation force exactly equals the inward gravitational force:

$$\frac{\sigma_e L_E}{4\pi c r^2} = \frac{G M_{\rm bh} m_p}{r^2}. \quad (21.12)$$

[6] It's natural that the gas should be ionized, since it's heated as it falls toward the black hole. The assumption that it's pure hydrogen is simply to make the calculation easier.

[7] As shown by equation (15.24), the cross-section goes as $1/m^2$, where m is the particle mass. Since electrons are less massive than protons by a factor $\sim 1/2000$, they have larger cross-sections by a factor ~ 4 million.

Note that the factors of r^2 cancel: the ratio of radiation force to gravitational force is independent of distance from the black hole. The Eddington luminosity for a black hole of mass $M_{\rm bh}$ is

$$L_E = \frac{4\pi G m_p c}{\sigma_e} M_{\rm bh} = 1.3 \times 10^{39}\,{\rm W}\left(\frac{M_{\rm bh}}{10^8 M_\odot}\right)$$
$$= 3.3 \times 10^{12} L_\odot \left(\frac{M_{\rm bh}}{10^8 M_\odot}\right). \tag{21.13}$$

If the luminosity is greater than L_E, then the ionized gas will be accelerated outward and accretion will cease. The existence of a maximum luminosity L_E leads to a maximum accretion rate for black holes:

$$\dot{M}_E = \frac{L_E}{\eta c^2} = 2 M_\odot\,{\rm yr}^{-1}\left(\frac{M_{\rm bh}}{10^8 M_\odot}\right)\left(\frac{\eta}{0.1}\right)^{-1}. \tag{21.14}$$

If you try to feed a black hole more rapidly than this, it spews the gas back out. It is sometimes useful to express the accretion rate in terms of the Eddington rate,

$$\dot{m} = \frac{\dot{M}}{\dot{M}_E}, \tag{21.15}$$

which we refer to as the **Eddington ratio**.

21.2.3 Accretion Disks

As discussed earlier in Section 18.3, infalling gas does not plummet directly through the Schwarzschild radius and into the black hole. If the gas has any net angular momentum, it will first pancake into an accretion disk. Let's suppose that such an accretion disk will be geometrically thin but optically thick, so that each point on the disk's surface radiates like a blackbody. Given these assumptions, consider a small patch of the accretion disk, with area dA, at a distance r from the black hole. The luminosity of the patch is

$$dL = (2\,dA)(\sigma_{\rm SB} T^4), \tag{21.16}$$

where T is the temperature of the patch; the factor of 2 enters because each surface of the patch is radiating. The mass of the patch is $dM = \Sigma\,dA$, where Σ is the mass surface density. Thus, the luminosity of the patch can also be written as

$$dL = \frac{2dM}{\Sigma}\sigma_{\rm SB} T^4. \tag{21.17}$$

The energy that is radiated away ultimately comes from the gravitational potential energy of the patch, which can be written in the form

$$dU = -\frac{GMdM}{r}, \tag{21.18}$$

where M is the mass of the central black hole. As viscosity causes the patch to slowly spiral inward, it loses gravitational potential energy at the rate

$$dL_{\rm pot} = \frac{GMdM}{r^2}\frac{dr}{dt} = \frac{GMdM}{r^2}v, \tag{21.19}$$

where v is the radial speed of the patch as it moves toward the black hole. If we assume that all the gravitational potential energy goes to heating the disk, we can combine equations (21.17) and (21.19) to find the temperature of the patch:

$$\frac{2\,dM}{\Sigma}\sigma_{\rm SB}T^4 \approx \frac{GMdM}{r^2}v, \tag{21.20}$$

or

$$T \approx \left[\frac{GM}{2\sigma_{\rm SB}r^2}\Sigma v\right]^{1/4}. \tag{21.21}$$

The above equation can be simplified if the accretion disk is axisymmetric and the mass infall rate $\dot{M}$ is constant with time. Under these conditions, mass conservation requires that

$$\dot{M} = 2\pi r\Sigma(r)v(r), \tag{21.22}$$

which is the two-dimensional equivalent of the three-dimensional continuity equation that we previously derived for a steady-state solar wind (equation 7.7).

Equations (21.21) and (21.22) can be combined to find the temperature profile of a steady-state, axisymmetric disk:

$$T(r) \approx \left[\frac{GM\dot{M}}{4\pi\sigma_{\rm SB}r^3}\right]^{1/4} \propto r^{-3/4}. \tag{21.23}$$

It is convenient to express the radius r in terms of the Schwarzschild radius (equation 18.54) and the mass accretion rate $\dot{M}$ in terms of the Eddington ratio (equation 21.15):

$$\dot{M} = \dot{m}\dot{M}_E = \frac{\dot{m}L_E}{\eta c^2} = \dot{m}\left(\frac{4\pi Gm_p}{\sigma_e\eta c}\right)M. \tag{21.24}$$

A little rearranging of equation (21.23) yields

$$\begin{aligned} T(r) &\sim \left(\frac{c^5 m_p}{8\sigma_{\rm SB}G\sigma_e}\right)^{1/4}\left(\frac{\dot{m}}{M}\right)^{1/4}\left(\frac{r}{r_{\rm Sch}}\right)^{-3/4} \\ &\sim 3\times10^5\,{\rm K}\,\dot{m}^{1/4}\left(\frac{M}{10^8 M_\odot}\right)^{-1/4}\left(\frac{r}{r_{\rm Sch}}\right)^{-3/4}. \end{aligned} \tag{21.25}$$

The continuum emission from quasars typically peaks in the ultraviolet, at an emitted wavelength of about 1000 Å. If this is the peak thermal emission from the accretion disk, Wien's law tells us that the temperature of the disk is about 3×10^4 K; inserting this temperature into equation (21.25) tells us that UV emission arises at

$$\left(\frac{r}{r_{\rm Sch}}\right) \sim 14 \left(\frac{\dot{m}}{\eta}\right)^{1/3} \left(\frac{M}{10^8 M_\odot}\right)^{-1/3}. \tag{21.26}$$

21.3 ▪ THE STRUCTURE OF AGNS AND UNIFIED MODELS

Seyfert galaxies, BL Lac objects, and radio galaxies seem to be something of a mixed bag, with different morphologies and spectra. However, it is possible to describe all of the various types of AGN in the context of **unified models**; the goal of unified models is to describe the broadest range of AGN phenomenology while using the fewest possible free parameters. The basic idea is that AGN structure is more flattened than spherical and that the visibility or strength of various components depends strongly on the inclination of the AGN axis relative to the line of sight.

We start by listing the major components of AGN, starting from the inside and working outward. At the heart of every AGN, there is a supermassive black hole. While this was suspected for many years based on the combination of the energetics arguments outlined earlier and the observed rapid flux variability (implying a compact emitting region), more recent spectroscopic evidence for supermassive black holes in AGNs comes in the form of the high-speed motions of stars and gas in galactic nuclei. The Schwarzschild radius of a $10^8 M_\odot$ black hole (equation 18.54) is $\sim 3 \times 10^{11}$ m ~ 2 AU.

Surrounding the black hole is an accretion disk that accounts for the ultraviolet and visible continuum emission of AGNs. Equation (21.25) tells us that the ultraviolet and visible emission from a $10^8 M_\odot$ black hole arises on a scale of 10^{12}–10^{13} m. The origin of X-rays is less well understood, but they seem to be produced in a hot corona that surrounds the accretion disk. The jets observed in some objects are thought to arise on this same physical scale. Some ionized gas is ripped from the accretion disk by electromagnetic fields, and spirals along magnetic field lines away from the disk, forming a jet. The accelerated electrons in the ionized gas emit synchrotron emission, accounting for the radio emission from the jet.

Proceeding outward, the size of the broad-line region is measured by timing the delay between flux variations of the ultraviolet and visible continuum and the response of the broad emission lines, a technique known as **reverberation mapping**; the time delay is simply due to the light travel time across the broad-line region. The size of the broad-line region scales with luminosity:

$$\left(\frac{R_{\rm blr}}{10^{15}\ {\rm m}}\right) \approx 0.26 \left(\frac{L_{\rm bol}}{10^{37}\ {\rm W}}\right)^{1/2} \tag{21.27}$$

It has been speculated that the broad-line region may be the outer part of the accretion disk structure.

The outer extent of the broad-line region seems to be defined by the **dust sublimation radius**, the closest point to the continuum source where graphite grains can survive the intense ultraviolet radiation emitted by AGNs; dust at smaller radii, where the equilibrium blackbody temperature exceeds ~ 1500 K, is simply vaporized. The existence of dust is important in unified models because dust supplies the source of opacity that can block our direct view of the inner regions from some directions. The sublimation radius marks the inner edge of dusty structure with a larger scale height than the inner regions. It is often referred to as the **obscuring torus**, or dusty torus (although evidence is accumulating that this region is not a doughnut-shaped structure but is rather the cool, dense part of a disk wind that arises from the accretion-disk structure itself).

The inner part of the narrow-line region is on about the same scale as the obscuring torus but can extend out to hundreds of parsecs. The morphology of the narrow-line region is often wedge-shaped or conical and along the axis of the black-hole/accretion-disk system. The narrow-line region gas is apparently just the interstellar medium of the host galaxy; interstellar gas that is not shielded from the central source by the obscuring torus is photoionized by the AGN ultraviolet radiation.

A schematic view of such a unified model is illustrated in Figure 21.9, which shows that the type of AGN you see from Earth depends on the orientation of the outer torus

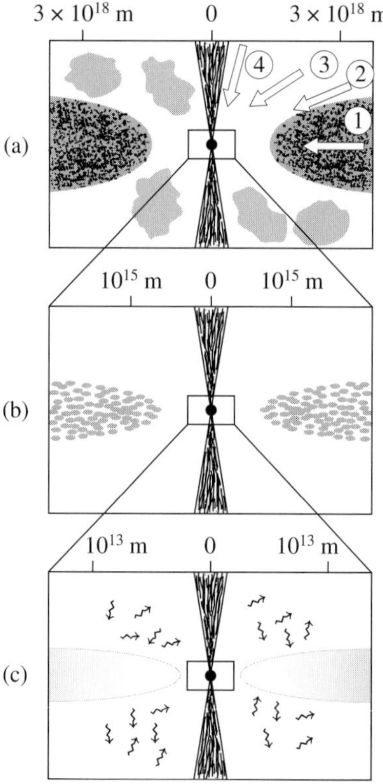

FIGURE 21.9 Cross-sectional view of an active galactic nucleus on three different scales. (a) An active galactic nucleus on a large scale. Vertical structures are jets of gas; in the equatorial plane is a torus of dusty clouds. (b) Expanded view of the central region. Small clouds of gas are concentrated near the equator. (c) Further expanded view. In the equatorial plane is a gas disk that is very bright in visible light. X-rays are emitted by hot gas near the disk.

of dusty clouds relative to the line of sight. If you look directly along the jet (vector 4 in Figure 21.9), you see primarily synchrotron emission from the jet. In this case, you see the featureless continuum spectrum characteristic of BL Lac objects. If you look at an angle close to the jet (vector 3), you get a good look at the accretion disk and the broad-line region, close to the event horizon where gas is moving very rapidly. In this case, you see broad emission lines from the rapidly moving gas, and the AGN is classified as a Seyfert 1 galaxy. If you look at an angle close to the disk (vector 1), the broad-line region is hidden by dust, and you can see only the narrow-line region, farther from the central black hole. In this case, you see narrow emission lines, and the AGN is classified as a Seyfert 2 galaxy. At some angles (vector 2), you are playing peekaboo with the edge of the obscuring torus.[8]

The key piece of evidence supporting this scenario is that some Seyfert 2 galaxies have clear broad-line region components visible in their *polarized* spectra. The light from the continuum source and the broad-line region is scattered into the observer's line of sight by electrons, which polarize the light they scatter, above the plane of the obscuring region.

It is also worth noting that while the majority of low-luminosity AGNs are Seyfert 2 galaxies, outnumbering Seyfert 1 galaxies by about three to one, "quasar 2" objects are extremely rare. Certainly this has something to do with the large dust sublimation radius for more luminous objects. If, for example, the scale height of the obscuring dust is independent of luminosity, then in more luminous quasars, the obscuring torus covers a smaller fraction of the sky as seen from the central source.

21.4 ▪ QUASARS OVER COSMIC HISTORY

Astronomers are fond of the saying "A telescope is a time machine." As we look farther out in space, we are looking farther back in time. When we look at 3C 273, 2 billion light years away, we are seeing it as it was 2 billion years ago. When we look at the quasar with the highest known redshift, $z = 6.4$, the relativistically correct equations tell us that we are looking at a quasar that is currently 27 billion light years away, and we are seeing it as it was 13 billion years ago.[9] We know from the measured masses of black holes today that quasars must have relatively short lives. The most luminous quasars, with $L \approx 3 \times 10^{14} L_\odot$, must have accretion rates $\dot{M} \approx 200 M_\odot\,\mathrm{yr}^{-1}$ to maintain that luminosity. The biggest supermassive black holes today have $M_{\rm bh} \sim 4 \times 10^9 M_\odot$. To grow to that mass at the accretion rate of the most luminous quasars would take a time

$$t \approx \frac{M_{\rm bh}}{\dot{M}} \approx \frac{4 \times 10^9 M_\odot}{200 M_\odot\,\mathrm{yr}^{-1}} \approx 20\,\mathrm{Myr}. \tag{21.28}$$

[8] In rare cases of extreme variability, AGNs appear to change from type 1 to type 2, or vice versa. Such extreme behavior can be understood in the context of a patchy obscuring medium.

[9] The $z = 6.4$ quasar is currently more than 13 billion light years away because the distance between us and it has been constantly stretching during the past 13 billion years. At the time the quasar emitted the light that we see today, it was only 3.7 billion light years away. The details of computing the distance to high-redshift objects are given in Section 23.4.

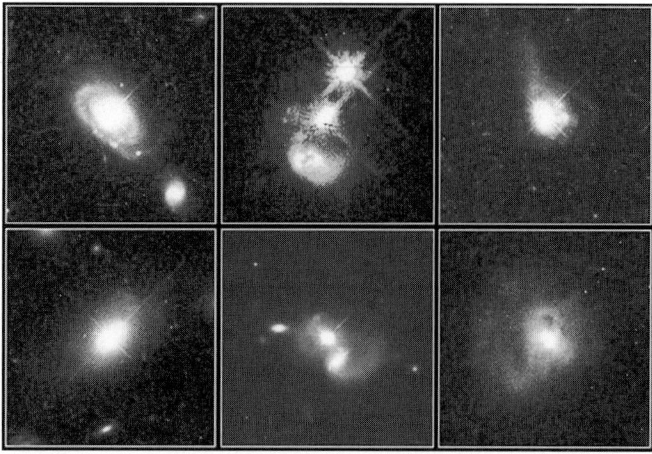

FIGURE 21.10 Host galaxies of six quasars ($d \sim 400 \to 900$ Mpc).

If the most luminous quasars maintained their luminosity for the age of the universe ($\sim$ 14 Gyr), they would have grown to $\sim 3 \times 10^{12} M_\odot$ by the present day.

When we compute the number density of quasars as a function of redshift, we find that there were many more quasars in the past than there are now. At the present moment, only one in a million bright galaxies is a quasar. When the universe was about 20% of its present age (that is, when it was $\sim$ 2.7 Gyr old), about one in a thousand bright galaxies was a quasar. The black holes that accreted gas and drove the luminosity of the quasars haven't gone away. Black holes can only grow more massive with time.[10] This means that the black holes that powered quasars are still around today, bigger and fatter than ever. They are just less luminous because they are being fed gas at a lower rate. Old quasars never die—they just go into hibernation when there's nothing to eat.

One reason why supermassive black holes accreted gas more rapidly in the past is hinted at when you look at the high-redshift galaxies that host quasars (Figure 21.10). Although some quasar hosts are fairly normal-looking spiral and elliptical galaxies (Figure 21.10a), most are peculiar-looking galaxies in the process of merging. Galaxy mergers were more common in the past than they are now—in an expanding universe, galaxies were closer together in the past. When galaxies merge, their gas clouds collide, lose angular momentum, and fall to the center of the merged galaxy, where the black hole(s) are ready to accrete them.[11]

[10] Unless they are extremely tiny, in which case Hawking radiation makes them evaporate.

[11] When two galaxies, each with a central, supermassive black hole, merge to form a new, larger galaxy, it is not known with certainty how long it takes for their two black holes to become one. They may just form a stable black hole binary at the center of the new galaxy.

21.5 ▪ PROBING THE INTERGALACTIC MEDIUM

Quasars are interesting not merely for their own sake, but because they act as probes of the **intergalactic medium**. Light from quasars hundreds of megaparsecs away must pass through hundreds of megaparsecs of intergalactic gas before it reaches us. Just as stellar spectra can show absorption lines from interstellar gas clouds (see Section 16.2), quasar spectra can show absorption lines from intergalactic gas. Because the gas clouds between us and the quasar are at a smaller distance d than the quasar (Figure 21.11), they have a smaller redshift z. Thus, the absorption lines are at shorter wavelengths than the corresponding emission line from the quasar itself. As an example, consider the Lyman alpha line of neutral hydrogen, which has a rest wavelength of $\lambda_0 = 1216$ Å. The quasar QSO 1425+6039 has a redshift of $z = 3.18$, so its Lyman alpha emission line peaks at a wavelength $\lambda = (1+z)\lambda_0 = (4.18)(1216\text{ Å}) = 5080$ Å, in the visible range of the spectrum, which makes it convenient for observation (Figure 21.12). The absorption lines from neutral hydrogen in the intervening atomic gas will be at wavelengths ranging from 1216 Å to 5080 Å.

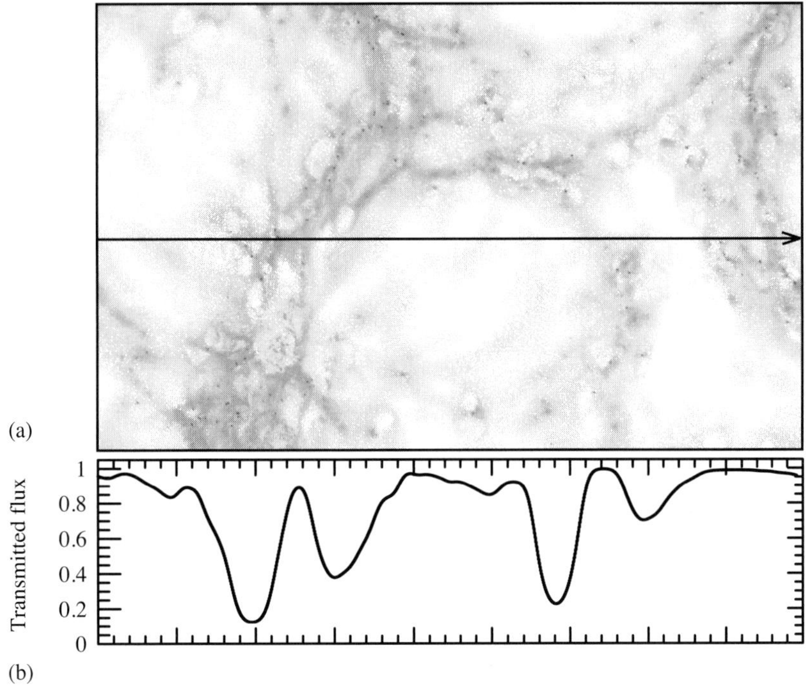

FIGURE 21.11 (a) An observer looks through the distribution of intergalactic gas, looking toward a quasar lying in the direction of the arrow. Darker regions indicate denser gas. (b) The resulting Lyman alpha absorption spectrum.

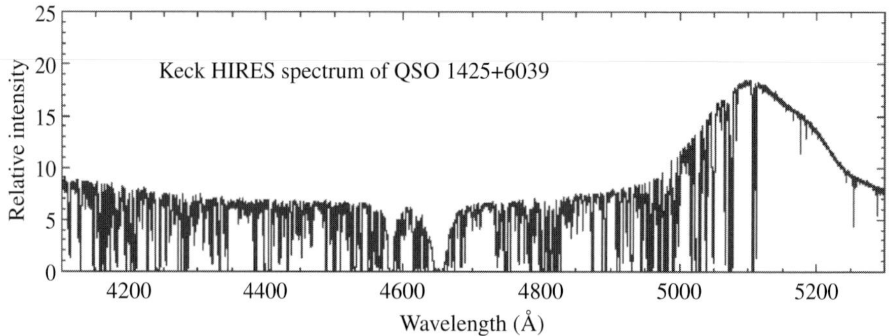

FIGURE 21.12 Spectrum of the $z = 3.18$ quasar, QSO 1425+6039. The broad emission line at $\lambda = 5080$ Å is the red-shifted Lyman α line from the quasar itself. The absorption lines at shorter wavelengths arise in neutral hydrogen gas between us and the quasar and constitute the "Lyman alpha forest." Note the damped Lyman α absorption line at $\lambda \approx 4650$ Å, corresponding to high-column-density gas at a redshift $z \approx 2.8$.

For high-redshift quasars like QSO 1425+6039, a great many individual absorption lines are seen. The thicket of Lyman alpha absorption lines is referred to as the "Lyman alpha forest." Some of the absorbers along the line of sight are extremely optically thick ($\tau > 10^4$) at the wavelength of Lyman alpha. Therefore, they produce Lyman alpha absorption lines that are saturated in the center and have Lorentz damping wings to either side (see Section 5.5). These high-density absorbers are called "damped Lyman alpha" systems. Much of the neutral gas in the universe at high redshifts is in damped Lyman alpha systems. By converting part of their gas into stars, damped Lyman alpha systems will evolve into galaxies of the type we see today.

The study of Lyman alpha systems is a topic of much interest, since it gives us an unfolding picture of how galaxies and other structures in the universe evolve. In particular, observations of quasar absorption lines (see Figure 21.12, for instance) are well explained by simulations such as the one shown in Figure 21.11, which includes the effects of gravity, gas dynamics, and ionization by the ultraviolet background light. In such simulations, it is found that much of the volume of intergalactic space contains low-density ionized gas. The observed Lyman alpha forest is produced by denser neutral gas that tends to lie along filaments. The damped Lyman alpha lines are produced by high-column-density neutral gas that seems to correspond to the gas-rich disks of young galaxies.

▪ APPENDIX: SUPERLUMINAL RADIO SOURCES

Beginning in the 1970s, the technique of very-long-baseline interferometry allowed the imaging of quasars at resolutions of one milliarcsecond or less. These high-resolution

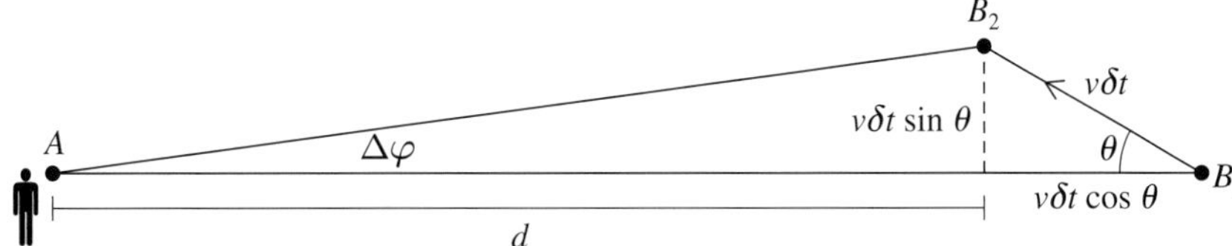

FIGURE 21.13 The observer at A makes two observations of a radio source, the first at position B_1 and the second when the radio source has moved to B_2.

radio images revealed that some radio-loud quasars have multiple components that are moving apart from each other with startlingly large proper motions. For instance, the quasar 3C 273 has a radio jet containing compact "knots" of emission. These knots are moving away from the base of the jet with a proper motion as large as $\mu'' \approx 7 \times 10^{-4}$ arcsec yr^{-1}. Given the distance to 3C 273 implied by its redshift, $d \approx 0.158(c/H_0) \approx 6.8 \times 10^8$ pc, this naïvely implies a tangential velocity (equation 19.22) of

$$v_t \approx 4.74(6.8 \times 10^8)(7 \times 10^{-4}) \text{ km s}^{-1} \approx 2 \times 10^6 \text{ km s}^{-1} \approx 7c. \tag{21.29}$$

This faster-than-light (or **superluminal**) tangential velocity led some astronomers to conclude that, in order to avoid violating special relativity, the jet of 3C 273 must be at less than 1/7 the distance implied by its redshift. However, it can be shown that *apparent* superluminal motion can result from relativistic motion close to the line of sight.

The conditions under which apparent superluminal motion can occur are illustrated in Figure 21.13. An observer is at point A. At a time t_1, a radio source (a hot blob of gas, for instance) is located at point B_1. At some later time t_2, the same radio source is at point B_2. In moving from position B_1 to B_2, the radio source has traveled at an angle θ relative to the observer's line of sight. If the source is moving at a speed v, the distance between B_1 and B_2 is $v\delta t$, where $\delta t \equiv t_2 - t_1$.

The photons emitted by the radio source at time t_1 and position B_1 are detected by the observer at A at a time

$$t_1' = t_1 + \frac{d + v\delta t \cos\theta}{c}, \tag{21.30}$$

where the second term is the light travel time from B_1 to A. The photons emitted by the radio source at time t_2 and position B_2 are detected at a time

$$\begin{aligned} t_2' &= t_2 + \frac{1}{c}\left[d^2 + v^2\delta t^2 \sin^2\theta\right]^{1/2} \\ &\approx t_2 + \frac{d}{c}, \end{aligned} \tag{21.31}$$

where the second term is the light travel time from B_2 to A, assuming $v\delta t \ll d$. The interval between the two observations is thus

$$\begin{aligned}\Delta t = t_2' - t_1' &= t_2 - t_1 - \frac{v\delta t \cos\theta}{c} \\ &= \delta t - \frac{v\delta t \cos\theta}{c} \\ &= \delta t(1 - \beta\cos\theta),\end{aligned} \tag{21.32}$$

where in the last step we have used the common definition $\beta \equiv v/c$. If the radio source is moving toward the observer ($\cos\theta > 0$), then the time between observations, $\Delta t = t_2' - t_1'$, will be shorter than the time between emissions, $\delta t = t_2 - t_1$. If the radio source is moving almost straight toward the observer at highly relativistic speeds, then the factor $\beta\cos\theta$ will be nearly equal to 1, resulting in $\Delta t \ll \delta t$.

During the time interval Δt between observations, the radio source will move through an angle $\Delta\phi = v\delta t \sin\theta/d$ as seen by the observer. Thus, the observer will compute a proper motion (in radians per unit time)

$$\begin{aligned}\mu &= \frac{\Delta\phi}{\Delta t} = \frac{v\delta t \sin\theta}{d}\frac{1}{\delta t(1-\beta\cos\theta)} \\ &= \frac{v\sin\theta}{d(1-\beta\cos\theta)}.\end{aligned} \tag{21.33}$$

We can multiply the proper motion by the distance d to get the tangential velocity v_t; then, dividing this by the speed of light we obtain

$$\beta_t = \frac{\mu d}{c} = \frac{v\sin\theta}{c(1-\beta\cos\theta)} = \frac{\beta\sin\theta}{1-\beta\cos\theta}. \tag{21.34}$$

It is left as an exercise for the reader to show that the maximum value of $\beta_t(\theta)$ occurs when the radio source is moving at an angle $\theta^{\max} = \cos^{-1}\beta$ relative to the line of sight; in that case, the value of β_t is

$$\beta_t^{\max} = \frac{\beta}{(1-\beta^2)^{1/2}}. \tag{21.35}$$

As $\beta \to 1$, $\beta_t^{\max}$ can become arbitrarily high. For $\beta_t^{\max} \geq 7$, required to explain the apparent superluminal motion in the jet of 3C 273, the true velocity of the jet material must be $\beta \geq (49/50)^{1/2} \approx 0.99$, and it must be moving at an angle $\theta \approx \cos^{-1}\beta \leq 8^\circ$ relative to the line of sight.

PROBLEMS

21.1 The quasar PDS 456 has a redshift $z = 0.184$ and an apparent magnitude $m_V = 14.0$.

(a) What is the distance to this quasar?
(b) What is its absolute magnitude, M_V?

21.2 At what redshift will the Lyα line ($\lambda_0 = 1215$ Å) be centered in the Johnson–Cousins U band? (Quasars will be particularly easy to discover at this redshift.)

21.3 The Eddington limit applies to stars as well as to accreting black holes and places an upper limit on their mass. Using the mass–luminosity relation for high-mass stars (see Section 13.6), determine the maximum mass of a star that is stable against disruption by radiation pressure.

21.4 Stellar-mass black holes in close binary systems can have hot accretion disks. These systems, sometimes called "microquasars," are bright X-ray sources. Explain quantitatively why microquasars are so luminous in X-rays. (Hint: equation (21.25) gives the temperature of an accretion disk.)

21.5 Starting from equation (21.27), demonstrate that the angular diameter of an AGN's broad-line region is proportional to the square root of the AGN's bolometric flux. The apparently brightest AGN is NGC 4151, with bolometric flux $F_{\rm bol} = 1.2 \times 10^{-12}$ W m^{-2} and redshift $z = 0.00332$. What is the angular size of the broad line region of NGC 4151, measured in arcseconds?

21.6 If the widths of the broad-emission lines in AGN spectra were due to thermal broadening, how hot would the gas have to be? On what grounds can we exclude the possibility of pure thermal broadening?

21.7 Suppose that you have a spectrum of the quasar 3C 273, as shown in Figure 21.4. You measure the width of the Hβ emission line to be 3500 km s^{-1} and the optical flux to be $F_{\rm opt} = 10^{-13}$ W m^{-2}. From these data, estimate the following:

(a) the bolometric luminosity of the quasar, assuming $L_{\rm bol} \approx 9L_{\rm opt}$
(b) the size of the Hβ-emitting region
(c) the mass of the central black hole
(d) the Eddington luminosity
(e) the Eddington ratio

21.8 For a quasar jet, show that the maximum value of $\beta_t(\theta)$ occurs when the radio source in the jet is moving at an angle $\theta^{\rm max} = \cos^{-1}\beta$ relative to the line of sight; then prove that equation (21.35) is correct.

21.9 Most quasars do not show damped Lyman alpha lines in their "Lyman alpha forest." However, by observing many quasars, we can estimate that the mean free path λ between damped Lyman alpha absorbers is $\sim 70{,}000\,\text{Mpc}$. Assuming that these systems are associated with luminous galaxies, which have a space density $n \approx 0.01\,\text{Mpc}^{-3}$, what does this say about the size of the atomic hydrogen disks of typical luminous galaxies?

22 Clusters and Superclusters

If the universe is a loaf of raisin bread (a metaphor that we adopted in Section 20.5), then the raisins are not uniformly distributed through the loaf. Speaking less metaphorically, galaxies are not uniformly distributed through space. Figure 22.1 is a plot of the distribution of galaxies in the northern sky with $B < 19$ mag. The plot is based on the Shane–Wirtanen catalog of galaxies. Shane and Wirtanen photographed the portion of the sky visible from Lick Observatory, then spent long hours counting the $\sim 10^6$ galaxies detected on the photographic plates. In the plot of the Shane–Wirtanen counts, we definitely see a nonuniform distribution. People tend to use the words "bubbly" or "spongy" when they describe the large-scale distribution of galaxies. The galaxies lie along walls or filaments, with particularly strong concentrations in the clusters where filaments meet.

The universe shows **hierarchical structure**; that is, it contains structure on a very wide range of length scales.

- **Stars** (typical diameter $d \sim 10^6$ km) are found largely in gravitationally bound objects called galaxies, containing $10^6 \to 10^{12}$ stars, plus gas, dust, and dark matter.
- **Galaxies** (typical diameter $d \sim 10$ kpc) are found largely in gravitationally bound objects called groups or clusters, containing $10 \to 10^4$ galaxies, plus gas and dark matter.
- **Groups and clusters** (typical diameter $d \sim 1$ Mpc) are found largely in currently collapsing objects called **superclusters**.

Superclusters have a maximum length $d \sim 100$ Mpc and are among the largest structures in the universe.

22.1 ▪ CLUSTERS OF GALAXIES

Groups and clusters are both gravitationally bound aggregations of galaxies, differing only in the number of galaxies that they contain. Groups contain fewer galaxies than clusters; typically, collections of fewer than 50 galaxies are called groups, but the boundary between a large group and a small cluster is not sharply defined.

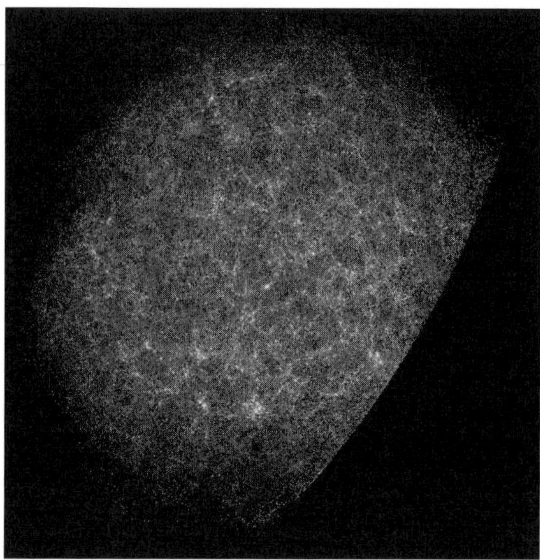

FIGURE 22.1 Distribution of galaxies in the Northern Galactic Hemisphere; the missing slice at lower right is inaccessible from Mount Hamilton, California, where Lick Observatory is located.

To start at home, our own galaxy is part of a group called the **Local Group**. The Local Group, mapped in Figure 22.2, contains at least 40 galaxies. The exact number of galaxies in the Local Group is not known, since most of the galaxies are inconspicuous dwarf spheroidal and dwarf irregular galaxies. Like most small clusters, the Local Group is irregularly shaped. Most of the galaxies clump around our own galaxy (labeled "Milky Way" in Figure 22.2) and M31 (also known as the Andromeda Galaxy). Our own galaxy and M31 contain most of the mass and luminosity in the Local Group. Third and fourth place go to M33 (an Sc galaxy) and the Large Magellanic Cloud. Most of the remaining galaxies are low-luminosity, low-surface brightness "fluff muffins."

The nearest cluster to the Local Group is the Virgo Cluster. The distance to the center of the Virgo Cluster is $d = 16$ Mpc, found using Cepheid stars as standard candles. The diameter of the Virgo Cluster is $D \approx 2$ Mpc; the caveat should be added that the Virgo Cluster is not particularly close to being spherical. Since the Virgo Cluster is both large and nearby (as clusters go), it subtends a large angle against the sky. It sprawls over a region $\sim 7°$ across, covering much of the constellation Virgo and stretching into Coma Berenices.

The four brightest galaxies in the Virgo Cluster (named M49, M60, M86, and M87) are all giant elliptical galaxies, swollen to a large size by cannibalizing smaller galaxies.[1] These four giant ellipticals each have an apparent magnitude $V \sim 9$ mag, so even in a

[1] As we've seen in Chapter 21, M87 is also a radio galaxy.

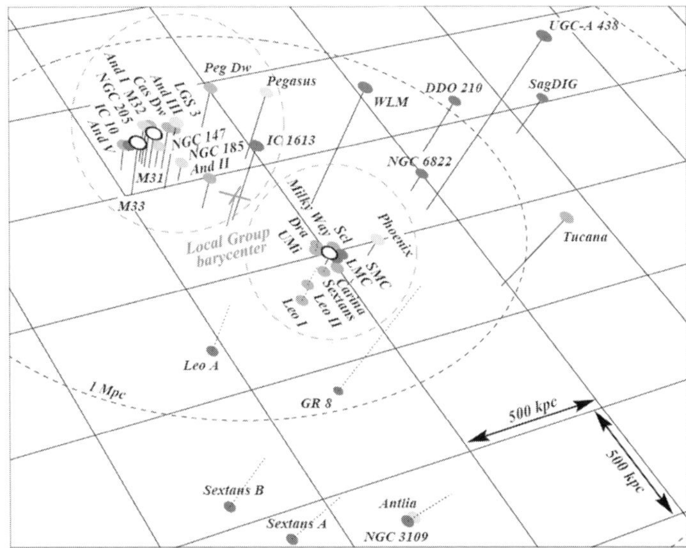

FIGURE 22.2 A map of the Local Group; note that the barycenter (a.k.a. the center of mass) is roughly midway between the Milky Way Galaxy and M31.

nearby cluster like Virgo, distance has rendered highly luminous galaxies invisible to the naked eye. There are several other Virgo-sized clusters within 60 Mpc of us (Abell 3655, Hydra, and Centaurus, for instance) as well as numerous smaller groups.

The nearest *rich* cluster of galaxies is the Coma Cluster, so called because it is in the constellation Coma Berenices, slightly north of the Virgo Cluster. A snapshot of the Coma Cluster (seen in the left panel of Color Figure 27) shows only the most luminous galaxies in the cluster. The two brightest galaxies in the cluster, NGC 4889 and NGC 4874, have an apparent magnitude $V \sim 13$ mag. Deeper images of the Coma Cluster reveal as many as 10,000 galaxies.

The distance to the Coma Cluster can be estimated from the Hubble law (see Section 20.5). The 100 brightest galaxies in the cluster have an average redshift of $\langle z \rangle = 0.0232$. This corresponds to a radial velocity

$$v_r = c\langle z \rangle = 6960 \text{ km s}^{-1}. \tag{22.1}$$

The distance to the Coma Cluster is, from the Hubble law (equation 20.30),

$$d_{\text{Coma}} = \frac{c}{H_0}\langle z \rangle = (4300 \text{ Mpc})(0.0232) = 100 \text{ Mpc}, \tag{22.2}$$

about six times the distance to the Virgo Cluster.

Knowing the distance, we can now estimate the absolute magnitude of the bright Coma galaxies, NGC 4889 and NGC 4874:

$$M_V = V - 5\log d + 5 \approx 13 - 5\log(10^8) + 5$$
$$\approx 13 - 40 + 5 \approx -22 \text{ mag}. \tag{22.3}$$

This is a magnitude brighter than M31 or the Milky Way, and comparable to the very brightest galaxies known. The luminosity of the Coma Cluster as a whole is estimated to be

$$L_B \approx 8 \times 10^{12} L_{B,\odot}. \tag{22.4}$$

Our own galaxy has $L_{B,\text{MW}} \approx 2 \times 10^{10} L_{B,\odot}$, so the luminosity of the Coma Cluster in the B band is equal to 400 times the luminosity of the Milky Way. Just as most of the visible light in a galaxy comes from a relatively few luminous stars, most of the visible light in a cluster comes from a relatively few luminous galaxies.

The mass of the Coma Cluster can be estimated from the virial theorem (equation 20.20):

$$M \approx 7.5 \frac{\sigma^2 r_h}{G}. \tag{22.5}$$

Now *galaxies*, not stars, are the individual masses whose line-of-sight velocity we measure. For the 100 brightest galaxies in the Coma Cluster, the dispersion in the radial velocities is

$$\sigma = 880 \text{ km s}^{-1} = 8.8 \times 10^5 \text{ m s}^{-1}. \tag{22.6}$$

The half-light radius of the cluster is

$$r_h = 1.5 \text{ Mpc} = 4.6 \times 10^{22} \text{ m}. \tag{22.7}$$

The virial theorem estimate of the mass is then

$$M \approx 7.5 \frac{(8.8 \times 10^5 \text{ m s}^{-1})^2 (4.6 \times 10^{22} \text{ m})}{6.67 \times 10^{-11} \text{ m}^3 \text{ s}^{-2} \text{ kg}^{-1}} \tag{22.8}$$
$$\approx 4 \times 10^{45} \text{ kg} \approx 2 \times 10^{15} M_\odot.$$

With a mass-to-light ratio $M/L_B \approx 250 M_\odot / L_{B,\odot}$, the Coma Cluster must be amply supplied with dark matter.

The mass of the Coma Cluster can also be estimated by looking at its X-ray emission (right panel of Color Figure 27). The X-ray emission, produced by hot intergalactic gas, looks smoothly distributed, which suggests that the hot gas is in hydrostatic equilibrium. From the X-ray spectrum, it is estimated that the average temperature of the hot gas is

$$\langle T_{\text{gas}} \rangle \approx 1 \times 10^8 \text{ K}, \tag{22.9}$$

and that the total amount of gas radiating X-rays is

$$M_{\text{gas}} \approx 2 \times 10^{14} M_\odot, \tag{22.10}$$

about 10% of the total mass of the cluster. Back when we were discussing stellar interiors (see Section 15.1), we noted that if a sphere of gas with radius R is in hydrostatic equilibrium, its central temperature must be

$$T_c \approx \frac{2GM\mu m_p}{Rk}, \tag{22.11}$$

if the pressure is described by a perfect gas law. If we substitute $R \sim 2r_h$ and $T_c \sim \langle T \rangle$, we can estimate the mass of the Coma Cluster:

$$M \sim \frac{r_h k \langle T \rangle}{G\mu m_p}. \tag{22.12}$$

With $\mu \approx 0.6$ (assuming a mix of ionized hydrogen and helium with a few metals), $r_h \approx 1.5$ Mpc, and $\langle T \rangle \approx 10^8$ K, the mass estimated from X-ray emission is

$$M \sim 10^{45}\,\text{kg} \sim 0.5 \times 10^{15} M_\odot. \tag{22.13}$$

Despite the crudity of our estimate, we get a mass in the same ballpark as our earlier mass estimate $M \approx 2 \times 10^{15} M_\odot$ from the virial theorem.[2]

22.2 ▪ WHEN GALAXIES COLLIDE!

The novel *When Worlds Collide*, by Wylie and Balmer, is a classic of science fiction. But how often do objects actually collide in the real universe? Consider a population of stars, each with radius R. One particular star is moving with a speed v relative to the average velocity of the stars in its vicinity. When its center comes within a distance $2R$ of the center of any other star, the two stars will collide. During a time t, the star sweeps out a cylindrical volume of length vt and radius $2R$; any star whose center lies within this cylinder will collide with the moving star. The volume of the cylinder is

$$V(t) = vt(4\pi R^2). \tag{22.14}$$

If the number density of stars is n, the average number of stars colliding with our moving star will be

$$N_\star(t) = nV(t) = nvt(4\pi R^2). \tag{22.15}$$

The mean time between collisions for any individual star, $t_\star$, will be roughly the time required to make $N_\star = 1$, that is,

$$t_\star \approx \frac{1}{nv(4\pi R^2)}. \tag{22.16}$$

Scaled to the properties of stars in the solar neighborhood,

[2] More sophisticated models of the X-ray emission yield $M = (1.3 \pm 0.5) \times 10^{15} M_\odot$ within 3 Mpc of the cluster's center.

$$t_\star \approx 5 \times 10^{10}\,\text{Gyr}\left(\frac{R}{R_\odot}\right)^{-2}\left(\frac{v}{30\,\text{km s}^{-1}}\right)^{-1}\left(\frac{n}{0.1\,\text{pc}^{-3}}\right)^{-1}. \tag{22.17}$$

During the Sun's lifetime of 4.6 Gyr, its probability of slamming into another star is only 1 in 10 billion; no wonder such a collision hasn't occurred. In the central parsec of our galaxy, where $n \sim 10^7\,\text{pc}^{-3}$ and $v \sim 200\,\text{km s}^{-1}$, the time between collisions is just $t_\star \sim 80$ Gyr, nearly a billion times shorter than in the solar neighborhood. Since $t_\star$ near the Galactic center is only 8 times the age of the Galaxy, the probability that a low-mass star will undergo a collision during its lifetime is no longer negligible.[3]

We can also ask how often *galaxies* collide with each other. Galaxies are much larger than stars but are also more widely spaced. Let's take, as an example, the Coma Cluster, which contains about 10,000 detectable galaxies. The typical size of these galaxies is

$$r \sim 3\,\text{kpc} \sim 1.3 \times 10^{11} R_\odot. \tag{22.18}$$

The typical relative speed, assuming an isotropic velocity dispersion, is

$$v \sim \sqrt{3}\sigma \sim \sqrt{3}(880\,\text{km s}^{-1}) \sim 1500\,\text{km s}^{-1}. \tag{22.19}$$

The average number density of galaxies, assuming that half the galaxies are inside the half-light radius, is

$$\begin{aligned} n &\sim \frac{N/2}{(4\pi/3)r_h^3} \sim \frac{5000}{(4\pi/3)(1.5\,\text{Mpc})^3} \\ &\sim 350\,\text{Mpc}^{-3} \sim 3.5 \times 10^{-16}\,\text{pc}^{-3}. \end{aligned} \tag{22.20}$$

By plugging the above values for r, v, and n into equation (22.17), we find that the average time between collisions for a galaxy in the Coma Cluster is

$$t_\star \sim 17\,\text{Gyr} \sim 1.2 H_0^{-1}. \tag{22.21}$$

Thus, the collision time is comparable to the Hubble time, and a typical galaxy in the Coma Cluster is as likely as not to undergo a collision. In a rich cluster, collisions between galaxies are the rule rather than the exception. In poor clusters and groups, the velocities and number densities are lower, but there is still a significant probability of collisions. Notice that when two galaxies collide, the individual stars within the galaxy do not collide with each other. Their cross-sections are just too tiny, even when you jack up the relative velocities to $v \sim 1500\,\text{km s}^{-1}$ and double the number density n of stars.

For every head-on collision, where the galaxies actually interpenetrate, there are many close encounters, where there is no overlap, but the tidal distortions of each galaxy are

[3] If two main sequence stars collide, they can merge to form a single massive star. There is evidence for such stellar merger remnants in the cores of globular clusters. Since globular clusters have ages of $t \sim 10$ Gyr or more, all stars with initial mass $M > 1M_\odot$ should have evolved off the main sequence. However, globular clusters are observed to have a few stars on the main sequence with $M > 1M_\odot$. These stars are called "blue stragglers," since they are hotter and bluer than expected for main sequence stars in an old cluster. Color Figure 28 shows the H–R diagram for the globular cluster M55, with its blue stragglers indicated.

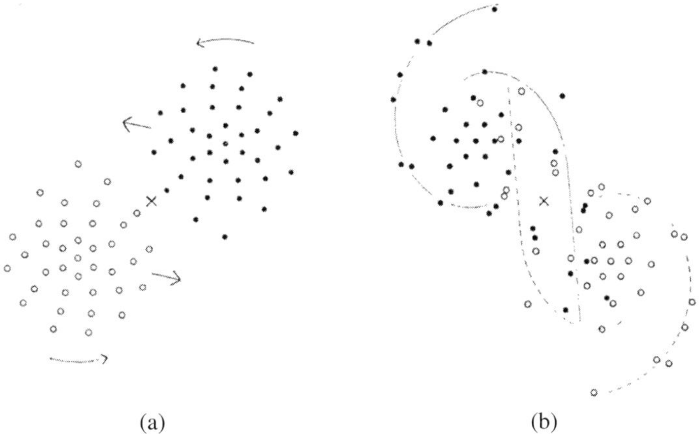

FIGURE 22.3 One of Holmberg's simulated galaxy encounters. (a) Before closest approach. (b) After closest approach, with tidal tails sketched in.

significant. Suppose two galaxies are on hyperbolic orbits about their mutual center of mass. Near the time of closest approach, each galaxy raises tidal bulges on the other galaxy. If the differential tidal force is strong enough to liberate loosely bound stars from a galaxy, then the stars from the tidal bulge closer to the center of mass will have a faster orbital speed (and thus will *lead* the main body of the galaxy), while stars from the tidal bulge farther from the center of mass will have a slower orbital speed (and thus will *lag* the main body of the galaxy).

Galaxy interactions can be modeled using computer simulations. In a typical n-body simulation, the galaxies are approximated as a distribution of n point masses, interacting according to an inverse square law gravitational force. The grandmother of all n-body simulations was performed by Erik Holmberg in 1941; he used an analog computer consisting of $n = 74$ light bulbs being pushed across a $3\text{ m} \times 4\text{ m}$ patch of floor covered with black paper.[4] Despite the simplicity of Holmberg's "computer," he was able to reproduce the tidal tails seen in interacting galaxies (Figure 22.3). More recent numerical simulations (for example, Figure 22.4) can contain $n > 10^8$ particles. It's the same physics, though. If you've seen one inverse square law, you've seen them all.

Many interacting galaxies can be seen in the real universe, particularly in rich clusters, where close encounters are more frequent. The Coma Cluster, for instance, contains a pair of galaxies known as "the Mice" because of their long tidal tails (Figure 22.5). The Mice, known also as NGC 4676 A and B, are a favorite system of astronomers doing computer models of galaxy encounters. The Mice can be modeled as a pair of identical

[4] Since the flux from each bulb was an inverse square law, just like gravity, the computed acceleration for each bulb was taken to be equal to the net flux of light at the bulb's position. At each time step, the velocity of each bulb was recomputed, and the bulb was moved the appropriate distance across the black paper.

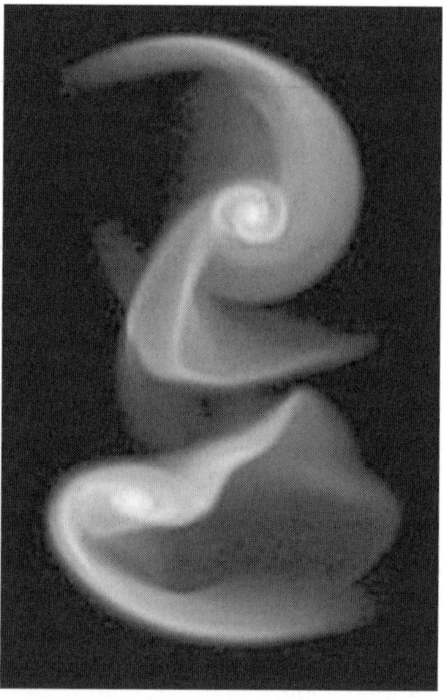

FIGURE 22.4 A recent simulated galaxy encounter.

FIGURE 22.5 The Mice ($d \approx 100$ Mpc).

spiral galaxies that have just passed pericenter in their encounter. ("Just," in this case, means that the closest approach was 1 or 2 hundred million years ago.)

The Cartwheel Galaxy, shown in Figure 22.6, is another interesting system involving interacting galaxies. It's located in the constellation Sculptor, about 130 Mpc away. The Cartwheel is a galaxy that recently had a smaller but denser galaxy zip through it at high speed—higher than the escape velocity from the Cartwheel Galaxy. The impulse provided by the high-speed intruder caused a circular "ripple" to run outward from the Cartwheel's center, similar to the ripple caused when a rock is dropped into a pool of

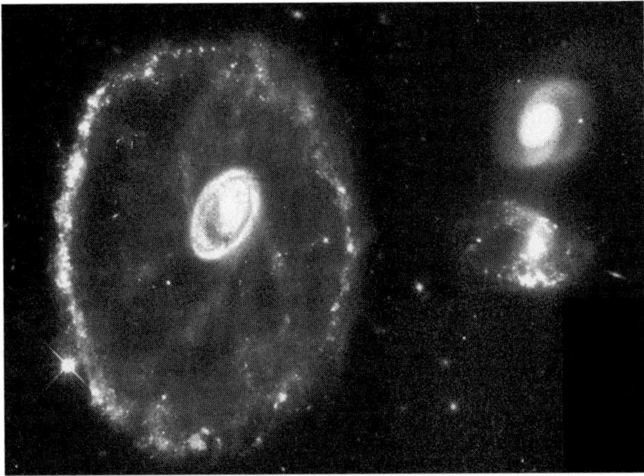

FIGURE 22.6 The Cartwheel and companions ($d \approx 130$ Mpc).

water. The circular ripple causes the Cartwheel Galaxy's ringlike appearance.[5] The outer ring of the Cartwheel is blue because the ripple running through the interstellar gas triggered star formation, creating hot, short-lived O and B stars.

If the relative speed of the galaxies is comparable to or less than the escape speed of the galaxies, then a nearly head-on encounter will result in the **merger** of the galaxies to form a single larger galaxy. Things to remember when galaxies collide:

- Individual stars do not collide with each other. There will not be a sudden bonanza of blue stragglers.
- Galaxy collisions are *inelastic*. That is, during the collision, some of the orbital kinetic energy of the galaxies is converted to internal energy, in the form of random motions of stars. Thus, a pair of galaxies that start out on hyperbolic orbits relative to each other can still end up as a bound system.
- Although stars don't collide, gas clouds do. A giant molecular cloud has $R_{\rm gmc} \sim 10\,{\rm pc} \sim 4 \times 10^8 R_\odot$. The large cross-section for molecular clouds means that the gas clouds will collide when two gas-rich galaxies pass through each other.

Thanks to the colliding gas clouds, merging galaxies are hotbeds of star formation. The class of galaxies known as **ultraluminous infrared galaxies**, or **ULIRGs**, are frequently found to be merging galaxies, in which large numbers of dust-enshrouded protostars produce copious amounts of far infrared light. Merging galaxies also produce lots of blue and ultraviolet light, thanks to the large numbers of O and B stars produced. The merging galaxies known as "the Antennae" (a tribute to their long tidal tails) are 14 Mpc away in

[5] The intruding galaxy is not necessarily either of the two galaxies on the right side of Figure 22.6. The high-speed culprit may already have fled the scene of the crime.

the constellation Corvus. The Antennae are very luminous in their central regions, both at blue wavelengths and at infrared wavelengths (Color Figure 29).

Galaxy mergers are like car crashes in that they are very effective at increasing entropy. Galaxies differ in the amount of entropy, or disorder, they contain. Spiral galaxies contain stars on orderly, one-way, nearly circular orbits; spirals are *low* in entropy. Elliptical galaxies contain stars on disorderly, randomly oriented orbits; ellipticals are *high* in entropy. When two neat, orderly compact cars collide, you don't get a neat, orderly SUV; you get a disordered heap of rubble. Similarly, when two neat, orderly spiral galaxies collide, you don't get a neat, orderly giant spiral; you get a disordered heap of stars (otherwise known as an elliptical galaxy).

The giant elliptical galaxies in the centers of rich clusters, like NGC 4889 and NGC 4874 in the middle of the Coma cluster, have grown to their present large size by the process of **galactic cannibalism**. The "cannibalism" refers to the fact that when a large galaxy merges with a small one, the large galaxy retains its identity, while the smaller one is disrupted. Sometimes "partially digested" cannibalized galaxies can be seen as multiple nuclei within a giant elliptical galaxy.

Our galaxy is a cannibal, as well, but only on a low level. It is currently tidally disrupting the Sagittarius dwarf galaxy and the Canis Majoris dwarf galaxy, which are disintegrating into long tidal streams that will eventually be mixed in with the halo of our galaxy. These minor encounters are trivial in comparison to our upcoming encounter with M31 (the Andromeda Galaxy). M31 is blueshifted with respect to the center of our galaxy. The relative velocity of the centers of the two galaxies is $v_r = -123\,\mathrm{km\,s^{-1}} = -126\,\mathrm{kpc\,Gyr^{-1}}$. The tangential velocity of M31 relative to our galaxy is unknown, thanks to the difficulty of measuring proper motions at large distances, but is likely to be small. Thus, in a time

$$t \sim \frac{d_{\mathrm{M31}}}{|v_r|} \sim \frac{700\,\mathrm{kpc}}{126\,\mathrm{kpc\,Gyr^{-1}}} \sim 6\,\mathrm{Gyr}, \tag{22.22}$$

M31 and our own galaxy will merge. Detailed computer simulations indicate that in ~ 3.2 Gyr, the two galaxies will have a close passage, inducing long tidal tails. The two distorted galaxies will fall back toward each other; about ~ 4.6 Gyr from now, they will have formed a large elliptical galaxy, surrounded by tidal debris. When the Sun becomes a red giant, it will be a member of an elliptical galaxy, not a spiral galaxy.

22.3 ▪ SUPERCLUSTERS AND VOIDS

In the hierarchy of structures in our universe, the average density of a star is greater than that of a galaxy. The average density of a galaxy is greater than that of a cluster. Finally, the average density of a cluster is greater than that of a supercluster. The superclusters consist of regions that are just now collapsing under their own gravity. For superclusters to be collapsing today, they must have a freefall time t_{ff} that is comparable to the age of the universe, which in turn is comparable to the Hubble time, H_0^{-1}. Using the value of the freefall time in terms of the average density ρ_0 of a structure (equation 17.6), we find that for superclusters,

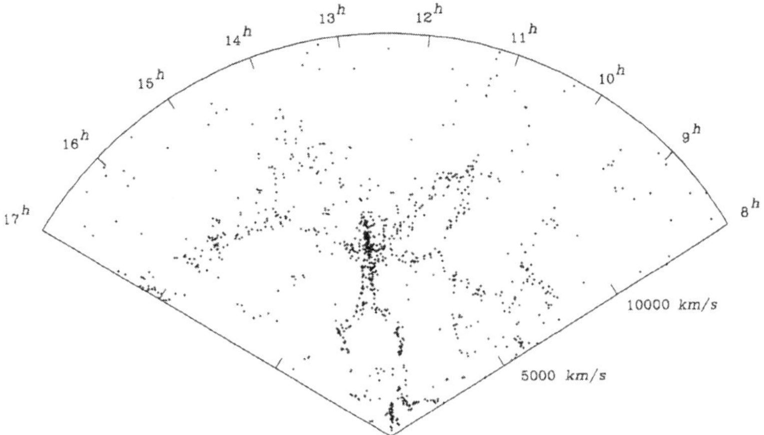

FIGURE 22.7 A slice of the universe: each dark dot represents a galaxy.

$$t_{\text{ff}} = \left(\frac{3\pi}{32G\rho_0} \right)^{1/2} \approx H_0^{-1}. \tag{22.23}$$

Thus, the average density of a supercluster must be

$$\rho_0 \approx \frac{3\pi H_0^2}{32G} \approx 2 \times 10^{-26}\,\text{kg m}^{-3} \approx 3 \times 10^{11} M_\odot\,\text{Mpc}^{-3}. \tag{22.24}$$

This is equivalent to 14 hydrogen atoms per cubic meter; even compared to the low-density coronal gas of the interstellar medium, the average density of a supercluster is not great.

The arrangement of clusters and groups into superclusters was first discovered by the astronomer Gerard de Vaucouleurs, who pointed out that the Local Group and the Virgo Cluster, along with other nearby groups and clusters, were arranged in a flattened "supercluster." This is now known as the Local Supercluster, or the Virgo Supercluster. More distant superclusters are most easily seen in three-dimensional maps of the universe, rather than in two-dimensional projections on the sky (such as Figure 22.1). A "redshift map" of the universe can be made by using the redshift z of a galaxy as a surrogate for its distance, since $d \propto z$ for small z. To make such a redshift map, you can start by measuring redshifts for galaxies in a long, narrow strip of the sky.

Among the earliest redshift maps of this kind were those produced in the 1980s by the CfA Redshift Survey.[6] A wedge of the universe from the CfA survey is shown in Figure 22.7. To make this redshift map, redshifts were measured for galaxies with right ascension $8^h < \alpha < 17^h$ and declination $26.5° < \delta < 32.5°$, down to a limiting apparent magnitude $B = 15.5$ mag. For the 1061 galaxies meeting these criteria, $v_r = cz$ was

[6] CfA stands for the Harvard–Smithsonian Center for Astrophysics.

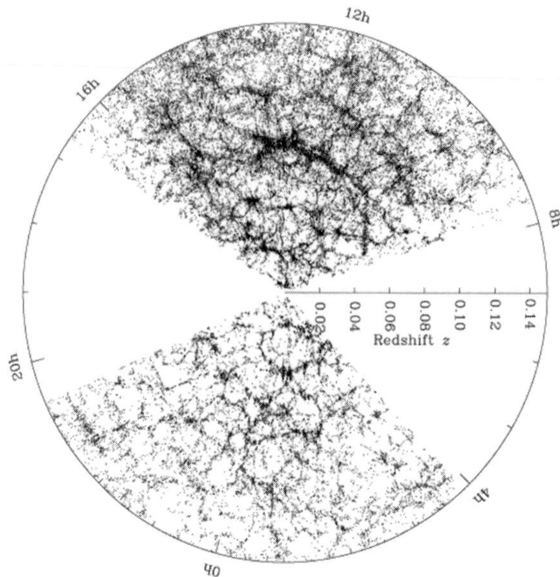

FIGURE 22.8 A bigger slice of the universe: each tiny dark dot represents a galaxy.

plotted versus right ascension. Note that dense regions in the redshift map tend to be elongated, while the underdense regions, or voids, are more nearly spherical (or more nearly circular, in the case of this thin slice).

A more recent redshift map, with a fainter apparent magnitude limit, is shown in Figure 22.8. This shows a slice from the Sloan Digital Sky Survey (SDSS). When the SDSS is complete, it will have provided redshifts for about a million galaxies in the Northern Galactic Hemisphere, down to a limiting apparent magnitude $m_r \approx 17.8$ mag.[7] This gives a detailed map of the galaxy distribution out to a redshift $z \sim 0.2$, corresponding to a distance $d \sim (c/H_0)z \sim 900$ Mpc. Thus, the Sloan Digital Sky Survey probes the universe to four times the distance of the CfA Redshift Survey, which reached to $z \sim 0.05$.

We learn something about the large-scale structure of the universe just from looking at redshift maps. Superclusters tend to be flattened pancakelike structures, or elongated filaments. Voids, by contrast, are more nearly spherical. Voids, by definition, have a very low density of bright galaxies. The density of nonstellar matter in voids is not always well determined. The transparency of intergalactic space places an upper limit on the density of dust: $\rho_{\rm dust} < 4 \times 10^{-30}\,{\rm kg\,m^{-3}}$. The limits on Lyman alpha absorption by neutral gas in voids places a stringent upper limit on the density of neutral hydrogen: $\rho_{\rm H} < 10^{-36}\,{\rm kg\,m^{-3}}$. However, the amount of ionized hydrogen and of dark matter in voids is not as well constrained.

[7] The Sloan r band is centered at a wavelength $\lambda \approx 6160$ Å and is roughly comparable to the Johnson R band.

Redshift maps should be used with the caveat that not all redshifts are due to the expansion of the universe. For instance, rich clusters in redshift maps show the "Finger of God" effect, which means that they are elongated in the radial direction. Figure 22.7 shows an example of a Finger of God. At the center of the figure is a structure that looks vaguely like a bowlegged stick figure. The elongated torso of the stick figure is actually the Coma Cluster. Note how the Coma Cluster is stretched out in the radial direction. The Finger of God effect received its name because the grossly elongated cluster is jokingly compared to God's finger pointing accusingly at the observer, while the Voice of God booms out, "You are WRONG!"

What the sinning observer has done wrong, in this case, is to assume that the redshift z measured for each galaxy in the cluster can be converted directly to a distance using the simple formula $d = (c/H_0)z$. In fact, if a cluster has a velocity dispersion σ along the line of sight, then some galaxies will have a radial velocity of approximately $+\sigma$ relative to the cluster's center of mass, while other galaxies will have a radial velocity of approximately $-\sigma$. Thus, two galaxies that are actually close to each other physically can have redshifts that differ by

$$\Delta z \sim \frac{2\sigma}{c}. \tag{22.25}$$

If we naïvely convert observed redshifts to distances using the Hubble law, we will conclude that the Finger of God has a length

$$\Delta d = \frac{c}{H_0}\Delta z \sim \frac{2\sigma}{H_0}. \tag{22.26}$$

The Coma Cluster, which has $\sigma = 880\,\text{km s}^{-1}$, appears to be stretched to a length of $\Delta d \sim 25\,\text{Mpc}$ in a redshift map, when its actual diameter is $d \sim 3\,\text{Mpc}$, an order of magnitude smaller.

Since the largest structures in the universe are superclusters and voids about 100 Mpc across, we expect that the region withing a few hundred megaparsecs of us, containing several superclusters and several voids, should be a fair sample of the universe at the present day. When we perform a census of galaxies in this region (out to $z \sim 0.05$), we can compute the number density of galaxies as a function of their luminosity. This function, known as the **luminosity function** of galaxies, is found to be well fitted by a power-law with an exponential cutoff:

$$\Phi(L) = \Phi_* \left(\frac{L}{L_*}\right)^\alpha \exp\left(-\frac{L}{L_*}\right)\frac{dL}{L_*}, \tag{22.27}$$

where $\Phi(L)dL$ is the number density of galaxies with luminosity in the range L to $L + dL$. The luminosity function of equation (22.27) is often called a Schechter function, after the astronomer who first applied it to the galaxy luminosity function. Schechter's own plot of the galaxy luminosity function is given in Figure 22.9. The exponential cutoff in the luminosity function occurs at

$$L_* \approx 2 \times 10^{10} L_\odot \approx L_{\text{MW}}. \tag{22.28}$$

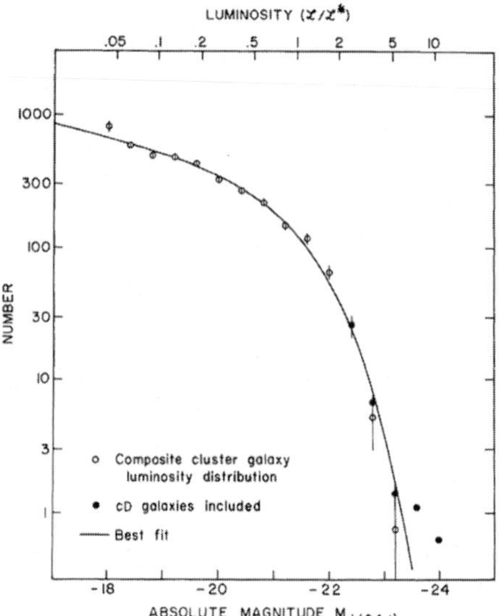

FIGURE 22.9 The number density of galaxies as a function of luminosity.

There are relatively few galaxies with luminosities greater than that of the Milky Way Galaxy, L_{MW}. Most of the scarce ultraluminous galaxies are fat "cannibals" in rich clusters of galaxies. At luminosities below the cutoff, the luminosity function is a power-law with index

$$\alpha \approx -1.2. \tag{22.29}$$

Thus, the luminosity function is weighted toward dim galaxies. Presumably, there is a cutoff at low luminosities as well as at high luminosities; otherwise, the total number density of galaxies,

$$n_{\text{gal}} = \int_0^\infty \Phi(L) dL, \tag{22.30}$$

would go to infinity as $L \to 0$. However, given the extreme difficulty of counting dwarf galaxies (we don't even know how many are in the Local Group, after all), how the luminosity function cuts off at low luminosities is not well determined.

The normalization of the luminosity function is

$$\Phi_* \approx 0.01 \, \text{Mpc}^{-3}. \tag{22.31}$$

The total luminosity density of galaxies (in $L_\odot \, \text{Mpc}^{-3}$) does not diverge. By integrating the luminosity function of equation (22.27) weighted by the luminosity, we find a total luminosity density[8]

$$\rho_L = \int_0^\infty L\Phi(L)dL = \Phi_* L_* \Gamma(2+\alpha) \approx 2.3 \times 10^8 L_\odot \text{ Mpc}^{-3}. \qquad (22.32)$$

This is equivalent to having a single 40-watt light bulb inside a sphere 1 AU in radius. By terrestrial standards, the universe is a poorly lit place.

PROBLEMS

22.1 Suppose you are an astronomer orbiting a star somewhere near the middle of the Virgo Cluster, and are observing the Local Group.

(a) What would be the angular size of the Local Group?
(b) What would be the angular size of the Milky Way Galaxy?
(c) What would be the apparent B magnitude of the Milky Way Galaxy?

22.2 Compute the mean apparent V magnitude of a $P = 100$ day Cepheid star in the Coma Cluster ($d = 100$ Mpc).

22.3 How would the "Finger of God" effect change if the dominant motion of clusters was rotation?

22.4 Using data from this chapter, estimate the time it takes a galaxy in the Coma Cluster to cross from one side of the cluster to the other. Does this result tell you anything about whether or not the cluster is gravitationally bound?

22.5 In a few billion years, as we note in this chapter, our galaxy and the Andromeda Galaxy will merge. Compute the expected number of collisions between stars when this occurs. (Assume that the typical star in each galaxy is an M dwarf, and that their average space density is equal to that of the solar neighborhood.)

22.6 The number of galaxies brighter than $m_B = 12$ mag is about 0.014 per square degree of the celestial sphere. Suppose that you wanted to observe every galaxy brighter than $m_B = 16$ mag in one-quarter of the sky. How many galaxies would you expect in such a sample?

[8] Since most of the light is contributed by the bright galaxies with $L \gtrsim L_*$, the uncertainty in the faint end of the luminosity function doesn't greatly affect our estimate of the luminosity density of the universe.

23 Cosmology

Cosmology is the study of the universe, or cosmos, regarded as a whole. Some questions addressed by cosmologists are What is the universe made of? Is it finite or infinite in spatial extent? Did it have a beginning at some time in the past? Will it come to an end at some time in the future?

In addition to dealing with Very Big Things, cosmology also deals with very small things. Early in its history, as we'll see later on, the universe was very hot in addition to being very dense, and interesting particle physics phenomena were occurring. Thus, a brief review of elementary particle physics will be useful as a preface to this chapter. For most particle physics applications, the electron volt ($1\,\text{eV} = 1.602 \times 10^{-19}\,\text{J}$) tends to be an inconveniently small unit of energy. Thus, particle physicists tend to measure energy in units of MeV (10^6 eV), GeV (10^9 eV), or TeV (10^{12} eV).

The most cosmologically important particles are listed in Table 23.1.[1] The objects that surround us in everyday life are made of protons, neutrons, and electrons. Protons and neutrons are both examples of **baryons**, where a baryon is defined as a particle made of three quarks.[2] A proton (p) contains two "up" quarks, each with a charge $q = +2/3$, and a "down" quark, with a charge of $q = -1/3$. A neutron (n) contains one "up" quark and two "down" quarks. A proton has a mass (or equivalently, a rest energy) that is 0.1% less than that of a neutron. A free neutron is unstable, decaying into a proton with a decay time of $\tau_n = 940$ s, about a quarter of an hour.

Electrons (e^-) are examples of **leptons**, a class of elementary particles that are not made of quarks.[3] The mass of an electron is small compared to that of a proton or neutron; the electric charge of an electron is equal in magnitude, but opposite in sign, to that of a proton. On large scales, the universe seems to be electrically neutral, with equal numbers of protons and electrons. The component of the universe made of atoms, molecules, and ions is called **baryonic matter**, since only the baryons contribute significantly to the mass density.

[1] Other particles, exotic by current standards, were abundantly present during the first few seconds of the universe. Since then, however, the particles we tabulate have been the most abundant.

[2] "Baryon" comes from the Greek root *barys*, meaning "heavy" or "weighty." A *barometer* measures the weight of the atmosphere, and the *barycenter* of the Local Group (see Figure 22.2) is the center of gravity, or center of mass.

[3] "Lepton" comes from the Greek root *leptos*, meaning "small" or "thin." In Greece, the euro cent (1/100 of a euro) is called a *lepton*, since it is the smallest coin minted.

TABLE 23.1 Particle Properties

Particle	Symbol	Rest energy (MeV)	Charge
proton	p	938.3	+1
neutron	n	939.6	0
electron	e^-	0.511	−1
neutrino	ν_e, ν_μ, ν_τ	$< 2 \times 10^{-6}$	0
photon	γ	0	0
dark matter	?	?	0

Neutrinos (ν) are also leptons. Neutrinos have no electric charge and interact with other particles only through the weak nuclear force or gravity. There are three types, or flavors, of neutrinos: electron neutrinos (ν_e), muon neutrinos (ν_μ), and tau neutrinos (ν_τ). Although recent experiments indicate that the different neutrino types have different masses, those masses must be small compared to the electron mass, with $m_\nu c^2 < 2$ eV being the approximate upper limit on the rest energy.

A particle known to be massless is the photon (γ). Unlike neutrinos, photons interact readily with electrons, protons, and neutrons. Although photons are massless, they have an energy $E = hc/\lambda$, where λ is the wavelength.

The most mysterious component of the universe is the dark matter. As discussed in Section 19.2, some of the dark matter may be baryonic (in the form of brown dwarfs or other dense, dim MACHOs). Some of the dark matter, but not much, is contributed by the lightweight neutrinos. It is likely that some of the dark matter is contributed by WIMPs, weakly interacting massive particles that are far more massive than neutrinos.

23.1 ▪ BASIC COSMOLOGICAL OBSERVATIONS

Observations of the universe around us have led cosmologists to adopt the **Hot Big Bang** model, which states that the universe has expanded from an initial hot and dense state to its current cooler and lower-density state, and that the expansion is continuing today. Several observations have contributed to the acceptance of the Hot Big Bang model. Many of these observations are recent and depend on sophisticated technology. However, the first observation on which the Hot Big Bang is based is ancient and requires nothing more sophisticated than your own eyes.

The first observation underpinning modern cosmology is this: **The night sky is dark.** When you go outside on a clear night and look upward, you see scattered stars on a dark background. The fact that the night sky is dark at visible wavelengths, rather than being uniformly bright with starlight, is known as Olbers's paradox, after the astronomer Heinrich Olbers, who wrote a paper on the subject in the year 1826.[4] Olbers was not

[4] Closer to home, Olbers is also known as the discoverer of the asteroids Pallas and Vesta.

actually the first person to think about Olbers's paradox; as early as 1576, Thomas Digges was worrying in print about the darkness of the night sky.

Why should the darkness of the night sky be paradoxical? First, consider the light from a single star of luminosity L at a distance r. The flux from the star is given by the inverse square law:

$$F = \frac{L}{4\pi r^2}. \tag{23.1}$$

The solid angle subtended by the star is also inversely proportional to its distance; if the star's radius is $R_\star$, its angular area (in steradians) is

$$d\Omega = \frac{\pi R_\star^2}{r^2}. \tag{23.2}$$

This means that the surface brightness $\Sigma_\star$ of the star (in watts per square meter per steradian) is independent of distance:

$$\Sigma_\star = \frac{F}{d\Omega} = \frac{L}{4\pi^2 R_\star^2} = 2.0 \times 10^7 \text{ W m}^{-2} \text{ ster}^{-1} \left(\frac{L}{L_\odot}\right)\left(\frac{R_\star}{R_\odot}\right)^{-2}. \tag{23.3}$$

For a Sun-like star, this corresponds to $\Sigma_\star \sim 0.5 \text{ mW m}^{-2} \text{ arcsec}^{-2}$. Since even nearby stars, like those of the Alpha Centauri system, have angular areas $d\Omega < 10^{-5} \text{ arcsec}^2$, any individual star other than the Sun will cover only a tiny fraction of the celestial sphere and contribute only a tiny flux here at Earth. But what if the universe stretches to infinity in all directions?

Let $n_\star$ be the average number density of stars in the universe, and let L and $R_\star$ be the average stellar luminosity and radius. Consider a thin spherical shell of radius r and thickness dr centered on the Earth (Figure 23.1). The total number of stars in the shell will be

$$dN_\star = n_\star 4\pi r^2 dr. \tag{23.4}$$

Since each star covers an angular area $d\Omega = \pi R_\star^2/r^2$, the fraction of the shell's area covered with stars will be

$$df = \frac{dN_\star d\Omega}{4\pi} = n_\star \pi R_\star^2 dr, \tag{23.5}$$

independent of the radius r of the shell. The fraction of the sky covered by stars within a distance r of us will then be

$$f = \int_0^r df = n_\star \pi R_\star^2 \int_0^r dr = n_\star \pi R_\star^2 r. \tag{23.6}$$

The coverage becomes complete when $f \approx 1$, corresponding to a distance

$$r_{\text{Olb}} \approx \frac{1}{n_\star \pi R_\star^2}. \tag{23.7}$$

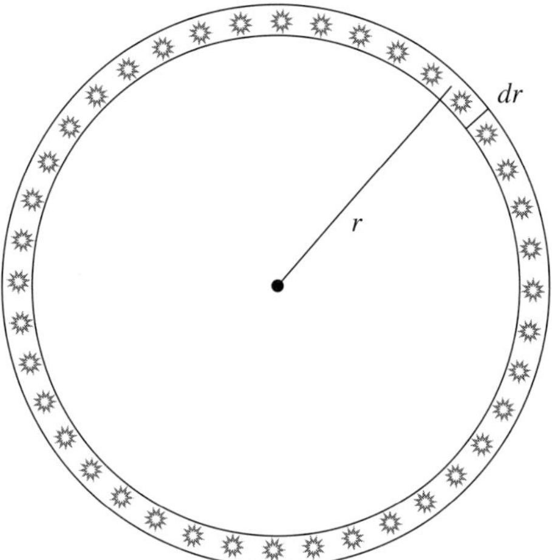

FIGURE 23.1 A star-filled spherical shell.

Thus, if the universe extends for a distance $r \geq r_{\rm olb}$, the sky must be uniformly bright, with a surface brightness equal to that of a typical star.

Obviously, the sky is *not* uniformly bright; at least one of the assumptions that went into our calculation must be wrong. One assumption we made was that $n_\star$ and R_* were independent of distance. This might be wrong. Distant stars might be smaller or less numerous than nearby stars.

A second assumption is that the universe is bigger than $r_{\rm olb}$. This might be wrong. If the universe stretches only to a distance $r_0 < r_{\rm olb}$, then the fraction of the night sky covered by stars will be

$$f \approx n_\star \pi R_\star^2 r_0 < 1, \tag{23.8}$$

and the average surface brightness of the sky will be

$$\Sigma_{\rm sky} = f\Sigma_\star \approx n_\star \pi R_\star^2 r_0 \frac{L}{4\pi^2 R_\star^2} \approx \frac{n_\star L r_0}{4\pi}. \tag{23.9}$$

This result will also be found if the universe is infinitely large but empty of stars beyond a distance r_0.

A third assumption, slightly more subtle, is that the universe is infinitely old. This might be wrong. If the universe has a finite age t_0, then the greatest distance we can see is $r_0 \approx ct_0$, and the average surface brightness of the sky will be

$$\Sigma_{\rm sky} \approx \frac{n_\star L c t_0}{4\pi}. \tag{23.10}$$

This result will also be found if the universe is eternally old but has contained stars only for a finite time t_0.

A fourth assumption made in computing the surface brightness is that the flux of stars is given by the inverse square law of equation (23.1). This might be wrong. The assumption that $F \propto r^{-2}$ follows directly from Euclid's laws of geometry. However, on large scales, the universe is under no obligation to be Euclidean. In some non-Euclidean geometries, the flux falls off more rapidly than an inverse square law.[5]

The darkness of the night sky caused astronomers to question many of their assumptions; an infinitely large, infinitely old, Euclidean universe can't stand up to close scrutiny.

The second observation on which modern cosmology is based is the Hubble law: **Galaxies show a redshift proportional to distance.** As noted in Section 20.5, the Hubble law is a natural consequence of homogeneous, isotropic expansion. If the expansion is perfectly homogeneous and isotropic, then the distance $r(t)$ between any two points can be written in the form

$$r(t) = a(t) r_0, \tag{23.11}$$

where $r_0 \equiv r(t_0)$ is the separation at the current time t_0, and $a(t)$ is a dimensionless function known as the **scale factor**. The homogeneity and isotropy of the expansion imply that $a(t)$ is not a function of position or direction but only of the time t. The distance between the two points will increase at the rate

$$v(r) = \dot{a} r_0 = \frac{\dot{a}}{a(t)}[a(t) r_0] = \frac{\dot{a}}{a(t)} r(t). \tag{23.12}$$

Thus, the velocity–distance relation takes the form of the Hubble law: $v(t) = H(t) r(t)$, where $H(t) \equiv \dot{a}/a$. The function $H(t)$ is called the **Hubble parameter**. Its value at the present day, $H_0 \equiv H(t_0)$, is called the **Hubble constant**.

Note how the Hubble law ties in with Olbers's paradox. If the universe is of finite age, $t_0 \sim H_0^{-1}$, then we expect that the **horizon distance**, the maximum distance from which light has had time to reach us, will be of the order $r_0 \sim c t_0 \sim c/H_0$. The luminosity density of starlight in the universe, computed in Section 22.3, is

$$n_\star L = \rho_L = 2.3 \times 10^8 L_\odot \,\mathrm{Mpc}^{-3}. \tag{23.13}$$

The average surface brightness of the sky should then be approximately

$$\begin{aligned} \Sigma_{\rm sky} &\sim \frac{n_\star L}{4\pi} \frac{c}{H_0} \sim \frac{(2.3 \times 10^8 L_\odot \,\mathrm{Mpc}^{-3})(4300\,\mathrm{Mpc})}{4\pi} \\ &\sim 8 \times 10^{10} L_\odot \,\mathrm{Mpc}^{-2}\,\mathrm{ster}^{-1} \sim 3 \times 10^{-8}\,\mathrm{W\,m^{-2}\,ster^{-1}}. \end{aligned} \tag{23.14}$$

When we compare this to the surface brightness of a Sun-like star,

$$\Sigma_\star \approx 2.0 \times 10^7 \,\mathrm{W\,m^{-2}\,ster^{-1}}, \tag{23.15}$$

[5] Of course, in other non-Euclidean geometries, the flux falls off less rapidly than an inverse square law, which will only increase the problem.

we find that $\Sigma_\star \sim 6 \times 10^{14} \Sigma_{\rm sky}$. Thus, for the entire sky to have a surface brightness as great as the Sun's, the universe would have to be 600 trillion times older than it is—*and* you'd have to keep the stars shining during all that time.

The primary resolution to Olbers's paradox is that the universe has a finite age. Stars beyond the horizon distance are invisible to us because their light hasn't had enough time to reach us. A secondary contribution to the darkness of the night sky is the redshift of distant light sources, close to the horizon, which reduces their flux as measured from Earth.

A third observation on which modern cosmology is based was made in the year 1965: **The universe is filled with a cosmic microwave background (CMB)**. The discovery of the CMB by Arno Penzias and Robert Wilson has entered cosmological folklore. Using a microwave antenna at Bell Labs, they discovered a slightly stronger signal than they expected from the sky. The extra signal was isotropic and constant with time. After removing all sources of noise that they could, they realized that they were truly detecting an isotropic background of microwave radiation. More recently, the *Cosmic Background Explorer (COBE)* satellite revealed that the CMB has a spectrum indistinguishable from a Planck function (equation 5.86) to a very high degree of accuracy:

$$I_\nu = \frac{2h\nu^3}{c^2} \frac{1}{e^{h\nu/kT_0} - 1}, \tag{23.16}$$

with a temperature

$$T_0 = 2.725 \pm 0.001\,\text{K}. \tag{23.17}$$

That is, the CMB is just what we would see if we were inside a hollow blackbody at a temperature of 2.725 K. The current energy density of the CMB is

$$u_0 = \frac{4\sigma_{\rm SB}}{c} T_0^4 = 4.17 \times 10^{-14}\,\text{J m}^{-3} = 0.260\,\text{MeV m}^{-3}, \tag{23.18}$$

where $\sigma_{\rm SB}$ is the Stefan–Boltzmann constant. The average energy of a single CMB photon, integrating over the complete Planck spectrum, is

$$\varepsilon_0 = 2.7kT_0 = 6.34 \times 10^{-4}\,\text{eV}. \tag{23.19}$$

The average photon energy ε_0 corresponds to a wavelength $\lambda_0 = hc/\varepsilon_0 \approx 2$ mm, in the microwave range of the electromagnetic spectrum (hence the name cosmic *microwave* background). The number density of CMB photons is

$$n_0 = \frac{u_0}{\varepsilon_0} = \frac{2.60 \times 10^5\,\text{eV m}^{-3}}{6.34 \times 10^{-4}\,\text{eV}} = 4.11 \times 10^8\,\text{m}^{-3}. \tag{23.20}$$

In an expanding Big Bang universe, cosmic background radiation arises naturally if the universe was initially very hot in addition to being very dense. Suppose the initial temperature was $T \gg 10^4$ K. At such high temperatures, the baryonic matter in the universe was completely ionized (Figure 23.2), and scattering of photons from the free electrons rendered the universe opaque. A dense, hot, opaque medium produces blackbody radiation, with a Planck spectrum. However, as the universe expanded, it

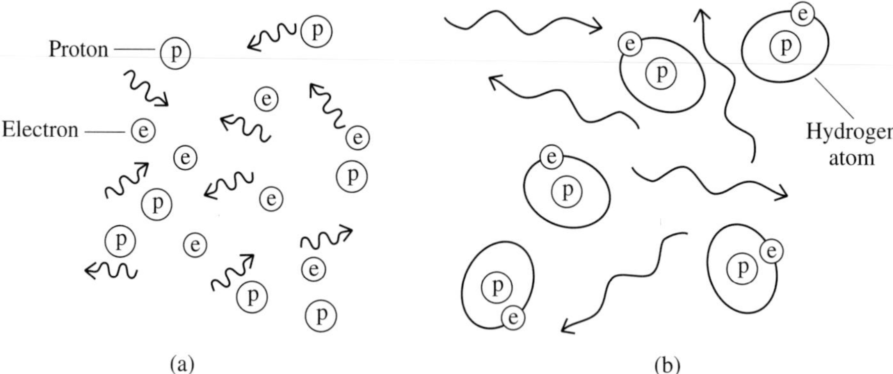

FIGURE 23.2 (a) The ionized, opaque universe before recombination. (b) The transparent universe after recombination.

cooled. When the temperature dropped to $T \sim 3000$ K, ions and free electrons combined to form neutral atoms. When the universe no longer contained a significant number of free electrons, the liberated blackbody photons started streaming through the universe, without further scattering.

At the time the universe became transparent, the temperature of the background radiation was $T \sim 3000$ K, about the temperature of an M star's photosphere. The temperature of the background radiation today is $T_0 = 2.725$ K, a factor of 1100 lower. Why has the background radiation cooled? It's a consequence of the expansion of the universe.

Consider a region of volume V that expands along with the universe, so that $V(t) \propto a(t)^3$, where $a(t)$ is the scale factor. The blackbody radiation within this volume can be thought of as a photon gas with energy density $u = (4\sigma_{SB}/c)T^4$ and pressure $P = u/3$. The photon gas within our volume obeys the first law of thermodynamics:

$$dQ = dE + PdV, \tag{23.21}$$

where dQ is the amount of heat flowing into or out of the volume, and dE is the change in the internal energy of the photon gas. In a homogeneous and isotropic universe, there is no flow of heat, since everything is at the same temperature; thus, $dQ = 0$. The first law of thermodynamics, applied to a gas in an expanding universe, then becomes

$$\frac{dE}{dt} = -P(t)\frac{dV}{dt}. \tag{23.22}$$

For the photons of the CMB, the internal energy is $E(t) = u(t)V(t) = (4\sigma_{SB}/c)T(t)^4V(t)$ and the pressure is $P(t) = (1/3)u(t) = (4\sigma_{SB}/3c)T(t)^4$. Equation (23.22) then becomes

$$\frac{4\sigma_{SB}}{c}\left(4T^3\frac{dT}{dt}V + T^4\frac{dV}{dt}\right) = -\frac{4\sigma_{SB}}{3c}T^4\frac{dV}{dt}, \tag{23.23}$$

or, with a little algebraic manipulation,

$$\frac{1}{T}\frac{dT}{dt} = -\frac{1}{3V}\frac{dV}{dt}. \tag{23.24}$$

Since $V(t) \propto a(t)^3$ as the universe expands, equation (23.24) can be rewritten in the form

$$\frac{1}{T}\frac{dT}{dt} = -\frac{1}{a}\frac{da}{dt}, \tag{23.25}$$

or

$$\frac{d}{dt}(\ln T) = -\frac{d}{dt}(\ln a). \tag{23.26}$$

This implies the simple relation $T(t) \propto a(t)^{-1}$; the temperature of the CMB drops as the universe expands. Note that it also implies $\varepsilon(t) \propto a(t)^{-1}$ for the average photon energy and $\lambda(t) \propto a(t)$ for the average photon wavelength. The background radiation has dropped in temperature by a factor of 1100 since the universe became transparent because the scale factor has grown by a factor of 1100 since then.

The observations we have noted so far—the dark night sky, the Hubble law, and the CMB—all fit neatly within the framework of the **Hot Big Bang** model for the universe, in which the universe was initially very hot and dense but has since cooled as it expanded. Although an exact treatment of how the universe expands requires knowledge of general relativity, many of the most important aspects of the expanding universe can be explained using purely Newtonian dynamics.

23.2 ▪ COSMOLOGY À LA NEWTON

Let's compute, using Newton's law of gravity and second law of motion, how the scale factor $a(t)$ depends on time. Consider a homogeneous sphere of matter, with fixed total mass M. The sphere is expanding (or contracting) homogeneously, so that its radius $r(t)$ is changing with time (Figure 23.3). Place a test mass, of infinitesimal mass m, at the surface of the sphere. The gravitational acceleration of the test mass will be

$$\frac{d^2r}{dt^2} = -\frac{GM}{r(t)^2}. \tag{23.27}$$

If we multiply each side of equation (23.27) by dr/dt and integrate over time, we find

$$\frac{1}{2}\left(\frac{dr}{dt}\right)^2 = \frac{GM}{r(t)} + k, \tag{23.28}$$

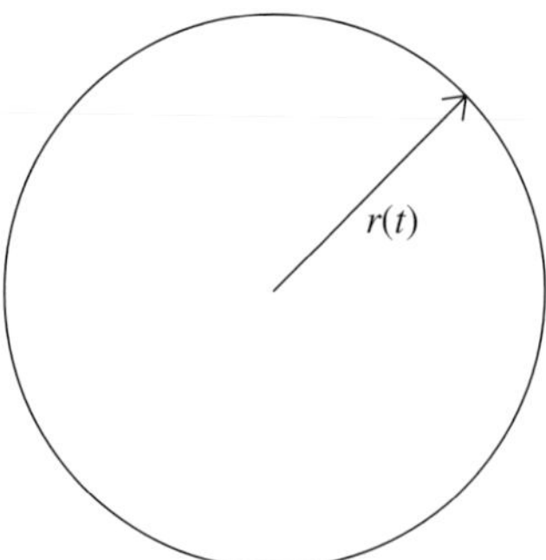

FIGURE 23.3 A sphere of fixed mass M and variable radius $r(t)$.

where k is the constant of integration. Equation (23.28) is an energy conservation statement. The sum of the kinetic energy per unit mass and the gravitational potential energy per unit mass is a constant (k) for a bit of mass at the sphere's surface.[6]

The future of an expanding, self-gravitating sphere falls into one of three classes, depending on the sign of the constant k. First, consider the case $k > 0$. In this case, the right-hand side of equation (23.28) is always positive. Therefore, the left-hand side of the equation never goes to zero, and the expansion continues forever. Second, consider the case $k < 0$. In this case, the right-hand side of equation (23.28) goes to zero at a maximum radius $r_{max} = GM/k$, and the expansion stops. However, at the maximum radius, the acceleration, given by equation (23.27), is still negative, so the sphere will then contract. Third and last, consider the case $k = 0$. This is the boundary case in which dr/dt asymptotically approaches zero as $t \to \infty$.

The three possible fates of an expanding sphere in a Newtonian universe are analogous to the three possible fates of a ball thrown upward from the Earth's surface. First, the ball can be thrown upward with a speed greater than the escape speed v_{esc}. In this case, the ball goes upward forever. Second, the ball can be thrown upward with a speed less than the escape speed. In this case, the ball reaches a maximum height, then falls back down. Third and last, the ball can be thrown upward with a speed exactly equal to v_{esc}. In this case, the speed of the ball asymptotically approaches zero as $t \to \infty$.

Equation (23.28), describing an expanding (or contracting) sphere, can be rewritten in such a way that it applies to a sphere of arbitrary radius and mass. The mass M, which is constant, can be written in the form

$$M = \frac{4\pi}{3}\rho(t)r(t)^3. \tag{23.29}$$

Since the expansion is isotropic about the center of the sphere, we can write

$$r(t) = a(t)r_0, \tag{23.30}$$

[6] Note also that the expansion velocity, dr/dt, enters equation (23.28) only as its square. This means that a contracting sphere ($dr/dt < 0$) is simply a time reversal of an expanding sphere ($dr/dt > 0$).

where $a(t)$ is the dimensionless scale factor and r_0 is the current radius of the sphere. Using these relations, equation (23.28) can be written in the form

$$\frac{1}{2}r_0^2\dot{a}^2 = \frac{4\pi}{3}Gr_0^2\rho(t)a(t)^2 + k, \tag{23.31}$$

or, dividing each side of the equation by $r_0^2a^2/2$,

$$\left(\frac{\dot{a}}{a}\right)^2 = \frac{8\pi G}{3}\rho(t) + \frac{2k}{r_0^2}\frac{1}{a(t)^2}. \tag{23.32}$$

The left-hand side of equation (23.32) is the square of the Hubble parameter, $H(t) \equiv \dot{a}/a$. Thus, we now have an equation that links the expansion rate of the universe to its mass density ρ. Equation (23.32) is called the **Friedmann equation**, after the Russian cosmologist Alexander Friedmann, who first found it (using a relativistically correct derivation) in the 1920s.

For a given value of the Hubble parameter, $H(t)$, there is a **critical mass density** $\rho_c(t)$ for which $k = 0$, and the universe is exactly on the boundary between eternally expanding ($k > 0$) and eventually recollapsing ($k < 0$). The value of the critical density is, from equation (23.32),

$$\rho_c(t) = \frac{3H(t)^2}{8\pi G}. \tag{23.33}$$

At the present moment in the real universe, $H_0 = 70$ km s^{-1} Mpc^{-1} and the value of the critical density is

$$\rho_{c,0} = \frac{3H_0^2}{8\pi G} = 9.2 \times 10^{-27}\,\text{kg m}^{-3} = 1.4 \times 10^{11} M_\odot\,\text{Mpc}^{-3}. \tag{23.34}$$

If the average density of the universe is greater than this value, then (if our Newtonian analysis is adequate) the universe will eventually collapse in a "Big Crunch." If the average density is less than or equal to this value, then it will expand forever in an increasingly tenuous "Big Chill." Is the average density greater than or less than $\rho_{c,0}$? It's not immediately obvious. Although $\rho_{c,0}$ is equivalent to a density of one hydrogen atom per 200 liters—much more tenuous than even the lowest density coronal gas in the interstellar medium—you must remember that most of the universe consists of very low density voids.

In a strictly Newtonian universe, the fate of the universe—Big Crunch or Big Chill—is determined solely by the ratio of the average mass density to the critical density. However, if a **cosmological constant** is present, then this ratio of densities no longer uniquely determines the ultimate fate of the universe. A cosmological constant is an entity that provides a positive acceleration ($\ddot{a} > 0$) to the expansion of the universe. The cosmological constant was introduced by Einstein in the context of general relativity. Since the Newtonian view is that gravity is always an attractive force ($\ddot{a} < 0$), it will be necessary for us to dabble in general relativity in order to understand the cosmological constant and the possibility of an accelerating universe.

23.3 ▪ COSMOLOGY À LA EINSTEIN

In Newton's view of the universe, space is static, unchanging, and Euclidean. In Euclidean, or "flat," space, all the axioms and theorems of plane geometry (as codified by Euclid around 300 BC) hold true. In Newton's view, an object with no net force acting on it moves through this Euclidean space with a constant velocity. However, when we look at real celestial objects (comets, planets, asteroids, and so forth) we find that their velocity is not constant; they move on curved lines with continuously changing speeds. Why is this? Newton would say, "Their velocities are changing because there is a force acting on them; the force called *gravity*."

Newton derived a useful formula for computing the gravitational force between two objects. Every object in the universe, said Newton, has a property that we may call the "gravitational mass." Let the gravitational mass of two spherical objects be m_g and M_g, and let the distance between their centers be r. The gravitational force acting between the objects is

$$F_{\rm grav} = -\frac{GM_g m_g}{r^2}, \tag{23.35}$$

where G is the Newtonian gravitational constant. The gravitational mass of an object is a nonnegative number, so the Newtonian gravitational force is always attractive, with $F_{\rm grav} \leq 0$. Newton also provided us with a useful formula that tells us how objects move in response to a force. Every object in the universe, said Newton, has a property that we may call the "inertial mass." If the inertial mass of an object is m_i, then if a net force F is applied to it, Newton's second law of motion tells us that its acceleration will be

$$a = F/m_i. \tag{23.36}$$

In equations (23.35) and (23.36), we use different subscripts to distinguish between the gravitational mass m_g and the inertial mass m_i. One of the fundamental principles of physics (a rather remarkable one, if you stop to think about it) is that the gravitational mass and the inertial mass of an object are identical:

$$m_g = m_i. \tag{23.37}$$

The equality of gravitational and inertial mass is known as the **equivalence principle**. The gravitational acceleration a of an object under the influence of a sphere of mass M_g will generally be

$$a = \frac{F_{\rm grav}}{m_i} = -\frac{GM_g}{r^2}\left(\frac{m_g}{m_i}\right). \tag{23.38}$$

If the equivalence principle didn't hold true, then different objects would fall at different rates in the Earth's gravitational field. The observation that $a = -9.8\ {\rm m\ s^{-2}}$ for all objects near the Earth's surface is supporting evidence that the equivalence principle holds true.

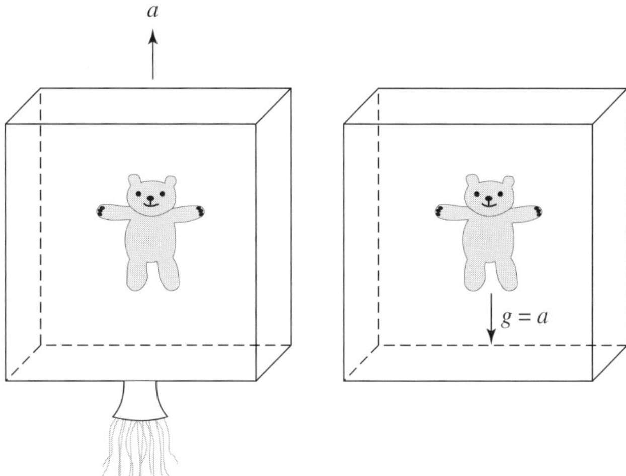

FIGURE 23.4 Equivalence principle (teddy bear version).

It is the equivalence principle that led Einstein to devise his theory of general relativity. To see why, let's do a thought experiment.[7] Suppose you wake up one morning to find that you've been sealed inside a small, opaque, soundproof box. You are so startled by this, you drop your teddy bear. Observing the falling bear, you find that it falls toward the floor with an acceleration $a = -9.8\ \mathrm{m\ s^{-2}}$. "Whew!" you say with relief. "At least I am still on the Earth's surface, and not being abducted by space aliens." At that moment, a window in the side of the box opens to reveal that you are in an alien spacecraft that is being accelerated at $a = 9.8\ \mathrm{m\ s^{-2}}$ by a rocket engine. When you drop a teddy bear, or any other object, in a small, sealed box, the equivalence principle allows two possible interpretations, illustrated in Figure 23.4: (1) The bear is moving at a constant velocity, and the box is being accelerated upward by a constant nongravitational force; or (2) The box is moving at a constant velocity (which may be zero), and the bear is being accelerated downward by a constant gravitational force. The observed behavior of the bear is the same in each case.

Now suppose you are still in the sealed box, being accelerated through space by a rocket at $a = 9.8\ \mathrm{m\ s^{-2}}$. You grab the flashlight you keep on the bedside table and shine a beam of light perpendicular to the acceleration vector (Figure 23.5). Since the box is accelerating upward, the path of the light will appear to you to be bent downward toward the floor, as the floor of the box accelerates upward to meet the photons. However, thanks to the equivalence principle, we can replace the accelerated box with a stationary box experiencing a constant gravitational acceleration. Since there's no way to distinguish between these two cases, we are led to the conclusion that the paths of photons will

[7] This thought experiment, as well as some other arguments in Chapters 23 and 24, are taken from *Introduction to Cosmology* (Barbara Ryden, 2003).

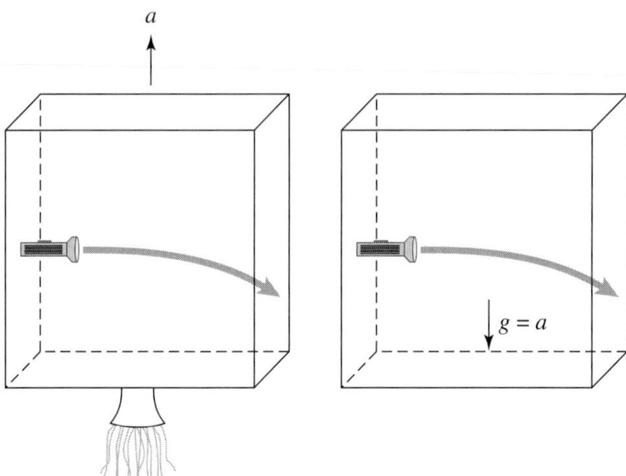

FIGURE 23.5 Equivalence principle (flashlight version).

be *curved* in the presence of a gravitational field. Gravity affects photons, Einstein concluded, even though they have no mass.

Contemplating the curved path of the photons, Einstein had another insight. A fundamental principle of optics is Fermat's principle, which states that light travels between two points along a path that minimizes the travel time.[8] In a vacuum, where the speed of light is constant, this translates into the requirement that light takes the shortest path between two points. In Euclidean space, the shortest distance between two points is a straight line. In the presence of gravity, however, the path taken by light in a vacuum is a *curved* line. This led Einstein to conclude that space is non-Euclidean.

The presence of mass, in Einstein's view, causes space to be curved. More broadly, in the theory of general relativity, mass and energy (which Newton thought of as very different things) are interchangeable, via the equation $E = mc^2$. Moreover, space and time (which Newton thought of as very different things) form a four-dimensional **spacetime**. A more complete summary of Einstein's viewpoint, then, is that the presence of mass-energy causes spacetime to be curved. This gives us a third way of thinking about the motion of the teddy bear in the box: (3) No forces are acting on the bear; it is simply following a **geodesic** in curved spacetime.[9]

In general, computing the curvature of spacetime is a complicated problem. Since the distribution of mass and energy is inhomogeneous on small scales, the curvature of space and time is also inhomogeneous, with strong curvature near black holes and neutron stars, and weak curvature in intergalactic voids. However, on scales bigger than 100 Mpc, the spatial distribution of mass and energy appears homogeneous and isotropic. Thus, we

[8] More precisely, Fermat's principle requires that the travel time be an extremum. Under most circumstances, the path minimizes travel time rather than maximizes it.

[9] The word "geodesic," in this context, is shorthand for "the shortest distance between two points."

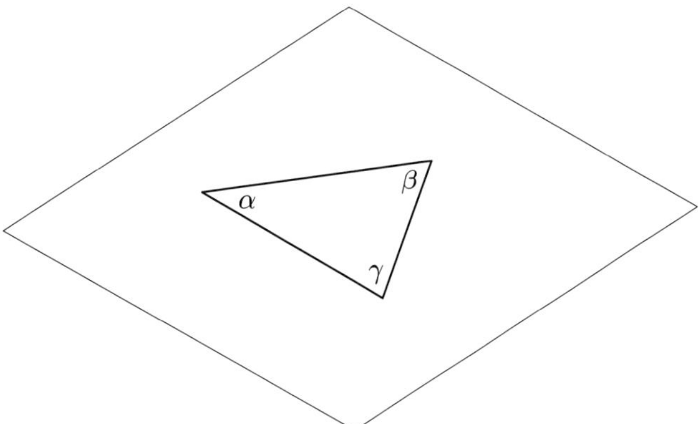

FIGURE 23.6 A flat, two-dimensional space (plane).

conclude that the curvature of space is also homogeneous and isotropic on large scales. The assumption of homogeneity and isotropy vastly simplifies the problem. There are only three basic geometries that space can have under such restrictive conditions. Since picturing the curvature of three-dimensional space is difficult, we'll start by considering the curvature of two-dimensional spaces, whose pictures can be neatly drawn on paper; later, we'll generalize to three dimensions.

First of all, space could be **flat**, or Euclidean. A picture of a flat two-dimensional space, otherwise known as a plane, is given in Figure 23.6. In flat space, all of Euclidean geometry holds true. For instance, in flat space, a geodesic is a straight line. If a triangle is constructed in flat space by connecting three points with geodesics, the angles at the vertices (α, β, and γ in Figure 23.6) must obey the relation

$$\alpha + \beta + \gamma = \pi, \tag{23.39}$$

when the angles are measured in radians. A plane has an infinite area,[10] and has no edge or boundary.

Another two-dimensional space with homogeneous, isotropic curvature is the surface of a sphere, as illustrated in Figure 23.7. On a sphere, a geodesic is a portion of a great circle.[11] If a triangle is constructed on the surface of a sphere by connecting three points with geodesics, the angles at its vertices (α, β, and γ in Figure 23.7) must obey the relation

$$\alpha + \beta + \gamma = \pi + A/r_c^2, \tag{23.40}$$

where A is the area of the triangle and r_c is the radius of the sphere. Spaces in which $\alpha + \beta + \gamma > \pi$ are called **positively curved** spaces. A sphere has a finite area, $4\pi r_c^2$, but no edge or boundary.

[10] Figure 23.6, of course, only shows a portion of a plane.

[11] If the Earth is approximated as a sphere, a line of constant longitude falls along a great circle. The equator is a great circle, but other lines of constant latitude are not.

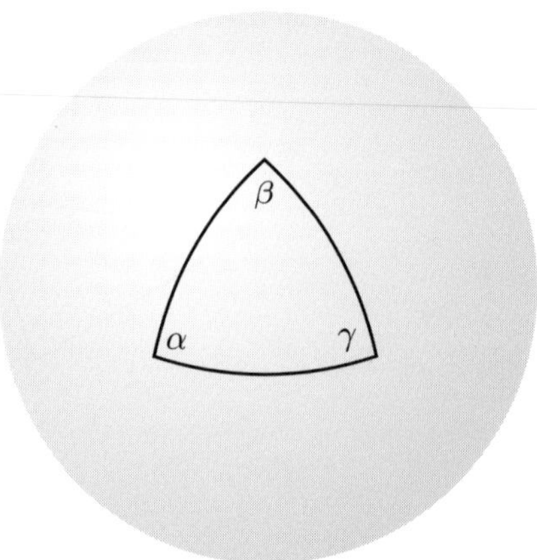

FIGURE 23.7 A positively curved, two-dimensional space (sphere).

In addition to flat spaces and positively curved spaces, there exist **negatively curved** spaces. An example of a negatively curved, two-dimensional space is the hyperboloid, or "saddle shape," shown in Figure 23.8. Consider a two-dimensional space of constant negative curvature, with radius of curvature r_c. If a triangle is constructed on this surface by connecting three points with geodesics, the angles at its vertices (α, β, and γ in Figure 23.8) must obey the relation

$$\alpha + \beta + \gamma = \pi - A/r_c^2, \tag{23.41}$$

where A is the area of the triangle. A surface of constant negative curvature has infinite area, just as a plane does.

If you want a two-dimensional surface to have homogeneous, isotropic curvature, only three cases fit the bill: it can be uniformly flat, it can have uniform positive curvature, or it can have uniform negative curvature. The same holds true for three-dimensional spaces. Thus, the curvature of homogeneous, isotropic space can be specified by just two numbers, κ and r_c. The number κ, called the **curvature constant**, is $\kappa = 0$ for flat space, $\kappa = +1$ for positively curved space, and $\kappa = -1$ for negatively curved space. If κ is not zero, then r_c, which has dimensions of length, is the **radius of curvature** of the space. Generally, $r_c(t)$ is a function of time, with $r_c(t) = a(t)r_{c,0}$ if the space is to remain homogeneous and isotropic.

So what is the curvature of the universe—positive, negative, or flat? As early as the year 1829, long before Einstein's parents were twinkles in his grandparents' eyes, the mathematician Nikolai Ivanovich Lobachevski, one of the founders of non-Euclidean geometry, proposed observational tests to determine the curvature of the universe. In principle, measuring the curvature is simple. Just draw a triangle, then measure its area A and the angles α, β, and γ at its vertices. From equations (23.39), (23.40), and (23.41), we know that

$$\alpha + \beta + \gamma = \pi + \frac{\kappa A}{r_{c,0}^2}, \tag{23.42}$$

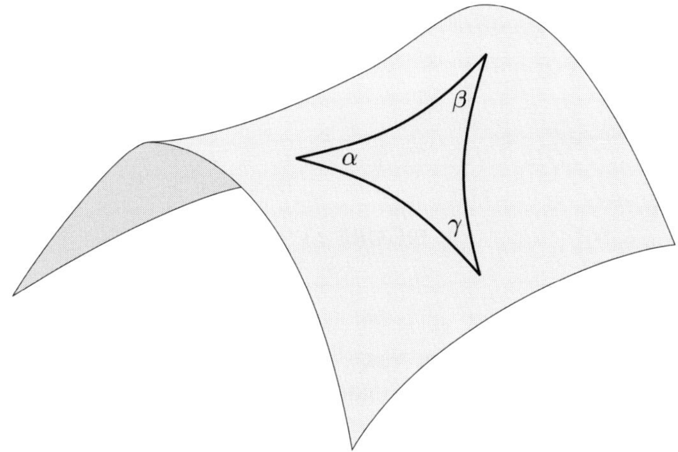

FIGURE 23.8 A negatively curved two-dimensional space (hyperboloid).

where κ is the curvature constant and $r_{c,0}$ is the present radius of curvature. Thus, we can compute

$$\frac{\kappa}{r_{c,0}^2} = \frac{\alpha + \beta + \gamma - \pi}{A}. \tag{23.43}$$

Unfortunately for this elegant plan, the deviation of $\alpha + \beta + \gamma$ from π radians is tiny unless the area of the triangle is comparable to $r_{c,0}^2$. Really, really big triangles are required.

We can conclude that if the universe is curved, with $\kappa = \pm 1$, the radius of curvature cannot be much smaller than the Hubble distance, $c/H_0 \approx 4300$ Mpc. To see why this is true, consider looking at a galaxy of diameter D that is at a distance d from the Earth (Figure 23.9). In a flat universe, in the limit $D \ll d$, we can use the small angle formula to compute the angular size α of the galaxy:

$$\alpha_{\text{flat}} = \frac{D}{d}, \tag{23.44}$$

where the angle α is in radians. However, in positively or negatively curved space, the angular size is no longer proportional to $1/d$.

In a space with uniform positive curvature, the angular size is

$$\alpha_{\text{pos}} = \frac{D}{r_{c,0}\sin(d/r_{c,0})} > \frac{D}{d}. \tag{23.45}$$

In a positively curved universe, the mass–energy content acts as a magnifying gravitational lens, making galaxies appear larger than they would in flat space. There are two interesting consequences of equation (23.45). First, the angular size blows up when $d = \pi r_{c,0}$; that is, when a galaxy is at a distance equal to half the circumference of the

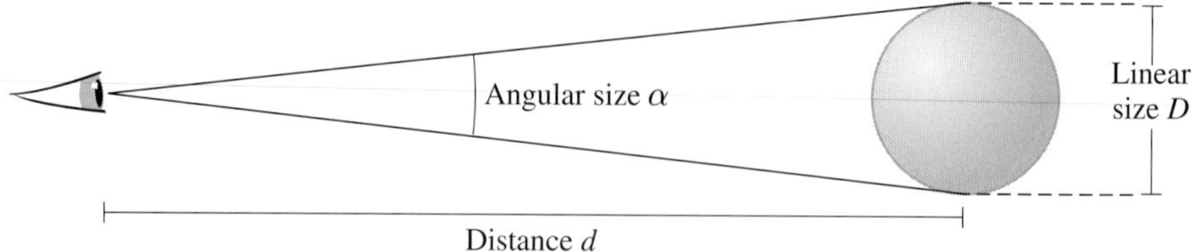

FIGURE 23.9 Angular size of a distant galaxy.

universe, it fills the entire sky.[12] No such enormous, sky-filling galaxies are seen. Second, since the universe has a finite circumference $C_0 = 2\pi r_{c,0}$, an object seen at a distance d will also be seen, with the same angular size, at a distance $d + C_0$, and at a distance $d + 2C_0$, and at a distance $d + 3C_0$, and so forth, ad nauseum. No such periodic galaxy images are seen. If the universe is positively curved, its radius of curvature must therefore be comparable to or greater than the Hubble distance.

In a space with uniform negative curvature, the angular size of a galaxy is

$$\alpha_{\rm neg} = \frac{D}{r_{c,0}\sinh(d/r_{c,0})} < \frac{D}{d}. \tag{23.46}$$

A negatively curved universe thus acts as a demagnifying lens.[13] If a galaxy is at a distance $d \gg r_{c,0}$, we can use the approximation $\sinh x \approx e^x/2$ when $x \gg 1$. With this approximation,

$$\alpha_{\rm neg} \approx \frac{2D}{r_{c,0}} e^{-d/r_{c,0}}. \tag{23.47}$$

In a negatively curved universe, objects at a distance much greater than the radius of curvature will appear exponentially tiny. Since galaxies are resolved in angular size, with $\alpha > 1$ arcsec, out to distances comparable to the Hubble distance, we conclude that if the universe is negatively curved, its radius of curvature must be comparable to or greater than the Hubble distance.

The conclusion of cosmologists, using geometrical arguments like the ones given above, is that the universe is consistent with being *flat* ($\kappa = 0$). Although we cannot rule out the possibility of slight positive or negative curvature, the radius of curvature in that case would be bigger than the Hubble distance, and would have negligible effects on the

[12] As a two-dimensional analogy, suppose that you were at the north pole of the Earth and a light source were at the south pole. If the light were constrained to follow great circles on the Earth's surface, it would flow along all the lines of longitude stretching away from the south pole and converge on your position at the north pole. No matter which way you turned, you would see the south pole beacon.

[13] Or a demagnifying rear view mirror: "Objects in mirror are closer than they appear."

small bit of the universe within a Hubble distance of us.[14] To make life simpler, we will assume, in many of the following equations, that the universe is perfectly flat.

23.4 ▪ METRICS OF SPACETIME

Astronomers study events that are widely spread out in space, and also widely spread out in time. Thus, it is useful for them to be able to compute the distance between two events in a four-dimensional spacetime. Computing the distance between two points in a flat, three-dimensional space is easy. If one point is at (x, y, z) and the other is at $(x + dx, y + dy, z + dz)$, the distance $d\ell$ between them is given by the formula

$$d\ell^2 = dx^2 + dy^2 + dz^2. \tag{23.48}$$

A formula such as equation (23.48) that gives the distance between two points is known as a **metric**. Equation (23.48) uses the convention, common among relativists, that $d\ell^2 = (d\ell)^2$, not $d(\ell^2)$; omitting the parentheses reduces visual clutter. The metric of flat space appears different when different coordinate systems are used. For instance, in spherical coordinates, the metric of flat space is

$$d\ell^2 = dr^2 + r^2(d\theta^2 + \sin^2\theta d\phi^2). \tag{23.49}$$

By extension, we can compute the four-dimensional spacetime distance between two events, one at (t, x, y, z) and the other at $(t + dt, x + dx, y + dy, z + dz)$. According to special relativity, the spacetime distance between these events is

$$\begin{aligned} d\ell^2 &= -c^2dt^2 + dx^2 + dy^2 + dz^2 \\ &= -c^2dt^2 + dr^2 + r^2(d\theta^2 + \sin^2\theta d\phi^2). \end{aligned} \tag{23.50}$$

The metric given in equation (23.50) is called the **Minkowski metric**, and the spacetime in which it holds true is called Minkowski spacetime. Note that the sign of the term involving time $(-c^2dt^2)$ is opposite to that of the terms involving the spatial coordinates.[15] The Minkowski metric applies only in the context of *special* relativity, which deals with the special case in which spacetime is not distorted by the presence of mass or energy. Thus, the Minkowski metric represents a static, empty, spatially flat universe.

In an expanding (or contracting) universe, the metric we use to measure spacetime distances is called the **Robertson–Walker metric**. If space is flat, then the Robertson–Walker metric takes the form

$$d\ell^2 = -c^2dt^2 + a(t)^2[dr^2 + r^2(d\theta^2 + \sin^2\theta d\phi^2)]. \tag{23.51}$$

[14] Similarly, a small bit of the Earth's curved surface is reasonably well described by a flat map. A flat map of the entire Earth results in distortions of size or shape (think of Greenland or Antarctica in a Mercator projection), but a flat map of Ohio doesn't have perceptible distortions.

[15] Some textbooks use the opposite sign convention: $d\ell^2 = c^2dt^2 - dx^2 - dy^2 - dz^2$. This is purely a formal convention and has no physical meaning. It's like the arbitrary pronouncement that electrons have negative charge and protons have positive; physics would be unchanged if we assigned + to electrons and − to protons.

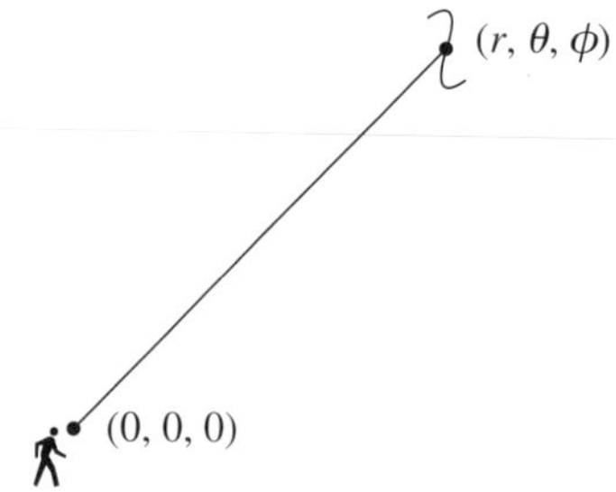

FIGURE 23.10 An observer looks at a galaxy.

Notice how the spatial component of the Robertson–Walker metric is scaled by the square of the scale factor $a(t)$. The time variable t in the Robertson–Walker metric is the **cosmic time**, which is the time measured by an observer who sees the universe expanding uniformly around him or her. The spatial variables (r, θ, ϕ) in the Robertson–Walker metric are the **comoving coordinates** of a point in space. If the expansion of the universe is perfectly homogeneous and isotropic, then the comoving coordinates of any point remain constant with time.[16]

Suppose you are observing a distant galaxy and want to know how far away it is. Since we are in an expanding universe, when we assign a distance ℓ between two objects (such as an astronomer and a galaxy), we must specify the time t at which that distance is correct. For convenience, let's set up a coordinate system in which you are at the origin and the galaxy is at a comoving coordinate position (r, θ, ϕ), as shown in Figure 23.10. The **proper distance** $\ell_p(t)$ between two points in space is the length of the geodesic between them when the cosmic time is fixed at the value t, and the scale factor is thus fixed at the value $a(t)$. The proper distance between an observer and a galaxy in a flat universe can be found by using the Robertson–Walker metric of equation (23.51) at fixed time t:

$$d\ell^2 = a(t)^2[dr^2 + r^2(d\theta^2 + \sin^2\theta d\phi^2)]. \tag{23.52}$$

Along the geodesic between the galaxy and observer, the angle (θ, ϕ) is constant, and thus

$$d\ell = a(t)dr. \tag{23.53}$$

The proper distance $\ell_p(t)$ is found by integrating over the radial comoving coordinate r:

$$\ell_p(t) = a(t)\int_0^r dr = a(t)r. \tag{23.54}$$

[16] Similarly, if the Earth were uniformly expanding or contracting with time, the latitude and longitude of any point would remain constant with time.

The normalization $a(t_0) = 1$ for the scale factor means that the comoving coordinate r is simply the current proper distance to the galaxy: $r = \ell_p(t_0)$.

Unfortunately, the proper distance $\ell_p(t_0)$ to a distant galaxy is impossible to measure, since we don't have gigaparsec-long tape measures that can be extended infinitely rapidly. As astronomers, we are condemned to a passive role; we learn what we can about the galaxy in Figure 23.10 by gathering up the photons that it emits. A photon that we collect at time t_0 was emitted at an earlier time $t_e < t_0$. Photons travel on geodesics through spacetime; more precisely, they travel on **null geodesics**. A null geodesic is a geodesic for which $d\ell = 0$ along every infinitesimal section of its path. Given equation (23.51), a photon must satisfy the relation

$$c^2 dt^2 = a(t)^2[dr^2 + r^2(d\theta^2 + \sin^2\theta d\phi^2)] \tag{23.55}$$

as it travels through an expanding, spatially flat universe. A photon traveling from the galaxy at (r, θ, ϕ) to an observer at the origin follows a beeline with θ and ϕ constant. This implies

$$c^2 dt^2 = a(t)^2 dr^2 \tag{23.56}$$

along every infinitesimal segment of the photon's radial path. Rearranging equation (23.56), we find

$$c\frac{dt}{a(t)} = dr, \tag{23.57}$$

in which the left-hand side depends only on t and the right-hand side depends only on r. Integrating along the photon's path,

$$c\int_{t_e}^{t_0} \frac{dt}{a(t)} = \int_0^r dr = r. \tag{23.58}$$

Since the comoving distance r is equal to the current proper distance $\ell_p(t_0)$, this implies that the proper distance is related to the scale factor by the relation

$$\ell_p(t_0) = c\int_{t_e}^{t_0} \frac{dt}{a(t)}. \tag{23.59}$$

In a static universe, where $a(t) = 1$ for all time, equation (23.59) states that the proper distance to a galaxy is equal to the speed of light times the photon's travel time: $\ell_p = c(t_0 - t_e)$. If the universe has been steadily expanding since t_e and t_0, then $a(t)$ was smaller in the past than it is now, and thus $\ell_p(t_0) > c(t_0 - t_e)$. In general, although the current proper distance $\ell_p(t_0)$ isn't something we can measure, it's something we can compute if we know $a(t)$.

Although we can't directly measure the current proper distance of a galaxy, there is a consolation prize; we can measure the galaxy's *redshift*. The redshift z tells us something useful: the scale factor $a(t_e)$ at the time the observed light was emitted. When we considered the cooling of the CMB, we learned that the wavelength of light expands along with the expansion of the universe: $\lambda(t) \propto a(t)$. This applies to all photons, not just CMB photons. If we observe a galaxy's emission line with wavelength λ_0 at time t_0, it was emitted with a shorter wavelength λ_e at an earlier time t_e. The relation between observed wavelength λ_0 and emitted wavelength λ_e is

$$\frac{\lambda_e}{a(t_e)} = \frac{\lambda_0}{a(t_0)}. \tag{23.60}$$

Using the definition of redshift,

$$z = \frac{\lambda_0 - \lambda_e}{\lambda_e}, \tag{23.61}$$

we find that the redshift is simply related to the scale factor at the time of emission:

$$1 + z = \frac{\lambda_e}{\lambda_0} = \frac{a(t_0)}{a(t_e)} = \frac{1}{a(t_e)}. \tag{23.62}$$

If we observe a quasar with $z = 6.4$, we are observing it as it was when the universe had a scale factor $a(t_e) = 1/7.4 = 0.135$.

The most distant objects we can see, in theory, are those for which the light emitted at time $t = 0$ is just now reaching us at $t = t_0$. The proper distance to such an object is called the **horizon distance**. In the limit $t_e \to 0$, equation (23.59) tells us that the current horizon distance is

$$\ell_{\rm hor}(t_0) = c \int_0^{t_0} \frac{dt}{a(t)}. \tag{23.63}$$

As an example, let's suppose that the scale factor is a power-law, with $a(t) = (t/t_0)^n$. If $n < 1$, the horizon distance is finite, with

$$\ell_{\rm hor}(t_0) = c \int_0^{t_0} \frac{dt}{(t/t_0)^n} = \frac{ct_0}{1-n}. \tag{23.64}$$

Since the Hubble constant is

$$H_0 = \left(\frac{\dot{a}}{a}\right)_{t=t_0} = \frac{n}{t_0}, \tag{23.65}$$

the horizon distance can also be written in the form

$$\ell_{\rm hor}(t_0) = \frac{n}{1-n}\frac{c}{H_0}, \tag{23.66}$$

when $0 < n < 1$. Thus, if we want to know the exact relation between the Hubble distance c/H_0 and the horizon distance, we need to know the functional form of $a(t)$.

23.5 ▪ THE FRIEDMANN EQUATION

In the context of general relativity, the form of $a(t)$ as well as the curvature constant κ and radius of curvature $r_{c,0}$ are dictated by Einstein's field equations. In general relativity, the field equations link the curvature of spacetime at any point to the energy density and pressure at that point.[17] The equation that links $a(t)$, κ, and $r_{c,0}$ is the **Friedmann equation**. We have already seen the Newtonian version of the Friedmann equation; it's the energy conservation equation for the expanding sphere (equation 23.32):

$$\left(\frac{\dot{a}}{a}\right)^2 = \frac{8\pi G}{3}\rho(t) + \frac{2k}{r_0^2}\frac{1}{a(t)^2}. \tag{23.67}$$

The relativistically correct form of the Friedmann equation is

$$\left(\frac{\dot{a}}{a}\right)^2 = \frac{8\pi G}{3c^2}u(t) - \frac{\kappa c^2}{r_{c,0}^2}\frac{1}{a(t)^2} + \frac{\Lambda}{3}. \tag{23.68}$$

Equation (23.68) is offered without proof. (A derivation should be done only by a highly trained relativist; please don't try this at home!)

Consider the changes made in going from the Newtonian form of the Friedmann equation to the relativistically correct form. First, the mass density ρ has been replaced by an energy density u. Relativistic particles, such as photons, have an energy $\varepsilon = hc/\lambda$ that contributes to the energy density. Not only do photons respond to the curvature of spacetime, they also contribute to it.

Second, in going from the Newtonian to the relativistic form, we make the substitution

$$\frac{2k}{r_0^2} \rightarrow -\frac{\kappa c^2}{r_{c,0}^2}. \tag{23.69}$$

In the Newtonian model, the constant k told us whether the universe was gravitationally bound ($k < 0$) or unbound ($k > 0$). In the relativistic model, the constant κ tells us whether the universe is positively curved ($\kappa > 0$) or negatively curved ($\kappa < 0$).

Third and last, in going from the Newtonian to the relativistic form, we add a new term, $\Lambda/3$, to the right-hand side of the equation. The Greek letter "Λ" is the symbol for the famous (or perhaps infamous) **cosmological constant**. The cosmological constant has a checkered history, going back to the year 1917, when Einstein published his first paper on the cosmological implications of general relativity. In a formal mathematical sense, Λ is a constant of integration resulting from solving Einstein's field equations, which are a set of differential equations. In addition, however, the cosmological component can be given a physical meaning.[18]

A close look at the Friedmann equation (eq. 23.68) shows that adding the Λ term is equivalent to adding a new component to the universe that has a constant energy density

$$u_\Lambda = \frac{c^2\Lambda}{8\pi G}. \tag{23.70}$$

[17] The field equations are the relativistic equivalent of Poisson's equation, which links the gravitational potential at any point to the mass density ρ at that point.

[18] Pedantic note: we are using the convention that Λ has units of 1/[time]^2. Other authors use a value of Λ that differs by a factor $1/c^2$, and thus has units of 1/[length]^2. Just a warning, in case you want to go browsing through the cosmological literature.

Thus, any component of the universe whose energy density is constant with time will play the part of a cosmological constant. One such component is the **vacuum energy**. In quantum physics, a vacuum is not a sterile void. The Heisenberg uncertainty principle allows particle/antiparticle pairs to spontaneously appear and then annihilate in an otherwise empty space. Just as there is an energy density u associated with real particles, there's an energy density $u_{\rm vac}$ associated with the virtual particles and antiparticles. The vacuum energy density $u_{\rm vac}$ is a small-scale quantum effect that is unaffected by the large-scale expansion of the universe; hence, $u_{\rm vac}$ remains constant as the universe expands. (Unfortunately, quantum field theory cannot tell us the expected numerical value of $u_{\rm vac}$.)

Let's rewrite the Friedmann equation in terms of the energy density of the universe, including the energy density u_Λ associated with the cosmological constant:

$$\left(\frac{\dot{a}}{a}\right)^2 = \frac{8\pi G}{3c^2}[u_r(t) + u_m(t) + u_\Lambda] - \frac{\kappa c^2}{r_{c,0}^2}\frac{1}{a(t)^2}. \tag{23.71}$$

We've subdivided the energy density into three categories. First, the **radiation density** u_r is the energy density contributed by relativistic particles, such as photons. Second, the **matter density** u_m is the energy density contributed by nonrelativistic particles such as protons, neutrons, electrons, and WIMPs. For nonrelativistic particles, $u_m = \rho_m c^2$. Finally, the **lambda density**, a.k.a. the vacuum density, is the constant energy density provided by the cosmological constant Λ.

The fact that our universe is flat (or very close to it) means that the total energy density is equal to the **critical energy density** (or very close to it). For perfect flatness ($\kappa = 0$),

$$u_r + u_m + u_\Lambda = u_c, \tag{23.72}$$

where the critical energy density is

$$u_c = \rho_c c^2 = \frac{3H(t)^2 c^2}{8\pi G}. \tag{23.73}$$

Since the Hubble parameter is currently $H_0 = 70\ \mathrm{km\ s^{-1}\ Mpc^{-1}}$, this translates into a current critical density

$$u_{c,0} = \frac{3H_0^2 c^2}{8\pi G} = 8.3 \times 10^{-10}\ \mathrm{J\ m^{-3}} = 5200\ \mathrm{MeV\ m^{-3}}. \tag{23.74}$$

This is one of the more fascinating results of general relativity. Because the universe is flat on large scales, we know the average energy density of the universe! Even if we don't know how much is contributed by each component, we know that the total must come to 5200 MeV $\mathrm{m^{-3}}$.[19]

Since the critical density $u_c(t)$ is vital to an understanding of the curvature and expansion of the universe, cosmologists frequently express the energy density of the

[19] That's the calorie content of a standard candy bar spread over a million cubic meters.

universe in terms of the dimensionless **density parameter**

$$\Omega(t) \equiv \frac{u(t)}{u_c(t)}. \tag{23.75}$$

If $\Omega < 1$, the universe is negatively curved; if $\Omega > 1$, the universe is positively curved. Saying "The universe is flat" is equivalent to saying "Omega equals one." By extension, we can write down a density parameter for each component of the universe:

$$\Omega_r(t) \equiv \frac{u_r(t)}{u_c(t)}, \qquad \Omega_m(t) \equiv \frac{u_m(t)}{u_c(t)}, \qquad \Omega_\Lambda(t) \equiv \frac{u_\Lambda}{u_c(t)}. \tag{23.76}$$

Knowing how the universe expands with time requires knowing how much energy density is in radiation, matter, and the cosmological constant today, and knowing how the energy density of radiation and matter evolves with time. (There are various exotic cosmologies that contain other components, like cosmic strings and domain walls and various types of dark energy, but for simplicity, we'll stick to a universe with just radiation, nonrelativistic matter, and a cosmological constant.)

PROBLEMS

23.1 Suppose that we smooth the Earth so that it's a perfect sphere of radius $R_\oplus = 6371$ km. If we then draw on its surface an equilateral triangle with sides of length $L = 1$ km, what will the sum of the interior angles be?

23.2 Imagine a universe full of regulation basketballs, each with mass $m_{\rm bb} = 0.62$ kg and radius $r_{\rm bb} = 0.12$ m.

(a) What number density of basketballs, $n_{\rm bb}$, is required to make the mass density equal to the current critical density, $\rho_{c,0} = 3H_0^2/(8\pi G)$?

(b) Given this density of basketballs, how far on average would you be able to see in any direction before your line of sight intersected a basketball?

(c) In fact, we can see galaxies at a distance $d \approx c/H_0 \approx 4300$ Mpc. Does the transparency of the universe on this length scale place useful limits on the number density of intergalactic basketballs?

23.3 Just as the universe has a cosmic microwave background dating back to the time when the universe was opaque to photons, it has a cosmic *neutrino* background dating back to the earlier time when the universe was opaque to neutrinos. The calculated number density of cosmic neutrinos is $n_\nu = 3.36 \times 10^8\ \text{m}^{-3}$.

(a) How many cosmic neutrinos are inside your body right now?

(b) What average neutrino mass, m_ν, would be required to make the mass density of cosmic neutrinos equal to the critical density $\rho_{c,0}$?

23.4 Suppose you are in a Newtonian universe whose density is equal to the critical density $\rho_{c,0}$. The scale factor $a(t)$ is implicitly given by the relation

$$\frac{\dot{a}^2}{a^2} = \frac{8\pi G\rho_{c,0}}{3}\frac{1}{a^3}.$$

(a) What is the functional form of $a(t)$, given the boundary condition $a = 1$ at $t = t_0$?
(b) What is t_0 in terms of the Hubble constant, H_0?
(c) In our universe, $H_0 = 70\ \mathrm{km\ s^{-1}\ Mpc^{-1}}$ and the oldest stars have an age $t_\star = 13\,\mathrm{Gyr}$. Are these two observations consistent with a Newtonian universe that has $\rho_0 = \rho_{c,0}$?

23.5 Prove that a redshifted blackbody is still a blackbody but at a temperature $T/(1+z)$.

24 History of the Universe

Knowing how the scale factor $a(t)$ grew in the past and predicting how it will change in the future is an important goal of cosmologists. The Friedmann equation tells us that the growth of the scale factor is related to the energy density of the universe. It is useful to divide the energy content into radiation (relativistic particles), matter (nonrelativistic particles), and a cosmological constant. This is because each of these components has an energy density with a different dependence on the scale factor.

A cosmological constant has an energy density u_Λ that is constant with time. To see how the energy density of radiation and matter behaves as the universe expands, consider a volume V that expands with the universe, so that $V(t) \propto a(t)^3$. If particles are neither created nor destroyed, then the number density of particles, n, is diluted by the expansion of the universe at the rate $n(t) \propto V(t)^{-1} \propto a(t)^{-3}$, as illustrated in Figure 24.1. The energy of the nonrelativistic particles is contributed entirely by their rest mass, $\varepsilon = mc^2$, which remains constant as the universe expands. Thus, for nonrelativistic particles, a.k.a. "matter," the energy density has the dependence

$$u_m(t) = n(t)\varepsilon = n(t)mc^2 \propto a(t)^{-3}. \tag{24.1}$$

The energy of relativistic particles, such as photons, has the dependence $\varepsilon(t) = hc/\lambda(t) \propto a(t)^{-1}$. Thus, for relativistic particles, a.k.a. "radiation," the energy density has the dependence

$$u_r(t) = n(t)\varepsilon(t) = n(t)hc/\lambda(t) \propto a(t)^{-4}. \tag{24.2}$$

Given the different rates of decrease for the energy density, we find that the total energy density u was contributed mainly by radiation at early times, when $a \ll 1$ (Figure 24.2). In the language of cosmologists, the early universe was "radiation dominated." If the universe has a positive cosmological constant Λ, then it becomes "lambda dominated" if it reaches a sufficiently large scale factor.

24.1 ▪ THE CONSENSUS MODEL

In recent years, cosmologists (ordinarily a contentious bunch) have found themselves approaching an approximate consensus on the curvature, contents, and age of the universe. The curvature is flat (or nearly so), implying that the energy density today is close to the

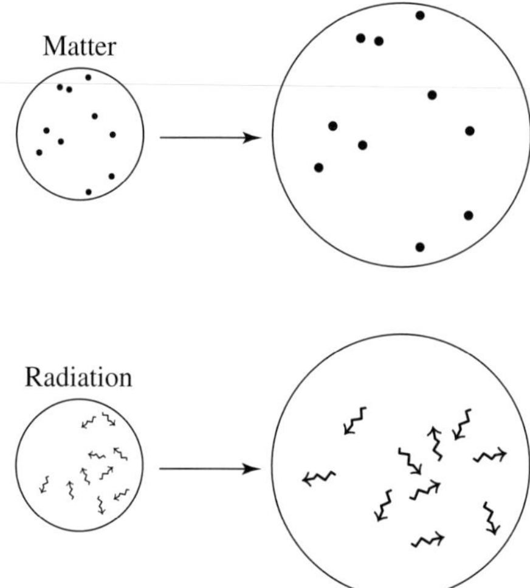

FIGURE 24.1 Dilution of nonrelativistic particles ("matter") and relativistic particles ("radiation").

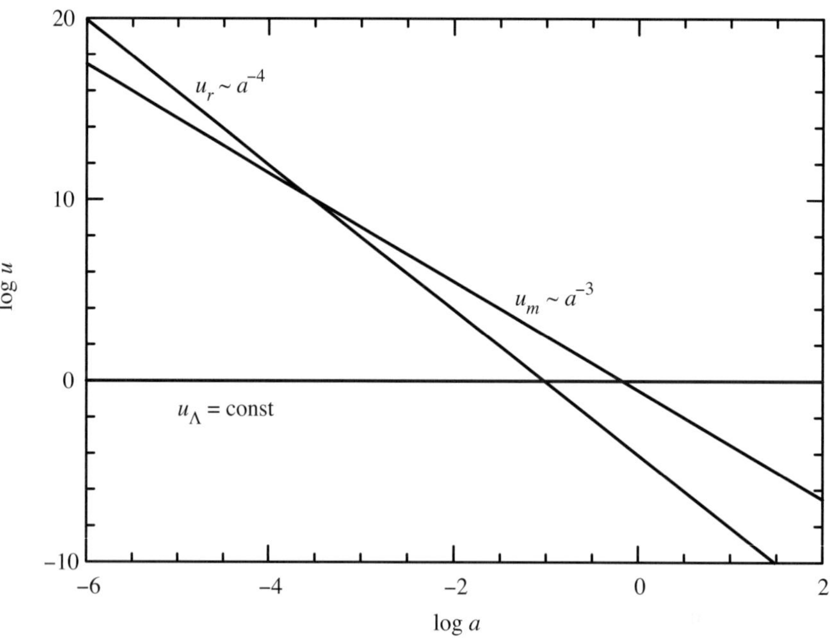

FIGURE 24.2 Dependence of energy density u on the scale factor $a(t)$.

critical density $u_0 \approx u_{c,0} \approx 5200\ \text{MeV m}^{-3}$. To see how this energy density is allocated among the different components, let's do a census of the universe.

Most of the energy density of *photons* is provided by the cosmic microwave background; although stars have been shining away for ~ 13 Gyr, starlight still provides less than 10% of the total photon energy of the universe.[1] The current energy density of the CMB, as computed in equation (23.18), is $u_{\text{cmb},0} = 0.260\ \text{MeV m}^{-3}$. The contribution of the CMB to the critical density is thus

$$\Omega_{\text{cmb},0} = \frac{u_{\text{cmb},0}}{u_{c,0}} = \frac{0.260\ \text{MeV m}^{-3}}{5200\ \text{MeV m}^{-3}} = 5.0 \times 10^{-5}. \tag{24.3}$$

The CMB is a relic of the time when the universe was hot and dense enough to be opaque to photons. If we extrapolate to earlier times and smaller scale factors, we reach a time when the universe was hot and dense enough to be opaque to *neutrinos*. Thus, there should be a cosmic neutrino background (CNB) analogous to the CMB. A detailed statistical mechanics calculation (of which we omit the details) reveals that the energy density of the CNB should be

$$u_{\nu,0} = 0.68 u_{\text{cmb},0} = 0.177\ \text{MeV m}^{-3}, \tag{24.4}$$

if the neutrinos are still traveling fast enough to qualify as relativistic particles today (that is, if the energy per particle, ε_ν, is much larger than the rest energy $m_\nu c^2$). The CNB has not yet been detected. The energy per neutrino is comparable to the energy per photon in the CMB—less than 10^{-3} eV. Detecting such low-energy neutrinos is not yet technically feasible.

If neutrinos are low enough in mass to be relativistic today, the present density parameter in radiation is

$$\Omega_{r,0} = \Omega_{\text{cmb},0} + \Omega_{\nu,0} = 1.68\Omega_{\text{cmb},0} = 8.4 \times 10^{-5}. \tag{24.5}$$

Thus, photons and neutrinos contribute a small fraction of the critical density today; about 1 part in 12,000. Most of the density must currently be provided by nonrelativistic matter and/or a cosmological constant (or by some other, hitherto unsuspected, component).

The energy density of the CMB has been measured with high precision; the energy density of the CNB has been computed using well-understood principles of physics. The energy density of matter is not as well determined. If we add together the mass of all the clusters of galaxies in our neighborhood, we find that the density of clustered matter is

$$\Omega_{\text{cluster},0} \approx 0.2. \tag{24.6}$$

This number doesn't include any smoothly distributed matter in the intercluster voids. The best estimate for the current density of nonrelativistic matter, using all available data, is

$$\Omega_{m,0} \approx 0.3. \tag{24.7}$$

[1] Thus, although photon number is not strictly conserved, as we assumed when computing $u_r \propto a^{-4}$, it's a better approximation than we might guess.

The best estimate for the current density of *baryonic* matter (that is, stuff made of protons, neutrons, and electrons) is

$$\Omega_{\text{bary},0} = 0.04. \tag{24.8}$$

(We see how this number is determined in Section 24.3.) The majority of the matter in the universe must consist of nonbaryonic dark matter, such as WIMPs.

The total mass density of baryonic matter today is

$$\rho_{\text{bary},0} = 0.04\rho_{c,0} = 3.7 \times 10^{-28}\,\text{kg m}^{-3}. \tag{24.9}$$

The number density of baryons is thus $n_{\text{bary},0} \approx \rho_{\text{bary},0}/m_p \approx 0.22\,\text{m}^{-3}$. This is much lower than the number density of photons. The photon-to-baryon ratio in the universe is approximately

$$\frac{n_{\text{cmb},0}}{n_{\text{bary},0}} \approx \frac{4.11 \times 10^8\,\text{m}^{-3}}{0.22\,\text{m}^{-3}} \approx 2 \times 10^9. \tag{24.10}$$

Baryons are badly outnumbered by photons, by a ratio of 2 billion to 1.

The available observational evidence has led cosmologists to a **Consensus Model of the Universe**. This model is flat and contains radiation, nonrelativistic matter, and a cosmological constant (a.k.a. Λ or "lambda," a.k.a. "vacuum energy," a.k.a. "dark energy"). Some of the current properties of the Consensus Model are listed in Table 24.1.

For the Consensus Model, with its mix of radiation, matter, and cosmological constant, the Friedmann equation (eq. 23.71) is

$$H(t)^2 = \frac{8\pi G}{3c^2}\left[\frac{u_{r,0}}{a(t)^4} + \frac{u_{m,0}}{a(t)^3} + u_\Lambda\right], \tag{24.11}$$

where $H(t) \equiv \dot{a}/a$. Dividing by H_0^2, and using the definition of the critical density (equation 23.73), we find that

$$\frac{H(t)^2}{H_0^2} = \frac{1}{u_{c,0}}\left[\frac{u_{r,0}}{a(t)^4} + \frac{u_{m,0}}{a(t)^3} + u_\Lambda\right], \tag{24.12}$$

TABLE 24.1 The Consensus Model

Component	Property
photons	$\Omega_{\gamma,0} = 5.0 \times 10^{-5}$
neutrinos	$\Omega_{\nu,0} = 3.4 \times 10^{-5}$
total radiation	$\Omega_{r,0} = 8.4 \times 10^{-5}$
baryonic matter	$\Omega_{\text{bary},0} = 0.04$
nonbaryonic dark matter	$\Omega_{\text{dm},0} = 0.26$
total matter	$\Omega_{m,0} = 0.30$
cosmological constant	$\Omega_{\Lambda,0} \approx 0.70$

or, in terms of the dimensionless density parameter Ω (equation 23.76),

$$\frac{H(t)^2}{H_0^2} = \frac{\Omega_{r,0}}{a(t)^4} + \frac{\Omega_{m,0}}{a(t)^3} + \Omega_\Lambda. \tag{24.13}$$

The Friedmann equation thus provides us with a differential equation for the scale factor $a(t)$:

$$\frac{da}{dt} = H_0 \left[\frac{\Omega_{r,0}}{a(t)^2} + \frac{\Omega_{m,0}}{a(t)} + \Omega_{\Lambda,0} a(t)^2 \right]^{1/2}. \tag{24.14}$$

Given values for H_0, $\Omega_{r,0}$, $\Omega_{m,0}$, and $\Omega_{\Lambda,0}$, equation (24.14) can be integrated to yield the scale factor as a function of time, given our usual normalization $a(t_0) \equiv 1$. Unfortunately, the solution of equation (24.14) doesn't have a simple analytic form. However, since the right-hand side of equation (24.14) is always positive for the Consensus Model, we can immediately predict that the universe will continue to expand forever. There is no Big Crunch for the Consensus Model.

Since the three components (radiation, matter, and Λ) have different dependences on scale factor, there will be long stretches in the history of the universe when one component dominates the energy density. At the moment, $u_\Lambda > u_{m,0} \gg u_{r,0}$. At an earlier time, and a smaller scale factor $a_{m\Lambda}$, the density of matter u_m was equal to u_Λ. This equality took place when

$$u_\Lambda = \frac{u_{m,0}}{a_{m\Lambda}^3}, \tag{24.15}$$

or

$$a_{m\Lambda} = \left(\frac{u_{m,0}}{u_\Lambda} \right)^{1/3} = \left(\frac{\Omega_{m,0}}{\Omega_{\Lambda,0}} \right)^{1/3} = \left(\frac{0.7}{0.3} \right)^{1/3} = 0.75. \tag{24.16}$$

When we observe a galaxy with redshift $z = 1/a_{m\Lambda} - 1 = 0.33$, we are looking back to a time when matter was equal in density to the cosmological constant.

If we go to earlier times, there was a scale factor a_{rm} at which the density of radiation u_r was equal to the density of matter u_m. This equality took place when

$$\frac{u_{m,0}}{a_{rm}^3} = \frac{u_{r,0}}{a_{rm}^4}, \tag{24.17}$$

or

$$a_{rm} = \frac{u_{r,0}}{u_{m,0}} = \frac{\Omega_{r,0}}{\Omega_{m,0}} = \frac{8.4 \times 10^{-5}}{0.3} = 2.8 \times 10^{-4}. \tag{24.18}$$

This scale factor corresponds to a redshift $z = 1/a_{rm} - 1 = 3600$. This is higher than the redshift at which the universe became transparent ($z \approx 1100$), so we cannot directly see the time of radiation–matter equality.

Early in the history of the universe, when the scale factor was small ($a \ll a_{rm} \approx 0.00028$), the universe was **radiation-dominated**. That is, the vast majority of the density was provided by photons and highly relativistic particles such as neutrinos. The Friedmann equation (eq. 24.14) in a radiation-dominated universe reduces to the form

$$\frac{da}{dt} = \frac{\Omega_{r,0}^{1/2} H_0}{a(t)}. \tag{24.19}$$

This equation has the solution

$$a(t) = [2\Omega_{r,0}^{1/2} H_0 t]^{1/2}, \tag{24.20}$$

as the reader can verify by substitution. Since $a(t) \propto t^{1/2}$ in the early universe, the horizon size

$$\ell_{\rm hor}(t_0) = \int_0^{t_0} \frac{dt}{a(t)} \tag{24.21}$$

does not diverge as $t \to 0$, and we live in a universe with a finite horizon. The acceleration in the early universe was negative:

$$\ddot{a} = -\frac{1}{4t^2} a(t) < 0, \tag{24.22}$$

indicating that the expansion of the early universe was slowed by gravity acting on photons and other relativistic particles.

At intermediate scale factors, when $a_{rm} \ll a \ll a_{m\Lambda}$, or $0.00028 \ll a \ll 0.75$, the universe was **matter-dominated**. That is, the majority of the density was provided by nonrelativistic particles, such as WIMPs, protons, and neutrons. During the matter-dominated era, the Friedmann equation (eq. 24.14) takes the simplified form

$$\frac{da}{dt} = \frac{\Omega_{m,0}^{1/2} H_0}{a(t)^{1/2}}, \tag{24.23}$$

which has the solution

$$a(t) = [\frac{3}{2}\Omega_{m,0}^{1/2} H_0 t]^{2/3}, \tag{24.24}$$

again verifiable by substitution. Since $a(t) \propto t^{2/3}$ during the matter-dominated era, the acceleration was

$$\ddot{a} = -\frac{2}{9t^2} a(t) < 0. \tag{24.25}$$

A matter-dominated universe, like a radiation-dominated universe, is decelerating.

In the future, when the scale factor becomes large ($a \gg a_{m\Lambda} \approx 0.75$), the universe will become **lambda-dominated**. When the cosmological constant is the only significant contributor to the energy density, the Friedmann equation (eq. 24.14) takes the form

$$\frac{da}{dt} = \Omega_{\Lambda,0}^{1/2} H_0 a(t). \tag{24.26}$$

This equation has an exponential solution:

$$a(t) \propto e^{Kt}, \tag{24.27}$$

where

$$K = \Omega_{\Lambda,0}^{1/2} H_0 = \frac{1}{16.7\ \text{Gyr}}. \tag{24.28}$$

When the cosmological constant takes over, the universe will expand exponentially, with an e-folding time of 16.7 Gyr. In the lambda-dominated universe, the Hubble parameter will be

$$H = \frac{\dot{a}}{a} = K = \Omega_{\Lambda,0}^{1/2} H_0. \tag{24.29}$$

The Hubble constant really will be constant with time. In addition, when the universe is lambda-dominated, the acceleration will be *positive*:

$$\ddot{a} = \Omega_{\Lambda,0} H_0^2 a(t) > 0. \tag{24.30}$$

In a relativistic universe, a cosmological constant $\Lambda > 0$ plays the role of a *repulsive* force in a Newtonian universe; that is, it causes the relative speed of any two points to increase with time.[2]

The Friedmann equation for the Consensus Model can be integrated numerically to find $a(t)$ for all times, not just those special epochs when a single component is dominant. The resulting scale factor is shown in Figure 24.3. Note that the transitions from radiation to matter domination, and from matter to lambda domination, are smooth and gradual.

With a complete knowledge of $a(t)$, the time corresponding to any scale factor can be computed. The scale factor of radiation–matter equality, $a_{rm} = 0.00028$, corresponds to a time

$$t_{rm} = 3.3 \times 10^{-6} H_0^{-1} = 47{,}000\ \text{yr}. \tag{24.31}$$

Despite the brevity of the radiation-dominated era, a lot of interesting physics was going on back then, and cosmologists have focused a great deal of attention on it. The scale factor of matter–lambda equality, $a_{m\Lambda} = 0.75$, corresponds to a time

$$t_{m\Lambda} 0.70 H_0^{-1} = 9.8\ \text{Gyr}. \tag{24.32}$$

This should be compared to the current age of the universe in the Consensus Model, which turns out to be

$$t_0 = 0.964 H_0^{-1} = 13.5\ \text{Gyr}. \tag{24.33}$$

[2] It can be shown that if a component of the universe has an energy density $u \propto a^{-n}$, then if $n < 2$, it will cause $\ddot{a} > 0$. In general, components that cause $\ddot{a} > 0$ are given the generic name of "dark energy." The "cosmological constant" is a special case of "dark energy," with $n = 0$.

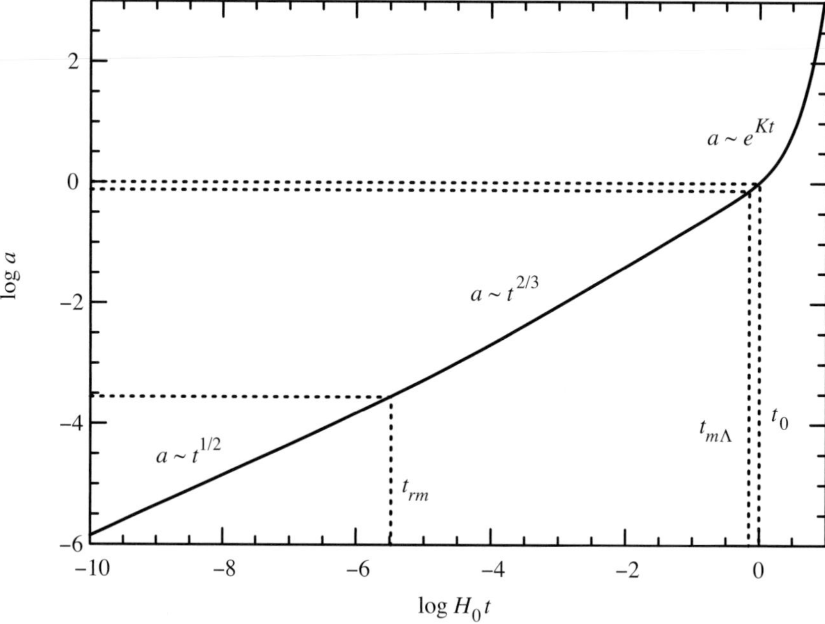

FIGURE 24.3 Scale factor as a function of time (in units of the Hubble time) for the Consensus Model.

What's amusing is that the period of deceleration (when the universe was dominated by radiation and matter) was almost exactly balanced by the later period of positive acceleration (when the universe was dominated by Λ). The net effect is that the age of the universe is nearly equal to H_0^{-1}, the naïve result you would get by assuming no acceleration at all.

With $a(t)$ known, other properties of the Consensus Model can be computed. For instance, Figure 24.4 shows the current proper distance, $\ell_p(t_0)$, to a galaxy with measured redshift z. The bold, solid line shows the results for the Consensus Model. For comparison, the dotted line shows the proper distance in a flat, matter-only universe and the dot-dash line shows the proper distance in a flat, lambda-only universe. As $z \to \infty$, the proper distance in the Consensus Model reaches a limiting value, $\ell_p(t_0) \to 3.24c/H_0$. Thus, the Consensus Model has a finite horizon distance,

$$\ell_{\rm hor}(t_0) = 3.24\frac{c}{H_0} = 3.12ct_0 = 14{,}000\ {\rm Mpc}. \tag{24.34}$$

In the matter-only universe, the horizon distance is $\ell_{\rm hor} = 2c/H_0$; in the lambda-only universe, the horizon distance is infinite.

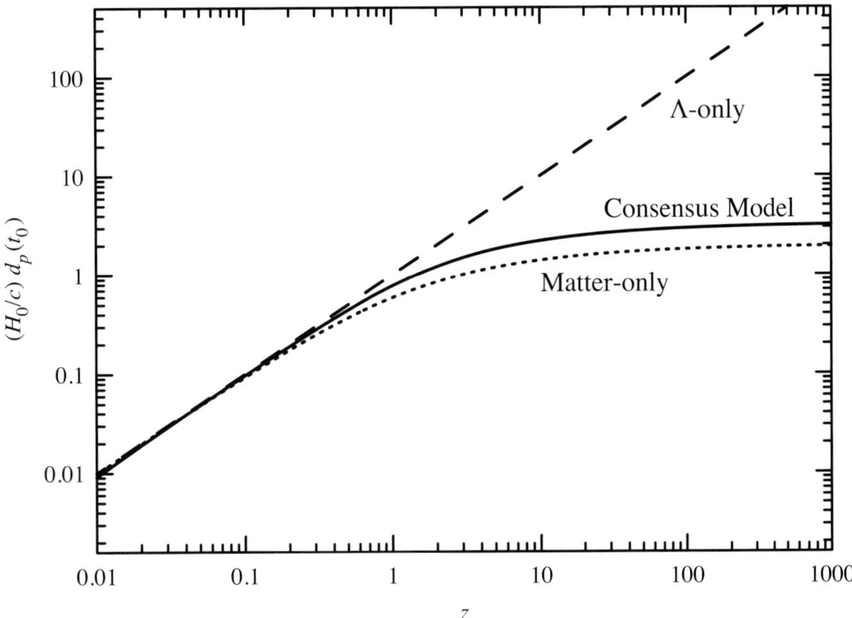

FIGURE 24.4 Current proper distance (in units of the Hubble distance) to a galaxy with redshift z.

24.2 ▪ THE ACCELERATING UNIVERSE

The Friedmann equation for the Consensus Model can be written in the form

$$\dot{a} = H_0 \left[\frac{\Omega_{r,0}}{a^2} + \frac{\Omega_{m,0}}{a} + \Omega_{\Lambda,0} a^2 \right]^{1/2} . \tag{24.35}$$

By taking the derivative with respect to t, then doing a bit of algebra, we find an equation for the second time-derivative of the scale factor:

$$\ddot{a} = H_0^2 \left[-\frac{\Omega_{r,0}}{a^3} - \frac{\Omega_{m,0}}{2a^2} + \Omega_{\Lambda,0} a \right] . \tag{24.36}$$

Note that on the right-hand side of equation (24.36), the terms involving radiation and matter are negative (they slow down the expansion), while the term involving Λ is positive (it speeds up the expansion). At present, $a(t_0) = 1$, so the acceleration of the expansion is

$$\ddot{a}_0 = H_0^2 [-\Omega_{r,0} - \Omega_{m,0}/2 + \Omega_{\Lambda,0}] = 0.55 H_0^2 \tag{24.37}$$

for the Consensus Model. The speeding up of the expansion is a remarkable—and in the context of Newtonian gravity, counterintuitive—result. What led cosmologists to embrace the accelerating universe? The conclusion was based largely on measuring the flux of standard candles at high redshifts.

Suppose we are looking at a standard candle, of known luminosity L, whose current proper distance is $r = \ell_p(t_0)$. In a static, flat universe, the observed flux would be given by an inverse square law:

$$F_{\text{static}} = \frac{L}{4\pi r^2}. \tag{24.38}$$

If the universe is expanding rather than static, the observed flux of the standard candle will be *lower* than this value, for two reasons.

First, the expansion of the universe causes the energy of each photon from the standard candle to decrease. The photon begins with an energy ε_e when it is emitted at time t_e. By the time we observe the photon at time t_0, its energy will have dropped to the value

$$\varepsilon_0 = \varepsilon_e \frac{a(t_e)}{a(t_0)} = \frac{\varepsilon_e}{1+z}, \tag{24.39}$$

where z is the measured redshift of the standard candle.

Second, the expansion of the universe will cause the time between photon detections to increase. If two photons are emitted in the same direction separated by a time interval δt_e, the proper distance between them will initially be $\delta r_e = c(\delta t_e)$. However, by the time we detect the two photons at the later time t_0, the proper distance between them will be stretched to $\delta r_0 = c(\delta t_e)(1+z)$, and we will detect them separated by a time interval $\delta t_0 = \delta t_e(1+z)$.

The net result of these two effects—lower energy photons and a longer time interval between photons—is that the observed flux f in an expanding (but spatially flat) universe will be

$$F_{\text{expand}} = \frac{L}{4\pi r^2(1+z)^2}. \tag{24.40}$$

Converting from fluxes to apparent magnitudes, we can also write down the observed apparent magnitude m in an expanding universe:

$$m = M + 5\log[r(1+z)] - 5, \tag{24.41}$$

where r is in parsecs. The distance modulus for a standard candle in an expanding (but spatially flat) universe is thus

$$m - M = 5\log r + 5\log(1+z) - 5, \tag{24.42}$$

where r is the current proper distance $\ell_p(t_0)$ to the standard candle. Consider the proper distances shown in Figure 24.4 for three different flat universes. Since the exponentially expanding, lambda-only universe has the largest proper distance r for a given redshift z, it will have the *faintest* standard candles at that redshift. Figure 24.5 shows the distance

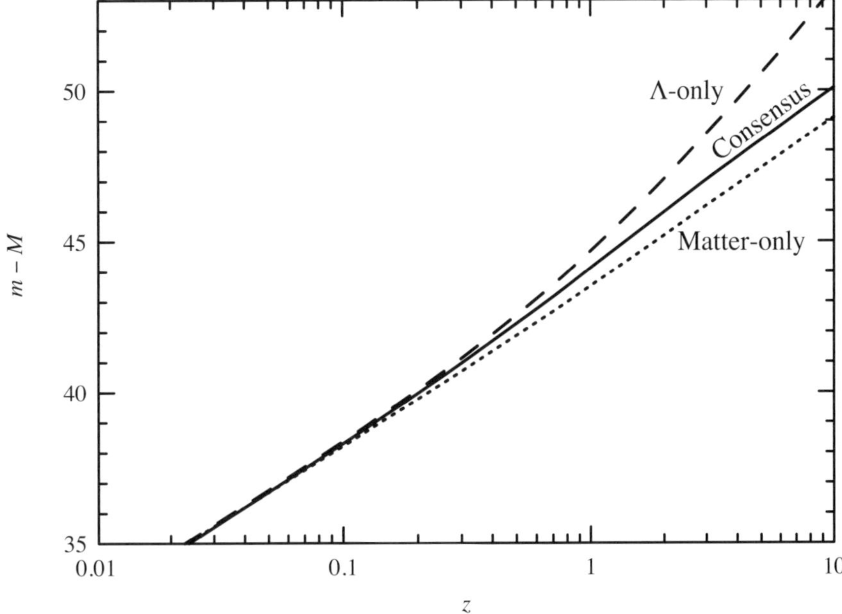

FIGURE 24.5 Distance modulus for an object with redshift z. The solid line represents the Consensus Model; the dotted line, a flat, matter-only universe; and the dot-dash line, a flat, lambda-only universe.

modulus for standard candles in our three different flat universes. At a very small redshift ($z \ll 1$), the distance modulus reduces to

$$m - M \approx 5 \log \left(\frac{c}{H_0} z \right) - 5 \approx 43.17 - 5 \log z, \tag{24.43}$$

regardless of the values of $\Omega_{m,0}$ and $\Omega_{\Lambda,0}$. It is only at $z > 1$ that the differences between models becomes large.

As an example, consider a type Ia supernova with an absolute magnitude $M = -20.0$ mag. If it is seen at a redshift $z = 1$, then its apparent magnitude in the Consensus Model will be $m = 24.1$ mag. Its apparent magnitude in the flat, lambda-only model will be $m = 24.7$ mag, 0.6 mag fainter than in the Consensus Model. Its apparent magnitude in the flat, matter-only model will be $m = 23.5$ mag, 0.6 mag brighter than in the Consensus Model.

Using the apparent magnitude of distant type Ia supernovae to distinguish among different models requires accurate photometry of apparently faint sources. It's difficult, but it can be done. Figure 24.6 shows the results from two different surveys of type Ia supernovae. The observational results are compared to three different models. In Figure 24.6a, the top line is the result expected in the Consensus Model; the bottom

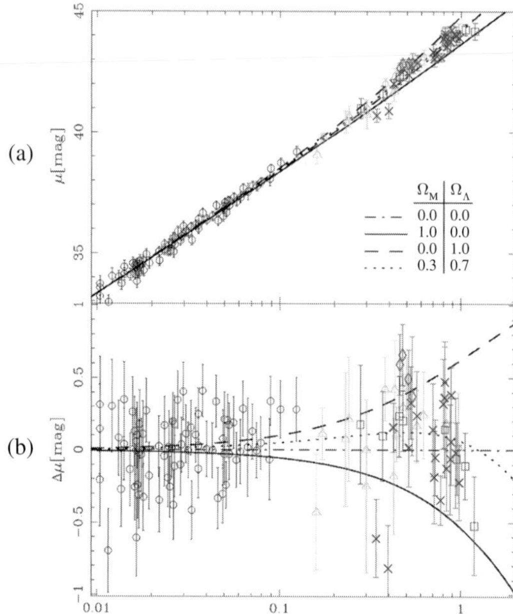

FIGURE 24.6 (a) Distance modulus versus redshift for type Ia supernovae. (b) Difference between the data and the predictions for an empty ($\Omega = 0$) universe.

line is the result for a flat, matter-only universe; and the middle line is for a negatively curved universe with $\Omega_{m,0} = 0.3$ and $\Lambda = 0$. The data are best fitted by the Consensus Model; this is better seen in Figure 24.6b, which shows the difference between the data and the predictions of the negatively curved $\Omega_{m,0} = 0.3$ model.

Instead of fitting just three models to the supernova data, we can ask more generally, What values of $\Omega_{m,0}$ and $\Omega_{\Lambda,0}$ give the best fits to the data.[3] After choosing values for $\Omega_{m,0}$ and $\Omega_{\Lambda,0}$, the relation between distance modulus and redshift can be computed, then compared to the supernova data. The results of fitting the model universes are shown in Figure 24.7. This is a rather busy plot that repays careful scrutiny. Since the radiation density is negligible, the criterion for flatness is $\Omega_m + \Omega_\Lambda = 1$, represented in Figure 24.7 by the dashed line running diagonally downward from left to right. Positively curved universes (labeled "Closed") lie above and to the right; negatively curved universes (labeled "Open") lie below and to the left. The solid line that runs diagonally upward from left to right divides universes with $\ddot{a}_0 > 0$ (labeled "Acceleration") from universes with $\ddot{a}_0 < 0$ (labeled "Deceleration"). Finally, the slightly curved line that runs nearly horizontally from ($\Omega_m = 0$, $\Omega_\Lambda = 0$) divides the Big Chill

[3] At $z < 1$, the role of radiation is negligible, so the supernova fluxes tell us nothing about the density of radiation, $\Omega_{r,0}$.

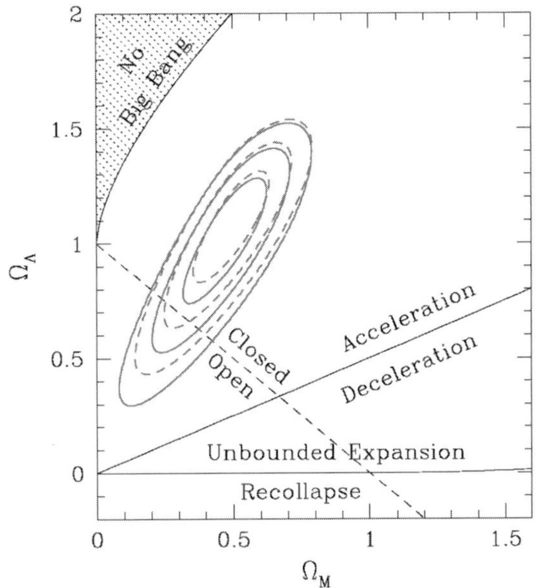

FIGURE 24.7 The values of $\Omega_{m,0}$ and $\Omega_{\Lambda,0}$ that best fit the relation between $m - M$ and z for type Ia supernovae. The solid and dashed lines represent two slightly different samples of supernovae.

universes (labeled "Unbounded Expansion") from the Big Crunch universes (labeled "Recollapse").[4]

The concentric ovals in Figure 24.7 show the region of parameter space that gives the best fit to the available supernova data. (The smallest, innermost oval gives the best fit, but the largest, outermost oval cannot be excluded at the 99.5% confidence level.) Decelerating universes can be strongly ruled out by the supernova data, as can Big Crunch universes. It is the supernova data that have led cosmologists to conclude that we live in a universe whose expansion is accelerating, leading to an exponentially chilly future for our universe.

Notice, however, that the supernova data cannot by themselves distinguish between positively curved, flat, or negatively curved universes. The curvature of the universe is constrained by looking at the angular size of distant objects, as outlined in Section 23.3. The most distant things we can see in the universe are hot and cold spots in the CMB. The angular size of these spots has been measured by the *Wilkinson Microwave Anisotropy Probe (WMAP)* and by ground-based and balloon-borne experiments. It is the preferred

[4] Your curiosity may be piqued by the wedge labeled "No Big Bang" in the upper left corner. These models, when extrapolated backward in time, have $\dot{a} = 0$ when $a > 0$; that is, they started their expansion in a state where the density was low compared to the extraordinarily high initial density we expect in a true Big Bang universe.

angular size of the structure in the CMB that provides the best evidence for the flatness of the universe ($\Omega \approx 1$). It is only when we combine the CMB results (the universe is flat) with the supernova results (the universe is accelerating) that we reach the Consensus Model, with $\Omega_m \approx 0.3$ and $\Omega_\Lambda \approx 0.7$.

If the cosmological constant is truly constant with time, then we face an accelerating future. The Local Supercluster will remain gravitationally bound (we don't have to worry about the Virgo Cluster and the Local Group being yanked apart), but more distant superclusters will move away from us with exponentially increasing velocity.

24.3 ▪ THE EARLY UNIVERSE

To understand the origins of the universe, we want to look as far back in time as possible. The oldest photons we see today are the photons of the CMB. As described in Section 23.1, when the initially hot and dense universe became sufficiently cool, protons and electrons combined to form neutral hydrogen atoms:

$$p + e^- \rightarrow H + \gamma. \tag{24.44}$$

At this time, the universe became transparent, since the photons of the cosmic background radiation no longer scattered off free electrons.

As we look outward in space, we look backward in time. Thus, we (and every other observer in the universe) are surrounded by a spherical **last scattering surface**, illustrated in Figure 24.8. The last scattering surface is where photons underwent their last scattering from a free electron before streaming freely through the newly transparent universe. The last scattering surface is the surface of the glowing, opaque ionized gas that filled the early universe.[5]

The universe became transparent, and photons underwent their last scattering, at a temperature $T_{\rm ls} = 3000$ K. The scale factor at the time of last scattering was $a_{\rm ls} = T_0/T_{\rm ls} = 2.725\ {\rm K}/3000\ {\rm K} = 9.1 \times 10^{-4}$, corresponding to a redshift $z_{\rm ls} = 1/a_{\rm ls} - 1 = 1100$. In the Consensus Model, the time of last scattering was $t_{\rm ls} = 2.5 \times 10^{-5} H_0^{-1} = 0.4$ Myr. Thus, the CMB gives us a glimpse of what the universe was like 400,000 years after the Big Bang.

At every point of the sky, the CMB has a blackbody spectrum. Although the average temperature of the CMB is $T_0 = 2.725$ K, the actual temperature varies slightly across the celestial sphere. Color Figure 30 shows a plot of the temperature of the CMB, as derived from *WMAP* data. The temperatures show a *dipole* distortion, with one hemisphere of the sky being blueshifted to higher temperatures, and the other hemisphere being redshifted to lower temperatures. This dipole distortion is simply a Doppler shift, caused by the motion of *WMAP* through space.[6] Once we subtract away the orbital motion of *WMAP* about the Sun ($v \approx 30\ {\rm km\ s^{-1}}$), the orbital motion of the Sun about the Galactic Center

[5] We can think of it as an inside-out photosphere, since the photosphere of a star is also the surface of a glowing, opaque, ionized gas.

[6] *WMAP* is at the Earth's L_2 point (see Figure 11.3), and not in a low Earth orbit.

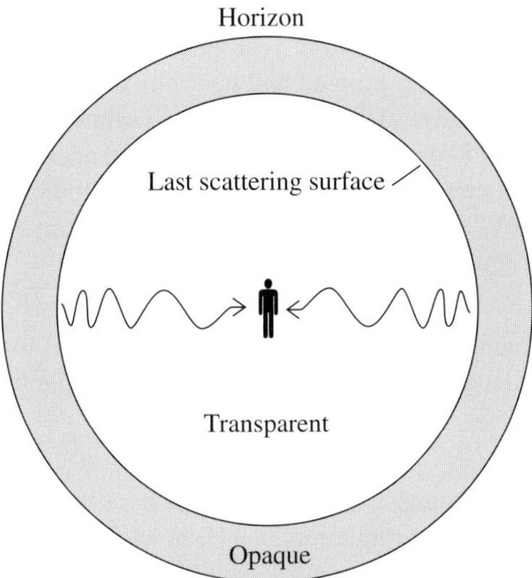

FIGURE 24.8 Observer surrounded by the "last scattering surface."

($v \approx 220$ km s^{-1}), and the orbital motion of the Galaxy relative to the center of mass of the Local Group ($v \approx 80$ km s^{-1}), we find that the Local Group is moving in the direction of Hydra, with a speed $v_{\rm lg} \approx 630$ km s^{-1}. Thus, the dipole distortion of the CMB is telling us about motion of the Local Group here and now (which is undeniably interesting but doesn't tell us directly about the early universe).

Color Figure 31 shows the remaining low-amplitude temperature fluctuations after the dipole Doppler distortion has been subtracted. The angular size of the hot and cold spots in this image are what cosmologists use to determine the curvature of space. The amplitude of the fluctuations is not large: typically, $\delta T/T \sim 10^{-5}$. The small temperature fluctuations result from small density fluctuations at the time of last scattering. A photon that happens to find itself in a dense region when the universe becomes transparent will lose energy as it climbs out of the gravitational potential well that is associated with the dense region, and will thus become redshifted to lower temperatures. Conversely, a photon that happens to be in a low-density region will be blueshifted to higher temperatures. The low-amplitude density fluctuations that were present at $t \approx 0.4$ Gyr have grown with time to the high-amplitude density fluctuations that we see at $t_0 \approx 13.5$ Gyr (superclusters, clusters, galaxies, etc.)

The opacity of the early universe draws a frustrating veil over the first 400 millennia of the history of the universe. Nevertheless, cosmologists can still deduce indirectly what was happening back then. For instance, we know that in the early universe, neutral hydrogen atoms couldn't exist because some of the cosmic background photons had energies larger than the hydrogen ionization energy ($\chi = 13.6$ eV). If we go farther back in time, we should reach a time at which bound atomic nuclei could exist because some of the cosmic background photons had energies larger than the nuclear binding energy (typically several MeV). Thus, just as there was a time when protons and electrons combined to form neutral hydrogen atoms (at $t \approx 0.4$ Myr), there must have been an earlier time when protons and neutrons combined to form atomic nuclei. This time is known as the era of **Big Bang nucleosynthesis (BBN)**.

Consider, for simplicity, a deuterium (D) nucleus. This is the simplest of all compound nuclei; it consists of a proton and neutron bound together with a binding energy $B = 2.22$ MeV. A gamma-ray photon with $\varepsilon > B$ can photodissociate deuterium:

$$D + \gamma \rightarrow p + n. \tag{24.45}$$

This reaction can run in the opposite direction, too; a proton and neutron can fuse to form a deuterium nucleus, with a gamma-ray photon carrying off the excess energy:

$$p + n \rightarrow D + \gamma. \tag{24.46}$$

Deuterium synthesis (equation 24.46) has obvious parallels to the radiative recombination of hydrogen (equation 24.44). In each case, two particles become bound together, with a photon carrying away excess energy. The most striking difference between the processes is the different energies involved. The photodissociation energy of deuterium is $B = 2.22\,\text{MeV} = (1.6 \times 10^5)(13.6\,\text{eV})$. The energy released when a deuterium nucleus is formed is 160,000 times the energy released when a neutral hydrogen atom is formed; thus, we expect the temperature at the time of nucleosynthesis to be 160,000 times greater than the temperature at the time of last scattering, when neutral hydrogen formed. This implies a nucleosynthesis temperature

$$T_{\rm nuc} = \frac{B}{\chi} T_{\rm ls} = (1.6 \times 10^5)(3000\,\text{K}) = 5 \times 10^8\,\text{K}. \tag{24.47}$$

In the Consensus Model, the universe had this temperature at an age $t_{\rm nuc} \sim 400\,\text{s} \sim 7$ min.[7]

Once a significant amount of deuterium forms, it can be converted to heavier nuclei. For instance, tritium (^{3}H) is made by the reaction

$$D + n \rightarrow {}^3\text{H} + \gamma. \tag{24.48}$$

Light helium (^{3}He) is made by the reaction

$$D + p \rightarrow {}^3\text{He} + \gamma. \tag{24.49}$$

Ordinary helium (^{4}He) can be made by reactions such as

$${}^3\text{H} + p \rightarrow {}^4\text{He} + \gamma \tag{24.50}$$

and

$${}^3\text{He} + n \rightarrow {}^4\text{He} + \gamma. \tag{24.51}$$

[7] This is a slight overestimate of the age; when Steven Weinberg entitled his book on BBN *The First Three Minutes*, he was using a more accurate calculation.

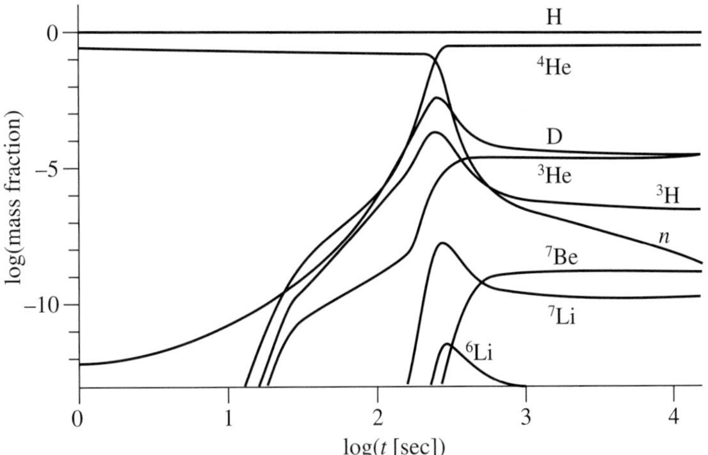

FIGURE 24.9 Mass fraction of nuclei (and free neutrons) during the time of BBN.

Once ^{4}He is reached, the orderly march of nucleosynthesis to larger atomic numbers hits a roadblock. There are no stable nuclei with atomic number 5. If we try to add a proton to ^{4}He to make ^{5}Li, it won't work; ^{5}Li is not a stable nucleus. If we try to add a neutron to ^{4}He to make ^{5}He, it won't work; ^{5}He is not a stable nucleus. We can make small amounts of lithium by the reactions

$$^{4}\mathrm{He} + D \to {}^{6}\mathrm{Li} + \gamma \tag{24.52}$$

and

$$^{4}\mathrm{He} + {}^{3}\mathrm{H} \to {}^{7}\mathrm{Li} + \gamma, \tag{24.53}$$

but then we hit another roadblock. There are no stable nuclei with atomic number 8. If we try to fuse two ^{4}He nuclei together to form ^{8}Be, it won't work; ^{8}Be is not a stable nucleus.[8]

In summary, BBN works rapidly and efficiently up to ^{4}He, but few nuclei heavier than helium are produced. The precise yields of the different elements and isotopes can be computed using a computer code that takes into account the cross-sections for the different nuclear reactions. Results of a typical BBN code are shown in Figure 24.9. At $t \approx 1$ s, almost all the baryons are in the form of free protons (labeled H in the figure) and free neutrons (labeled n).

[8] As you may recall from Section 15.3, the instability of ^{8}Be is the main reason why the triple alpha process in stars requires such high temperature and density; a ^{4}He nucleus must be slammed into the ^{8}Be nucleus during the brief interval before it falls apart.

Because protons have a lower rest energy than neutrons, the laws of statistical mechanics state that protons will be more numerous than neutrons in the early universe. By $t \approx 100$ s, when nucleosynthesis kicks into high gear, there are seven protons for every neutron in the universe. Consider a representative group of two neutrons and 14 protons. The two neutrons swiftly combine with two of the protons to form a single ^{4}He nucleus, leaving 12 lonely protons left over.[9] At $t \approx 10^4\,\mathrm{s} \approx 3\,\mathrm{hr}$, the temperature has dropped too low for further nuclear reactions and the epoch of BBN is over. At this point, the mass fraction of hydrogen is $X \approx 12/16 \approx 0.75$ and the mass fraction of helium is $Y \approx 4/16 \approx 0.25$. Only tiny amounts of elements other than ^{1}H and ^{4}He are present.

A basic prediction of BBN is that helium contributed 25% of the baryon density even before the first generation of stars began to pollute the universe with heavy elements. Observations of gas and stars reveal that hydrogen is invariably mixed with helium. The helium mass fraction of the Sun is $Y = 0.250$, but the Sun is contaminated by helium formed in earlier generations of stars. When we look at interstellar gas that hasn't been run through the stellar mill, the helium mass fraction can be as low as $Y = 0.24$, but not any lower. This is in good agreement with the predictions of BBN.

24.4 ▪ THE VERY EARLY UNIVERSE

So far we have accentuated the positive when discussing the Hot Big Bang universe in general, and the Consensus Model in particular. However, the standard Hot Big Bang scenario, in which the universe was dominated by radiation at early times, has a pair of problems that have puzzled cosmologists. These are the **flatness problem** and the **horizon problem**. Let's examine the flatness problem first.

The curvature of the universe is related to its energy content by the Friedmann equation,

$$H(t)^2 = \frac{8\pi G}{3c^2}u(t) - \frac{\kappa c^2}{r_{c,0}^2}\frac{1}{a(t)^2}. \tag{24.54}$$

If we divide each side by $H(t)^2$, we can rewrite the Friedmann equation in the form

$$1 = \Omega(t) - \frac{\kappa c^2}{r_{c,0}^2}\frac{1}{a(t)^2 H(t)^2}. \tag{24.55}$$

If the density parameter is exactly equal to 1, then the universe is perfectly flat. At the present moment, the observational results are consistent with the limits

$$|1 - \Omega_0| \leq 0.1. \tag{24.56}$$

[9] The solitary life of the protons ends 400,000 years later, when they find electron sidekicks and become neutral hydrogen atoms.

Why should the value of the density parameter be so close to 1 today? We might just shrug and say, "It's a coincidence." However, when you extrapolate the value of $\Omega(t)$ back into the past, the closeness of Ω to 1 becomes more difficult to dismiss as a coincidence.

Equation (24.55) tells us

$$|1 - \Omega(t)| \propto \frac{1}{a(t)^2 H(t)^2}. \quad (24.57)$$

During the matter-dominated era, $a(t) \propto t^{2/3}$ and $H(t) \equiv \dot{a}/a \propto t^{-1}$. Thus, during the matter-dominated era, the difference between Ω and 1 grew at the rate

$$|1 - \Omega(t)|_m \propto t^{2/3} \propto a(t). \quad (24.58)$$

During the radiation-dominated era, $a(t) \propto t^{1/2}$ and $H(t) \equiv \dot{a}/a \propto t^{-1}$. Thus, during the radiation-dominated era, the difference between Ω and 1 grew at the rate

$$|1 - \Omega(t)|_r \propto t \propto a(t)^2. \quad (24.59)$$

If we extrapolate back to the time of BBN ($t_{\rm nuc} \sim 3$ min), we compute that the deviation of Ω from one was

$$|1 - \Omega(t_{\rm nuc})| \leq 10^{-14}. \quad (24.60)$$

At the time deuterium and helium were forming, the density of the universe was equal to the critical density with an accuracy of 1 part in 100 trillion. Our very existence depends on the astonishingly close match between the actual density and the critical density in the early universe. If, for instance, the deviation of Ω from 1 at the time of Big Bang nucleosynthesis had been 1 part in 100 thousand instead of one part in 100 trillion, the universe would have collapsed in a Big Crunch or expanded to a low-density Big Chill after only a few years. In either case, galaxies, stars, planets, and cosmologists would not have had time to form.

The flatness problem is simply the statement that Ω was very, very close to 1 in the early universe. It would be satisfying if we could find a physical mechanism for flattening the universe early in its history, rather than invoking a highly implausible coincidence.

The horizon problem is simply the statement that the universe is nearly homogeneous and isotropic on large scales. Why is this a problem? To see why large-scale homogeneity and isotropy is puzzling in the standard Hot Big Bang scenario, consider two antipodal points on the last scattering surface, as shown in Figure 24.10. Since the last scattering of CMB photons took place long ago ($t_{\rm ls} \approx 0.4$ Myr $\approx 3 \times 10^{-5} t_0$), the current proper distance to the last scattering surface is only slightly smaller than the horizon distance. In the Consensus Model, the last scattering surface is at a distance $\ell_p = 0.98\ell_{\rm hor}$ from us. Thus, two antipodal points on the last scattering surface are currently separated by a distance $1.96\ell_{\rm hor}$. Since the two points are farther apart than the horizon distance, they are not in causal contact. That is, they haven't had time to send messages to each other. In particular, they haven't had time to come into thermal equilibrium with each other. Nevertheless, the two points have the same temperature, once the dipole distortion is subtracted, to within 1 part in 10^5.

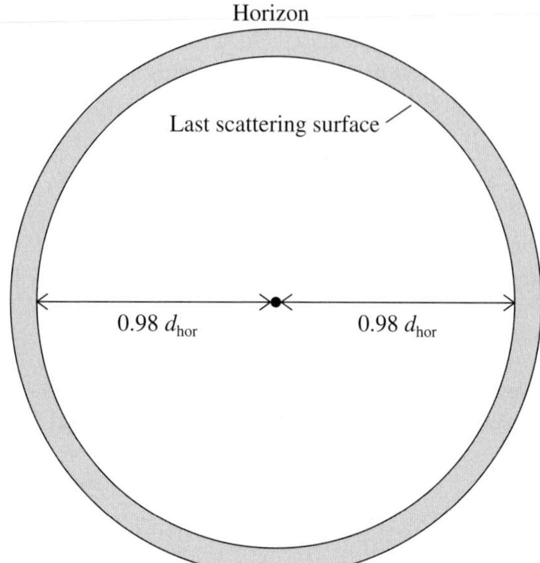

FIGURE 24.10 The distance between antipodal points on the last scattering surface.

How can two points that haven't had time to swap information be so nearly identical in their properties? It would be satisfying if we could find a physical mechanism for homogenizing the universe early in its history, rather than invoking a highly implausible coincidence.

The first satisfying solution to the flatness and horizon problems was provided by Alan Guth, who put forward the **inflationary theory** in 1981. In a cosmological context, "inflation" is the statement that there was a very early period when the acceleration of the expansion was positive ($\ddot{a} > 0$). As usually implemented, inflationary theory supposes that the universe was temporarily dominated by a cosmological constant Λ_i very much larger than the cosmological constant Λ present today.

When the universe is dominated by a cosmological constant, it expands exponentially (equation 24.27):

$$a(t) \propto e^{H_i t}, \qquad (24.61)$$

where

$$H_i = \left(\frac{8\pi G u_\Lambda}{3c^2} \right)^{1/2}. \qquad (24.62)$$

To see how inflation can solve the flatness and horizon problems, suppose that the universe had a period of exponential growth in the middle of its early radiation-dominated phase. For simplicity, let's suppose the exponential expansion switched on instantaneously at a time t_i, and lasted until some later time t_f, when it switched off instantaneously. In this simple case, the scale factor grows during the inflationary era ($t_i < t < t_f$) by a factor

$$\frac{a(t_f)}{a(t_i)} = e^N, \tag{24.63}$$

where N, the number of e-foldings of inflation, is

$$N \equiv H_i(t_f - t_i). \tag{24.64}$$

If the duration of inflation, $t_f - t_i$, was long compared to the Hubble time during inflation, H_i^{-1}, then N was large and the growth of the scale factor was exponentially huge.

For concreteness, let's take one popular model for inflation. According to Grand Unified Theories of particle physics, or GUTs, there was a phase transition that took place at a time $t_{\rm GUT} \approx 10^{-35}$ s, when the strong nuclear force separated from the electroweak force. In the GUT model of inflation, exponential growth began at the GUT time, $t_i \approx t_{\rm GUT} \approx 10^{-35}$ s, with a Hubble parameter $H_i \approx t_{\rm GUT}^{-1} \approx 10^{35}\,{\rm s}^{-1}$, and lasted for $N \sim 100$ e-foldings. In the GUT model, the growth in scale factor during inflation was

$$\frac{a(t_f)}{a(t_i)} \sim e^{100} \sim 10^{43}, \tag{24.65}$$

all happening in a time $\sim 100 t_{\rm GUT} \sim 10^{-33}$ s.

How does inflation resolve the flatness problem? In an exponentially expanding universe, equation (24.57) can be written in the form

$$|1 - \Omega(t)| \propto \frac{1}{a(t)^2 H(t)^2} \propto \frac{1}{e^{2H_i t}} \propto e^{-2H_i t}. \tag{24.66}$$

If we compare Ω at the beginning of inflation ($t = t_i$) to Ω at the end of inflation ($t = t_f = t_i + N/H_i$), we find

$$|1 - \Omega(t_f)| = e^{-2N} |1 - \Omega(t_i)|. \tag{24.67}$$

If the universe were strongly curved prior to inflation, with

$$|1 - \Omega(t_i)| \sim 1, \tag{24.68}$$

then 100 e-foldings of inflation would flatten it like the proverbial pancake, and then some:

$$|1 - \Omega(t_f)| \sim e^{-200} \sim 10^{-87}. \tag{24.69}$$

The current limits on the density parameter, $|1 - \Omega_0| \leq 0.1$, imply that $N > 60$ if inflation took place at the GUT time.

How does inflation resolve the horizon problem? Consider the entire universe directly visible to us today, that is, the region bounded by the surface of last scattering (Figure 24.10). Currently, the proper distance to the surface of last scattering is

$$\ell_p(t_0) = 0.98 \ell_{\rm hor}(t_0) = 14{,}000 \text{ Mpc}. \tag{24.70}$$

If inflation ended at a time $t_f \sim 10^{-33}$ s, this corresponds to a scale factor $a_f \sim 8 \times 10^{-27}$ in the Consensus Model. Thus, immediately after inflation, the portion of the universe visible to us today was crammed into a sphere of radius

$$\begin{aligned} \ell_p(t_f) &= a_f \ell_p(t_0) \\ &\sim (8 \times 10^{-27})(1.4 \times 10^4 \text{ Mpc}) \sim 10^{-22} \text{ Mpc} \sim 4 \text{ m}. \end{aligned} \tag{24.71}$$

Immediately after inflation, all the mass–energy destined to become the hundreds of billions of galaxies we see today was contained in a sphere a few yards in radius. This may boggle your mind. If so, be prepared for additional boggling. If there were $N \sim 100$ e-foldings of inflation, then prior to the inflationary epoch, the currently visible universe was contained in a sphere of radius

$$\ell_p(t_i) \sim e^{-100} \ell_p(t_f) \sim 10^{-43} \text{ m}. \tag{24.72}$$

What matters for the solution of the horizon problem is not that this distance is small (which it certainly is!) but that it is smaller than the horizon distance at t_i, the start of inflation. Since the universe was radiation-dominated before inflation, the preinflationary scale factor was $a(t) = a_i (t/t_i)^2$, and the horizon distance at t_i was

$$\ell_{\text{hor}}(t_i) = c a_i \int_0^{t_i} \frac{dt}{a_i (t/t_i)^2} = 2ct_i \sim 6 \times 10^{-27} \text{ m}, \tag{24.73}$$

assuming that inflation began at the GUT time, $t_i \sim 10^{-35}$ s. This horizon distance is over 16 orders of magnitude bigger than the size of the currently visible universe at time t_i. Thus, everything we see today had plenty of time to swap photons back and forth prior to inflation and come to thermal equilibrium.

The detailed particle physics behind inflation is beyond the scope of this book. The usual driving mechanism behind inflation involves a scalar field being caught in a "false vacuum state" for a finite length of time. A false vacuum state is one for which the potential energy of the field is not the global minimum. It takes some length of time for the scalar field to transit to the global minimum of the potential (the true vacuum state). During the time of transition, the energy of the scalar field plays the role of a cosmological constant. A scalar field in a false vacuum state is sometimes compared to a supercooled liquid. Freezing would lower the energy of the supercooled liquid, but until some disturbance initiates the freezing, it temporarily remains in the higher-energy, liquid state. When the freezing finally occurs, the latent heat of fusion is released and warms the surroundings. Similarly, when a scalar field goes from a false vacuum to the true vacuum, the energy released in going from a higher to lower potential energy warms up the universe, returning the temperature of the universe to what it was before ~ 100 e-foldings of inflation chilled it down.

It is tempting to extrapolate the scale factor back to $t = 0$, and $a = 0$, representing an infinite density singularity. One shortcoming of general relativity, however, is that it doesn't take quantum effects into account. A complete "quantum gravity" theory has not yet been devised. However, it is speculated that time is quantized in units of the **Planck time**

$$t_{\rm Pl} = \left(\frac{G\hbar}{c^5}\right)^{1/2} = 5 \times 10^{-44}\ \text{s}, \tag{24.74}$$

and that talking about times earlier than the Planck time may not be physically meaningful. Although invoking quantum gravity prevents us from having to contemplate infinitely dense initial conditions, the properties at $t \approx t_{\rm Pl}$ were fairly mind-boggling in themselves. At $t \approx t_{\rm Pl}$, the number density of particles would have been $n \sim 10^{104}\ \text{m}^{-3}$, and the average particle energy would have been $E \sim 10^{28}$ eV; that's an energy comparable to the kinetic energy of a cruising passenger jet, concentrated in a single elementary particle. From this incredibly dense, hot state evolved the complex universe we see around us today.

PROBLEMS

24.1 (a) Given that the current scale factor is $a(t_0) = 1$, at what scale factor did the temperature of the cosmic background radiation equal the temperature of the Sun's photosphere?
(b) At what scale factor did it equal the temperature of the Sun's center?
(c) If the current mass density of the universe is equal to $0.3\rho_{c,0}$, what was the mass density of the universe when the temperature was equal to that of the Sun's center? Compare this mass density to the average density of the Sun.

24.2 Explicitly calculate the redshifts for the following:
(a) The universe goes from radiation-dominated to matter-dominated.
(b) The universe goes from matter-dominated to dark-energy-dominated.

24.3 At the time this problem was written, the highest-redshift quasar known was CFHQS J2329-0301, which has a redshift $z = 6.43$.
(a) What was the scale factor a of the universe at the time the quasar light we are observing now left the quasar?
(b) How old was the universe at the time the light left the quasar?
(c) What is the distance modulus of the quasar?

24.4 Suppose that star formation stops today, everywhere in the universe.
(a) At what time $t_{\rm die}$ will the last stars die out?
(b) What will be the scale factor $a(t_{\rm die})$ at that time?

24.5 Estimate how high the temperature of the universe must be for proton–proton pair production to occur. What was the approximate age of the universe when it had cooled enough for proton–proton pair production to cease?

A Astronomical Data

TABLE A.1 Physical Constants

Name	Symbol	Value	Units
Gravitational constant	G	6.673×10^{-11}	$m^3\,kg^{-1}\,s^{-2}$
Permittivity of the vacuum	ϵ_o	8.854×10^{-12}	$C^2\,N^{-1}\,m^{-2}$
Permeability of the vacuum	μ_o	$4\pi \times 10^{-7}$	W m
Elementary charge	e	1.602×10^{-19}	C
Speed of light in vacuum	c	2.998×10^{8}	$m\,s^{-1}$
Planck constant	h	6.626×10^{-34}	J s
Reduced Planck constant	$\hbar \equiv h/2\pi$	1.055×10^{-34}	J s
Boltzmann constant	k	1.381×10^{-23}	$m^2\,kg\,s^{-2}\,K^{-1}$
Stefan–Boltzmann constant	σ_{SB}	5.670×10^{-8}	$W\,m^{-2}K^{-4}$
Thomson cross-section	σ_e	6.652×10^{-29}	m^2
Proton mass	m_p	1.673×10^{-27}	kg
Electron mass	m_e	9.109×10^{-31}	kg

TABLE A.2 Astronomical Constants

Name	Symbol	Value	Units
Mass of Earth	$M_{\oplus}$	5.974×10^{24}	kg
Mass of Sun	$M_{\odot}$	1.989×10^{30}	kg
Mass of Moon		7.36×10^{22}	kg
Equatorial radius of Earth	$R_{\oplus}$	6378	km
Equatorial radius of Sun	$R_{\odot}$	6.955×10^{5}	km
Equatorial radius of Moon		1737	km
Mean density of Earth		5515	kg m^{-3}
Mean density of Sun		1408	kg m^{-3}
Mean density of Moon		3346	kg m^{-3}
Luminosity of Sun	$L_{\odot}$	3.839×10^{26}	W
Effective temperature of Sun		5778	K
Hubble constant	H_o	70 ± 5	$\text{km s}^{-1}\,\text{Mpc}^{-1}$
Light-year		9.461×10^{12}	km
Astronomical unit	AU	1.496×10^{8}	km
Parsec	pc	3.086×10^{13}	km

TABLE A.3 The Planets

Name	Mean radius $(R_\oplus)$[a]	Mass $(M_\oplus)$[b]	Rotation period (days)	Orbital semimajor axis (AU)	Orbital eccentricity	Orbital period (years)
Planets						
Mercury	0.383	0.0553	58.6	0.387	0.2056	0.241
Venus	0.950	0.8150	−243.0	0.723	0.0068	0.615
Earth	1.000	1.0000	0.997	1.000	0.0167	1.000
Mars	0.532	0.1074	1.026	1.524	0.0934	1.881
Jupiter	10.97	317.8	0.414	5.203	0.0484	11.86
Saturn	9.14	95.16	0.444	9.537	0.0539	29.45
Uranus	3.98	14.50	−0.718	19.19	0.0473	84.02
Neptune	4.18	17.20	0.671	30.07	0.0086	164.8
Dwarf Planets						
Ceres	0.075	0.00016	0.378	2.767	0.0795	4.599
Pluto	0.188	0.00220	−6.387	39.45	0.2502	247.9
Haumea[c]	0.11	0.00070	0.163	43.13	0.1950	283.3
Makemake	0.12	∼ 0.0007	unknown	45.43	0.1612	306.2
Eris	0.20	0.00280	∼ 1	67.90	0.4362	559.6

a. $R_\oplus = 6371$ km (Note: this table uses mean radius rather than equatorial radius.)
b. $M_\oplus = 5.974 \times 10^{24}$ kg
c. Haumea is ellipsoidal due to its rapid rotation $(0.15R_\oplus \times 0.12R_\oplus \times 0.08R_\oplus)$.

TABLE A.4 Major Satellites in the Solar System[a]

Name	Mean radius (km)	Mass (10^{20} kg)	Orbital semimajor axis (10^3 km)	Orbital period (days)
Earth				
Moon	1737	734.8	384.4	27.32
Mars				
Phobos	11.1	1.066×10^{-4}	9.378	0.3189
Deimos	6.2	0.148×10^{-4}	23.46	1.2624
Jupiter				
Amalthea	83	0.021	181.4	0.4982
Io	1822	893.2	421.7	1.769
Europa	1561	480.0	670.9	3.551
Ganymede	2631	1482	1070	7.155
Callisto	2410	1076	1883	16.69
Himalia	85	0.07	11460	250.6
Saturn				
Janus	89	0.0190	151.5	0.6947
Mimas	198	0.3751	185.5	0.9424
Enceladus	252	1.079	237.9	1.370
Tethys	533	6.176	294.6	1.888
Dione	562	10.96	377.4	2.737
Rhea	764	23.07	527.1	4.518
Titan	2575	1346	1222	15.94
Hyperion	135	0.0559	1481	21.28
Iapetus	736	18.06	3561	79.32
Phoebe	107	0.0829	12960	−550.6
Uranus				
Puck	81	0.029	86.00	0.7618
Miranda	236	0.66	129.4	1.413
Ariel	579	12.9	191.0	2.520
Umbriel	585	12.2	266.0	4.144
Titania	789	34.2	435.9	8.706
Oberon	761	28.8	583.5	13.46
Neptune				
Galatea	88	0.037	61.95	0.4287
Larissa	97	0.049	73.55	0.5547
Proteus	210	0.50	117.6	1.122
Triton	1353	213.9	354.8	−5.877
Nereid	170	0.31	5514	360.1
Pluto				
Charon	604	15	17.54	−6.387

a. Includes all natural satellites of terrestrial planets, plus all satellites of Jovian and dwarf planets with mean radius > 80 km.

TABLE A.5 MK Spectral Types (Main Sequence Stars)

Type	M_V	$B - V$	T_{eff}	BC	$M/M_\odot$ [a]	$R/R_\odot$ [b]	$\log(g/g_\odot)$ [c]
Main Sequence (luminosity class V)							
O5	−5.7	−0.33	42,000	−4.40	60	12	−1.5
B0	−4.0	−0.30	30,000	−3.16	17.5	7.4	−1.4
B5	−1.2	−0.17	15,200	−1.46	5.9	3.9	−1.00
A0	+0.65	−0.02	9790	−0.30	2.9	2.9	−0.7
A5	+1.95	+0.15	8180	−0.15	2.0	1.7	−0.4
F0	+2.7	+0.30	7300	−0.09	1.6	1.5	−0.3
F5	+3.5	+0.44	6650	−0.14	1.4	1.3	−0.2
G0	+4.4	+0.58	5940	−0.18	1.05	1.1	−0.1
G2	+4.7	+0.63	5790	−0.20	1.00	1.00	0.0
G5	+5.1	+0.68	5560	−0.21	0.92	0.92	−0.1
K0	+5.9	+0.81	5150	−0.31	0.79	0.85	+0.1
K5	+7.35	+1.15	4410	−0.72	0.67	0.72	+0.25
M0	+8.8	+1.40	3840	−1.38	0.51	0.60	+0.35
M5	+12.3	+1.64	3170	−2.73	0.21	0.27	+1.0
M8					0.06	0.10	+1.2

a. $M_\odot = 1.989 \times 10^{30}$ kg
b. $R_\odot = 6.955 \times 10^{5}$ km
c. $g_\odot = 275$ m s^{-2}

TABLE A.6 MK Spectral Types (Giants and Supergiants)

Type	M_V	$B - V$	T_{eff}	BC	$M/M_\odot$ [a]	$R/R_\odot$ [b]	$\log(g/g_\odot)$ [c]
Giants (luminosity class III)							
B0					20	15	−2.2
B5					7	8	−0.95
A0					4	5	−1.5
G0					1.0	6	−2.4
G5	+0.9	+0.86	5050	−0.34	1.1	10	−3.0
K0	+0.7	+1.00	4660	−0.50	1.1	15	−3.5
K5	−0.2	+1.50	4050	−1.02	1.2	25	−4.1
M0	−0.4	+1.56	3690	−1.25	1.2	40	−4.7
M5	−0.3	+1.63	3380	−2.48			
Supergiants (Luminosity class I)							
O5					70	30	−2.6
B0					25	30	−3.0
B5	−6.2	−0.10	13,600	−0.95	20	50	−3.8
A0	−6.3	−0.01	9980	−0.41	16	60	−4.1
A5	−6.6	+0.09	8610	−0.13	13	60	−4.2
F0	−6.6	+0.17	7460	−0.01	12	80	−4.6
F5	−6.6	+0.32	6370	−0.03	10	100	−5.0
G0	−6.4	+0.76	5370	−0.15	10	120	−5.2
G5	−6.2	+1.02	4930	−0.33	12	150	−5.3
K0	−6.0	+1.25	4550	−0.50	13	200	−5.8
K5	−5.8	+1.60	3990	−1.01	13	400	−4.1
M0	−5.6	+1.67	3620	−1.29	13	500	−7.0
M5	−5.6	+1.80	2880	−3.47			

a. $M_\odot = 1.989 \times 10^{30}$ kg
b. $R_\odot = 6.955 \times 10^5$ km
c. $g_\odot = 275$ m s^{-2}

TABLE A.7 25 Closest Stars

Star	α_{2000}	δ_{2000}	m_V	Spectral Type	Distance (pc)
Sun	varies	varies	−26.72	G2 V	4.848×10^{-6}
Proxima Cen	14 29 43.0	−62 40 46	11.09	M5.5 V	1.301
α Cen A	14 39 36.5	−60 50 02	0.01	G2 V	1.338
α Cen B	14 39 35.1	−60 50 14	1.34	K0 V	1.338
Barnard's Star	17 57 48.5	+04 41 36	9.53	M4 V	1.828
Wolf 359	10 56 29.2	+07 00 53	13.44	M6 V	2.386
Lalande 21185	11 03 20.2	+35 58 12	7.47	M2 V	2.542
Sirius A	06 45 08.9	−16 42 58	−1.43	A1 V	2.631
BL Ceti	01 39 01.3	−17 57 01	12.54	M5.5 V	2.676
UV Ceti	01 39 01.3	−17 57 01	12.99	M6 V	2.676
Ross 154	18 49 49.4	−23 50 10	10.43	M3.5 V	2.968
Ross 248	23 41 54.7	+44 10 30	12.29	M5.5 V	3.165
ϵ Eri	03 32 55.8	−09 27 30	3.73	K2 V	3.226
Lacaille 9352	23 05 52.0	−35 51 11	7.34	M1.5 V	3.293
Ross 128	11 47 44.4	+00 48 16	11.13	M4 V	3.348
EZ Aqr A	22 38 33.4	−15 18 07	13.33	M5 V	3.454
EZ Aqr B	22 38 33.4	−15 18 07	13.27	M	3.454
EZ Aqr C	22 38 33.4	−15 18 07	14.03	M	3.454
Procyon	07 39 18.1	+05 13 30	0.38	F5 IV-V	3.496
61 Cyg A	21 06 53.9	+38 44 58	5.21	K5 V	3.496
61 Cyg B	21 06 55.3	+38 44 31	6.03	K7 V	3.496
GJ 725 A	18 42 46.7	+59 37 49	8.90	M3 V	3.534
GJ 725 B	18 42 46.9	+59 37 37	9.69	M3.5 V	3.534
GX And	00 18 22.9	+44 01 23	8.08	M1.5 V	3.564
GQ And	00 18 22.9	+44 01 23	11.06	M3.5 V	3.564

TABLE A.8 25 Apparently Brightest Stars

Star	α_{2000}	δ_{2000}	m_V	Spectral Type	Distance (pc)
Sun	varies	varies	−26.72	G2 V	4.848×10^{-6}
Sirius A	06 45 08.9	−16 42 58	−1.43	A1 V	2.631
Canopus	06 23 57.1	−52 41 44	−0.62	F0 Ib	96
Arcturus	14 15 39.7	+19 10 57	−0.05	K2 IIIp[a]	11.3
α Cen A	14 39 36.5	−60 50 02	0.01	G2 V	1.338
Vega	18 36 56.3	+38 47 01	0.03	A0 Vvar[b]	7.76
Capella	05 16 41.4	+45 59 53	0.08	M1 III	12.9
Rigel	05 14 32.3	−08 12 06	0.18	B8 Ia	240
Procyon	07 39 18.1	+05 13 30	0.38	F5 IV-V	3.496
Betelgeuse	05 55 10.3	+07 24 25	0.45	M2 Ib	130
Achernar	01 37 42.8	−57 14 12	0.45	B3 Vp	44
β Cen	14 03 49.4	−60 22 23	0.61	B1 III	160
Altair	19 50 47.0	+08 52 06	0.76	A7 IV-V	5.14
α Cru	12 26 35.9	−63 05 57	0.77	B0.5 IV	98
Aldeberan	04 35 55.2	+16 30 34	0.87	K5 III	20.0
Spica	13 25 11.6	−11 09 41	0.98	B1 V	80
Antares	16 29 24.5	−26 25 55	1.06	M1 Ib	190
Pollux	07 45 19.0	+28 01 34	1.16	K0 IIIvar	10.3
Fomalhaut	22 57 39.0	−29 37 20	1.17	A3 V	7.69
Deneb	20 41 25.9	+45 16 49	1.25	A2 Ia	1000
β Cru	12 47 43.3	−59 41 20	1.25	B0.5 III	110
α Cen B	14 39 35.1	−60 50 14	1.34	K0 V	1.338
Regulus	10 08 22.3	+11 58 02	1.36	B7 V	23.8
Adhara	06 58 37.6	−28 58 20	1.50	B2 II	130
Castor	07 34 35.9	+31 53 18	1.58	A2 V	15.8

a. p = peculiar
b. var = variable

Bibliography

Beatty, J. K., Petersen, C. C., and Chaikin, A., eds. 1999, *The New Solar System*, 4th ed. (Cambridge, MA: Sky Publishing; Cambridge: Cambridge University Press).

Binney, J., and Merrifield, M. 1998, *Galactic Astronomy* (Princeton, NJ: Princeton University Press).

Binney, J., and Tremaine, S. 2008, *Galactic Dynamics*, 2nd ed. (Princeton, NJ: Princeton University Press).

Carroll, B. W., and Ostlie, D. A. 2007, *An Introduction to Modern Astrophysics*, 2nd ed. (San Francisco: Addison-Wesley).

Cox, A. N., ed. 2000, *Allen's Astrophysical Quantities*, 4th ed. (New York: Springer-Verlag).

Frank, J., King, A., and Raine, D. 2002, *Accretion Power in Astrophysics*, 3rd ed. (Cambridge: Cambridge University Press).

Irwin, J. A. 2007, *Astrophysics: Decoding the Cosmos* (Chichester: Wiley).

Karttunen, H., Kröger, P., Oja, H., Poutanen, M., and Donner, K. J., eds. 2003, *Fundamental Astronomy*, 4th ed. (Berlin: Springer-Verlag).

Kundt, W. 2001, *Astrophysics: A Primer* (Berlin: Springer-Verlag).

Kutner, M. L. 2003, *Astronomy: A Physical Perspective* (Cambridge: Cambridge University Press).

Landstreet, J. D. 2003, *Physical Processes in the Solar System* (London, ON: Keenan and Darlington).

Maoz, D. 2007, *Astrophysics in a Nutshell* (Princeton, NJ: Princeton University Press).

Maran, S., ed. 1992, *The Astronomy and Astrophysics Encyclopedia* (New York: Van Nostrand Reinhold).

Osterbrock, D. E., and Ferland, G. J. 2006, *Astrophysics of Gaseous Nebulae and Active Galactic Nuclei*, 2nd ed. (Sausalito, CA: University Science Books).

Peebles, P. J. E. 1993, *Principles of Physical Cosmology* (Princeton, NJ: Princeton University Press).

Peterson, B. M. 1997, *An Introduction to Active Galactic Nuclei* (Cambridge: Cambridge University Press).

Prialnik, D. 2000, *An Introduction to the Theory of Stellar Structure and Evolution* (Cambridge: Cambridge University Press).

Ryden, B. 2003, *Introduction to Cosmology* (San Francisco: Addison-Wesley).

Schwarzschild, M. 1958, *Structure and Evolution of the Stars* (Princeton, NJ: Princeton University Press).

Shu, F. H. 1982, *The Physical Universe* (Mill Valley, CA: University Science Books).

Sparke, L. S., and Gallagher, J. S. 2007, *Galaxies in the Universe: An Introduction*, 2nd ed. (Cambridge: Cambridge University Press).

Spitzer, L. 1978, *Physical Processes in the Interstellar Medium* (New York: Wiley).

Swihart, T. L. 1992, *Quantitative Astronomy: Topics in Astrophysics* (Englewood Cliffs, NJ: Prentice Hall)

Weymann, R. J., Swihart, T. L., Williams, R. E., Cocke, W. J., Pacholczyk, A. G., and Felten, J. E. 1976, *Lecture Notes on Introductory Theoretical Astrophysics* (Tucson, AZ: Pachart).

Zeilik, M., and Gregory, S. A. 1998, *Introductory Astronomy and Astrophysics*, 4th ed. (Brooks/Cole).

Credits

PHOTO AND ART CREDITS

Chapter 1

Figure 1.6 Peter Michaud (Gemini Observatory)/AURA/NSF

Figure 1.14 Frank Zullo/SPL

Chapter 2

Figure 2.6 Royal Astronomical Society/SPL

Chapter 4

Figure 4.2(a) Jeff Newbery

Figure 4.2(b) Jeff Newbery

Figure 4.15(a) Photo Researchers, Inc.

Figure 4.15(b) Photo Researchers, Inc.

Figure 4.15(c) Thierry Legault/Eurelios/SPL

Figure 4.19 Fred Espenak/NASA

Chapter 6

Figure 6.1 Bettmann/Corbis

Figure 6.5 Colin Eberhardt

Figure 6.11 Bradley Peterson

Figure 6.13 Starizona

Figure 6.18 Large Binocular Telescope Observatory

Chapter 7

Figure 7.1 Big Bear Solar Observatory

Figure 7.3 Big Bear Solar Observatory

Figure 7.4 SOHO-EIT Consortium, ESA, NASA

Figure 7.5 Stefan Seip

Figure 7.11 Juan Carlos Casado

Figure 7.12 G.E. Hale, F. Ellerman, S. F. Nicholson, & A. H. Joy, "The Magnetic Polarity of Sun-Spots," 1919, *The Astrophysical Journal,* v. 49, p. 153 (Plate VII, panels A & B)

Figure 7.14 E. W. Maunder, "Note the Distribution of Sun-Spots in Heliographic Latitude," 1904, *Monthly Notices of the Royal Astronomical Society,* v. 64, p. 747 (plate 16)

Figure 7.15 Big Bear Solar Observatory

Figure 7.16 SOHO/EIT, ESA, NASA

Figure 7.17 Science Applications Internal Corporation

Chapter 8

Figure 8.3 M.J. McCaughrean, H. Chen, J. Bally, E. Erickson, R. Thompson, M. Rieke, G. Schneider, S. Stolovy, & E. Young "High-Resolution Near-Infrared Imaging of the Orion 114-426 Silhouette Disk," 1998, *The Astrophysical Journal Letters,* v. 492, p. L157 (figure 2)

Figure 8.4 NASA/JHU APL/Carnegie Institution of Washington

Chapter 9

Figure 9.8 NASA

Figure 9.9 NASA/Apollo 17

Figure 9.10 NASA/NSSDC

Figure 9.12 NASA/Apollo 15

Chapter 10

Figure 10.2 NASA/JHU APL/Carnegie Institution of Washington

Figure 10.3 NASA/JPL

Figure 10.5 NASA/MOLA Science Team

Figure 10.6 NASA/MOLA Science Team

Figure 10.7 NASA/JPL/ASU

Figure 10.9(a) NASA/JPL/Malin Space Science Systems

Figure 10.9(b) NASA/JPL/Malin Space Science Systems

Figure 10.10 USGS Astrogeology Research Program

Figure 10.11 W. B. Hubbard, T. Guillot, M. S. Marley, A. Burrows, J. I. Lunine, & Saumon, D. S. "Comparative Evolutin of Jupiter and Saturn," 1999, Planetary & Space Science, v. 47, p. 1175, fig. 2

Figure 10.14 NASA Headquarters

Figure 10.15 A. Tayfun Oner

Figure 10.16 NASA/JPL/ASU

Figure 10.17 NASA/JPL/DLR

Figure 10.18 NASA/JPL/Space Science Institute

Figure 10.19 NASA/ESA/L. Sromovsky & P. Fry (U. Wisconsin), H. Hammel (Science Space Institute), & K. Rages (SETI Institute)

Figure 10.20 NASA/JPL

Figure 10.22 NASA/JPL/Space Science Institute

Figure 10.23 NASA/JPL/Space Science Institute

Chapter 11

Figure 11.4 Dr. R. Albrecht, ESA/ESO Space Telescope European Coordinating Facility/NASA

Figure 11.6 NASA/JPL/UMD

Figure 11.7 NASA/A. Feild (Space Telescope Science Institute)

Figure 11.9 Yuri Beletsky (ESO)

Figure 11.11 Daniel R Suchy/Kansas Geological Survey

Figure 11.12 David Parker/SPL

Chapter 12

Figure 12.1(a) NASA/JPL

Figure 12.1(b) NASA/JPL/Space Science Institute

Figure 12.1(c) NASA/JPL

Figure 12.2(a) NASA/Lunar Orbiter 4

Figure 12.2(b) NASA/JPL

Figure 12.2(c) NASA/JPL

Chapter 13

Figure 13.3 Andrea Dupree (Harvard-Smithsonian CfA), Ronald Gilliland (STScI), NASA and ESA

Figure 13.5 NASA/SAO/CXC

Figure 13.6 D. Benset & J. L. Duvent

Chapter 14

Figure 14.1 W. W. Morgan, P Keenan, & E. Kellman

Figure 14.2(a) H. N. Russell, 1914, Nature, v. 93, n. 2323, p. 252, fig. 1

Figure 14.2(b) J. Kovalevsky, 1998, Ann. Rev. Astron. Astrophys., v. 36, p. 99, fig. 9

Chapter 15

Figure 15.8 SOI/MDI, Stanford Lockheed Institute for Space Research

Chapter 16

Figure 16.2 Spoon, H. W. W., et al., 2004, Ap. J. Supp., v.154, p.184, fig. 1

Figure 16.4 NASA/UC Berkley/SpaceDev

Chapter 17

Figure 17.1 T.A.Rector (NOAO/AURA/NSF) and Hubble Heritage Team (STScI/AURA/NASA)

Chapter 18

Figure 18.1 J. L. Provencal, H. L. Shipman, E. Hog, & P. Thejl, 1998, *The Astrophysical Journal,* 494, 759

Figure 18.2 M. A. Barstow, H. E. Bond, J. B. Holberg, M. R. Burleigh, I. Hubeny, & D. Koester, 2005, *Monthly Notices of the Royal Astonomical Society,* 362, 1134

Figure 18.5 V. Burwitz, F. Haberl, R. Neuhauser, P. Predehl, J. Trumper, & V. E. Zavlin, 2003, *Astronomy & Astrophysics,* 399, 1109

Figure 18.7 Frank Zullo/Science Source

Figure 18.8 NASA

Figure 18.9 UC Regents/Lick Observatory

Figure 18.10 WIYN/NOAO/NSF

Chapter 19

Figure 19.1 Steward Observatory, The University of Arizona

Figure 19.2 "On the Construction of the Heavens," William Herschel, 1785, Philosophical Transactions of the Royal Society of London, v. 75, p. 213

Figure 19.3 NOAO/AURA/NSF

Figure 19.4 Large Binocular Telescope Observatory

Figure 19.5 ESO

Figure 19.6 NASA/JPL-Caltech/R. Hurt (SSC-Caltech)

Figure 19.16 "Rotation Effects, Interstellar Absorption, and Certain Dynamical Constants of the Galaxy Determined from Cepheid Variables," Alfred H. Joy, 1939, *The Astrophysical Journal,* v. 89, p. 356

Figure 19.17 "Galactic Kinematics of Cepheids from Hipparcos Proper Motions," Michael Feast & Patricia Whitelock, 1991, *Monthly Notices of the Royal Astronomical Society,* v. 291, p. 683

Figure 19.19 "Massachusetts—Stony Brook Galactic Plane CO Survey—The Galactic Disk Rotation Curve," D. P. Clemens, 1985, *The Astrophysical Journal,* v. 295, p. 422

Figure 19.20 "Three-dimensional Distribution of the ISM in the Milky Way Galaxy: I. The HI Disk," Hiroyuki Nakanishi & Yoshiaki Sofue, 2003, Publications of the Astronomical Society of Japan, v. 55, p. 191

Figure 19.21 "The Interstellar Environment of Our Galaxy," Katia M. Ferriere, 2001, Reviews of Modern Physics, v. 73, p. 1031

Figure 19.22 "Performance of the Canada-France-Hawaii Telescope Adaptive Optics Bonnette," F. Rigaut, D. Salmon, R. Arsenault, J. Thomas, O. Lai, D. Rouan, J.P. Veran, P. Gigan, D. Crampton, J.M. Fletcher, J. Stilburn, C. Boyer, & P. Jagourel, 1998, Publications of the Astronomical Society of the Pacific, v. 110, p. 152

Figure 19.23 "Stellar Orbits Around the Galactic Center Black Hole," A. M. Ghez, S. Salim, S. D. Hornstein, A. Tanner, J. R. Lu, M. Morris, E. E. Becklin, & G. Duchêne, 2005, *The Astrophysical Journal,* v. 620, p. 744

Chapter 20

Figure 20.1 NASA/ESA

Figure 20.2(a) Princeton University Press

Figure 20.2(b) Princeton University Press

Figure 20.2(c) Princeton University Press

Figure 20.3(a) Paul B. Eskridge, et al. 2002, *The Astrophysical Journal Supplement*

Figure 20.3(b) Paul B. Eskridge, et al. 2002, *The Astrophysical Journal Supplement*

Figure 20.3(c) Paul B. Eskridge, et al. 2002, *The Astrophysical Journal Supplement*

Figure 20.4(a) Paul B. Eskridge, et al. 2002, *The Astrophysical Journal Supplement*

Figure 20.4(b) Paul B. Eskridge, et al. 2002, *The Astrophysical Journal Supplement*

Figure 20.4(c) Paul B. Eskridge, et al. 2002, *The Astrophysical Journal Supplement*

Figure 20.5 John Kormendy/University of Texas at Austin

Figure 20.6 Fred Espenak

Figure 20.7 David Malin/Anglo-Australian Observatory

Figure 20.8 John Hibbard/NRAO

Figure 20.9 "The Supermassive Black Hole of M87 and the Kinematics of its Associated Gaseous Disk," F. Macchetto, A. Marconi, D. J. Axon, A. Capetti, W. Sparks, & P. Crane, 1997, *the Astrophysical Journal,* v. 489, p. 579

Figure 20.10 John Kormendy/University of Texas

Figure 20.11 "A Relation Between Distance and Radial Velocity Among Extra-Galactic Nebulae," Edwin Hubble, 1929, Proceedings of the National Academy of Sciences, v. 15, p. 168

Chapter 21

Figure 21.1 Barbara Ryden

Figure 21.3 Sloan Digital Sky Survey

Figure 21.5(a) Misty Bentz—MDM Observatory

Figure 21.5(b) Misty Bentz—MDM Observatory

Figure 21.5(c) Misty Bentz—MDM Observatory

Figure 21.6 "The Surface Brightness of the Nebulosity in BL Lacertae," T. D. Kinman, 1975, *The Astrophysical Journal Letters,* 197, L49

Figure 21.7 "The inner radio structure of Centaurus A—Clues to the origin of the jet X-ray emission," J. O. Burns, E. D. Feigelson, & E. J. Schreier, 1983, *The Astrophysical Journal,* v. 273, p. 128

Figure 21.8 NASA/The Hubble Heritage Team (STScI/AURA)

Figure 21.10 John Bahcall (IAS)/Mike Disney (U. Wales)/NASA

Figure 21.12 W. L. W. Sargent/California Institute of Technology

Chapter 22

Figure 22.1 "New Reduction of the Lick Catalog of Galaxies," M. Seldner, B. Siebers, J. Groth, & P. J. E. Pebbles, 1977, *The Astrophysical Journal,* v. 82, p. 249

Figure 22.2 "The Stellar Content of the Local Group," 1999, IAU Symp. 192, eds. P. Whitelock & R. Cannon (Provo: ASP), 17-38

Figure 22.3 "On the Clustering Tendencies among the Nebulae. II. A Study of Encounters Between Laboratory Models of Stellar Systems by a New Integration Procedure," Erik Holmberg, 1944, *The Astrophysical Journal,* v. 94, p. 385

Figure 22.4 John Dubinski/University of Toronto

Figure 22.5 NASA, H. Ford (JHU), G. Illingworth (UCSC/LO), M.Clampin (STScI), G. Hartig (STScI), the ACS Science Team, and ESA

Figure 22.6 Kirk Borne (STScl)/NASA

Figure 22.7 "A Slice of the Universe," V. de Lapparent, M. J. Geller, & J. P. Huchra. 1986, *The Astrophysical Journal (Letters),* v. 302, p. L1

Figure 22.8 Sloan Digital Sky Survey

Figure 22.9 "An Analytic Expression for the Luminosity Function for Galaxies," 1976, *The Astrophysical Journal,* v. 203, p. 297

Chapter 24

Figure 24.6 "Hubble Space Telescope and Ground-based Observations of Type la Supernova at Redshift 0.5: Cosmological Implications," 2006, Alexander Clocchiatti et al., *The Astrophysical Journal,* v. 642, p. 1

Figure 24.7 "Hubble Space Telescope and Ground-based Observations of Type la Supernova at Redshift 0.5: Cosmological Implications," 2006, Alexander Clocchiatti et al., *The Astrophysical Journal,* v. 642, p. 1

Color Plates

Color Figure 1 Pekka Parviainen/Photo Researchers

Color Figure 2 Richard Pogge

Color Figure 3 Richard Pogge

Color Figure 4 Fred Espenak/Photo Researchers

Color Figure 5 Voyager 2 Team/NASA

Color Figure 5 NASA Headquarters

Color Figure 5 NASA/JPL

Color Figure 6 Image Science & Analysis Laboratory/ NASA Johnson Space Center

Color Figure 7 Lunar & Planetary Institute

Color Figure 8 Lunar & Planetary Institute

Color Figure 9 NASA/JPL/DLR

Color Figure 10 NASA/JPL/Univ. of Arizona

Color Figure 11 NASA/JPL/USGS

Color Figure 12 NASA/J. C. Casado

Color Figure 13 Nigel Sharp/NOAO/NSO/Kitt Peak FTS/AURA/NSF

Color Figure 14 NASA/Robert Hurt

Color Figure 15 NASA/ESA/M. Robberto/HST Orion Treasury Project Team

Color Figure 16 Hubble Heritage Team/AURA/STScL/ NASA

Color Figure 17 T. Dame (CfA, Harvard) et. Al., Columbia 1.2-m Radio Telescope

Color Figure 18 Nigel Sharp/NOAO/AURA/NSF

Color Figure 19 2MASS/J. Carpenter/M. Skrutskie/R. Hurt

Color Figure 20 F. Yusef-Zadeh, M. R. Morris, D. R. Chance/NRAO/AUI/NSF

Color Figure 21 K. Y. Lo, M. J. Clausen/NRAO/AUI/ NSF

Color Figure 22 NASA/ESA/S. Beckwith (STScl)/ HUDF Team

Color Figure 23 NASA/CXC/SAO/PSU/CMU/N. A. Sharp/AURA/NOAO/NSF/MERLIN/VLA

Color Figure 24 R. L. Davies, et al. 2001, *The Astrophysical Journal Letters,* 548, 33

Color Figure 25 R. L. Davies, et al. 2001, *The Astrophysical Journal Letters,* 548, 33

Color Figure 26 H. Ford, R. Harms, Z. Tsvetanov, A. Davidson, G. Kriss, R. Bohlin, G. Hartig, L. Dressel, A. K. Kohhar, B. Margon

Color Figure 27 S.L. Snowden/NASA/GSFC

Color Figure 28 B. J. Mochejska/J. Kaluzny

Color Figure 29 NASA/JPL-Caltech/Z. Wang (Harvard-Smithsonian CfA)

Color Figure 30 C. L. Bennett et al., 2001, *The Astrophysical Journal Supplement,* 148, 1

Color Figure 31 WMAP Science Team

Index